Software Solutions

for

Engineers

and

Scientists

Software Solutions for Engineers and Scientists

Julio Sanchez
Maria P. Canton

CRC Press
Taylor & Francis Group
Boca Raton   London   New York

CRC Press is an imprint of the
Taylor & Francis Group, an informa business

CRC Press
Taylor & Francis Group
6000 Broken Sound Parkway NW, Suite 300
Boca Raton, FL 33487-2742

© 2008 by Taylor & Francis Group, LLC
CRC Press is an imprint of Taylor & Francis Group, an Informa business

No claim to original U.S. Government works
Printed in the United States of America on acid-free paper
10 9 8 7 6 5 4 3 2 1

International Standard Book Number-13: 978-1-4200-4302-0 (Hardcover)

Library of Congress Cataloging-in-Publication Data

Sanchez, Julio, 1938-
    Software solutions for engineers and scientists / authors, Julio Sanchez and Maria P. Canton.
      p. cm.
    Includes bibliographical references and index.
    ISBN 978-1-4200-4302-0 (alk. paper)
    1. Engineering--Data processing. 2. Science--Data processing. I. Canton, Maria P. II. Title.

TA345.S31535 2008
620'.0285--dc22
                                        2007034749

Visit the Taylor & Francis Web site at
http://www.taylorandfrancis.com

and the CRC Press Web site at
http://www.crcpress.com

Table of Contents

Preface   xxiii

PART I — TECHNIQUES AND CODE

Chapter 1 — Computer Number Systems

Chapter Summary   3
1.0 Counting   3
    1.0.1 The Tally System   3
    1.0.2 Roman Numerals   4
1.1 The Origins of Our Number System   5
    1.1.1 Number Systems for Digital-Electronics   6
    1.1.2 Positional Characteristics   7
    1.1.3 Radix or Base   7
1.2 Types of Numbers   8
    1.2.1 Whole Numbers   8
    1.2.2 Signed Numbers   8
    1.2.3 Rational and Irrational Numbers   8
    1.2.4 Real and Complex Numbers   9
1.3 Radix Representations   10
    1.3.1 Decimal Versus Binary Numbers   10
    1.3.2 Octal and Hexadecimal Numbers   11
1.4 General Theory of Counting   12
1.5 Assembly Language Conversion Routines   14
    1.5.1 Binary-to-ASCII-Decimal Conversion   15
    1.5.2 Binary-to-Hexadecimal Conversion   18
    1.5.3 Decimal-to-Binary Conversion   19
1.6 C++ Conversion Routines   23
    1.6.1 C++ Binary-to-ASCII Conversions   23
    1.6.2 C++ ASCII-to-Binary Conversions   25

Chapter 2 — Numeric Data in Memory

Chapter Summary   31
2.0 Electronic-Digital Machines   31
2.1 Storage of Numerical Data   32
    2.1.1. Word Size   32
2.2 Integer Encodings   33
    2.2.1 Sign-Magnitude Representation   34
    2.2.2 Radix Complement Representation   35
2.3 Encoding of Fractional Numbers   39
    2.3.1 Fixed Point Representations   40

2.3.2 Floating-Point Representations — 41
2.4 Standardized Floating-Point Encodings — 42
    2.4.1 ANSI/IEEE 754 Single Format — 42
    2.4.2 Floating-Point Exponent in ANSI/IEEE 754 Single Format — 43
    2.4.3 Floating-Point Significand in ANSI/IEEE 754 Single Format — 44
    2.4.4 Decoding Floating-Point Numbers — 45
2.5 Binary-Coded Decimals (BCD) — 49
    2.5.1 Floating-Point BCD — 49
2.6 BCD Conversions — 51
    2.6.1 BCD Conversion Functions — 51

Chapter 3 — Machine Arithmetic

Chapter Summary — 55
3.0 Intel Microprocessors — 55
    3.0.1 CPU Flags — 56
3.1 Logical Instructions — 57
    3.1.1 Logical AND — 58
    3.1.2 Logical OR — 59
    3.1.3 Logical XOR — 59
    3.1.4 Logical NOT — 60
3.2 Arithmetic Instructions — 60
    3.2.1 Signed and Unsigned Arithmetic — 60
    3.2.2 Operations on Decimal Numbers — 61
3.3 Auxiliary and Bit Manipulation Instructions — 63
    3.3.1 Bit Shift and Rotate Instructions — 63
        Bit Shift Instructions — 63
        Bit Rotate Instructions — 66
        Double Precision Shift Instructions — 67
        Shift and Rotate Addressing Modes — 68
    3.3.2 Comparison, Bit Scan, and Bit Test Instructions — 68
        Signed and Unsigned Conditional Jumps — 70
    3.3.3 Increment, Decrement, and Sign Extension Instructions — 71
    3.3.4 486 and Pentium Proprietary Instructions — 72
        BSWAP — 72
        XADD — 73
        CMPXCHG and CMPXCHG8B — 74
3.4 CPU Identification — 74

Chapter 4 — High-Precision Arithmetic

Chapter Summary — 79
4.0 Applications of BCD Arithmetic — 79
    4.0.1 ANSI/IEEE 854 Standard — 80
4.1 Algorithms for BCD Arithmetic — 81
4.2 Floating-Point BCD Addition — 82
4.3 Floating-Point BCD Subtraction — 82
4.4 Floating-Point BCD Multiplication — 83
4.5 Floating-Point BCD Division — 85
4.6 C++ BCD Arithmetic Functions — 86
4.6 High-Precision BCD Arithmetic — 90

Chapter 5 — Floating-Point Hardware

Chapter Summary 95
5.0 A Mathematical Coprocessor 95
5.1 Intel Math Units 96
    5.1.1 Math Unit Applications 97
    5.1.2 Math Unit Limitations 98
    5.1.3 Processor/Coprocessor Interface 99
    5.1.4 Math Unit Versions 100
        8087 101
        80287 101
        80387 102
    5.1.5 The Numeric Unit in 486 and Pentium CPU 102
5.2 Detecting and Identifying the Math Unit 102
5.3 ANSI/IEEE 754 Standard 106
    5.3.1 Numeric Data Encoding 107
    5.3.2 Rounding 109
    5.3.3 Interval Arithmetic 110
    5.3.4 Treatment of Infinity 110
    5.3.5 Not a Number (NaN) 112
        Signaling and Quiet NaNs 112
    5.3.6 Exceptions 113
        Invalid Operation Exception 113
        Division by Zero Exception 114
        Overflow Exception 114
        Underflow Exception 115
        Inexact Result Exception 116

Chapter 6 — Floating Point Data and Conversions

Chapter Summary 117
6.0 Math Unit Data Formats 117
    6.0.1 Binary Integers 118
    6.0.2 Decimal Integers 119
    6.0.3 Binary Reals 120
6.1 Special Encodings for Reals 121
6.2 Low-Level Numeric Data in Memory 123
    6.2.1 Initializing Data with the DW Directive 124
    6.2.2 Initializing Data with DD and DQ Directives 124
    6.2.3 Initializing Data with the DT Directive 125
    6.2.4 Memory Image of the Special Encodings 125
    6.2.5 Operating on Memory Variables 126
6.3 High-Level Numeric Data 127
6.4 Numeric Data Conversions 127
    6.4.1 Data Conversion in ANSI/IEEE 754 128
    6.4.2 Conversion Requirements in ANSI/IEEE 754 128
    6.4.3 FPU_INPUT Procedure 130
        Calculating 10^y 131
        Post-Conversion Operations 132
    6.4.4 FPU_OUTPUT Procedure 132
    6.4.5 ASCII-to-Exponential Conversion 133
6.5 High-Level Interface Functions 133

Chapter 7 — Math Unit Architecture and Instruction Set

Chapter Summary   137
7.0 Math Unit Internal Organization   137
    7.0.1 Math Unit Register Stack   137
    7.0.2 Math Unit Control Register   140
        Error Conditions   141
    7.0.3 Math Unit Status Register   143
    7.0.4 The Environment Area   150
        Tag Word Register   152
        Instruction and Data Pointers   154
    7.0.5 Math Unit State Area   156
7.1 Math Unit Instruction Patterns   158
    7.1.1 Register Operands   158
    7.1.2 Memory Operands   159
7.2 Math Unit Instruction Set   159
    7.2.1 Data Transfer Instructions   160
    7.2.2 Nontranscendental Instructions   161
        Basic Arithmetic   161
        Scaling and Square Root   162
        Partial Remainder   163
        Update of the Partial Remainder   166
        Manipulating the Encoding   168
    7.2.3 Comparison Instructions   171
    7.2.4 Transcendental Instructions   172
        Transcendental Algorithms   174
    7.2.5 Constant Instructions   177
    7.2.6 Processor Control Instructions   178

Chapter 8 — Transcendental Primitives

Chapter Summary   181
8.0 Developing Math Unit Software   181
8.1 Exponential Functions   182
    8.1.1 Calculation of Powers   183
        Logarithmic Approximation of Exponentials   184
        Binary Powering   186
        Exponent Factoring   190
        Applications   198
        Mixed Methods   198
8.2 Math Unit Trigonometry   199
    8.2.1 Angular Conversions   199
    8.2.2 Range-Scaling Operations   201
        Reduction to the Unit Circle   201
        Reduction to the First Octant   202
    8.2.3 Trigonometric Functions   204
        Calculating Tangent, Sine, and Cosine   205
        Trigonometric Arcfunctions   207
8.4 Logarithms   211
    8.4.1 Calculating Natural and Common Logarithms   211
    8.4.2 Calculating Antilogarithms   212
8.5 C++ Interface to Transcendentals   213

Chapter 9 — General Mathematical Functions

Chapter Summary 215
9.0 Calculator Operations 215
    9.0.1 Calculating Hyperbolic Functions 216
    9.0.2 Factorial 220
    9.0.3 Ordering Numeric Data 221
    9.0.4 Calculating the Modulus 224
    9.0.5 Integer and Fractional Parts 225
    9.0.6 Solving Triangles 227
9.1 Quadratic Equations 229
9.2 Imaginary and Complex Numbers 232
    9.2.1 Operations in Complex Arithmetic 233
    9.2.2 Real and Complex Roots of a Quadratic Equation 240
    9.2.3 Polar and Cartesian Coordinates 244

Chapter 10 — Financial Calculations

Chapter Summary 249
10.0 Interest Calculations 249
    10.0.1 Simple Interest 249
    10.0.2 Compound Interest 251
        Future Value at Compound Interest 252
        Exponentials in Financial Formulas 253
        Present Value at Compound Interest 254
        Effective Rate 256
10.1 Amortization 257
    10.1.1 Periodic Payment Calculations 257
    10.1.2 Add-On Interest 259
    10.1.3 Annual Percentage Rate 261
10.2 Annuities 263
    10.2.1 Annuity Future Value 263
    10.2.2 Annuity Present Value 265
    10.2.3 Annuity Due 266
    10.2.4 Sinking Fund 269
    10.2.5 Number of Compounding Periods 271
10.3 Numerical Errors in Financial Calculations 272
    10.3.1 Conversion Errors 273
    10.3.2 Representation Errors 273
    10.3.3 Precision and Computation Errors 274
        Cancellation Error 275
10.4 Financial Software 275

Chapter 11 — Statistical Calculations

Chapter Summary 277
11.0 About Statistical Data 277
    11.0.1 Data-Type-Flexible Coding 278
11.1 Data Manipulation Primitives 278
    11.1.1 Common Summations 278
    11.1.2 Sorting 280
11.2 Counting Techniques 281

11.2.1 Permutations   282
11.2.2 Combinations   283
11.2.3 Binomial Probability   285
11.3 Measures of Central Tendency   286
11.3.1 Mean   287
11.3.2 Median   287
11.3.3 Midrange   288
11.3.4 Mode   289
11.3.5 Weighted Measures of Central Tendency   292
11.4 Measures of Dispersion   294
11.4.1 Range   294
11.4.2 Variance   295
11.4.3 Standard Deviation   296
11.5 Normal Distribution   297
11.5.1 Normal Curve   297
Standard Normal Curve   298
11.5.2 Calculating f(x)   299
11.5.3 Probability in Normal Distribution   301
11.6 Linear Correlation and Regression   304
11.6.1 Linear Correlation Coefficient   305
11.6.2 Linear Regression Analysis   307

Chapter 12 — Interpolation, Differentiation, and Integration

Chapter Summary   311
12.0 Interpolation   311
12.0.1 Linear Interpolation   313
12.0.2 Lagrange Interpolation   316
12.0.3 Least-Squares Interpolation   319
Linear Models   319
Calculating the Error Sum   321
Least-Squares Linear Interpolation Function   321
Non-Linear Models   323
12.1 Numerical Differentiation   327
12.2 Numerical Integration   332
12.2.1 Integration by the Trapezoidal Rule   333
12.2.2 Integration by Simpson's Rule   336

Chapter 13 — Linear Systems

Chapter Summary   341
13.0 Linear Equations   341
13.0.1 Systems of Linear Equations   342
13.0.2 Matrix Representations of Linear Systems   344
13.1 Numeric Data in Matrix Form   345
13.1.1 Matrices in C++   345
13.1.2 Locating a Matrix Entry   348
13.2 Operations on Matrix Entries   349
13.2.1 Vectors   350
13.2.2 Vector-by-Scalar Operations in C++   350
13.2.3 Low-Level Vector-by-Scalar Operations   352
13.2.4 Matrix-by-Scalar Operations   357

13.2.5 Matrix-by-Matrix Operations	359
Matrix Addition	360
Matrix Multiplication	363
13.3 The Solution of a Linear System	368
13.3.1 Gauss-Jordan Elimination	368
13.3.2 Errors in Gaussian Elimination	370
13.4 A Gauss-Jordan Algorithm	371
13.5 Solution of a Linear System	372

Chapter 14 — Solving and Parsing Equations

Chapter Summary	375
14.0 Function Mapping	375
14.1 Developing a Parser	377
14.2 Evaluating User Equations	379
14.2.1 Equation Grammar	379
14.2.2 Equation Syntax	380
14.2.3 Symbol Table and Numeric Data	381
14.3 An Equation-Solving Algorithm	382
14.3.1 The _EVALUATE Procedure	382
14.3.2 _CALCULATE_Y Procedure	386

Chapter 15 — Neural Networks

Chapter Summary	389
15.0 Reverse-Engineering the Brain	389
15.0.1 The Biological Neuron	389
15.0.2 The Artificial Neuron	391
15.0.3 Artificial Neural Networks	394
15.1 The Network as a Classifier	396
15.1.1 Multiple-Node Networks	396
15.1.2 Software Model for Neural Nets	398
15.2 The Perceptron	398
15.2.1 Perceptron as a Classifier	399
15.2.2 Perceptron Learning	400
15.2.3 Training the Perceptron	402
15.2.4 A Perceptron Function	403
15.3 The Adaline	405
15.3.1 Widrow-Hoff Learning	405
15.3.2 A Neuron for Adaline	406
15.4 Improving the Classification Function	409
15.4.1 Calculating the Error Sum	409
15.4.2 Improving Perceptron Results	410
15.5 Backpropagation Networks	412
15.5.1 Nonlinear Neurons	413
15.5.2 Backpropagation Algorithm	414
15.5.3 Software Model for Backpropagation	415
15.5.4 Executing the Network	417
15.5.5 A Backpropagation Trainer	418

PART II — APPLICATION DEVELOPMENT

Chapter 16 — The C++ Language on the PC

Chapter Summary   423
16.0 Introducing C++   423
    16.0.1 Evolution of C++   423
    16.0.2 Advantages of the C++ Language   424
    16.0.3 Disadvantages of the C++ Language   424
16.1 PC Implementations of C and C++   425
16.2 Flowcharts and Software Design   425
16.3 The C++ Console Application   427

Chapter 17 — Event-Driven Programming

Chapter Summary   431
17.0 Graphical Operating Systems   431
17.1 Enter Windows   432
    17.1.1 Text-based and Graphical Programs   432
    17.1.2 Graphics Services   434
17.2 Programming Models   434
    17.2.1 Event-Driven Programs   434
        The Event Manager   436
        The Event Handler   436
    17.2.2 Event Types   436
        System Events   436
        Control Events   437
        Program Events   437
    17.2.3 Event Modeling   437
17.3 File Structure of a Windows Program   438
    17.3.1 Source Files   438
    17.3.2 Library Files   439
    17.3.3 Resource Files   440
    17.3.4 Make Files   440
    17.3.5 Object Files   441
    17.3.6 Executable Files   442
    17.3.7 Dynamic Linking   444

Chapter 18 — The Window Program Components

Chapter Summary   447
18.0 "Hello, World"   447
18.1 Naming Conventions   449
18.2 Constants and Handles   451
    18.2.1 Windows Handles   452
18.3 Visual Elements   453
    18.3.1 The Main Window   453
    18.3.2 Controls   454
    18.3.3 Other Visual Components   455
18.4 Programming Style   456
    18.4.1 Commented Headers   456

18.4.2 Assertions Notation   457
    ASSERT   458
    INV   458
    PRE and POST   458
    FCTVAL   458
18.4.3 Programming Templates   458

Chapter 19 — A First Windows Program

Chapter Summary   463
19.0 Preliminary Steps   463
19.1 The Program Project   464
   19.1.1 Creating a Project   464
19.2 Elements of a Windows Program   468
   19.2.1 WinMain()   468
      Parameters   469
   19.2.2 Data Variables   470
   19.2.3 WNDCLASSEX Structure   471
   19.2.4 Registering the Windows Class   475
   19.2.5 Creating the Window   476
   19.2.6 Displaying the Window   480
   19.2.7 The Message Loop   480
19.3 The Window Procedure   481
   19.3.1 Windows Procedure Parameters   482
   19.3.2 Windows Procedure Variables   483
   19.3.3 Message Processing   483
      WM_CREATE Message Processing   484
      WM_PAINT Message Processing   484
      WM_DESTROY Message Processing   485
   19.3.4 The Default Windows Procedure   485
19.4 The WinHello Program   486
   19.4.1 Modifying the Program Caption   486
   19.4.2 Displaying Text in the Client Area   487
   19.4.3 Creating a Program Resource   489
   19.4.4 Creating the Icon Bitmap   490
19.5 WinHello Program Listing   493

Chapter 20 — Text Display

Chapter Summary   497
20.0 Text in Windows   497
20.1 The Client Area   498
20.2 Device and Display Contexts   498
   20.2.1 The Display Context   499
   20.2.2 Display Context Types   500
   20.2.3 Window Display Context   502
20.3 Mapping Modes   502
   20.3.1 Screen and Client Area   503
   20.3.2 Viewport and Window   504
20.4 Programming Text Operations   505
   20.4.1 Typefaces and Fonts   507
   20.4.2 Text Formatting   508

20.4.3 Paragraph Formatting   511
20.4.4 The DrawText() Function   515
20.5 Text Graphics   518
20.5.1 Selecting a Font   518
20.5.2 Drawing with Text   523

Chapter 21 — Keyboard and Mouse Programming

Chapter Summary   525
21.0 Keyboard Input   525
21.1 Input Focus   526
21.1.1 Keystroke Processing   527
21.1.2 Determining the Key State   529
21.1.3 Character Code Processing   530
21.1.4 Keyboard Demonstration Program   531
21.2 The Caret   534
21.2.1 Caret Processing   535
21.2.2 Caret Demonstration Program   535
21.3 Mouse Programming   538
21.3.1 Mouse Messages   539
21.3.2 Cursor Location   541
21.3.3 Double-Click Processing   542
21.3.4 Capturing the Mouse   543
21.3.5 The Cursor   543
21.4 Mouse and Cursor Demonstration Program   546

Chapter 22 — Graphical User Interface Elements

Chapter Summary   549
22.0 Window Styles   549
22.1 Child Windows   550
22.1.1 Child Windows Demonstration Program   552
22.2 Basic Controls   554
22.2.1 Communicating with Controls   558
22.2.2 Controls Demonstration Program   563
22.3 Menus   566
22.3.1 Creating a Menu   568
22.3.2 Menu Item Processing   569
22.3.3 Shortcut Keys   570
22.3.4 Pop-Up Menus   571
22.3.5 The Menu Demonstration Program   573
22.4 Dialog Boxes   573
22.4.1 Modal and Modeless   574
22.4.2 The Message Box   574
22.4.3 Creating a Modal Dialog Box   576
22.4.4 Common Dialog Boxes   578
22.4.5 The Dialog Box Demonstration Program   581
22.5 Common Controls   581
22.5.1 Common Controls Message Processing   582
22.5.2 Toolbars and ToolTips   583
22.5.3 Creating a Toolbar   584
22.5.4 Standard Toolbar Buttons   589

22.5.5 Combo Box in a Toolbar 591
22.5.6 ToolTip 592

Chapter 23 — Drawing Lines and Curves

Chapter Summary 597
23.0 Drawing in a Window 597
23.1 The Redraw Responsibility 598
    23.1.1 The Invalid Rectangle 599
    23.1.2 Screen Updates On-Demand 600
    23.1.3 Intercepting WM_PAINT 600
23.2 The Graphics Device Interface 602
    23.2.1 Device Context Attributes 603
    23.2.2 DC Info Demonstration Program 607
    23.2.3 Color in the Device Context 610
23.3 Graphic Objects and GDI Attributes 611
    23.3.1 Pens 611
    23.3.2 Brushes 613
    23.3.3 Foreground Mix Mode 615
    23.3.4 Background Modes 617
    23.3.5 Current Pen Position 617
    23.3.6 Arc Direction 618
23.4 Pixels, Lines, and Curves 618
    23.4.1 Pixel Operations 619
    23.4.2 Drawing with LineTo() 620
    23.4.3 Drawing with PolylineTo() 621
    23.4.4 Drawing with Polyline() 621
    23.4.5 Drawing with PolyPolyline() 622
    23.4.6 Drawing with Arc() 623
    23.4.7 Drawing with ArcTo() 624
    23.4.8 Drawing with AngleArc() 624
    23.4.9 Drawing with PolyBezier() 626
    23.4.10 Drawing with PolyBezierTo() 629
    23.4.11 Drawing with PolyDraw() 629
    23.4.12 Pixel and Line Demonstration Program 633

Chapter 24 — Drawing Solid Figures

Chapter Summary 635
24.0 Closed Figures 635
24.1. Closed Figure Elements 636
    24.1.1 Brush Origin 636
    24.1.2 Object Selection Macros 638
    24.1.3 Polygon Fill Mode 638
    24.1.4 Creating Custom Brushes 640
24.2 Drawing Closed Figures 642
    24.2.1 Drawing with Rectangle() 643
    24.2.2 Drawing with RoundRect() 643
    24.2.3 Drawing with Ellipse() 644
    24.2.4 Drawing with Chord() 645
    24.2.5 Drawing with Pie() 646
    24.2.6 Drawing with Polygon() 647

24.2.7 Drawing with PolyPolygon()   649
24.3 Operations on Rectangles   650
    24.3.1 Drawing with FillRect()   650
    24.3.2 Drawing with FrameRect()   652
    24.3.3 Drawing with DrawFocusRect()   652
    24.3.4 Auxiliary Operations on Rectangles   653
    24.3.5 Updating the Rectangle() Function   659
24.4 Regions   660
    24.4.1 Creating Regions   661
    24.4.2 Combining Regions   664
    24.4.3 Filling and Painting Regions   666
    24.4.4 Region Manipulations   667
    24.4.5 Obtaining Region Data   670
24.5 Clipping Operations   671
    24.5.1 Creating or Modifying a Clipping Region   672
    24.5.2 Clipping Region Information   675
24.6 Paths   676
    24.6.1 Creating, Deleting, and Converting Paths   678
    24.6.2 Path-Rendering Operations   679
    24.6.3 Path Manipulations   681
    24.6.4 Obtaining Path Information   684
24.7 Filled Figures Demo Program   685

Chapter 25 — Displaying Bit-Mapped Images

Chapter Summary   687
25.0 Raster and Vector Graphics   687
25.1 The Bitmap   688
    25.1.1 Image Processing   689
    25.1.2 Bitblt Operations   690
25.2 Bitmap Constructs   691
    25.2.1 Windows Bitmap Formats   691
    25.2.2 Windows Bitmap Structures   691
    25.2.3 The Bitmap as a Resource   691
25.3 Bitmap Programming Fundamentals   693
    25.3.1 Creating the Memory DC   693
    25.3.2 Selecting the Bitmap   693
    25.3.3 Obtaining Bitmap Dimensions   694
    25.3.4 Blitting the Bitmap   695
    25.3.5 A Bitmap Display Function   697
25.4 Bitmap Manipulations   698
    22.4.1 Hard-Coding a Monochrome Bitmap   699
    25.4.2 Bitmaps in Heap Memory   701
    25.4.3 Operations on Blank Bitmaps   706
    25.4.4 Creating a DIB Section   708
    25.4.5 Creating a Pattern Brush   713
25.5 Bitmap Transformations   714
    25.5.1 Pattern Brush Transfer   714
    25.5.2 Bitmap Stretching and Compressing   715
25.6 Bitmap Demonstration Program   719

PART III — PROJECT ENGINEERING

Chapter 26 — Fundamentals of Systems Engineering

Chapter Summary   723
26.0 What Is Software Engineering   723
26.0.1 The Programmer as an Artist   725
26.1 Software Characteristics   725
    26.1.1 Software Qualities   726
        Correctness   726
        Reliability   726
        Robustness   727
        Efficiency   727
        Verifiability   728
        Maintainability   728
        User Friendliness   728
        Reusability   729
        Portability   729
        Other Properties   729
    26.1.2 Quality Metrics   730
26.2 Principles of Software Engineering   730
    26.2.1 Rigor   731
    26.2.2 Separation of Concerns   733
    26.2.3 Modularization   734
    26.2.4 Abstraction and Information Hiding   735
    26.2.5 Malleability and Anticipation of Change 736
    26.2.6 Maximum Generalization   736
    26.2.7 Incremental Development   737
26.3 Software Engineering Paradigms   738
    26.3.1 The Waterfall Model   738
    26.3.2 Prototyping   740
    26.3.3 The Spiral Model   742
    26.3.4 A Pragmatic Approach   743
26.4 Concurrent Documentation   744
    26.4.1 Objections and Excuses   745
    26.4.2 Advantages of Good Documentation   745

Chapter 27 — Description and Specification

Chapter Summary   747
27.0 System Analysis Phase   747
    27.0.1 The System Analyst   748
    27.0.2 Analysis and Project Context   749
27.1 The Feasibility Study   750
    27.1.1 Risk Analysis   751
        Risk Identification   753
        Risk Estimation   753
        Risk Assessment   753
        Risk Management   754
    27.1.2 Risk Analysis in a Smaller Project   754
    27.1.3 Cost-Benefit Analysis   755
27.2 Requirements Analysis and Specification   757

27.2.1 The Requirements Analysis Phase   757
        Customer/User Participation   758
        The Virtual Customer   758
27.2.3 The Specifications Phase   758
        The Software Specifications Document   760
27.3.4 Formal and Semiformal Specifications   761
27.3.5 Assertions Notation   762
        ASSERT   762
        INV   762
        PRE and POST   763
27.4 Tools for Process and Data Modeling   764
    27.4.1 Data Flow Diagrams   765
            Event Modeling   767
    27.4.2 Entity-Relationship Diagrams   769

Chapter 28 — The Object-Oriented Approach

Chapter Summary   773
28.0 History and Chronology   773
28.1 Object-Oriented Fundamentals   774
    28.1.1 Problem-Set and Solution-Set   775
    28.1.2 Rationale of Object Orientation   776
28.2 Classes and Objects   776
    28.2.1 Classes and Data Abstraction   777
    28.2.2 Classes and Encapsulation   778
    28.2.3 Message Passing   778
    28.2.4 Inheritance   779
    28.2.5 Polymorphism   780
            Abstract Classes   780
28.4 A Notation for Classes and Objects   781
28.5 Example Classification   783
28.6 When to Use Object Orientation   786
    28.6.1 Operational Guidelines   786

Chapter 29 — Object-Oriented Analysis

Chapter Summary   789
29.0 Elements of Object-Oriented Analysis   789
    29.0.1 Modeling the Problem-Domain   790
    29.0.2 Defining System Responsibilities   790
    29.0.3 Managing Complexity   791
            Abstraction   791
            Encapsulation   791
            Inheritance   791
            Message Passing   791
29.1 Class and Object Decomposition   792
    29.1.1 Searching for Objects   796
    29.1.2 Neat and Dirty Classes   796
29.2 Finding Classes and Objects   797
    29.2.1 Looking at Class Associations   797
            Gen-Spec Structures   798
            Multiple Inheritance in Gen-Spec Structures   800

	Whole-Part Structures	800
	Compound Structures	802
29.2.2	Looking at Mechanisms and Devices	802
29.2.3	Related Systems	803
29.2.4	Preserved Data	803
29.2.5	Roles Played	803
29.2.6	Operational Sites	804
29.2.7	Organizational Units	804
29.3	Testing Object Validity	804
29.3.1	Information to Remember	805
29.3.2	Object Behavior	805
29.3.3	Multiple Attributes	805
29.3.4	Multiple Objects	805
29.3.5	Always-Applicable Attributes	806
29.3.6	Always-Applicable Methods	806
29.3.7	Objects Relate to the Problem Domain	806
29.3.8	Derived or Calculated Results	807
29.4	Subsystems	807
29.4.1	Subsystems as Modules	808
	Subsystem Cohesion	809
	Subsystem Coupling	809
29.5	Attributes	809
29.5.1	Attribute Identification	810
29.5.2	Attributes and Structures	811
29.6	Methods or Services	812
29.6.1	Identifying Methods	812
	Object States	812
	Required Services	812
29.7	Instance Connections	814
29.7.1	Instance Connection Notation	814
29.8	Message Connections	814
29.8.1	Message Connection Notation	814
29.9	Final Documentation	815

PART IV — APPENDICES

Appendix A — C++ Math Unit Programming

Summary	819
AA.0 Programming the Math Unit	819
AA.1 MASM Sources in C++ Programs	820
AA.1.1 Sample Code	821

Appendix B — Accuracy of Exponential Functions

Summary	829
AB.0 Accuracy Calculations	829

Appendix C — C++ Indirection

Summary	833

AC.0 Indirection in C++   833
    AC.0.1 Pointer Phobia   833
AC.1 Indirect Addressing   836
AC.2 Pointer Variables   837
    AC.2.1 Dangling Pointers   838
    AC.2.2 Pointers to Variables   838
        Pointer Variable Declaration   838
        Pointer Variable Assignment   838
        Pointer Variable Dereferencing   839
AC.3 Pointers to Arrays   839
AC.4 Pointers to Structures   840
AC.5 Pointer Arithmetic   842
AC.6 Pointers to Void   843
    AC.6.1 Programming with Pointers to Void 845
AC.7 Reference Variables   847
AC.8 Dynamic Memory Allocation in C++   848
    AC.8.1 Dynamic Data   849
    AC.8.2 The *New* and *Delete* Operators 850
AC.9 Pointers to Functions   852
    AC.9.1 Simple Dispatch Table   854
    AC.9.2 Indexed Dispatch Table   856
AC.10 Compounding Indirection   858

Appendix D — Multiple File Programs

Summary   861
AD.0 Partitioning a Program   861
    AD.0.1 Class Libraries   862
    AD.0.2 Public and Private Components   862
    AD.0.3 Object-Oriented Class Libraries   863
AD.1 Multifile Support in C++   864
AD.2 Multilanguage Programming   864
    AD.2.1 Naming Conventions   865
    AD.2.2 Calling Conventions   866
    AD.2.3 Parameter-Passing Conventions   867
    AD.2.4 Difficulties and Complications   867
AD.3 Mixing Low- and High-Level Languages   868
    AD.3.1 C++ to Assembly Interface   869
AD.4 Sample Interface Programs   869
    AD.4.1 Microsoft C-to-Assembly Interface 871
        Return Values in Microsoft C++ Compilers   874
    AD.4.2 Borland C++ to Assembly Interface 875
    AD.4.3 Using Interface Headers   877

Appendix E — The MATH32 Library

Summary   879

Appendix F — Windows Structures

Summary   883

Bibliography 895

Index 903

Preface

Our book was conceived as a programming toolkit and a problem-solving resource for professional engineers and scientists who take on the role of programmers. The book's original idea was based on the fact that engineers and scientists often need to develop software to suit their particular needs, but hardly ever are they "super-programmers" at the start. On the other hand, the scientist/engineer who becomes an improvised programmer is often the originator of major software products: who knows better what the software must accomplish than the person who is an expert in the field of application and the intended user of the product? Who can better design and code the bridge-building program than the bridge builder?

At the same time, scientists and engineers are intellectuals, trained in the sciences, with good mathematical backgrounds, and who already know the many complexities involved in scientific and technological calculations. They constitute a group of readers who do not accept black boxes or unexplained methodologies. Furnishing software and development tools for them requires finding a balance that, on one hand, does not ignore their scientific, technical, and mathematical competence, and on the other one, provides practical shortcuts that avoid unnecessary complications. We have attempted to achieve this balance in our book.

The software requirements of applications intended for engineering and scientific uses are almost always computational. The typical scientific and engineering program has a strong number-crunching component. But the computational capabilities of conventional programming languages are limited and often not well documented. For example, the mathematical functions that are part of the C++ language barely include the common trigonometric and logarithmic functions. An engineering application that requires calculating a statistical function, or that performs basic differentiation or integration, cannot be easily developed in this or most other programming languages. Consequently, the scientist/engineer software developer is forced to search for or develop algorithms and methods to perform the required calculations, often ending up with untested, undocumented, and unreliable routines. Providing these numerical tools has been our primary objective in this book.

But it does not end with the tool, algorithm, or even the canned routine that solves a computational problem. The engineering or scientific application must execute in a modern computer and do so with professional appearance and functionality. In a very few cases is a minimal, text-based program sufficient for the purpose at hand. To be effective, the program must execute in a graphical operating system such as Windows; therefore the engineer/scientist programmer must also have access to these skills.

Furthermore, the scientist/engineer programmer must also be capable of analyzing and designing the application using state-of-the-art tools and technology. The "make it up as you go along" approach does not work for digging a mountain tunnel or for crafting a computer program. This means that project engineering skills will also be necessary, as they are for any other engineering or scientific enterprise.

Book Structure

The requirements just listed account for the three principal parts of the book. Part I, titled *Techniques and Code*, starts at the most elementary level: number systems, storage of numerical data, and basic machine arithmetic. A discussion of binary-coded decimal numbers and the development of high-precision arithmetic routines in floating-point follows. Then come chapters on the Intel math unit architecture, data conversions, and the details of math unit programming. With this framework established, the remainder of the first part is devoted to developing routines useful in engineering and scientific code. These routines include general mathematics, financial calculations, statistics, differentiation and integration, linear systems, equation parsing, and topics in artificial intelligence.

Part II is titled *Application Development*. Starting with Chapter 16 we cover the actual implementation of a C++ program on the PC. The first chapter of Part II is an introduction of the PC implementations of C++, flowcharting as a program development tool, and the use of Microsoft's Developer Studio in creating a text-based program, called a Console Application. The remaining chapters of Part II are a tutorial on Windows programming. The intention of this part is to provide essential skills that will allow the reader to create a professional quality program while avoiding topics and details that have little interest for the engineer/scientist programmer.

Part III, titled *Project Engineering*, covers topics quite familiar to the scientist/engineer programmer since a software product is created following engineering principles that are similar to those used in any engineering endeavor. The first chapter of Part III is an overview of the software engineering field and describes its most common qualities, principles, and paradigms. This is followed by a discussion of the description and specification of software projects. The remainder of this part is about the object-oriented approach to software development and the use of object-orientation as an analysis tool. Object-oriented programming has proven itself in the past 20 years and the scientist/engineer programmer should have some familiarity with the possible uses of this development model.

The final part of the book contains several appendices that either cover topics of general interest in the field of scientific programming or provide useful tools for the developer. Included in the topics covered is an overview of math unit programming, the accuracy of exponential functions, the use of pointers and indirection in C++, and the use of multiple files in applications. Also in Part IV is an appendix listing of all the primitives contained in the MATH32 library furnished in the book's software package and an appendix that lists the Windows structures mentioned in the text.

Languages and Environment

The principal programming language used in the book is C++, but some primitive functions are coded in Intel Assembly Language. This is necessary since the Intel math unit, which is the number-crunching engine on the PC, can only be programmed in low-level code. However, all machine-level routines discussed in the text are accessible from C++ by means of furnished interface functions. This ensures that readers not fluent in Assembly Language can still understand the material and put the software to work.

Microsoft's Visual C++ and MASM were used in developing the software. The text is compatible with Visual C++ versions 5, 6, and 7. The code adopts the flat, 32-bit memory model, which makes it compatible with both Visual C++ Console Applications and Windows programs.

Software On-line

The book includes a software package, furnished on-line, with all the sample programs developed in the text. Also included are the source and object files for the 32-bit numerical library named MATH32.LIB, and for an MS DOS numerical library named MATH16.LIB. The procedures in the MATH32 library are listed in Appendix E. The demonstration programs test and exercise the primitives and functions developed in the text and serve as an illustration on how to use these resources in the reader's own code. The publisher's web site is:

www.taylorandfrancis.com

Julio Sanchez

Maria P. Canton

Part I

Techniques and Code

Chapter 1

Computer Number Systems

Chapter Summary

One of the fundamental applications of a computer system is the processing and storing of numeric data, sometimes called "number crunching." In order to perform more efficient digital operations on numeric data, mathematicians have devised systems and structures that differ from those used traditionally. This chapter presents the background material necessary for understanding and using the number systems and numeric data storage structures employed in digital computers. The material includes low- and high-level integer conversion routines.

1.0 Counting

The origin and the fundamental application of a number system is counting. Imagine a primitive hunter using his or her fingers to show other members of the tribe how many mammoth were spotted during a scouting trip. By this simple scheme the hunter is able to transmit a unique type of information. The information the hunter wishes to convey does not relate to the species, size, or color of the animals, but to their numbers. This is possible because our minds are able to isolate the notion of "oneness" from other properties of objects.

The most primitive method of counting consists of using a common object to represent the degrees of oneness. The hunter uses one finger to represent each mammoth. Alternatively, the hunter could have resorted to pebbles, sticks, lines on the ground, or scratches on the cave wall to show how many units there were of the object, independently of any other characteristic or attribute.

1.0.1 The Tally System

The first number system probably started from notches on a stick or scratches on a cave wall. In the simplest form, sometimes called a *tally system*, each scratch, notch, or line represents an object. The method is so simple and intuitive that we continue resorting to it, even after mastering the intricacies of the calculi. Tallying is based on a one-to-one correspondence between objects and their representation. The hunter counts the mammoth hiding in the ravine by making notches on a stick, at the rate of one notch for each mammoth. In order to share this information with other members

3

of the tribe, the hunter can later transfer the notches on a stick to lines or scratches on the cave wall. Tallying requires no knowledge of quantity and no elaborate symbols. Had there been 12 mammoth in the ravine the cave wall would have appeared as follows:

$$| | | | | | | | | | | |$$

The next step in the evolution of the tally system probably consisted of grouping the marks to help visualize the number of objects. Since we have five fingers on each hand, the 12 mammoth may have been grouped as follows:

$$| | | | |\quad | | | | |\quad | |$$

A primitive mathematical genius added one final sophistication to the tally system. By drawing one tally line diagonally the visualization is further improved, as in this familiar style:

$$\cancel{||||}\qquad\cancel{||||}\qquad ||$$

1.0.2 Roman Numerals

By observing the early Roman numerals we can see how a simple graphical tally evolved into a symbolic numeric representation. The first five digits were encoded with the symbols

I, II, III, IIII, and V

The Roman symbol V is conceivably a simplification of the tally encoding using a diagonal line to complete the grouping. Table 1.1 shows the decimal value of the Roman symbols.

Table 1.1
Symbols in the Roman Numeration System

ROMAN	DECIMAL
I	1
V	5
X	10
L	50
C	100
D	500
M	1000

The Roman numeral system is based on an add-subtract rule whereby the elements of a number, read left-to-right, are interpreted as follows:

1. If the value of the numeral to the right is equal to or larger than the numeral to its left, its value is added to the previous total.

2. If the value of the numeral to the right is larger, its value is subtracted from the previous total. For example:

```
III = I + I + I = 3
 IV = V - I     = 4
 XL = L - X     = 40
```

The decimal 1994 is represented in Roman numerals as follows:

```
MCMXCIV = M + (C - M) + (X - C) + (I - V)
        = 1000 + (1000 - 100) + (100 - 10) + (5 - 1)
        = 1000 + 900 + 90 + 4
        = 1994
```

The uncertainty in the positional value of each digit, the absence of a symbol for zero, and the fact that some numbers require either one or two symbols (I, IV, V, IX, and X) complicates the rules of arithmetic using Roman numerals.

1.1 The Origins of Our Number System

The one element of our civilization which has transcended all cultural and social differences, is a system of numbers. While mankind is yet to agree on the most desirable political order, on generally acceptable rules of moral behavior, or on a universal language, the Hindu-Arabic numerals have been adopted by practically all the nations and cultures of the world. There must be something extraordinary about this number system for it to achieve such general acceptance.

By the year 800 A.D. the Arabs were using a ten-symbol positional system of numbers which included the special symbol for 0. This system (later called the Hindu-Arabic numerals) was introduced into Europe during the 8th century, probably through Spain. Pope Sylvester II, who had studied the Hindu-Arabic numbers in Spain, was the first European scholar known to adopt and teach them. The Latin title of the first book on the subject of "Indian numbers" is *Liber Algorismi de Numero Indorum*. The author is the Arab mathematician al-Khuwarizmi.

In spite of the evident advantages of this number system its adoption in Europe took place only after considerable debate and controversy. Many scholars of the time still considered that Roman numerals were easier to learn and more convenient for operations on the abacus. The supporters of the Roman numeral system, called *abacists*, engaged in intellectual combat with the *algorist*, who were in favor of the Hindu-Arabic numerals as described by al-Khuwarizmi. For several centuries abacists and algorists debated about the advantages of their systems, with the Catholic church often siding with the abacists. This controversy explains why the Hindu-Arabic numerals were not accepted into general use in Europe until the beginning of the 16th century.

It is often said that the reason for there being ten symbols in the Hindu-Arabic numerals is related to the fact that we have ten fingers. However, if we make a one-to-one correlation between the Hindu-Arabic numerals and our fingers, we find that the last finger must be represented by a combination of two symbols (10). Also, one Hindu-Arabic symbol (0) cannot be matched to an individual finger. In fact, the decimal system of numbers, as used in a positional notation that includes a zero digit, is a refined and abstract scheme which should be considered one of the greatest achievements of human intelligence. We will never know for certain if the

Hindu-Arabic numerals are related to the fact that we have ten fingers, but its profoundness and usefulness clearly transcends this biological fact.

The most significant feature of the Hindu-Arabic numerals is the presence of a special symbol (0) which by itself represents no quantity. Nevertheless, the special symbol 0 is combined with the other ones. In this manner the nine other symbols are reused to represent larger quantities. Another characteristic of decimal numbers is that the value of each digit depends on its position in a digit string. This positional characteristic, in conjunction with the use of the special symbol 0 as a placeholder, allow the following representations:

$$
\begin{array}{rcl}
1 &=& \text{one} \\
10 &=& \text{ten} \\
100 &=& \text{hundred} \\
1000 &=& \text{thousand}
\end{array}
$$

The result is a counting scheme where the value of each symbol is determined by its column position. This positional feature requires the use of an almost-magical symbol (0), which does not correspond to any unit-amount, but is used as a place-holder in multi column representations. We must marvel at the intelligence, abstraction capability, and even the sense of humor of the mind that conceived a counting system that has a symbol that represents nothing. We must also wonder about the evolution of mathematics, science, and technology had this system not been invented. One intriguing question is whether a positional counting system that includes the zero symbol is a natural and predictable step in the evolution of our mathematical thought, or whether its invention was a stroke of genius that could have been missed for the next two thousand years.

1.1.1 Number Systems for Digital-Electronics

The computers built in the United States during the early 1940s operated on decimal numbers. However, in 1946, von Neumann, Burks, and Goldstine published a trend-setting paper titled *Preliminary Discussion of the Logical Design of an Electronic Computing Instrument*, in which they state:

> *"In a discussion of the arithmetic organs of a computing machine one is naturally led to a consideration of the number system to be adopted. In spite of the long-standing tradition of building digital machines in the decimal system, we must feel strongly in favor of the binary system for our device."*

In their paper, von Neumann, Burks, and Goldstein also consider the possibility of a computing device that uses binary-coded decimal numbers. However, the idea is discarded in favor of a pure binary encoding. The argument is that binary numbers are more compact than binary-coded decimals. Later in this book you will see that binary-coded decimal numbers (called BCD) are used today in some types of computer calculations.

The use of the binary number system in digital calculators and computers was made possible by previous research on number systems and on numerical representations, starting with an article by G.W. Leibniz published in Paris in 1703. Researchers concluded that it is possible to count and perform arithmetic operations using any set of symbols as long as the set contains at least two symbols, one of which must be zero.

In digital electronics the binary symbol 1 is equated with the electronic state ON, and the binary symbol 0 with the state OFF. The two symbols of the binary system can also represent conducting and non-conducting states, positive or negative, or any other bi-valued condition. It was the binary system that presented the Hindu-Arabic decimal number system with the first challenge in 800 years. In digital-electronics two steady states are easier to implement and more reliable than a ten-digit encoding.

1.1.2 Positional Characteristics

All modern number systems, including the decimal, the hexadecimal, and the binary systems, are positional. It is the positional feature that is used to determine the total value of a multi-digit representation. For example, the digits in the decimal number 4359 have the following positional weights:

```
4 3 5 9
| | | |_____  units
| | |_____  ten units
| |_____  hundred units
|_____  thousand units
```

The total value is obtained by adding the column weights of each unit:

```
    4000 ——— 4 thousand units
     300 ——— 3 hundred units
+     50 ——— 5 ten units
       9 ——— 9 unit
    ----
    4359
```

or also

$$4 * 10^3 + 3 * 10^2 + 5 * 10^1 + 9 * 10^0 = 4359$$

Recall that $10^1 = 10$ and $10^0 = 1$.

1.1.3 Radix or Base

In any positional number system the weight of each column is determined by the total number of symbols in the set, including zero. This is called the *base* or *radix* of the system. The base of the decimal system is 10, the base of the binary system is 2, and the base of the hexadecimal system is 16. The positional value or weight (P) of a digit in a multi digit number is determined by the formula

$$P = d \times B^c$$

where d is the digit, B is the base or radix, and c is the zero-based column number, starting from right to left. Notice that the increase in column weight from right to left is purely conventional. You could easily conceive a number system in which the column weights increase in the opposite direction. In fact, in the original Hindu notation the most significant digit was placed at the right.

In radix-positional terms a decimal number can be expressed as a sum of digits expressed by the formula

$$\sum_{i=0}^{n} d_i \times 10^i \quad \text{for } 0 \leq d_i \leq 9 \ (d_i \text{ an integer})$$

The summation formula for a binary radix, positional representation is as follows

$$\sum_{i=0}^{n} b_i \times 2^i \quad \text{for } b_i = 0 \text{ or } 1$$

where i is the system's range and n is its limit.

1.2 Types of Numbers

By the adoption of special representations for different types of numbers the usefulness of a positional number system can be extended beyond the simple counting function.

1.2.1 Whole Numbers

The digits of a number system, called the positive integers or natural numbers, are an ordered set of symbols. The notion of an ordered set means that the numerical symbols are assigned a predetermined sequence. A positional system of numbers also requires a special digit, named zero. The special symbol 0, by itself, represents the absence of oneness, or nothing, and thus is not included in the set of natural numbers. However, 0 assumes a cardinal function when it is combined with other digits, for instance, 10 or 30. The whole numbers are the set of natural numbers, including the number zero.

1.2.2 Signed Numbers

A number system also represents direction. We generally use the + and – signs to represent opposite numerical directions. The typical illustration for a set of signed numbers is as follows:

```
-9  -8  -7  -6  -5  -4  -3  -2  -1   0  +1  +2  +3  +4  +5  +6  +7  +8  +9
negative numbers   <-           zero          -> positive numbers
```

The number zero, which separates the positive and the negative numbers, has no sign of its own. Although in some binary encodings, which are discussed later in this book, we end up with a negative and a positive zero.

1.2.3 Rational and Irrational Numbers

A number system also represents parts of a whole. For example, when a carpenter cuts one board into two boards of equal length we can represent the result with the fraction ½; the fraction ½ represents 1 of the two parts which make up the object.

Rational numbers are those expressed as a ratio of two integers, for example, 1/2, 2/3, 5/248. Note that this use of the word *rational* is related to the mathematical concept of a ratio, not to reason.

The denominator of a rational number expresses the number of potential parts. In this sense 2/5 indicates two of five possible parts. There is no reason why the number 1 cannot be used to indicate the number of potential parts, for example 2/1, 128/1. In this case the ratio $x/1$ indicates x elements of an undivided part. Therefore, it follows that $x/1 = x$. The implication is that the set of rational numbers includes the integers, since an integer can be expressed as a ratio by using a unit denominator.

But not all non-integer numbers can be written as an exact ratio of two integers. The discovery of the first *irrational number* is usually associated with the investigation of a right triangle by the Greek mathematician Pythagoras (approximately 600 BC). Here again, irrational refers to not-a-ratio, not to "unreasonable." The Pythagorean theorem states that in any right triangle the square of the longest side (hypotenuse) is equal to the sum of the squares of the other two sides.

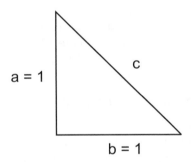

For this triangle, the Pythagorean theorem states that

$$a^2 + b^2 = c^2$$
$$2 = c^2$$
$$2 = c * c$$
$$c = \sqrt{2}$$

Therefore, the length of the hypotenuse in a right triangle with unit sides is a number, that when multiplied by itself, gives 2. This number (approximately 1.414213562) cannot be expressed as the exact ratio of two integers. Other irrational numbers are $\sqrt{3}$, $\sqrt{5}$, as well as the mathematical constants π and e.

1.2.4 Real and Complex Numbers

The set of numbers that includes the natural numbers, the whole numbers, and the rational and irrational numbers is called the *real numbers*. Most common mathematical problems are solved using real numbers. However, during the investigation of squares and roots we notice the following interesting peculiarity

```
+2  *  +2  =  4
-2  *  -2  =  4
```

Since the square of a positive number is positive and the square of a negative number is also positive, there can be no real number whose square is negative. Therefore, $\sqrt{-2}$ and $\sqrt{-4}$ do not exist in the real number system. But mathematicians of the 18th century extended the number system to include operations with roots of negative numbers. They defined the *imaginary unit* as

```
        i² = -1
or,  i = √-1
```

This gives rise to a new set of numbers, called *complex numbers*, that consist of a real part and an imaginary part. One of the uses of complex numbers is in finding the solution of a quadratic equation. Complex numbers are also useful in vector analysis, graphics, and in solving many engineering, scientific, and mathematical problems.

1.3 Radix Representations

The *radix* of a number system is the number of symbols in the set, including zero. In this manner, the radix of the decimal system is 10, and the radix of the binary system is 2. For certain applications, the decimal system is not ideal. Computing machines are based on electronic circuits that can be in one of two stable states. Therefore, a number system based on two symbols is better suited for computer work, since each state can be represented by a digit.

1.3.1 Decimal Versus Binary Numbers

The binary system of numbers uses only two symbols, 1 and 0. It is the simplest possible set of symbols with which we can count and perform arithmetic. Most of the difficulties in learning and using the binary system is a result of this simplicity. Figure 1.1 shows sixteen groups of four electronic cells each in all possible combinations of two states.

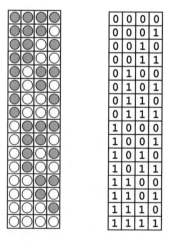

Figure 1.1 *Electronic Cells and Binary Numbers*

Note that the representation in binary numbers matches the physical state of each electronic cell. If we think of each cell as a miniature light bulb, then the binary number 1 can be used to represent the state of a charged cell (light ON) and the binary number 0 to represent the state of an uncharged cell (light OFF).

1.3.2 Octal and Hexadecimal Numbers

Binary numbers are convenient in digital-electronics; however, one of their drawbacks is the number of symbols required to encode a large value. For example, the number 9134 is represented in four decimal digits. However, the binary equivalent 10001110101110 requires 14 digits. By the same token, large binary numbers are difficult to remember.

One possible way of compensating for these limitations of binary numbers is to use individual symbols to represent groups of binary digits. For example, a group of three binary numbers allow eight possible combinations. In this case we can use the decimal digits 0 to 7 to represent each possible combination of three binary digits. This grouping of three binary digits gives rise to the octal number system, shown in the following table:

```
       binary          octal
       0  0  0            0
       0  0  1            1
       0  1  0            2
       0  1  1            3
       1  0  0            4
       1  0  1            5
       1  1  0            6
       1  1  1            7
```

The octal encoding serves as a shorthand representation for groups of 3-digit binary numbers. Imagine a digital-electronic device built with two-state elements, grouped in multiples of 3. In this case octal numbers provide a convenient way for representing the binary data stored in this device. For instance, if the device's memory consisted of groups of twelve cells, four octal digits could be used to represent any stored value.

Hexadecimal numbers (base 16) are used for representing values encoded in four binary digits. Since there are only ten decimal digits, the hexadecimal system borrows the first six letters of the alphabet (A, B, C, D, E, and F). The result is a set of sixteen symbols, as follows:

```
       0  1  2  3  4  5  6  7  8  9  A  B  C  D  E  F
```

Most modern computers are designed with memory cells, registers, and data paths in multiples of four binary digits. Table 1.2 lists some common units of memory storage.

Table 1.2

Units of Memory Storage

UNIT	BITS	HEX DIGITS	HEX RANGE
Nibble	4	1	0 to F
Byte	8	2	0 to FF
Word	16	4	0 to FFFF
Doubleword	32	8	0 to FFFFFFFF
Quadword	64	16	0 to FFFFFFFFFFFFFFFF

In most digital-electronic devices memory addressing is organized in multiples of four binary digits. Here again, the hexadecimal number system provides a convenient way to represent addresses. Table 1.3 lists some common memory addressing units and their hexadecimal and decimal range.

Table 1.3

Units of Memory Addressing

UNIT	DATA PATH IN BITS	ADDRESS RANGE DECIMAL	ADDRESS RANGE HEXADECIMAL
1 paragraph	4	0 to 15	0-F
1 page	8	0 to 255	0-FF
1 kilobyte	16	0 to 65,535	0-FFFF
1 megabyte	20	0 to 1,048,575	0-FFFFF
4 gigabytes	32	0 to 4,294,967,295	0-FFFFFFFF

1.4 General Theory of Counting

Mathematics originates in our intuitive sense of oneness. If you could not distinguish between one object and two objects, as a characteristic unrelated to the properties of the objects themselves, you could not count. Without counting there could not be arithmetic, algebra, or calculus. Counting is an abstract process, which implies the following assumptions:

1 The counter uses a set of symbols, which are successively assigned to the units being counted.

2. There must be a first symbol in this set. The counting process always begins with this symbol.

3. The remaining symbols in the set must be ordered; 2 follows 1, 3 follows 2, and so forth.

4. The set of symbols must be finite: there is a last symbol in the set.

Usually counting proceeds beyond the last symbol of the set. If not, we could only count up to nine units in the decimal system. Counting beyond the last symbol in the set requires additional rules. One counting scheme, known as the *positional system*, aligns the digits in columns that are assigned different weights. In this scheme a special digit named *zero* is used as a place-holder. Counting in a zero-based positional implies the following additional rule:

5. There is a special symbol, called zero, which by itself represents no units.

Counting starts with the first symbol of the set and proceeds along the set of symbols until the last symbol of the set. In the decimal system we count

```
1
2
3
4
5
6
7
8
9
```

We can continue counting, past the last symbol of the set, using the properties of a positional, zero-based system. If column weight increases from right to left (as is the case in our common systems) we count past the last symbol in the set by setting the right-most column to zero, then bumping the column to the left to the next symbol of the set. According to this scheme, the number following 9 is 10.

Because we are so familiar with decimal counting, these rules may appear trivial. However, the rules apply to any zero-based positional system, such as binary, octal, or hexadecimal. Counting in binary has the peculiarity that the first symbol of the set (1) is also the last symbol of the set. The rules of counting can be applied to any number system, for example:

The first unit is represented with the first symbol of the set

binary	decimal	octal	hexadecimal
1	1	1	1

We proceed along the set of symbols until we reach the last symbol of the set

binary	decimal	octal	hexadecimal
1	9	7	F

At this point we turn the right-most column to zero and bump the column to the left to the next or first symbol of the set

binary	decimal	octal	hexadecimal
10	10	10	10

Note that in different numerical bases the same symbols represent different amounts. For this reason it is meaningless to say ten binary, or ten octal, or ten hexadecimal. Ten is the name given to the digit combination 10 in the decimal system. Since digit combinations in systems with other bases are unnamed, we are forced to use one-zero binary, or one-zero octal, or one-zero hexadecimal. The same applies to the decimal designations of hundred, thousand, and their combinations.

By the same token, it is incorrect to say one thousand hexadecimal, since the word thousand implies the decimal number system.

In summary: counting in a zero-based positional number system proceeds at the right-most column by indexing through the set of symbols in the corresponding set. When the right-most column is at the last symbol of the set, it is changed to 0 and the next column to the left is bumped to the next symbol of the set. This reset-and-bump sequence, which takes place when a column reaches the last symbol of the set, continues until we reach a column that is not at the last symbol of the set. If we represent the bump with a # sign, then the process of counting in decimal numbers can be described as follows

```
  1 .... first symbol of the set
  2 .... next symbol of the set
  3 .... proceed along the set of symbols
  .
  .
  .
  9 .... until the last symbol of the set
  #
 10 .... set right-most column to zero and bump
```

In the binary system the counting proceeds as follows

```
  1 .... the first symbol of the set is also the last
  #
 10 .... set right-most column to zero and bump
 11 .... proceed along the set of symbols until the last
  #
  0 .... reset right-most column and bump
  #
 00 .... reset next column and bump
100 .... column not at the last symbol of the set
101 .... proceed with the right-most column
  #
  0 .... reset right-most and bump
110 .... column not at the last symbol of the set
```

Notice that when all columns are holding the last symbol of the set, counting can only proceed by opening a new column to the left. Also notice that the symbol combination that follows this condition is the same in all bases, for example, after

```
binary     decimal     octal     hexadecimal
  111         999        777          FFF
```

the next number is

```
binary     decimal     octal     hexadecimal
 1000        1000       1000         1000
```

1.5 Assembly Language Conversion Routines

We use decimal numbers in our everyday life because they meaningfully represent common units used in the real world. To state that a certain historical event took place in the year 7C6 hexadecimal would convey little information to the average person. However, in computer systems based on two-state electronic cells binary representations are more convenient. Also note that hexadecimal and octal numbers are handy shorthand for representing groups of binary digits.

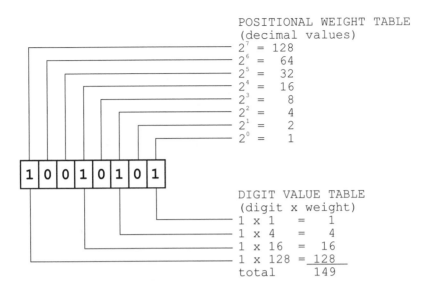

Figure 1.2 *Example of Binary to ASCII Decimal Conversion*

Numerical conversions between positional systems of different radices are based on the number of symbols in the respective sets and on the positional value (weight) of each column. But methods used for manual conversions are not always suitable for machine conversions, as we will see in the forthcoming sections.

1.5.1 Binary-to-ASCII-Decimal Conversion

To manually convert a binary number to its decimal equivalent we take into account the positional weight of each binary digit, as shown in Figure 1.2.

The positional weight table in Figure 1.2 lists the decimal value of each binary column. These weights are powers of the system's base (2 in the binary system). In the digit value table, also in Figure 1.2, the decimal values of the binary columns holding a 1 digit are added. The sum of the weights of all the one-digits in the operand is the decimal equivalent of the binary number. In this case 10010101 binary = 149 decimal.

The method shown in Figure 1.2, although useful in manual conversions, is not a good algorithm for computer conversions. Modern digital computers contain machine instructions to perform integer division. Using integer division by 10 provides a simpler and more effective conversion. The remainder of each division is the decimal digit of the result, while the quotient becomes the new value to be divided. A quotient of 0 indicates that the conversion has concluded. Figure 1.3 is a flowchart of a low-level binary-to-decimal conversion. The algorithm for the processing in Figure 1.3 can be written as follows:

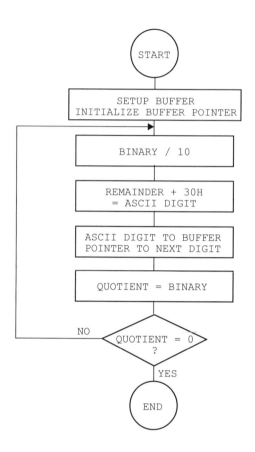

Figure 1.3 *Flowchart for a Binary to ASCII Decimal Conversion*

1. Set up and initialize a buffer to hold the ASCII decimal digits of the result. Set up the buffer pointer to the right-most digit position of the result.

2. Obtain the remainder of the value divided by 10.

3. Add 30H to remainder digit to convert to ASCII representation.

4. Store remainder digit in buffer and index the buffer pointer to the preceding digit.

5. Quotient of division by 10 becomes the new binary value.

6. End conversion routine if quotient is equal to 0. Otherwise, continue at step 2.

PROGRAMMER'S NOTEBOOK

The ASCII character set was designed so that the decimal digits are in the range 30H to 39H. Converting binary-to-ASCII and ASCII-to-binary is a simple matter of adding or subtracting 30H.

In coding an 80x86 assembly language conversion routine for binary-to-ASCII-decimal we start by creating a buffer to hold the result. The buffer size depends on the maximum number of ASCII decimal digits that are expected. For example, if the binary can have up to 8 bits, then the ASCII decimal buffer must be capable of storing three characters, since the highest value would be 255 decimal. By the same token, if the binary can have up to 16 bits then the ASCII decimal buffer must be capable of storing five characters, since in this case the maximum decimal value would be 65,535. The code first clears the ASCII buffer by entering the necessary blank characters. This step is necessary because the conversion ends as soon as there is a zero quotient. Therefore, if the buffer is not previously cleared, the result could be polluted by garbage from a previous conversion.

```
;*****************************************
;   definitions for 32-bit flat model
;*****************************************
                .486
                .MODEL flat
                .CODE

_BIN_TO_ASC10    PROC
; Procedure to convert a 16-bit binary number into 5 ASCII decimal
; digits
; On entry:
;          AX = binary number
;          EDI --> 5 byte ASCII buffer
; On exit:
;          Buffer holds ASCII decimal digits
;**********************|
;   clear ASCII buffer |
;**********************|
        MOV     ECX,5               ; Five digits to clear
CLEAR_5:
        MOV     BYTE PTR [EDI],20H       ; Clear digit
        INC     EDI                 ; Bump pointer
        LOOP    CLEAR_5             ; Repeat for 5 digits
        DEC     EDI                 ; Adjust buffer pointer to digit
        MOV     ECX,10              ; Decimal divisor to CX
;**********************|
;  obtain ASCII decimal |
;        digit          |
;**********************|
;
GET_ASC10:
        MOV     DX,0                ; Clear for word division
        DIV     CX                  ; Perform division AX/CX
; Quotient is in AX and remainder in DL
; Convert decimal to ASCII decimal
        ADD     DL,30H              ; Add 30H to bring to ASCII range
;**********************|
;     store digit       |
;  bump buffer pointer  |
;**********************|
        MOV     BYTE PTR [EDI],DL        ; Store digit in buffer
        DEC     EDI                 ; Buffer pointer to next digit
; Note: the binary quotient is left in AX by the DIV CX instruction
;**********************|
;   is quotient = 0 ?   |
;**********************|
```

```
              CMP     AX,0              ; Test for end of binary
              JNZ     GET_ASC10        ; Continue if not 0
    ;***********************|
    ;   end of conversion   |
    ;***********************|
              CLD
              RET

_BIN_TO_ASC10        ENDP
```

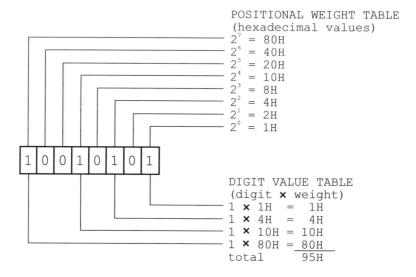

POSITIONAL WEIGHT TABLE
(hexadecimal values)
2^7 = 80H
2^6 = 40H
2^5 = 20H
2^4 = 10H
2^3 = 8H
2^2 = 4H
2^1 = 2H
2^0 = 1H

DIGIT VALUE TABLE
(digit ✗ weight)
1 ✗ 1H =   1H
1 ✗ 4H =   4H
1 ✗ 10H = 10H
1 ✗ 80H = 80H
total   95H

Figure 1.4 *Example of Binary to ASCII Hexadecimal Conversion*

1.5.2 Binary-to-Hexadecimal Conversion

The method described in Section 1.5.1 for a binary to ASCII decimal conversion can be adapted to other radices by representing the positional weight of each binary digit in the number system of choice. In the case of a binary to ASCII hexadecimal conversion the positional weight of each binary digit is a hexadecimal value. Figure 1.4 shows the conversion of the binary value 10010101 into hexadecimal by using the corresponding positional weights.

   The 80x86 assembly language procedure to convert from binary to ASCII hexadecimal is similar to the binary-to-ASCII-decimal routine (BIN_TO_ASC10) listed in Section 1.5.1. In the case of the conversion into ASCII hexadecimal digits the buffer need only hold four ASCII characters, since a 16-bit binary cannot exceed the value FFFFH. In either case, the divisor for obtaining the digits is the base of the number system. Therefore, the binary-to-hex routine uses the value 16. The following procedure converts a binary number to ASCII hexadecimal.

```
_BIN_TO_ASC16      PROC
; Procedure to convert a 16-bit binary number into 4 ASCII
; hexadecimal digits
```

```
;
; On entry:
;          AX = binary number
;          EDI --> 4 byte ASCII buffer
; On exit:
;          Buffer holds ASCII decimal digits
;*********************|
;   clear ASCII buffer  |
;*********************|
        MOV      ECX,4                ; Four digits to clear
CLEAR_4:
        MOV      BYTE PTR [EDI],20H      ; Clear digit
        INC      EDI                  ; Bump pointer
        LOOP     CLEAR_4              ; Repeat for 4 digits
        DEC      EDI                  ; Adjust buffer pointer to last
                                      ; digit
        MOV      ECX,16               ; Hexadecimal divisor to CX
;*********************|
;   obtain ASCII hex   |
;        digit         |
;*********************|
GET_ASC16:
        MOV      DX,0                 ; Clear for word division
        DIV      CX                   ; Perform division AX/CX
; Quotient is in AX and remainder in DL
; Convert decimal to ASCII
; Test for digit range 0 to 9
        CMP      DL,9
        JA       IS_LETTER            ; Digit is Hex letter
        ADD      DL,30H               ; Add 30H to bring to ASCII range
        JMP      STORE_DIGIT
IS_LETTER:
        ADD      DL,55                ; Convert to ASCII letter
;*********************|
;     store digit      |
;  bump buffer pointer |
;*********************|
STORE_DIGIT:
        MOV      BYTE PTR [EDI],DL       ; Store digit in buffer
        DEC      EDI                  ; Buffer pointer to next digit
; Note: the binary quotient is left in AX by the DIV CX instruction
;*********************|
;   is quotient = 0 ?  |
;*********************|
        CMP      AX,0                 ; Test for end of binary
        JNZ      GET_ASC16            ; Continue if not 0
;*********************|
;   end of conversion  |
;*********************|
        CLD
        RET
_BIN_TO_ASC16      ENDP
```

1.5.3 Decimal-to-Binary Conversion

Longhand conversion of decimal into binary can be performed by using the positional weights to find the binary 1-digits and then subtracting this positional weight from the decimal value. The process is shown in Figure 1.5.

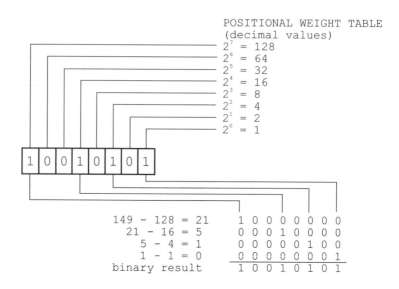

Figure 1.5 *Example of Decimal to Binary Conversion*

In the example of Figure 1.5 we start with the decimal value 149. Since the highest power of 2 smaller than 149 is 128, which corresponds to bit 7, we set bit 7 and perform the subtraction:

```
149 - 128 = 21
```

At this point the highest positional weight smaller than 21 is 16, which corresponds to bit 4. Therefore we set bit 4, and perform the subtraction:

```
21 - 16 = 5
```

The remaining steps in the conversion can be seen in Figure 1.5. The conversion is finished when the result of the subtraction is 0.

Suppose that the user enters a numerical value in the form of a string of ASCII decimal, octal, or hexadecimal digits. In order for the processor to perform simple arithmetic operations on these integers, load the user values into machine registers or dedicated memory locations. However, methods suited for manual conversion do not always make a good computer algorithm. Figure 1.6 shows two decimal-to-binary conversion algorithms that are suited for machine coding.

Using the first method of Figure 1.6, the individual decimal digits are multiplied by their corresponding positional values. The final result is obtained by adding all the partial products. This method is used frequently, however, it has the disadvantage that a different multiplier is used during each iteration (1, 10, 100, 1000, et cetera). The second method in Figure 1.6 starts with the high-order ASCII-decimal digit. The calculations consist of multiplying an accumulated value by 10. Initially, this accumulated value is set to 0. After multiplication by 10, the value of the digit is added to the accumulated value. The following algorithm is based on the second method in Figure 1.6, shown on the following page.

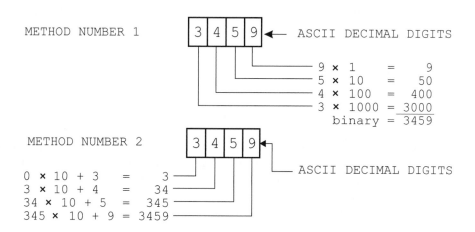

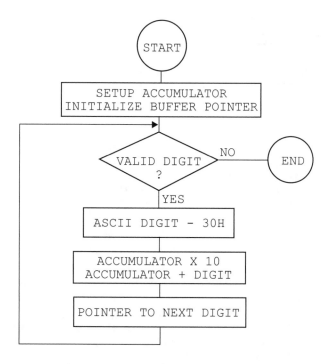

Figure 1.6 *Machine Conversion of ASCII Decimal to Binary.*

Figure 1.7 is a flowchart of the conversion algorithm.

Figure 1.7 *Flowchart for ASCII to Machine Register Conversion*

1. Set up and initialize to zero a storage location for holding the value accumulated during conversion. Set up a pointer to the highest order ASCII digit.

2. Test the ASCII digit for a value in the range 0 to 9. End of routine if the ASCII digit is not in this range.

3 Subtract 30H from ASCII decimal digit.

4. Multiply accumulated value by 10.

5. Add digit to accumulated value.

6. Increment the pointer to the next digit and continue at step 2.

   The listed procedure, named ASCII_TO_EDX, performs the necessary processing for loading a 5-digit ASCII number into the EDX register. The conversion routine is based on the flowchart shown in Figure 1.7.

```
_ASCII_TO_EDX        PROC
; Convert an ASCII number to binary and store in EDX register
;
; On entry:
;         ESI > start of 10-digit ASCII buffer holding decimal
;               string in the range 0 to 4,294,967,295
; On exit:
;         EDX = binary number
;
;
;***********************|
;  initialize registers |
;***********************|
         MOV     EDX,0              ; Clear accumulator register
         MOV     ECX,10             ; Load multiplier into ECX
;***********************|
; test digit for range  |
;        0 to 9         |
;***********************|
DIGIT_TO_ACC32:
         MOV     AL,[ESI]           ; Get ASCII digit
         CMP     AL,'0'             ; Test for lower limit
         JB      EXIT_ASC32         ; Exit if less than 0
         CMP     AL,'9'             ; Test for higher limit
         JA      EXIT_ASC32         ; Exit if larger than 9
;***********************|
; ASCII digit to binary |
;***********************|
         SUB     AL,30H             ; ASCII to decimal
;***********************|
;    accumulator x 10   |
;       + digit         |
;***********************|
         MOV     AH,0               ; Clear high byte in AX
         PUSH    AX                 ; Save digit in stack
         MOV     EAX,EDX            ; Previous product to EAX
         MUL     ECX                ; Perform EAX = EAX * 10
         MOV     EDX,EAX            ; Move product to accumulator
         MOV     EAX,0              ; Clear 32 bits
         POP     AX                 ; Restore decimal digit
         ADD     EDX,EAX            ; Add digit to accumulator
;***********************|
; pointer to next digit |
;***********************|
         INC     ESI                ; Bump pointer
```

```
        JMP     DIGIT_TO_ACC32  ; Continue
EXIT_ASC32:
        RET
_ASCII_TO_EDX    ENDP
```

1.6 C++ Conversion Routines

Conversion routines in high-level languages are more compact but less efficient than in assembler. C and C++ programmers often rely on library routines to perform these conversions. However, at times we must manually perform conversions. In the following sections we develop C++ integer conversion routines equivalent to the low-level ones previously listed. Two types of routines are considered: those that convert a binary value stored in a primitive variable into a string of ASCII digits, and those that convert a string of ASCII digits into a binary value.

1.6.1 C++ Binary-to-ASCII Conversions

C++ library functions often perform conversions. For example, when you use one of the stream classes to display a data primitive, the corresponding function performs the conversion automatically. For example:

```
int aNumber = 12345;
. . .
cout << "\nThe value is: " << aNumber;
```

In this case the cout stream converts the binary value stored in the variable aNumber into a displayable string of ASCII digits, and sends this string to the standard output device. But there are times in which we need to perform the conversion ourselves. The logic for a high-level routine to convert a binary value to a string of ASCII digits is similar to the one developed in Section 1.5.1. In C++ programming the processing logic can be described as follows:

1. Declare and initialize a buffer of type char to hold the ASCII decimal digits of the result.

2. Obtain the remainder of the binary value divided by 10.

3. Add 0x30 to remainder digit to convert to ASCII.

4. Store remainder digit in buffer at the rightmost position.

5. Index the buffer pointer to the preceding digit.

6. Obtain the new binary value by dividing the previous binary by 10.

7. Continue processing at step 2 while the binary value is greater than zero.

The following function, named Bin2AscDec(), converts the value stored in a C++ primitive of type unsigned long, into a string of ASCII digits. The digits are placed in a buffer passed by the caller. The conversion routine is as follows:

```
void Bin2AscDec( char digits[], unsigned long value)
// Function to convert a binary value into a string
// of ASCII decimal digits
// Pre:
//    1. digits[] is a char array for storing the ASCII
//       decimal digits
//    2. value is the binary number stored in a variable
//       of type unsigned long
```

```
//     3. Array is of sufficient size for storing all
//        ASCII digits. If not, the returned result is
//        invalid.
// Post:
//     Returns a string of ASCII decimal digits in the
//     argument array
{
   int size = 0;
   // Calculate array size
   while(digits[size])
      size++;
   while(value > 0 && size > 0)
   {
    digits[size-1] = (value % 10) + 0x30;// Obtain and store
                                        // low-order digit
    // Note:
    //      size is the number of elements in the digits
    //      array. The storage offset is size - 1

    value = value / 10;          // New operand
    size --;                     // Update digit position
   }
    return;
}
```

With slight modifications we can change the function Bin2AscDec() into one that produces a string of ASCII hexadecimal digits. In order to adapt the function to hexadecimal we must use the base of the hex number system (16) to calculate the remainder and the new binary value. In addition, the binary-to-ASCII-hex routine must examine the value of the resulting digit: if the value is smaller than 10, then the ASCII hex digit is obtained by adding 0x30, as in the case of the ASCII-decimal conversion. If the digit value is 10 or larger, then we must add a different constant in order to obtain one of the alpha hex digits, in the range A through F. To obtain an upper-case hex digit, the constant to be added is 0x37. The constant is 0x57 for lower-case hex digits. The following function, named Bin2AscHex() converts a binary value in a primitive of type unsigned long, into a string of ASCII hex digits.

```
void Bin2AscHex( char digits[], unsigned long value)
// Function to convert a binary value into a string
// of ASCII hexadecimal digits
// Pre:
//     1. digits[] is a char array for storing the ASCII
//        hex digits
//     2. value is the binary number stored in a variable
//        of type unsigned long
//     3. Array is of sufficient size for storing all
//        ASCII hex digits. If not, the returned result is
//        invalid.
//
// Post:
//     Returns a string of ASCII hexadecimal digits in the
//     argument array
{
   int size = 0;                   // Elements in array
   unsigned int hexDigit;
   unsigned int factor = 0x37;    // Change to 0x57 to obtain
                                  // lower-case digits in the
```

```
                                    // range a to f
  // Calculate array size
  while(digits[size])
     size++;
  while(value > 0 && size > 0)
  {
   hexDigit = value % 16;
   // Convert and store digits in the range 0 to 9
   if(hexDigit < 10)
     digits[size-1] = hexDigit + 0x30;    // Obtain and store
                                          // low-order digit
   else
     digits[size-1] = hexDigit + factor;
   // Note:
   //      If size is the number of elements in the digits
   //      array, then the storage offset is (size - 1)
   value = value / 16;                    // New operand
   size --;                               // Update digit position
  }
  return;
}
```

1.6.2 C++ ASCII-to-Binary Conversions

Another set of conversion routines is necessary in order to parse an integer represented in a string of ASCII digits into the corresponding binary value. Each number system requires a different conversion routine. The most common cases are strings of binary, decimal, and hexadecimal digits.

A high-level routine to convert a string of ASCII binary digits into a binary value is based on the weight of each binary digit. The weight of the first eight binary digits is shown in Figure 1.2. In fact, the positional weight is calculated by raising 2 to the corresponding digit position. For example, binary digit number 13 has a decimal weight of 2^{13}.

A reasonable algorithm for an ASCII-binary-to-binary conversion is to examine the string of binary digits right-to-left, that is, starting with the lowest-order digit. If the binary digit is 1, then the digit weight is added in an accumulator. The weight of the low-order bit is 1. The next digit weight is calculated by doubling the previous one. The routine continues examining digit-by-digit until it reaches the end of the binary digit string. The algorithm is shown in the flowchart of Figure 1.8.

The following function, named AscBin2Bin() receives as an argument a string of ASCII binary digits, stored in an array of type char. The digit string is converted into binary and returned to the caller in a variable of type unsigned long. The routine allows the use of spaces and dashes as separators in the string of binary digits.

```
unsigned long AscBin2Bin( char digits[])
// Function to convert string of binary digits to a binary
// value
//
// Pre:
//    digits[] is a char array of binary digits.
//    The array can contain dashes used as
//    separators.
```

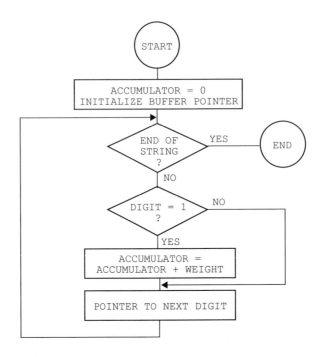

Figure 1.8 *Conversion of an ASCII-Binary String into Binary*

```
// Post:
//    Returns the binary value of the digits
//    in a variable of type unsigned long
{
    int size = 0;                      // Array size
    unsigned long binWeight = 1;       // Binary digit weight
    unsigned long accumulator = 0;     // Temp value
    unsigned int bitOffset;            // Offset in bit array
    unsigned int digitPos = 0;         // Relative digit position

    // Calculate array size
    while(digits[size])
       size++;
    // Routine starts at the least significant digit, which
    // has a weight of 1. The weight of other digits to the
    // left of the first one is 2 * the weight of the
    // previous bit
    for(int counter = 0; counter < size; counter++)
    {
       bitOffset = (size - 1) - digitPos;
       if(digits[bitOffset] == ' ' || digits[bitOffset] == '-')
       {
          digitPos++;
          continue;
       }
    // Test for 1-digit and add weight to accumulator
    if(digits[bitOffset] == '1')
        accumulator = accumulator + binWeight;
    // Update digit weight
```

```
      binWeight *= 2;
      // Update digit position in string
      digitPos++;
      }
   return accumulator;
}
```

Another useful high-level routine is one to convert a string of decimal digits into a binary value. In C++ the digit string is typically stored in an array of type char. The processing routine obtains the digits, one by one, from the array, starting with the least significant digit. The ASCII digit is converted to binary by subtracting 0x30 from its value. The result is multiplied by the weight of the digit's position and added to an accumulator. When the routine concludes, the accumulator contains the binary equivalent of the digit string. The algorithm is shown in the flowchart of Figure 1.9.

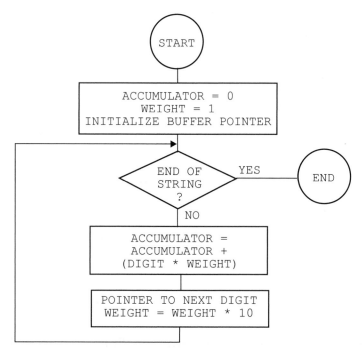

Figure 1.9 *Conversion of an ASCII-Decimal String into Binary*

The following function, named AscDec2Bin() receives as an argument a string of ASCII decimal digits, stored in an array of type char. The decimal digit string is converted into binary and returned to the caller in a variable of type unsigned long. The routine allows using commas as separators in the string of ASCII decimal digits.

```
unsigned long AscDec2Bin( char digits[])
// Function to convert string of decimal digits to a binary
// value
// Pre:
//    digits[] is a char array of decimal digits.
```

```
//     The array can contain commas as separators.
// Post:
//     Returns the binary value of the digits
//     in a variable of type unsigned long
//
{
    int size = 0;                      // Array size
    unsigned long decWeight = 1;       // Decimal digit weight
    unsigned long accumulator = 0;     // Temp value
    unsigned int bitOffset;            // Offset in bit array
    unsigned int digitPos = 0;         // Relative digit position
    char thisDigit;                    // Storage for digit

    // Calculate array size
    while(digits[size])
        size++;

    // Routine starts at the least significant digit, which
    // has a weight of 1. The weight of other digits to the
    // left of the first one is 10 * the weight of the
    // previous one
    for(int counter = 0; counter < size; counter++)
    {
      bitOffset = (size - 1) - digitPos;
      if(digits[bitOffset] == ',')
    {
        digitPos++;
        continue;
    }
    // ASSERT:
    //       Valid decimal digit in the range 0 to 9
    // Convert ASCII digit to binary
    thisDigit = digits[bitOffset] - 0x30;
    // Accumulate digit weight
    accumulator = accumulator + (thisDigit * decWeight);
    // Update digit weight
    decWeight *= 10;
    // Update digit position in string
    digitPos++;
    }
    return accumulator;
}
```

A similar routine allows converting a string of ASCII hexadecimal digits into binary and storing the results in a primitive variable of type unsigned long. The logic requires minor variations since we must convert each ASCII hex digit into binary before multiplying by the corresponding digit weight. This part of the routine must take into account that the binary value of a numeric hex digit (range 0 to 9) is obtained by subtracting 0x30 from the value of the ASCII character. With the alphabetic hex digits (range A through F) the conversion requires subtracting 0x37 if the digit is upper-case, or 0x57 if the digit is lower-case. In the hexadecimal conversion routine the digit weight is calculated by multiplying by 16, instead of by 10. The function AscHex2Bin() performs the processing.

```
unsigned long AscHex2Bin( char digits[])
// Function to convert string of hex digits to a binary
// value
// Pre:
```

```
//     digits[] is a char array of hex digits.
// Post:
//     Returns the binary value of the digits
//     in a variable of type unsigned long
// Note:
//     The function allows the standard C++ hex
//     format characters (0x) in the digit string.
//     Hex alpha characters (A through F) can be
//     upper- or lower-case.
//
{
    int size = 0;                       // Array size               //
    unsigned long hexWeight = 1;        // Digit weight
    unsigned long accumulator = 0;      // Temp value
    unsigned int offset = 0;            // Offset in array
    unsigned int digitPos = 0;          // Relative digit position
    unsigned int digitValue;            // Binary digit value

    // Obtain the size of the array
    while(digits[size])
      size++;

    // Routine starts at the least significant digit, which
    // has a weight of 1. The weight of other digits to the
    // left of the first one is 16 * the weight of the
    // previous bit
    for(int counter = 0; counter < size; counter++)
    {
      offset = (size - 1) - digitPos;
      if(digits[offset] == 'x')
      {
         digitPos++;
         continue;
      }
    // ASSERT:
    //         Valid hex digit in the range 0 to F
    // Convert ASCII digit to binary

    if(digits[offset] < 'A')
       digitValue = digits[offset] - 0x30;
    else
    {
      // Test for upper case hex digit
      if(digits[offset] < 'G')
         digitValue = digits[offset] - 0x37;
      else
         digitValue = digits[offset] - 0x57;
    }
    // Accumulate digit weight
    accumulator = accumulator + (digitValue * hexWeight);
    // Update digit weight
    hexWeight *= 16;
    // Update digit position in string
    digitPos++;
    }
  return accumulator;
}
```

The function AscHex2Bin(), listed previously, allows the presence of the standard C++ hex formatting characters (0x) in the digit string. Also, the hex alphabetic characters (A through F) can appear in upper- or lower-case.

SOFTWARE ON-LINE

The source code for the C++ conversion routines developed in this chapter can be found in the file INTCOV.H, contained in the project folder CHAPTER1\INTEGER CONVERSIONS in the book's on-line software. The code for the Assembly Language conversion routines are in the UN32_1.ASM module of the MATH32 library.

Chapter 2

Numeric Data in Memory

Chapter Summary

In this chapter we discuss several encodings and storage formats used in the computer representation of numeric data. We also continue developing routines for converting numeric data to-and-from the various encodings. Here, the conversion routines are to manipulate binary-coded decimals. Numeric data conversion routines are one of the building blocks required for developing mathematical software. BCD arithmetic routines are the topic of Chapter 4. The material covered in this chapter does not include the manipulation of binary floating-point representations, such as those used in the Pentium math unit. This is one of the topics of Chapters 5 and 6.

2.0 Electronic-Digital Machines

Efforts at mechanizing arithmetic can be traced from the abacus, slide rule, mechanical calculators, and punch card machines, to today's digital-electronic calculators and computers. The work of John von Neumann at Princeton's Institute for Advanced Study and Research marks the first major breakthrough in the design and construction of a digital-electronic calculating machine. Von Neumann's efforts, which began in 1945, produced in 1951 the first machine that used electronic circuits to hold and store data, to perform calculations, and to store its own instructions. In von Neumann's design, data and instructions are stored in a common memory area. A consequence of this architecture is that a program can easily modify stored data and even its own instructions.

The calculating power of the first von Neumann computer was approximately 2000 operations per second, while previous electro-mechanical predecessors were capable of performing only 3 or 4 operations per second. Today's machines can execute 1 billion instructions per second. The first commercial digital-electronic computer was manufactured by Remington Rand Incorporated and named UNIVAC I (Universal Automatic Computer). UNIVAC I used vacuum tubes, mercury delay lines, magnetic storage, and punched tape technology. The computers of its generation filled entire buildings and consumed enough electricity to power a small neighborhood. Technological advances and miniaturization techniques have reduced the cost and size of computing machinery. A modern PC costing less than $1500 easily exceeds the storage capacity and calculating speed of a multi-million dollar mainframe of a few years ago.

2.1 Storage of Numerical Data

In order to perform arithmetic operations, the computing machine must be capable of storing and retrieving numerical data. The numerical data items are stored in standard formats, designed to minimize space and optimize processing efficiency. Historically, numeric data was stored in data structures devised to fit the characteristics of a specific machine, or the preferences of its designers. It was in 1985 that the Institute of Electrical and Electronics Engineers (IEEE) and the American National Standards Institute (ANSI) formally approved mathematical standards for digital computers.

The electronic and physical mechanisms used for storing computer data have evolved with technology. One common feature of many microelectronic devices, from the transistor to the VLSI chip, is that they are based on the properties of a type of material called a *semi-conductor*. Integrated circuit technology has made possible the combining of transistors, resistors, and capacitors, and their electronic linkages, on the same piece of semi-conductor material.

Electronic technology is based on bi-stable components, which explains why the binary number system has been almost universally adopted. Data stored in processor registers, in magnetic media, in optical devices, or in punched tape, is usually encoded in binary. For this reason, the programmer and the operator can often ignore the physical characteristics of the storage medium. In other words, the bit pattern 10010011 can be encoded as holes in a strip of paper tape, as magnetic charges on a Mylar-coated disk, as positive voltages in an integrated circuit memory cell, or as minute craters on the surface of the CD. In all cases this bit pattern represents the decimal number 147.

Binary encodings must take into account, not only numeric values, but symbols and formats used in mathematical notation. For example, representing signed numbers requires a convention for encoding the plus and minus sign. Another case is the representation of decimal fractions, which requires special protocols and encoding schemes that facilitate storage and speed up calculations and other data processing operations. In all cases, numeric data encodings have been designed to minimize storage requirements and optimize processing.

2.1.1. Word Size

In electronic devices the bi-stable states are represented by a binary digit, or bit. Circuit designers group several individual cells to form a unit of storage that holds several bits. In a particular machine the basic unit of data storage is called the *word size*. Table 2.1 lists the word size of some historical computer systems.

Table 2.1

Computer Word Size in Historical Systems

COMPUTER FAMILY	WORD SIZE
TRS 80 Microcomputers (Z80 processor) Apple Microcomputers (6502 processor	8 bits
IBM PC, XT, AT, PCjr, Model 30, 50, 60 and other microcomputers equipped with Intel 8086, 8088 or 80286 CPU DEC PDP 11	16 bits
IBM PS/2 Model 55 SX, Model 70, Model 80 and other microcomputers equipped with the Intel 80386, 486, or PENTIUM CPU IBM 360/370 series IBM 303X, 308X series DEC VAX 11 Prime computers	32 bits
DEC 10 UNIVAC Honeywell	36 bits
CDC 6000 CDC 7000 CDC CYBER series	60 bits

In the PC the smallest unit of storage individually addressable is 8 bits (one byte). Individual bits are not directly addressable and must be manipulated as part of larger units of data storage.

The machine's word size determines the units of data storage, the machine instruction size, and the units of memory addressing. PCs equipped with the Intel 8086, 8088, or 80286 CPU have 16-bit wide registers, transfer data in 8 and 16-bit units to memory and ports, and address memory using a 16-bit base (segment register) and 16-bit pointers (offset register). Since the data registers in these CPUs are 16-bits wide, the largest value that can be held in a register is 11111111 11111111 binary, or 65,535 decimal. PCs that use the Intel 80386, 486, and Pentium CPU have 32-bit internal registers and a flat address space that is 32-bits wide. In these machines the word size is 32 bits. For compatibility reasons some operating systems and application code use the 80386, 486, and Pentium microprocessors in a mode compatible with their 16-bit predecessors.

2.2 Integer Encodings

The integers are the set of whole numbers, which can be positive or negative. The integer digits are located one storage unit apart and do not have a decimal point. The computer storage of unsigned integers is in a straightforward binary encoding. Since the smallest addressable unit of storage in the PC is one byte, the CPU logic pads with leading zeros numbers that are smaller than one byte. Figure 2.1 is a representation of an integer number stored electronically in a computer cell.

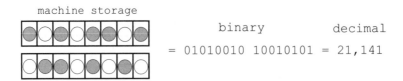

Figure 2.1 *Representation of an Unsigned Integer*

In Chapter 1 we discussed the conversion of unsigned binary numbers into ASCII decimal and hexadecimal values and develop the corresponding low-level and high-level routines.

2.2.1 Sign-Magnitude Representation

Representing signed numbers requires a special convention in order to differentiate positive from negative magnitudes. The most generally accepted scheme is to devote one bit to represent the sign. A common signed number storage format sets the high-order bit to indicate negative magnitudes and clears it to indicate positive magnitudes and zero. In this scheme the decimal numbers 93 and – 93 are represented as follows:

```
01011101 binary = 93 decimal
11011101 binary = -93 decimal

|

|—————— sign bit
```

Notice that the left-most digit is set for a negative number and clear for a positive one. This way of designating negative numbers, called a *sign-magnitude* representation, corresponds to the conventional way in which we write signed numbers longhand. That is, we precede the number by its sign. Sign-magnitude representation has the following characteristics:

1. The absolute value of positive and negative numbers is the same.

2. Positive from negative numbers can be distinguished by examining the high-order bit. If this bit is 1, then the number is negative. Otherwise, the number is positive.

3. There are two possible representations for zero, one negative (10000000B) and one positive (00000000B).

One limitation of the sign-magnitude representation is that the logic required to perform addition is different from that for subtraction. While this is not insurmountable, there are other numeric representations (discussed later in this chapter) which arithmetic operations are performed in the same way. A related limitation of sign-magnitude representation is that, in some cases, it is necessary to take into account the magnitude of the operands in order to determine the sign of the result. Finally, the presence of an encoding for negative zero reduces the numerical range of the representation and is, for most practical uses, an unnecessary complication.

The limitations of the sign-magnitude format can be seen in the complicated rules required for the addition of signed numbers. Assuming two signed operands, labeled x and y, the following rules must be observed for performing addition:

1. If x and y have the same sign, they are added directly and the result is given the common sign of the addends.

2. If the absolute value of x is larger than or equal to the absolute value of y, then y is subtracted from x and the result is given the sign of x.

3. If the absolute value of y is larger than the absolute value of x, then x is subtracted from y and the result is given the sign of y.

4. If both x and y are − 0, then the sum is 0.

The rules for subtracting numbers in sign-magnitude form are even more complicated.

2.2.2 Radix Complement Representation

Complementary numbers originated in an effort to simplify the mechanized addition and subtraction of signed numbers. Arithmetic complements arise during subtraction. In general, the radix complement of a number is defined as the difference between the number and the next integer power of the base that is larger than the number. In decimal numbers the radix complement is called the *ten's complement*. In the binary system the radix complement is called the *two's complement*. For example, the radix complement of the decimal number 89 (ten's complement), is calculated as follows:

```
        100   = higher power of 10
    -    89
        ____
         11   = ten's complement of 89

    a = 602
    b = 353
```

Consider the operation x = a − b, as follows:

```
         602
       - 353
        ____
    x =  249
```

Notice that in the process of performing longhand subtraction we had to perform two borrow operations. Now consider that the radix complement (ten's complement) of 353 is

```
    1000 - 353 = 647
```

Using complements we can reformulate subtraction as the addition of the ten's complement of the subtrahend, as follows

```
         602
       + 647
        ____
        1249
       |_____ discarded digit
```

The result is adjusted by discarding the digit that *overflows* the number of digits in the operands.

In regards to longhand decimal arithmetic there is probably little advantage in replacing subtraction with ten's complement addition. The additional work of calculating the ten's complement cancels out any other possible benefit. However, in binary arithmetic the use of radix complements entails significant computational advantages, principally because a binary machine can calculate complements very rapidly.

The two's complement of a binary number is obtained in the same manner as the ten's complement of a decimal number, that is, by subtracting the number from an integer power of the base that is larger than the number. For example, the two's complement of the binary number 101 is:

```
  1000B  =   2³ = 8 decimal (higher power of 2)
-  101B  =        5 decimal
  -----      ----------
   011B  =        3 decimal
```

By the same token, the two's complement of 10110B is calculated

```
  100000B  =   2⁵ = 32 decimal (higher power of 2)
-  10110B  =        22 decimal
  -------      ----------
   01010B          10 decimal
```

You can perform the binary subtraction of 11111B minus 10110B by finding the two's complement of the subtrahend, adding the two operands, and discarding any overflow digit, as follows:

```
      11111B  =   31 decimal
  +   01010B  =   10 decimal (two's complement of 22)
      -------
      101001B
discard_____I
       01001B  =    9 decimal (31 minus 22 = 9)
```

In addition to the radix complement representation (called ten's complement in the decimal system and two's complement in the binary system) there is a *diminished radix representation* that is often useful. This encoding, sometimes called the *radix-minus-one* form, is created by subtracting 1 from an integer power of the base that is larger than the number, then subtracting the operand from this value. In the decimal system the diminished radix representation is sometimes called the *nine's complement*. This is due to the fact that an integer power of ten, minus one, results in one or more 9-digits. In the binary system the diminished radix representation is called the *one's complement*. The nine's complement of the decimal number 76 is calculated as follows:

```
     100  = next highest integer power of 10

      99  = 100 minus 1
 -    76
 --------
      23  = nine's complement of 89
```

The one's complement of a binary number is obtained by subtracting from an integer power of the base that is larger than the number, minus one. For example, the one's complement of the binary number 101 (5 decimal) can be calculated as follows:

```
    1000B  =   2³ = 8 decimal

     111B  =  1000B minus 1 =  7 decimal
 -   101B                       5 decimal
 -------                     ----------
     010B  =                    2 decimal
```

Notice that the one's complement can be obtained by changing every 1 binary digit to a 0 and every 0 binary digit to a 1. In the above example 010B is the one's complement of 101B. In this context the 0 binary digit is often said to be the complement of the 1 binary digit, and vice versa. Most modern computers contain an instruction that inverts all the digits of a value by transforming all 1 digits into 0, and all 0 digits into 1. The operation is also known as a logical or arithmetic negation. In the 80x86 CPU family the NOT instruction generates the one's complement of the operand.

An interesting fact is that the two's complement can be derived by adding one to the one's complement of a number. Therefore, instead of calculating

```
     100000B
 -    10110B
    ---------
      01010B
```

we can find the two's complement of 10110B as follows

```
    10110B  = number
    01001B  = change 0 to 1 and 1 to 0 (one's complement)
 +      1B    then add 1
 --------
    01010B  = two's complement
```

This simple algorithm provides a convenient way of calculating the two's complement in a machine equipped with a logical NOT, or a complement instruction. However, in the Intel 80x86 family this is unnecessary because these processors have a NEG instruction which calculates the two's complement directly.

A third way of calculating the two's complement is subtracting the operand from zero and discarding the overflow. The following code fragment illustrates the use of the NOT and NEG instructions in obtaining the one's and two's complement.

```
; Use of the 80x86 NOT instruction to obtain the one's complement
; of a byte operand
      MOV    AL,22              ; AL is loaded with 00010110
```

```
        NOT     AL                      ; Inversion of 00010110
                                        ;             is 11101001 = 233

; Use of the 80x86 NEG instruction to form the two's complement
; of a byte operand
        MOV     AL,22                   ; AL is loaded with 00010110
        NEG     AL                      ; Negation of 00010110
                                        ;             is 11101010 = 234
```

We have mentioned that the radix complement of a number is defined as the difference between the number and an integer power of the base that is larger than the number. Following this rule, we calculate the radix complement of the binary number 10110 as follows

```
        100000B   =   2⁵ = 32 decimal
    -    10110B   =        22 decimal
        --------              -----------
         01010B              10 decimal
```

However, the 80x86 NEG instruction calculates the two's complement of a byte operand as follows

```
        100000000B   =   2⁸ = 256 decimal
    -    00010110B   =         22 decimal
        -----------              ------------
         11101010B               234 decimal
```

The difference is due to the fact that in the longhand method we have used the next higher integer power of the base compared to the value of the subtrahend (in this case 100000B) while the CPU uses the next higher integer power of the base compared to the operand's word size, which can be 8 or 16 bits. In the above example the operand's word size is 8 bits and the next highest integer power of 2 is 100000000B. In either case, the results from two's complement subtraction are valid as long as the minuend is an integer power of the base that is *larger than* the subtrahend.

For example, to perform the binary subtraction of 00011111B (31 decimal) minus 00010110B (22 decimal) we can find the two's complement of the subtrahend and add, discarding any overflow digit, as follows.

```
            00011111B   =   31 decimal
        +   11101010B   =   234 decimal (two's complement of 22)
            ---------
            100001001B
discard____|
            00001001B   =    9 decimal (31 minus 22 = 9)
```

One advantage of complementing to the word size is that the high-order bit can be used to detect the sign of the number. This means that two's complement number obtained using the 80x86 NEG instruction always has the high-order bit set.

Another advantage of the two's complement notation is that there is no representation for negative 0. Although there are some mathematical uses for a signed zero notation, in most cases this distinction is unnecessary. While both the two's complement and the one's complement schemes can be used to implement binary arithmetic, computer designers usually prefer the two's complement.

2.3 Encoding of Fractional Numbers

In Chapter 1 we saw that, in a positional number system, the weight of each integer digit can be determined by the formula:

$$P = d * B^c$$

where d is the digit, B is the base or radix, and C is the zero-based column number, starting from right to left. Therefore, the value of a multidigit positive integer to n digits can be expressed as a sum of the digit values:

$$d_n * B^n + d_{n-1} * B^{n-1} + d_{n-2} * B^{n-2} + \ldots + d_0 * B_0$$

where d_i (i = 0...n) is the value of the digit and B is the base or radix of the number system. This representation can be extended to represent fractional values. Recalling that

$$x^{-n} = \frac{1}{x^n}$$

we can extend the sequence to the right of the radix point, as follows:

$$. + d_{n-1} \times B^{-1} + dn_{-2} \times B^{-2} \ldots$$

In the decimal system the value of each digit to the right of the decimal point is calculated as 1/10, 1/100, 1/1000, and so on. The value of each successive digit of a binary fraction is the reciprocal of a power of 2, hence the sequence: 1/2, 1/4, 1/8, 1/16, et cetera. Figure 2.2 shows the positional weight of the integer and the fractional digits in a binary number.

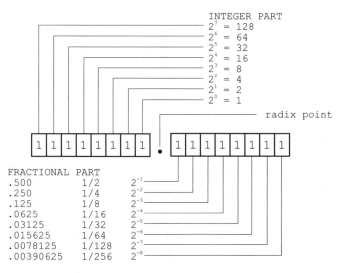

Figure 2.2 *Positional Weights in a Binary Fraction*

In Chapter 1 you used the positional weight of the binary digits to convert a binary number to its decimal equivalent. You can follow a similar method to convert the fractional part of a binary number. Using the decimal equivalents shown in Figure 2.2 you can convert the binary fraction .10101 to a decimal fraction as follows

```
                                        .1 0 1 0 1
                                         |   |   |
         .500    _____|   |   |
         .125    _____|   |
         .03125  _____|

         _____
         .65625
```

2.3.1 Fixed Point Representations

The encoding and storage of real numbers in binary form presents several difficulties. The first one is related to the position of the radix point. Since there are only two symbols in the binary set, and both are used to represent the numerical value of the number, there is no other symbol available for representing the radix point. To represent the positive decimal number 58.125 you could use one data element to encode the integer part, and another one for the fractional part, as follows:

```
         58 = 111010B
         .125 = .001B
```

While in the decimal system you can write 58.125, the lack of a binary radix point makes it impossible to explicitly encode binary fractions. This means that we must find other ways for determining the position of the radix point in a binary encoding. One possible scheme is to assume that the radix point is located at a fixed position in the representation. Figure 2.3 shows one possible convention for storing a real number in a fixed point binary format.

```
          binary                    decimal

= 0o111010 00100000 = 58.125
                 ▲
         _____|_____
         implied binary point
```

Figure 2.3 *Binary Fixed-Point Representation*

In Figure 2.3 we assumed that the binary point is positioned between the eighth and the ninth digit of the encoding. When using a binary fixed point representation of a real number, whatever distribution of digits is selected for the integer and the fractional part of the representation must be maintained in all cases. This is the greatest limitation of the fixed point formats. Suppose we want to store the value 312.250. This number can be represented in binary as follows:

```
         312 = 100111000
         .250 = .01
```

The total number of binary digits required for the binary encoding is 11. The number can be physically stored in a 16-digit structure (as the one in Figure 2.3) with five cells to spare. However, since the fixed point format adopted in Figure 2.3 assigns 8 cells to represent the integer part of the number, 312.250 cannot be encoded because the integer part (312) requires 9 binary digits. In spite of this limitation, the fixed point format was the only one used in early computers.

2.3.2 Floating-Point Representations

Contrary to fixed-point scheme, floating-point representations do not use a fixed position for the radix point. The idea of separately encoding the position of the radix point originated in what is usually called *scientific notation*. In standard scientific notation a number is written as a base greater than or equal to 1 and smaller than 10, multiplied by a power of 10. For example, the value 310.25 in scientific notation is written:

$$3.1025 \times 10^2$$

A number encoded in scientific notation has a real part and an exponent part. Using the terminology of logarithms these two parts are sometimes called the *mantissa* and the *characteristic*. The following simplification of scientific notation is often used in computer work:

$$3.1025 \text{ E2}$$

In the computer version of scientific notation the multiplication symbol and the base are assumed. The letter E, which is used to signal the start of the exponent part of the representation, accounts for the name *exponential form*. Numbers smaller than 1 can be represented in scientific notation or in exponential form by using negative powers. For example, the number .0004256 can be written:

$$4.256 \times 10^{-4}$$

or as

$$4.256 \text{ E-4}$$

In computer arithmetic this notation is often called a *floating-point* representation to indicate that the radix point floats according to the value of the exponent. Floating-point representations provide a more efficient use of the computer's storage space. For example, the numerical range of the fixed point encoding shown in Figure 2.3 is from 255.99609375 to 0.00390625 (refer to Figure 2.2). To improve this range we can re-assign the 16 bits of storage so that 4 bits are used for encoding the exponent and 12 bits for the fractional part, which is called the *significand*. In this case the encoded number appears as follows:

```
0000 000000000000
---- ------------
   |       |_____  12-bit fractional part
   |
   |_____ 4-bit exponent part
```

Assume that we were to use the first bit of the exponent to indicate the sign of the exponent, then the range of the remaining 3 digits would be 0 to 7. Notice that the sign of the exponent indicates the direction in which the decimal point is to be moved, which is unrelated to the sign of the number. In this case the fractional part (or significand) could hold values in the range 1,048,575 to 1. The combined range of exponent and significand allows representing decimal numbers in the range 4095 to 0.00000001, which considerably exceeds the range in the same storage space in fixed point format.

2.4 Standardized Floating-Point Encodings

Both the significand and the exponent of a floating-point number can be stored either as an integer, or in sign magnitude, or radix complement representations. Furthermore, the number of bits assigned to each field can vary according to the range and the precision required. For example, the computers of the CDC 6000, 7000, and CYBER series used a 96-digit significand with an 11-digit exponent, while the PDP 11 series used 55-digit significands and 8-digit exponents in their extended precision formats.

Historical variations, incompatibilities, and inconsistencies in floating-point formats created a need to standardize these elements. In March and July 1985, the Computer Society of the Institute of Electric and Electronic Engineers (IEEE) and the American National Standards Institute (ANSI) approved a standard for binary floating-point arithmetic (ANSI/IEEE Standard 754-1985). This standard established four formats for encoding binary floating-point numbers. Table 2.2 summarizes the characteristics of these formats.

Table 2.2
ANSI/IEEE Floating Point Formats

PARAMETER	SINGLE	SINGLE EXTENDED	DOUBLE	DOUBLE EXTENDED
total bits	32	≥43	64	≥79
significand bits	24	≥32	53	≥64
maximum exponent	+127	≥+1023	+1023	≥+16383
minimum exponent	−126	≥+1022	−1022	≥+16382
exponent width	8	≥11	11	≥15
exponent bias	+127	----	+1023	----

2.4.1 ANSI/IEEE 754 Single Format

Figure 2.4 shows the ANSI/IEEE floating-point single format.

Figure 2.4 *ANSI/IEEE Floating-Point Single Format*

Note that the rudimentary floating-point encoding developed in Section 2.3.2 allows the representation of positive real numbers, but not of negative numbers. In this case the sign of the exponent, which determines the direction in which the radix point must be moved, should not be confused with the sign of the number. If a floating-point encoding is to allow the representation of signed numbers it must devote one binary digit to encode the number's sign. In the IEEE 754 single format in Figure 2.4 the high-order bit represents the sign of the number. A value of 1 indicates a negative number.

2.4.2 Floating-Point Exponent in ANSI/IEEE 754 Single Format

The exponent part of a binary floating-point number represents the integer power of the base with which the significand must be multiplied. As mentioned in Section 2.4, the exponent can be stored in integer, sign magnitude, or radix complement representations. All these forms have been successfully used in various computer hardware and software systems. The ANSI/IEEE 754 standard for floating-point arithmetic establishes that the exponent be stored in *biased* form, although in some formats (single extended and double extended) the bias is not defined (see Table 2.2).

The word bias, in this context, means a constant value that is added to the exponent. The term *excess-n* notation has also been used to designate a biased encoding. The constant usually added to create the bias form is approximately one half the numerical range of the exponent field. For example, in the IEEE single format devotes 8 digits for the exponent field (see Figure 2.4). The numerical range of 8 binary digits is 0 to 255 decimal. One half of this range is approximately 127. Adding the constant 127 to all positive exponents places them in the range 127 to 255. The lower half of the range (1 to 126) is used for negative exponents. A 0 value in the exponent field is reserved to encode zero and denormals. Denormals are a special type of numbers discussed in the following paragraphs. Table 2.3 shows the values of the exponent and the biased representation in the ANSI/IEEE single format for floating-point numbers.

Notice in Table 2.3 that the exponent value 00000000B is used to represent zero and *denormal* numbers. Denormals, or *denormalized* numbers, occur when the exponent of the number is too small to represent in the corresponding floating-point format. On the other hand, the exponent 11111111B is used to encode numbers that are too large for the single format, or to represent error conditions. The exponent range 00000001B to 11111110B (decimal values 1 to 254) is used to represent *normal* numbers, that is, numbers that are within the range of the format.

In ANSI/IEEE 754 floating-point formats the high bit of the exponent field does not encode the sign, as is the case in the sign-magnitude form. Instead, the ANSI/IEEE single format uses a bias 127 scheme, mentioned previously, to represent negative and positive exponents. This means that negative exponents are in the range 1 to 127 (see Table 2.3) and positive exponents are in the range 128 to 254. In contrast with fixed point conventions, the high bit of the exponent is set to indicates a positive exponent, and is zero to indicate negative exponent. The main advantage of a biased exponent is that the numbers can be compared bitwise, from left to right, to determine the larger one. The number's true exponent is obtained by subtracting the bias.

Table 2.3

Interpretation of Exponent in the ANSI/IEEE Single Format

BIASED EXPONENT	SIGN OF NUMBER	TRUE EXPONENT	SIGNIFICAND	CLASS
0000 0000	+	---	00 ... 00	positive zero
	−	---	00 ... 00	negative zero
			11 ... 11	
			to	
			00 ... 01	denormals
0000 0001	---	-126	00 ... 00	normals
to		to	to	
0111 1111		0	11 ... 11	
1000 0000	---	1	00 ... 00	normals
to		to	to	
1111 1110		127	11 ... 11	
1111 1111	+	---	00 ... 00	+ infinity
	−		00 ... 00	− infinity
	−		10 ... 00	indefinite
	−		00 ... 01	
			to	
			11 ... 11	not-a-number

2.4.3 Floating-Point Significand in ANSI/IEEE 754 Single Format

The last field of a floating-point representation is known by several names: fractional part, mantissa, characteristic, or significand. We prefer the word significand because it is the one most commonly used the literature. Like the exponent, the significand can be stored as an in integer, or in sign magnitude or radix complement representations.

A floating-point binary number is said to be in *normalized form* when the first digit of its significand is 1. An un-normalized binary floating-point number can be normalized by successively shifting the digits of the significand to the left, while simultaneously subtracting one from the exponent. This process is continued until the high-order bit of the significand is a binary 1. The process does not change the value of the number, since shifting the significand bits to the left effectively multiplies the number by 2, while subtracting one from the exponent divides the number by 2. Note that the value of a number does not change if it multiplied and divided by the same value.

Notice that normalization applies to the entire encoded number since it requires adjustments of both the exponent and the significand. Therefore, it is not correct to speak of a normalized significand or a normalized mantissa, we should rather refer to the significand of a normalized floating-point number.

One advantage of the normalized form is that the significand contains a maximum number of significant bits. However, addition and subtraction of floating-point numbers require that both operands have the same exponent. Therefore, before performing these operations it is often necessary to shift the significand digits to the right or to the left so that the exponents are equal.

The ANSI/IEEE standard for floating-point arithmetic takes advantage of the fact that a normalized significand of a binary floating-point starts with a 1-digit. In the single and double precision formats this leading bit of the significand is assumed, therefore increasing by one bit the range of the representation. Not so in the extended formats, in which the digit must be explicitly coded. Notice that this assumption is not valid if the exponent is all zeros. A zero exponent and a non-zero significand indicate a denormal, as shown in Table 2.3. Also, the use of an implicit bit makes necessary a special representation for zero (see Table 2.3). This special zero must be handled separately during arithmetic operations.

2.4.4 Decoding Floating-Point Numbers

The formats in the ANSI/IEEE 754 standard for binary floating-point arithmetic were designed to provide maximum storage capacity and processing efficiency. For example, the exponent in the ANSI/IEEE single format, stored in biased from, takes up 8 bits; however, these 8 bits do not fall on a byte boundary. The exponent bit take up 7 bit positions in the high-order byte, and 1 bit position in the next byte, as shown in Figure 2.4. In the same ANSI/IEEE single encoding the significand takes up seven bits of the second byte as well as the third and fourth bytes. The sign of the number is the high-order bit of the high-order byte. Figure 2.5 shows the number 127.375 stored in the floating-point single format.

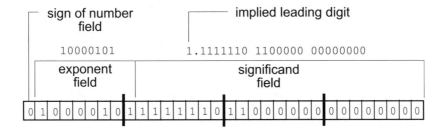

Figure 2.5 *Encoding of the Number 127.375 in IEEE Single Format*

The encoding in Figure 2.5 is interpreted as follows:

```
sign of number = 0 (positive)

biased exponent = 10000101B = 133 decimal

real exponent = 133 - bias = 133 - 127 = 6

significand = 1.1111110 11000000 00000000 (adding explicit digit)
```

significand is adjusted by moving the radix point 6 places to the right:

```
new significand = 1111111.01100...000
```

The significand bits are interpreted as follows::

```
integer part = 1111111 = 127
fractional part = .01100..00 = .375
bit value: 11111110-11000000-00000000
            |-------|----------------|
                |            |
                |            |_____ fractional part
                |_____ integer part
number: 127.375
```

Computer systems based on the Intel 80x86 family of microprocessors store data items in memory with the low byte at the lowest memory address. This ordering scheme is called the *little-endian* format. In graphs and illustrations floating-point numbers are usually laid out with the high-order bit to the left (as in Figures 2.4 and 2.5). However, in machines that use little-endian format, the bytes are actually stored in memory in reverse order. Figure 2.6 shows the memory image of the number 127.375 in the little-endian byte ordering scheme.

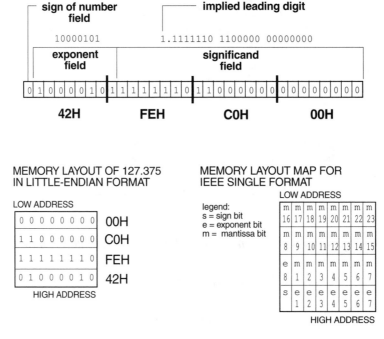

Figure 2.6 *Memory Storage of 127.375 in ANSI/IEEE Single Format*

C++ variables of type float are encoded in IEEE single format, while variables of type double are in double format. The C++ function named DecodeSingle(), listed on the next page, decodes a C++ variable of type float. The results are stored in three variables of int type: one for the sign, a second one for the exponent, and the third int one stores the number's significand. The DecodeSingle() function allows C++ code to examine the sign, exponent, and significand and to access these elements for conversions or arithmetic operations.

PROGRAMMER'S NOTEBOOK

Some C++ compilers (including Microsoft's Visual C++) allow typing assembly language code directly into the C++ source file. This development mode, called inline assembly, is useful in optimizing routines. Also, it is the only way in which a C++ program can access the Pentium math units. Although there are limitations to inline assembly, it is a powerful tool. In Visual C++, inline assembly is accessed by means of the _asm keyword, as shown in the function that follows this note.

```
void DecodeSingle(float single, int *sig, int *exp, int *signf )
{
_asm
{
// C++ function to separate the elements of a floating point number
// stored in memory in IEEE single format
// Pre:
//        Parameter single is an IEEE floating point single
// Post:
//        Parameter sig is the number's sign field:
//                1 = negative sign
//                0 = positive sign
//        Parameter exp is  8-bit exponent in bias 127 form
//        Parameter signf is 24-bit significand
//                Implicit significand bit is set if exponent
//                is not zero

        LEA     ESI,single
;*********************|
;      decode sign    |
;*********************|
        ADD     ESI,3               ; Set pointer to last byte
        MOV     AH,BYTE PTR[ESI]         ; Get byte into AH
        AND     AH,10000000B    ; Mask off exponent bits
        MOV     CL,7            ; Counter to shift 7 bits
        SHR     AH,CL           ; Shift sign bit to right-most bit
;*********************|
;    decode exponent  |
;*********************|
        MOV     AL,[ESI]        ; Get 7 high bits of exponent
        SHL     AL,1            ; Shift left one bit
        DEC     ESI             ; Pointer to byte at offset 2
        MOV     BH,[ESI]        ; Get byte into BH
        TEST    BH,10000000B    ; Test for exponent bit 8 set
        JZ      OK_EXPONENT     ; Exponent OK if bit not set
        OR      AL,00000001B    ; Set low exponent bit
;*********************|
;  decode significand |
;*********************|
OK_EXPONENT:
; Test for a non-zero exponent so set implicit significand bit
        CMP     AL,0            ; Is exponent zero
        JE      IMPLICIT_0      ; Implicit bit is zero with a
                                ; zero exponent. Do not set
;*********************|
;    explicit 1 bit   |
;*********************|
```

```
        OR      BH,10000000B    ; Set BH high bit
IMPLICIT_0:
        DEC     ESI             ; Point to byte at offset 1
        MOV     BL,[ESI]        ; Load into BL
        DEC     ESI             ; Point to byte at offset 0
        MOV     CH,[ESI]        ; Load into CH
;**********************|
;  Data to variables   |
;**********************|
; Move IEEE format data from registers to variables
; Start with exponent, in AL
        MOV     EDX,EAX
        AND     EDX,0FFH        ; Preserve AL
        MOV     EDI,exp
        MOV     [EDI],EDX
; Store sign, in AH
        MOV     EDX,EAX
        AND     EDX,0FF00H      ; Preserve AH
        SHR     EDX,8
        MOV     EDI,sig
        MOV     [EDI],EDX
; Store significand (in BH, BL, and CH)
        MOV     EDX,EBX
        AND     EDX,0FFFFH      ; Clear high-order bits
        SHL     EDX,8           ; Make space for low order bits
        AND     ECX,0FF00H      ; Preserve CH
        SHR     ECX,8           ; Move to low order bits
        OR      EDX,ECX
        MOV     EDI, signf
        MOV     [EDI],EDX
}
}
```

SOFTWARE ON-LINE

The DecodeSingle() C++ function is located in the BCD12 Rtns.cpp module found in the \Sample Code\Chapter02\BCD12 Conversions folder in the book's on-line software. A low-level version of this routine, named _DECODE_SINGLE is found in the Un32_1 module of the MATH32 library.

Notice that the function DecodeSingle(), listed previously, sets the explicit bit of the significand only in the case in which the exponent is not zero. This is necessary in order to maintain the correct encoding for positive and negative zero and for denormals, as shown in Table 2.3. Also notice that the function returns the decimal value of the exponent and the significand fields as stored in the ANSI/IEEE single precision format. These field values do not usually coincide with the number's decimal representation. For example, if we use the DecodeSingle() function on the value 127.375 the result is as follows:

```
Sign: 0 (positive number)
Exponent (bias 127): 133 = 10000101b
Real exponent (133-127) = 6
Significand: 16,695,296 = 11111110-11000000-00000000b
```

Notice that these values coincide with those shown in Figure 2.5.

The conversion of ANSI/IEEE 754 floating-point encodings into ASCII decimal numbers and vice versa is discussed in more detail in Chapter 6.

2.5 Binary-Coded Decimals (BCD)

Binary floating-point encodings are usually considered the most efficient format for storing numerical data in a digital computer. Other representations are also used in computer work. *Binary-coded decimal* (BCD) is a representation of decimal digits in binary form. There are two common ways of storing BCD digits. One is known as the *packed* BCD format and the other one as *unpacked*. In the unpacked format one BCD digit is represented in one byte of memory storage. In packed form two BCD digits are encoded per byte of storage. The unpacked BCD format does not use the four high-order bits of each byte, which is wasted storage space. On the other hand, the un-packed format facilitates conversions and arithmetic operations on some machines. Figure 2.7 shows the memory storage of a packed and unpacked BCD number.

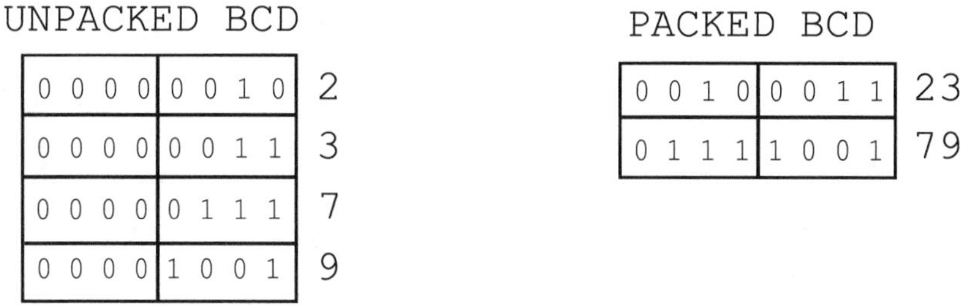

```
UNPACKED BCD                      PACKED BCD

0 0 0 0 | 0 0 1 0   2            0 0 1 0 | 0 0 1 1   23

0 0 0 0 | 0 0 1 1   3            0 1 1 1 | 1 0 0 1   79

0 0 0 0 | 0 1 1 1   7

0 0 0 0 | 1 0 0 1   9
```

Figure 2.7 *Packed and Unpacked BCD*

2.5.1 Floating-Point BCD

Binary-coded decimal representations and BCD arithmetic have not been explicitly described in a formal standard. Each machine or software package stores and manipulates BCD numbers in a unique and often incompatible way. The Intel mathematical coprocessors, as well as the math unit of the 486 and Pentium, provide a *packed decimal format* (which is a BCD representation) as well as two opcodes for performing BCD conversions. But this BCD format is not considered in the ANSI/IEEE standard for binary floating-point arithmetic.

The Pentium math unit packed decimal format is a sign-magnitude BCD representation in 18 digits. It serves well for the purposes at hand, mainly conversions and input-output operations. However, since this is an integer format its general use is very limited. For the routines and examples in this book we have developed two floating-point BCD representations, which we call them the BCD12 and BCD20 formats. The BCD12 format is shown in Figure 2.8 and Table 2.4.

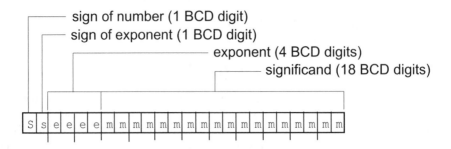

sign of number (1 BCD digit)
sign of exponent (1 BCD digit)
exponent (4 BCD digits)
significand (18 BCD digits)

S s e e e e m m m m m m m m m m m m m m m m m m m m

Figure 2.8 *Map of the BCD12 Format*

Table 2.4

Field Structure of the BCD12 Format

FIELD CODE	FIELD NAME	BITS WIDE	BCD DIGITS	RANGE
S	sign of number	4	1	0 – 1 (BCD)
s	sign of exponent	4	1	0 – 1 (BCD)
e	exponent	16	4	0 – 9999
m	significand	72	18	0 – 99..99 (18 digits)

Format size   96 (12 bytes)

Notes:
  1. The significand is scaled (normalized) to a number in the range 1.00..00 to 9.99..99.
  2. The encoding for the value zero (0.00..00) is a special case.
  3. The special value FH in the sign byte indicates an invalid number.

The BCD12 format requires 12 bytes of storage. The format is based on the BCD arithmetic features of the 80x86 and those in the Pentium math unit. The format is described as follows:

1. The sign of the number (S) is encoded in the left-most packed BCD digit. Therefore, the first 4 bits are either 0000B (positive number) or 0001B (negative number).

2. The sign of the exponent is represented in the 4 low-order bits of the first byte. This means that the sign of the exponent is also encoded in one packed BCD digit. As is the case with the sign of the number field, the sign of the exponent is either 0000B (positive exponent) or 0001B (negative exponent).

3. The following 2 bytes encode the exponent in 4 packed BCD digits. The decimal range of the exponent is 0000 to 9999.

4. The remaining 9 bytes are devoted to the significand field, consisting of 18 packed BCD digits. Positive and negative numbers are represented with a significand nor-

malized to the range 1.00...00 to 9.00...99. The decimal point following the first significand digit is implied. The special value 0 has an all-zero significand.

5. The special value FF hexadecimal in the number's sign byte indicates an invalid number.

The BCD12 and BCD20 formats, as is the case in all BCD encodings, do not make ideal use of the available storage space. In the first place, each packed BCD digit requires 4 bits, which in binary could serve to encode 6 additional combinations. At a byte level the wasted space is 100 encodings (BCD 0 to 99) out of a possible 256 (0 to FFH). The sign of the exponent field is wasteful since only one binary digit is actually required for storing the sign. Regarding efficient use of storage, BCD formats cannot compete with floating-point binary encodings. The advantages of BCD representations are a greater ease of conversions into decimal forms, and, in the case of the 80x86 CPU, the possibility of using of the processor's BCD arithmetic instructions.

The BCD20 format, discussed in detail in Section 4.6, is identical to BCD12 except that it allows representing a 34-digit significand, thus expanding the precision of the BCD12 encoding.

SOFTWARE ON-LINE

The file BCD12.h located in the folder Sample Code\Chapter04\BCD12 Arithmetic, in the book's on-line software, contains conversion an arithmetic functions for the BCD12 format. The file BCD20.h, in Sample Code\Chapter04\BCD20 Arithmetic, contains conversion and arithmetic routines for the BCD20 format.

2.6 BCD Conversions

In order to manipulate BCD numbers in C++ we first need to develop conversion routines. A number encoded in the BCD12 format, which was developed earlier in this chapter, can be stored in a C++ array of char. Since the BCD12 format requires 12 bytes of storage, the space is allocated as follows:

```
char BCDNum[12];
```

Notice that the BCD encodings developed in this book are not supported in C or C++. You cannot use C or C++ functions to display values stored in BCD12 or BCD20 formats, since the output will be meaningless.

The conversion routines for BCD arithmetic were developed in Microsoft Visual C++ Version 6. The executables that exercise the routines and demonstrate their use are console applications (DOS programs). However, the sources are in Win32 format and can be called from Windows applications.

2.6.1 BCD Conversion Functions

The functions listed in this section provide a C++ interface to the low-level conversion procedures in the MATH32 library. The first function, named AsciiToBcd12(), converts a string of ASCII decimal digits into a BCD number in BCD12 format. The second function, named Bcd12ToAscStr() converts a number in BCD12 format into a string of ASCII digits. Both functions use inline assembly in order to access the low-level procedures in the MATH32 library.

```
void AsciiToBcd12(char ascNum[], char bcdNum[])
{
// Conversion of an ASCII decimal number into BCD12 format
// On entry:
//          parameter ascNum[] is an ASCII decimal string,
//          terminated in a NULL byte or control code
//               Note: the caller's buffer may contain the $ sign
//                     and commas as formatting symbols. The string
//                     terminator (00H) may be preceded by a CR
//                     (0DH) or LF (0AH) control codes. If the
//                     characters E or e are in the string, the
//                     procedure assumes that the string is in
//                     ASCII exponential form. The code assumes that
//                     exponential strings are normalized.
//
//          parameter bcdNum[] is 12-byte variable to hold the number
//          encoded in BCD12 format
//
// Sample input strings:
//          123.223344
//          -0000.000002345
//          1.273E-22
//          -1.44556677 E230
//
// On exit:
//          parameter bcdNum[] contains the input number
//          formatted in BCD12 format, as follows:
//
//           S s e e e e m . m m m m m m m m m m m m m m m m m
//                         |
//                         |____ implicit decimal point
//
//      S = sign of number (1 BCD digit)
//      s = sign of exponent (1 BCD digit)
//      e = exponent (4 BCD digits)
//      m = normalized significand (18 BCD digits)
//      . = implicit decimal point between the first and second
//          significand digits
_asm
{
        MOV     ESI,ascNum      // Pointer to source
        MOV     EDI,bcdNum      // Pointer to results buffer
        CALL    ASCII_TO_BCD12
}
return;
}

void Bcd12ToAscStr(char ascBuf[], char bcdBuf[])
{
// Convert a number in BCD12 format into an ASCII string
// On entry:
//          parameter bcdBuf[] is a 12-byte BCD12 buffer
//                    holding number
//          parameter ascBuf[] is a 31-byte ASCII buffer
```

```
//          to hold ASCII string
// On exit:
//          ascBuf[] holds ASCII string
//
// Operation:
//          Numbers with exponent < 10 are converted to decimal
//          format. If exponent is > 10 the number is converted to
//          exponential format
//
// BCD12 format:
//          S s e e e e m . m m m m m m m m m m m m m m m m m
//                       |
//                       |____ implicit decimal point
//
//     S = sign of number (1 BCD digit)
//     s = sign of exponent (1 BCD digit)
//     e = exponent (4 BCD digits)
//     m = normalized significand (18 BCD digits)
//         (in positive and negative numbers the first significand
//          digit must be non-zero)
//     . = implicit decimal point between the first and second
//         significand digits
//
_asm
{

        MOV     ESI,bcdBuf      // Pointer to BCD
        MOV     EDI,ascBuf      // Pointer to result
        CALL    BCD12_TO_ASCSTR
}
return;
}
```

Chapter 3

Machine Arithmetic

Chapter Summary

This chapter is about arithmetic operations on the Intel central processors, including operations on ASCII decimal and binary code decimal numbers. Chapter topics relate to the characteristics of the various Intel Central Processing Units and to the logical, arithmetic, and bit manipulation instructions. It includes a description of the instructions most used in performing calculations in decimal arithmetic. The material is intended as background for the processing routines developed later in the book, as well as a short refresher course in 80x86 assembler.

3.0 Intel Microprocessors

The central processor used in all PC models is a microprocessor designed and manufactured by Intel Corporation. The original IBM PC, released in 1981, used the Intel 8088 CPU. Later various versions and upgrades of the 8088 chip were used in other IBM and IBM-compatible machines. In summary: the 8088 CPU is found in the PC, PC XT, and PCjr. The 8086 is used in the PS/2 Model 25 and Model 30; the 80286 in the PC AT, the PS/2 Model 50 and Model 60, and in some PS/1 machines. The 80386 is found in the PS/2 Model 70, Model 70P, and Model 80, as well as in IBM-compatible machines of that same generation. The 486 was first offered as an upgrade for the PS/2 Model 70 A-21, and as standard equipment in the Model 56, Model 57, Model 76, Model 90, and Model 95. The first IBM machines equipped with the Pentium CPU are the PS/2 Model 9595 Server and the Value Point Models 6384-189 and 6384-199. Hundreds of models of IBM-compatible microcomputers also use the Pentium, Pentium Pro, and Pentium III, Pentium IV, and P5. Since all of these processors, from the 8088 to the P5, share many common features, at times we will refer to them generically as the Intel 80x86 family. We mention the specific microprocessors when referring to features that are unique to a particular chip.

The numerical instructions of the Intel 80x86 microprocessors are usually classified into three general groups:

1. Logical instructions. These are called the Boolean operators in the Intel literature. The group includes instructions with the mnemonics AND, NOT, OR, and XOR. They perform the logical functions that correspond to their names.

2. Arithmetic instructions. This group of instructions performs integer addition, subtraction, multiplication, and division of signed and unsigned binary and binary coded decimal numbers.

3. Auxiliary and bit manipulation instructions. This group includes logical and mathematical instructions to shift and rotate bits, to compare operands, to test individual binary digits, and to perform various auxiliary operations.

These three instruction groups are discussed in detail starting in Section 3.1.

3.0.1 CPU Flags

All 80x86 CPUs are equipped with a special register that reflects the current status of the microprocessor. This register has been called the status word, the Status register, the Flags register, and the E flags register. The status register contains bits that are meaningful during the execution of logic and arithmetic operations. These bits sometimes called flags. Figure 3.1 shows the structure of the Status register as well as the name and position of the status bits or flags.

Figure 3.1 *CPU Status Flags*

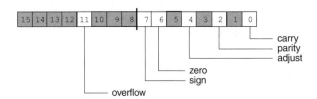

The 80286 CPU introduced several new control flags. The Status register was expanded to 32 bits in the 80386 CPU and renamed the E flags register. However, the flags used for arithmetic and logical operations, usually called the status flags, remained unchanged. Their function is as follows:

1. The carry flag is set if there has been a carry or a borrow out of the high-order bit of the operand after the execution of an instruction that affects this flag. The carry bit is often used to detect an overflow condition while performing arithmetic operations with operands that can exceed the size of the processor's registers.

2. The parity flag is set if the low-order byte of the result contains an even number of one bits (parity even) and cleared otherwise. This flag was provided for compatibility with the Intel 8080 and 8005 processors. This instruction, which is occasionally useful in data transmission and reception routines, finds little use in machine arithmetic.

3. The adjust flag, also called the auxiliary carry or half-carry flag, is set if there has been a carry or a borrow out of the low-order nibble (bits 0 to 3) of the operand. The adjust flag is used internally by several 80x86 instructions that make adjustments required in BCD arithmetic.

4. The zero flag is set if the result of an instruction that affects this flag is zero. Because not all instructions affect the zero flag, it cannot be used as an infallible test for an empty register. For example, the instruction

  MOV AX,0

loads the AX register with a value of zero but the zero flag is not affected. The logical NOT is another 80x86 instruction that does not affect the zero flag. Consequently, the following test for the zero flag is not valid.

```
MOV   AX,0FFFFH ; AX is all 1 bits

NOT   AX        ; Now AX = 0

JZ    PROGRAM_LABEL    ; Since NOT does not affect the

               ; zero flag, this jump may fail
```

5. The sign flag is set if the high-order bit of the operand is set after the execution of a CPU instruction. In signed representations the high bit is set if the number is negative, and clear if it is positive. Therefore, this flag can be used to determine the sign of the result when using signed operands. It is usually ignored when operating with unsigned numbers, since, in this case, the high-order bit is part of the numeric value.

6. The overflow flag indicates a signed positive number that is too large to represent in the format, or a signed negative number that is too small. Signed arithmetic routines test this condition to detect overflow.

  Table 3.1 lists the 80x86 logical, arithmetic, and bit manipulation instructions and their action on the status flags.

3.1 Logical Instructions

In the Intel literature the logical instructions include the Boolean operators, the TEST instruction, and the shifts and rotates. We prefer to include the shifts and rotates in a different category. Therefore we consider that the logical instructions are AND, OR, NOT, XOR, and TEST. Their action is as follows:

1. AND, OR, and XOR logically combine each bit in the source operand with the corresponding bit in the destination operand. The result does not affect the neighboring bits.

2. The NOT opcode inverts all bits in the destination operand.

  These actions explain the term bitwise operation sometimes used to describe the action of the logical instructions.

Table 3.1

Instructions that Affect the Status Flags

OPCODE	CARRY	PARITY	FLAG ADJUST	ZERO	SIGN	OVERFLOW
AAA	M	?	T/M	?	?	?
AAD	?	M	?	M	M	?
AAM	?	M	?	M	M	?
AAS	M	?	T/M	?	?	?
ADC	T/M	M	M	M	M	M
ADD	M	M	M	M	M	M
AND	0	M	?	M	M	0
CBW	X	X	X	X	X	X
CLC	0	X	X	X	X	X
DAA	T/M	M	T/M	M	M	?
DAS	T/M	M	T/M	M	M	?
DEC	X	M	M	M	M	M
DIV	?	?	?	?	?	?
IDIV	?	?	?	?	?	?
IMUL	M	?	?	?	?	M
INC	X	M	M	M	M	M
MUL	M	?	?	?	?	M
NEG	M	M	M	M	M	M
NOT	?	?	?	?	?	?
OR	0	M	?	M	M	0
RCL/RCR 1	T/M	X	X	X	X	M
RCL/RCR count	T/M	X	X	X	X	M
ROL/ROR 1	M	X	X	X	X	M
ROL/ROR count	M	X	X	X	X	?
SAL/SAR/SHL/SHR 1	M	M	?	M	M	M
SAL/SAR/SHL/SHR count	M	M	?	M	M	?
SBB	T/M	M	M	M	M	M
STC	1	?	?	?	?	?
SUB	M	M	M	M	M	M
TEST	0	M	?	M	M	0
XOR	0	M	?	M	M	0

Legend:
  T = instruction tests the status bit
  M = instruction modifies the status bit
  0 = instruction clears the status bit
  1 = instruction sets the status bit
  ? = instruction's action on the status bit is undefined
  X = instruction does not affect the status bit

3.1.1 Logical AND

The AND opcode performs a logical AND of the two operands. This action determines that a bit in the result is set if and only if the corresponding bits are set in both operands. A frequent use of the AND instruction is to clear one or more bits without affecting the remaining ones. This is possible because ANDing with a 0 bit always clears the result bit, while and ANDing with a 1 bit preserves the value of the first operand.

We can use the AND operation to isolate the 4 low-order bits as follows:

```
               hexadecimal              binary
                  34                  0011 0100
         AND      0F                  0000 1111
               _____      _____

                  04                  0000 0100
```

The second operand, in this case 0FH, is called a mask. The AND operation preserves, in the other operand, the bits that are 1 in the mask, while clearing the mask bits that are 0. Consequently, the mask 00001111B clears the four high-order bits and preserves the original value of the four low-order bits.

3.1.2 Logical OR

The OR opcode performs the logical inclusive OR of two operands. The action of the OR operation is that a bit in the result is set if and only if one or both of the corresponding bits in the operands are set. A frequent use for the OR is to selectively set one or more bits in a memory location or a machine register. The action takes place because ORing with a 1 bit always sets the result bit, while ORing with a 0 bit preserves the original value in the other operand.

For example, to set the high-order bit (bit number 7) we can OR with a 1 bit, as follows

```
               hexadecimal              binary
                  34                  0011 0100
         OR       80                  1000 0000
               ____                 _____

                  B4                  1011 0100
```

The OR operation sets the bits that are 1 in the mask and preserves the bits that are 0.

3.1.3 Logical XOR

The XOR opcode performs the logical exclusive OR of the two operands. The action of the XOR opcode is that a bit in the result is set if and only if the corresponding bits in the operands have opposite values. This explains why XORing a value with itself always generates a zero result, since, in this case, all the bits necessarily have the same value. On the other hand, XORing with a 1-bit inverts the value of the other operand, since 0 XOR 1 is 1 and 1 XOR 1 is 0. This toggling action of XORing with a 1 bit generates identical bitwise results as the NOT operation. By selecting the value of an XOR mask the programmer can control which bits of the operand are inverted and which are preserved.

You can invert the four high-order bits of an operand by XORing with a mask that has these bits set. If the four low-order bits of the mask are clear, then the original values of these bits in the original operand are preserved in the result. For example:

```
               hexadecimal              binary
                  55                  0101 0101
         XOR      F0                  1111 0000
               _____        _____

                  A5                  1010 0101
```

In this example, the XOR operation inverted the bits that are 1 in the mask and preserved the bits that are 0. In other words, the XOR mask 11110000B inverts the four high-order bits and preserves the value of the low-order bits in the original operand.

3.1.4 Logical NOT

While the AND, OR, and XOR operations require two operands, the NOT instruction acts on a single value. Its action is consistent with a Boolean NOT function, which converts all 1-bits to 0 and all 0-bits to 1. Arithmetically, the result is the one's complement of the original value (see Section 2.2.2). The NOT instruction is useful in obtaining the two's complement representation by performing the logical NOT, and then adding one to the results. In the 80x86 the NOT and increment manipulation is unnecessary since the two's complement can be obtained directly through the use of the NEG instruction.

3.2 Arithmetic Instructions

The 80x86 is not arithmetically powerful. The chip's designers did not consider it important to provide extensive arithmetic functions in the 8086, since the CPU could be optionally complemented with a mathematical coprocessor chip, originally called the 8087. The mathematical coprocessor chip was later integrated into the floating-point unit of the 486 DX and the Pentium. The 80x87 mathematical coprocessor and the math unit of the Pentium are discussed in detail, starting in Chapter 5.

The mathematical instructions in the 80x86 processors can operate on 8-, 16-, and 32-bit operands, although 32-bit operations are available only on the 80386, 486, and Pentium. The largest unsigned decimal number representable in 8 bits is 255, in 16 bits is 65,535, and in 32 bits is 4,294,967,295. Since signed representations require one bit for encoding the sign, the numerical range of signed numbers is by approximately one half of the range of unsigned numbers.

The limit determined by the largest value that can be held in a single register does not imply that machine arithmetic is restricted to this range. It is possible to develop multi-digit arithmetic routines that extend the numeric range to any value desired. In fact, the 80x86 arithmetic instructions were specifically designed to support multi-digit operations.

3.2.1 Signed and Unsigned Arithmetic

Some 80x86 arithmetic instructions can be used with either signed or unsigned operands. Such is the case with addition and subtraction, in which the opcodes ADD, ADC, SUB, and SBB can be used with signed or unsigned integers. This is possible because the instructions update the different flags in the Status register in a way that allows the results to be interpreted as either signed or unsigned.

For example, the carry flag (CF) and the zero flag (ZF) are used in operations with unsigned integers. In this case the code can test CF to determine if the addition or subtraction operation exceeded the limits of the destination operand. In single-digit routines this condition (carry flag set) typically indicates that the result is invalid, since it does not fit in the designated storage. In multi-digit routines for unsigned operands the carry flag indicates a bit that must be carried to the next digit. The 80x86 instruction set includes two opcodes, ADC (add with carry) and SBB (subtract with borrow), which take into account the status of the carry bit.

80x86 addition and subtraction opcodes serve for both signed and unsigned operands. The Intel mnemonics for signed and unsigned multiplication and division are somewhat confusing since the opcode names do not precisely represent the actual operation performed by the instruction. The instructions labeled IMUL (integer multiply) and IDIV (integer divide) are used for operations with signed numbers, while MUL (multiply) and DIV (divide) are used for unsigned operations. The designation IMUL and IDIV, in which the prefix letter "I" is usually associated with the word integer, seems to imply that fractional multiplication and division are also available, which is not the case. The 80x86 programmer must translate this use of the word integer to mean signed.

In summary: the 80x86 CPUs support addition, subtraction, multiplication, and division of signed and unsigned integers. In the case of addition and subtraction the same opcodes are used with signed and unsigned operands. Code must determine if result is signed or unsigned by interpreting the high-order bit and evaluating the status flags. In the case of multiplication and division, there are separate opcodes for signed and unsigned operations. The letter I is used as a prefix (IMUL and IDIV) in the opcodes for signed operations.

3.2.2 Operations on Decimal Numbers

Although the processors of the Intel 80x86 family are binary machines, their instruction set includes operation codes for binary-coded decimal arithmetic. In Chapter 2 you saw that BCD numbers can be stored in packed or unpacked form. In packed form two BCD digits are stored per byte. The low-order BCD digit takes up bits 0 to 3 and the high-order BCD digit takes up bits 4 to 7. Unpacked BCD digits, on the other hand, are stored one digit per byte. In the case of unpacked BCD numbers the high-order nibble is not used. The packed and unpacked binary coded decimal formats are shown in Figure 2.7.

Here again, Intel terminology can be somewhat confusing, since it refers to packed BCD numbers as decimal numbers. The word decimal, as used in the opcodes DAA (decimal adjust after addition) and DAS (decimal adjust after subtraction), actually refers to operations on packed BCD digits. DAA is used to adjust the result after adding two valid packed BCD digits in the AL register. This instruction corrects the result so that the sum is also stored in packed BCD form. When using the DAS and DAA opcodes, the carry flag (CF) is set if there is a carry out of the high digit, that is, if the sum is larger than the value 99H. The following code fragment illustrates the operation of the DAA instruction.

```
; Code fragment to illustrate the packed BCD adjustment performed
; by the DAA instruction
        MOV     AL,27H          ; AL now holds BCD digits 27
        MOV     AH,35H          ; AL holds the BCD digits 35
```

```
        ADD     AL,AH           ; AL = AL + AH
; At this point the sum in AL is the value 5CH
        DAA
; After adjustment AL = 62H
```

In interpreting the above code fragment recall that BCD digits are nibble-based values. Thus, they can be represented by hex digits in the range 0 to 9. The DAA instruction adjusts the contents of AL to two packed BCD digits after an addition operation.

Unpacked binary coded decimal digits are sometimes called ASCII numbers in Intel literature. This is an unconventional interpretation of the term ASCII, which is actually a standard character encoding. The instruction AAA (ASCII adjust after addition) corrects the contents of the AL register after the addition of two un-packed BCD digits. The adjustment always implies zeroing the 4 high-order bits of the result, since this bit pattern is necessary in the unpacked BCD representation. If the sum exceeds the value 9, the carry flag is set and AH is incremented. The following code fragment illustrates the action of the AAA instruction.

```
; Action of AAA instruction in the addition of two unpacked
; BCD numbers
        MOV     AH,0            ; Clear AH
        MOV     AL,7H           ; AL = unpacked BCD 7
        MOV     CL,3H           ; CL = unpacked BCD 3
        ADD     AL,CL           ; AL now holds 0AH
        AAA                     ; Adjust result
; AH = 01H and AL = 00H. Carry flag is set
```

As can be seen in the above fragment, the instruction AAA (ASCII adjust after addition) modifies the AL register after an addition operation of two unpacked BCD digits so that the result is correctly represented. The four high-order bits of the result are zeroed. If a borrow was required, AH is incremented and the carry flag is set.

The instruction AAM (ASCII adjust after multiplication) corrects the product after the multiplication of two unpacked BCD numbers. The high-order digit is found in AH and the low-order digit in AL. If the digits used in the multiplication are valid, unpacked BCD numbers, then there can be no overflow, since the prod-uct of two unpacked BCD digits cannot exceed the destination storage. The following fragment illustrates the action of the AAM instruction.

```
; Action of AAM instruction in the multiplication of two
; unpacked BCD numbers
        MOV     AH,0            ; Clear AH
        MOV     AL,9H           ; AL = unpacked BCD 9
        MOV     CL,8H           ; CL = unpacked BCD 8
        MUL     CL              ; AL now holds 48H
        AAM                     ; Adjust result
; AH = 07H and AL = 02H
```

AAD (ASCII adjust before division) modifies the contents of the dividend in AL so that the result of a division is a valid, unpacked, binary coded decimal digit.

Notice that division of unpacked BCD numbers on the 80x86 requires that the AH register be zero before executing the DIV instruction. After the execution of AAD, the AL register holds the quotient digit and the AH holds the remainder. The high-order nibbles of AH and AL are zero. Also notice that the AAD instruction must be executed before performing the arithmetic operation, in contrast with the other three adjustment instructions used in unpacked BCD arithmetic. The following fragment illustrates the action of the AAD instruction.

```
; Action of AAD instruction in the division of two unpacked
; BCD numbers
        MOV     AH,0          ; Clear AH
        MOV     AL,9H         ; AL = unpacked BCD 9
        MOV     CL,4H         ; CL = unpacked BCD 4
        ADD                   ; Adjust dividend for division
        DIV     CL            ; Divide 9/4 = 2, 1 remainder
; AH = 01H (remainder) and AL = 02H (quotient)
```

3.3 Auxiliary and Bit Manipulation Instructions

We include in this group all the 80x86 opcodes that are related to the logical and mathematical instructions or that serve to support them. In the Intel classification these instructions are contained in the arithmetic and in the bit manipulation group. The instruction group can be subdivided as follows:

1. Bit shift and rotate instructions, including the double precision shift opcodes introduced with the 80386.

2. Comparison instructions. This sub-group includes CMP (compare) and TEST opcodes and the bit scan and bit test opcodes introduced with the 80386 CPU.

3. Type conversion and pointer handling instructions. This group includes the opcodes to perform sign extension operations as well as the INC (increment) and DEC (decrement) opcodes. INC and DEC are typically used in pointer arithmetic.

4. New arithmetic opcodes introduced with the 486 and Pentium, namely: BSWAP (byte swap), XADD (exchange and add), CMPXCHG (compare and exchange), and CMPXCHG8B (compare and exchange 8 bytes).

3.3.1 Bit Shift and Rotate Instructions

The opcodes in the shift and rotate group perform bit manipulations that are often useful in BCD and binary arithmetic routines. These bit handling operations can also be used to implement multiplication and division by bit shifting. Although, in programming the 80x86 CPUs, the multiplication and division opcodes, mentioned earlier in this chapter, are usually preferred.

Bit Shift Instructions

Shift instructions transpose to the left or right all the bits in the operand. In 80x86 systems the operand can be a processor register or a memory variable. For example, after a right shift operation all the bits in the value 01110101B (75H) are moved one position to the right, resulting in the value 00111010B (3AH). Notice that on a right shift the right-most bit disappears. The 80x86 opcode SHR (shift logical right) performs the operation described above. In this opcode the right-most bit is moved into the carry flag. Figure 3.2 shows the action of the 80x86 shift instructions.

The 80x86 opcodes for performing a bit shift to the left are SHL (shift logical left) and SAL (shift arithmetic left). Notice that SHL and SAL are different mnemonics for the same operation (see Figure 3.2). In SHL and SAL it is the left-most bit of the operand that is moved into the carry flag.

The terms logical and arithmetic, as used in the SHL and SAL opcodes, reflect a potential problem associated with shifting bits in a signed representation. The problem is that negative numbers in two's complement form always have the high bit set. Therefore, when the bits of a two's complement number are shifted, the sign bit can change unpredictably. For this reason, in left-shift operations of signed operands the sign bit is moved into the carry flag. After performing the shift, software can test the carry flag and make the necessary adjustments.

On the other hand, in a right-shift operation the sign bit is moved from bit number 7 to bit number 6, and a zero bit is introduced into the sign bit position. This action makes all signed numbers positive. In order to make possible shift operations of signed numbers the 80x86 instruction set has a separate opcode for the right-shift of signed numbers. The SAR opcode (shift arithmetic right) preserves the sign bit (bit number 7) while shifting all other bits to the right. This action can be seen in the diagram for the SAR instruction in Figure 3.2. Note that, in the SAR instruction, the left-most bit (sign bit) is both preserved and shifted. For example, the value 10000000B becomes 11000000B after executing the SAR operation. This action is sometimes called a sign extension operation.

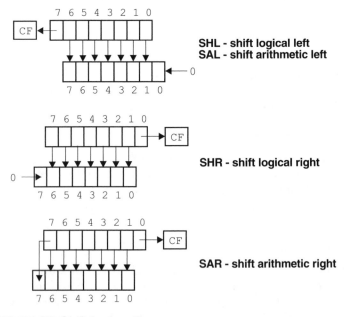

Figure 3.2 *80x86 Bit Shift Instructions*

The 8-bit microprocessors that preceded the 80x86 family (such as the Intel 8080, the Zilog Z80, and the Motorola 6502) did not include multiplication and division instructions. In these chips multiplication and division had to be performed by software. One approach to multiplication was through repeated addition. Occasionally this approach is still useful. The following code fragment illustrates multiplication by repeated addition using 80x86 code.

```
; Multiplication of AL * CX using repeated addition
      MOV     AH,0            ; Clear register used to
                             ; accumulate sum
      MOV     AL,10           ; Load multiplicand
      MOV     CX,6            ; Load multiplier
MULTIPLY:
      ADD     AH,AL           ; Add AL to sum in AH
      LOOP    MULTIPLY
; AH now holds product of 10 * 6
```

An often-used method for performing fast multiplication and division operations is by shifting the bits of the operand. This method is based on the positional properties of the binary number system. In the binary number scheme the value of each digit is a successive power of 2 (see Chapter 1). Therefore, by shifting all digits to the left, the value 0001B (1 decimal) successively becomes 0010B (2 decimal), 0100B (4 decimal), and 1000B (8 decimal).

A limitation of binary multiplication by means of bit shift operations is that the multiplier must be a power of 2. If not, then the software must shift by a power of 2 that is smaller than the multiplier and add the multiplier as many times as necessary to complete the product. For example, to multiply by 5 we can shift left twice and add once the value of the multiplicand.

A more practical approach can be based on the same algorithm used in longhand multiplication. For example, the multiplication of 00101101B (45 decimal) by 01101101B (109 decimal) can be expressed as a series of products and shifts, in the following manner:

```
                        0 0 1 0 1 1 0 1 B = 45 decimal
            times       0 1 1 0 1 1 0 1 B = 109 decimal
                        -------------------------------
                          0 0 1 0 1 1 0 1
                        0 0 0 0 0 0 0 0
                      0 0 1 0 1 1 0 1
                    0 0 1 0 1 1 0 1
                  0 0 0 0 0 0 0 0
                0 0 1 0 1 1 0 1
              0 0 1 0 1 1 0 1
            0 0 0 0 0 0 0 0
            -------------------------------------------
            0 0 1 0 0 1 1 0 0 1 0 1 0 0 1 B = 4905 decimal
```

The actual calculations using this method of binary multiplication are quite simple, since the product by a 0 digit is zero and the product by a 1 digit is the multiplicand itself. The multiplication routine simply tests each digit in the multiplier. If the digit is 1, the multiplicand is shifted left and added into an accumulator. If the digit is 0, then the bits are shifted but the addition is skipped.

Shift-based multiplication routines were quite popular in processors that were not equipped with a multiplication instruction. In the case of the 80x86 there seems to be little use for multiplication routines based on bit shifts, since the processor is capable of performing efficient multiplications internally. For this reason, 80x86 programmers find little practical use for the SAR and SAL opcodes in developing arithmetic routines, although these opcodes are still useful for other bit manipulations.

Bit Rotate Instructions

The 80x86 rotate instructions also shift the bits in the operand to the left or right. The difference between the shift and the rotate is that in the rotate the bit shifted out is either re-introduced at the other end of the operand or is stored in the carry flag. The ROL opcode (rotate left) shifts the bits to the left while the high-order bit is cycled back to the low-order bit position, as well as stored in the carry flag. The ROR opcode operates in a similar manner, except that the action takes place left-to-right. In both instructions, ROL and ROR, the carry flag is used to store the recycled bit, which can be conveniently tested by the software. Figure 3.3 shows the action of the 80x86 rotate instructions.

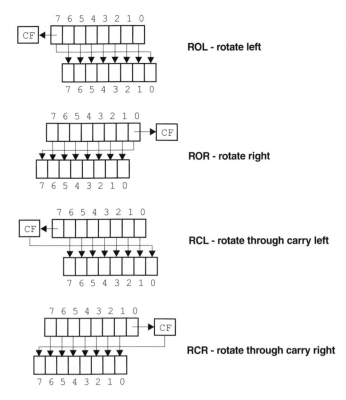

Figure 3.3 *80x86 Bit Rotate Instructions*

Two rotate instructions, RCL (rotate through carry left) and RCR (rotate through carry right), use the carry flag as a temporary storage for the bit that is shifted out. This action can be seen in the diagrams of Figure 3.3. Note that the bit shifted out is not recovered at the other end of the operand until the instruction is re-executed. It is also interesting that by repeating the rotation as many times as there are bits in the destination operand the rotate instructions preserve the original value. This requires rotating a byte-size operand 8 times, a word-size operand 16 times, and so on.

Double Precision Shift Instructions

The 386 introduced two new opcodes for performing bitwise operations on long bit strings. These opcodes have the mnemonic SHLD (double precision shift left) and SHRD (double precision shift right). The instructions are also available in the 486 and the Pentium.

The double precision shift instructions SHLD and SHRD require 3 operands. For example:

```
SHLD      AX,BX,12
```

The left-most operand (AX) is the destination of the shift. The right-most operand (12) is the bit count. The middle operand (BX) is the source. The bits in the source operand are moved into the destination operand, starting with the sources' high order bits. Source and destination must be of the same size, for example, if the destination is a word-size register then the source has to be a word size register or memory variable. By the same token, if the destination is a doubleword register or memory location then the source must also be 32-bits wide. Either source or destination may be a memory operand, but at least one of them must be a machine register. The count operand can be an immediate byte or the value in the CL register. The limit of the shift count is 31 bits. The following code fragment shows a double precision bit shift.

```
; Demonstration of the action preformed by the double precision
; shift left (SHLD)
        MOV     EAX,3456H        ; One operand to destination
        MOV     EBX,10000000H    ; Source operand
        SHLD    EAX,EBX,4        ; Shift left EAX digits 4 bits
                                 ; and introduce EBX bits into
                                 ; EAX bits vacated by the shift
; At this point:
;       EAX = 34561
;       EBX = 10000000 (unchanged)
```

The most common used of the SHLD and SHRD instructions is in manipulating long bit strings. For example, you can overlay a memory variable with a register value, as in the following code fragment using inline assembly:

```
int var1;
main()
{
_asm
{
        MOV    EBX,12300000H      ; Source operand
```

```
        SHLD   var1,EBX,12
// ASSERT:
//      VAR1 = 123H
}
}
```

In the above code fragments notice that the SHLD instruction has been used to shift 4 packed BCD digits. The digit shift is accomplished by selecting a bit count that is a multiple of 4, since each digit takes up 4 bits. In this manner a bit count of 8 would have shifted 2 packed BCD digits. Also notice that the source register is unchanged by the double precision shift.

Shift and Rotate Addressing Modes

The addressing modes for shift and rotate opcodes have undergone several changes in the different microprocessors of the 80x86 line. In the 8086 and 8088, shift and rotate can use a count in the CL register or the number 1 as an immediate operand. Later processors allow an 8-bit immediate operand. The following code fragment illustrates the valid addressing modes in each case.

```
; Shift and rotate addressing modes in the 8086 and 8088 chips
      SHL    AL,1        ; Shift left 1 bit position
      MOV    CL,4        ; Shift count to CL
      SHL    AL,CL       ; Shift left 5 bit positions
       .
       .
       .
; Shift and rotate addressing modes in the 80286, 80386, 486,
; and Pentium, in which an 8-bit immediate operand can be specified
; directly
      SHR    AX,3        ; Shift right 3 bits
       .
       .
       .
; In the 80386, 486, and Pentium the shift and rotate opcodes allow
; a 32-bit register operand as a destination, for example
      SHL    EBX,4       ; Shift EBX 4 bits
       .
       .
       .
```

3.3.2 Comparison, Bit Scan, and Bit Test Instructions

The CMP (compare) instruction changes the flags as if a subtraction had taken place but does not change the value of the operands. The action can be described as setting the Status register as if the source operand had been subtracted from the destination. The instruction is typically followed by a conditional jump. The following code fragment shows the use of CMP in determining the relative value of an operand in a machine register.

```
; Use of CMP to determine if BX > AX, BX < AX, or BX = AX
; Code assumes that the values in AX and BX are unsigned binary
      CMP    AX,BX       ; Simulate AX minus BX
      JA     AX_ABOVE    ; Go if AX > BX
      JB     AX_BELOW    ; Go if AX < BX
; At this point AX = BX
       .
```

```
        .
        .
        .
; Entry point for AX > BX
AX_ABOVE:
        .
        .
        .
; Entry point for AX < BX
AX_BELOW:
        .
        .
        .
```

The TEST instruction performs a logical AND and updates the flags without changing the operands. If a TEST instruction is followed by JNZ, the jump is taken if there are matching 1-bits in both operands. The following code fragment shows the use of the TEST opcode.

```
; Use of TEST to determine if bit 7 of the AL register is set
      TEST    AL,10000000B    ; ANDing AL and binary mask
      JNZ     HIGH_BIT_SET    ; Go if AL bit 7 = 1
; At this point AL bit 7 = 0
        .
        .
        .
; Entry point for AL bit 7 set
HIGH_BIT_SET:
        .
        .
        .
```

The 80386 CPU introduced several new bit manipulating instructions that allow more elaborate bit scanning and testing. The BSF (bit scan forward) opcode scans the source operand low-to-high and stores, in the destination operand, the bit position of the first 1-bit found. If all bits of the source operand are 0, then the zero flag is set, otherwise the zero flag is cleared. BSR (bit scan reverse) performs the same test but starting at the high-order bit position. Both instructions require word or doubleword operands; byte operands are not allowed. The following code fragment shows the operation of BSF.

```
; Use of the BSF and BSR instructions to determine the number of
; the first bit set in the source operand.
      MOV     AX,10001000B    ; Right-to-left first bit
                              ; set is number 3
      BSF     BX,AX           ; AX bit number into BX
; At this point BX = 03 since the first bit set is in bit
; position number 3 when read low-to-high. Zero flag is clear
      BSR     CX,AX           ; AX bit number into CX
                              ; read high-to-low
; At this point CX = 07 since bit number 7 of AX is the first
; bit set when read high-to-low. Zero flag is clear
```

The bit test opcodes BT (bit test), BTS (bit test and set), BTR (bit test and reset), and BTC (bit test and complement) were also introduced with the 386 processor. All of these opcodes copy the value of a specified bit into the carry flag. The code can later include a JC or JNC instruction to direct execution according to the state of the carry flag. In addition, the bit tested can be modified in the destination operand: BTS sets the tested bit, BTR clears the tested bit, and BTC complements the tested bit. The following code fragment shows the action of these opcodes.

```
; Use of BT, BTS, BTR, and BTC opcodes to test and manipulate
; bits according to their position
      MOV    AX,10001000B    ; Set value in operand
      BT     AX,3            ; Test AX bit 3
; Carry flag is set since AX bit 3 is set. AX is not changed
      BTS    AX,0            ; Test AX bit 0
; Carry flag is clear since AX bit 0 is not set
; AX = 10001001B since the instruction sets the specified bit
      BTR    AX,7            ; Test AX bit 7
; Carry flag is set since AX bit 7 is set
; AX = 00001001B since bit 7 is reset (cleared) by BTR
      BTC    AX,1            ; Test AX bit 1
; Carry flag is clear since bit 1 is cleared
; AX = 00001011B since bit 1 is toggled (complemented) by BTC
```

Signed and Unsigned Conditional Jumps

The 80x86 provides two categories of conditional jump opcodes: one for operating on integers and one for operating on signed numbers in two's complement form. For example, JA (jump if above) and JB (jump if below) assume that the operands are unsigned integers while JG (jump if greater) and JL (jump if less) assume that the operands are signed numbers in two's complement format. Table 3.2 shows the 80x86 conditional jump instructions according to their signed or unsigned interpretation.

Notice in Table 3.2 that the conditional jump instructions that assume signed operands use the sign and the overflow flag to determine their action. The sign flag is clear when the result of the operation is a binary positive number, that is, one in which the high bit is 0. The sign flag is set if the result of the previous operation is a binary negative number, that is, one in which the high bit is set. On the other hand, unsigned arithmetic routines usually ignore the sign flag since the high-order bit of unsigned binary numbers is interpreted as value. The overflow flag indicates a signed positive number that is too large to represent in the format, or a signed negative number that is too small. In signed arithmetic this flag indicates an overflow, however, it is usually ignored when operating on unsigned binary numbers.

Several jump instructions in Table 3.2 are based on the parity flag, namely: JNP (jump if no parity), JPO (jump if parity odd), JP (jump if parity), and JPE (jump if parity even). This flag is set if the low-order eight bits of the result contain an even number of 1-bits (parity even) and cleared otherwise. This flag was provided for compatibility with the Intel 8080 and 8005 processors. Although the parity flag can be used to assure the integrity of data transmissions, it has no application in arithmetic or logic routines.

Table 3.2

x86 Conditional Jumps

MNEMONIC	FLAG ACTION	DESCRIPTION
CONDITIONAL JUMPS THAT ASSUME UNSIGNED OPERANDS		
JA	(CF or ZF) = 0	jump if above
JNBE		jump if not below or equal
JAE	CF = 0	jump if above or equal
JNB		jump if not below
JNC		jump if no carry
JB	CF = 1	jump if below
JNAE		jump if not above or equal
JC		jump if carry set
JBE	(CF or ZF) = 1	jump if below or equal
JNA		jump if not above
JE	ZF = 1	jump if equal
JZ		jump if zero
JNE	ZF = 0	jump if not equal
JNZ		jump if not zero
JNP	PF = 0	jump if no parity
JPO		jump if parity odd
JP	PF = 1	jump if parity
JPE		jump if parity even
CONDITIONAL JUMPS THAT ASSUME SIGNED OPERANDS		
JG	((SF xor OF) or ZF) = 0	jump if greater
JNLE		jump if not less or equal
JGE	(SF xor OF) = 0	jump if greater or equal
JNL		jump if not less
JL	(SF XOR OF) = 1	jump if less
JNGE		jump if not greater or equal
JLE	((SF xor OF) or ZF) = 1	jump if less or equal
JNG		jump if not greater
JNO	OF = 0	jump if no overflow
JNS	SF = 0	jump if positive (no sign)
JO	OF = 1	jump if overflow
JS	SF = 1	jump if negative (sign set)

Legend:
CF = carry flag   ZF = zero flag   PF = parity flag
SF = sign flag   OF = overflow flag

3.3.3 Increment, Decrement, and Sign Extension Instructions

The INC (increment) instruction adds 1 to the value of the destination while the DEC (decrement) instruction subtracts 1. INC and DEC are often used in manipulating pointers although they find occasional application in arithmetic routines, mainly in adjusting after overflow or underflow conditions. Both instructions assume that the operand is an unsigned integer, therefore they do not affect the carry flag. For this reason, when operating with signed magnitudes it is preferable to use the ADD and SUB instructions.

The 80x86 instruction set also includes several opcodes whose action is often described as performing a sign extension of the source operand. CBW (convert byte to word) converts a signed byte in two's complement form into a signed word, also in two's complement. The source is always the AL register and the destination is AX. The conversion is performed by copying the most significant bit of AL into all AH bits. Therefore the signed value 0083H is converted into FF83H, hence the use of the term sign extension to describe its action. The opcode CWD (convert word to doubleword) performs the same conversion regarding a word in AX to a doubleword in DX:AX.

The 80386 processor introduced two new sign extension instructions designed to operate on 32-bit and 64-bit operands. CWDE (convert word to doubleword extended) converts a signed 16-bit number in AX into a signed 32-bit number in EAX. The CDQ (convert doubleword to quadword) assumes a two's complement number in EAX and converts it into a signed 64-bit integer in EDX:EAX. The sign extension opcodes are useful in performing signed multiplication and division when one of the operands is in a different format than the destination. The following code fragment is a demonstration of the use of the CBW instruction.

```
; Use of CBW to multiply a signed word operand in BX by a
; signed byte in AL
        MOV     BX,-1234    ; Load byte multiplier
        MOV     AL,-104     ; Load multiplicand (98H)
        CBW                 ; Convert to word
; At this point AX holds FF98H (signed byte converted to word)
        IMUL    BX          ; -1234 * -104
; Result of -1234 * -104 is 128,336. The product is stored
; in DX:AX as 0001:F550H
```

3.3.4 486 and Pentium Proprietary Instructions

The 486 and Pentium processors introduced 4 new instructions that are related to arithmetic processing; these are: BSWAP (byte swap), XADD (exchange and add), CHPXCHG (compare and exchange), and CMPXCHG8B (compare and exchange 8 bytes).

BSWAP

The BSWAP instruction reverses the byte order in a 32-bit machine register. One use of BSWAP is in converting data between the little endian and the big endian formats. In this sense it is possible to use BSWAP to reverse the order of unpacked decimal digits loaded from a memory operand into a 32-bit machine register. For example: assume four unpacked decimal digits are stored in a memory operand with the least significant digit in the lowest order location, as would be the case in a conventional BCD format. When these digits are loaded into a machine register by means of a MOV instruction their order would be reversed. The following code simulates this situation.

```
DATA    SEGMENT
FOUR_DIGS   DB      01H,02H,03H,04H
DATA    ENDS
```

If these digits are now loaded into a 32-bit machine register, typically by means of a pointer register, their order would be reversed, as shown in the following fragment.

```
      LEA    SI,FOUR_DIGITS     ; Pointer to unpacked BCD
      MOV    EAX,DWORD PTR [SI] ; Load EAX using pointer
; EAX = 04030201H
```

At this point the unpacked BCD digits are reversed in the EAX register. In a Pentium machine the situation can be easily corrected by means of the BSWAP instruction. The instruction would reverse the bytes in EAX, as follows

```
      BSWAP  EAX                     ; Swap bytes in EAX
; EAX = 01020304H
```

Figure 3.4 shows the action of the BSWAP instruction.

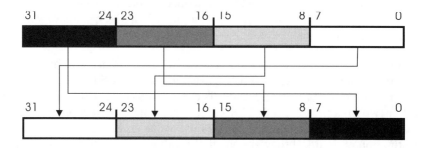

Figure 3.4 *Action of the 486 BSWAP Instruction*

In a 386 CPU reversing the byte order in a 32-bit register requires several XCHG (exchange) operations. The following procedure simulates the BSWAP in a 80386 machine.

```
BSWAP_EAX       PROC    NEAR
; Simulate the 486 BSWAP EAX instruction on a 386 machine
; Comments assume that on entry EAX = 0403 0201H
; After byte inversion EAX will hold 0102 0304H
;
      PUSH    EBX               ; Save EBX in stack
      MOV     EBX,EAX           ; Copy EAX in EBX
      SHR     EBX,16            ; Shift high word into low word
; At this point:
;             EAX = 0403 0201H
;             EBX = 0000 0403H
      XCHG    AH,AL             ; EAX = 0403 0102H
      SHL     EAX,16            ; EAX = 0102 0000H
      XCHG    BH,BL             ; EBX = 0000 0304H
      OR      EAX,EBX           ; EAX = 0102 0304H
      POP     EBX               ; Restore EBX
      RET
BSWAP_EAX       ENDP
```

XADD

The 486 XADD (exchange and add) instruction requires a source operand in a machine register and a destination operand, which can be a register or a memory variable. When XADD executes, the source operand is replaced with the destination and the destination is replaced with the sum of both original operands. The main purpose of this instruction is to provide a multiprocessor mechanism whereby several CPUs can execute the same loop.

CMPXCHG and CMPXCHG8B

The 486/Pentium CMPXCHG (compare and exchange) opcode requires three operands. The source must be a machine register. The destination can be either a machine register or a memory variable. The third operand is the accumulator, which can be either AL, AX, or EAX. If the value in the destination and the accumulator are equal then CMPXCHG replaces the destination operand with the source. In this case the zero flag (ZF) is set. Otherwise, the destination operand is loaded into the accumulator. In either case the flags are set as if the destination operand had been subtracted from the accumulator. Intel documentation states that CMPXCHG is primarily intended for manipulating semaphores.

The Pentium processor includes a version of the compare and exchange opcode with the mnemonic CMPXCHG8B (compare and exchange 8 bytes). Like CMPXCHG, CMPXCHG8B requires three operands. The destination must be a memory variable. The other two operands are a 64-bit (8 byte) value in EDX:EAX and a 64-bit value in ECX:EBX. When the instruction executes the value in EDX:EAX is compared with the destination operand. If they are equal, the value in ECX:EBX is then stored in the destination. In this case the zero flag is set. If they are not equal then the destination is loaded into EDX:EAX. In this case the zero flag is cleared. Intel documentation states that CMPXCHG8B is also intended for manipulating semaphores.

3.4 CPU Identification

Software often needs to determine on which version of the CPU the program is running in order to use or bypass one or more instructions or to select among available features. For example, previously we developed a procedure named BSWAP_EAX, which simulates the 486/Pentium BSWAP of the EAX register on a 386 machine. In order to develop code that can execute in any machine environment it is possible to create several alternative processing routes. A CPU test function can be called to determine which processing branch is required.

In later versions of the 486 CPU, Intel introduced an instruction named CPUID. This instruction can be used to obtain information about the vendor, as well as the CPU family, model, and stepping mode. The information returned by the instruction depends on the value passed in the EAX register. If CPUID is executed with 0 in EAX, then the instruction returns in EAX the highest input parameter that it can understand. For a Pentium family processor the smallest value returned in EAX is 1. Also in this case the EBX, EDX and ECX registers may contain a string that identifies the CPU vendor. If the Pentium is made by Intel Corporation, the string is "GenuineIntel." Other vendors may provide a different identification string.

If the CPUID instruction is executed with a value of 1 in EAX, then it returns additional CPU information. Other values can also be loaded in EAX according to the CPU processor family. Table 3.3 lists the values returned by several implementations of the CPUID instruction.

Table 3.3

Information Returned by CPUID Instruction

EAX VALUE	INFORMATION PROVIDED
0H	EAX = maximum input understood by CPUID EBX = "Genu" (756E6547H) EDX = "inel" (49656E69H) ECX = "ntel" (6C65746EH)

1H	EAX = version (type, family, model, and stepping ID)
	EBX = brand index
	EDX = feature information:
	    Bit: description
	    0 math unit on chip
	    1 Virtual 8086 mode enhancements
	    2 debugging extensions
	    3 page size extensions
	    ... other information according to CPU version

2H	EAX-EBX-ECX-EDX = cache and TLB information
3H	ECX-EDX = Processor serial number

The following function, named IdCpu(), tests for five different CPU options used in IBM microcomputers: 8086/8088, 80286, 80386, 486, and Pentium. If the CPU is a Pentium then the CPUID instruction is executed with a value of 0 in EAX to test for a "GenuineIntel" signature. If the signature is "GenuineIntel" then the CPUID instruction is executed a second time with a value of 1 in EAX. When execution returns to the caller the variables passed as an argument hold a CPU identification code. If the processor was a Pentium made by Intel, then a second variable contains the version information.

```
void IdCpu(int *CPUtype, int *Cid)
{
_asm
{
// Function to determine the CPU in a PC
// Post:
// Parameter CPUtype as follows:
//          1 if CPU is 8086 or 8088
//          2 if CPU is 80286
//          3 if CPU is 80386
//          4 if CPU is 486
//          5 if CPU is Pentium
// Parameter Cid contains the CPU identification code
// if processor id string is 'GenuineIntel'
// Bits are as follows:
//      xxxxxxxx xxxxxxxx xxxxxxxx xxxxxxxx <= bits
//      |<-- ignored -------->|--| |---|--|
//                             |    | |   |____[3-0] stepping ID
//                             |    |_____[7-4] model number
//                             |_____[11-8] family
// Otherwise Cid is unchanged

;**********************|
;    test for 8086/8088  |
;**********************|
```

```
; Bits 12 to 15 in the flag register are always set in the 8086
; and 8088 CPU
        PUSHF                       ; Flag register to stack
        POP     AX                  ; Store flags in AX
        AND     AX,0FFFH            ; Clear bits 12 to 15
        PUSH    AX                  ; AX to stack
        POPF                        ; and to flags register
        PUSHF                       ; Flags to stack
        POP     AX                  ; and to AX for reading
        AND     AX,0F000H           ; Preserve bits 12 to 15
        CMP     AX,0F000H           ; Test for bits set
        JNE     TEST_286            ; Go if bits not set
; At this point processor is a 8086 or 8088
        MOV     AX,1                ; Return code
        MOV     DX,0
        JMP     ID_EXIT             ; Exit
;***********************|
;     test for 80286    |
;***********************|
; Bits 12 to 15 in the flag register are always clear in the Intel
; 80286 CPU
TEST_286:
        PUSHF                       ; Flag register to stack
        POP     BX                  ; Store flags in BX
        OR      BX,0F000H           ; Make sure bit field is set
        PUSH    BX                  ; To stack
        POPF                        ; And to flag register
        PUSHF                       ; Flags to stack
        POP     AX                  ; And to AX
        AND     AX,0F000H           ; Clear all other bits
        JNZ     TEST_386            ; Go if bits not clear
; At this point processor is an 80286
        MOV     AX,2                ; Return code
        MOV     DX,0
        JMP     ID_EXIT             ; Exit
;***********************|
;     test for 80386    |
;***********************|
; Bit 18 of the E flags register was introduced in the 486 CPU
; This bit cannot be set in the 80386
TEST_386:
        PUSHFD                      ; 32-bits E flags to stack
        POP     EAX                 ; Flags to EAX
        OR      EAX,40000H          ; Make sure bit 18 is set
        PUSH    EAX                 ; New flags to stack
        POPFD                       ; An to E flags register
        PUSHFD                      ; Back to stack
        POP     EAX                 ; And to EAX
        AND     EAX,40000H          ; Clear all except bit 18
        JNZ     TEST_486            ; Go if bit 18 is clear
; At this point processor is a 80386
        MOV     AX,3                ; Return code
        MOV     DX,0
        JMP     ID_EXIT             ; Exit
;***********************|
;     test for 486      |
;***********************|
; Bit 21 (ID flag) of the E flags register cannot be set in the
; 486
TEST_486:
```

```
        PUSHFD                      ; 32-bits E flags to stack
        POP       EAX               ; Flags to EAX
        OR        EAX,200000H       ; Make sure bit 21 is set
        PUSH      EAX               ; New flags to stack
        POPFD                       ; An to E flags register
        PUSHFD                      ; Back to stack
        POP       EAX               ; And to EAX
        AND       EAX,200000H       ; Clear all except bit 21
        JNZ       IS_PENTIUM        ; Go if bit 21 is clear
; At this point processor is a 486
        MOV       EAX,4               ; Return code
        MOV       EDX,0
        JMP       ID_EXIT             ; Exit
;**********************|
;  processor is PENTIUM |
;**********************|
IS_PENTIUM:
;**********************|
;    use CPUID         |
;**********************|
        MOV       EAX,0
        CPUID
        CMP       EBX,'uneG'
        JE        IS_INTEL
        MOV       EAX,5             ; Is Pentium type
        JMP       ID_EXIT           ; but not Intel
IS_INTEL:
        MOV       EAX,1
        CPUID
; At this point:
; EAX = contains the following information
//      xxxxxxxx xxxxxxxx xxxxxxxx xxxxxxxx <= bits
//      |<-- ignored -------->|--| |---|--|
//                            |    |   |   |____[3-0] stepping ID
//                            |    |   |____[7-4] model number
//                            |____[11-8] family
;
        MOV       EDI,Cid
        MOV       [EDI],EAX
        MOV       EAX,5             ; Pentium code
ID_EXIT:
        AND       EAX,0FH           ; Clear all other bits
        MOV       EDI,CPUtype
        MOV       [EDI],EAX
}
}
```

SOFTWARE ON-LINE

The Id CPU() function is found in the file Id CPU.h located in the folder Sample Code\Chapter02\Id CPU in the book's on-line software. The program Id CPU.cpp, also in this folder, calls the Id CPU() function and interprets the results.

Chapter 4

High-Precision Arithmetic

Chapter Summary

This chapter is about the algorithms and functions used in performing fundamental arithmetic operations on packed BCD numbers. We develop C++ interface functions for multi-digit BCD addition, subtraction, multiplication, and division. The chapter concludes with the development of high-precision BCD-arithmetic functions that allow manipulating numbers with 34 significant digits.

4.0 Applications of BCD Arithmetic

The Intel mathematical coprocessor and math units are indeed powerful calculating tools. These devices store and manipulate floating-point numbers according to the formats defined in the ANSI/IEEE 754 standard. C and C++ use these standards in representing floating point numbers. The C/C++ float type corresponds to ANSI/IEEE single format and the C/C++ double type to ANSI/IEEE double format.

Table 2.2 shows that the significand in the ANSI/IEEE 754 double format is 53 binary digits wide, to which we must add an implicit 1-bit. The largest decimal significand allowed in 54 bits is 720,575,940,379,277,743, which makes it possible to represent up to 18 significant digits. This precision is sufficient for many mathematical applications; however, in science, business, and technology we occasionally need to represent numbers of more than 18 significant digits. When this is the case, the programmer must take on the task of encoding numeric values and performing the necessary calculations.

One option for representing numeric values and performing calculations to higher precision than ANSI/IEEE 754 is BCD arithmetic. The main disadvantages of BCD arithmetic on the main CPU, compared to floating-point calculations using the math unit, is that BCD code executes much slower and that encodings take up more space. The one major advantage of developing BCD arithmetic routines is that the precision of the calculations is not limited by the design of Intel floating-point hardware. Numeric operations on the floating-point units, such as the math unit of the Pentium and the MMX, must be performed in the specific numeric data formats that are built into the hardware. We have seen that, with present day floating-point hardware, the maximum numeric precision of the result is of 18 significant digits. The use of floating-point BCD arithmetic is an option when designing routines that are capable of mathematical calculations to any desired precision.

Another consideration that, on occasions, favors the use of BCD arithmetic relates to round-off errors. The math unit is a binary machine and decimal numbers must be converted to binary before processing. After the calculations have concluded, the results must be converted back to decimal numbers for output. The binary-to-decimal and decimal-to-binary conversions often introduce errors, since many decimal numbers cannot be exactly represented in binary. BCD arithmetic, on the other hand, is decimal arithmetic. In BCD arithmetic no conversion errors are introduced.

In developing the BCD arithmetic routines that are the topic of this chapter we continue using the BCD12 format that was introduced in Chapter 2. However, the BCD12 format is limited to numbers with 18 significant digits, which is approximately the same precision of the Intel floating-point hardware. To make possible high-precision BCD arithmetic we need a wider numeric format. At the end of the chapter we present the BCD20 format, which allows representing numbers to 34 significant digits. The processing of BCD20 numbers is similar to that of BCD12; therefore BCD20 routines are not listed in the text. These functions can be found in the bcd20math.cpp module that is furnished in the book's CD ROM.

4.0.1 ANSI/IEEE 854 Standard

On March 12, 1987, the Standards Board of the Institute of Electrical and Electronic Engineers approved the IEEE Standard for Radix-Independent Floating-Point Arithmetic. This project was sponsored by the Technical Committee on Microprocessors and Microcomputers of the IEEE Computer Society. The document was approved by the American National Standards Institute (ANSI) on September 10, 1987.

It is stated in the Foreword that the purpose of this standard is "to generalize ANSI/IEEE 754-1985 Standard for Binary Floating-Point Arithmetic, to remove dependencies on radix and word length." ANSI/IEEE 854 applies to BCD arithmetic as well as to binary, decimal, octal, or floating-point arithmetic in any other radix. However, ANSI/IEEE 854 does not specify formats for floating-point numbers or encodings of integers or strings representing decimal numbers. Therefore BCD and ASCII formats, such as the BCD12 and BCD20, used in the examples in this chapter, need not comply with any specific sizes or other requirements.

Furthermore, compliance or incompliance with the standard is not determined at the level of the core routines, such as those developed in the remainder of this chapter, but by how the results obtained from the core routines are handled by the hardware and software. In other words, since compliance with ANSI/IEEE 854 is determined at the implementation level, no statement of compliance or incompliance can be made about routines, procedures, sub-programs, or any component part of a software or hardware product.

Notice that Standard 854 was directly derived from ANSI/IEEE 754, which makes both standards quite similar.

4.1 Algorithms for BCD Arithmetic

Computer algorithms for multi-digit arithmetic on binary coded decimal numbers are often derived from longhand methods. These are the traditional grade-school algorithms for longhand addition, subtraction, multiplication, and division. However, the calculating routines can take advantage of certain facilities that are available in a digital machine. In addition, the particular encoding used in representing the numerical values can serve to facilitate or to hinder the actual calculations. Finally, the algorithms and routines should include error processing to identify illegal values, such as a zero divisor, and perform the necessary rounding operations on the results in order to ensure accuracy. The following points apply to the BCD arithmetic routines presented in this chapter:

1. The BCD arithmetic routines receive input in numbers coded and stored in floating-point BCD12 and BCD20 formats. This means that the processing algorithms are based on the floating-point exponential representation used in the BCD12 and BCD20 encodings.

2. The routines calculate results to double the number of significand digits of the input format, plus a possible carry. That is, the BCD12 routines calculate to 37 binary coded decimal digits, and the BCD20 routines to 69 binary coded decimal digits. These results are rounded and returned in the BCD12 or BCD20 format of the operands, respectively. Doubling the precision during calculations ensures that the significant digits of the formats are maintained in multiplication and division.

3. While the BCD12 and BCD20 formats store digits in packed form, the arithmetic routines unpack these digits prior to performing numerical calculations. One reason for this practice is that the Intel CPUs do not contain instructions for multiplication and division of packed BCD operands. In order to maintain uniform processing all operations are performed on unpacked digits.

4. The same rounding procedure is used by all BCD arithmetic routines. Rounding takes place to the nearest even number.

5. Some functions use a common scratchpad area for temporary calculations and for volatile data. No effort was made at optimizing the use of this scratchpad space. Temporary buffers and local variables were chosen to make the routines easy to develop and understand, rather than to save a few bytes of memory.

6. The routines do not save the caller's machine registers except for those used as pointers to the passed data.

7. The exponents are stored as 4 packed BCD digits in both the BCD12 and BCD20 formats. The packed BCD exponent is converted to biased form during processing. This conversion operation is performed by the function EXP_2_BIAS. Since the range of the exponent in the BCD12 and the BCD20 formats is –9999 to +9999, the bias value of 10000 was chosen as a mid-range approximation. The convenience of a biased exponent in performing numerical calculations was discussed in Chapter 2.

8. The BCD arithmetic routines are compatible with all Intel 486 and Pentium CPUs used in the PC.

9. The functions use a flat, 32-bit address space that is characteristic of the Win32 convention. The functions were developed using Visual C++ version 6.0 as Win32 console applications. However, the source modules can also be used by Windows programs.

10. All BCD arithmetic functions (add, subtract, multiply, and divide) take three parameters in the respective BCD12 or BCD20 format. The first two parameters are the operands, and the third one is used to return the result of the calculations.

The description of the functions, in the following sections, refer to the BCD12 format. The BCD20 format is described at the end of this chapter. For each function in BCD12 arithmetic there is a corresponding one in BCD20.

4.2 Floating-Point BCD Addition

The function SignAddBcd12(), listed in Section 4.6, performs the signed addition of two floating-point numbers encoded in BCD12 format. The processing assumes that the BCD12 number has been normalized so that there are no leading zeros in the significand, except for the encoding of the value 0. The implicit decimal point is located between the first and second significand digits. The BCD12 encoding is described in Chapter 2.

The algorithm for BCD addition is shown in the flowchart of Figure 4.1. The logic for the operation z = x+y can be described as follows:

1. If the addends (x and y) have the same sign, the significands are added and the sum is given the sign of the addends.

2. If the addends have unequal signs, the significand of the addend with the smaller absolute value is subtracted from the absolute value of the larger significand and the result is given the sign of the addend with the larger absolute value.

3. The exponent of the sum is the exponent of the addend with the larger absolute value. The operations performed on the significands may require adjusting the exponent in order to maintain a normalized result.

The sum of the significands is rounded to 18 significant digits. If the difference between exponents exceeds the final number of digits (18), then the addition of the significands will not affect the result. This case, which is labeled the trivial case, is illustrated in the code and is handled separately by the routine.

4.3 Floating-Point BCD Subtraction

The function named SignSubBcd12(), listed in Section 4.6, performs the signed subtraction of two floating-point numbers encoded in BCD12 format. Algebraic subtraction is performed by reversing the sign of the subtrahend and adding the operands.

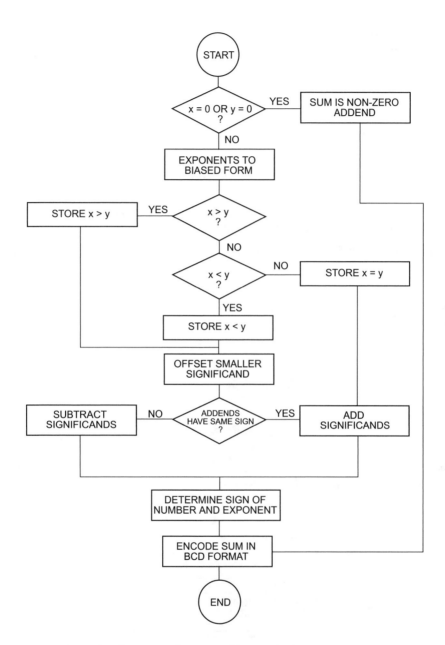

Figure 4.1 *Flowchart for Signed BCD Addition*

4.4 Floating-Point BCD Multiplication

The function SignMulBcd12(), listed in Section 4.6, performs the signed multiplication of two floating-point numbers encoded in BCD12 format. Processing assumes, as in addition and subtraction, that the BCD12 encoding has been normalized so that there are no leading zeros in the significand, except if the number is 0.

If the multiplication operation is represented as $z = x \cdot y$ then the algorithm can be described as follows:

1. If one of the factors is zero (x or y) then the product is zero.

2. If the factors have equal signs the product is positive, if they have unequal signs the product is negative.

3. The exponent of the product is the sum of the exponents of the multiplicand and the multiplier.

4. The significand of the product is the significand of the multiplicand times the significand of the multiplier.

5. The operations performed on the significands may require adjusting exponents in order to maintain a normalized result.

Figure 4.2 is a flowchart of the processing performed by the SignMulBcd12() function.

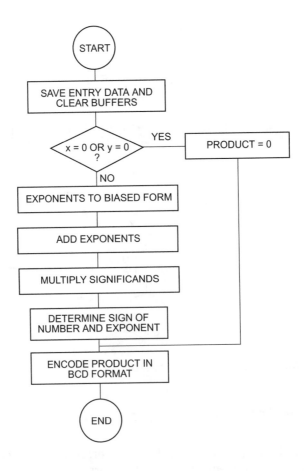

Figure 4.2 *Flowchart for Signed BCD Multiplication*

4.5 Floating-Point BCD Division

The function SignDivBcd12(), listed in Section 4.6, performs the signed division of two floating-point numbers encoded in BCD12 format. Here again, the processing assumes that the BCD12 encoding has been normalized so that, in the representation of non-zero values, there are no leading zeros in the significand.

Figure 4.3 is a flowchart of BCD division. If the division operation is in the form z = x / y, then the algorithm can be described as follows:

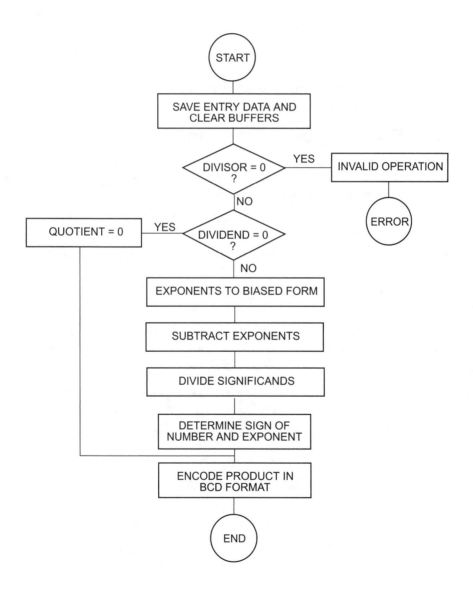

Figure 4.3 *Flowchart for Signed BCD Division*

1. If the dividend is zero (x = 0) the quotient is zero.

2. Division by zero is not defined, therefore a zero divisor (y = 0) is an invalid operation. In this case the first byte of the BCD result is set to FF hexadecimal. This special encoding is detected by the BCD conversion routines and handled as an invalid operand.

3. If the elements x and y have equal signs, the quotient is positive. If they have unequal signs, the quotient is negative. This rule for the sign of the result is the same as the one used in the multiplication algorithm.

4. The exponent of the quotient is the difference between the exponent of the dividend and the exponent of the divisor.

5. The significand of the quotient is the significand of the dividend divided by the significand of the divisor.

6. The operations performed on the significands may require adjusting the exponents in order to maintain a normalized result.

4.6 C++ BCD Arithmetic Functions

This section contains the listing of the C++ functions for BCD arithmetic. Each function provides an interface with the low-level procedures that perform the actual calculations. The following functions are listed:

1. SignAddBcd12() performs signed addition of two floating-point BCD numbers encoded in BCD12 format.

2. SignSubBcd12() performs signed subtraction of two floating-point BCD numbers encoded in BCD12 format.

3. SignMulBcd12() performs signed multiplication of two floating-point BCD numbers encoded in BCD12 format.

4. SignDivBcd12() performs signed division of two floating-point BCD numbers encoded in BCD12 format.

```
//***********************************************************************
//                         BCD12 arithmetic
//***********************************************************************
void SignAddBcd12(char bcd1[], char bcd2[], char result[])
{
// Addition of two signed BCD numbers stored in BCD12 floating point
// format
// Operation:
//           z = x + y
//           where x, y, and z are signed, floating point numbers
// On entry:
//           bcd1[] =  Addend (element x in z = x + y)
//           bcd2[] =  Augend (element y in z = x + y)
//           result[] is 12-byte storage area for result in BCD12 format
// Note: the code assumes that the BCD12 numbers are in normalized
//       form, that is, that there are no leading zeros in the
//       significand
// On exit:
//           result = Sum (element z in z = x + y)
//
```

```
// This routine operates on two numbers encoded in BCD12 format
// as follows:
//
//          Ss e e e e m . m m m m m m m m m m m m m m m m m m
//                       |
//                       |____ implicit decimal point
//
//    S = sign of number (1 BCD digit)
//    s = sign of exponent (1 BCD digit)
//    e = exponent (4 BCD digits)
//    m = normalized significand (18 BCD digits)
//        (first significand digit must be non-zero)
//    . = implicit decimal point between the first and second
//        significand digits
//***********************************************************************
//
// BCD signed addition algorithm:
// CASE 1:
//    If x and y have the same sign, the absolute values are
//    added and the result has the common sign
// CASE 2:
//    If x and y have different signs, the smaller value is
//    subtracted from the larger value and the result has the sign
//    of the larger
//***********************************************************************
//
// Routine operations
// CASE 1 and 2:
// A. The input elements are tested for zero values. If one element
//    is zero the result is the value of the other element
// B. The packed significands in BCD12 format are unpacked and moved
//    into work buffers located in the code segment
// C. The unpacked significands are aligned in the work buffers SIG_L
//    (for the significand of the number with the larger absolute
//     value) and SIG_S (for the significand of the number with the
//     smaller absolute value)
// CASE 1 (x and y have the same sign)
//    SIG_R will hold the significand of the sum
//    Addition operation:
//    SIG_L =     D D D . . . D D D 0 0 0 0 0 0 0 0 0 0 0 0 0 0 0
// +  SIG_S =     0 0 0 0 d d d . . . d d d 0 0 0 0 0 0 0 0 0 0 0
//            -------------------------------------------------
//    SIG_R = C s s s s s s s s s s s s s 0 0 0 0 0 0 0 0 0 0 0
//        legend:
//            D = digits in the larger significand
//            d = digits in the smaller significand
//            s = digits in the sum
//            C = possible carry digit in the sum significand
//                (in this case the exponent must be adjusted)
//
// CASE 2 (x and y have different signs)
//    SIG_R will hold the significand of the difference
//    Subtraction operation:
//    SIG_L =     D D D . . . D D D 0 0 0 0 0 0 0 0 0 0 0 0 0 0 0
// -  SIG_S =     0 0 0 0 d d d . . . d d d 0 0 0 0 0 0 0 0 0 0 0
//            -------------------------------------------------
//    SIG_R = B s s s s s s s s s s s s s 0 0 0 0 0 0 0 0 0 0 0
//        legend:
//            B = possible borrow digit in the difference
// TRIVIAL CASE:
```

```
//     If the difference between exponents is larger than the number
//     of significand digits, then the aligned significands will be
//     as follows:
//       SIG_L =     D D D . . . D D D 0 0 0 0 0 0 0 0 0 0
//       SIG_S =     0 0 0 0 0 0 0 0 0 0 d d d . . . d d d
//                   ---------------------------------------
//       SIG_R =   0 D D D . . . D D D 0 ? ? ? . . . ? ? ?
//     This means that the result (rounded to the format's significand
//     size) will equal the significand of the larger number. Therefore
//     the addition or subtraction operation would be trivial
// D. The exponent of the sum is the exponent of the element with
//     the larger absolute value, adjusted according to the operations
//     performed on the significands

_asm
{
; Store entry variables
        MOV     ESI,bcd1
        MOV     EDI,bcd2
        MOV     EBX,result
        CALL    SIGN_ADD_BCD12
}
return;
}

void SignSubBcd12(char bcd1[], char bcd2[], char result[])
{
// Subtraction of two signed BCD numbers stored in BCD12 floating
// point format
// Operation:
//          z = x - y
//          where x, y, and z are signed, floating point numbers
// On entry:
//          bcd1[] = Minuend (element x in z = x - y)
//          bcd2[] = Subtrahend (element y in z = x - y)
//          result[] is 12-byte storage area for difference (element z)
// Note: the code assumes that the BCD12 numbers are in normalized
//       form, that is, that there are no leading zeros in the
//       significand
// On exit:
//          result[] = difference (element z in z = x - y)
// Operation:
//    Processing is based on the algebraic principle of changing the
//    sign of the subtrahend and proceeding as in addition
//

_asm
{
; Store entry variables
        MOV     ESI,bcd1
        MOV     EDI,bcd2
        MOV     EBX,result
        CALL    SIGN_SUB_BCD12
}
return;
}

void SignMulBcd12(char bcd1[], char bcd2[], char result[])
{
// Multiplication of two signed BCD numbers stored in BCD12 floating
```

```
// point format
// Operation:
//              z = x * y
//              where x, y, and z are signed, floating point numbers
// On entry:
//          bcd1[] is multiplicand (element x in z = x * y)
//          bcd2[] is multiplier (element y in z = x * y)
//          result[] is 12-byte storage area for product in BCD12 format
// Note: the code assumes that the BCD12 numbers are in normalized
//       form, that is, that there are no leading zeros in the
//       significand
// On exit:
//          result[] = product (element z in z = x * y)
//*****************************************************************
// BCD signed multiplication algorithm
// A. If one of the factors is zero, the product is zero
//     If neither factor is zero then:
// B. If the factors have equal signs the product is positive
//     If the factors have unequal sings the product is negative
// C. The significand of the product is the product of the
//     significands of the factors
// D. The exponent of the product is the sum of the exponents of
//     the factors
//*****************************************************************
//
_asm
{
; Store entry variables
          MOV     ESI,bcd1
          MOV     EDI,bcd2
          MOV     EBX,result
          CALL    SIGN_MUL_BCD12
}
return;
}


void SignDivBcd12(char bcd1[], char bcd2[], char result[])
{
// Division of two signed BCD numbers stored in BCD12 floating point
// format
// Operation:
//              z = x / y
//              where x, y, and z are signed, floating point numbers
// On entry:
//          bcd1[] is dividend (element x in z = x / y)
//          bcd2[] is divisor (element y in z = x / y)
//          result[] is a 12-byte storage area for quotient
// Note: the code assumes that the BCD12 numbers are in normalized
//       form, that is, that there are no leading zeros in the
//       significand
// On exit:
//          result[] = quotient (element z in z = x / y)
//          carry set if divisor equals zero
//          carry clear if divisor not equal zero
//
//*****************************************************************
// BCD signed division algorithm
// A. If the dividend is zero the quotient is zero
//     If the divisor is zero the operation is undefined
//     If neither element is zero then:
```

```
// B. If the factors have equal signs the quotient is positive
//    If the factors have unequal sings the quotient is negative
// C. The significand of the quotient is the quotient resulting from
//    dividing the significand of the dividend by the significand of
//    the divisor
// D. The exponent of the quotient is the difference between the
//    exponent of the dividend minus the exponent of the divisor
//******************************************************************

_asm
{
; Store entry variables
            MOV         ESI,bcd1
            MOV         EDI,bcd2
            MOV         EBX,result
            CALL        SIGN_DIV_BCD12
}
return;
}
```

SOFTWARE ON-LINE

The C++ functions for the BCD12 arithmetic are found in the file BCD12.h located in the folder Sample Code\Chapter04\BCD12 Arithmetic in the book's on-line software. The project BCD12 Arithmetic exercises and tests the low-level procedures located in the Un32_2 module of the MATH32 library.

4.6 High-Precision BCD Arithmetic

One of the advantages of BCD arithmetic on the main CPU is that there is practically no limit to the numerical precision. In Chapter 2 we developed the BCD12 format with 18 significant digits and an exponent in the range –9999 to 9999. Although BCD12 is suitable for explaining BCD arithmetic operations, it is limited to 18 significant digits. 18-digits is approximately the same precision obtained with the Intel floating-point hardware, which is explained starting in Chapter 5. In order to perform high-precision arithmetic in the CPU we need a more extensive BCD format. The BCD20 format allows representing 34 significant digits and uses the same sign encoding and exponent range as BCD12. The structure of the BCD20 format is shown in Figure 4.4 and in Table 4.1 .

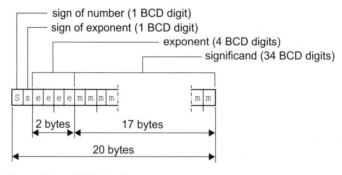

Figure 4.4 *Map of the BCD20 Format*

Table 4.1

Field Structure of the BCD20 Format

CODE	FIELD NAME	BITS WIDE	BCD DIGITS	RANGE
S	sign of number	4	1	0 – 1 (BCD)
s	sign of exponent	4	1	0 – 1 (BCD)
e	exponent	16	4	0 – 9999
M	significand	136	34	0 – 99..99 (34 digits)

	Format size .... 160 (20 bytes)			

Notes:

1. The significand is scaled (normalized) to a number in the range 1.00..00 to 9.99..99.
2. The encoding for the value zero (0.00..00) is a special case.
3. The special value FFH in the sign byte indicates an invalid number.

The BCD20 format requires 20 bytes of storage. The format is described as follows:

1 The sign of the number (S) is encoded in the left-most packed BCD digit. Therefore, the first 4 bits are either 0000B (positive number) or 0001B (negative number).

2. The sign of the exponent is represented in the 4 low-order bits of the first byte. The sign of the exponent is also encoded in one packed BCD digit. As is the case with the sign of the number field, the sign of the exponent is either 0000B (positive exponent) or 0001B (negative exponent).

3. The next 2 bytes encode the exponent in 4 packed BCD digits. The decimal range of the exponent is 0000 to 9999. The actual exponent is stored in bias 10000 form.

4. The remaining 17 bytes are devoted to the significand field, consisting of 34 packed BCD digits. Positive and negative numbers are represented with a significand normalized to the range 1.00...00 to 9.00...99. The decimal point following the first significand digit is implied. The special value 0 has an all-zero significand.

5. The special value FF hexadecimal in the number's sign byte indicates an invalid number.

As with the BCD12 format, the BCD20 format does not make ideal use of the available storage space, however, the numerical precision of 34 digits doubles that of the BCD12 and of the double precision format of the ANSI/IEEE 754 standard. BCD20 arithmetic and conversion functions are found in the book's on-line software package.

To the programmer, BCD20 arithmetic allows operating on numeric values with valid results up to 34 significant digits. The processing provided by C and C++ is limited to the double precision format of ANSI/IEEE 754. During input or processing any value that exceeds this precision is automatically truncated.

For example, in C++ programming you may attempt to define a variable of type double to 19 significant digits. The C++ compiler rounds-off the entered value to the maximum precision supported, which can never exceed that of ANSI/IEEE 754. The following small program shows the results:

```
#include <iostream.h>
#include <iomanip.h>
#include <stdio.h>

double largeDouble = 1.2233445566778877665544; // Initialized to 23
                                               // digits

int main()
{

    cout < "\ndouble defined as: 1.2233445566778877665544";
    cout < setprecision(50) < setw(50);
    cout < "\ndisplayed          : " < largeDouble;
    cout < "\n\n";
    return 0;
}
```

Notice that we have used iomanip operators to set the precision and the displayed width to 50 digits. When the above program executes in Visual C++ 6.0, output is as follows:

```
double defined as: 1.2233445566778877665544
displayed:         1.22334455667789
```

The compiler has rounded-off the displayed result to 15 significant digits. Although, internally, values in double format are stored to 17 or 18 digits precision, several digits defined in the initialization string have been lost in the operand. We should mention that Visual C++, as well as most C and C++ compilers, contain compile-time switches and options that allow changing the precision and rounding of floating-point operands. However, the resulting precision can never be higher than supported by the adopted formats.

Using BCD20 format and arithmetic you can represent and manipulate numbers up to 34 significant digits without loss of precision. The following short program shows an example.

```
#include <iostream.h>
#include <iomanip.h>
#include "bcd20math.h"
#include <stdio.h>

// Source numbers for tests (to 33 significant digits)
char asc1[] = "1.22334455667788771122334455667711";
char asc2[] = "1.11111111111111111111111111111111";
// BCD20 data
char num1[20];
char num2[20];
char bcdResult[20];
// ASCII data
char ascResult[52];

int main()
{
```

```
// BCD20 addition
AsciiToBcd20(asc1, num1);
AsciiToBcd20(asc2, num2);

SignAddBcd20(num1, num2, bcdResult);
Bcd20ToAscStr(bcdResult, ascResult);

cout < "\nFirst addend  : " < asc1;
cout < "\nSecond addend : " < asc2;
cout < "\nSum           :" < ascResult < "\n\n";

return 0;
}
```

When the above program executes, the results are as follows:

```
First addend  : 1.22334455667788771122334455667711
Second addend : 1.11111111111111111111111111111111
Sum           : 2.33445566778899882233445566778822
```

As previously mentioned, the price paid for the higher precision that can be obtained with BCD arithmetic is much slower processing and less efficient storage formats.

SOFTWARE ON-LINE

The C++ functions for the BCD20 arithmetic are found in the file BCD20.h located in the folder Sample Code\Chapter04\BCD20 Arithmetic in the book's on-line software. The project BCD20 Arithmetic exercises and tests the low-level procedures located in the Un32_3 module of the MATH32 library.

Chapter 5

Floating-Point Hardware

Chapter Summary

This chapter presents an introduction to mathematical coprocessor hardware in general, and in particular to the Intel Floating-point Units: 8087, 80287, 80387, 487 SX, and the math unit of the 486 DX and Pentium. The chapter also includes a discussion of the ANSI/IEEE 754 Standard for Binary Floating-point Arithmetic which is closely related to the Intel hardware components listed above.

5.0 A Mathematical Coprocessor

The CPU used in all IBM and IBM-compatible microcomputers manufactured to date is an Intel microprocessor of the 80x86 family. The first microprocessor of this family was the 8086, released in mid-1978. In 1981 IBM made public its first desktop computer, called the IBM Personal Computer, which was equipped with the Intel 8088 CPU, a version of the 8086 chip. Both the 8086 and 8088 were conceived and designed as general-purpose devices. The mathematical instructions in the 8086/8088 are limited to the fundamental arithmetic operations on signed and unsigned binary integer numbers and on unsigned integer decimals. The supported operations are addition, subtraction, multiplication, division of binary and binary coded decimal numbers, as well as numerical conversions between these formats. Although the more recent CPUs of the 80x86 family (80386, 486, and Pentium) are capable of operating on larger numbers than the original ones, the arithmetic instruction set of the 80x86 family has remained basically unchanged on the various processor implementations.

Perhaps the most important limitation of the 8086 and its descendants is their inability to operate on fractional numbers; a fact that did not go unnoticed to its original designers. Bill Pohlman, the 8086 project manager, defined a floating-point extension to the chip and implemented an interface for a coprocessor. In addition to the mathematical extension, coprocessors have been used to assist the main CPU in performing other specialized tasks, such as graphics, text and data manipulations, communications, and multimedia. Intel coprocessors include the 8089 input/output channel processor for data operations, the 82586 coprocessor for communications, the 82730 text processor, an entire family of mathematical coprocessors, and the Multimedia Extension, called the MMX.

Previously, floating-point hardware had been used mainly in many mainframe and mini-computers. The first implementation of mathematical coprocessor technology was in the IBM 704 in 1953. At that time, large computing machines usually included the floating-point hardware. In mini-computers the floating point hardware was usually furnished as an option. The Intel mathematical coprocessor was the first implementation of floating-point hardware in a microprocessor.

Although the Intel mathematical coprocessors are the best known and most frequently used in the PC, they are not the only one. The Weitek chip set and processors were considerably faster than the Intel math units, although much less powerful, and offered a different approach to mathematical calculations. Some PCs of the time provided support for the Weitek processors.

5.1 Intel Math Units

The first mathematical coprocessor for the 8086 and 8088, named the 8087, was introduced by Intel in 1980. The original design work was the work of John Palmer and Bruce Ravenel. In the preface to their book on the 8087, titled *The 8087 Primer*, (see Bibliography) Palmer and Morse give extensive credit to Prof. William Kahan, of the University of California, Berkeley. Others associated with the 8087 are Jean Claude Cornet, who directed the 8086 project, John Bayliss and Bob Koehler, who share a patent for the functional partitioning of processor functions. Rafi Nave managed the chip's design at Intel Israel.

The 8087 is also known as the math unit, the numeric data processor (NDP), the numeric data coprocessor, the math coprocessor, and the numeric processor extension, or NPX. Later in this Chapter we list the versions of the math coprocessor that correspond to the various Intel processors used in the PC. The 486 DX and the Pentium include the floating-point operations of the NDP in their own instruction set. This functional area is referred to as the floating-point unit or math unit. The names "math coprocessor" and "math unit" are used interchangeably throughout the book to refer generically to all members of the Intel family of mathematical coprocessors, including the math unit of the 486 and Pentium.

Originally, the math unit was not standard equipment in the PC, although most systems included an empty, wired socket for its optional installation. Two exceptions are the IBM PCjr and the PC Convertible, which have no coprocessor socket. Installation of the coprocessor consisted of pushing an 8087, 80287, or 80387 chip into this socket. Some of the earlier hardware also required changing the position of a mechanical switch. While in others the initialization included a software program to log-on the newly installed coprocessor.

Because the math unit was originally an optional device, it was sometimes imitated in software by a program called a coprocessor or 8087 emulator. The emulator software allowed programmers to use math unit instructions even if a math unit was not physically present in the system. The only difference between using the math unit hardware or a correctly coded emulator program was that execution took longer with the emulator than with the coprocessor. Emulators were made available by Intel Corporation and other sources.

The math unit programmer accesses eight coprocessor registers, each of which is 80 bits wide. The registers are located in a stack structure and can be addressed explicitly or implicitly. In scientific and technical applications the 80x87 chip is typically programmed to use the long real format. Numbers in this format can range from 4.19E–307 to 1.67E+308. Coprocessor operations are carried out in an expanded numeric representation, called the temporary real format. Numbers represented in this format can range from 3.4E–4932 to 1.2E+4932. The additional precision of the temporary real format serves to absorb possible errors that occur during computation and round off operations. In business and financial applications the math unit can process decimal numbers of up to 18 digits without rounding. Exact integer arithmetic, particularly useful in graphics applications, can be performed on numbers as large as 2.0E+18.

5.1.1 Math Unit Applications

The math unit processes and stores numerical data encoded in seven data formats. However, all internal calculations are performed in an 80-bit data format that allows representation of 19 significant decimal digits. The maximum precision available to the user is in the long real format, which encodes 17 to 18 significant decimal digits. The chip's processing capability includes the following operations:

1. Data transfer from memory into the processor's registers and from processor registers into memory. These transfers take place in one of the seven data formats recognized by the math unit. Conversion from ASCII into the processor's internal formats and vice versa must be executed in external software.

2. Arithmetic operations on integers and floating-point numbers.

3. Square roots, scaling, absolute value, remainder, sign change, and extraction of the integer and fractional parts of a number.

4. Direct loading of the constants 0, 1, p, and several logarithmic primitives.

5. Comparison and testing of internal processor operands.

6. Calculation of trigonometric functions and of primitives from which other functions can be obtained.

7 Calculation of several exponential bases and transcendental bases.

8. Control instructions to initialize the processor, to set internal operational modes, to store and restore the processor's registers and status, and to perform other housekeeping and auxiliary functions.

Intel states that the math unit improves the execution speeds of mathematical calculations by a factor of 10 to 100, compared with equivalent processing performed by 80x86 software. The math unit extends the functions of the main CPU by adding an instruction set of approximately 70 instructions, as well as the eight specialized registers for numerical operands.

A system containing an Intel math unit is capable of loading, storing, and exchanging all supported numeric data types. The system can perform basic arithmetic operations, including the calculation of square roots, scaling, finding the integer part and the absolute value of a number, and changing its sign. Comparison operations permit examining, comparing, and testing the numeric operands in the registers or in memory. Transcendental instructions allow determining the tangent, arctangent, and several basic logarithmic functions. The constants 1, 0, p, and several logarithmic primitives can be loaded directly as operands. Finally, several processor control instructions allow changing the machine's control and status words, initializing the processor or the emulator, storing and loading the coprocessor environment, enabling and disabling interrupts, and clearing the error exception flags.

5.1.2 Math Unit Limitations

One difficulty encountered in programming floating-point operations on the math unit is that data must be entered into the coprocessor registers using one of the chip's internal data formats. This means that the user's input, typically in the form of a string of ASCII decimal numbers, must be converted by external software into one of the formats supported by the hardware. By the same token, the result of floating-point calculations performed in the math unit must be converted from the internal formats into an ASCII decimal representation that can be interpreted by the user. In Chapter 6 we develop suitable conversion routines.

Another limitation relates to the fact that the math unit does not always output mathematical functions in a form that can be directly used by the software. For example, the only trigonometric result that can be obtained in 8087 and 80287 systems is the tangent of an angle in the range 0 to p/4 radians. User input must be scaled to this range and the other trigonometric functions must be obtained from this tangent by external software. Similar situations apply in the calculation of logarithms, roots (other than the square root), and powers. Although these limitations were partly corrected in the 387, programs that use the math unit still depend on considerable external processing for input and output as well as in scaling and other manipulations.

When the 8087 and 80287 chips were released (1980) the ANSI/IEEE 754 Standard for Binary Floating-Point Arithmetic had not yet been approved. Although the developers of the 8087 were committed to complying with the standard, and had been involved in the standard's development, it was not possible for them to know in advance all the details of its final version. For this reason several elements of the original coprocessor were later in disagreement with the floating-point standard. During the development of the 387, its designers had to introduce modifications in order to ensure that the new version of the math unit would comply with all the provisions of ANSI/IEEE 754. These changes are the cause of minor incompatibilities between the 387 and its predecessors. The result is that code written for the 8087 or the 80287 can execute with variations in a 387 or in the math unit of the 486 or the Pentium.

5.1.3 Processor/Coprocessor Interface

The main CPU and the math unit behave as a single entity that combines the instruction set and processing capabilities of both chips. To the programmer, the CPU/math unit combination appears as a single device. Both devices use the same clock generator, system bus, and interface components. The 486 and the Pentium include the 80x86 and the FPU instruction sets, in addition to some new instructions.

In both cases, instructions for the central processor and the math unit, are intermixed in memory in the instruction stream. The first 5 bits of the opcode identify a coprocessor or math unit escape sequence (bit code 11011xxx). This bit pattern identifies the CPU ESC (escape) operation code. All instructions that match these first 5 bits are executed by the math unit. The CPU distinguishes between escape instructions that reference memory and those that do not. If the instruction contains a memory operand, the CPU performs the address calculations on behalf of the coprocessor. On the other hand, if the escape instruction does not contain a memory operand, then the CPU ignores it and proceeds with the next instruction in line.

Processor/coprocessor synchronization was pioneered by Intel with the 8087. The original 8086/8087 design allows a central processor and a coprocessor to execute simultaneously. Originally, a BUSY pin in math unit was connected to a TEST pin in the CPU. The math unit's BUSY pin is set high whenever the coprocessor is executing an instruction. The CPU's TEST pin, upon receiving a WAIT (or FWAIT) instruction, forces the central processor to cease execution until the coprocessor has finished. However, the processor/coprocessor synchronization is implemented differently in the 8087 than in the 80287 and 80387 hardware and the math unit of the 486 and the Pentium.

The 8086 must not present a new instruction to the math unit while it is still executing the previous one. This is guaranteed by inserting a WAIT instruction either before or after every coprocessor ESC opcode. If the WAIT follows the ESC, then the 8086 does nothing while the coprocessor is executing. Most assemblers insert a WAIT instruction before the coprocessor ESC opcode in order to allow concurrent processing by the CPU and the math unit. In this case, the CPU can continue executing its own code until it finds the next ESC in the instruction stream.

Nevertheless, it is possible that if the WAIT precedes the ESC, the CPU will access a memory operand before the coprocessor has finished acting on it. If this situation can happen, the programmer must detect it and insert an additional WAIT. The alternative mnemonic FWAIT is usually preferred in this case, since some emulator libraries do not recognize the WAIT opcode. The following code fragment shows a typical circumstance that requires the insertion of an FWAIT instruction.

```
FSTCW      CTRL_WORD ; Store control word in memory
FWAIT                ; Force the CPU to wait for NDP
                     ; to finish before
  MOV  AX,CTRL_WORD  ; recovering the control word
                     ; into the AX register
```

Synchronization requirements are different in 80286/80287, 80386/80387 systems, and in the math unit of the 486 and the Pentium. The 80286 and 80386 CPU automatically check that the coprocessor has finished executing the previous instruction before sending the next one. For this reason, unlike the 8087, the 80287 and 80387 do not require the WAIT instruction for synchronization. However, the possibility of both processors accessing the same memory operand simultaneously also exists in 80287/80387 systems and must be prevented as previously described for the 8087.

In conclusion, programs intended for 8087 systems must follow 8087 synchronization requirements. However, some 80287 assemblers (such as Intel's ASM286) omit the FWAIT opcode. Other assemblers (such as Microsoft's MASM version 5.0 and later) have options that allow the FWAIT instructions to be automatically inserted or not inserted. In either case, code in which the ESC instructions are not accompanied by a CPU FWAIT do not execute correctly in 8087 systems. If the assembler program used to generate the machine code does not automatically insert the FWAIT instruction preceding each coprocessor escape, then the programmer has to manually insert the FWAIT opcode in the source file if the code is to execute correctly in an 8087 system. In this case, another option is to develop a set of macro instructions that automatically include FWAIT.

Some math unit processor control instructions have an alternative mnemonic that instructs the assembler not to prefix the instruction with a CPU FWAIT. This mnemonic form is characterized by the letters FN, signifying NO WAIT, for example, FINIT/FNINIT and FENI/FNENI. The no-wait form should be used only if CPU interrupts are disabled and the math unit is set up so that it cannot generate an interrupt that would precipitate an endless wait. In all other cases, the normal version of the instruction should be preferred.

5.1.4 Math Unit Versions

Three versions of the mathematical coprocessor have been released by Intel. The original 8087 chip was intended for use with the 8086 and the 8088 and is also compatible with the 80186 and 80188. The 80287 is the version designed to function with the 80286 CPU, and the 80387 for the 80386 central processor (see Table 5.1).

Table 5.1
Intel Processors, Coprocessors, and Math Units

CPU	MATH COPROCESSOR
8086 8088 80186 80188	8087
80286	80287
80386 486 SX	80287 or 80387
486 DX Pentium	Built-in math unit

8087

8087 is Intel's designation for the original mathematical coprocessor chip. The chip was first offered to the public in 1980. It was developed simultaneously with the ANSI/IEEE proposed standard for binary floating-point arithmetic, which was not finalized until 1984. This explains the minor differences between the 8087 chip and the standard; in most cases the difference consists of the 8087 exceeding the standard's requirements.

80287

The 80287, sometimes called the 287, was introduced in 1983. The 80287 is the version of the Intel mathematical coprocessor designed for the 80286 CPU. The 80287 extends numerical coprocessing to the protected-mode, multitasking environment supported by the 80286 CPU. When multiple tasks execute in the 80287, they receive the memory management and protection features of the central processor. According to Intel, the performance of the 80287 chip is 41 to 266 times that of equivalent software routines. The 80287 is also compatible with the 80386 CPU.

The internal architecture and instruction set of the 80287 are almost identical to those of its predecessor. Most programs for the 8087 execute unmodified in the 80287 protected mode, except for the handling of numeric exceptions. The following are the major differences between the 80287 and the 8087:

1. The 80286 uses a dedicated line to signal processing errors to the CPU. This signal does not pass through the system's interrupt controller.

2. The 8087 instructions for enabling and disabling interrupts, FENI/FNENI and FDISI/FNDISI, serve no purpose and are not implemented in the 80287. The opcodes are ignored by the processor.

3. The 80287 instruction opcodes are not saved when executing in protected mode, but exception handlers can retrieve these opcodes from memory.

4. While the address of the ESC instruction saved by the 8087 does not include leading prefixes (such as segment overrides), the 80287 does include them.

5. The FSETPM instruction, used to enable 80287 protected-mode operation, has no equivalent in the 8087.

6. The FSTSW and FNSTSW instructions in the 80287 allow the AX register as a destination operand. Writing the status word to a processor register optimizes conditional branching.

8087 instructions must be preceded by an FWAIT instruction to synchronize processor and coprocessor. This opcode is automatically generated by most assemblers. The FWAIT instruction is not required for the 80287, which has an asynchronous interface with the main processor. For this reason, reassembling programs intended for the 80287 exclusively may result in a more compact code that executes slightly faster (see Section 5.1.3). On the other hand, this code does not execute on 8087 systems.

80387

The Intel 80387, sometimes called the 387, is a mathematical coprocessor intended for the 80386 central processing unit. The 80387 supports all 8087 and 80287 operations and instructions. Programs developed for the 8087 or the 80287 generally execute unmodified on the 80387. A version of the 80387 designated the 487 SX is compatible with the 486 SX chip. The 487 SX is functionally identical to the 80387, therefore it is not discussed separately.

The 80387 conforms with the final version of the ANSI/IEEE 754 standard for binary floating-point arithmetic, approved in 1985. This has made necessary the following changes in coprocessor behavior:

1. Automatic normalization of denormalized operands.

2. Affine interpretation of infinity. Note that the 8087 and 80287 support both affine infinity and projective infinity.

3. Unordered compare instructions, which do not generate an invalid operation exception if one operand is a NAN.

4. A partial remainder instruction that behaves as expected by the ANSI/IEEE 754 standard. The 80387 version of the FPREM instruction is named FPREM1.

The 80387 instructions FUCOM, FUCOMP, and FUCOMPP differ from the previous FCOM, FCOMP, and FCOMPP instructions in that they do not generate an invalid operation exception if one of the operands is tagged as not-a-number (NAN). The 80387 instruction set has been expanded with the opcodes FSIN, to calculate sines, FCOS, to calculate cosines, and FSINCOS, to calculate both sine and cosine functions simultaneously. This last instruction can be followed by a division operation to directly obtain tangents and cotangents.

The operand range of the instructions FPTAN, FPATAN, F2XM1, and FSCALE was expanded. This expansion simplifies the calculation of some trigonometric and transcendental functions.

5.1.5 The Numeric Unit in 486 and Pentium CPU

The Intel 486 (DX) and Pentium include the mathematical coprocessor functions as part of the central processor. According to Intel, the numeric functions and floating-point instructions of the 486 (DX) and Pentium CPU are identical to those of the 80387 mathematical coprocessor, as described in Section 5.1.4. Therefore no specific discussion of the mathematical unit of the 486 (DX) and Pentium CPU is required.

5.2 Detecting and Identifying the Math Unit

You have seen the variations in the operation and instruction set of the various versions of the coprocessor and the math unit of the 486 and Pentium. Starting with the Pentium, the math unit became part of the CPU. In the future, programmers may be able to assume that math unit hardware is always available on a PC. However, at the present time, software may still need to determine on which device a program is running in order to use or bypass one or more instructions, or to select among several processing branches.

The function IdMathUnit(), listed below, tests for the presence of a coprocessor or math unit. If one is detected, code identifies one of the following implementations: 8087, 80287, 80387, 486 math unit or 486/80387 system, or Pentium math unit. When execution returns to the caller, the int variable passed as a parameter contains a code that identifies which version of the coprocessor, if any, is installed on the host machine.

```
void IdMathUnit(int *userCode)
{
// Local data
unsigned short CONTROL_87 = 0; // Storage for control word
unsigned short STATUS_87 = 0;  // Storage for status word

_asm
{
; Determine if there is a mathematical coprocessor installed and
; if it is an 8087, 80287, 80387, or the math unit of a 486 or
; Pentium
; On exit:
;          userCode = 0 if no coprocessor present
;                     1 if 8087
;                     2 if 80287
;                     3 if 80387
;                     4 if math unit of 486 or 486SX / 487 system
;                     5 if math unit of Pentium
;****************************************************************

        FNINIT              ; Initialize coprocessor (if present)
; Note that the no-wait form must be used to prevent a wait
; forever condition if a coprocessor is not present
        MOV     AX,5A5AH        ; Value to set in status word
        MOV     STATUS_87,AX
        FNSTSW  STATUS_87       ; Store status word
        MOV     AX,STATUS_87    ; Read status into AX
        CMP     AL,0            ; Test for no status bits
        JE      CHK_CONTROL     ; Go if 0
; At this point no math unit is detected in system
        MOV     AX,0            ; Return code for no coprocessor
        JMP     EXIT_FPU
; A secondary test is based on the math unit control word
CHK_CONTROL:
        FNSTCW  CONTROL_87      ; Store control word
        MOV     AX,CONTROL_87   ; Read into AX
        CMP     AL,0            ; Test for no status bits
        AND     AX,103FH        ; Bit mask is
                                ; AH = 0001 0000
                                ; AL = 0011 1111
        CMP     AX,003FH        ; Test for AL bits unchanged
        JE      SYS_80X87       ; 80x87 is present
; At this point no math unit is detected in system
        MOV     AX,0            ; Return code for no coprocessor
        JMP     EXIT_FPU
;********************|
;  coprocessor present |
;********************|
```

```
SYS_80X87:
; The first test for the coprocessor type can be based on the CPU
; type installed in the host
; determine the CPU type, as follows:
; AX =       1 if CPU is 8086 or 8088
;            2 if CPU is 80286
;            3 if CPU is 80386
;            4 if CPU is 486
;            5 if CPU is Pentium
;**********************|
;   test for 8086/8088 |
;**********************|
; Bits 12 to 15 in the flag register are always set in the 8086
; and 8088 CPU
        PUSHF                   ; Flag register to stack
        POP     AX              ; Store flags in AX
        AND     AX,0FFFH        ; Clear bits 12 to 15
        PUSH    AX              ; AX to stack
        POPF                    ; and to flags register
        PUSHF                   ; Flags to stack
        POP     AX              ; and to AX for reading
        AND     AX,0F000H       ; Preserve bits 12 to 15
        CMP     AX,0F000H       ; Test for bits set
        JNE     TEST_286        ; Go if bits not set
; At this point processor is a 8086 or 8088
        MOV     AX,1            ; Return code
        JMP     ID_EXIT         ; Exit
;**********************|
;     test for 80286   |
;**********************|
; Bits 12 to 15 in the flag register are always clear in the Intel
; 80286 CPU
TEST_286:
        PUSHF                   ; Flag register to stack
        POP     BX              ; Store flags in BX
        OR      BX,0F000H       ; Make sure bit field is set
        PUSH    BX              ; To stack
        POPF                    ; And to flag register
        PUSHF                   ; Flags to stack
        POP     AX              ; And to AX
        AND     AX,0F000H       ; Clear all other bits
        JNZ     TEST_386        ; Go if bits not clear
; At this point processor is an 80286
        MOV     AX,2            ; Return code
        JMP     ID_EXIT         ; Exit
;**********************|
;     test for 80386   |
;**********************|
; Bit 18 of the E flags register was introduced in the 486 CPU
; This bit cannot be set in the 80386
TEST_386:
        PUSHFD                  ; 32-bits E flags to stack
        POP     EAX             ; Flags to EAX
        OR      EAX,40000H      ; Make sure bit 18 is set
        PUSH    EAX             ; New flags to stack
        POPFD                   ; An to E flags register
        PUSHFD                  ; Back to stack
        POP     EAX             ; And to EAX
        AND     EAX,40000H      ; Clear all except bit 18
        JNZ     TEST_486        ; Go if bit 18 is clear
```

```
; At this point processor is a 80386
        MOV     AX,3            ; Return code
        JMP     ID_EXIT         ; Exit
;**********************|
;     test for 486     |
;**********************|
; Bit 21 (ID flag) of the E flags register cannot be set in the
; 486
TEST_486:
        PUSHFD                  ; 32-bits E flags to stack
        POP     EAX             ; Flags to EAX
        OR      EAX,200000H     ; Make sure bit 21 is set
        PUSH    EAX             ; New flags to stack
        POPFD                   ; An to E flags register
        PUSHFD                  ; Back to stack
        POP     EAX             ; And to EAX
        AND     EAX,200000H     ; Clear all except bit 21
        JNZ     IS_PENTIUM      ; Go if bit 21 is clear
; At this point processor is a 486
        MOV     AX,4            ; Return code
        JMP     ID_EXIT         ; Exit
;**********************|
;  processor is PENTIUM |
;**********************|
IS_PENTIUM:
        MOV     AX,5            ; Return code
ID_EXIT:
; At this point AL holds CPU ID code as follows:
;           AL = 1 if FPU is 8087
;           AL = 2 if FPU is 80287
;           AL = 3 if FPU is 80287 or 80387 since the 80287 can
;                be installed in an 80386 system
;           AL = 4 if FPU is 486 math unit
;           AL = 5 if FPU is Pentium math unit
;
; If return code is not 3 then there is no uncertainty regarding
; the FPU type
        CMP     AX,3            ; Test for undefined case
        JE      IS_386          ; Go if 386 CPU
; If not 386, return with coprocessor code in AL
        JMP     EXIT_FPU
;**********************|
;    80287 or 80387    |
;        system        |
;**********************|
; The test for an 80287 system is based on the state of the
; condition code bits after FINIT. In the 80387 C0 to C3 bits
; are zero after FINIT, in the 80287 they contain the previous
; value
; Note: at this time the wait form of math unit opcodes can be used
;        since the presence of a coprocessor is certain
IS_386:
        FINIT                   ; Make sure stack is EMPTY
        FXAM                    ; C0 and C3 are set after FXAM if
                                ; the stack top is EMPTY
        FINIT                   ; In an 80287 C0 and C3 will remain
                                ; set after the second FINIT
        FSTSW   STATUS_87       ; Status word to memory
        FWAIT                   ; Synchronize CPU and FPU
        MOV     AX,STATUS_87    ; Status word to AX register
```

```
        AND     AH,01000001B  ; Clear all bits except C3
                              ; and C0
        JNZ     TYPE_287      ; Go if bits not clear
;********************|
;     80387 or 486   |
;********************|
        MOV     AX,3          ; Return code for 80387
        JMP     EXIT_FPU
;********************|
;        80287       |
;********************|
TYPE_287:
        MOV     AX,2          ; Return code for 80287
EXIT_FPU:
        AND     EAX,0FFFFH
        MOV     ESI,userCode
        MOV     [ESI],EAX
}
}
```

SOFTWARE ON-LINE

The IdMathUnit() function is found in the ID Math Unit.h file located in the folder Sample Code\Chapter05\Id Math Unit in the book's on-line software.

5.3 ANSI/IEEE 754 Standard

During the early years of computing the implementation of floating-point mathematics in mainframe and mini computers was based on proprietary data formats and processing routines. Dewar and Smosma in their book titled *Microprocessors: A Programmer's View*, (see Bibliography), refer to a style of numerical programming prevalent during this period, which they call a "hope for the best" method. They cite numerous examples of this style including a Cray supercomputer which was unable to perform exact divisions by 2, a Honeywell machine in which the precision guard bits would disappear unexpectedly, and a case in which multiplication by 1.0 could cause an overflow. By the late 1970's a movement for the development of a standard that would bring some order to this chaotic situation had gained support in academia and in the computer professional societies.

The first actual suggestion of a computer mathematics standard was an article entitled *A Proposed Standard for Binary Floating-Point Arithmetic*, published in the SIGNUM Newsletter of the Association for Computing Machinery (ACM) in October, 1979. The article is signed by Jerome T. Coonen, William Kahan, John F. Palmer, Tom Pittman, and David Stevenson. John Palmer was at the time associated with Intel, and together with Ravenel and Nave, holds a patent for the invention of the 8087.

The second publication was in an article titled *An Implementation Guide to a Proposed Standard for Floating-Point Arithmetic*, by Jerome T. Coonen, which appeared in *Computer Magazine*, January, 1980. To solicit public comments, draft 8.0 of the standard was published in the March 1981 edition of *Computer*. The standard was approved by the IEEE Standards Board on March 21, 1985. Approval by the American National Standards Institute took place July 26, 1985. The final version was a product of the Floating-Point Working Group of the Microprocessor Standards Subcommittee of the IEEE Computer Society, presided by David Stevenson. The current version of the standard is designated as the ANSI/IEEE Standard for Binary Floating-Point Arithmetic, ANSI/IEEE Standard 754-1985. A more general standard was published in 1987, under the designation of ANSI/IEEE Standard 854 for Radix-Independent Floating-Point Arithmetic.

The Foreword to ANSI/IEEE 754 states that the intent of the standard is to promote the portability of numeric software, to provide a uniform environment for programs, and to encourage the development of better, safer, and more sophisticated mathematical code. Among the specific refinements of ANSI/IEEE 754 are the diagnosis of anomalies at execution time, the improved handling of exception conditions, and the implementation of interval arithmetic. In addition, the standard provides for standard elementary functions, very high precision calculations, and the use of algebraic symbolism in numerical operations.

According to the standard, a system in compliance with the ANSI/IEEE 754 Standard can be implemented in hardware, in software, or in both. Conformance to the standard is not determined by the internal properties of a system, but by the user's perception. ANSI/IEEE 754 specifically indicates that if a hardware product requires additional software to comply with the provisions of the standard, it should not be stated, in general terms, that it conforms. The Standard lists the following operations as specifically included:

1. Floating-point numeric formats.

2. The arithmetic operations of addition, subtraction, multiplication, division, square root, remainder, and compare.

3. Conversions between integer and floating-point, between the various floating-point formats, and between decimal strings and floating-point formats.

4. The handling of errors and exceptions.

The following topics are specifically excluded from the standard:

1. Decimal and integer formats.

2. Interpretation of the sign and the significand fields in non-numeric encodings (NaNs).

3. Binary to decimal and decimal to binary conversion of numbers encoded in the Standard's extended formats.

The development of the original 8087 was closely linked with the ANSI/IEEE Standard for Floating-Point Arithmetic. However, not until the 80387 did an Intel numeric processor fully comply with its terms.

5.3.1 Numeric Data Encoding

ANSI/IEEE 754 Standard for Binary Floating-Point Arithmetic defines four floating-point encodings divided into two groups. The first group is called the basic group and the second group is called the extended group. The basic formats are specified in detail by the standard, while for the extended formats the standard lists only the minimum requirements. Both groups have a single and a double precision encoding. Table 5.2 shows the requirements for encodings in these four formats.

Table 5.2

Numeric Data Encoding in ANSI/IEEE 754

| | SINGLE | | DOUBLE | |
	BASIC	EXTENDED	BASIC	EXTENDED
significand bits	24	≥32	53	≤64
maximum exponent	+127	≥+1023	+1023	≥+16383
minimum exponent	−126	≤+ 1022	- 1022	≤−16382
exponent bias	+127	unspecified	+1023	unspecified
exponent bits	8	≥11	11	≥15
total bits	32	≤43	64	≤79

Each binary encoding in the ANSI/IEEE 754 Standard contains three elements or fields:

1. The first field is the most significant bit and is used to encode the sign of the number. A 1-bit represents a negative number and a zero bit a positive number.

2. The second field is used for encoding the exponent of the number in biased form. The biased encoding makes it unnecessary to store the exponent sign. An exponent smaller than the bias is in the negative range. An exponent larger than the bias is in the positive range. The exponent is zero if it is equal to the bias.

3. The third field is called the significand, or the fraction field. In ANSI/IEEE formats this field has an implied 1-bit to the left of an also implied binary point. However, the standard validates encodings in the extended format in which the significand's leading bit is explicitly represented.

Figure 5.1 shows the bit structure and fields in the single and double formats of the ANSI/IEEE 754 Standard.

ANSI/IEEE 754 Standard leaves considerable freedom regarding encodings in the extended formats. The extended formats are defined as having a minimum number of parameters and an unspecified exponent bias, as shown in Table 5.2. In relation to the extended formats, the standard states that the developer may encode values redundantly, and reserve bit strings for purposes not described. ANSI/IEEE 754 requires that all implementations support the single format and recommends that at least one extended format be implemented for the widest basic format used. This means that an implementation that supports the single basic format, should also have an extended single encoding. By the same token, an implementation that supports the basic double format, should also have an extended double. The intention of this recommendation is that the extended formats be used for storing intermediate results with more precision than the format used for the result. This scheme serves to improve computational accuracy.

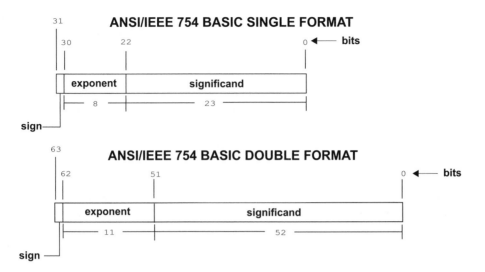

Figure 5.1 *ANSI/IEEE 754 Single and Double Formats*

5.3.2 Rounding

Rounding (or rounding-off) is the process of adjusting a numerical value so that it fits in a particular format. In general, the purpose of rounding operations is to reduce the error that arises from the loss of one or more digits. For example, the number 27,445.89 can be reduced to an integer value by truncating it to 27,445 or by rounding to 27,446. In this case the rounded value is a more accurate representation of the original number than the one obtained by chopping-off the last two digits.

ANSI/IEEE 754 requires that implementations provide the following rounding modes:

1. Round to nearest. This should be the default rounding mode. In this mode the result is the nearest representable value. The standard also describes how rounding is to take place when the result is equally near two representable values. This case, sometimes called the halfway case, occurs when rounding decimal numbers in which the last non-zero digit is 5.

   For example, regarding the number 128.500 the arbitrary rounding rule often taught in high-school is to round up. The above value would be rounded to the integer 129. An alternative rounding mode is called round to nearest even. In rounding the value 20,000.50 to an integer value there are two equally near options: 20,001 and 20,000. In the rounding to the nearest even mode the number 20,000 is preferred since it is an even number. Binary representations can be easily rounded to nearest even result by selecting the value in which the least significant bit is zero. Note that this method is also valid regarding binary coded decimals. The BCD rounding procedure named ROUND_SIGF, discussed in Chapter 4, determines if a BCD digit is odd or even by testing the least significant bit.

2. Round to positive infinity. In this rounding mode the result is rounded to the next highest representable value. This rounding mode is sometimes called rounding up.

3. Round to negative infinity. In this rounding mode the result is rounded to the next lowest representable value. This rounding mode is sometimes called rounding down.

4. Truncate. According to the definition at the beginning of this section, truncation is not considered a rounding mode. Truncation, also called chopping or chopping-off, consists in discarding the non-representable portion and disregarding its value. The chop-off operation is sometimes used in generating an integer result from a fractional operand.

5.3.3 Interval Arithmetic

The possibility of selecting rounding to positive infinity or negative infinity (round-up and round-down) allows the use of a technique known as interval arithmetic. Interval arithmetic is based on executing a series of calculations twice: once rounding up and once rounding down. This allows the determination of the upper and lower bounds of the error. Using interval arithmetic, it is possible, in many cases, to certify that the correct result is a value not larger than the result obtained while rounding up, and no smaller than the result obtained while rounding down. This places the exact result within a certain boundary.

Although ANSI/IEEE 754 does not specifically mention interval arithmetic, it does require directed rounding modes. Interval arithmetic can be a powerful numerical tool, although there are exceptional cases in which these results are not valid, as demonstrated by Dewar and Smosna in their book *Microprocessors: A Programmer's View* (see Bibliography). Not all mathematical calculations can be subject to interval analysis. The fundamental rules are as follows:

1. The operation must consist of multiple steps.

2. At least one intermediate result in the calculations must be subject to rounding.

3. The value zero should not be in the error range, that is, both results must have the same sign. The subsequent possibility of division by zero or a by a very small number introduce other potential problems that are not evident in interval arithmetic.

4. The calculations should not be, in themselves, a method for approximating results. Compounded approximations render invalid intervals.

5.3.4 Treatment of Infinity

The concept of infinity arises in relation to the range of a system of real numbers. One approach, called a projective closure, describes infinity as an unsigned representation for very small or very large numbers. When projective infinity is adopted, the symbol ∞ is used to represent a number that is either too small or too large to be encoded in the system.

An alternative approach, called affine closure, recognizes the difference between values that exceed the number system by being too large (+∞) or too small (−∞) to be represented. Figure 5.2 represents graphically the projective and the affine methods for the closure of a number system.

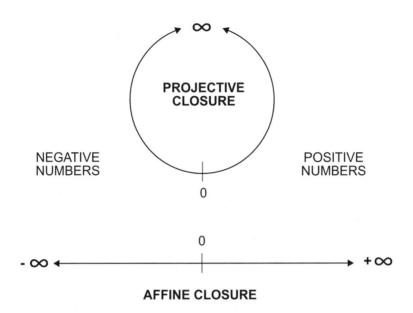

Figure 5.2 *Representations of Infinity*

According to the ANSI/IEEE 754, infinity must be interpreted in the affine sense. That is, any representable finite number x shall be located

$$-\infty \ (x) \ +\infty$$

This requirement contradicts the behavior of the 8087 and 80287 coprocessors, which default to a projective closure. Also the Intel literature for these chips recommends the use of the projective closure over the affine and states that, "there are occasions when the sign (of infinity) may in fact represent misinformation." The Intel 80387 mathematical coprocessor, which was released after the ANSI/IEEE 754, conforms with the standard's treatment of infinity. In the 80387 it is no longer possible for the programmer to select an infinity mode, since only affine closure is supported.

The Standard also provides that arithmetic operations with one or more infinity operands must be exact. Nevertheless, certain operations on infinity are considered invalid and generate the corresponding exception, specifically:

1. Addition (or subtraction) of infinities with opposite signs, for instance, $(-\infty) + (+\infty)$.

2. Multiplication of $0 \cdot \infty$.

3. Division of ∞ / ∞.

4. Remainder operations in the form x REM y, when $x = \infty$.

5. When infinity is created by the overflow of a finite operand.

The practical consequence of the affine treatment of infinities in ANSI/IEEE 754 is that when infinity forms part of an arithmetic operation the results are algebraically valid, since the correct sign is preserved.

Notice in Table 5.2 that in the basic encoding of the single format the exponent (bias 127) is in the range +127 to –126. This leaves unrepresented the exponent values 0H and FFH. Also, in the encoding of the double format, the exponent values 0H and 7FFH are unrepresented. These values are intentionally left unused by the standard so that they are available for encoding un-normalized numbers, infinity, and non-numeric values that represent invalid operations.

In the single and double formats, infinity is represented with an exponent of all one bits, and a significand of all zeros in the fractional portion. In other words, in encodings that use an implicit 1-bit in the significand, the significand for an infinity appears as all zeros. On the other hand, in encodings with an explicit 1-bit, infinity is represented as 100..00. In ANSI/IEEE 754 affine treatment of infinity is achieved by setting the number's sign bit for negative infinity and clearing it for positive infinity.

5.3.5 Not a Number (NaN)

ANSI/IEEE 754 requires that a number that exceeds the capacity of the destination format due to overflow or underflow, including the representable denormals, be replaced with the special encoding for infinity. Thereafter infinity arithmetic generates exact and valid results on these operands and no exception is signaled, except for the special conditions listed in Section 5.3.4.

On the other hand, certain operations generate results that are absurd, un-representable, or mathematically undefined; for example, attempts to perform division by zero, to multiply $0 \times \infty$, and to calculate the square root of a negative number. In these cases the standard provides a special encoding to represent results that are classified as Not a Number (NaN). The general pattern is an exponent of all 1-bits (as in the encoding for infinity) and a non-zero fractional portion of the significand. Note that the NaN encoding is easily differentiated from the infinity encoding because infinity requires all zeros in the fractional portion of the significand. Since these are the only requirement of the standard for a NaN encoding, implementations are free to use variations of the non-zero significand to represent different types of NaNs.

Signaling and Quiet NaNs

ANSI/IEEE 754 also requires the support of two different types of NaNs: signaling NaNs and quiet NaNs. The difference between them is that when a signaling NaN appears as an operand in an arithmetic calculation it forces the generation of an error exception. Quiet NaNs, on the other hand, will silently propagate signaling no error.

In ANSI/IEEE 754, signaling NaN and quiet NaN encodings are left to the implementor's discretion. The Standard does mention that signaling NaNs are typically used in representing un-initialized variables, complex infinity encodings, and other particular enhancements of the implementation. The signaling mechanism provides an automatic method for detecting an attempt to use these values as numeric operands. It is left to the implementation whether the simple copying of a signaling NaN generates an error.

According to ANSI/IEEE 754, signaling NaNs are not to be propagated by the system. If the result of an operation is a NaN it should be represented as a quiet NaN, even if both operands are signaling NaNs. Since the only requirement of the NaN encoding is that the exponent be formed with all one-bits and that the significand be non-zero, there is an abundant number of possible NaN combinations, even in the smaller formats. For example, in the single format (see Figure 5.1), which uses a 23-bit significand, there are over 8 million possible encodings for positive NaNs, and as many for negative NaNs. How these encodings are assigned to the various signaling and non-signaling NaNs is also left to the implementation.

The first Intel mathematical coprocessor to recognize and implement signaling and quiet NaNs is the 80387. In this chip, a leading one-bit in the fractional portion of the significand indicates a quiet NaN, and a zero bit in this position indicates a signaling NaN. In the Intel documentation for the 80387, 486, and Pentium, signaling NaNs are called SNaNs and quiet NaNs are called QNaNs. The Intel chips convert a SNaN to a QNaN by changing this high bit. Leaving the other bits unchanged provides a simple way for preserving the encoded diagnostic information.

5.3.6 Exceptions

The ANSI/IEEE 754 Standard requires the identification and signaling of five different error conditions, as follows:

1. Invalid operation

2. Division by zero

3. Overflow

4. Underflow

5. Inexact result

The signaling of an exception condition is performed by setting a flag, executing a trap routine, or both. The default response is to bypass the trap routine. The trap (which is different for each exception condition) transfers control to the user's error handler. The implementation must provide a different error flag for each exception.

Invalid Operation Exception

According to ANSI/IEEE 754 the following conditions generate an invalid operation exception:

1. An operation on a signaling NaN.

2. Addition or subtraction operations in which one or both operands are infinities.

3. Multiplication of $0 \cdot \infty$.

4. Division of 0/0 or ∞/∞.

5. Remainder operation, in the form x REM y, in which $x = 0$ or $y = \infty$.

6. The square root of a negative number.

7. Conversion operations from binary floating-point formats into integer or decimal formats that produce a result that cannot be faithfully represented.

8. Comparison operations in which one or both operands are NaNs.

Division by Zero Exception

This exception occurs when the divisor is zero and the dividend is non-zero. According to the standard the result is encoded as infinity. Note that the operation 0/0, which generates an invalid exception, is not considered a division by zero.

Overflow Exception

Table 5.2 shows the exponent encodings in the ANSI/IEEE 754 basic single format, which ranges from -126 to $+127$. Since this exponent is bias 127, the maximum absolute exponent is the decimal value 254 (11111110B) and the minimum absolute exponent is the decimal value 1 (00000001B). The exponent encodings of 0 (00000000B) and -127 (11111111B) are not used in representing real numbers in the basic single format. An analysis of the valid exponents in the other formats confirms that the exponent digit value of 00..00B and of 11..11B are also not part of the legal range assigned for the representation of real numbers.

This approach is based on the fact that any computer representation of real numbers is necessarily limited to a certain range. Numbers approach the limits of this range as they become very large or very small. The overflow condition takes place whenever a real number exceeds the representable range by becoming too large. In ANSI/IEEE 754 basic and extended formats, the maximum representable values have an exponent in the form 11..10B and a significand of 11..11B. Regarding positive real numbers, adding the smallest possible value to this encoding generates a number that exceeds the representable range (overflow).

The Standard requires that, when an overflow condition is detected, an exception be signaled and a special encoding be entered as a result of the operation. There are four possible variations of actual result, depending on the selected rounding mode, as follows:

1. If round to nearest is selected, the result of an overflow is encoded as an infinity with the sign of the intermediate result.

2. If the truncate mode is selected, the result of an overflow is represented with the format's encoding for the largest representable number.

3. If round to negative infinity is selected, the result of a positive overflow is represented with the format's encoding for the largest representable value and a negative overflow with the encoding for $-\infty$.

4. If round to positive infinity is selected, the result of a negative overflow is represented with the format's encoding for the smallest representable value and a positive overflow with the encoding for $+\infty$.

Note that in ANSI/IEEE 754 overflow is always abrupt (also called a sudden overflow). Because of the limitations in the representation of real numbers there are no provisions for gradual overflow. The result of the overflow of a positive number results in $+\infty$ or in the larger representable positive real, while the overflow of a negative number results in $-\infty$ or in the smallest representable negative real. Which action is taken depends on the selected rounding mode.

Underflow Exception

Overflow conditions take place as the absolute value of a number becomes very large. Underflow, on the other hand, takes place as the absolute value of a number becomes very small, in other words, as its value approximates zero. One method of handling numbers that approximate zero is to make them equal zero. This operation, sometimes called flush to zero, has been frequently used as a simple solution to the problem of underflow. But this sudden underflow presents some peculiar problems. For example, in the equation

$$(x-y)+y = x$$

if y is a sufficiently large number, then the portion (x–y) could suddenly underflow to zero, therefore

$$(0)+y = y$$

and

$$y = x$$

instead of

$$(x-y)+y = x$$

which is the expected result.

You have seen that, according to the provisions of ANSI/IEEE 754, overflow conditions are handled by abruptly converting the result to an infinity, or to the largest representable real. Which method is adopted depends on the rounding mode in effect. In order to avoid the dangers of sudden underflow, the standard requires using a special un-normalized representation of real numbers, called denormals.

In order to understand gradual underflow you must recall that a floating-point representation is said to be normalized when the first digit of the significand is non-zero. Normalization is designed to preserve the maximum number of significand digits and, therefore, the precision of the stored value. You can deduce that the smallest representable number in either format is encoded with an exponent pattern of 00..01B and a significand of 00..00B. Gradual underflow is based in the use of a special encoding for real numbers (the so-called denormals) which are characterized by an exponent in the form 00..00B and a denormalized significand. This representation, easily identified by an exponent containing all zero bits, allows representing numbers smaller than the smallest one that could be encoded using a normalized significand. Gradual underflow is made possible at the expense of precision. As the significand becomes denormalized, the number of its significant digits diminishes.

The ANSI/IEEE 754 requires the use of denormalized representations as well as the gradual underflow of very small numbers. The standard describes two correlated events that can contribute to underflow. One is the creation of a representable number which is yet so small that it may generate an error exception. An example is the overflow condition that could result from dividing by a very small operand. The second event is the loss of accuracy that results from representing very small numbers by denormalizing the significand.

Inexact Result Exception

The inexact results can occur from many arithmetic operations performed on valid operands. For example, the division operations 1/3, 1/7 and 1/9 cannot be exactly represented in binary form. This exception, sometimes called the precision exception, is designed as a warning that the rounded result of the previous operation cannot be exactly represented. In most computational situations this is the most frequent exception, and also the error condition that is most often ignored.

Chapter 6

Floating Point Data and Conversions

Chapter Summary

This chapter presents the formats and standards used in storing numerical data in the Intel math units. Supported data types include integer and real binary numbers and binary-coded decimals. The chapter also includes the development of conversion routines for input of user data into the math unit, and for converting math unit numeric data into strings of user-readable ASCII digits.

6.0 Math Unit Data Formats

In Chapter 5 we examined the floating-point storage formats prescribed by the ANSI/IEEE 754 Standard for Binary Floating-Point Arithmetic. In the following sections we refer to the encoding of numeric data as performed by the Intel math units. We have mentioned that the 80387 and the math unit of the 486 and the Pentium are the only Intel numeric data processors that are in full compliance with the provisions of ANSI/IEEE 754. Although all Intel math units store and manipulate numeric data in identical form, in this discussion we use the terminology that appears in the Intel documentation for the 80387, 486, and the Pentium. The following differences should be noted:

1 The numeric data type designated as short real in the 8087 and 80287 documentation corresponds to the single precision data type of the 80387, 486, and Pentium.

2. The numeric data type designated as long real in the 8087 and 80287 documentation correspond to the 80387, 486, and Pentium double precision format.

3. The packed decimal data type that appears in the 8087 and 80287 documentation correspond to the packed BCD data type in the 80387, 486, and Pentium.

4. The temporary real data type in the 8087 and 80287 correspond to the extended precision data type of the 80387, 486, and Pentium.

The Intel math units operate on three classes of numbers: binary integers, decimal integers, and binary real numbers. Table 6.1 lists the fundamental characteristics and the available encodings for each of these classes of numbers.

Table 6.1

Math Unit Formats for Numerical Data

NAME	BITS	RANGE	ENCODING		
BINARY INTEGERS: 　word integer	16	10^4	POSITIVES AS BINARY INTEGERS NEGATIVES IN TWO'S COMPLEMENT		
short integer	32	10^9	POSITIVES AS BINARY INTEGERS NEGATIVES IN TWO'S COMPLEMENT		
long integer	64	10^{18}	POSITIVES AS BINARY INTEGERS NEGATIVES IN TWO'S COMPLEMENT		
DECIMAL INTEGERS:			sign bits	exponent bits	significand bits
packed BCD	80	10^{18}	1	NO	72
BINARY REAL NUMBERS:			sign bits	exponent bits	significand bits
single 　precision	32	$10^{\pm38}$	1	8	23
double 　precision	64	$10^{\pm308}$	1	11	52
extended 　precision	80	$10^{\pm4932}$	1	15	64

6.0.1 Binary Integers

In the Intel math unit binary integer numbers can be stored in three formats. All three have identical structure but different capacity. The word integer format (see Table 6.1) occupies two bytes, the short integer format takes up a doubleword, and the long integer format a quadword. In all three formats the most significant bit encodes the sign of the number. In accordance with the general convention, a sign bit of 1 represents a negative number and a sign bit of 0 a positive number. Positive numbers are stored in pure binary form. Negative numbers are represented in two's complement form. Figure 6.1 shows the general structure of the binary integer formats.

The designers of the math unit provided some compatibility between the binary integer data types and the data types used by the 80x86 CPU family. A value stored as a math unit word integer (16 bits) can be loaded into an 80x86 16-bit register directly from memory. All 80x86 arithmetic instructions can operate on numbers encoded in this format. The short integer format can be loaded directly into a 32-bit register in the 80386, 486, and Pentium. However, a value stored in the short integer format cannot be loaded directly into a 8086/8088 or 80286 register.

Programs that deal with small integers can make good use of this compatibility. For example, a graphics application that operates on pixel addresses in the range 0 to ±32,767 can store screen pixel data in word integer format. The pixel addresses can then be loaded directly into the math unit for the necessary mathematical calculations and later stored back in the same memory variables. Since the math unit includes instructions to load and store integers in these formats, no data conversion operations are necessary. At the same time, graphics display routines can access data in memory variables in order to load from and store into

80x86 registers. This data sharing results in a considerable increase in performance since it facilitates calculations, which are easier to perform on the math unit, while display routines and other graphics manipulations can continue to be executed by 80x86 code.

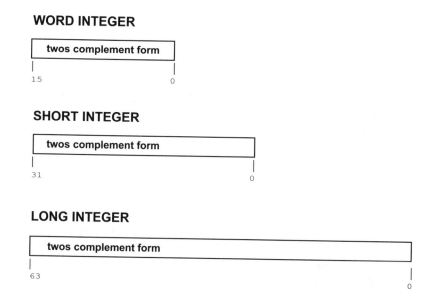

Figure 6.1 *Math Unit Binary Integer Formats*

6.0.2 Decimal Integers

The decimal integer data type contains a single storage format, designated as packed BCD in Table 6.1. In the packed BCD format the high-order bit encodes the sign of the number. A 1-bit encodes a negative number and a 0-bit a positive number, in the conventional manner. This contrasts with the binary integer formats in which negative numbers are encoded using the two's complement form. There is no exponent field in the decimal integer format since the format is limited to integer values. The significand consists of 2 binary coded decimal digits packed into each byte. A total of 18 significand digits are represented in the 72-bit significand. Figure 6.2 shows the bit map of the decimal integer encoding.

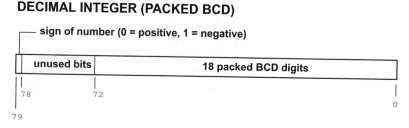

Figure 6.2 *Math Unit Decimal Integer Format*

The memory image of a number encoded as a decimal integer consists of 10 bytes (80 bits), with the high-order byte located at the lowest numbered memory address. The high-order byte encodes the sign of the number in its most significant bit. The remaining 7 bits in the sign byte are unused. The next 9 bytes hold 18 packed binary coded decimal digits, at the rate of two digits per byte. An 80-bits packed BCD value can be conveniently stored in a tenbyte memory variable. Most assembler programs, including Microsoft's MASM, Borland's Turbo Assembler, and Intel's ASMx86, encode a tenbyte variable declared with the DT directive as a number in decimal integer format.

We should emphasize that, in contrast with the BCD12 and BCD20 formats developed in previous chapters, the math unit decimal integer format is not a floating-point encoding. The decimal integer representation is limited to integer numbers. Perhaps the most important application of this format is in converting floating-point binary numbers into ASCII decimals. This use of the decimal integer format is shown later in this chapter.

6.0.3 Binary Reals

Rigorously speaking, the term real numbers is a mathematical designation for the set that includes all rational and irrational numbers; however, this term is also used to designate a number that can be represented in signed, floating-point form. The three math unit binary real encodings can be seen in Table 6.1. Figure 6.3 shows the general structure of the real number formats.

All three real number formats have the following fields:

1. The sign bit field, which is the most significant bit in the encoding, represents the sign of the number. A one bit in the sign field indicates a negative number and a zero bit indicates a positive number.

2. The exponent field encodes the position of the significand's binary point. The exponent encoding is in bias form. Therefore, if the value of the exponent is less than the bias, then the exponent is negative. If the exponent is greater than the bias, then it is positive. If the exponent is equal to the bias, it is zero.

3. The significand field encodes the number's significant digits as a binary fraction. Normal numbers (see Chapter 5) have an exponent in the range 11..10 to 00..01 and the significand is a binary fraction in the form 1.xx..xx. The number of digits in the fractional part of the significand changes in the different formats (see Table 6.2). The integer digit of the significand is implicit in the single and double precision formats but is explicitly coded in the extended precision format.

Internally, the math unit stores all numbers in the extended precision format of the binary real encoding. Numbers encoded in the remaining six formats (see Table 6.1), including the single and double precision real, exist only in memory. When a number is loaded from a memory variable into the math unit, it is automatically converted into the extended precision format. Therefore, all internal computations performed by the coprocessor are carried out in extended precision. The additional exponent and significand bits of this format provide a safety net to absorb computational errors that can result from conversions and rounding operations and during overflow and underflow. This extended precision format can also be used for the storage of constants and intermediate results. However, following the guidelines of the ANSI/IEEE 754, Intel literature recommends that the extended format not be used to increase the accuracy of computations, since this practice compromises the safety margin that it was designed to provide. Table 6.2 shows the format parameters for the 80x87 binary real numbers.

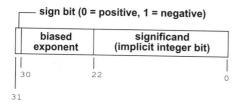

SINGLE PRECISION BINARY REAL

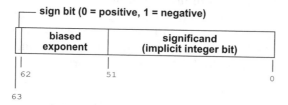

DOUBLE PRECISION BINARY REAL

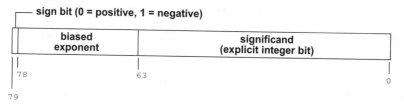

EXTENDED PRECISION BINARY REAL

Figure 6.3 *Math Unit Real Formats*

6.1 Special Encodings for Reals

In Chapter 5 we mentioned that certain types of numbers are treated separately by ANSI/IEEE 754; these are: infinities, NaNs, zeros, normal, and denormals. The Intel math units implement and recognize these special encodings as well as others not mentioned in the standard. Notice that the special encodings apply only to the real formats named single, double and extended precision and not to the binary or decimal integers.

Table 6.2

Floating-Point Binary Formats

	SINGLE PRECISION	DOUBLE PRECISION	EXTENDED PRECISION
sign bit	1	1	1
exponent bits	8	11	15
maximum exponent	+127	+1023	+16383
minimum exponent normal numbers)	−126	−1022	−16382
exponent bias	+127	+1023	+16383
significand bits	23	52	64
significand fractional bits (precision)	23	52	63
explicit binary point	NO	NO	YES
total bits	32	64	80

We have mentioned that, in the single and double precision formats, the significand is encoded with an implied 1- bit. This scheme, required by ANSI/IEEE 754, allows gaining one significant digit of precision. However, the standard does not specify the exact format for extended precision representations. In the case of extended precision, the implementor is free to use or not to use an implicit integer bit in the significand. Table 6.3 shows the bit fields for the math unit's special encodings in the single precision and double precision real formats.

Table 6.3

Special Encodings in Single and Double Precision

DESIGNATION	SIGN	BIT PATTERNS EXPONENT	SIGNIFICAND
ZERO			
positive zero	0	00..00	00..00
negative zero	1	00..00	00..00
INFINITY			
positive infinity	0	11..11	00..00
negative infinity	1	11..11	00..00
DENORMALS			
positive denormals	0	00..00	11..11 to 00..01
negative denormals	1	00..00	11..11 to 00..01
NANS			
indefinite	1	11..11	10..00
positive signaling NaNs	0	11..11	01..11 to 00..01
negative signaling NaNs	1	11..11	01..11 to 00..01
positive quiet NaNs	0	11..11	11..11 to 10..00
negative quiet NaNs	1	11..11	11..11 to 10..00
NORMALS			
positive normal numbers	0	11..10 to 00..01	11..11 to 00..00
negative normal numbers	1	11..10 to 00..01	11..11 to 00..00
single precision	1 bit	8 bits	23 bits
double precision	1 bit	11 bits	52 bits

The designers of the original 8087 mathematical coprocessor opted for an explicit 1 bit in the significand in the extended precision format. This practice has been preserved in all subsequent versions of the math unit. One reason for having an explicit 1-bit in the extended precision format is it facilitates mathematical calculation routines.

The implicit 1-bit scheme used in the single and double formats does not apply to denormal and zero encodings, which must always have a leading zero bit in the significand. Therefore, when using an implicit 1-bit storage scheme, the software must perform additional tests on the exponents to rule out the special cases in which the implicit bit must be zero. Table 6.4 shows the bit fields for the math unit special encodings in the extended precision real formats.

6.2 Low-Level Numeric Data in Memory

We have mentioned that most popular assembler programs, including Microsoft's MASM, Borland's Turbo Assembler, and Intel's ASMx86, provide assistance for storing numeric data in the math unit formats. This is a convenient way for creating and initializing memory variables and constants, which can then be read directly into the mathematical coprocessors. Fortunately, all major assembler programs mentioned above operate almost identically regarding the memory storing of numeric data. Table 6.5 lists the assembler directives and their corresponding math unit data formats.

Table 6.4

Special Encodings in Extended Precision

DESIGNATION	SIGN	BIT PATTERNS EXPONENT	SIGNIFICAND
ZERO			
positive zero	0	00..00	0.00..00
negative zero	1	00..00	0.00..00
INFINITY			
positive infinity	0	11..11	1.00..00
negative infinity	1	11..11	1.00..00
DENORMALS			
positive denormals	0	00..00	0.11..11 to 0.00..01
negative denormals	1	00..00	0.11..11 to 0.00..01
NANS			
indefinite	1	11..11	1.10..00
positive signaling NaNs	0	11..11	1.01..11 to 1.00..01
negative signaling NaNs	1	11..11	1.01..11 to 1.00..01
positive quiet NaNs	0	11..11	1.11..11 to 1.00..01
negative quiet NaNs	1	11..11	1.11..11 to 1.00..01
NORMALS			
positive normal numbers	0	11..10 to 00..01	1.11..11 to 1.00..00
negative normal numbers	1	11..10 to 00..01	1.11..11 to 1.00..00
Field size	1 bit	15 bits	64 bits

6.2.1 Initializing Data with the DW Directive

We mentioned that the math unit word integer data type is the only numeric format that is entirely compatible with all 80x86 CPUs, and that the short integer format is compatible with the 80386, 486, and Pentium, but not with its predecessors. The DW assembler directive can be used to allocate and initialize two bytes of storage, thus creating a variable compatible with the math unit word integer data type. This compatibility between the CPU word and the math unit word integer formats can be quite useful in applications that perform numeric operations on small integer numbers, as previously mentioned.

The following code fragment instructs the assembler to initialize the named variables in the word integer format:

```
; Initialize word variables to numeric values
TEN          DW    10       ; 000AH
TWO          DW    2        ; 0002H
MINUS_2      DW    -2       ; FFFEH (two's complement of 2)
MINUS_10     DW    -10      ; FFF6H (two's complement of 10)
TEN_THOU     DW    10000    ; 2710H
MAX_INT      DW    0FFFFH   ; FFFFH
```

Table 6.5

Assembler Directives for Initializing Numeric Data in Memory

DIRECTIVE	VERBAL EQUIVALENT	BINARY INTEGER	DECIMAL	REAL
DW	define word	word integer	NO	NO
DD	define doubleword	short integer	NO	single precision
DQ	define quadword	long integer	NO	double precision
DT	define tenbyte	NO	packed BCD	extended precision

6.2.2 Initializing Data with DD and DQ Directives

In Table 6.5 we see that the DD (define doubleword) and the DQ (define quadword) directives can generate numbers encoded as a binary integer or as a single precision real. This makes it necessary for the code to inform the assembler which format is to be created. The simplest way to identify the two possible formats is to enter single precision floating-point values with an explicit decimal point. The following code lines illustrate this operation.

```
; Initialize double and quadword variables to numeric values
SHORT_INT       DD      12345     ; Initialize short integer
LONG_INT        DQ      56789     ; Initialize long integer
SINGLE_REAL     DD      123.45    ; Initialize single precision
SINGLE_IN_HEX   DD      4332200H  ; Input of hexadecimal digits
DOUBLE_REAL     DQ      56789.0   ; Initialize double precision
DOUBLE_EXP      DQ      -1.3E12   ; Input in exponential notation
```

Notice that the presence of a decimal point determines in which format is the memory variable initialized. For instance, the value 56789 is stored as a long integer, while the value 56789.0 determines the creation of a double precision real. The difference in storage format must be taken into account when loading a variable into a coprocessor register, since the FLD opcode is used to load real and the FILD instruction to load integers. Also note that input can be entered in exponential notation, as in the case of the variable DOUBLE_EXP in the previous code sample.

6.2.3 Initializing Data with the DT Directive

In Table 6.5 we see that the DT (define tenbyte) directive can generate a number encoded as a decimal or as an extended precision real. Also in this case we instruct the assembler to generate a real number by entering an explicit decimal point or by using exponential notation. If no decimal point is present, the assembler encodes the number in packed BCD format. The creation of tenbyte variables is shown in the following code fragment.

```
; Initialize tenbyte variables to numeric values
PACKED_DEC        DT        123456789     ; Packed BCD
EXTENDED_PREC     DT        123456789.0   ; Extended precision real
EXTENDED_EXP      DT        1.23E8        ; Exponential input
EXTENDED_IN_HEX   DT        7F6ABC749318049EFF3FH
                                          ; Hex input
```

Here again the program must take into account the variable format when loading it into the coprocessor. The FLD opcode is used to load an extended precision real and the FBLD opcode to load a packed BCD variable.

6.2.4 Memory Image of the Special Encodings

In the math unit encodings for single precision and double precision binary real numbers the first digit of the significand is an implicit 1-bit (see Figure 6.3). This encoding scheme, which is required by ANSI/IEEE 754, has the advantage of doubling the numeric range of the representation. However, the implicit 1-bit makes it difficult to create a memory image of some of the special encodings, since the hardware will automatically interpret that the significand is preceded by a 1-bit. On the other hand, in the extended precision format all significand digits are explicitly encoded. Therefore it is a relatively simple matter to create any of the special encodings in extended precision format, as shown in the following code fragment.

```
; Special encodings in the extended precision format
ZERO_POS        DT      0000000000000000000000H
ZERO_NEG        DT      8000000000000000000000H
DENORMAL_POS    DT      00000000000000000000FFH
DENORMAL_NEG    DT      80000000000000000000FFH
NAN_POS         DT      7FFF80000000000000000FFH
NAN_NEG         DT      0FFFF80000000000000000FFH
INFINITY_POS    DT      7FFF8000000000000000000H
INFINITY_NEG    DT      0FFFF8000000000000000000H
;                        |  |                   |
;                        |--|----------------|
;      sign and exponent —^        ^————— significand
```

Notice that some of the above encodings are an arbitrary selection of one of many valid combinations of sign/exponent/significand that are legal in the format. For example, the significand field of a denormal number in extended precision format can range from 7F..FFH to 00..01H. The value selected in the preceding code fragment (00..FFH) is just one of many legitimate ones.

Software that must preserve special encodings in memory variables should use the tenbyte units of the extended precision format, as shown in this section. This is due to the fact that the smaller fields and the implicit 1-bit of the other real encodings could force a change of the representation as it is stored in memory by the NDP.

6.2.5 Operating on Memory Variables

Assembler programs automatically keep track of the storage format assigned to each floating-point variable. For example, the FLD instruction can be used to load single, double, and extended precision variables and the FILD opcode to load word, short, and long integers. In each case the assembler checks for a valid data type and generates the corresponding opcode. In this manner, the instruction:

```
FLD        REAL_VAR
```

assembles with no error and executes correctly if the variable REAL_VAR is a single, double, or extended precision real. By the same token, the instruction:

```
FILD       INTEGER_VAR
```

assembles and executes correctly if the variable INTEGER_VAR is a word, short, or long integer.

By exception it may be desirable to use an operand that does not explicitly indicate its type. For instance, if the register ESI points to a single precision real variable, simply coding FADD [ESI] may not work correctly, since the assembler has no way of telling if the variable pointed at by ESI is single or double precision. The solution is to inform the assembler of the variable type by using a variable type override operator. If, as in the above example, the ESI register points to a single precision variable stored in a doubleword of memory, the instruction can be coded in the form:

```
FADD       QWORD PTR [ESI]
```

Table 6.6 lists the type operators for the various numeric data variables.

Table 6.6

Type Override Operators for Numeric Data

DATA VARIABLE	STORAGE	OVERRIDE OPERATOR
word integer	DW	WORD PTR
short integer single precision	DD	DWORD PTR
long integer double precision	DQ	QWORD PTR
packed BCD extended precision	DT	TBYTE PTR

6.3 High-Level Numeric Data

Data formats in a particular implementation of a high-level language must match the machine-level data formats. But a high-level language does not have to implement all low-level formats. In C/C++ the numeric data formats are compiler-dependent. Visual C++ Version 6.0 defines the data types shown in Table 6.7.

Table 6.7

Visual C++ Numeric Data Types

TYPE NAME	BYTES	OTHER NAMES	RANGE OF VALUES
int	2/4	signed, signed int	system dependent
unsigned int	2/4	unsigned	system dependent
__int8	1	char, signed char	−128 to 127
__int16	2	short, short int, signed short int	−32,768 to 32,767
__int32	4	signed, signed int	−2,147,483,648 to 2,147,483,647
__int64	8	none	−9,223,372,036,854,775,808 to 9,223,372,036,854,775,807
char	1	signed char	−128 to 127
unsigned char	1	none	0 to 255
short	2	short int, signed short int	−32,768 to 32,767
unsigned short	2	unsigned short int	0 to 65,535
long	4	long int, signed long int	−2,147,483,648 to 2,147,483,647
unsigned long	4	unsigned long int	0 to 4,294,967,295
float	4	none	3.4E +/− 38 (7 digits)
double	8	none	1.7E +/− 308 (15 digits)
long double	10	none	1.2E +/− 4932 (19 digits)

Note that the long double data type (80-bit, 10-byte precision) is implemented as a double (64-bit, 8-byte precision) in Windows 95/NT and later.

Also that int and unsigned int types have the size of the system word. In MS DOS and Win16 it is 2 bytes, in 32-bit operating systems the size is 4 bytes. Code that must ensure portability should rely on the sized integer types: __int8, __int16, __int32, and __int64. Also notable is that Visual C++ provides no equivalent data type for the 10-byte packed BCD format.

6.4 Numeric Data Conversions

In developing numerical software it is usually necessary to have some way for entering data into the math unit, and for displaying the results of computations. High-level language programmers can use the languages' input and output functions for this purpose. This is possible because C and C++ functions take care of converting user input into ANSI/IEEE integer or floating point formats. In the test programs developed for this book and contained in the book's CD ROM we often use the C++ cin() function to obtain keyboard data and the cout() function to display messages and results of computations.

However, cin and cout, as well as the corresponding C input and output functions, are software black boxes. The programmer has little control about how the conversion is performed and little feedback regarding possible errors or loss of precision. In the following sections we develop procedures for entering data into the math unit registers and for recovering data from math unit registers into user-readable ASCII strings. We also develop low-level and high-level interface routines to these core procedures.

6.4.1 Data Conversion in ANSI/IEEE 754

ANSI/IEEE 754 and 854 require that an implementation provide conversions between decimal strings and encoded floating-point numbers. However, the math unit does not contain conversion facilities, nor are conversion routines usually furnished with the hardware. The lack of conversion routines is one reason why the Intel math unit cannot be considered as a stand-alone implementation of ANSI/IEEE 754 Standard. Although the lack of full hardware conversion facilities should not, by itself, be considered a violation of the standard, since ANSI/IEEE 754 and 854 provide that an implementation can be realized in hardware, in software, or in a combination of both. The standards state that "It is the environment the programmer or user of the system sees that conforms or fails to conform to this standard." However, both standards also state that hardware components that require software support "shall not be said to conform apart from this software."

In Section 6.2 we saw that it is relatively easy for assembler code to initialize variables and to define memory constants and integer values which can be loaded into the math unit. But there is no simple or convenient way of converting ASCII decimal numbers entered by the user into math unit real formats and vice versa. Over the years Intel has made available software support packages for each of its math units. These products, named the 80x87 Numeric Support Libraries, include conversion routines, software emulators, libraries of common and complex-number functions, and error handlers. The principal problem with the 80x87 Numeric Support Libraries is that they are designed to be used with Intel's ASMx86 line of assemblers. The code is not easily portable to other high- or low-level development systems.

6.4.2 Conversion Requirements in ANSI/IEEE 754

ANSI/IEEE 754 Standard lists the requirements for performing binary to decimal and decimal to binary conversions. These can be summarized as follows:

1. The implementation must provide conversions between all the supported internal formats and at least one decimal format. If the number to be converted is expressed as

$$\pm M \times 10^{\pm N}$$

where M and N are integers, then the decimal conversion ranges for the single and double floating-point formats are as shown in Table 6.8.

Table 6.8
Conversion Ranges in ANSI/IEEE 754 Standard

FORMAT	DECIMAL TO BINARY		BINARY TO DECIMAL	
	MAX M	MAX N	MAX M	MAX N
Single	10^9-1	99	10^9-1	53
Double	$10^{17}-1$	999	$10^{17}-1$	340

2. When the integer M in Table 6.7 lies outside the specified range, the implementor may change the significant digits after the ninth one in the single format and after the seventeenth one in the double format. This change usually consists of replacing the corresponding digit with 0.

3. Conversions should be rounded according to the active rounding mode as specified by the Standard. In rounding to the nearest even the error should not exceed 0.47 units of the least significant digit. In all other rounding modes the error should not exceed 0.47 ulps (units in the last place). Table 6.9 lists the correct rounding range for decimal conversions.

Table 6.9
Correct Decimal Rounding Ranges in ANSI/IEEE 754 Standard

FORMAT	DECIMAL TO BINARY		BINARY TO DECIMAL	
	MAX M	MAX N	MAX M	MAX N
Single	10^9-1	13	10^9-1	13
Double	$10^{17}-1$	27	$10^{17}-1$	27

4. Conversions should be monotonic, that is, an increase in the value of the operand shall not decrease the value of the converted string. Monotonicity is usually related to the accuracy of the conversion routine.

5. When the rounding to nearest mode is enabled, the conversion from binary to decimal and back to binary should always result in the same number, as long as the decimal string is carried to the maximum precision shown in Table 6.7.

6. When converting a decimal number to a binary format the conversion routine must detect overflow and underflow. If so, the code must perform the actions specified by the standard. The required action consists of trapping to an error handler, possibly with an adjusted exponent, and the result is rounded to the destination's precision. If the result is too far outside the range to allow this adjustment, then a quiet NaN should be delivered instead.

The designer of a conversion routine that aims at conforming with ANSI/IEEE 754 Standard must pay careful attention to some of the above requirements. Regarding code that executes in the Intel math units, the use of the extended precision format aides in assuring the required accuracy, since the standard does not require conversions to exceed the double precision format for reals.

The MATH16 and MATH32 libraries in the book's CD ROM contains low-level conversion routines developed by the authors. FPU_INPUT converts an ASCII decimal string entered by the user into a binary floating-point and loads the number into the math unit stack top register. FPU_OUTPUT stores the FPU stack top register as an ASCII decimal number. A support procedure named ASCII_TO_EXP is also provided. This last routine converts an ASCII decimal number into exponential (scientific) notation. The operation is useful because FPU_INPUT assumes that the ASCII decimal number is already in exponential form. All three routines are contained in the UN16_4 module of the MATH16 library and the UN32_4 module of the MATH32 library in the book's CD ROM. Notice that the 32-bit versions of these procedures have a leading underscore in the procedure names in order to make them accessible to C++ code. In referring to these procedures we exclude the leading underscore symbol.

6.4.3 FPU_INPUT Procedure

This procedure loads a decimal floating-point number in exponential notation into the math unit stack top register. It is also an indirect means for converting a decimal input, in exponential form, into any one of the math unit integer or floating-point real formats. This is possible due to the fact that the contents of the math unit stack top can be stored in memory as an integer (FIST, FISTP, FBSTP opcodes) or as a floating-point real (FST and FSTP opcodes). The following code fragment shows this operation.

```
        .DATA
ASCII_DEC       DB      28 DUP(00H) ; Storage for decimal number
                                    ; in exponential form
SINGLE_PREC     DD      ?           ; Storage for single precision
                                    ; real number
        .CODE
        .
        .
        .

        LEA     EDX,ASCII_DEC       ; Pointer to stored decimal
        CALL    FPU_INPUT           ; Conversion and load procedure
; FPU stack top register now has input number in extended
; precision real format
        FST     SINGLE_PREC         ; Store stack top as a single
                                    ; precision real
    .
    .
    .
```

The FPU_INPUT procedure reports to the caller that the operation took place successfully by clearing the carry flag and loading the value 0 in the AL register. If the carry flag is set, then AL holds one of the following error codes:

AL = 1 if the input was not formatted in exponential notation. In this case the special encoding NAN is loaded into ST(0).

AL = 2 if the exponent exceeds the valid range error. If the excess constitutes an overflow, the special encoding for positive infinity is loaded into ST(0). If an underflow, then the special encoding for negative infinity is in ST(0).

AL = 3 if the caller's string contains one or more invalid characters. In this case the special encoding for NAN is loaded into ST(0).

The conversion algorithm can be described as follows:

1. The number, represented by a string of ASCII digits in exponential notation, is first separated into its exponent and significand fields. The exponent is stored as a binary value and the integer significand as a string of packed BCD digits.

2. The code tests the magnitude of the exponent for values that exceed the range of the format. If so, a special encoding (+ or – infinity) is loaded and an input error is flagged to the caller. Invalid characters in the caller's string determine that the encoding for NaN be loaded into the FPU and invalid input is flagged to the caller.

3. The BCD integer significand is then loaded into the math unit and divided by 10^{17} in order to scale it to a number in the range

$$0> \text{significand} >10$$

4. The constant 10 is raised to the power of the unsigned integer exponent (y). The significand is multiplied by 10^y if the exponent is positive and divided by 10^y if it is negative. The result is left in ST(0).

Calculating 10^y

The algorithm for loading a string of ASCII digits into the math unit stack top register requires evaluating 10^y, where y is the exponent of the number to be converted. Several approaches to this calculation are found in the literature. Perhaps the most common approach is to use logarithms. Problems that result from logarithmic approximation and other methods for calculating exponential functions are discussed in Chapter 8. Regarding conversion routines, the use of a logarithmic approximation for the calculation of 10^y introduces a considerable error. When operating close to the limits of the math unit temporary real format, the error can propagate to the 12th significand bit of the result. For example, the conversion of the ASCII value 1.0 E+4922 using a logarithmic approximation for calculating 10^{4922} produces the number

```
9.99999999999999502 E+4921.
```

In this case the error in the logarithmic approximation of 10^y has contaminated the result to a point where it exceeds the error allowed by ANSI/IEEE 754 Standard. The problem can be fixed by using a more accurate approximation of the value 10^y. The conversion routine used by the _FPU_INPUT procedure, which is based on a table of constants, calculates the value 10^y accurately to the 63rd significand bit. In this case, the ASCII string 1.0 E+4921 produces the number

```
1.00000000000000000 E+4921
```

which is exact to the lowest order decimal digit.

Post-Conversion Operations

The FPU_INPUT performs all calculations using the math unit extended precision format for reals. Since ANSI/IEEE 754 Standard does not require conversions beyond the precision of the double format, the use of extended precision could seem wasteful and unnecessary. In reality, the use of extended precision provides a simple and efficient way for insuring that the conversion meets the requirements of ANSI/IEEE 754 Standard. All that is necessary is that the code store the value returned by the FPU_INPUT routine in a double precision variable and then reload ST(0) from this storage. The store/reload sequence will automatically insure the following requirements of ANSI/IEEE 754.

1. Rounding to the double precision real format will be performed according to the setting of the rounding control field of the math unit control word.

2. Overflow and underflow exceptions will be correctly encoded and a trap handler will receive control if the exception response bits of the control word are set accordingly.

3. The higher accuracy of the extended format insures monotonicity of the result stored in the double precision variable.

In addition, the use of extended precision in the conversion routines guarantees that the error tolerance of 0.47 ulps for rounding to the nearest and 1.47 ulps for all other rounding modes are not exceeded during processing.

6.4.4 FPU_OUTPUT Procedure

The FPU_OUTPUT procedure converts the value in the math unit stack top register into an ASCII decimal number formatted in exponential notation. _FPU_OUTPUT also provides an indirect means for converting a binary number stored in memory in one of the FPU formats into an ASCII decimal in exponential form. The following code fragment shows the operations required.

```
        .DATA
;
STORED_SINGLE   DD      4332200H    ; Single precision real
OUTPUT_BUF      DB      25 DUP(00H) ; Storage for decimal number
                DB      0H
        .CODE
        .
        .
        .
        FLD     STORED_SINGLE       ; Load single precision real
                                    ; into FPU stack
        LEA     EDX,OUTPUT_BUF
        CALL    _FPU_OUTPUT
; The buffer OUTPUT_BUF now has a decimal string of ASCII digits
; that represent the value of the stored single precision real
```

The _FPU_OUTPUT procedure returns with the carry flag clear if the conversion was successful and the number in ST(0) is classified normal. If not, the carry flag is set and the AL register is as follows:

AL = 1 if the number in ST(0) is a positive or negative denormal or if its format is unsupported. In this case the caller's buffer contains the string: "denormal."

AL = 2 if the number in ST(0) is a positive infinity. In this case the caller's buffer contains the string: "+infinity."

AL = 3 if the number in ST(0) is a negative infinity. In this case the caller's buffer contains the string: "–infinity."

AL = 4 if the number in ST(0) is a positive or negative NAN. In this case the caller's buffer contains the string: "NAN (not-a-number)."

AL = 5 if ST(0) is tagged empty. In this case the caller's buffer contains the string: "EMPTY."

Signed zeros are reported as follows:

0.00000000000000000 E+0 represents a positive zero

–0.00000000000000000 E+0 represents a negative zero.

SOFTWARE ON-LINE

The low-level conversion procedures FPU_INTPUT and FPU_OUTPUT mentioned in the previous sections are found in the Un32_4 module of MATH32.LIB, in the book's on-line software. Support procedures are located in the Un32_1 module.

6.4.5 ASCII-to-Exponential Conversion

The input conversion routine FPU_INPUT requires that the input value be in exponential form. The ASCII decimal string must be entered in a format similar to scientific notation. For example, the value –128.765 must be entered in the form –1.28765 E+2. The procedure named ASCII_TO_EXP converts a string of ASCII decimal numbers into exponential notation. The ASCII_TO_EXP procedure also detects and bypasses a number already in exponential form. This makes it possible to pass strings in either ASCII decimal or exponential form. If the string is not in exponential notation it is converted. If it is already in exponential form it is copied to the output buffer.

6.5 High-Level Interface Functions

The folder Sample Code\Chapter6\Mu Conversions in the book's CD ROM contains the file Un32_4.h. This header contains three C++ functions that provide access to the conversion procedures mentioned in the previous sections. The project Mu Conversions demonstrates and exercises the following conversion functions:

1. FpuInput() provides access to the _FPU_INPUT procedure in MATH32.lib.

2. FpuOutput() provides access to the _FPU_OUTPUT procedure in MATH32.lib.

3. AsciiToExp() provides access to the _ASCII_TO_EXP procedure in MATH32.lib.

Following is a listing of the functions.

```
int FpuInput(char number[])
{
// On entry:
//     parameter number[] is a user-defined array holding an
//     ASCII decimal number formatted in exponential or
//     decimal notation.
//
//     Exponential notation as follows:
//
//                    sm.mmmmmmmmmmmmmmmmmm ESeeee
// where
// s = sign of number (blank = +)
// m = up to 18 significand digits. Extra digits are ignored
// . = decimal point following the first significand digit.
//     If no decimal point in string, it is assumed
// E = explicit letter E (or e) to signal start of exponent
// S = + or - sign of exponent. If no sign, positive is assumed
// e = up to four exponent digits. The numerical value of the
//     exponent cannot exceed +/- 4932
// One or more spaces can be used to separate the last digit of
// the significand and the start of the exponent
// Examples of input:
//                    1.781252345E-1
//                    -3.14163397  E+0
//                    1.2233445566778899 e1387
//                    11.2233
//                    -0.004455
// On exit:
//            User's value is loaded into math unit ST(0) register
//            Previous values in the stack are pushed down one
//            register
//            returned value = 0 (no error)
//
//         b. If returned value not zero, returned value indicates
//            the following errors
//            1    if no E or e symbol in the caller's string
//                 NAN encoding is loaded into ST(0)
//            2    if exponent exceeds valid range
//                 Overflow - positive infinity is loaded into ST(0)
//                 Underflow - negative infinity is loaded into ST(0)
//            3    if there is an invalid character in caller's
//                 string. NAN encoding is loaded into ST(0)
//
//         AX is destroyed. All other registers are preserved.
//         The caller's NDP environment is preserved, except
//         the stack top register.

int rtnError = 0;
_asm
{
     MOV      EDX,number
     CALL     FPU_INPUT
; Capture error code
     MOV      rtnError,EAX
}
   return rtnError;
}

void AsciiToExp(char ascStr[], char expStr[])
{
```

```
// Checks for the ASCII input string and converts it to
// exponential form
// On entry:
//          ascStr[]  is an ASCII decimal string, terminated in a NULL
//                    byte, which holds the caller's ASCII decimal
//                    input
//              Note: The caller's buffer may contain the $ sign
//                    and commas as formatting symbols. The string
//                    terminator (00H) may be preceded by a CR
//                    (0DH) or LF (0AH) control codes. If the
//                    characters E or e are in the string, the
//                    procedure assumes that the string is already
//                    in exponential notation
//
//          expStr[]  is a 27-byte unformatted output buffer which will
//                    be filled with the output string formatted in
//                    exponential form
//
// On exit:
//          expStr[]  is the caller's output buffer with the input number
//                    reformatted in exponential form, as follows:
//
//                    sm.mmmmmmmmmmmmmmmmmm ESeeee
// where
// s = sign of number (blank = +)
// m = 18 significand digits
// . = decimal point following the first significand digit
// E = letter E signals the start of the exponent field
// S = + or - sign of exponent
// e = up to four exponent digits
//
// NOTE:
//      This function can be used to make sure that the input string
//      passed to the FpuInput() function is formatted in exponential
//      notation.
_asm
{
    MOV     ESI,ascStr
    MOV     EDI,expStr
    CALL    ASCII_TO_EXP

}
  return;
}


int FpuOutput(char userBuf[])
{
//          Converts the value in the top register of the FPU
//          stack into an ASCII decimal number in scientific
//          notation
//
// On entry:
//          userBuf[] is the unformatted caller's buffer area
//          with minimum space for 25 bytes of storage
//
//          ST(0) holds value to be converted
//
// On exit:
//          Return value = 0
//          ST(0) holds value to converted
```

```
//          The caller's buffer will contain the ASCII decimal
//          representation of the number at the FPU stack top
//          register, formatted in scientific notation, as follows:
//
//                    sm.mmmmmmmmmmmmmmmmmm ESeeee
// where
// s = sign of number (blank = +)
// m = 18 significand digits
// . = decimal point following the first significand digit
// E = explicit letter E to signal start of exponent
// S = + or - sign of exponent
// e = up to four exponent digits
// Examples of output:
//                    1.78125234500000000 E-12
//                   -3.14163397000000000 E+0
//                    1.22334455667788998 E+1388
//      Return value = 1
//      Invalid value in top of stack
//
// Other number types return the following 12-character text strings
// in the caller's buffer
//                    + denormal
//                    - denormal
//                    + NAN
//                    - NAN
//                    + infinity
//                    - infinity
//                    unsupported
//                     EMPTY
//      Signed zeros are reported as follows:
//                    0.00000000000000000 E+0 (positive zero)
//                   -0.00000000000000000 E+0 (negative zero)
//
//      The caller's math unit environment is preserved
//********************************************************************
int rtnError = 0;
_asm
{
    MOV     EDX,userBuf
    CALL    FPU_OUTPUT
    JNC     FPU_EXIT
    // Error
    MOV     rtnError,1
FPU_EXIT:
}
    return rtnError;
}
```

SOFTWARE ON-LINE

The program project FP Conversions demonstrates the C++ interface functions. The project files are found in the folder Sample Code\Chapter06\FP Conversions, all in the book's on-line software.

Chapter 7

Math Unit Architecture and Instruction Set

Chapter Summary

This chapter describes the fundamental architecture of the Intel math units and discusses its math unit instruction set. Topics include programming methods and techniques especially suited for coding math unit routines as well as the operation and characteristics of the individual opcodes. The material serves as background for the programming algorithms and routines developed in later chapters.

7.0 Math Unit Internal Organization

Intel math units have many common elements in their internal architecture. Structurally, the math units are divided into a control unit (CU) and a numeric execution unit (NEU). Usually, these structures are invisible to the programmer, who perceives the math unit as consisting of the addressable data areas and a dedicated instruction set. The following are the math unit's visible data areas:

1. A stack of eight operational registers

2. A Control register

3. A Status register

4. A tag word, consisting of 2 tag bits for each stack register

5. An instruction pointer

6. A data pointer

7.0.1 Math Unit Register Stack

The numeric execution unit of the FPU contains eight internal registers for operational data. Many instructions allow these registers to be addressed explicitly, that is, according to their designation in the ST(i) form. Figure 7.1 represents the math unit register stack.

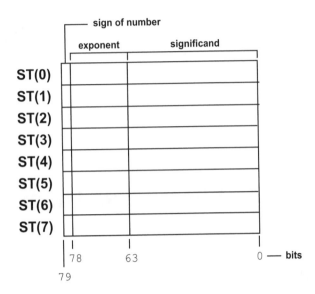

Figure 7.1 *Math Unit Register Stack*

The math unit's internal structure for holding operational data can be viewed as a stack-like arrangement of eight individually addressable units. Each register stores a binary real number in extended precision format (see Figure 6.3). Many math unit instructions allow addressing the stack registers explicitly, that is, according to the ST(i) designation, where i is a positive integer in the range 0 to 7. The left side of Figure 7.1 shows the explicit register names. Note that the register at the top of the stack is designated simply as ST, although most assemblers also allow the expression ST(0). However, there are exceptional cases in which the designation ST(0) is required.

The programmer can also address the math unit numeric data registers as a pure stack structure. As is the case in the main CPU, math unit load and push operations first decrement the top-of-stack pointer and then store the value in the new stack top register. Store and pop instructions perform the reverse action. The math unit register stack grows downward (toward lower-numbered register) in a similar manner as a memory stack. This mode of operation is consistent with the description for a last-in-first-out (LIFO) data structure.

Math unit instruction encodings must take into account that registers can be addressed explicitly (using the register names shown in Figure 7.1) or implicitly (as a stack). However, not all math unit operations can address the stack registers explicitly. For example, the square root instruction (FSQRT) does not allow an explicit register designation since its action is limited to the Stack Top register. Several other arithmetic instructions also do not allow explicit register designations, such as FABS (get absolute value) and FCHS (change sign). The same applies to the instructions that load constants onto the math unit, such as FLDPI, FLDZ, and FLD1, as well as to some transcendental instructions.

For this reason it is not always possible to code numeric programs that use a pure register model of the math unit. It is usually better to think of the math unit's numeric data registers as a stack structure in which the programmer has limited possibilities for addressing individual elements. In all cases it is up to the programmer to keep track of the registers in use.

The math unit register stack is sometimes represented as a circular structure equipped with a pointer that automatically wraps around to its first element, as shown in Figure 7.2. The pointer to the current top-of-stack register corresponds with a 3-bit field in the Control Word register known as the *stack top pointer*.

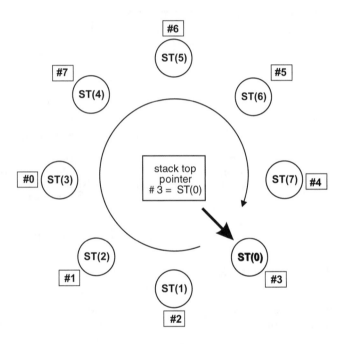

Figure 7.2 *Visualization of the Math Unit Stack*

In Figure 7.2 the designation of each stack register is indicated in two different ways. The values inside rectangles represent the register's physical number. The ST(i) designation, shown inscribed in circles, indicates the register's relative position in the stack. While the physical ordering of the registers does not change, their positions in the stack does. This position, which is stored in 3-bit field of the status word, signals which physical register is currently at the top of the stack. In Figure 7.2 physical register number 3 is at the top of the stack, therefore it is designated as ST(0). If, at this time, a new data item is pushed onto the math unit, it is placed in physical register number 3. The top-of-stack pointer is immediately decremented and ST(0) becomes physical register number 2. Notice that, conventionally, the math unit stack grows downward, that is, toward smaller numbered physical registers.

7.0.2 Math Unit Control Register

The programmer can modify math unit operation by means of the Control register. Originally, the Control register allowed changing the way the coprocessor handles rounding, infinity, and precision control. Because the final version of the ANSI/IEEE 754 Standard for Binary Floating-Point Arithmetic established affine closure as the only method for handling infinity, the 80387 and the math unit of the 486 and Pentium no longer have an infinity control function. Nevertheless, programs that modify the infinity control bit do not generate an error in the 80387, 486, and Pentium. Figure 7.3 is a bitmap of the math unit Control register.

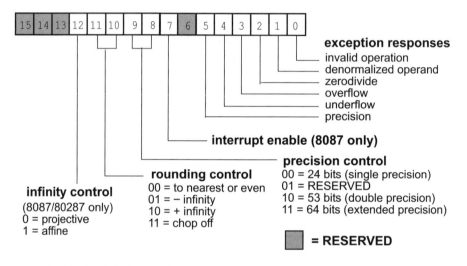

Figure 7.3 *Math Unit Control Register Bitmap*

The rounding control field of the Control register determines the active numerical rounding mode. Rounding to the nearest even number is probably the most used setting, since it is the least biased one. Numerical rounding was discussed in Chapter 5 in relation to the ANSI/IEEE 754 Standard. All the rounding modes required by the standard are implemented in the math unit.

Intel documentation states that the precision control field is provided mainly for compatibility with earlier generations of numerical processors. These bits control internal precision of the significand during the arithmetic operations FADD, FSUB, FMUL, FDIV, and FSQRT. The default setting is to the maximum precision, which is a 64-bit significand and corresponds with the extended precision real format. Occasionally math unit software reduces the precision of the calculations to prevent spurious rounding of excessively precise results. An example of this manipulation can be found during the calculation of the exponent in the FPU_OUTPUT procedure mentioned in Chapter 6.

In the 8087, bit number 7 of the Control register (see Figure 7.3) is used as a general mask to enable or disable all interrupts. This bit is not used (reserved) in the 80287, 80387, and in the math unit of the 486 and the Pentium. In 8087 systems if this bit is set, all interrupts are masked from the central processor.

Error Conditions

Bits 0 to 5 determine the math unit response to six error conditions that can originate during numeric processing. If the corresponding mask bit in the control word is set (exception masked) the math unit takes a default internal action. In this case processing continues uninterrupted. On the other hand, if the corresponding exception response mask bit is clear (exception unmasked) a software exception handler coded by the user receives control on every occurrence of the corresponding error. For example, a division by zero operation can generate the following actions:

1 If the division by zero exception is masked (bit number 2 of the control word = 1) the math unit returns an infinity encoding. The sign of this infinity is the result of XORing the signs of the operands. Numeric processing then continues undisturbed.

2. If the division by zero exception is unmasked (bit number 2 of the control word = 0) an interrupt is generated. This interrupt can be used to invoke a software error handler.

The masked responses for the exception conditions are designed to provide sufficient and adequate action for most programming situations. However, an application sometimes requires that a specific action be taken in response to an exception. In order for your program to gain control, the exception flag for the corresponding exception bit must be clear (mask = 0). This ensures that an interrupt will be generated, and intercepted by the error handler routine. In 8087 systems bit 7 of the Control register (the interrupt enable bit) must also be clear (bit = 0) for the interrupt to take place.

The actual vector for the error exception interrupt has changed in the different versions of the Intel math units. In the original IBM Personal Computer, the 8087 interrupt line was wired to the Non Maskable Interrupt (NMI), which was vectored to interrupt 2. Therefore, the address of the exception handler for such a machine must be located at the interrupt 2 slot in the vector table. Because this vector was shared with system error conditions, such as a memory parity error, the error handler had to check the 8087 Status register to make sure that the interrupt was caused by a numeric condition. If not, the handler could return control to the system's NMI routine.

The IBM PC AT was the first commercial machine to use the 80286 CPU and the 80287 coprocessor. The design of the AT makes use of two 8259 interrupt controllers, instead of the single one used in the original Personal Computer. In the AT the coprocessor interrupt is linked to request line number 13 (IRQ13) of the second 8259A. The handler routine is vectored through interrupt 75H, but the BIOS initialization points the NMI handler to this vector. This redirection is also used in the IBM microcomputers of the PS/2 line. For this reason, an MS DOS exception handler written for the original Personal Computer will also work on the AT and some models of the PS/2 line, but not in all IBM-compatible machines, and certainly not in Windows or other operating systems.

The following error conditions are determined by the exception mask bits. The descriptions correspond to the action of the 80387 coprocessor and the math unit of the 486 and the Pentium.

1. *Invalid operation.* This error usually results from loading a non-empty register, popping an empty register, or from a stack overflow. Undefined operations can also generate an invalid operation error, for example, a division of zero by zero or an attempt to calculate the square root of a negative number. Any arithmetic operation on a signaling NaN also generates this exception, as well as arithmetic and transcendental instructions in which one of the operands is infinity. ANSI/IEEE 754 Standard lists eight conditions that generate this exception. The invalid operation exception is the one most often vectored to a software interrupt handler.

2. *Overflow.* Overflow occurs when a number exceeds the largest representable value in the format. ANSI/IEEE 754 and 854 Standards state that the masked response to an overflow must be determined by the rounding mode currently in effect. The math unit responses to a masked overflow condition can be seen in Table 7.1. These responses are in accordance with the requirements of ANSI/IEEE 754 and 854 Standards.

   If the overflow exception is not masked, the math unit response depends on whether the results are to be stored in memory or in the register stack. If the unmasked overflow exception is from a store instruction (memory destination) then nothing is stored and the operand is left in the stack. If the destination is the stack, then the result is scaled to approximately the mid-range of the format. This action is consistent with the provisions of ANSI/IEEE 754.

Table 7.1

Math Unit Masked Response to Overflow

ROUNDING MODE	SIGN OF RESULT	RESULT
to nearest	+	$+\infty$
	−	$-\infty$
to $-\infty$	+	largest positive encoding
	−	$-\infty$
to $+\infty$	+	$+\infty$
	−	largest negative encoding
truncate	+	largest positive encoding
	−	largest negative encoding

3. *Underflow.* Underflow occurs when a number is too small for the normal range of the destination format. The concept of a *tiny* number is sometimes used with non-zero values with an exponent too small to represent in the format. In Chapter 5 we mentioned that ANSI/IEEE 754 requires that underflow conditions be handled gradually by progressively denormalizing the significand. In this sense a tiny number is a denormal. Tiny or denormal numbers can be easily identified by a zero exponent and a significand that has a zero integer bit, which can be represented explicitly or implicitly.

ANSI/IEEE 754 states that two correlated events contribute to underflow. The first one is the generation of a result that is considered to be tiny and the second one is the loss of accuracy resulting from this "tininess." The math unit response to underflow depends on whether the exception is masked or unmasked. If the exception is masked, the underflow condition is signaled only if the result is both tiny and inexact. In this case the calculated value is either a denormal or zero. If the underflow exception is not masked, it is signaled when the result is a denormal (tiny), regardless of loss of accuracy.

4. *Division by zero*. This exception results from an attempt to divide a non-zero number by zero, since the operation 0/0 generates an invalid exception. The math unit detects division by zero during the operations FDIV, FIDIV, FYL2X, and FXTRACT. The masked response is an infinity, which is signed according to the result of XORing the signs of the operands; except in FXTRACT, which returns ST = 0 and ST(1) = –INFI. If the exception is unmasked, the handler is invoked with the original operands.

Notice that the coprocessor's division by zero exception is unrelated to the divide by zero error of the central processor. A division by zero on the math unit does not generate an interrupt 0 error, as is the case in the main CPU.

5. *Denormal operand*. This exception is not specifically described in ANSI/IEEE 754 or 854. The error occurs if an operation is attempted when one or both operands are denormals. The idea of this exception is to inform of a condition that can affect the accuracy of the calculations.

If the denormal exception is masked, the 80387 and the math unit of the 486 and the Pentium automatically normalize the result. This contrasts with the behavior of the 8087 and the 80287 which generate a denormal in this case. This difference is due to changes in the final version of ANSI/IEEE 754.

6. *Precision*. This error indicates loss of numeric precision. The precision exception is named *inexact* in ANSI/IEEE 754 (see Chapter 5). The most frequent cause of a loss of precision is an internal rounding operation. Conversion operations can also generate a precision error. For instance, the decimal fraction 1/5 cannot be exactly represented in binary numbers.

7.0.3 Math Unit Status Register

The *Status register* is used to store the present state of the math unit hardware. This register can be inspected by the program through the instruction FSTSW. FSTSW stores the status word in a memory variable from where it can be accessed by code. The 80287, 80387, and the math unit of the 486 and the Pentium also allow transferring the contents of the Status register directly into the main processor's AX register. Figure 7.4, on the following page, is a bitmap of the Status register.

Bits 0 to 5 in the Status register reflect the error conditions corresponding to precision, underflow, overflow, division by zero, denormalized operand, and invalid operation. These error conditions match the exceptions conditions described previously and the response bit shown in Figure 7.3. If an exception condition has occurred, the corresponding bit in the Status register is set. Although the name of the exceptions are the same as the names of the masks, the action in the Status register is independent of the setting of the exception masks of the Control register.

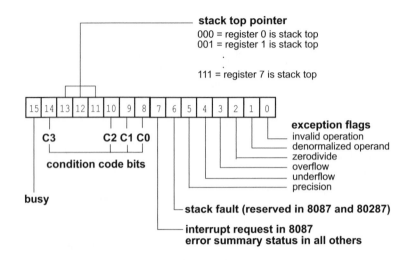

Figure 7.4 *Math Unit Status Register Bitmap*

The math unit exception bits in the Status register are said to be *sticky*. Once an error has occurred, the corresponding error flag remains set until explicitly re-set by a FCLEX (clear exceptions) instruction or until the coprocessor is re-initialized. The instructions FLDENV, FSAVE, and FRSTOR also clear the exception bits.

Bit 6 of the Status register is reserved in 8087 and 80287 systems. In the 80387 and the math unit of the 486 and the Pentium, bit 6 is designated the *stack fault* (SF) bit. In these devices the stack fault bit serves to identify invalid operations caused by a stack overflow or underflow. If SF is set, then bit 9 (labeled C1 in Figure 7.4) is set if the error was caused by overflow, and cleared if the error is due to stack underflow.

Bit 7 of the Status register is called the interrupt request bit in 8087 systems. Starting with the 80287, Intel literature refers to this bit as the *exception* or *error summary status* bit. In either case, this bit is set if any of the Status register exceptions (bits 0 to 5) are set. A software exception handler can test this bit to determine if an error took place.

Bits number 14, 10, 9, and 8 (see Figure 7.4) are called the *condition code* bits. This bit field of the Status register has been compared to the flags of the central processor. The state of the condition code bits is determined by the computational operations performed by the math unit and their setting reflects the outcome of these operations. Table 7.2 is a summary of the values assigned to the condition code bits by the various math unit instructions that operate on these bits and the interpretation of the resulting bit patterns.

The condition code bits can be tested by loading the status word into a processor register. In 80287, 80387, and the math unit of the 486 and the Pentium, this operation can be executed directly by coding

```
FSTSW    AX
```

To ensure compatibility with the 8087 it is necessary to first transfer the coprocessor's Status register to a memory variable, from which the value can be loaded into a machine register. At this point the code can branch to the corresponding routine by testing the state of the condition code bits.

A processing scheme can be devised by which the condition codes serve to index into a jump table which, in turn, directs execution to the corresponding processing branch. The jump table can be designed so that the various routines are ordered according to the absolute value of the condition codes after FXAM. Since each offset in the jump table is a doubleword (four bytes) the condition code bits must be manipulated so that they represent multiples of 4. This is accomplished by relocating bits C0 to C3 to bit positions 2, 3, 4, and 5 respectively. At this time a jump instruction can be encoded taking into account the offset of the jump table plus the value stored in the register that holds the relocated condition codes. For example, if the condition codes in the pattern 00xx xx00 are in the EBX register, and the jump table is at the code segment label named EXAM_TBL, the jump can be encoded as follows:

```
JMP     EXAM_TBL [EBX]
```

Table 7.2

Math Unit Condition Code Bits

INSTRUCTION	CONDITION CODE BITS C3	C2	C1	C0	INTERPRETATION
FXAM	0	0	0	0	Unsupported
	0	0	1	0	Unsupported
	0	1	0	0	+ Normal
	0	1	1	0	– Normal
	1	0	0	0	+ 0
	1	0	1	0	– 0
	1	1	0	0	+ Denormal
	1	1	1	0	– Denormal
	0	0	0	1	+ NaN
	0	0	1	1	– NaN
	0	1	0	1	+ Infinity
	0	1	1	1	– Infinity
	1	?	?	1	Empty
FUCOM	0	0	?	0	ST > source
FUCOMP	0	0	?	1	ST < source
FUCOMPP	1	0	?	0	ST = source
	1	1	?	1	Unordered
FCOM-FCOMP	0	0	?	0	ST > source
FCOMPP-FICOM	0	0	?	1	ST < source
FICOMP	1	0	?	0	ST = source
	1	1	?	1	ST or source undefined
FTST	0	0	?	0	ST is positive and nonzero
	0	0	?	1	ST is negative and nonzero
	1	0	?	0	ST is zero (+ or –)
	1	1	?	1	ST is not comparable

The following procedure, named NUM_AT_ST0, uses the condition codes to access a jump table as described in the preceding paragraph.

```
NUM_AT_ST0         PROC USES ebx ecx edx esi edi
; Diagnose the value at math unit stack top register
; On entry:
;               ST(0) holds number to be tested
;               EDI -> 12-byte buffer supplied by caller
; On exit:
;               Caller's buffer contains a NULL terminated
;               string defining the value at ST(0)
;               AL = error code, as follows:
;
;                      AL:    string by EDI:
;                       0.    '+ normal    ',0H
;                       1.    '- normal    ',0H
;                       2.    '+ denormal  ',0H
;                       3.    '- denormal  ',0H
;                       4.    '+ zero      ',0H
;                       5.    '- zero      ',0H
;                       6.    '+ NAN       ',0H
;                       7.    '- NAN       ',0H
;                       8.    '+ infinity  ',0H
;                       9.    '- infinity  ',0H
;                      10.    'EMPTY       ',0H
;                      11.    'unsupported',0H
;
;****************************|
;           table data       |
;****************************|
; Jump table for processing routines
;                               |--> CODE LABELS
;                               |
;                               |--------------|; CC bits = value
EXAM_TBL           DD          UNSUPPORTED     ; 0 0 0 0 = 0
                   DD          POS_NAN         ; 0 0 0 1 = 1
                   DD          UNSUPPORTED     ; 0 0 1 0 = 2
                   DD          NEG_NAN         ; 0 0 1 1 = 3
                   DD          POS_NORMAL      ; 0 1 0 0 = 4
                   DD          POS_INFINITY    ; 0 1 0 1 = 5
                   DD          NEG_NORMAL      ; 0 1 1 0 = 6
                   DD          NEG_INFINITY    ; 0 1 1 1 = 7
                   DD          POS_ZERO        ; 1 0 0 0 = 8
                   DD          EMPTY           ; 1 0 0 1 = 9
                   DD          NEG_ZERO        ; 1 0 1 0 = 10
                   DD          EMPTY           ; 1 0 1 1 = 11
                   DD          POS_DENORMAL    ; 1 1 0 0 = 12
                   DD          EMPTY           ; 1 1 0 1 = 13
                   DD          NEG_DENORMAL    ; 1 1 1 0 = 14
                   DD          EMPTY           ; 1 1 1 1 = 15
;
; Error message table. Each entry is 12 bytes
; Strings terminate in NULL byte
ERR_MESS_TBL       DB          '+ normal    ',0H
                   DB          '- normal    ',0H
                   DB          '+ denormal  ',0H
                   DB          '- denormal  ',0H
                   DB          '+ zero      ',0H
                   DB          '- zero      ',0H
```

```
                    DB          '+ NAN        ',0H
                    DB          '- NAN        ',0H
                    DB          '+ infinity ',0H
                    DB          '- infinity ',0H
                    DB          'EMPTY        ',0H
                    DB          'unsupported',0H

; Storage for Status word
STATUS_WORD     DW          ?
;
;**************************|
;     get condition codes  |
;**************************|
; Get condition codes for stack top register and store for later
; reference
        FXAM                            ; Examine condition codes
        FWAIT                           ; For 8087 compatibility
        FSTSW   STATUS_WORD             ; Store status word in memory
        FWAIT
;**************************|
;   store condition codes  |
;         in BX            |
;**************************|
;   AH bits: 7  6  5  4  3  2  1  0
;                 3           2  1  0   condition code bits
;                 |_____|__|__|_____ Condition code settings:
        MOV     BX,STATUS_WORD          ; Status word to BX
; Clear all non-condition code bits
; Note: shifts are coded with the operand in CL so that the code
;       will be compatible with 8086/8088//80286 systems
        AND     BH,01000111B            ; Mask to clear bits
        SHR     BX,6                    ; C0-C2 now 000x xx00 in BL
        SAL     BH,5                    ; C3 now 00x0 0000 in BH
        OR      BL,BH                   ; C0-C3 now 00xx xx00 in BL
        XOR     BH,BH                   ; Clear high byte
;**************************|
;   jump based on CC bits  |
;**************************|
        AND     EBX,0FFFFH              ; Clear garbage bits
        JMP     EXAM_TBL [EBX]          ; Jump using CC bits
;********************************|
;    destination labels for jumps    |
;********************************|
; Each jump label sets the corresponding error code in AX and
; exits to the epilogue routine
UNSUPPORTED:
        MOV     AX,11
        JMP     EXIT_RTN
POS_NAN:
        MOV     AX,6
        JMP     EXIT_RTN
NEG_NAN:
        MOV     AX,7
        JMP     EXIT_RTN
POS_NORMAL:
        MOV     AX,0
        JMP     EXIT_RTN
POS_INFINITY:
        MOV     AX,8
        JMP     EXIT_RTN
```

```
NEG_NORMAL:
        MOV     AX,1
        JMP     EXIT_RTN
NEG_INFINITY:
        MOV     AX,9
        JMP     EXIT_RTN
POS_ZERO:
        MOV     AX,4
        JMP     EXIT_RTN
EMPTY:
        MOV     AX,10
        JMP     EXIT_RTN
NEG_ZERO:
        MOV     AX,5
        JMP     EXIT_RTN
POS_DENORMAL:
        MOV     AX,2
        JMP     EXIT_RTN
NEG_DENORMAL:
        MOV     AX,3
;************************|
;   move message to caller's |
;       buffer and exit      |
;************************|
; At this point AX holds the error code number
; The value in AX is multiplied by 12 to obtain the offset
; of the corresponding error message
; Message is then moved to the caller's buffer by EDI
EXIT_RTN:
        PUSH    AX                  ; Save error code
        LEA     ESI,ERR_MESS_TBL    ; Offset of table
        MOV     CL,12               ; Length of each message
        MUL     CL                  ; AX = offset of correct message
        ADD     SI,AX               ; Add to table offset
; At this point:
;       ESI -> 12 byte number type message
;       EDI -> caller's buffer with 12 bytes minimum space
        MOV     ECX,12              ; Counter for 12 bytes
TRANSFER_12:
        MOV     AL,[ESI]            ; Get message character
        MOV     [EDI],AL            ; Place in caller's buffer
        INC     ESI                 ; Bump buffer pointers
        INC     EDI
        LOOP    TRANSFER_12
        POP     AX
        CLD
        RET
NUM_AT_ST0      ENDP
```

SOFTWARE ON-LINE

The procedure NUM_AT_ST0 is used by the FpuOutput() function developed in Chapter 6. The NUM_AT_ST0 procedure is found in the Un32_4 module of the MATH32 library, in the book's on-line software.

The condition code bits can be transferred to the processor's Flags register directly. A conditional branch instruction can then be executed according to the 80x86 flags. Transferring the status bits (in AX) to the Flags register is preformed with the SAHF instruction. This works because all four condition code bits are located in the high-order byte (AH). Figure 7.5 shows the migration of the condition code bits from AX to the 80x86 Flags register.

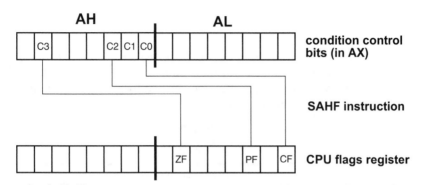

Figure 7.5 *Condition Code Bits Migration During SAHF*

When using this technique the actual branching possibilities are limited by the conditional jump opcodes available in the CPU. For example, since there is no conditional jump instruction that takes place if both the zero and the carry flag are set, there is no way of coding a conditional jump based on the state of bits C0 and C3. The following code fragment shows the processing necessary for comparing two numbers stored in memory and directing execution to one of four routines according to the result of the compare operation.

```
; Code assumes that two floating-point numbers are stored at
; the variables labeled FP_NUM1 and FP_NUM2
        FINIT                   ; Initialize FPU
        FLD     FP_NUM1         ; Number 1 to ST(0)
        FCOMP   FP_NUM2         ; Compare to number 2 and pop
        FSTSW   STATUS_WORD     ; Status word to memory
        FWAIT                   ; Wait for store to complete
        MOV     AX,STATUS_WORD  ; Load in AX
        SAHF                    ; AH to Flags register
; At this point:
;           carry flag = condition code C0
;          parity flag = condition code C2
;            zero flag = condition code C3
        JP      INVALID         ; C2 is set if compare invalid
        JC      ONE_IS_SMALLER  ; C0 set if ST(0) <- source
        JZ      BOTH_EQUAL      ; C3 set if ST(0) <- source
; If execution drops number 1 is bigger than number 2
        .
        .
        .
BOTH_EQUAL:
        .
        .
        .
```

```
ONE_IS_SMALLER:

        .

        .

        .

INVALID:

        .

        .

        .
```

Table 7.3 shows the conditional jump instructions and their relation to the CPU flags. These flags can be loaded from the condition codes by the SAHF instruction, therefore the transfer condition is determined by the relative magnitudes of the ST register and the source operand.

Table 7.3

80x86 Conditional Jumps

OPCODE	FLAGS	TRANSFER CONDITION
JB	CF = 1	Taken if ST < source or not comparable
JBE	(CF or ZF) = 1	Taken if ST ≤ source or not comparable
JA	(CF or ZF) = 0	Taken if ST > source
JAE	CF = 0	Taken if ST ≥ source
JE	ZF = 1	Taken if ST = source or not comparable
JNE	ZF = 0	Taken if ST ≠ source

The field formed by the Status register bits 11, 12, and 13, shown in Figure 7.4, holds the value of the physical stack register that is presently at the stack top. This field corresponds to the center box in Figure 7.2. In the 8087 and 80287, bit 15 of the Status register indicates if the coprocessor is idle or if it is executing an instruction or signaling an exception. In the 80387 and the math unit of the 486 and the Pentium, this bit is a copy of bit 7 (error summary bit) of the Status register and is provided for compatibility with previous versions of the coprocessor.

7.0.4 The Environment Area

The math unit *environment* is a 14-byte or a 28-byte data area according to the memory model used by the program. Programs that use a 16-bit address space store the math unit environment in a 14-byte memory area, while programs that use a 32-bit flat address space require a 28-byte area. The environment area holds fundamental information regarding the present state of the math unit. To inspect the environment area a program must first store its image in memory. The math unit instruction FSTENV (store environment) can be used for this purpose. Figure 7.6 shows the registers and data areas included in the environment area in the 16- and 32-bits memory models.

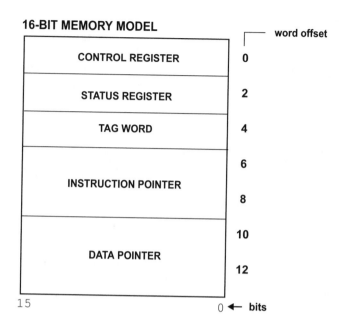

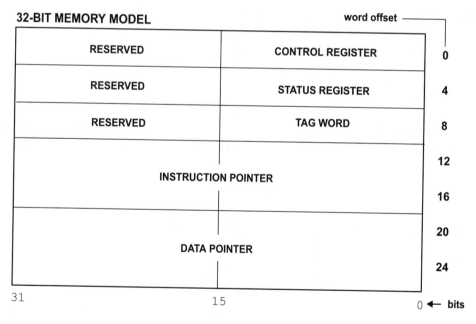

Figure 7.6 *Math Unit Environment Area in 16- and 32-bit Memory Models*

In addition to the math unit control and Status registers, discussed in preceding sections, the environment contains two other elements of interest to the programmer: the *tag word* and the *instruction and data pointers*. These last two elements are sometimes called the *exception pointers*.

Tag Word Register

The Tag register, or Tag Word register, is located in the math unit environment area (see Figure 7.6). The tag word contains a two-bit field for each one of the eight Numeric Data registers in the register stack. The two-bit codes offer a summary of the contents of each stack register. The Tag Word register is shown in Figure 7.7.

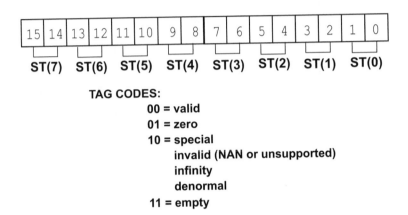

TAG CODES:
00 = valid
01 = zero
10 = special
     invalid (NAN or unsupported)
     infinity
     denormal
11 = empty

Figure 7.7 *Math Unit Tag Word Register Structure*

The tag codes are used internally by the processor in optimizing performance, but they are also accessible to software. Programs can read the tag word to obtain information regarding the contents of the stack registers. In reading the tag code it is usually necessary to inspect the stack top pointer field of the Status register in order to determine which tag corresponds to the desired stack register. The following procedure named GET_TAG read and returns the tag code for any desired math unit register, as well as a text string identifying the register as valid (tag 00), zero (tag 01), special (tag 10), or empty (tag 11).

```
            .486
            .MODEL flat
            .DATA
; Storage for environment variables in 32-bit format
ENVIRO_FPU      DD      0       ; FPU control word
STATUS_FPU      DD      0       ; FPU status word
TAG_WORD        DD      0       ; FPU tag word
; Storage for instruction and data pointers
                DD      0       ; Instruction pointer
                DD      0
                DD      0       ; Data pointer
                DD      0
; Temporary storage for requested register number
STACK_REG       DB      0
                DW      0

;
; Error message table. Each entry is 8 bytes
TAG_MESS_TBL    DB      'Valid  ',0H
                DB      'Zero   ',0H
                DB      'Special',0H
```

```
                    DB        'EMPTY ',0H

;*********************************************************************
;                              CODE
;*********************************************************************
        .CODE

_GET_TAG        PROC USES esi edi ebx ebp
; Procedure to return tag code corresponding to the requested
; math unit register
; An 8-character string with the tag text is passed to the caller
; On entry:
;          EDI -> caller's text buffer with a minimum of 8
;                 free bytes for message storage
;          AL = register stack desired (modulo 8)
;
; On exit:
;          caller's buffer (by EDI) contains tag string
;          BX = tag code for requested register

; Save caller's requested stack register number
        MOV       STACK_REG,AL          ; For later use
;
;*********************|
;   store environment |
;*********************|
        FSTENV    ENVIRO_FPU     ; Environment to memory
        FWAIT                    ; Wait for end of memory access
        MOV       EBX,TAG_WORD   ; Tag word into BX
        MOV       EAX,STATUS_FPU ; Status word to AX
;
;*********************|
;   isolate TOS field |
;*********************|
        AND       AH,00111000B   ; Mask off all other bits
        MOV       CL,3           ; Count of 3 bits to shift
        SHR       AH,CL          ; Shift TOS bits right 3 times
; AH now holds the number of the physical register presently
; mapped as the stack top (ST(0))
; The tag word is mapped as follows:
;   xx xx xx xx | xx xx xx xx
;   -- -- -- --   -- -- -- --
;    |  |  |  |    |  |  |  |
;    7  6  5  4    3  2  1  0  <- physical register
;
; To find the physical register that corresponds with the FPU
; stack register requested by the caller it is necessary to
; add the caller's register number to the current stack top
; and subtract 8 is the sum exceeds register 7. This calculates
; the modulus 8 address of the requested stack register
        MOV       AL,STACK_REG   ; Caller's requested stack
                                 ; register number
        ADD       AL,AH          ; AL holds sum
        CMP       AL,7           ; Test for limit
        JNA       SUM_IN_MOD_8   ; Go if less
        SUB       AL,8           ; Convert to modulus 8
SUM_IN_MOD_8:
; Unit mask is 00000000 00000011B (0003H)
        MOV       DX,0003H       ; Mask to DX
        MOV       CL,AL          ; Register number to CL
```

```
        ADD     CL,CL           ; Double number to get shift count
        SHL     DX,CL           ; Shift mask bits left
        AND     BX,DX           ; Mask off all other tag bits
        SHR     BX,CL           ; Shift unmasked tag bits right
;
;***************************|
;   move message to caller's |
;       buffer and exit      |
;***************************|
; At this point BX holds the tag code
        MOV     AX,BX           ; Tag code to AX
; The value in AX is multiplied by 8 to obtain the offset
; of the corresponding tag code text message
; Message is then moved to the caller's buffer by DS:DI
        LEA     ESI,TAG_MESS_TBL    ; Offset of table
        MOV     CL,8            ; Length of each message
        MUL     CL              ; AX -> offset of correct message
        ADD     ESI,EAX         ; Add to table offset
; At this point:
;       ESI -> 8-byte number type message
;       EDI -> caller's buffer with 8 bytes minimum space
        MOV     ECX,8           ; Counter for 8 bytes
TRANSFER_8:
        MOV     AL,[ESI]        ; Get message character
        MOV     [EDI],AL        ; Place in caller's buffer
        INC     ESI             ; Bump buffer pointers
        INC     EDI
        LOOP    TRANSFER_8
; End of processing
        CLD
        RET
_GET_TAG        ENDP
```

SOFTWARE ON-LINE

The GET_TAG procedure is found in the Un32_4 module of the MATH32 library, in the book's on-line software.

The contents of the Stack Top register can be determined more precisely using the FXAM or FTST instructions and interpreting the resulting condition code bits, as described in Section 7.0.3.

Instruction and Data Pointers

The *Instruction and Data Pointer* registers are part of the math unit environment (see Figure 7.6). These two registers are jointly called the *exception pointers*. After each floating-point instruction is executed, the math unit automatically saves its operation code and address, as well as the operand's address if one was contained in the instruction. This data, which is saved internally in the math unit, can be examined by storing the environment in memory. The operation of saving and inspecting the environment is shown in the GET_TAG procedure listed previously.

The information provided by the instruction and the data pointers is often used by exception handler routines to identify the instruction that generated an error.

In the 80287, 80387, and the math unit of the 486 and the Pentium, the storage formats for the instruction and data pointers depend on the operating mode as well as the memory model. In the real mode the value stored is in the form of a 20-bit physical address and an 11-bit math unit opcode. In protected mode the value stored is the 32-bit virtual address of the last coprocessor instruction. The 8087 stores this data as in the real mode mentioned above. Figure 7.8 is a map of the data stored in the exception pointers while the processor is operating in 16-bit real mode.

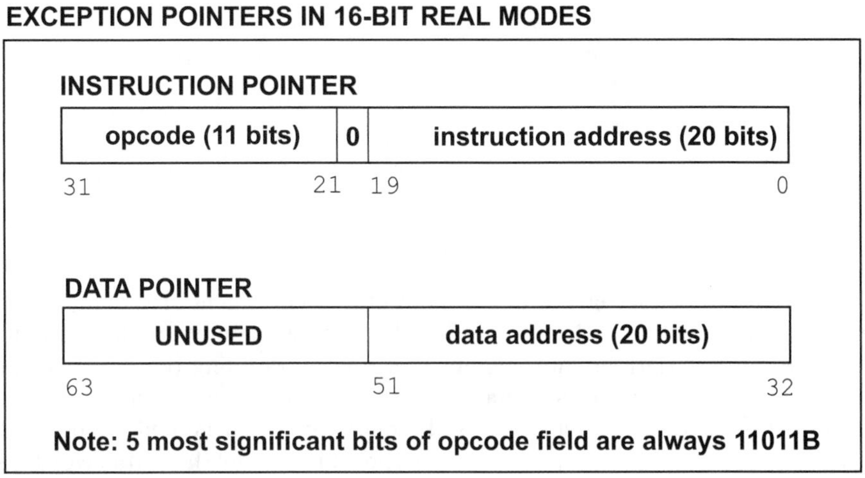

EXCEPTION POINTERS IN 16-BIT REAL MODES

INSTRUCTION POINTER

opcode (11 bits)	0	instruction address (20 bits)

31   21   19   0

DATA POINTER

UNUSED	data address (20 bits)

63   51   32

Note: 5 most significant bits of opcode field are always 11011B

Figure 7.8 *Exception Pointers Memory Layout*

Notice that on the 8087 the instruction address saved in the environment area does not include a possible segment override prefix. This was changed in the 80287 so that the address pointer includes a possible segment override. A portable error handler routine would have to take this difference into account.

As shown in Figure 7.6, the location of the exception pointers within the environment area changes according to the memory model. In the 16-bit model the instruction pointer is at word offset 6 from the start of the environment area and the data pointer at word offset 10. In the flat 32-bit memory model the instruction pointer is at word offset 12 and the data pointer at word offset 20. The following code fragment shows how the various data elements of the math unit environment area can be defined in the 32-bit memory model.

```
        .486
        .MODEL flat
        .DATA
;
; Storage for environment variables in 32-bit memory model
ENVIRO_FPU      DD      0       ; FPU control word - 4 bytes
STATUS_FPU      DD      0       ; FPU status word  - 4 bytes
TAG_WORD        DD      0       ; FPU tag word     - 4 bytes
INST_POINTER    DD      0       ; Instruction ptr  - 8 bytes
                DD      0
DATA_POINTER    DD      0       ; Data pointer     - 8 bytes
```

```
                DD      0
;                                                   =========
;                                  total ..........  28 bytes
```

In the 16-bit memory model the various areas can be defined as follows:

```
        .486
        .MODEL medium
        .DATA
;
; Storage for environment variables in 32-bit memory model
ENVIRO_FPU      DW      0       ; FPU control word - 2 bytes
STATUS_FPU      DW      0       ; FPU status word  - 2 bytes
TAG_WORD        DW      0       ; FPU tag word     - 2 bytes
INST_POINTER    DD      0       ; Instruction ptr  - 4 bytes
DATA_POINTER    DD      0       ; Data pointer     - 4 bytes
;                                                   =========
;                                  total ..........  14 bytes
```

The different memory layout of the math unit environment area compromises the portability of applications that execute in the various memory models. Applications must take these variations into account not only in defining the memory map, but also in coding CPU instructions that access the stored data. In the preceding code fragments the various data elements of the math unit environment are defined using variables of different sizes. For example, in the 16-bit model the status word is stored in a word variable, while in the 16-bit model it is stored in a doubleword variable. The coding for retrieving the status word into a 16-bit register could be as follows:

```
        MOV     AX,STATUS_FPU
```

while in a 32-bit model program the code would have to be changed to:

```
        MOV     EAX,STATUS_FPU
        AND     EAX,0FFFFH      ; Clear un-used bits
```

7.0.5 Math Unit State Area

The coprocessor *state area* is a data area that holds the environment area plus the eight registers in the math unit stack. Since the state area includes the environment, its size changes according to the memory model. In the 16-bit model the state area consists of 94 bytes, while in the 32-bit flat model it requires 108 bytes. The difference of 14 bytes is the difference in size of the environment area in the two models, as discussed in the previous section.

The math unit instruction set contains the FSAVE instruction that stores the state area in memory. The FRSTOR instruction serves to reload a saved state into the math unit. Figure 7.9 is a map of the data stored in the state area.

System and application software usually save the coprocessor state whenever they wish to clean up the math unit for a new task. In a multitasking environment this can occur at every context or task switch. In addition, an interrupt service routine or an exception handler saves the math unit state in order to use the coprocessor for its own calculations; later the math unit is restored to its original contents.

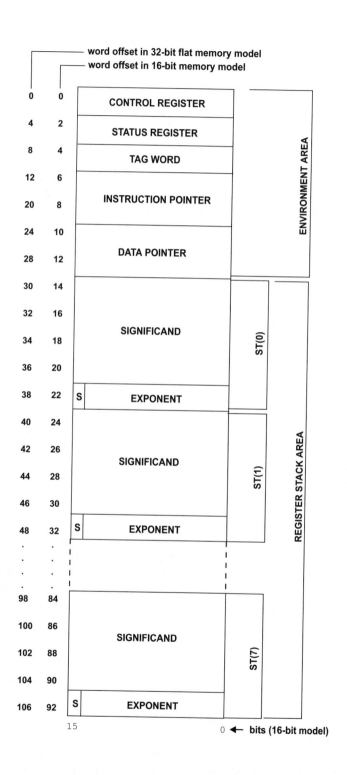

Figure 7.9 *Memory Map of Math Unit State Area*

7.1 Math Unit Instruction Patterns

You have seen that the math unit Data registers seem to share the characteristics of explicit storage units and that of a stack structure. Another feature of the math unit is that its instruction set can access memory operands using all the memory addressing modes of the central processor. This is due to the fact that the CPU performs all address calculations on behalf of the math unit. The result is an abundance of math unit operand patterns that are suitable for most programming situations.

A useful coding style is to use the comment area to keep track of the state of the math unit register stack. In this book we often use this notation style, although text space limitations often force the use of abbreviations that may be somewhat cryptic. In the code fragments listed in the following section we labeled three columns with the designations of the first three stack registers: ST, ST(1), and ST(2). Thus, the comment field is a snapshot of a portion of the math unit stack after the instruction executes. Examples of this coding style are found in the following sections.

7.1.1 Register Operands

Some math unit instructions can be coded using explicit Numeric Data register operands, for example:

```
                        ; |    ST     |    ST(1)    |    ST(2)    |
; Initialize processor
        FINIT           ; |   EMPTY   |    EMPTY    |    EMPTY    |
; Perform operations
        FLD1            ; |    1.0    |    EMPTY    |    EMPTY    |
        FLDZ            ; |    0.0    |     1.0     |    EMPTY    |
        FLDPI           ; |  3.1415.. |     0.0     |     1.0     |
        FADD    ST,ST(2);|  4.1415.. |     0.0     |     1.0     |
        FADD    ST(1),ST;|  4.1415.. |   4.1415..  |     1.0     |
```

In this listing the FADD instructions specifically designate which stack registers must be added, and which register holds the sum. Another type of FPU opcodes automatically pop the stack after each instruction executes. The mnemonic for these instructions end with the letter "P" (pop), for example, FADDP. The following code fragment illustrates the action of FADDP instruction.

```
                        ; |    ST     |    ST(1)    |    ST(2)    |
; Initialize processor
        FINIT           ; |   EMPTY   |    EMPTY    |    EMPTY    |
; Perform operations
        FLD1            ; |    1.0    |    EMPTY    |    EMPTY    |
        FLDPI           ; |  3.1415.. |     1.0     |    EMPTY    |
        FADDP   ST(1),ST;|  4.1415.. |    EMPTY    |    EMPTY    |
```

In the preceding fragment notice that the stack is popped after the instruction executes. Therefore, the destination operand cannot be the Stack Top register; if this were the case, the sum would be destroyed. Consequently it is illegal to code

```
        FADDP   ST,ST(i)
```

The math unit instruction set makes it possible to not designate registers explicitly, but the implicit encoding can sometimes produce unexpected results. For example

```
                          ; |    ST    |   ST(1)   |   ST(2)   |
; Initialize processor
        FINIT             ; |  EMPTY   |   EMPTY   |   EMPTY   |
; Perform operations
        FLD1              ; |   1.0    |   EMPTY   |   EMPTY   |
        FLDPI             ; | 3.1415.. |    1.0    |   EMPTY   |
        FADD              ; | 4.1415.. |   EMPTY   |   EMPTY   |
```

Notice that the FADD instruction is used with implicit operands. In this case the stack registers ST(1) and ST are added and the stack is then popped. The action of coding FADD with no operands is the same as coding FADDP ST(1),ST. However, it may seem reasonable that the implicit opcode mode would resemble the form FADD ST,ST(1), rather than its actual action.

7.1.2 Memory Operands

The math unit can access numeric data stored in memory using any of the five CPU addressing modes: direct, register indirect, base, indexed, and based indexed addressing. A difference between processor and coprocessor memory addressing is that math unit opcodes that reference memory have a single operand. For instance, it is possible to load a memory variable into any of the processor's general purpose registers

```
MOV     AX,MEM_VALUE_1    ; First variable to AX
MOV     BX,MEM_VALUE_2    ; Second variable to BX
MOV     DX,MEM_VALUE_1    ; First variable to DX
```

However, the two-operand format is not valid in the math unit instruction set. This is due to the fact that, if the instruction is a load (FLD, FILD, or FBLD) the destination is always the Stack Top register (ST), while if the operation is a store, the source is assumed to be in the Stack Top register. In instructions that perform calculations, a memory operand is always a source. For example

```
FLD     SINGLE_PREC       ; Memory variable to ST
FST     DOUBLE_PREC       ; ST stored in memory variable
FADD    LONG_INT          ; ST = ST + memory variable
  .
  .
  .
LEA     BX,DOUBLE_PREC    ; Set pointer to memory variable
FADD    QWORD PTR [BX]    ; ST = ST + variable -> [EBX]
```

7.2 Math Unit Instruction Set

The math unit instruction set is classified into six groups according to their operation. The groups of instructions are named data transfer, arithmetic, comparison, transcendental, constant, and processor control. In the following sections we present a brief description of the instructions in each of these groups.

7.2.1 Data Transfer Instructions

The *data transfer instructions* are used to move numeric data between stack registers, and between registers and memory. Any of the seven math unit data types can be read from a memory storage into the Stack Top register. The math unit automatically converts the numeric data into the extended precision format as it is loaded into the register stack. The data transfer instructions automatically update the Tag register. Separate instructions are provided for loading and storing real, integer, and packed binary coded decimal numbers. The FI prefix identifies the integer load and store instructions and the FB prefix the packed BCD transfers.

The FST (store real) instruction transfers the stack top to the destination operand, which can be a memory variable or another stack register. However, FST can only be used to store the stack top into a single or double precision real variable. FSTP (store real and pop) must be used to store into a memory destination in extended precision real format. Constants, special encodings, temporary results, and other operational data that could affect the precision of the final result should always be stored in extended precision format. On the other hand, final results should not be represented in the extended format since this defeats its purpose, which is absorbing rounding and computational errors.

The store opcodes that end in the letter "P" pop the stack after the data transfer is executed. The encoding FSTP ST(0) pops the stack without a data transfer, effectively discarding the contents of ST(0). Table 7.4 describes the nine opcodes related to math unit data transfer instructions.

Table 7.4

Math Unit Data Transfer Instructions

MNEMONICS	OPERATION	EXAMPLES
	TRANSFER OF REAL NUMBERS	
FLD	Load real memory variable or stack register onto stack top. Value is converted to extended real format	FLD SINGLE_REAL FLD DOUBLE_REAL FLD EXENDED_REAL FLD ST(2)
FST	Store stack top in another stack register or in a real memory variable. Rounding is according to RC field of control word. Coding FLD ST(0) duplicates the stack top	FST ST(3) FST SINGLE_REAL FST DOUBLE_REAL
FSTP	Store stack top in another stack register or in a real memory variable and pop stack. Rounding is according to RC field in control word.	FSTP ST(2) FSTP SINGLE_REAL FSTP DOUBLE_REAL FSTP EXTENDED_REAL

(continues)

Table 7.4

Math Unit Data Transfer Instructions (continued)

MNEMONICS	OPERATION	EXAMPLES
	TRANSFER OF REAL NUMBERS	
FXCH	Swap contents of stack top and another stack register. If no explicit register, ST(1) is used	FXCH ST(2) FXCH
	INTEGER TRANSFERS	
FILD	Load word, short or long integer to stack top. Loaded number is converted to extended real	FILD WORD_INTEGER FILD SHORT_INTEGER FILD LONG_INTEGER
FIST	Round stack top to integer. Rounding is according to the RC field in the control word. FIST stores in integer memory variable. FISTP (see below) must be used to store a long integer	FIST WORD_INTEGER FIST SHORT_INTEGER
FISTP	Round stack top to integer, per RC field in the status word, store in variable and pop stack	FISTP WORD_INTEGER FISTP SHORT_INTEGER FISTP LONG_INTEGER
	TRANSFER OF PACKED BCD	
FBLD	Load packed BCD to stack top	FBLD PACKED_BCD
FBSTP	Store stack top as a packed BCD integer and pop stack. Non-integers are rounded before storing	FBSTP PACKED_BCD

7.2.2 Nontranscendental Instructions

The math unit *nontranscendental instructions* provide the basic arithmetic operations required by ANSI/IEEE 754. These are: addition, subtraction, multiplication, division, and remainder. In addition, the math unit instruction set includes several other operations not required by the standard, such as the calculation of square roots, rounding, scaling, partial remainder, change of sign, and the extraction of exponent and significand. In the original Intel literature the nontranscendental instructions were called the arithmetic instructions.

Basic Arithmetic

The fundamental arithmetic instructions that perform addition, subtraction, multiplication, and division are straightforward and uncomplicated. Addition and multiplication are commutative, that is, the result is independent of the order of the operands. In order to extend this symmetry to all fundamental arithmetic operations, the math unit provides opcodes for reversing the operands of subtraction and division. Furthermore, there are separate operand modes for performing integer and real arithmetic. Table 7.5 lists the operand options for the math unit nontranscendental instructions that perform basic arithmetic.

In Table 7.5 notice that if no explicit operand is present in the mnemonic, the math unit operates as a pure stack machine. In this case the source operand is assumed to be in ST and the destination in ST(1). After performing the calculation the result is stored in ST(1) and the stack is popped, effectively replacing both operands with the result. Perhaps a more reasonable way of implementing a classical stack operation is to use an operand in the form ST(1),ST and the pop mnemonic form of the opcode (see Table 7.5). For example, in the instruction

```
FADDP     ST(1),ST
```

the sum of ST and ST(1) is placed in ST(1) and the stack is popped. The result is the same as coding FADD with no operand but the action of the instruction is more clearly expressed by the explicit encoding.

Table 7.5

Operand Modes for Arithmetic Instructions

INSTRUCTION TYPE	MNEMONIC FORMAT	OPERAND DESTINATION,SOURCE	SAMPLE CODING
implicit (pop stack)	F*opcode*	{ST(1),ST}	FADD
registers (explicit)	F*opcode*	ST(i),ST or ST,ST(i)	FADD ST,ST(1)
register (explicit and pop)	F*opcode*P	ST(i),ST	FADDP ST(2),ST
memory (real number)	F*opcode*	{ST},MEM_VAR	FADD MEM_VAR
memory (integer number)	FI*opcode*	{ST},MEM_INT	FIADD MEM_INT

F*opcode*:	ACTION:
ADD	destination <= destination + source
SUB	destination <= destination − source
SUBR	destination <= source - destination
MUL	destination <= destination · source
DIV	destination <= destination / source
DIVR	destination <= source / destination

Legend: Braces { } indicate implicit operands

Scaling and Square Root

The FSQRT instruction calculates the square root of the number in ST(0). Intel documentation states that the algorithm used in the calculation of the square root insures that the FSQRT instruction executes faster than ordinary division. At the time of the introduction of the 8087 this level of square root calculation performance had no precedent in commercial floating-point hardware. The result of the square root is accurate to within one-half of the last significand digit, which is the same precision obtained by the add, subtract, multiply, and divide operations.

The FSCALE (scale) opcode is designed to provide a fast multiplication and division by integral powers of 2. The operation interprets the value in ST(1) as an exponent and adds its value to the exponent field of the number in ST. This action can be expressed as

$$ST <= ST \cdot 2^{ST(1)}$$

For example, if the value in ST(1) is the integer 3, then the FSCALE instruction performs

$$ST <= ST \cdot 2^3$$
$$ST <= ST \cdot 8$$

If ST = 1 then FSCALE calculates a power of 2. Negative powers of the value in ST(1) indicates a subtraction of the exponent, which results in effectively dividing the operand in ST by the power of 2 in ST(1). The following fragment shows the processing for quickly and accurately obtaining p/4, a constant sometimes used in argument reduction prior to the calculation of trigonometric functions.

```
        .DATA
;
NEG_TWO   DW    -2   ; Storing of constant -2

        .CODE
    .
    .
    .
                        ; |    ST     |   ST(1)   |    ST(2)   |
                        ; |  EMPTY    |   EMPTY   |   EMPTY    |
        FILD    NEG_TWO ; |   -2      |   EMPTY   |   EMPTY    |
        FLDPI           ; |   PI      |    -2     |   EMPTY    |
        FSCALE          ; |  PI/4     |    -2     |   EMPTY    |
        FSTP    ST(1)   ; |  PI/4     |   EMPTY   |   EMPTY    |
; At this point ST(0) holds PI/4
```

In the 8087 and 80287 the scaling factor, in ST(1), must be an integer in the range ±32767. However, there is no limit to the scaling factor in the 80387 and the math unit of the 486 and the Pentium. In the newer machines, if the value in ST(1) is not an integer, it is chopped to the nearest integer before it is added to the exponent of ST. In order to ensure that the scaling factor is an integer, it is a good programming practice to define it in an integer variable and load it into the math unit by means of the FILD instruction, as in the preceding fragment.

Partial Remainder

The FPREM (partial remainder) instruction performs modulo division of ST by ST(1). In this case the modulus is assumed to be in ST(1). Like FSCALE, the FPREM instruction allows no explicit operands. FPREM produces an exact result, therefore the precision exception does not occur and the rounding field of the control word has no effect.

FPREM allows implementing operations of *finite algebra* and *modular arithmetic* on the math unit. These operations, sometimes referred to as *clock arithmetic*, are based on closed number systems which wrap around to the first number in the set. For example, consider a 12-hour clock showing the present time as 2 o'clock. The clock time 54 hours later is calculated as follows:

$$54 / 12 = 4 \text{ (remainder 6)}$$

$$2 + 6 = 8 \text{ o'clock}$$

In clock arithmetic the new time is obtained by adding, to the present time, the remainder of dividing the operand (54) by the clock modulus (12). In this case we can say that we have performed modulo 12 division of 54, which is 6.

Notice that if you use conventional division to calculate the remainder, the rounding of the operands could compromise the precision of the result. For example, the trigonometric functions (sine, cosine, tangent, etc.) are known to be periodic over the range 2p radian. Therefore,

$$\sin(x+2np) = \sin(x)$$

by the same token

$$\sin(x-2np) = \sin(x)$$

where n is an integer and x is the angle in radians. For this reason, any value of x can be reduced to the unit circle by calculating

$$y = x - (\text{remainder } (x / 2p))$$

Since $0 \le y \le 2p$ then we can also state that $\sin(x) = \sin(y)$. However, if this remainder is calculated using conventional division, as in the formula

$$r = \frac{x}{2\pi}$$

$$y = r - (\text{integer part of } r)$$

then we can see that the round off error makes r approximate an integer and y approximate 0 as x becomes very large. Therefore the trigonometric identities

$$\sin^2 x + \cos^2 x = 1$$

$$2 \sin x \cos x = \sin^2 x$$

will not hold for all arguments. For this reason the ANSI/IEEE 754 Standard requires that all implementations include an *exact remainder* operation that can be used, among other operations, in the calculation of accurate argument reductions.

The exact remainder can also be calculated by performing successive subtractions of the modulus until the difference is smaller than the modulus. The difficulty with this method is that with large operands and small moduli the calculation could require a large number of subtractions, tying-up the math unit for a long time. Since interrupts can take place only after an instruction has concluded, the long latency of a single-step remainder calculation could compromise system integrity. For this reason, the designers of the original 8087 provided this function in the form that they called a *partial remainder*. At the most 64 subtractions are performed in each execution of the instruction. Notice that the limit of 64 subtractions was chosen so that the FPREM instruction would never be slower than the FDIV instruction. If after 64 subtractions of the modulus a true remainder has not been obtained, its present value (partial remainder) is stored at ST(0) and execution concludes with condition code bit C2 set. On the other hand, if a true remainder is obtained (one that is smaller than the modulus) the instruction concludes with condition code bit C2 cleared. The operation of the FPREM instruction is shown in the following pseudo-code:

```
REPEAT:
    FPREM
    if bit C2 = 1 go to REPEAT
```

Software can detect the result of FPREM by storing the machine status word in a memory variable, inspecting the C2 bit, and re-executing the instruction until bit C2 is cleared. Since the partial remainder is left in ST(0) and the modulus in ST(1) no stack manipulation is required inside the loop. Alternatively, the code can compare the values in ST and ST(1). If ST > ST(1) then the FPREM instruction must be repeated.

In the calculation of the remainder the quotient keeps track of the number of subtractions of the modulus. For example

$$54 / 12 = 4 \text{ (remainder 6)}$$

In terms of clock arithmetic, the quotient (4 in this case) expresses the number of full circles completed by the hour hand. Trigonometric functions have a periodic interval of p/4 radians, which is one eighth of the unit circle. This value can be used as a modulus for argument reduction of angles that exceed p/4 radian. This relationship is shown in Figure 7.10.

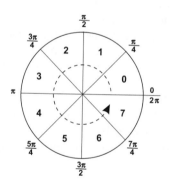

Figure 7.10 *Octants in the Unit Circle*

If argument reduction to the first octant (octant 0 in Figure 7.10) were performed by conventional division, we could examine the integer portion of the quotient, modulo 8, to determine the octant in which the original angle was located. The FPREM instruction does not report the complete value of the quotient obtained in the modular division operation. However, it does report the three low-order bits of the integer quotient when the execution has produced a true remainder. These three bits are located in the condition codes C1 (bit 0), C3 (bit 1), and C0 (bit 2). Condition code bit C2 is not used for this, since it is cleared if the reduction is complete and set otherwise. The interpretation of the condition code bits after FPREM can be seen in Table 7.6.

Table 7.6

Interpretation of Condition Codes Bits after FPREM

| CONDITION CODES | | | | INTERPRETATION |
C2	C0	C3	C1	
1	?	?	?	Incomplete reduction. More FPREM iteration are required. ST(0) holds partial remainder
0	?	?	?	Complete reduction. ST(0) holds true remainder
				Interpretation of C0, C3, and C1:
0	0	0	0	Angle in octant 0
0	0	0	1	Angle in octant 1
0	0	1	0	Angle in octant 2
0	0	1	1	Angle in octant 3
0	1	0	0	Angle in octant 4
0	1	0	1	Angle in octant 5
0	1	1	0	Angle in octant 6
0	1	1	1	Angle in octant 7

Programmers working with the original 8087 discovered that the condition code bits were not always reported correctly after FPREM. Therefore the evaluation of these bits to determine the octant of the original angle was not reliable. For this reason the argument reduction routines written for the 8087 and 80287 had to work around this bug by not using the condition code bits in determining the octant of the original angle. Palmer and Morse state in their book *The 8087 Primer* (see Bibliography) when referring to the octant interpretation of the condition code bits that "none of Intel's floating-point library routines use this feature." In Chapter 8, in the context of calculating trigonometric functions with the math unit, we present a routine that performs argument reduction to modulus p/4 and determines the octant without using the condition code bits.

Update of the Partial Remainder

When the final version of ANSI/IEEE 754 Standard was released in 1985 its requirements regarding the calculation of the partial remainder were different from those implemented in the FPREM instruction. ANSI/IEEE 754 states that the remainder function is defined by the formula

$$r = a - b \times q$$

where a is the argument, b is the modulus, and q is the nearest integer to the exact value of a/b. In other words, the standard requires that the quotient be rounded to the nearest integer. Furthermore, it also states that when the quotient is exactly halfway between two numbers it is rounded to an even value. This rounding mode, usually called rounding to the nearest even, is considered the least biased.

The actual implementation of the partial remainder function by the FPREM instruction differs from the standard in that FPREM requires that the sign of the remainder be the same as the sign of the argument. Also that the quotient is obtained by chopping off to the next smaller integer instead of by rounding to the nearest even one. Finally, in FPREM the magnitude of the remainder must be smaller than the modulus. Figure 7.11 is a graph of the FPREM and FPREM1 functions.

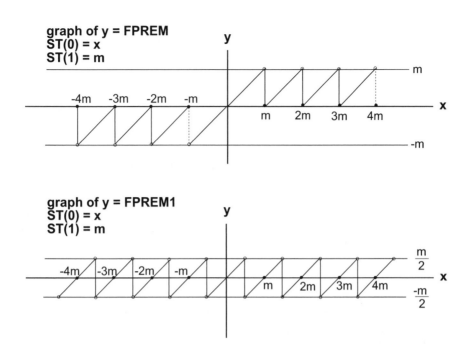

Figure 7.11 *Graph of FPREM and FPREM1 Instructions*

Notice in Figure 7.11 that the remainder obtained with FPREM is always positive if the argument (in this case x) is positive, and the remainder is negative otherwise. This constraint can cause undesirable results. The first problem is that the range of the remainder is doubled for any given value of the modulus. The second one is that the remainder is not periodic, therefore, we cannot expect it to remain unchanged if a constant is added to the argument. Both of these effects tend to defeat the intended purpose of the exact remainder function as described in ANSI/IEEE 754. Another difference in the operation of FPREM and FPREM1 is that in FPREM the magnitude of the remainder is always less than the modulus, while in FPREM1 the remainder is always less than one half the modulus.

All of the above considerations determine that FPREM cannot usually be replaced by FPREM1 without introducing other modifications in the code. For example, in a conventional argument reduction, the use of FPREM1 could introduce a negative remainder that, under some conditions, would not be acceptable. To correct this unexpected result after FPREM1, the code can test for a negative value in ST(0). If this is the case, the modulus can be added once to ST(0) to convert the remainder to a positive range. If positive, ST(0) is left unchanged.

Regarding the use of the remainder functions in the reduction of the arguments of trigonometric function, in the 80387 and the math unit of the 486 and the Pentium this reduction is usually unnecessary, since these math units have a considerably expanded operand range. Specifically: the valid operand range in the 8087 and 80287 is an angle between 0 and p/4 radian while in the 80387 and the math unit of the 486 and the Pentium this range is between 0 and 2^{64} radian. Considering that 2^{64} is approximately 1.84×10^{19}, it can be seen that the new range will be sufficient for most practical calculations.

Manipulating the Encoding

Several nontranscendental instructions allow transforming the value stored in ST(0) by manipulating elements of the floating-point encoding. The manipulations include rounding the value at the stack top to an integer, extracting the exponent and the significand, converting the value at ST(0) to a positive number, and complementing its sign.

FRNDINT (round to integer) rounds the stack top element to an integer value, which is left in ST. The rounding takes place according to the value stored in the rounding control field of the math unit control word (see Figure 7.3).

FXTRACT (extract exponent and significand) breaks down the number at the stack top into its exponent and significand fields. The exponent is stored in ST(1) and the significand in ST. Notice that this conversion refers to the actual binary exponents and significands in extended precision format and not to its decimal equivalents. For example, suppose that the number 178.125 is stored in ST, as follows:

```
ST(0):
    exponent field = 4006H
significand field = B220..00H
```

after performing FXTRACT

```
ST(1) (holds exponent of 178.125):
    exponent field = 4001H
 significand field = E000..00H
ST(0) (holds significand of 178.125):
    exponent field = 3FFFH
 significand field = B220..00H
```

The FXTRACT instruction is designed to be used in conjunction with FBSTP (store packed BCD and pop) in performing numeric conversions from the math unit binary format into BCD and ASCII. Nevertheless, the actual conversion routines usually require additional manipulations of the exponent and the significand fields. In fact, conversion routines often find it easier to decompose exponent and significand by operating on separate copies of the original value, as is the case in the procedure named FPU_OUTPUT mentioned in Chapter 6.

Two instructions are available for manipulating the sign of the value in ST(0). FABS (absolute value) makes the Stack Top register a positive number. FCHS (change sign) complements the sign bit of the number at ST, in fact reversing sign. Table 7.7 lists and describes the nontranscendental instructions.

Table 7.7

Math Unit Nontranscendental Instructions

MNEMONICS	OPERATION	EXAMPLES
	ADDITION AND SUBTRACTION	
FADD	Add source to destination with results in destination. ST can be doubled by coding: FADD ST,ST(0)	FADD ST,ST(2) FADD SINGLE_REAL FADD DOUBLE_REAL FADD
FADDP	Add and pop stack.	FADDP ST(2),ST
FIADD	Add integer in memory to stack top with sum in the stack top	FIADD WORD_INTEGER FIADD SHORT_INTEGER
FSUB	Subtract source from destination with difference in destination.	FSUB ST,ST(3) FSUB ST(1),ST FSUB SINGLE_REAL FSUB DOUBLE_REAL FSUB
FSUBP	Subtract source from destination with result in destination and pop stack	FSUBP ST(2),ST
FSUBR	Subtract destination from source with difference in destination. Reverse subtraction	FSUBR ST,ST(1) FSUBR ST(3),ST FSUBR SINGLE_REAL FSUBR DOUBLE_REAL FSUBR
FSUBRP	Subtract destination from source with difference in destination and pop stack	FSUBRP ST(3),ST
	ADDITION AND SUBTRACTION	
FISUB	Subtract integer memory variable from stack top. Difference to the stack top	FISUB WORD_INTEGER FISUB SHORT_INTEGER
FISUBR	Subtract stack top from integer memory variable. Difference to stack top	FISUBR WORD_INTEGER FISUBR SHORT_INTEGER
	MULTIPLICATION AND DIVISION	
FMUL	Multiply reals. Destination by source with product in destination.	FMUL ST,ST(2) FMUL ST(1),ST FMUL SINGLE_REAL FMUL DOUBLE_REAL FMUL
FMULP	Multiply reals and pop stack. (See FMUL)	FMULP ST(2),ST

(continues)

Table 7.7

Math Unit Nontranscendental Instructions (continued)

MNEMON9CS	OPERATION	EXAMPLES
FIMUL	Multiply integer memory variable by the stack top. Product in stack top	FIMUL WORD_INTEGER FIMUL SHORT_INTEGER
FDIV	Normal division. Divide stack top by the source operand and place quotient in the destination. If no explicit destination ST is assumed	FDIV ST,ST(2) FDIV ST(4),ST FDIV SINGLE_REAL FDIV DOUBLE_REAL FDIV
FDIVR	Reverse division. Divide source operand by the stack top and place quotient in destination. If no explicit destination ST is assumed	FDIVR ST,ST(2) FDIVR ST(3),ST FDIVR SINGLE_REAL FDIVR DOUBLE_REAL FDIVR
FDIVP	Divide destination by source with quotient in destination and pop stack (see FDIV)	FDIVP ST(3),ST
FDIVRP	Divide source by destination with quotient in destination and pop stack (see FDIVR)	FDIVRP ST(4),ST
FIDIV	Divide stack top by integer variable. Quotient in stack top.	FIDIV WORD_INTEGER FIDIV SHORT_INTEGER
FIDIVR	Divide integer memory variable by stack top. Quotient in stack top.	FIDIVR WORD_INTEGER FIDIVR WORD_INTEGER

OTHER ARITHMETIC OPERATIONS

FSQRT	Calculate square root of stack top Square root of −0 = −0	FSQRT
FSCALE	Scale variable. Add scale factor, integer in ST(1), to exponent of ST. Provides fast multiplication (division if scale is negative) by powers of 2. Range of factor is −32767 ≤ ST(1) < 32767 in 8087 And 80287. No limit in 80387 and later	FSCALE
FPREM	Partial remainder. Performs modulo division of the stack top by ST(1), producing an exact result Sign is unchanged. Formula used: Part. rem. = ST − ST(1) · quotient Result is exact. Unsigned remainder < modulus.	FPREM

(continues)

Table 7.7

Math Unit Nontranscendental Instructions (continued)

MNEMONICS	OPERATION	EXAMPLES
FPREM1	Calculates IEEE compatible partial 80387 remainder. See FPREM. Differs from FPREM in how the quotient ST/ST(1) is rounded. Result is exact. Signed remainder < (modulus/2)	
FRNDINT	Round the stack top to an integer according to the setting of the control word	FRNDINT
FXTRACT	Decompose stack top into exponent and significand. The exponent is found in ST(1) and the significand in ST	FXTRACT
FABS	Calculate absolute value of ST Positive values are unchanged Negative values are changed to positive	FABS
FCHS	Change sign of stack top element	FCHS

7.2.3 Comparison Instructions

The *comparison instructions* compare numerical data stored in the stack registers and report the results in the Status register. The FSTSW (store status word) instruction can be used to transfer the condition codes to memory so that they can be tested by the code. The interpretation of the condition codes for the different comparison instructions can be seen in Table 7.2.

Several operand modes are recognized by the compare opcodes. The various formats can be seen in Table 7.8, on the following page.

When ANSI/IEEE 754 was released in 1985 it contained requirements for the compare operation, not all of which were met by the compare instructions as implemented in the 8087 and 80287 processors. Specifically, the Standard requires that signaling NaNs raise the invalid operation exception, but that quiet NaNs do not. This is not the case in the 8087 and 80287 in which any NaN produces and invalid operation. This behavior was corrected in the 80387 by introducing three new compare opcodes, named the un-ordered compares. These are FUCOM (unordered compare), FUCOMP (unordered compare and pop), and FUCOMPP (unordered compare and pop twice).

The procedure named NUM_AT_ST0, listed in Section 7.0.3, demonstrates the use of the FXAM instruction in identifying the contents of the math unit stack registers. Table 7.9 lists and describes the comparison instructions.

Table 7.8

Operand Modes for Compare Instructions

INSTRUCTION TYPE	MNEMONIC FORMAT	OPERAND DESTINATION,SOURCE	SAMPLE CODING
Implicit	F*opcode*	{ST,ST(1)}	FCOM
Registers (explicit)	F*opcode*	ST,ST(i)	FCOM ST,ST(2)
Register (explicit and pop)	F*opcode*P	ST,ST(i)	FCOMP ST,ST(2)
Register (explicit and pop twice)	F*opcode*PP	ST,ST(i)	FCOMPP ST,ST(2)
Memory (real number)	F*opcode*	{ST},MEM_VAR	FCOM MEM_REAL
Memory (integer number)	Fi*opcode*	{ST},MEM_INT	FICOM MEM_INT
Unordered (pop once or twice)	FU*opcode*[PP]	ST,ST(i)	FUCOM ST,ST(2)

Legend: Braces { } indicate implicit operands

7.2.4 Transcendental Instructions

The *transcendental instructions* perform the calculations necessary for obtaining trigonometric, logarithmic, hyperbolic and exponential functions. The instructions are designed to do the necessary core work. They are normally used in computational routines that include processing to reduce the input to the range of the instruction and to scale the results. The transcendental instructions require that the operands be in ST or in ST and ST(1) and return the result in ST. All trigonometric transcendentals assume operands in radian measure.

In the 8087 and 80287 the scope and operand range for the trigonometric transcendentals was limited. For this reason the calculation routines had to include prologue code to scale the operand to this range and to determine its octant. In the 8087 and 80287 only two operations were available: FPTAN (partial tangent) to calculate the tangent of an angle in the range 0 to p/4 radian, and FPATAN (partial arctangent) to calculate the arc function. All other trigonometric functions had to be obtained from these primitives.

Table 7.9

Math Unit Comparison Instructions

MNEMONICS	OPERATION	EXAMPLES	
FCOM	Compare stack top with source operand (stack register or memory). If no source, ST(1) is assumed. Condition codes are set.	FCOM FCOM FCOM FCOM	ST(2) SINGLE_REAL DOUBLE_REAL
FCOMP	Compare stack top with source and pop stack (see FCOM).	FCOMP FCOMP FCOMP FCOMP	ST(2) SINGLE_REAL DOUBLE_REAL
FCOMPP	Compare stack top with ST(1) and pop stack twice. Both operands are discarded	FCOMPP	
FICOM	Compare integer in memory with stack top	FICOM FICOM	WORD_INT SHORT_INT
FICOMP	Compare integer in memory with stack top and pop stack. Stack top element is discarded. Condition codes are set	FICOMP WORD_INT FICOMP SHORT_INT	
FUCOM (80387)	Unordered compare. Operates like FCOM except that no invalid operation if one operand is a NaN.	FUCOM FUCOM FUCOM FUCOM	ST(2) SINGLE_REAL DOUBLE_REAL
FUCOMP (80387)	Unordered compare and pop. Like FCOMP except that no invalid operation if one operand is a NaN.	FUCOMP FUCOMP ST(2) FUCOMP SINGLE_REAL FUCOMP DOUBLE_REAL	
FUCOMPP (80387)	Unordered compare and pop twice. Operates like FCOMPP except that no invalid operation if one a NaN.	FUCOMPP	
FTST	Compare stack top with 0.0 and set condition codes	FTST	
FXAM	Examine stack top and report type of object in ST in condition codes (see Table 7.2)	FXAM	

The 80387 introduced several new transcendental instructions to simplify the calculations of trigonometric functions, and expanded the operand range of the existing ones. The new opcodes are FSIN, to calculate sines, FCOS, to calculate cosines, and FSINCOS, to calculate both sine and cosine functions simultaneously. In the 80387 and the math unit of the 486 and the Pentium, the operand range for all trigonometric functions is from 0 to 2^{63} radians. Since 2^{63} is approximately 9.22×10^{18}, many number crunching routines can perform the calculations without any preliminary range testing or argument reduction.

It has been documented by Intel that in the 80387 and the math unit of the 486 and the Pentium, argument reduction to the first octant is performed internally using a higher precision constant for the modulus p/4 than can be represented externally. For this reason, it is undesirable to use argument reduction routines designed for the 8087 and the 80287 when developing code that will be used exclusively in the 80387 or the math unit of the 486 and the Pentium. The calculation of trigonometric functions is discussed in Chapter 8.

The logarithmic transcendental primitives are FYL2X (y times log base 2 of x) and FYL2XP1 (y times log base 2 of x plus 1). Both instructions use a binary radix. Logarithms to other bases are calculated by means of the formula

$$\log_b(x) = \log_b(2) \cdot \log_2(x)$$

Because the above formula requires it, a multiplication operation is built into the math unit opcodes FYL2X and FYL2XP1. The calculation of logarithms is discussed in Chapter 8.

Table 7.10 lists and describes the transcendental instructions.

The Intel math units contain a single transcendental instruction for exponentiation, named F2XM1 (2 to the x minus 1), although the FSCALE instruction can be used to raise 2 to an integer power. In the 8087 and 80287 the argument for the F2XM1 instruction has to be in the range 0 to 1/2. In the 80387 and the math unit of the 486 and the Pentium the argument was expanded to the range –1 to +1. The fundamental exponentiation function required in high-level programming languages and general number-crunching is the operation y^x. Exponentiation routines, including one to obtain y^x, are developed in Chapter 8.

All transcendental instructions assume that the arguments are both valid and in range. Denormals, unnormals, infinities, and NaNs are considered invalid. Some functions accept a zero operand while for other functions zero is out-of-range. It is important for the code to certify the validity and range of the operand since invalid or out-of-range values produce an undefined result without signaling an exception.

Transcendental Algorithms

Up to 1993 Intel Corporation had not published much information regarding the algorithms used internally by the math unit in the calculation of transcendentals or of other primitives and functions. Palmer and Morse in their book *The 8087 Primer* (see Bibliography) do mention that in the original 8087 the transcendentals were obtained using a variation of the CORDIC (COordinated Rotation DIgital Computer) algorithm first published in 1971 (see Bibliography). The modification of the CORDIC consisted in reducing the size of the table of constants necessary for the calculations and using a rational approximation toward the end of the processing.

Table 7.10

Math Unit Transcendental Instructions

MNEMONICS	OPERATION	EXAMPLES
FCOS (80387)	Calculates cosine of stack top and returns value in ST. $\lvert ST \rvert < 2^{63}$. Input in radians	FCOS
FSIN (80387)	Calculates sine of stack top and returns value in ST. $\lvert ST \rvert < 2^{63}$. Input in radian	FSIN
FSINCOS (80387)	Calculates sine and cosine of ST. SIne appears in ST and cosine in ST(1). $\lvert ST \rvert < 2^{63}$. Input in radians. Tangent = Sine/Cosine	FSINCOS
FPATAN	Partial arctangent. Calculates ARCTAN m= (Y/X), X is ST and Y is ST(1). X and Y must observe $0 < Y < X < +\infty$. Stack is popped. X and Y are destroyed. 1 in radians. The result has the sign of ST(1) and must be < B	FPATAN
FPTAN	Partial tangent. Calculates Y/X = TAN m at ST, must be in the range $0 \leq m < p/4$. Y is returned in ST and X in ST(1). mis destroyed. Input in radians. Result is in the range $\lvert 0 \rvert < 2^{63}$	FPTAN
FYL2X	Calculates Z = log base 2 of X. X is the value at ST and Z in ST(1). Stack is popped and Y is found in ST. Operands must be in the range $0 < X < \infty$ and $-\infty < Y < +\infty$	FYL2X
FYL2XP1	Calculates Z = log base 2 of (X+1). X is in ST and must be in the range $0 < \lvert X \rvert < (1-\sqrt{2}/2)$. Y is in ST(1) and must be in the range $-\infty < Y < \infty$. Stack is popped and Z is found in ST	FYL2XP1
F2XM1	Calculates $Z = 2^x - 1$. X is in ST and must be in the range $0 \leq x \geq 0.5$ radian. The result replaces x in ST	F2XM1

In 1993 Intel published the *Pentium Processor User Manual* (see Bibliography). Volume 3 of this work, titled *Architecture and Programming Manual* contains appendix G, *Report on Transcendental Functions*. This appendix includes a summary discussion on the algorithms used in the calculation of the transcendentals. On this subject the Intel book mentions an alternative to the CORDIC, which is called a polynomial-based algorithm, described by Cody and Waite in their book *Software Manual for the Elementary Functions* (see Bibliography). The transcendental algorithms used by the Pentium are described as midway between the CORDIC and the polynomial-based method. In the case of the Pentium, a table of functions stored in ROM is used to shorten the calculations required by the polynomial-based method.

In the past, table-driven polynomial algorithms have been used in mathematical software packages. The method is well described by Tang in two articles published in the *ACM Transactions on Mathematical Software* (see Bibliography). The innovation of the Pentium is implementing these algorithms in hardware. The advantages mentioned by Intel relate to the following elements:

Accuracy. This element is measured in units of last place error or ulps. The error in ulps is defined by the formula

$$u = \left| \frac{f(x) - F(x)}{2^{k-63}} \right|$$

where f(x) is the exact value of the function, F(x) is the computed value, and k is an integer such that

$$1 \le 2^{-k} f(x) < 2$$

According to Intel, the worst case error in the calculation of transcendental functions in the Pentium processor is of 1 ulp when rounding to the nearest mode and of 1.5 ulps in all other rounding modes. This degree of precision represents an improvement of 2 to 3 ulps regarding the 486 math unit. No information has been provided by Intel regarding the comparative accuracy of other math units.

Monotonicity. This attribute refers to a function whose value always changes in the same direction as the argument. In other words, if the argument is larger, the function is also larger, and vice versa. In this case the monotonicity results from the accuracy of the calculations. The Pentium documentation guarantees that the transcendental functions are monotonic over their entire domain.

Proof of Correctness. The algorithm used in the calculation of the functions makes possible a rigorous and straightforward error analysis. The Intel document mentioned at the start of the section includes a verification summary for each of the functions calculated by the Pentium.

Performance. Intel documentation states that the transcendental algorithms used in the Pentium lead to higher performance. Typical values range from 54 to 115 clock cycles.

7.2.5 Constant Instructions

The math unit *constant instructions* are used to load numerical values that are commonly needed in mathematical calculations. All the constant instructions operate on the Stack Top register. The instructions in this group are a convenience, since these and other constants can be created and loaded from memory variables, as described in Chapter 6. Advantages of using internal constants is that they simplify programming and improve execution speed. The constants are loaded as if they were defined in the extended precision format. This insures that they are accurate to approximately 19 decimal places. Table 7.11 lists and describes the math unit constant instructions.

Table 7.11

Math Unit Constant Instructions

MNEMONICS	OPERATION	EXAMPLES
FLDLG2	Load logarithm base 10 of 2 on stack top. Constant is accurate to 64 bits (approximately 19 digits) $\text{Log}_{10}2 = 0.30102...$	FLDLG2
FLDLN2	Load logarithm base e of 2 on stack top. Constant is accurate to 64 bits (approximately 19 digits) $\text{Log}_e2 = 0.69315...$	FLDLN2
FLDL2E	Load logarithm base 2 of e on stack top. Constant is accurate to 64 bits (approximately 19 digits) $\text{Log}_2e = 1.44268...$	FLDL2E
FLDL2T	Load logarithm base 2 of 10 on stack top. Constant is accurate to 64 bits (approximately 19 digits) $\text{Log}_210 = 3.32192...$	FLDL2T
FLDPI	Load p on the stack top. Constant is accurate to 64 bits (approximately 19 digits) Value is 3.14159...	FLDPI
FLDZ	Load zero on the stack top. Constant is accurate to 64 bits (approximately 19 digits)	FLDZ
FLD1	Load +1.0 on the stack top. Constant is accurate to 64 bits (approximately 19 digits)	FLD1

7.2.6 Processor Control Instructions

Like the constant instructions, the *processor control instructions* perform no numerical calculations. Their purpose is to set up the processor for a desired mode of operation, to read its state during computations, and to make adjustments in the stack registers.

An alternative mnemonic form (NO WAIT) is provided for use in routines that must execute under circumstances where timing can be a critical factor. By using the NO WAIT form the programmer forces the assembler not to prefix the processor control opcode with the normal wait. The special mnemonic is identified by the letter N, for example, FINIT and FNINIT. In addition, the NO WAIT form ignores unmasked numeric exceptions. The no wait form is also required in code that cannot assume that a math unit is available in the system. In the absence of a math unit, the wait mnemonic could cause the machine to hang up. This coding method is shown in the ID_FPU procedure listed in Chapter 5. The processor control instructions appear in Table 7.12.

Table 7.12

Math Unit Processor Control Instructions

MNEMONICS	OPERATION	EXAMPLES
FCLEX	Clear exception flags, exception	FCLEX
FNCLEX	status, and busy flag in the status word	FNCLEX
FDECSTP	Decrement stack top pointer field in the status word. If field = 0 then it will change to 7. The effect is to rotate the stack	FDECSTP
FDISI FNDISI (8087)	Disable interrupts by setting mask. No action in 80287 and 80387	FDISI FNDISI
FENI FNENI (8087)	Enable interrupts by clearing the mask in the control register. No action in 80287 and 80387	FENI FNENI
FFREE	Change tag of destination register to EMPTY	FFREE ST(2)
FINCSTP	Add one to the stack top field in the status word. If field = 7 then it will change to 0. The effect is to rotate the stack	FINCSTP
FINIT	Initialize processor. Control word	FINIT

(continues)

Table 7.12

Math Unit Processor Control Instructions (continued)

MNEMONICS	OPERATION	EXAMPLES
FNINIT	is set to 3FFH. Stack registers are tagged EMPTY. Exception flags are cleared. All exceptions are masked. Rounding set to nearest Even. Precision set to 64-bits. Register number 0 is stack top	FNINIT
FLDCW	Load memory variable (word) into the control register	FLDCW CTRL_WORD
FLDENV	Load 14-byte environment from memory storage area. The environment should have been previously saved by FSTENV	FLDENV MEM_14
FNOP	Floating point no operation.	FNOP
FRSTOR	Restore state from 94-byte memory area previously written by a FSAVE or FNSAVE	FRSTOR MEM_94
FSAVE FNSAVE	Save state (environment and stack registers) to a 94-byte area in memory	FSAVE MEM_94
FSETPM (80287)	Sets protected mode addressing for 80287 systems. Interpreted as FNOP in 80387	FSETPM
FSTCW FNSTCW	Store control register in a memory variable (word)	FSTCW CTRL_WORD
FSTENV FNSTENV	Store 14-byte environment into memory storage area. See FLDENV	FSTENV MEM_14
FSTSW FNSTSW	Store status register in memory variable (word)	FSTSW STAT_WORD
	In the 80387, 486 and Pentium it is possible to code FSTSW AX	
FWAIT	Alternate mnemonics for WAIT. Must be used with Intel emulators	FWAIT

Chapter 8

Transcendental Primitives

Chapter Summary

In this chapter we discuss the design and development of primitive routines for calculating exponential, trigonometric, and logarithmic functions on the Intel math unit. These routines will perform the fundamental calculation of transcendental functions required in a typical floating-point package, a mathematical application, or a high-level language.

8.0 Developing Math Unit Software

Programming of the Intel math unit is not always a simple or intuitive task. In addition to the data conversion difficulties mentioned in Chapter 6, the following possible sources of problems must be taken into consideration:

1. The trigonometric functions are not directly available in all math unit implementations. On the 8087 and 80287 only a partial tangent (FPTAN) can be obtained, and its range is limited to an angle of 0 to p/4 radians. In these math units all other trigonometric functions must be derived from this partial tangent. Software must also reduce the input argument to a valid range. The 80387 coprocessor introduced new trigonometric instructions to calculate sine and cosine directly and expanded the operand range of the partial tangent instruction. However, the code will not run in 8087 and 80287 systems if these new opcodes are used.

2. Only one instruction, FPATAN, is provided for the calculation of inverse trigonometric functions. Arc-sine, arc-cosine, arc-tangent, as well as the arc functions of their reciprocals, must be calculated from a partial arc-tangent function.

3. The two logarithmic opcodes operate on a binary radix. Additional processing is necessary to obtain logarithms to other bases, such as the natural and common logs. Similar manipulations are necessary in the calculation of antilogarithms.

4. The instruction F2XM1 raises 2 to the x power and subtracts one. In the 8087 and 80287 the range of the exponent (x) must be a positive number between 0 and 0.5. Although

181

the exponent range was increased in the 80387, there is no single math unit instruction that raises a base to an arbitrary power.

5. A program containing math unit instructions will almost certainly hang-up if it executes in a machine that is not equipped with floating-point hardware. Although today a PC without a math unit is a rare occurrence, the problem cannot be totally ignored. The solution is a product called a floating-point emulator. The ideal emulator consists of a set of software routines that exactly imitate the hardware component in systems not equipped with the chip. However, in the 8087, emulator software cannot operate on the same opcodes used by the hardware component. The math unit opcodes must be replaced or patched with the opcodes recognized by the emulator. Math unit software emulators and support routines are usually not included with assemblers and development packages.

6. The math unit is a binary machine. Although it operates on integers, floating-point, and BCD data types, a typical numerical application uses mostly floating-point representations. This means that programs often require input and output in some form of user-readable ASCII encodings. Conversion routines to and from the math unit internal formats are not trivial. If incorrectly coded they can affect the result of calculations.

Some of these problems have already been addressed. In Chapter 5 we developed the function IdMathUnit() which can be used to identify the various implementations of the mathematical coprocessor. In Chapter 6 we presented routines which allow converting numeric data in ASCII into the math unit formats and vice versa. In this chapter we develop routines for performing fundamental operations in the calculation of exponential, trigonometric, and logarithmic functions.

8.1 Exponential Functions

Exponential functions, such as the calculation of 10^y, e^y, and x^y are essential operations in most mathematical and floating-point packages. In addition, many compilers and interpreters include an exponential operation which allows the calculation of powers and roots, although certain common powers and roots, such as squares, cubes, and square roots, are often provided separately. However, Palmer and Morse (1984, 105) mention that "the most difficult elementary function to compute that routinely appears in high-level languages is x^y." One of the reasons for the computational difficulty is that the expression x^y can represent a power or a root according to the value of the exponent. For example, x^4 represents the operation of multiplying by x by itself 4 times since

$$x \times x \times x \times x = x^4$$

On the other hand, $x^{\frac{1}{4}}$ represents the operation of extracting the 4th root of x

$$x^{\frac{1}{4}} = \sqrt[4]{x}$$

Computationally speaking, the functions are entirely different, however, a routine to calculate x^y can be used to calculate $x^{1/y}$ by virtue of the following identity:

$$x^{2\frac{1}{2}} = x^2 \times x^{\frac{1}{2}}$$
$$= x^2 \times \sqrt{x}$$

By the same token, a mixed exponent is interpreted as having an integer and a fractional part, for example:

$$y^{\sqrt{x}} = \frac{1}{x^y}$$

As you will see later in this chapter, the calculation of integer powers is easier and more accurate than the calculation of fractional powers. By factoring the exponent into an integer and a fractional component we can make the exponential calculation more accurate.

The Intel math units do not provide a specific opcode for the general calculation of exponentials, as would be convenient for directly calculating 10^y, e^y, or x^y. The instruction F2XM1 calculates 2 to the x and subtracts 1 from the result. The reason for subtracting 1 is to improve accuracy for values of x close to 0. In the 8087 and 80287 the operand range is limited to $0 = < x = < 0.5$. In the 80387 and the math unit of the 486 and the Pentium the operand range was expanded to $-1 < x < 1$. However, it has been documented that the error magnitude increases very rapidly as the operand approaches |1|.

Although the $2^x - 1$ function provided by F2XM1 does not allow direct calculations of powers of other bases, it can be combined with logarithmic instructions (discussed later in this chapter) to obtain an approximation of common exponentials, since

$$10^y = 1 + (2^{y*\log2(10)-1}) = 1 + \text{F2XM1}(y*\log_2(10))$$
$$e^y = 1 + (2^{y*\log2(e)-1}) = 1 + \text{F2XM1}(y*\log_2(e))$$
$$x^y = 1 + (2^{y*\log2(x)-1}) = 1 + \text{F2XM1}(y*\log_2(x)).$$

In Section 7.3.4 and 7.3.5 we saw that $\log_2(10)$ is obtained with FLDL2T, $\log_2(e)$ with FLDL2E, and $\log_2(x)$ with FYL2X. Therefore the formulas can be expressed, using math unit opcodes, as follows

$$10^y = 1 + \text{F2XM1}(x*\text{FLDL2T})$$
$$e^y = 1 + \text{F2XM1}(x*\text{FLDL2E})$$
$$x^y = 1 + \text{F2XM1}(y*\text{FYD2X}(y,x)).$$

8.1.1 Calculation of Powers

Integer powers are often required in floating-point packages, high-level languages, and many general-purpose applications. The rationale for not having a power function is that powers can be readily obtained through logarithms. Computer algorithms for the logarithmic approximation of the power functions are found in Koren (1993,

164–167), Brassard and Bratley (1988, pages 128–132), and many others. Implementations for the Intel math unit have been listed by Bradley (1984, pages 218–219), Startz (1985, pages 194–196), and Intel (1990, 20-18, 20-19). One variant of the log approximation algorithm obtains low-order powers from a tabulated list while the larger exponents are approximated through logs (Intel 1990, 20-18, 20-19).

The main objection to the logarithmic method for the calculation of exponential functions is its low accuracy. Palmer and Morse (1984, 105) state that in the design of the original 8087 chip it was necessary to provide a 64-bit field in the internal format to insure that the logarithmic evaluation of functions would be precise to 53 significant bits. Tang (1989) has analyzed the source and magnitude of the error in logarithmic approximations of exponenentials and proposed table-driven implementations that improve accuracy.

The error generated by the straight logarithmic approximation of powers is often tangible. For example, one method for converting real numbers represented in ASCII digits into one of the binary floating-point formats established in the ANSI/IEEE 754 requires the evaluation of 10^y, where y is the exponent of the input number (this problem was discussed in Section 6.3.2). The power of 10 is used by the routine in normalizing the significand, which is multiplied by 10^y if the exponent is signed positive, and divided by 10^y if the exponent is negative. However, if the conversion routines use a logarithmic method for obtaining 10^y, the resulting error can propagate to the 12th significand bit (see Section 6.3.2).

Logarithmic Approximation of Exponentials

In spite of their inaccuracy, logarithmic methods are often used since they provide a simple way for obtaining functions with integer, fractional, or mixed exponents. By the same token, the same routine serves to calculate powers and roots. The following low-level procedure allows the logarithmic approximation of xy.

```
        .DATA
; Constant defined in single precision format
ONE_HALF        DD      0.5     ; 1/2 in single precision
; Storage for math unit controls
STATUS_WW       DW      0       ; Machine status word
CONTROL_WW      DW      0       ; Control word
ROUND_DOWN      DW      177FH   ; Control word for round down
        .CODE
_X_TO_Y_BYLOG   PROC
; Raise a number (y) to an arbitrary power (x)
; On entry:
;       ST(0) = exponent
;       ST(1) = base
; On exit:
;       ST(0) = x**y
; The FPU exponential function x**y is implemented by the following
; identity:
;       x**y = 2**(y*Log2(x))
; Algorithm to raise 2 to an arbitrary power::
;       2**x = (2**i)*(2**f)
;       where:
;           i = the nearest integer smaller than x
;           f = the fraction x - i
```

```
; In calculations:
;            if f > 1/2 then
;               2**f = 2**((f-1/2) + 1/2)
;                    = 2**(f-1/2) * 2**(1/2)
;                    = 2**(f-1/2) * (square root of 2)
; This manipulation allows reducing f to the range of the instruction
; F2XM1. 2**i is calculated using FSCALE
        FXCH                    ; Invert x and y
;                       |   ST(0)   |   ST(1)   |   ST(2)
                        ;    y      |     x     |   EMPTY   |
        FYL2X           ;  y*LOG2(x) |   EMPTY   |
;*********************|
;   set to round down |
;*********************|
        FSTCW   CONTROL_WW   ; Store control word in memory
        FLDCW   ROUND_DOWN   ; Install new control word
; At this point the 80x87 is set to round down
; This ensures that i is smaller than x
;                       |   ST(0)   |   ST(1)   |   ST(2)
                        ;    x      |   EMPTY   |
        FLD     ST(0)   ;    x      |     x     |   EMPTY   |
        FRNDINT         ;    i      |     x     |   EMPTY   |
        FLDCW   CONTROL_WW   ; Restore original control word
        FSUB    ST(1),ST;    i      |     f     |   EMPTY   |
; f = x - i, that is, the fractional part of x
        FXCH            ;    f      |     i     |   EMPTY   |
        FLD     ONE_HALF
                        ;   1/2     |     f     |     i     |
        FXCH            ;    f      |    1/2    |     i     |
; Scale f to the range of F2XM1, which is 0 <= x => 1/2
        FPREM           ; REM(f/(1/2) |   1/2   |     i
; If the value of f => 0.5 then the quotient of FPREM is 1
; and bit C1 is set. Otherwise x is < 1/2
        FSTSW   STATUS_WW    ; Store status word
        FWAIT           ;    f      |    1/2    |     I     |
        FSTP    ST(1)   ;    f      |     i     |   EMPTY   |
        F2XM1           ;  2**(f)-1 |     i     |   EMPTY   |
; Since F2XM1 calculates 2^(f)-1, 1 must be added to the
; result
        FLD1            ;    1      | 2**(f)-1  |     I     |
        FADDP   ST(1),ST;  2**f     |     i     |   EMPTY   |
        MOV     AH,BYTE PTR STATUS_WW +1
                        ; Move status bits into AH
        TEST    AH,00000010B    ; Test bit C1
        JZ      CASE_2
; If f was => 1/2 then multiply by square root of 2
        FLD1            ;    1      |   2**f    |     I     |
        FADD    ST,ST(0);    2      |   2**f    |     I     |
        FSQRT           ;  2**(1/2) |   2**f    |     I     |
        FMULP   ST(1),ST;2**f*2**(1/2)|    i    |   EMPTY   |
CASE_2:
;                       |   2**f    |     i     |   EMPTY   |
        FSCALE          ;  2(f)*2(i) |     i     |   EMPTY   |
        FSTP    ST(1)   ;   2**x    |   EMPTY   |
EXTI_XTOYL:
        CLD
        RET
_X_TO_Y_BYLOG    ENDP
```

Notice that the _X_TO_Y_BYLOG procedure must separate the integer and the fractional part of the exponent to scale the operand to the range of the F2XM1 instruction. In order to make the code compatible with the 8087 and 80287, the fractional part of the exponent must be tested for a value > 0.5. If y > 0.5 the fractional element of the exponent (2^f) is factored as follows

$$2^f = 2^{(f^{\frac{1}{2}}+\frac{1}{2})}$$

$$= 2^{f^{\frac{1}{2}}} \times 2^{\frac{1}{2}}$$

$$= 2^{f^{\frac{1}{2}}} \times \sqrt{2}$$

Although this manipulation is not strictly necessary in 80387 systems and in the math unit of the 486 and Pentium, it serves to avoid values of the exponent close to 1, in which range errors have been documented to increase rapidly.

SOFTWARE ON-LINE

The procedure _X_TO_Y_BYLOG is found in the un32_5 module of the MATH32 library furnished in the book's on-line software. The C++ interface function named XtoYByLog() is in located in the Chapter8/Test Un32_5 project folder.

Binary Powering

Several non-logarithmic algorithms have been described for evaluating x^y when y is a positive integer. Knuth (1981, 2:441–466) discusses in detail what he calls the "S-and-X binary method" for exponentiation. This method is also examined by Gonnet and Baeza-Yates (1991, 240–242) under the name of binary powering.

The binary powering algorithm computes an integer power by raising the base to half the exponent and squaring the result. If the exponent is odd, the previous product is also multiplied by the base. Knuth describes the algorithm by letting the letter S represent the operations of squaring the previous product and the letter X represent the operation of multiplying the previous product by the base (x). The fundamental rule of binary powering is that every 1-bit of the exponent is replaced by the letters SX and every 0-bit by the letter S. For example, in performing x^{25} by binary powering we proceed as follows

$$25 = 11001 \text{ binary}$$

$$= \text{SX SX S S SX}$$

the first SX operation is now eliminated, leaving

$$\text{SX S S SX}$$

which means that we must successively compute $(x^2 * x)$, x^2, x^2, and $(x^2 * x)$. If $x = 2$ the iterations of the calculation of 225 by binary powering are

$$x^2 * x = 8$$

$$x^2 = 64$$

$$x^2 = 4096$$

$$x^2 * x = 33,554,432$$

An algorithm suitable for implementing binary powering in the 80x86 and math unit systems can be devised in terms of left-shift and bit test operations, as in the following pseudo-code description for calculating 10^y.

```
ALGORITHM 10_TO_Y_BY_BP

    constant BIT_SIZE = bit size of exponent (y)

Function 10_TO_Y(y)

    OP_COUNTER = BITS_SIZE

    REM ** skip leading zeros in n

    WHILE LEFTMOST BIT OF N   1

       OP_COUNTER = OP_COUNTER - 1

       SHIFT LEFT N

    REM ** DISCARD LEFTMOST LEADING BIT

    SHIFT LEFT N

    OP_COUNTER = OP_COUNTER - 1

    WHILE OP_COUNTER   0

       IF LEFTMOST BIT OF N = 1 THEN

          N = N * N * 10

       ELSE

          N = N * N

END Function 10_TO_Y()
```

In the algorithm 10_TO_Y_BY_BP the constant BITS_SIZE holds the maximum number of binary digits of the exponent. For instance, if the exponent is stored in a 16-bit register then BITS_SIZE = 16. The following procedure performs the calculation of xy by binary powering. Recall that binary powering requires an integer exponent, therefore the method cannot be used in the extraction of roots. In other words, binary powering cannot be used in the evaluation of functions with rational or mixed exponents.

The following low-level procedure performs x^y by binary powering.

```
            .DATA
; Storage for entry values
X_VALUE          DT      0
Y_VALUE          DW      0
OP_COUNT         DB      0           ; Iterations counter

         .CODE
_X_TO_Y_BYBP      PROC  USES esi edi ebx ebp
; Calculation of x**y by binary powering
; On entry:
;          ST(0) = exponent
;          ST(1) = base
; On exit:
;          ST(0) = x**y
;
; Algorithm (based on D. Knuth)
;       1. Determine maximum number of binary digits in
;          exponent. This is intial value of operations counter
;       2. Skip leading zeros in exponent decrementing operations
;          counter
;       3. Skip first 1-digit decrementing operations counter
;       4. Test leftmost binary digit of exponent
;          If 1
;              Square previous product and multiply by x
;              Goto step 5
;          If 0
;              Square previous product
;       5. Shift left exponent bits
;          Decrement operations counter
;              If not zero then step 4
;       6. End
;
;***************************|
; move data to work variable |
;   init operations counter  |
;***************************|
;                                    |   y   |   x   | EMPTY |
;                                    |   x   | EMPTY |
        FISTP     Y_VALUE      ;   x   | EMPTY |
        MOV       DX,Y_VALUE   ; Exponent to loop register
        MOV       OP_COUNT,16  ; Initialize operations counter
; DX hold exponent
;*************************|
;        special case      |
;        test for x**0     |
;*************************|
; By the rules of exponents any base to the 0 power = 1
        CMP       DX,0         ; Test for 0 exponent
        JNE       TEST_CASE_E1 ; Go if not zero
; At this point we have detected x**0
        FSTP      ST(0)        ; Clear stack
        FLD1                   ; Load 1
        JMP       EXIT_XTOYP
;*************************|
;        special case      |
;        test for x**1     |
;*************************|
; By the rules of exponents any base to the 1 power = base
TEST_CASE_E1:
        CMP       DX,1         ; Test for 1 exponent
        JNE       SUPRESS_0S   ; Go if not zero
```

```
; At this point we have detected x**1
; Base is already in ST(0)
;                                    |   x    |  EMPTY  |
        JMP     EXIT_XTOYP
;**************************|
;     suppress leading 0s         |
;    decrementing ops counter     |
;**************************|
SUPRESS_0S:
        TEST    DX,8000H         ; Is bit 15 set?
        JNZ     SUPRESS_LAST     ; Go if set
; Bit 15 is not set
        DEC     OP_COUNT         ; Adjust shift counter
        SAL     DX,1             ; Shift bits left
        JMP     SUPRESS_0S
; At this point all leading 0 bits have been eliminated
; OP_COUNT has been decremented accordingly
;**************************|
;     skip first 1-bit            |
;    decrement ops counter        |
;**************************|
SUPRESS_LAST:
        SAL     DX,1             ; Shift out this 1-bit
        DEC     OP_COUNT         ; Adjust shift counter
        FLD     ST(0)            ;    x    |    x    |  EMPTY   |
;**************************|
;     test leftmost bit           |
; if 1, square and multiply |
; if 0, square only               |
;**************************|
TEST_BIT:
        TEST    DX,8000H          ; Is bit 15 set?
        JNZ     SQUARE_AND_MUL10         ; Go if set
; Square only
        FMUL    ST,ST(0)         ; ST**2   |   10   |  EMPTY  |
        JMP     NEXT_OP
SQUARE_AND_MUL10:
        FMUL    ST,ST(0)         ; ST**2   |   10   |  EMPTY  |
        FMUL    ST,ST(1)         ; ST*10   |   10   |  EMPTY  |
;**************************|
;  shift exponent bits and  |
;   test for end of counter  |
;**************************|
NEXT_OP:
        SAL     DX,1             ; Shift all bits left
        DEC     OP_COUNT
; Test for end of processing
        JNZ     TEST_BIT          ; Continue if not zero
;**************************|
;      end of routine            |
;**************************|
        FSTP    ST(1)            ;  x**y    |  EMPTY  |
EXIT_XTOYP:
        RET
_X_TO_Y_BYBP    ENDP
```

The number of multiplications required in binary powering can be expressed in terms of the number of 0-digits (μ) and of 1-digits (v) of the binary value of the exponent as follows

$$Q(n) = \mu(n) + 2(v(n) - 1). \tag{I}$$

For example, in determining the number of multiplications for calculating 10249 we proceed in this manner

$$n = 249 = 11111001 \text{ binary,}$$

$$\mu(n) = 2 \text{ (number of 0-digits),}$$

$$v(n) = 6 \text{ (number of 1-digits),}$$

$$Q(n) = 2 + (6 - 1) = 7 \text{ multiplications.}$$

Binary powering offers a compact, accurate, and easy-to-implement method for calculating integer powers.

Exponent Factoring

An exponential function often required by compilers and applications is the evaluation of the integral power of a constant. This case can be represented as C^y where C is a constant and y a positive integer. In Chapter 6 we examined how the calculation of 10^y is used in an ASCII to binary conversion routine for normalizing the significand. At that time we discussed how the accuracy in the calculation of 10^y can affect the result of the conversion.

The accuracy of a computer system, sometimes called machine epsilon, or e_{mach}, is defined as the difference between the significands of a number x_0 and the next larger representable number x_1. In the math unit extended precision format, machine epsilon is the binary value of the 64th digit of the significand. This makes e_{mach} the smallest error value representable in a particular machine.

So far we have seen exponentials obtained by logarithmic methods and by binary powering. In the method described in this section the integral exponent of a power function is factored in such a way that allows the use of table values in the evaluation of the functions. The main advantage of the factoring method is accuracy of the result, which in some variants of the algorithm approximates e_{mach}. This high accuracy results from the fact that the table values are pre-defined as memory constants to the maximum representable precision.

The method of exponent factors is best represented using the notation of finite algebra. In this chapter we use the following expressions

y (MOD x) = the remainder of y/x,

INT(y/x) = the integer quotient of y/x.

The original notion for the algorithm stems from a simple application of the laws of exponents. For example

$$c^{456} = (c^{100})^4 \times (c^{10})^5 \times (c)^6$$

During computations of the above example the use of the constant C^{100} and C^{10} each saves a minimum of 99 multiplications. Calculating 10^{456} by brute force requires 455 multiplications. However, using 10^{100} and 10^{10} as factors the total number of multiplications is reduced to 15 (4 + 5 + 6). Perhaps the most important feature of this method is that the constants (C^{100} and C^{10} in the above case) can be stored in memory to machine epsilon precision. The use of high-precision constants reduces the cumulative error of the calculation. In other words, exponent factoring diminishes the cumulative error by decreasing the number of multiplications and by confining the multiplicative error to each place-value digit.

We introduce the general notation for exponent factoring by means of an example case in which we have predefined 4 place-value factors. Later in this chapter we generalize the notation to include any number of exponent factors to any set of predetermined values.

In the following example the initial values for the 4 exponent factors are the place-values 1000, 100, 10, and 1 of a 4-digit exponent. The following terms are used in the calculation of C^y:

$$F_3 = 1000 \qquad I_3 = INT(y/F_3) \qquad R_3 = y \, MOD(I_3 * F_3) \quad (II)$$

$$F_2 = 100 \qquad I_2 = INT(R_3/F_2) \qquad R_2 = R3 \, MOD(I_2 * F_2)$$

$$F_1 = 10 \qquad I_1 = INT(R_2/F_1) \qquad R_1 = R2 \, MOD(I_1 * F_1)$$

$$F_0 = 1 \qquad I_0 = INT(R_1/F_0).$$

Based on these factors, the place-value products for the expansion of C^y can be stated as

$$C^y = C(F_3 * I_3) * C(F_2 * I_2) * C(F_1 * I_1) * C(F_0 * I_0)$$

or as

$$C^y = (CF_3)I_3 * (CF_2)I_2 * (CF_1)I_1 * (CF_0)I_0. \qquad (III)$$

For the calculation of 10^{2456} the factor list is

C = 10

y = 2456

$$F_3 = 1000 \qquad I_3 = INT(2456/1000) \qquad R_3 = 2456 \, MOD(2*1000)$$
$$= 2 \qquad\qquad = 456$$
$$F_2 = 100 \qquad I_2 = INT(456/100) \qquad R_2 = 456 \, MOD(4*100)$$
$$= 4 \qquad\qquad = 56$$
$$F_1 = 10 \qquad I_1 = INT(56/10) \qquad R_1 = 56 \, MOD(5*10)$$
$$= 5 \qquad\qquad = 6$$
$$F_0 = 1 \qquad I_0 = INT(6/1)$$
$$= 6.$$

We can now apply formula (III):

$$10^{2456} = (10^{1000})^2 * (10^{100})^4 * (10^{10})^5 * (10)^6.$$

Notice that the total number of multiplications required in the calculation is the sum $I_3 + I_2 + ... + I_0$.

The exponent factoring algorithm for calculating C^y is not exactly applicable to the case x^y. In the latter case the required number of memory constants would be too large for practical application. However, the method can be modified to allow solving x^y by introducing an additional step in which the required factors are calculated and stored for later use. This additional step determines that the x^y variant of the algorithm shows lower comparative accuracy and lower performance than the C^y variants. The procedure named _TEN_TO_Y_BYFAC listed contained in the Un32_5 module of the MATH32 library calculates 10^y by exponent factoring. The following procedure, named _X_TO_Y_BYFAC, calculates any integer power of x by the same method.

```
            .DATA
// Data for procedure
X_TO_1000       DT      0
X_TO_100        DT      0
X_TO_10         DT      0
; Storage for entry values
X_VALUE         DT      0

            .CODE
_X_TO_Y_BYFAC    PROC  USES esi edi ebx ebp
; Exact calculation of x**y by exponent factoring according
; to the following factor list:
;
; F3 = 1000    I3 = INT(y/F3)         R3 = y MOD(I3*F3)
; F2 = 100     I2 = INT(R3/F2)        R2 = R3 MOD(I2*F2)
; F1 = 10      I1 = INT(R2/F1)        R1 = R2 MOD(I1*F1)
; F0 = 1       I0 = INT(R1/F0)
;
; Exponent factorization formula:
;
; 10**y = (10**F3)**I3 *  (10**F2)**I2 *  (10**F1)**I1 *  (10**F0)**I0
;         ------------     ------------     ------------     ------------
;              |               |                |                |
;              |               |                |                |
;            STAGE 3         STAGE 2          STAGE 1          STAGE 0
; partial      |               |                |                |
; products    S3              S2               S1               S0
;         ---------------------------------------------------------------
;                      e x e c u t i o n    s t a g e s
; On entry:
;          ST(0) = exponent
;          ST(1) = base
; On exit:
;          ST(0) = x**y
;
; During operations:
;          X_VALUE holds base (extended precision real)
;          Y_VALUE holds exponent (word integer)
;
;***************************|
;  move data to work variables |
;***************************|
;                              |   y   |   x   | EMPTY  |
;
```

```
        FISTP   Y_VALUE              ;    x    |  EMPTY  |  EMPTY  |
        FSTP    X_VALUE              ;  EMPTY  |
;**************************|
;       special case       |
;       test for x**0      |
;**************************|
; By the rules of exponents any base to the 0 power = 1
        MOV     CX,Y_VALUE           ; Exponent to loop register
        CMP     CX,0                 ; Test for 0 exponent
        JNE     NOT_ZERO_EXP         ; Go if not zero
; At this point we have detected x**0
        FLD1                         ; Load 1
        JMP     EXIT_XTOYF
;**************************|
;    test for y < 0        |
;**************************|
; By the rules of exponents any base to the 0 power = 1
NOT_ZERO_EXP:
        CMP     CX,0                 ; Test for 0 exponent
        JNL     NOT_EXP_NEG          ; Go if not negative
; Routine returns 0 for a negative exponent
        FLDZ                         ; Load 0
        JMP     EXIT_XTOYF
;**************************|
; calculate factors:       |
;           y**1000,       |
;           y**100,        |
;       and y**10          |
;**************************|
; These factors are needed in stages 3, 2, and 1 of the calculations
; Only the factors actually required are obtained
; Test for EXP > 1000
; First load 1 for the case that exp < 1000
NOT_EXP_NEG:
        FLD1                         ;    1    |  EMPTY  |  EMPTY  |
        CMP     CX,10                ; Is exp > 10
        JA      GET_FACTORS          ; Go if exponent > 10
; Load two more 1s for final product calculation
        FLD1
        FLD1                         ;    1    |    1    |    1    |
        JMP     STAGE_0              ; Go directly to last stage
;**************************|
;       calculate factors  |
;**************************|
; If the power is > 10 then all three factors are calculated and
; stored
GET_FACTORS:
        FSTP    ST(0)                ; EMPTY-- |
; Calculate y**1000. At this point CX is > 1000
        MOV     DX,CX                ; Save counter in DX
;**************************|
;       calculate x**1000  |
;       in 3 stages        |
;**************************|
; First stage: x**10
;                                    | ST(0) |  ST(1) |  ST(2) |
        FLD     X_VALUE              ;    x    | EMPTY-- |
        FLD1                         ;    1    |    x    | EMPTY  |
        MOV     CX,10                ; Multiplier
GET_10X:
```

```
        FMUL    ST,ST(1)            ;  prod   |    x    | EMPTY  |
        LOOP    GET_10X             ;  x**10  |    x    | EMPTY  |
        FSTP    ST(1)               ;  x**10  | EMPTY   |
        FLD     ST(0)               ;  x**10  | x**10   | EMPTY  |
; Store factor x**10
        FSTP    X_TO_10
        FLD1                        ;    1    | x**10   | EMPTY  |
        MOV     CX,10               ; Multiplier for second loop
GET_100X:
        FMUL    ST,ST(1)            ;  prod   | x**10   | EMPTY  |
        LOOP    GET_100X            ;  x**100 | x**10   | EMPTY  |
        FSTP    ST(1)               ;  x**100 | EMPTY   | EMPTY  |
        FLD     ST(0)               ;  x**100 | x**100  | x**10  |
; Store factor x**100
        FSTP    X_TO_100
; At this point ST(0) = x**100
        FLD1                        ;    1    | x**100  | EMPTY  |
        MOV     CX,10               ; Multiplier for third loop
GET_1000X:
        FMUL    ST,ST(1)            ;  prod   | x**100  | EMPTY  |
        LOOP    GET_1000X           ;  x**1000| x**100  | EMPTY  |
        FSTP    ST(1)               ;  x**100 | EMPTY   |
; Store factor x**1000
        FSTP    X_TO_1000
; All three factors are stored. Stack is empty
        MOV     CX,DX               ; Exponent to CX
;
;****************|
;    STAGE 3     |
;*****************************|
;    calculate (X**F3)**I3     |
;*****************************|
; Test for EXP > 1000
; First load 1 for the case that exp < 1000
        FLD1                        ;    1    | EMPTY-- |
        CMP     CX,1000             ; Is exp > 1000
        JNA     STAGE_2             ; Go if not above than 1000
        FSTP    ST(0)               ; EMPTY-- |
; At this point CX is > 1000
; Load 1.0 E1000 in ST(0)
        FLD     X_TO_1000           ;   S3    | EMPTY-- |
        FLD1                        ;    1    |   S3    | EMPTY  |
;************************|
; calculate INT(y/1000)  |
;************************|
; The division by 1000 is performed in 2 steps in order to avoid
; a divide overflow
        MOV     DX,CX               ; Save original exponent in DX
        MOV     AX,CX               ; Exponent to AX
        MOV     BX,100              ; First divisor
        DIV     BL                  ; Exponent / 100
        MOV     AH,0                ; Clear remainder of byte division
        MOV     BX,10               ; Second divisor
        DIV     BL                  ; First dividend / 10
        MOV     AH,0                ; Clear remainder of byte division
        MOV     CX,AX               ; Copy divisor to CX
        MOV     BX,1000             ; Base for this stage
;***************************|
; calculate  (x**F3)**I3    |
;***************************|
```

```
                ; At this point CX = y MOD 100
                MUL_STAGE_3:
                        FMUL       ST,ST(1)           ;  prod    |   S3   | EMPTY   |
                        SUB        DX,BX              ; Exponent - 1000 to calculate
                                                     ; remainder
                        LOOP       MUL_STAGE_3
                        FSTP       ST(1)              ;   S3    | EMPTY   |
                ; DX holds REM (y MOD 1000)
                        MOV        CX,DX              ; Copy remainder to CX
                ; Test for exponent exact multiple of 1000 (no remainder)
                        CMP        CX,0               ; In this case CX = 0
                        JNE        STAGE_2            ; Go if not
                ;************************|
                ;  quick exit if REM = 0  |
                ;*********************************************|
                ;    multiply exponent factors:              |
                ;       (y**F3)**I3 ----> S3 ----->   ST(3)   |
                ;*********************************************|
                ;                             |   S3   | EMPTY   |
                        JMP        EXIT_XTOYF      ; Return if exponent is exact
                                                  ; multiple of 1000
                ;
                ;****************|
                ;    STAGE 2     |
                ;*****************************|
                ;    calculate (x**F2)**I2    |
                ;*****************************|
                ; Test for EXP > 100
                ; First load 1 for the case that exp < 100
                ; At this point:               |  ST(0)  |  ST(1)  | ST(2)  |
                ;                              |   S3    | EMPTY   |
                ;                              |  or 1   |
                STAGE_2:
                ; Test for EXP > 100
                ; First load 1 for the case that exp < 100
                        FLD1                      ;   1    |   S3    | EMPTY   |
                        CMP        CX,100         ; Is exp > 100
                        JNA        STAGE_1        ; Go if not above than 100
                        FSTP       ST(0)          ; EMPTY-- |
                ; At this point CX is > 100
                ; Load x**100 factor
                        FLD        X_TO_100       ; x**100  |   S3    | EMPTY   |
                        FLD1                      ;   1     | x**100  |   S3    |
                ;************************|
                ; calculate INT(y/100)  |
                ;************************|
                        MOV        DX,CX          ; Save original exponent in DX
                        MOV        AX,CX          ; Exponent to AX
                        MOV        BX,100         ; Divisor
                        DIV        BL             ; Exponent / 100
                        MOV        AH,0           ; Clear reminder
                        MOV        CX,AX          ; Copy divisor to CX
                ;
                ;************************|
                ; calculate (y**F2)**I2  |
                ;************************|
                ; At this point CX = y MOD 100
                MUL_STAGE_2:
                        FMUL       ST,ST(1)       ;  prod   | x**100  |   S3    |
                        SUB        DX,BX          ; Exponent - 100 to calculate
```

```
                                     ; remainder
        LOOP    MUL_STAGE_2
        FSTP    ST(1)                ;    S2   |    S3   | EMPTY   |
; DX holds REM (y MOD 100)
        MOV     CX,DX                ; Copy remainder to CX
; Test for exponent exact multiple of 100 (no remainder)
        CMP     CX,0                 ; In this case CX = 0
        JNE     STAGE_1              ; Go if not
;************************|
;   quick exit if REM = 0   |
;****************************************************|
;    multiply exponent factors:                     |
;        (y**F3)**I3 ----> S3 ----->    ST(3)        |
;        (y**F2)**I2 ----> S2 ----->    ST(2)        |
;****************************************************|
;                               |   S2   |   S3   | EMPTY   |
        FMUL    ST,ST(1)
        FSTP    ST(1)                ;  x**y   | EMPTY   |
        JMP     EXIT_XTOYF           ; Exponent is exact multiple of 100
;
;****************|
;    STAGE 1        |
;****************************|
;    calculate (x**F1)**I1       |
;****************************|
; Test for EXP > 10
; First load 1 for the case that exp < 10
; At this point:               |  ST(0)  |  ST(1)  |  ST(2)  |
;                              |   S2   |   S3   | EMPTY   |
;                              |  or 1   |  or 1   |
; CX = REM (y MOD 100)
; Test for EXP > 10
; First load 1 for the case that exp < 10
STAGE_1:
        FLD1                         ;    1    |    S2   |    S3   |
        CMP     CX,10                ; Is exp > 10
        JNA     STAGE_0              ; Go if not above than 10
        FSTP    ST(0)                ;    S2   |    S3   | EMPTY   |
; At this point CX is > 10
        FLD     X_TO_10              ; x**10   |    S2   |    S3   |
        FLD1                         ;    1    |  x**10  |    S2   |..
;
;******************************|
;   calculate (x**F1)**I1          |
;******************************|
; CX = REM (y MOD 100)
        MOV     DX,CX                ; Save remainder in DX
        MOV     AX,CX                ; New exponent to AX
        MOV     BX,10                ; Divisor
        DIV     BL                   ; Exponent / 25
        MOV     AH,0                 ; Clear reminder
        MOV     CX,AX                ; Copy divisor to CX
; CX = INT(y/100)
MUL_STAGE_1:
        FMUL    ST,ST(1)             ; prod   | x**10   |   S2    |..
        SUB     DX,BX                ; Exponent - 10 to calculate
                                     ; remainder
        LOOP    MUL_STAGE_1
        FSTP    ST(1)                ;   S1   |   S2   |   S3   |
; DX holds remainder
```

```
        MOV     CX,DX              ; CX is new iteration counter
; Test for exponent exact multiple of 10
        CMP     CX,0               ; In this case CX = 0
        JNE     STAGE_0            ; Go if not
;***********************|
;  quick exit if REM = 0  |
;*******************************************************|
;    multiply exponent factors:                        |
;        (y**F3)**I3 ----> S3 ----->    ST(3)           |
;        (y**F2)**I2 ----> S2 ----->    ST(2)           |
;        (y**F1)**I1 ----> S1 ----->    ST(1)           |
;*******************************************************|
;                              |   S1    |   S2    |   S3   |..
        FMUL    ST,ST(1)
        FSTP    ST(1)          ;   PP    |   S3    | EMPTY  |
        FMUL    ST,ST(1)
        FSTP    ST(1)          ;  x**y   | EMPTY   |
        JMP     EXIT_XTOYF
;
;****************|
;     STAGE 0        |
;*******************************|
;    calculate (y**F0)**I0       |
;*******************************|

; At this point:               |  ST(0) |  ST(1) |  ST(2) |
;                              |   S1    |   S2    |   S3    |
;                              |  or 1   |  or 1   |  or 1   |
; CX = INT(y/10)
STAGE_0:
        FLD     X_VALUE        ;    x    |   S1    |   S2    |..
        FLD1
                               ;    1    |    x    |   S2    |..
MUL_STAGE_0:        FMUL    ST,ST(1)
                               ;   S0    |    x    |   S1    |..
        LOOP    MUL_STAGE_0
        FSTP    ST(1)          ;   S0    |   S1    |   S2    |..
;*******************************************************|
;    multiply exponent factors:                        |
;        (y**F3)**I3 ----> S3 ----->    ST(3)           |
;        (y**F2)**I2 ----> S2 ----->    ST(2)           |
;        (y**F1)**I1 ----> S1 ----->    ST(1)           |
;        (y**F0)**I0 ----> S0 ----->    ST(0)           |
;*******************************************************|
;                              |   S0    |   S1    |   S2    |..
        FMUL    ST,ST(1)
        FSTP    ST(1)          ;   PP    |   S2    |   S3    |
        FMUL    ST,ST(1)
        FSTP    ST(1)          ;   PP    |   S3    | EMPTY   |
        FMUL    ST,ST(1)
        FSTP    ST(1)          ;  x**y   | EMPTY   |
EXIT_XTOYF:
        CLD
        RET
;
_X_TO_Y_BYFAC   ENDP
```

Applications

Each of the methods for the calculation of exponential functions, described previously, has advantages and disadvantages:

1. Logarithmic approximations are relatively compact and efficient. Additionally, the same routine can be directly applied to the calculation of powers and roots. Their main disadvantage is the low accuracy of the result.

2. Binary powering provides high accuracy and high performance, but the method applies only to the calculation of integer powers.

3. Exponent factoring methods provide very high accuracy for the case C^y and also acceptable performance. However, implementation costs are higher than by the other methods. For the case x^y exponent factoring offers less advantages over binary powering than for C^y.

Mixed Methods

It is possible to design implementations that combine two different methods of obtaining exponential functions. For example, a routine can use binary powering or exponent factoring to evaluate the integer portion of the exponent, and logarithmic approximation to evaluate the fractional portion. This mode of implementation improves accuracy by limiting the log approximation error to the fractional component. The following procedure uses the mixed approach:

```
_X_TO_Y_MIXED     PROC   USES esi edi ebx ebp
; Mixed method of obtaining x**y
;     Integer part is calculated by binary powering
;     Fractional part is calculated by log approximation
; On entry:
;         ST(0) = exponent
;         ST(1) = base (must be integer)
; On exit:
;         ST(0) = x**y
;
; This routine calls the procedures named X_TO_Y_BYBP and
; X_TO_Y_BYLOG
;

; First calculate integer part using binary powering
                              ;   exp    |    base   |
          FSTP    EXP_ELEMENT ;   base   |   EMPTY   |
          FSTP    BASE_ELEMENT; EMPTY    |
          FLD     EXP_ELEMENT ;   exp    |    base   |
          FRNDINT             ; int(exp) |    base   |
          FLD     BASE_ELEMENT;   base   | int(exp)  |
          FXCH                ; int(exp) |    base   |
          CALL    _X_TO_Y_BYBP
                              ; x** int y|   EMPTY   |
; Now calculate fractional part using log approximation
          FLD     EXP_ELEMENT ;   exp    | x**int y|
          FLD     ST(0)       ;   exp    |    exp    | x**int y |...
          FRNDINT             ; int(exp) |    exp    | x**int y |...
          FSUBP   ST(1),ST    ; frc(exp) | x**int y|   EMPTY   |
          FLD     BASE_ELEMENT;   base   | frc(exp)| x**int y |...
          FXCH                ; frc(exp) |    base   | x**int y |...
          CALL    _X_TO_Y_BYLOG
```

```
                              ;x**(frc y)| x**int y|  EMPTY   |
        FMUL    ST,ST(1)      ;  x**y     | x**int y|  EMPTY   |
        FSTP    ST(1)         ;  x**y     | EMPTY   |
; ST(0) holds x**y
        CLD
        RET
_X_TO_Y_MIXED   ENDP
```

The procedure _X_TO_Y_MIXED is used in calculating exponentials for the financial primitives developed in Chapter 10.

8.2 Math Unit Trigonometry

In the original 8087 and 80287 mathematical coprocessors all trigonometric functions were implemented by means of two elementary instructions: FPTAN to calculate the partial tangent, and FPATAN to calculate the partial arc-tangent. With the 80387 Intel introduced major changes on the trigonometric instructions. These changes, which were discussed in detail starting in Section 7.3.4, can be summarized as follows:

1. The FPTAN (partial tangent) instruction was modified to accept an input angle (r) with absolute value in the range $0 \pounds r \pounds 2^{63}$. In the 8087 and 80287 the input angle must be in the range $0 \pounds r \pounds p/4$ radians. Note that 2^{63} is approximately 9.22 E+18, which means that FPTAN in the 80387 can be considered to have a practically unlimited range for most technical applications.

2. Three new trigonometric opcodes, FSIN, FCOS, and FSINCOS, have been added to the math unit instruction set. These instructions, which have no equivalent in the 8087 and 80287 hardware, are designed to facilitate the direct calculation of sine and cosine functions.

3. The FPATAN instruction (partial arc-tangent), used in calculating inverse trigonometric functions, has an unrestricted operand range in the 80387. In the 8087 and 80287 the absolute value in ST(0) must be smaller than the one in ST(1).

The modifications and additions introduced in the 80387 coprocessor and preserved in the math unit of the 486 and the Pentium can considerably simplify the coding and improve the performance of trigonometric routines, as shown later in this chapter. However, programs that use the new features do not operate correctly in systems equipped with the 80287 or 8087. The fact that the 80386 central processor is compatible with the 80287 and the 80387, introduces an additional complication. In the following sections we present trigonometric primitives that are compatible with all Intel 80387 or the math unit of the 486 and the Pentium. The MATH16 library, furnished in the book's CD ROM, contains trigonometric routines that use the instruction set and processing capabilities of the original Intel coprocessors.

8.2.1 Angular Conversions

Instructions for calculating trigonometric functions and arcfunctions require radian measure. While radians are a standard representation in many technical, engineering, and scientific fields, non-technical applications often express angles in decimal degrees. The procedures listed in this section perform conversions from radians to degrees and vice-versa. Both conversion formulas use the constant 180. This value is stored in memory in the extended format to ensure maximum precision.

Due to the binary architecture of the math unit, the use of decimal constants introduces the possibility of conversion errors, which can sometimes give rise to unexpected results. For example, the trigonometric functions of angles that lie exactly at the octant or quadrant boundaries take special values. In this manner, the tangent of an angle of p or 2p radians is 0 and the tangent of an angle of p/2 or 2p/3 radians is infinity. Now suppose that you were to use a conversion routine to obtain the radian measure of an angle of 270 decimal degrees and then calculate the tangent of this value. In this case small errors in the conversion of decimal degrees to radians could cause the expected value (infinity) not to be produced. If such were the case, the fault would not lie in the calculation of the tangent, but in the conversion from decimal degrees to radians.

The following low-level procedures converts degrees to radians and radians to degrees.

```
        .DATA
; Constant used in conversion
ONE_80          DT       180.0

        .CODE

;************************************
;        degrees to radians
;************************************
;
_DEG_TO_RAD      PROC
; Convert entry value in ST(0) to radian measure
;
; On entry:
;         ST(0) = value in degrees
; On exit:
;         ST(0) = value converted to radians
;
; Formula:
;         r = d * (pi/180)
;
; where r = angle in radians, d = angle in degrees
;
;                           ;   ST(0)   |   ST(1)   |   ST(2)
;                           ;     d     |   EMPTY   |   EMPTY   |
        FLD     ONE_80
                            ;    180    |     d     |
        FLDPI               ;    pi     |    180    |     d     |
        FDIV    ST,ST(1);  pi/180   |    180    |     d     |
        FSTP    ST(1)   ;  pi/180   |     d     |
        FMUL    ST,ST(1);     r     |     d     |
        FSTP    ST(1)   ;     r     |   EMPTY   |

; ST(0) now holds entry value in radians
        CLD
        RET
_DEG_TO_RAD      ENDP
```

```
;************************************
;          radians to degrees
;************************************
_RAD_TO_DEG        PROC

; Convert entry value in ST(0) to degrees
;
; On entry:
;          ST(0) = value in radians
; On exit:
;          ST(0) = value converted to degrees
;
; Formula:
;          d = r * (180/pi)
;
; where r = angle in radians, d = angle in degrees
;
;                        ;   ST(0)   |   ST(1)   |   ST(2)
;                        ;     d     |   EMPTY   |
        FLDPI           ;     r     |    pi     |   EMPTY   |
        FLD     ONE_80
                        ;    180    |    pi     |    r      |...
        FDIV    ST,ST(1);   180/pi  |    pi     |    r      |...
        FSTP    ST(1)   ;   180/pi  |    r      |   EMPTY   |
        FMUL    ST,ST(1);     d     |    r      |   EMPTY   |
        FSTP    ST(1)   ;     d     |   EMPTY   |
; ST(0) now holds entry value in degrees
        CLD
        RET
_RAD_TO_DEG        ENDP
```

8.2.2 Range-Scaling Operations

In Chapter 7 we saw that FPTAN calculates the partial tangent of an angle in ST(0), expressed in radian measure. In the 8087 and 80287 the input angle must be less than p/4 radians. This limitation of the input argument makes it necessary that a general-purpose routine to calculate functions of any angle perform some preliminary testing, scaling, and reduction operations.

Reduction to the Unit Circle

The first reduction operation is based on the periodicity identities, which state that any trigonometric function of an angle > 2p radians (360 degrees) is equal to the function of this angle minus 2B radians. An angle in the range 0 to 2p radians is often said to be in the unit circle. The arithmetic reduction to the unit circle consists of successive subtractions of the constant 2B until the result is smaller than 2B. In Section 7.3.2. we discussed how the FPREM and FPREM1 instructions are used to perform modular arithmetic. The following procedure named CIRCLE_SCALE performs argument reduction to 2p radians. The procedure is found in the Un32_6 module of the MATH32 library in the book's CD ROM.

```
        .DATA
STATUS_WW         DW      0       ; Storage for status word

        .CODE
_CIRCLE_SCALE      PROC
; Reduce angle in ST(0) (r) to a value less than 2pi radians
```

```
; by calculating   n = r (MOD 2pi)
; The code assumes that a word-size variable, named STATUS_WW
; has been previously defined in the code segment.

;                          ;    ST(0)    |    ST(1)    |    ST(2)
                           ;      r      | EMPTYEMPTY  |
     FLDPI                 ;     pi      |      r      |   EMPTY   |
     FADD    ST,ST(0);     2pi      |      r      |   EMPTY   |
     FXCH                  ;      r      |     2pi     |   EMPTY   |

UNIT_CIRC:
     FPREM            ; REM(r/2pi)  |      2pi     |   EMPTY   |
     FSTSW   STATUS_WW    ; Store status bits in memory
     FWAIT                    ; Delay until end of store
     MOV     AH,BYTE PTR STATUS_WW+1
                         ; Move status bits into AH
     AND     AH,00000100B  ; Mask off all except C2
     JNZ     UNIT_CIRC     ; Repeat if C2 not clear
;                          ;    ST(0)    |    ST(1)    |    ST(2)
     FSTP    ST(1)    ;      n      |    EMPTY    |
; ST(0) now holds an angle in the unit circle (n)
     CLD
     RET
_CIRCLE_SCALE      ENDP
```

Reduction to the First Octant

The reason why the argument of the original FPTAN instruction is limited to p/4 radians is that all trigonometric functions are periodic. If the octant location of the original angle is known, software can use the fundamental identities to determine which function is required in each particular case. Table 8.5 lists the four formulas that are applicable for calculating the tangent of an angle according to its octant position.

Table 8.5

Octant Formulas for the Tangent Function of Angle r

OCTANT	CONDITION CODES C3	C1	FORMULA
0 and 4	0	0	tan (r)
1 and 5	0	1	1/tan (p/4− (r (MOD p/4)))
2 and 6	1	0	−1/tan (r)
3 and 7	1	1	− tan (p/4 − (r (MOD p/4)))

We have seen that FPREM stores the three least-significant bits of the integer quotient in the condition code bits labeled C3, C1, and C0. These bits are significant only if the value used as a modulus is p/4, which seems to indicate that FPREM was designed with trigonometric scaling in mind. Table 7.6 shows the setting of the condition code bits after FPREM. However, Palmer and Morse (1984, 101), referring to the 8087, express doubts regarding the validity of the condition code settings after an FPREM instruction. They state:

"… it turned out to be very difficult to implement such condition-code settings correctly, and nobody has been able to prove that the 8087 condition codes do indeed return the correct octant in all cases."

A few lines later the authors state:

"So what was thought to be a useful feature early in the design of the 8087, was realized later not to be necessary and is possibly not even implemented reliably."

However, the doubts regarding the validity of the setting of the condition code bits after FPREM refer to cases in which the instruction performs more than 63 subtraction iterations. When FPREM concludes in a single scaling cycle, the condition code bits have been proven to be reliable. We can use this fact to develop a work-around to the original bug by scaling to the unit circle before attempting to scale to the first octant. In this manner the condition code setting that results from scaling to the unit circle can be ignored, while the ones that result from the second scaling operation are known to be reliable, since no more than 8 reductions are necessary in this step.

The following procedure named OCTANT_SCALE is designed to reduce an angle to the first octant while preserving the validity of the condition code bits.

```
        .DATA
STATUS_WW       DW      0       ; Storage for status word

        .CODE
_OCTANT_SCALE   PROC
; Scale an input angle (in radian measure) to the first octant
; and report the octant of the original angle
; On entry:
;       ST(0) = positive angle in the unit circle
; On exit:
;       ST(0) = input scaled to pi/4 radians (first octant)
;       ST(1) = pi/4
;       AH = condition codes C3, C1, and C0 in their original
;            bit positions
;***************************|
;   reduce to unit circle   |
;***************************|
;                       ;   ST(0)   |   ST(1)   |   ST(2)
                        ;     r     |   EMPTY   |
        FLDPI           ;    pi     |     r     |   EMPTY   |
        FADD    ST,ST(0);    2pi    |     r     |   EMPTY   |
        FXCH            ;     r     |    2pi    |   EMPTY   |
SCALE_TO_UNIT:
        FPREM           ; r = MOD(2pi)|   2pi   |   EMPTY   |
        FSTSW   STATUS_WW    ; Store status bits in memory
        FWAIT                        ; Delay until end of store
        MOV     AH,BYTE PTR STATUS_WW+1
                             ; Move status bits into AH
        AND     AH,00000100B    ; Mask off all except C2
        JNZ     SCALE_TO_UNIT   ; Repeat if C2 not clear
;                       ;    ST(0)   |   ST(1)   |   ST(2)
```

```
        FSTP    ST(1)    ;   r (scaled  |   EMPTY   |   EMPTY   |
;                        |   to 2pi)  |
; ST(0) now holds an angle in the unit circle
;***************************|
;       reduce to pi/4       |
;***************************|
        FCLEX
        FLD1                     ;    1   |    r   | EMPTY   |
        FADD    ST(0),ST         ;    2   |    r   | EMPTY   |
        FCHS                     ;   -2   |    r   | EMPTY   |
        FLDPI                    ;   pi   |   -2   |   r    |
        FSCALE                   ;  pi/4  |   -2   |   r    |
        FSTP    ST(1)            ;  pi/4  |    r   | EMPTY   |
        FXCH                     ;    r   |  pi/4  | EMPTY   |
        FPREM                    ; MOD(pi/4)|  pi/4  | EMPTY   |
        FSTSW   STATUS_WW        ; storage for status word
; At this point ST(0) contains an angle in the first octant
; The FPREM instruction will produce a valid remainder in one
; execution if the input is < 2pi radians
        MOV     AH,BYTE PTR STATUS_WW+1
                            ; Move status bits into AH
        AND     AH,01000011B   ; Mask off unused bits
;                        |    ||_____ C0
;                        |    |_____ C1
;                        |_____ C3
; FPU stack on exit
;                              |r (scaled |   pi/4   | EMPTY   |
;                              | to pi/4) |
        CLD
        RET
_OCTANT_SCALE    ENDP
```

Notice that the _OCTANT_SCALE procedure listed above leaves the ratio p/4 in ST(1). This is convenient if the scaling operation is to be followed by the calculation of a trigonometric function, since the ratio p/4 is required in several common formulas (see Table 8.5). If the procedure is to be used for other purposes, the calling code should FSTP ST(1) to discard the value p/4 from the math unit stack.

SOFTWARE ON-LINE

The function _OCTANT_SCALE is found in the Un32_6 module of the MATH32 library in the book's software on-line.

8.2.3 Trigonometric Functions

In Chapter 7 we saw that the 80387 introduced several important changes in the trigonometric instructions, which are also present in the math unit of the 486 and the Pentium. These changes consist in an expanded operand range and the introduction of three new opcodes: FSIN, FCOS, and FSINCOS. The result is that trigonometry in the 80387 and the math unit of the 486 and the Pentium is much easier to implement that in its predecessors. In addition, the resulting code image is smaller and execution time is faster than in the 8087 or 80287. However, before using 80387-specific code the programmer should make certain that the routines will not be run in the older versions of the math unit, since the resulting code is likely to fail.

In the 80387 and the math unit of the 486 and the Pentium, the calculation of the tangent function is done with the FPTAN instruction. The 80387 version of FPTAN has an expanded operand range such that the magnitude of the input angle in radians cannot exceed 263. Considering that 263 is approximately 9.22 E+18, we can see that the expanded range is indeed generous: in fact, for many practical applications it can be considered unlimited. In any case, code can scale the input to the unit circle in order to make the function valid over the entire range of the extended precision format. The procedure named _CIRCLE_SCALE listed in Section 8.2.2 is suited to this purpose.

In the sections that follow we develop primitive routines for calculating the tangent, sine, and cosine functions using code that is compatible with the 80387 and the math unit of the 486 and Pentium. Routines to calculate these functions in the 8087 and 80287 are found in the MATH16 library in the book's CD ROM.

Calculating Tangent, Sine, and Cosine

In the 80387 and the math unit of the 486 and Pentium the calculating of the tangent, sine, and cosine functions is almost trivial. The FPTAN opcode returns the tangent, FSIN the sine, and FCOS the cosine. All three instructions accept an argument in radians in the range 0 to 263. Although this range is quite extensive, it may still be a good idea to scale the angle to the first quadrant. The procedures _TANGENT, _SINE, and _COSINE, in the Un32_6 module of the MATH32 library, perform the calculations. The first one is listed below.

```
_TANGENT        PROC
; Compute tangent of an angle in radians in the 80387 or the math
; unit of the 486 and Pentium. (no restrictions on range or sign)
;
; On entry:
;        ST(0) = an angle in radian (can be positive or negative)
; On exit:
;        ST(0) = value of tangent
;* * * * * * * * * * * * * * * * * * * * * * * * * * * * * * * * * * * * * * * * * * * * * * * * * * * * * * * * *
;
        CALL    _CIRCLE_SCALE
        FPTAN                           ;   1    | tan r   |  EMPTY   |
        FSTP    ST(0)                   ; tan r  | EMPTY   |
        CLD
        RET
_TANGENT        ENDP
```

The FSINCOS opcode, introduced in the 80387, calculates both the sine and the cosine. The cosine is returned in ST(1) and the sine in ST. Applications that need both functions can save time using FSINCOS than with the FSIN and FCOS operations.

The co-functions, co-tangent, secant, and co-secant are derived from the primitives for tangent, sine, and cosine by applying the fundamental identities

$$cot(r) = \frac{1}{tan(r)}$$

$$sec(r) = \frac{1}{cos(r)}$$

$$csc(r) = \frac{1}{sin(r)}$$

The following procedure, named _COTANGENT, calculates the function by obtaining the reciprocal of the tangent function.

```
_COTANGENT          PROC
; Compute co-tangent of an angle in radians
; (no restrictions on range or sign)
; On entry:
;        ST(0) = an angle in radians (can be positive or negative)
; On exit:
;        ST(0) = value of co-tangent
;
; Processing is based on the identity
;               cot (r) = 1 / tan(r)
;
        FPTAN                       ;   1    |   tan r  | EMPTY   |
        FDIV    ST,ST(1)            ; 1/tan  |   tan r  | EMPTY   |
        FSTP    ST(1)               ;  cot   |  EMPTY   |
        CLD
        RET
;
_COTANGENT          ENDP
```

Since the inverse trigonometric functions are not periodic, scaling to the unit circle is not required in these cases. Also notice that since FPTAN leaves the scale factor 1 in ST(0), the calculation of the cotangent is a simple matter of dividing ST/ST(1). In the case of the secant and cosecant functions the value 1 have to be loaded onto ST(0), as in the procedure _COSECANT listed below.

```
_COSECANT           PROC
; Compute secant of an angle in radians
; (no restrictions on range or sign)
; On entry:
;        ST(0) = an angle in radian (can be positive or negative)
; On exit:
;        ST(0) = value of secant
;
; Processing is based on the identity
;               cosc (r) = 1 / cos(r)
        FCOS                        ; cos r  |  EMPTY   |
        FLD1                        ;   1    |  cos r   | EMPTY   |
        FDIV    ST,ST(1)            ; 1/cos  |  cos r   | EMPTY   |
        FSTP    ST(1)               ;  cosc  |  EMPTY   |
        CLD
        RET
;
_COSECANT           ENDP
```

SOFTWARE ON-LINE

The procedures _COTANGENT, _SECANT, and COSECANT are found in the
Mu32_6 module of the MATH32 library, in the book's on-line software.

Trigonometric Arcfunctions

While there are four instructions (FPTAN, FSIN, FCOS, and FSINCOS) for calculating
trigonometric functions, the single instruction FPATAN is used in obtaining the inverse functions. Since the inverse trigonometric functions are not periodic, the calculation does not require scaling.

Calculating the arc-tangent function requires considering several cases:

1. If the value of the argument is positive, then processing must consider three possibilities:

   a. If value = 1, then arctan (1) = p/4

   b. If value > 1, then ATAN (x) = p/2 – ATAN (1/x)

   c. If value < 1, then use the FPATAN instruction

2. If the value of the argument is negative, then

   arctan(–x) = – arctan(x)

3. If the value of the argument is zero, then

   arctan (0) = 0

The following procedure, named _ARC_TANGENT shows the necessary processing.

```
_ARC_TANGENT      PROC
; Find the angle (in radians whose tangent is in ST(0)
; On entry:
;          ST(0) = tangent of an angle (in radians)
;                  Must be a normal number
; On exit:
;          ST(0) = arc tangent of entry value in ST(0)
;
; Entry values and identities:
; 1. If ST(0) is positive, proceed as follows:
;    1a. If ST(0) = 1 then arctan (1) = pi/4
;    1b. If ST(0) > 1 then ATAN (x) = pi/2 - ATAN (1/x)
;    1c. If ST(0) < 1 then in range for FPATAN instruction
; 2. If ST(0) is negative then arctan(-x) = - arctan(x)
; 3. If ST(0) = 0 then arctan (0) = 0
;
; The BH register is used to store the sign of the original
; value. If BH = 0 original input was positive. If BH = 1
; original input was negative

         MOV      BX,0               ; Clear BX. Default sign is +
;                           ;     ST(0)    |     ST(1)    |     ST(2)
                            ;       r      |    EMPTY     |    EMPTY    |
         FXAM               ; Examine stack top register
```

```
        FSTSW    STATUS_WW
        FWAIT
; Relevant condition codes:
; C3      C2      C1      C0
; 0       1       0       0    if  ST(0) = positive normal
; 0       1       1       0    if  ST(0) = negative normal
; 1       0       ?       0    if  ST(0) = zero
;
        MOV      AH,BYTE PTR STATUS_WW+1
        AND      AH,01000111B    ; Mask off all unused bits
        CMP      AH,00000110B    ; Bit code for negative
        JNE      NON_NEGATIVE    ; Argument is not negative
;********************|
;   negative argument   |
;********************|
; Change sign of ST(0) and store original negative sign
        MOV      BH,1            ; Code for negative input
        FCHS                     ; Change sign of ST(0)
        JMP      CASE_1  ; Continue as if input was positive
NON_NEGATIVE:
        AND      AH,01000011B    ; Mask off C3 bit
        CMP      AH,01000000B    ; Test for zero
        JNE      CASE_1          ; Argument is not zero
;********************|
;     zero argument     |
;********************|
; The arc tangent of zero is zero
;                        ;    ST(0)    |    ST(1)    |    ST(2)
;                        ;     r       |   EMPTY     |
        FSTP     ST(0)   ; EMPTY       |
        FLDZ             ;     0.0     |   EMPTY     |
        JMP      EXIT_ATAN
;********************|
;   positive normal     |
;     argument          |
;********************|
CASE_1:
;                        ;    ST(0)    |    ST(1)    |    ST(2)
;                        ;     r       |   EMPTY     |
        FLD1             ;     1.0     |     r       |   EMPTY   |
        FCOM             ; Compare angle with 1 to determine if
                         ; r < 1, r > 1, or r = 1
        FSTSW    STATUS_WW
        FWAIT
; Condition codes after FCOM:
; C3      C2      C0
; 0       0       0    if  1 > r
; 0       0       1    if  1 < r
; 1       0       0    if  1 = r
;
        MOV      AH,BYTE PTR STATUS_WW+1
        AND      AH,01000011B    ; Mask off all unused bits
        JZ       R_LESS_1        ; Go if r < 1
        CMP      AH,00000001B    ; Test for r > 1
        JE       R_GREATER_1     ; Go if r > 1
;********************|
; case 1A            |
;          ST(0) = 1 |
;********************|
; If ST(0) = 1 the arc tangent ST(0) = pi/4
```

```
;                              ;    ST(0)    |    ST(1)    |    ST(2)
;                              ;     1.0     |     r      |   EMPTY    |
        FADD     ST(0),ST;           2.0     |     r      |   EMPTY    |
        FADD     ST(0),ST;           4.0     |     r      |   EMPTY    |
        FLDPI            ;            pi      |    4.0     |     r      |...
        FDIV     ST,ST(1);          pi/4     |    4.0     |     r      |...
        FSTP     ST(1)   ;          pi/4     |     r      |   EMPTY    |
        JMP      SIGN_RTN
;*********************|
; case 1B             |
;          ST(0) > 1  |
;*********************|
R_GREATER_1:
;                              ;    ST(0)    |    ST(1)    |    ST(2)
;                              ;     1.0     |     r      |   EMPTY    |
        FXCH             ;             r     |    1.0     |   EMPTY    |
        FPATAN           ;  arctan(r) |    1.0     |   EMPTY    |
        FLD1             ;             1     | arctan(r)  |    1.0     |
        FADD     ST,ST(0);            2     | arctan(r)  |    1.0     |
        FLDPI            ;            pi     |    2.0     | arctan(r)  |..
        FDIV     ST,ST(1);          pi/2     |    2.0     | arctan(r)  |..
        FSTP     ST(1)   ;          pi/2     | arctan(r)  |    1.0     |
        FSUBRP   ST(1),ST;         angle     |    1.0     |   EMPTY    |
        JMP      SIGN_RTN
;*********************|
; case 1C             |
;          ST(0) < 1  |
;*********************|
R_LESS_1:
;                              ;    ST(0)    |    ST(1)    |    ST(2)
;                              ;     1.0     |     r      |   EMPTY    |
        FPATAN           ;         angle     |     r      |   EMPTY    |
;*********************|
;     sign routine    |
;*********************|
; Change sign of angle if original tangent was negative
SIGN_RTN:
        CMP      BH,1               ; BH = 1 if negative input
        JNE      EXIT_ATAN          ; Go if not negative
        FCHS
EXIT_ATAN:
        CLD
        RET
_ARC_TANGENT     ENDP
```

Calculating the arc-sine and the arc-cosine is based on the following identities:

$$\arcsin(r) = \text{arc-tangent} \frac{r}{1 - r^2}$$

$$\arccos(r) = \text{arc-tangent} \sqrt{\frac{r^2}{r}}$$

The following procedures calculate the arc-sine and arccosine functions.

```
_ARC_SINE        PROC
; Find the angle (in radians) whose sine is in ST(0)
```

```
; On entry:
;           ST(0) = sine of an angle (in radians)
;                   Must be a normal number in the range 0 < ST < 1
;                   this corresponds to an angle in the first
;                   quadrant
; On exit:
;           ST(0) = arc sine of entry value in ST(0)
;                   expressed in radians
; Formula:
;           arcsin (r) = arc-tangent (r / SQR(1 - r^2))
;
; Legend: x2 = x**2
;         y2 = y**2
;         SQR = square root
;                               ;     ST(0)    |    ST(1)    |     ST(2)
;                               ;      r       |   EMPTY     |
        FLD     ST(0)   ;       r       |    r        |   EMPTY   |
        FMUL    ST,ST(1);      r^2      |    r        |   EMPTY   |
        FLD1            ;       1       |   r^2       |    r      |
        FSUB    ST,ST(1);     1-r^2     |   r^2       |    r      |
        FSQRT           ; SQR(1-r^2)    |   r^2       |    r      |
        FSTP    ST(1)   ; SQR(1-r^2)    |    r        |  EMPTY    |
; FPATAN calculates arc-tangent ST(1)/ST
        FPATAN          ; arcsin (r)    |  EMPTY      |
        CLD
        RET
_ARC_SINE       ENDP


_ARC_COSINE     PROC
; Find the angle (in radians) whose cosine is in ST(0)
; On entry:
;           ST(0) = cosine of an angle (in radians)
;                   Must be a normal number in the range 0 < ST < 1
;                   this corresponds to an angle in the first
;                   quadrant (0 < angle < pi/2)
; On exit:
;           ST(0) = arc cosine of entry value in ST(0)
; Formula:
;           arccos (r) = arc-tangent (SQR (1 - r^2) / r)
;
; Legend: x2 = x**2
;         y2 = y**2
;         SQR = square root
;                               ;     ST(0)    |    ST(1)    |     ST(2)
;                               ;      r       |   EMPTY     |
        FLD     ST(0)   ;       r       |    r        |   EMPTY   |
        FMUL    ST,ST(1);      r^2      |    r        |   EMPTY   |
        FLD1            ;       1       |   r^2       |    r      |
        FSUB    ST,ST(1);     1-r^2     |   r^2       |    r      |
        FSQRT           ; SQR(1-r^2)    |   r^2       |    r      |
        FSTP    ST(1)   ; SQR(1-r^2)    |    r        |  EMPTY    |
        FXCH            ;      r        | SQR(1-r^2)  |  EMPTY    |
; FPATAN calculates arc-tangent ST(1)/ST
        FPATAN          ; arcsin (r)    |  EMPTY      |
        CLD
        RET
_ARC_COSINE     ENDP
```

The hyperbolic functions and arc functions are discussed in Chapter 9.

8.4 Logarithms

The calculations of logarithmic functions on the Intel math unit is much simpler than the calculation of trigonometric or exponentials. In Chapter 7 we saw that the math units contain two logarithmic instructions (FYL2X and FYL2XP1) and allows the loading of four logarithmic constants. The instructions related to logarithms operate as follows:

1. FYL2X calculates the base 2 logarithm of the positive operand in ST(0) and multiplies this log by the value in ST(1). The result appears in ST(0) and the stack is popped. This instruction provides a simple means for calculating logarithms to other bases, since

$$\log_b(x) = \log_b 2 * \log_2 x.$$

For example, if x is in ST(0) and log102 is in ST(1), FYL2X performs the following calculation:

$$ST(0) = \log_2 x * \log_{10} 2 = \log_{10} x$$

2. FYL2XP1 adds 1 to the value in ST(0), calculates the base 2 logarithm of this number, and multiplies this result by the value in ST(1). This type of operation is useful in calculating logarithms of values very close to 1, as is required in some financial formulas, since it improves the accuracy of the results.

3. Four load instruction allow entering logarithmic constants into the Stack Top register, these are:

FLDL2T - Load $\log_2 10$ to ST(0)

FLDL2E - Load $\log_2 e$ to ST(0)

FLDLG2 - Load $\log 10_2$ to ST(0)

FLDLN2 - Load $\log_e 2$ to ST(0)

8.4.1 Calculating Natural and Common Logarithms

Routines for calculating logarithms to bases other than 2 make use of the simple relations listed in Section 8.4. The algorithms are quite straightforward. The actual computations are shown in the procedures named _LOG_10 and _LOG_E, listed below, and located in the Un32_5 module of the MATH32 library in the book's CD ROM.

```
_LOG_E              PROC
; Calculate the natural log (base e)
; On entry:
;         ST(0) = positive number
; On exit:
;         ST(0) = ln ST(0) (Natural log)
;
;                             |    ST(0)   |    ST(1)   |    ST(2)
;                          ;       x      |   EMPTY    |
          FLDLN2          ;    LOGe(2)    |      x     |   EMPTY    |
          FXCH            ;       x       |   LOGe(2)  |   EMPTY    |
          FYL2X           ;    LOGe(2)*   |
                          ;    LOG2(x)    |   EMPTY    |
          CLD
```

```
        RET
_LOG_E              ENDP
;
_LOG_10             PROC
; Calculate the common log (Base 10)
; On entry:
;         ST(0) = positive number
; On exit:
;         ST(0) = LOG ST(0) (Common log)
;
;                         |   ST(0)    |   ST(1)    |   ST(2)
;                         ;     x      |   EMPTY    |
        FLDLG2           ;  LOG10(2)   |     x      |  EMPTY    |
        FXCH             ;     x       |  LOG10(2)  |  EMPTY    |
        FYL2X            ;  LOG10(2)*  |
                         ;  LOG2(x)    |   EMPTY    |

        CLD
        RET
;
_LOG_10             ENDP
```

8.4.2 Calculating Antilogarithms

By definition, the antilogarithm is the inverse function of the logarithm. Antilogarithms to any base can be generalized in the following equation:

$$alog_b(x) = b^x$$

The following equations represent the natural antilog (base e) and the common antilog (base 10):

$$alog_{10}(x) = 10^x$$

$$alog_e(x) = e^x$$

The following procedures, named _ALOG_10 and _ALOG_E, calculate these antilogarithms. They share a common routine, named _ALOG_2 which is used to raise the logarithm base 2 of a number to an arbitrary power.

```
_ALOG_10            PROC
; Calculate the common antilogarithm of the number in ST(0)
; On entry:
;         ST(0) holds a positive logarithm base 10
; On exit:
;         ST(0) holds the antilogarithm of the number
;
;                         |   ST(0)    |   ST(1)    |   ST(2)
;         FLDL2T          ;  log2(10)  |     x      |  EMPTY    |
        CALL    _ALOG_2
                          ;  alog10(x) |   EMPTY    |

        CLD
        RET
_ALOG_10            ENDP


;
_ALOG_E             PROC
; Calculate the common antilogarithm of the number in ST(0)
; On entry:
;         ST(0) holds a positive logarithm base 10
; On exit:
```

```
;              ST(0) holds the antilogarithm of the number
;
;                           |    ST(0)    |    ST(1)    |    ST(2)
          FLDL2E        ;     log2(10)  |      x      |   EMPTY    |
          CALL    _ALOG_2
                        ;   alog2(x)    |   EMPTY    |
          CLD
          RET

_ALOG_E              ENDP


_ALOG_2              PROC
; Core routine to calculate antilog base e and base 10
; On entry:
;         ST(0) = logarithm base 2 of a number (N)
;         ST(1) = positive power to which N must be raised (x)
; On exit:
;         ST(0) = N**x
;
;                           |    ST(0)    |    ST(1)    |    ST(2)
;                           |   log2(N)   |      x      |   EMPTY    |
          FMULP   ST(1),ST; LOG2(N)*x=z  |   EMPTY    |
; Set control word to round down
          FSTCW   CONTROL_WW   ; Store caller's control word
          FLDCW   ROUND_DOWN   ; Temporary control word for
                               ; rounding down
          FLD1          ;     1       |      z      |   EMPTY    |
          FCHS          ;    -1       |      z      |   EMPTY    |
          FLD     ST(1) ;     z       |     -1      |     z      |
          FRNDINT       ; i = int(z)  |     -1      |     z      |
; The value i is the closest integer smaller than z
          FLDCW   CONTROL_WW   ; Restore caller's control word
          FXCH    ST(2) ;     z       |     -1      |     i      |
          FSUB    ST,ST(2);    f      |     -1      |     i      |
          FSCALE        ;    f/2      |     -1      |     i      |
; FSCALE using the value -1 decrements the binary exponent of
; ST(0), which is the same as dividing by 2
          F2XM1         ;2**(f/2) -1  |     -1      |     i      |
          FSTP    ST(1) ;2**(f/2)- 1  |      i      |   EMPTY    |
          FLD1          ;     1       |2**(f/2) - 1|      i      |
          FADDP   ST(1),ST;  2**(f/2) |      i      |   EMPTY    |
          FMUL    ST,ST(0);   2**f    |      i      |   EMPTY    |
          FSCALE        ; 2**f * 2**i |      i      |   EMPTY    |
          FSTP    ST(1) ;    N**x     |   EMPTY    |
          CLD
          RET
_ALOG_2              ENDP
```

8.5 C++ Interface to Transcendentals

The header file Un32_5.h contains the C++ interface functions to the power and logarithmic primitives developed in this chapter. The prototypes are as follows:

```
unsigned long TenToYByFac(int);   // 10^y by exponent factoring
double XtoYByFac(double, int);    // x^y by exponent factoring
double XtoYByBP(double, double);  // x^y by binary powers
double XtoYByLog(double, double); // x^y by logarithms
double XtoYMixed(double, double); // x^y by logs and binary power
double LogE(double);              // Natural logarithm
```

```
double Log10(double x);        // Common logarithm
double AlogE(double x);        // Natural antilogarithm
double Alog10(double x);       // Common antilogarithm
double Alog2(double x);        // Base 2 logarithm
```

The header file Un32_6.h contains the C++ interface functions to the trigonometric primitives. The prototypes are as follows:

```
double DegToRad(double);       // Degrees to radians
double RadToDeg(double);       // Radians to degrees
double Tangent(double);        // Tangent
double Sine(double);           // Sine
double Cosine(double);         // Cosine
double CoTangent(double);      // Cotangent
double Secant(double);         // Secant
double CoSecant(double);       // Cosecant
double ArcSine(double);        // Arc-sine
double ArcCosine(double);      // Arc-cosine
double ArcTangent(double);     // Arc-tangent
```

SOFTWARE ON-LINE

The projects Chapter08\Math Primitives and Chapter08\Trig Primitives contain tests for the primitives, as well as for the corresponding C++ the interface functions. The files are found in the book's on-line software.

Chapter 9

General Mathematical Functions

Chapter Summary

In this chapter we present an assortment of numerical routines and primitives often required in floating-point packages, numerical applications, graphics, and high-level languages. These include the so-called calculator operations (such as factorials and hyperbolic functions), solution of quadratic equations, operations on complex numbers, and conversions between polar and Cartesian coordinates.

9.0 Calculator Operations

In addition to fundamental arithmetic and to the transcendental primitives developed in Chapter 8, numerical applications often require other scalar operations such as those found in more-or-less powerful electronic calculators. Some of these are executed by the math unit by means of a single instruction. For example, the square root is calculated by the FSQRT instruction. Others require the transcendental primitives developed in Chapter 8, binary/ASCII conversions, or other support routines from preceding chapters.

In the presentation of calculator-like operations we have excluded those that can be implemented with one or two math unit opcodes, such as addition, subtraction, multiplication, division, changing the sign, absolute value, rounding to integer, and calculating square root. In these cases the programming is so simple and straightforward that it is not necessary to develop a dedicated procedure.

The routines in this chapter are core primitives and they are intended as examples. Many of the routines perform no checking for valid arguments or for errors in the result. Error-handling is an application-dependent topic. The level and refinement of error detection code depends on the application requirements. In coding the primitives we assumed that the implementor will provide error handlers in separate code, or will incorporate them into the procedures if and when they are needed. In all routines code assumes that the FPU rounding and precision control are in the default setting. Notice that this assumption may not be acceptable in some applications.

9.0.1 Calculating Hyperbolic Functions

Many scientific and engineering problems require the evaluation of hyperbolic functions. These functions, called the hyperbolic tangent (tanh), hyperbolic sine (sinh), and hyperbolic cosine (cosh), have similar properties as the circular trigonometric functions, except that the hyperbolic functions are related to the hyperbola rather than to the circle. However, the hyperbolic functions cannot be readily derived from their trigonometric counterparts. Instead they are expressed by the following formulas

$$\sinh(x) = \frac{e^x - e^{-x}}{2}$$

$$\cosh(x) = \frac{e^x + e^{-x}}{2}$$

$$\tanh(x) = \frac{e^x - e^{-x}}{e^x + e^{-x}}$$

or also

$$\tanh(x) = \frac{\sinh(x)}{\cosh(x)}$$

Evaluating the hyperbolic formulas requires the calculation of e^x and e^{-x}. However, the value of e is not one of the constants that can be directly loaded into the math unit. One possible solution is to load the logarithm base 2 of e and then call a routine that calculates the base 2 antilogarithm, such as the procedure ALOG_2 listed in Chapter 8. Perhaps a more direct approach is to define e as a memory constant, possibly to extended precision. The following procedure, named E_TO_Y calculates an exponential of the constant e, defined as a memory constant.

```
                    .DATA
; Definition of the constant e
E                   DT      4000ADF85458A2BB4A9BH

                    .CODE
_E_TO_Y             PROC
; Procedure for obtaining e^y in support of hyperbolic function
; calculations
;
; On entry:
;           ST(0) = y
; On exit:
;           ST(0) = e^y
;
            FLD     E                     ;   e   |   y   |  EMPTY  |
            FXCH                          ;   y   |   e   |  EMPTY  |
            CALL    _X_TO_Y_BYLOG         ;  e^y  |  EMPTY  |
            RET
_E_TO_Y             ENDP
```

Notice that the procedure _E_TO_Y uses the exponential primitive
_X_TO_Y_BYLOG developed in Chapter 8. The logarithmic calculation of the exponential is necessary in order to allow signed and fractional exponents. Equipped with a routine that calculates e^x and e^{-x}, the calculation of the hyperbolic functions is a matter of applying the formulas listed previously. The following procedure calculates the hyperbolic sine.

```
_SINH    PROC
; Calculate hyperbolic sine using formula
;        sinh(x) = (e^x - e^-x)/2
;
; On entry:
;        ST(0) = x
; On exit:
;        ST(0) = sinh(x)
;
;                               ;    x    |   EMPTY   |
; First obtain e^-x
         FLD     ST(0)          ;    x    |    x    |  EMPTY  |
         FCHS                   ;   -x    |    x    |  EMPTY  |
         CALL    E_TO_Y         ;   e^-x  |    x    |  EMPTY  |
         FXCH                   ;    x    |   e^-x  |  EMPTY  |
         CALL    E_TO_Y         ;   e^x   |   e^-x  |  EMPTY  |
         FSUB    ST,ST(1)       ; (e^x -
                                ;   e^-x) |   e^-x  |  EMPTY  |
         FSTP    ST(1)          ;(e^x ..) | EMPTY - |  EMPTY  |
         FLD1                   ;    1    |(e^x ..) |  EMPTY  |
         FADD    ST,ST(0)       ;    2    |(e^x ..) |  EMPTY  |
         FDIV    ST(1),ST       ;    2    | sinh (x)|  EMPTY  |
         FSTP    ST(0)          ; sinh(x) |  EMPTY  |
         CLD
         RET
_SINH    ENDP
```

SOFTWARE ON-LINE

Low-level procedures for calculating the remaining hyperbolic arc functions are in the Un32_6 module of the MATH32 library. The program Math Primitives CHAPTER08\Math Primitives contains the C++ interface functions to the hyperbolic primtives. The files are found in the book's on-line software.

The C++ interface functions to the hyperbolic procedures in the Un32_6 module of the MATH32 library are as follows:

```
double Sinh(double angle)
// Compute hyperbolic sine of an angle in radians using the
// 80387 or the math unit of the 486 and Pentium.
// (no restrictions on range or sign)
//
// On entry:
//       parameter angle is an angle in radians
//       (can be positive or negative)
// On exit:
//       returns hyperbolic sine of angle
```

```
{
double result;
_asm
{
    FLD      angle
    CALL     SINH
    FSTP     result
}
return result;
}

double Cosh(double angle)
// Compute hyperbolic cosine of an angle in radians using the
// 80387 or the math unit of the 486 and Pentium.
// (no restrictions on range or sign)
//
// On entry:
//       parameter angle is an angle in radians
//       (can be positive or negative)
// On exit:
//       returns hyperbolic cosine of angle
{
double result;
_asm
{
    FLD      angle
    CALL     COSH
    FSTP     result
}
return result;
}

double Tanh(double angle)
// Compute hyperbolic tangent of an angle in radians using the
// 80387 or the math unit of the 486 and Pentium.
// (no restrictions on range or sign)
//
// On entry:
//       parameter angle is an angle in radians
//       (can be positive or negative)
// On exit:
//       returns hyperbolic tangent of angle
{
double result;
_asm
{
    FLD      angle
    CALL     TANH
    FSTP     result
}
return result;
}

double ArcSinh(double trigFun)
// Computes the angle in radians corresponding to a hyperbolic
// sine function, using the 80387 or the math unit of the 486
// and Pentium.
// (no restrictions on range or sign)
//
// On entry:
```

```
//        parameter trigFun is a hyperbolic sine function
// On exit:
//        returns the angle in radians
{
double result;
_asm
{
    FLD        trigFun
    CALL       ARC_SINH
    FSTP       result
}
return result;
}


double ArcCosh(double trigFun)
// Computes the angle in radians corresponding to a hyperbolic
// cosine function, using the 80387 or the math unit of the 486
// and Pentium.
// (no restrictions on range or sign)
//
// On entry:
//        parameter trigFun is a hyperbolic cosine function
// On exit:
//        returns the angle in radians
//
{
double result;
_asm
{
    FLD        trigFun
    CALL       ARC_COSH
    FSTP       result
}
return result;
}


double ArcTanh(double trigFun)
// Computes the angle in radians corresponding to a hyperbolic
// tangent function, using the 80387 or the math unit of the 486
//  and Pentium.
// (no restrictions on range or sign)
//
// On entry:
//        parameter trigFun is a hyperbolic sine function
// On exit:
//        returns the angle in radians
//
{
double result;
_asm
{
    FLD        trigFun
    CALL       ARC_TANH
    FSTP       result
}
return result;
}
```

SOFTWARE ON-LINE

The interface functions are contained in the file Calc Primitives.h which is located in the folder Sample Code\Chapter09\Calc Primitives, in the book's on-line software.

9.0.2 Factorial

The factorial of a number is the product of all positive integers less than or equal to the number, for example:

$$5! = 5 \cdot 4 \cdot 3 \cdot 2 \cdot 1 = 120$$

Computer algorithms for the calculation of factorials are often recursive. The recursive definition of the factorial function is as follows:

$$0! = 1$$

$$n! = n \cdot (n-1)! \quad - \text{for } n > 0$$

In C++ code a recursive factorial function can be coded as follows:

```
int Factorial(int n)
{
   if(n == 0)
      return 1;
   else
      return n * Factorial(n - 1);
}
```

The code assumes that the input value is a positive integer.

In math unit programming, factorials can be calculated by using a processor register for counting the multiplication operations. Since the highest factorial that can be calculated in extended precision format is 1754!, it is safe to use a word-size register to count the number of iterations. The following procedure is a non-recursive factorial calculation using the math unit.

```
            .486
            .MODEL flat
            .DATA
FAC_ARGUMENT    DW      0         ; Storage for factorial argument

            .CODE
_FACTORIAL      PROC
; Calculation of the factorial function
; On entry:
;           ST(0) = x (x is an integer)
; On exit:
;           ST(0) = x!
;
            FRNDINT                   ; Make sure argument is integer
            FABS                      ; and positive
            FIST    FAC_ARGUMENT      ; Store argument in memory
            MOV     CX,FAC_ARGUMENT   ; Load into counter register
            CMP     CX,0              ; Test for 0!
            JE      FACTORIAL_ZERO    ; Special case is 0!
```

```
        DEC     CX              ; Number of factors is 1 less
        JNZ     FAC_OK
; At this point operation is 0! or 1!
FACTORIAL_ZERO:
        FSTP    ST(0)           ; Clear stack
        FLD1                    ;    1    |  EMPTY -- |
        JMP     FAC_EXIT
;                               ;    x    |  EMPTY -- |
FAC_OK:
        FLD1                    ;    1    |    x    |  EMPTY - |
        FXCH                    ;    x    |    1    |  EMPTY - |
        FLD     ST(0)           ;    x    |    x    |    1     |
DO_FACTORS:
        FSUB    ST,ST(2)        ;  x - 1  |  pfact  |    1     |
        FXCH                    ;  pfact  |  x - 1  |    1     |
        FMUL    ST,ST(1)        ;  pfact  |  x - 1  |    1     |
        FXCH                    ;  x - 1  |  pfact  |    1     |
        LOOP    DO_FACTORS      ; Repeat
        FSTP    ST(0)           ;   x!    |    1    |  EMPTY - |
        FSTP    ST(1)           ;   x!    |  EMPTY -- |
FAC_EXIT:
        CLD
        RET
_FACTORIAL
```

C++ access to the _FACTORIAL procedure is as follows:

```
long Factorial(int value)
// Calculate factorial of argument
// On entry:
//        parameter value holds x
// On exit:
//        returns x!
{
long fac;
_asm
{
    FILD    value
    CALL    FACTORIAL
    FISTP   fac
}
return fac;
}
```

SOFTWARE ON-LINE

The _FACTORIAL procedure is found in the Un32_7 module of the MATH32 library. The C++ interface function named Factorial() is located in the file Calc Primitives.h which is found in the folder Sample Code\Chapter09\Calc Primitives, all in the book's on-line software.

9.0.3 Ordering Numeric Data

Software often needs to determine which of two data items is the larger one. In one version of this operation the data items are ordered in the math unit stack. In another version the larger data item is returned to the caller and the smaller one is discarded. The Un32_7 module of the MATH32 library contains a low-level procedure named ORDER that organizes two items in the math unit stack so that the larger one is in ST(0) and the smaller one in ST(1). Another procedure in this module, named MAX, selects the larger of two elements and discards the smaller one. Finally, a procedure named MIN preserves the smaller element and discards the larger one.

```
_ORDER    PROC
; Find larger of two values, x and y
;
; On entry:
;          ST(0) = x
;          ST(1) = y
; On exit:
;          ST(0) = the larger of x and y
;          ST(1) = the smaller of x and y
;
          FCOM                      ;    x    |    y    |  EMPTY  |
          FSTSW    STATUS_WW        ; Store status word
          FWAIT
          MOV      AH,BYTE PTR STATUS_WW + 1
          SAHF                      ; AH to 80x86 flags register
          JA       STACK_OK         ; Go if x > y
; At this point ST(0) =< ST(1)
          FXCH                      ;    y    |    x    |  EMPTY  |
STACK_OK:
          CLD
          RET
_ORDER    ENDP


_MAX      PROC
; Find larger of two values, x and y
;
; On entry:
;          ST(0) = x
;          ST(1) = y
; On exit:
;          ST(0) = the larger of x and y
;          ST(1) = empty
;
          FCOM                      ;    x    |    y    |  EMPTY  |
          FSTSW    STATUS_WW        ; Store status word
          FWAIT
          MOV      AH,BYTE PTR STATUS_WW + 1
          SAHF                      ; AH to 80x86 flags register
          JA       MAX_IS_X         ; Go if x > y
; At this point ST(0) =< ST(1)
          FXCH                      ;    y    |    x    |  EMPTY  |
MAX_IS_X:
          FSTP     ST(1)            ; Scrap ST(1)
          CLD
          RET
_MAX      ENDP


;
_MIN      PROC
; Find smaller of two values, x and y
;
```

```
; On entry:
;           ST(0) = x
;           ST(1) = y
; On exit:
;           ST(0) = the smaller of x and y
;           ST(1) = empty
;
        FCOM                            ;   x    |    y    |  EMPTY  |
        FSTSW     STATUS_WW             ; Store status word
        FWAIT
        MOV       AH,BYTE PTR STATUS_WW + 1
        SAHF                            ; AH to 80x86 flags register
        JB        MIN_IS_X              ; Go if x > y
; At this point ST(0) =< ST(1)
        FXCH                            ;   y    |    x    |  EMPTY  |
MIN_IS_X:
        FSTP      ST(1)                 ; Scrap ST(1)
        CLD
        RET
_MIN    ENDP
```

The C++ interface functions to the low-level procedures listed previously are as follows:

```
void Order(double x, double y)
// Place the larger value in ST(0) and the smaller one in
// ST(1)
// On entry:
//        parameters x and y contain values
// On exit:
//        ST(0) holds larger value
//        ST(1) holds smaller value
{
_asm
{
    FLD       x
    FLD       y
     CALL     ORDER
}
return;
}

void InitMU(void)
{
// Re-initialize and clear the math unit
_asm
{
    FINIT
}
}

double Max(double x, double y)
// Return the larger value
// On entry:
//        parameters x and y contain values
// On exit:
//        returns larger
{
double value;
```

```
_asm
{
    FLD      x
    FLD      y
     CALL    MAX
    FSTP     value
}
return value;
}

double Min(double x, double y)
// Return the smaller value
// On entry:
//          parameters x and y contain values
// On exit:
//          returns smaller
{
double value;
_asm
{
    FLD      x
    FLD      y
     CALL    MIN
    FSTP     value
}
return value;
}
```

SOFTWARE ON-LINE

The functions named Order(), Max(), Min() are contained in the file Calc Prim-itives..h which is located in the folder Sample Code\Chapter09\Calc Primitives in the book's on-line software.

9.0.4 Calculating the Modulus

The fundamental operation of finite algebra, called the modulus, consists of finding the remainder that results from the division of two arguments. Some confusion in terminology exists because the divisor argument has sometimes also been called the modulus. To avoid this predicament in the following discussion we use the word modulus for the result of the operation. Therefore if

$$r = \text{remainder of } x/y$$

then also

$$r = x \ (\text{MOD } y).$$

The following low-level procedure calculates the modulus function of two arguments.

```
      .CODE
_MODULUS           PROC
; Find x (MOD y), which is the remainder of x/y
;
; On entry:
;          ST(0) = x (argument)
```

```
;               ST(1) = y (modulus)
;
; On exit:
;               ST(0) = x (MOD y)

;                                       ;  ST(0)    |   ST(1)    |   ST(2)  |
;                                       ;   x       |    y       |  EMPTY   |
REPEAT_MOD:
        FPREM                           ; REM(x/y)  |    y       |  EMPTY   |
        FSTSW   STATUS_WW               ; Store status bits in memory
        FWAIT                           ; Delay until end of store
        MOV     AH,BYTE PTR STATUS_WW+1
                        ; Move status bits into AH
        AND     AH,00000100B            ; Mask off all except C2
        JNZ     REPEAT_MOD              ; Repeat if C2 not clear
        FSTP    ST(1)                   ; x (MOD y) | EMPTY   |
        CLD
        RET
_MODULUS        ENDP
```

The C++ interface function to the _MODULUS procedure is as follows:

```
int Modulus(int x, int y)
// Return the value x(MOD y)
// On entry:
//        parameters x and y contain values
// On exit:
//        returns x modulo y
{
int modulo;
_asm
{
    FILD    y
    FILD    x
    CALL    MODULUS
    FISTP   modulo
}
return modulo;
}
```

The Modulus() function is found in the Un_32.h file which is located in the folder CHAPTER9\TEST UN32_7 in the book's CD ROM.

9.0.5 Integer and Fractional Parts

Rational and irrational numbers must often be separated into their integer and fractional parts. Apparently, this operation could be easily accomplished using the FRNDINT math unit opcode to obtain the integer part, which can then be subtracted from the original number to produce the fractional part. However, FRNDINT rounds to an integer according to the current setting of the rounding control field for the control word. Therefore, if the rounding control is in the default mode (round to nearest even) the opcode can fail to correctly produce the integer part. For example, if ST(0) holds the number 125.66 and rounding control is to nearest even, then FRNDINT produces 126, which is not the integer part of 125.66.

The problem is easily solved by forcing the math unit to round down or to chop. In this case it is also necessary to save to caller's control word so that it can be cor-

rectly restored. The following procedure resolves the integer and fractional parts
of a number in ST(0).

```
              .DATA
ROUND_DOWN   DW        177FH    ; Control word for round down
                                ; operations
CONTROL_WW   DW        0        ; Storage for caller's control word

              .CODE
_INT_FRAC    PROC
; Separate integer and fractional parts a number in ST(0)
;
; On entry:
;          ST(0) = n
;
; On exit:
;          ST(0) = integer part of n
;          ST(1) = fractional part of n
;*********************|
;   set to round down |
;*********************|
        FSTCW   CONTROL_WW   ; Store control word in memory
        FLDCW   ROUND_DOWN   ; Install new control word
; At this point the math unit is set to round down
;                                   |  ST(0)  |   ST(1)  |   ST(2)
                                    ;    n    |  EMPTY - |
        FLD     ST(0)               ;    n    |    n     | EMPTY  |
        FRNDINT                     ;    i    |    n     | EMPTY  |
        FLDCW   CONTROL_WW   ; Restore original control word
        FSUB    ST(1),ST            ;    i         f     | EMPTY  |
        CLD
        RET
_INT_FRAC          ENDP
```

Since _INT_FRAC produces both the integer and the fractional parts of the operand, high-level interface routines can be developed that use _INT_FRAC to return either of these elements. The C++ functions IntPart() and FracPart() perform
these operations. The code is as follows:

```
double FracPart(double num)
// Returns the fractional part of a float
// On entry:
//       parameter num holds operand
// On exit:    returns fractional part of num
//
{
double frac;
_asm
{
    FLD     num
    CALL    INT_FRAC
    FXCH
    FSTP    frac
}
return frac;
}

int IntPart(double num)
```

```
// Returns the integer part of a float
// On entry:
//          parameter num holds operand
// On exit:
//          returns integer part of num
//
{
int part;
_asm
{
     FLD     num
     CALL    INT_FRAC
     FISTP   part
}
return part;
}
```

SOFTWARE ON-LINE

The functions IntPart() and FracPart() are contained in the file Calc Primitives.h which is located in the folder Sample Code\Chapter09\Calc Primitives in the book's on-line software.

9.0.6 Solving Triangles

One of the theorems of Euclidean geometry, usually attributed to the Greek mathematician Pythagoras, states the relationship between the sides of a right triangle, as shown in Figure 9.1.

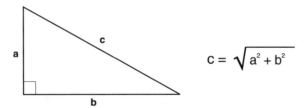

$$c = \sqrt{a^2 + b^2}$$

Figure 9.1 *Pythagoras' Theorem*

The following procedure allows solving for the hypotenuse of a right triangle when the dimensions of the two sides are known.

```
_PYTHAGORAS       PROC
; Solve Pythagoras' equation: c = SQRT(x^2 + y^2)
;
; On entry:
;          ST(0) = x
;          ST(1) = y
;
; On exit:
;          ST(0) = SQRT (x^2 + y^2)
;
;                              ;  ST(0)   |   ST(1)   |   ST(2)  |
;                              ;    x     |     y     |   EMPTY  |
```

```
        FMUL    ST(0),ST        ;    x^2      |      y      |   EMPTY   |
        FXCH                    ;     y       |     x^2     |   EMPTY   |
        FMUL    ST(0),ST        ;    y^2      |     x^2     |   EMPTY   |
        FADDP   ST(1),ST        ; y^2 + y^2 | EMPTY   |
        FSQRT                   ;     c       |  EMPTY   |
        CLD
        RET
_PYTHAGORAS     ENDP
```

The C++ interface function is as follows:

```
double Pythagoras(double a, double b)
// Return the square root of the sum of 2 squares
// On entry:
//          parameters a and b input values
// On exit:
//          returns square root of a^2 + b^2
{
double c;
_asm
{
    FLD     a
    FLD     b
    CALL    PYTHAGORAS
    FSTP    c
}
return c;
}
```

The solution of a right triangle is often required in terms of one side and the adjacent angle. The tangent function can be applied to this case, as shown in Figure 9.2.

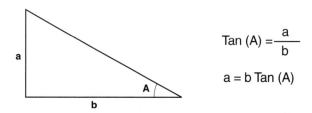

$$\text{Tan}(A) = \frac{a}{b}$$

$$a = b\,\text{Tan}(A)$$

Figure 9.2 *Side-Angle Formula for Right Triangle*

The following procedure finds the opposite side of a right triangle in terms of a side and the adjacent angle. The processing assumes that the angle is entered in degrees. The conversion from degrees to radians and the calculation of the tangent function is performed by calling the procedures developed in Chapter 8.

```
_SOLVE_FOR_A        PROC
; Solve a right triangle in terms of one side and the adjacent
; angle (A) using the formula
;
;           a = b * Tan A
;
; On entry:
```

```
;            ST(0) = angle A (in degrees)
;            ST(1) = side b
;
; On exit:
;            ST(0) = side a
;
;                                      ;  ST(0)  |  ST(1)  |   ST(2)  |
;                                      ;   A     |    b    |  EMPTY   |
; Convert degrees to radians
        CALL     _DEG_TO_RAD   ;  A (rads)|    b    |  EMPTY   |
        CALL     _TANGENT      ;  Tan A   |    b    |  EMPTY   |
        FMULP    ST(1),ST      ;  b * Tan A|  EMPTY  |
        CLD
        RET
_SOLVE_FOR_A    ENDP
```

The C++ interface function is coded as follows:

```
double SolveForA(double angle, double sideb)
// Solve a right triangle in terms of one side and the adjacent
// angle using the formula
//                      a = b * Tan A
// On entry:
//       parameter angle and sideb hold triangle data
// On exit:
//       returns side a

{
double sidea;
_asm
{
    FLD      sideb
    FLD      angle
    CALL     SOLVE_FOR_A
    FSTP     sidea
}
return sidea;
}
```

SOFTWARE ON-LINE

The functions Pythagoras() and SolveForA() are contained in the file Calc Primitives.h which is located in the folder Sample Code\Chapter09\Calc Primitives in the book's on-line software.

9.1 Quadratic Equations

The general quadratic equation is expressed by the formula:

$$ax^2 + bx + c = 0$$

where the coefficients a, b, and c are constants. The solution a quadratic equation can be attempted by applying the standard quadratic formula:

$$x = \frac{-b \pm \sqrt{b^2 - 4ac}}{2a}$$

The solution set can have two equal real roots, two different real roots, or no real roots. The following procedure, named QUAD_FORM_REAL, calculates the real roots, if they exist, of a quadratic equation by solving the quadratic formula.

```
                    .DATA
; For the quadratic procedures
CONST_A          DT      0        ; Storage for constant a
CONST_B          DT      0        ; Storage for constant b
CONST_C          DT      0        ; Storage for constant c
DISCRIMINANT     DT      0        ; Storage for discriminant
                    .CODE
_QUAD_FORM_REAL    PROC
; Solution of a quadratic equation in the form:
;
;                      ax^2 + bx + c = 0
;
; by means of the formula
;
;                    -b + SQRT [(b^2 - 4ac)]
;          x(1) =   EMPTY   EMPTY   EMPTY ----
;                   _       2a
;
;                    -b - [SQRT (b^2 - 4ac)]
;          x(2) =   EMPTY   EMPTY   EMPTY ----
;                              2a
;
; where x(1) and x(2) are two possible roots and a, b, and c are
; constants
;
; On entry:
;                   ST(0) = c
;                   ST(1) = b
;                   ST(2) = a
; On exit:
;                   ST(0) = x(1)
;                   ST(1) = x(2)
;
;                                 |  ST(0)  |  ST(1) |  ST(2)
;                                 ;    c    |    b   |    a    |
;***************************|
;     store constants       |
;***************************|
        FSTP    CONST_C         ;    b    |    a    |  EMPTY  |
        FSTP    CONST_B         ;    a    |  EMPTY  |
        FSTP    CONST_A         ;  EMPTY - |
;***************************|
;  evaluate the discriminant |
;     SQRT (b^2 - 4ac)       |
;***************************|
        FLD     CONST_C         ;    c    |  EMPTY  |
        FLD     CONST_A         ;    a    |    c    |  EMPTY  |
        FMULP   ST(1),ST        ;    ac   |  EMPTY  |
        FLD1                    ;    1    |    ac   |  EMPTY  |
```

```
        FADD    ST(0),ST    ;    2     |    ac    |   EMPTY    |
        FADD    ST(0),ST    ;    4     |    ac    |   EMPTY    |
        FMULP   ST(1),ST    ;   4ac    |  EMPTY   |
        FLD     CONST_B     ;    b     |   4ac    |   EMPTY    |
        FMUL    ST(0),ST    ;   b^2    |   4ac    |   EMPTY    |
        FXCH                ;   4ac    |   b^2    |   EMPTY    |
        FSUB                ; b^2-4ac  |  EMPTY   |
        FSQRT               ;   dscr   |  EMPTY   |
; Store discriminant
        FSTP    DISCRIMINANT ;  EMPTY - |
;**************************|
;  evaluate the root x(1)  |
;**************************|
        FLD     CONST_A     ;    a     |  EMPTY   |
        FLD1                ;    1     |    a     |   EMPTY    |
        FADD    ST(0),ST    ;    2     |    a     |   EMPTY    |
        FMULP   ST(1),ST    ;   2a     |  EMPTY   |
        FLD     CONST_B     ;    b     |   2a     |   EMPTY    |
        FCHS                ;   -b     |   2a     |   EMPTY    |
        FLD     DISCRIMINANT ;  dscr   |   -b     |    2a     |
        FADDP   ST(1),ST    ; -b + dscr|   2a     |   EMPTY    |
        FDIVR               ;   x(1)   |  EMPTY   |
; ST(0) holds root x(1)
;**************************|
;  evaluate the root x(2)  |
;**************************|
        FLD     CONST_A     ;    a     |   x(1)   |
        FLD1                ;    1     |    a     |   x(1)    |
        FADD    ST(0),ST    ;    2     |    a     |   x(1)    |
        FMULP   ST(1),ST    ;   2a     |   x(1)   |   EMPTY   |
        FLD     CONST_B     ;    b     |   2a     |   x(1)    |
        FCHS                ;   -b     |   2a     |   x(1)    |
        FLD     DISCRIMINANT ;  dscr   |   -b     |    2a     |
        FSUBP   ST(1),ST    ; -b - dscr|   2a     |   x(1)    |
        FDIVR               ;   x(2)   |   x(1)   |   EMPTY   |
        FXCH                ;   x(1)   |   x(2)   |   EMPTY   |
        CLD
        RET
_QUAD_FORM_REAL   ENDP
```

The above procedure has been designed to show the processing as clearly as possible and to make minimum use of the math unit stack. It is possible to improve the efficiency of the calculations by retaining all data elements in the math unit, instead of using memory variables. The values 4 and 2 are obtained by the routine by manipulating the constant 1. Performance could be further improved by storing these values as integer constants.

The C++ interface function to the _QUAD_FORM_REAL procedure is as follows:

```
void QuadFormReal(double a, double b, double c,\
     double *x1, double *x2)
// Solution of a quadratic equation in the form:
//
//                     ax^2 + bx + c = 0
//
// by means of the formula
//
//
```

```
//                      -b + SQRT[(b^2 - 4ac)]
//            x(1) = ----------------------
//                             2a
//
//                      -b - [SQRT(b^2 - 4ac)]
//            x(2) = ----------------------
//                             2a
//
// where x(1) and x(2) are two possible roots and a, b, and c are
// constants
//
// On entry:
//          parameters a, b, and c contain equation coefficients
// On exit:
//          parameters x1 and x2 contain the roots, if the
//          equation has real roots
//
{
double r1, r2;
_asm
{
        FLD     a
        FLD     b
        FLD     c
        CALL    QUAD_FORM_REAL
        FSTP    r1
        FSTP    r2
}
*x1 = r1;
*x2 = r2;
return;
}
```

SOFTWARE ON-LINE

The QuadFormReal() function is found in the Calc Primitives.h file located in the folder Sample Code\Chapter09\Calc Primitives in the book's on-line software.

9.2 Imaginary and Complex Numbers

The Intel math units have no provisions for handling imaginary or complex numbers. Complex arithmetic must be implemented in software. Since there is no explicit way of representing the imaginary unit (i), the calling routine must be aware of which stack register contains an implicit imaginary number. For the purpose of the discussion that follows, the reader should recall that the imaginary unit is defined as

$$i = \sqrt{-1}$$

therefore,

$$\sqrt{-x} = i\sqrt{x}$$

Also recall that complex numbers consist of a real and an imaginary part united by a plus or a minus sign. They are usually represented as

$$a + bi$$

$$a - bi$$

where a is the real part and bi is the imaginary part. A pure imaginary number (bi) can be represented as a complex number in the form

$$0 + bi.$$

This representation allows unifying the treatment of both imaginary and complex numbers.

9.2.1 Operations in Complex Arithmetic

Arithmetic operations of addition, subtraction, multiplication, division, powers, roots, and absolute value have been defined for complex numbers. The following formulas are pertinent to complex arithmetic.

1. Complex addition and subtraction

$$(a + bi) + (c + di) = (a + c) + i(b + d)$$

2. Complex multiplication

$$(a + bi)(c + di) = (ac - bd) + i(ad + bc)$$

3. Complex division

$$\frac{a+bi}{c+di} = \frac{(ac+bd)+i(bc-ad)}{c^2+d^2}$$

The magnitude or absolute value of a complex number represents the distance between the point that defines the complex number and the origin of the coordinate plane, as shown in the Argand diagram in Figure 9.3.

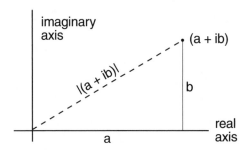

Figure 9.3 *Absolute Value (Magnitude) of a Complex Number*

4. The formula for determining the absolute value is the Pythagorean expression

$$|x| = \sqrt{a^2 + b^2}$$

The following low-level procedures perform the fundamental arithmetic operations on complex numbers.

```
                    .DATA
; Data for complex arithmetic
PART_A          DT      0          ; Storage for a
PART_B          DT      0          ; Storage for bi
PART_C          DT      0          ; Storage for c
PART_D          DT      0          ; Storage for di


                    .CODE
_ADD_CMPLX          PROC
; Addition of two complex numbers (a + bi and c + di) by means of
; the formula:
;
;       (a + bi) + (c + di) = (a + c) + i(b + d)
;
; On entry:
;           ST(3) = a
;           ST(2) = b
;           ST(1) = c
;           ST(0) = d
; On exit:
;           ST(0) = (a + c) = real part
;           ST(1) = (b + d) = imaginary part
;
;                                      | ST(0) | ST(1) | ST(2) | ST(3) |
;                                      |   a   |   b   |   c   |   d   |
;***************************|
;       store inputs        |
;***************************|
        FSTP        PART_D            ;   bi  |   c   |   di  | EMPTY |
        FSTP        PART_C            ;   c   |   di  | EMPTY |
        FSTP        PART_B            ;   d   | EMPTY |
        FSTP        PART_A            ; EMPTY |
;***************************|
;    calculate (a + c)      |
;***************************|
        FLD         PART_A            ;   a   |EMPTY- |
        FLD         PART_C            ;   c   |   a   | EMPTY |
        FADD                          ; a + c |EMPTY- |
;***************************|
;    calculate i(b + d)     |
;***************************|
        FLD         PART_B            ;   b   | a + c | EMPTY |
        FLD         PART_D            ;   d   |   b   | a + c | EMPTY |
        FADD                          ; b + d | a + c | EMPTY |
        FXCH                          ; a + c | b + d | EMPTY |
                                      ;   |       |___ imaginary part
                                      ;   |_____ real part

        CLD
        RET
_ADD_CMPLX          ENDP
;
;
_SUB_CMPLX          PROC
_MUL_CMPLX          PROC
; Multiplication of two complex numbers (a + bi and c + di) by
; means of the formula:
```

```
;
;           (a + bi)(c + di) = (ac - bd) + i(ad + bc)
;
; On entry:
;           ST(3) = a
;           ST(2) = b
;           ST(1) = c
;           ST(0) = d
; On exit:
;           ST(0) = (ac - bd) = real part
;           ST(1) = (ad + bc) = imaginary part
;
;                                      | ST(0) | ST(1) | ST(2) | ST(3) |
;                                      |   a   |  bi   |   c   |  di   |
;****************************|
;       store inputs         |
;****************************|
        FSTP      PART_D        ;   bi  |   c   |  di   | EMPTY |
        FSTP      PART_C        ;   c   |  di   | EMPTY | EMPTY |
        FSTP      PART_B        ;   di  | EMPTY | EMPTY | EMPTY |
        FSTP      PART_A        ; EMPTY | EMPTY | EMPTY | EMPTY |
;****************************|
;    calculate (ac - bd)     |
;****************************|
;                             ; EMPTY |
        FLD       PART_A        ;   a   | EMPTY |
        FLD       PART_C        ;   c   |   a   | EMPTY |
        FMULP     ST(1),ST      ;   ac  | EMPTY |
        FLD       PART_B        ;   b   |   ac  | EMPTY-|
        FLD       PART_D        ;   d   |   b   |  ac   |
        FMULP     ST(1),ST      ;   bd  |   ac  | EMPTY |
        FSUB                    ; ac-bd | EMPTY |
;****************************|
;    calculate i(ad + bc)    |
;****************************|
;                             ; ac-bd |
        FLD       PART_A        ;   a   | ac-bd | EMPTY |
        FLD       PART_D        ;   a   |   d   | ac-bd |
        FMULP     ST(1),ST      ;   ad  | ac-bd | EMPTY |
        FLD       PART_B        ;   b   |   ad  | ac-bd | EMPTY |
        FLD       PART_C        ;   c   |   b   |  ad   | ac-bd |
        FMULP     ST(1),ST      ;   bc  |   ad  | ac-bd | EMPTY |
        FADD                    ; ad+bc | ac-bd | EMPTY |
        FXCH                    ; ac-bd | ad+bc | EMPTY |
                                ;   |        |___ imaginary part
                                ;   |_____ real part
        CLD
        RET
_MUL_CMPLX        ENDP
;
;
_DIV_CMPLX        PROC
; Division of two complex numbers (a + bi and c + di) by means of
; the formula:
;
;       (a + bi)              (ac + bd) + i(bc - ad)
;        EMPTY -      =       EMPTY  EMPTY  EMPTY -
;       (c + di)                  c**2 + d**2
;
; On entry:
```

```
;           ST(3) = a
;           ST(2) = b
;           ST(1) = c
;           ST(0) = d
; On exit:
;           ST(0) = (ac + bd)/(c**2 + d**2) = real part
;           ST(1) = (bc - ad)/(c**2 + d**2) = imaginary part
;
;                                      | ST(0) | ST(1) | ST(2) | ST(3) |
;                                      |   a   |  bi   |   c   |  di   |
;****************************|
;       store inputs         |
;****************************|
        FSTP    PART_D        ;   bi  |   c   |  di   | EMPTY |
        FSTP    PART_C        ;   c   |  di   | EMPTY | EMPTY |
        FSTP    PART_B        ;   di  | EMPTY | EMPTY | EMPTY |
        FSTP    PART_A        ; EMPTY | EMPTY | EMPTY | EMPTY |
;****************************|
;    calculate (ac + bd)     |
;****************************|
;                            ; EMPTY |
        FLD     PART_A        ;   a   | EMPTY |
        FLD     PART_C        ;   c   |   a   | EMPTY |
        FMULP   ST(1),ST      ;  ac   | EMPTY |
        FLD     PART_B        ;   b   |  ac   | EMPTY |
        FLD     PART_D        ;   d   |   b   |  ac   |
        FMULP   ST(1),ST      ;  bd   |  ac   | EMPTY |
        FADD                  ; ac+bd | EMPTY |
;****************************|
;    calculate i(bc - ad)    |
;****************************|
;                            ; ac+bd |
        FLD     PART_B        ;   b   | ac+bd | EMPTY |
        FLD     PART_C        ;   c   |   b   | ac+bd | EMPTY |
        FMULP   ST(1),ST      ;  bc   | ac+bd | EMPTY |
        FLD     PART_A        ;   a   |  bc   | ac+bd | EMPTY |
        FLD     PART_D        ;   d   |   a   |  bc   | ac+bd |
        FMULP   ST(1),ST      ;  ad   |  bc   | ac+bd | EMPTY |
        FSUB                  ; bc-ad | ac+bd | EMPTY |
;****************************|
;    calculate c**2 + d**2   |
;****************************|
;                            | bc-ad | ac+bd | EMPTY |
        FLD     PART_C        ;   c   | bc-ad | ac+bd | EMPTY |
        FMUL    ST(0),ST      ;  c**2 | bc-ad | ac+bd | EMPTY |
        FLD     PART_D        ;   d   |  c**2 | bc-ad | ac+bd |
        FMUL    ST(0),ST      ;  d**2 |  c**2 | bc-ad | ac+bd |
        FADD                  ; c2+d2 | bc-ad | ac+bd | EMPTY |
;****************************|
; calculate imaginary part   |
; (bc - ad)/(c**2 + d**2)    |
;****************************|
;                            | ST(0) | ST(1) | ST(2) | ST(3) |
;                            | c2+d2 | bc-ad | ac+bd | EMPTY |
        FXCH                  ; bc-ad | c2+d2 | ac+bd | EMPTY |
        FLD     ST(1)         ; c2+d2 | bc-ad | c2+d2 | ac+bd |
        FDIV                  ; bc-ad |
                              ;/c2*d2 | c2+d2 | ac+bd | EMPTY |
;****************************|
;    calculate real part     |
```

```
; (ac + bd)/(c**2 + d**2)    |
;************************|
        FXCH    ST(2)               ; ac+bd | c2+d2 | img p | EMPTY |
        FXCH    ST(1)               ; c2+d2 | ac+bd | img p | EMPTY |
        FDIV                        ; ac+bd | bc-ad |
                                    ;/c2+d2 |/c2+d2 | EMPTY |
                                    ;   |          |___ imaginary part
                                    ;   |_____ real part
        CLD
        RET
_DIV_CMPLX      ENDP
;
;
_ABS_CMPLX      PROC
; Find the absolute value of a complex number using the formula:
;
;           |a + bi| = SQRT(a**2 + b**2)
;
; On entry:
;        ST(0) = a
;        ST(1) = b
; On exit:
;        ST(0) = |a + bi|
;                                    | ST(0) | ST(1) | ST(2) | ST(3) |
;                                    |   a   |   b   |
        FMUL    ST(0),ST            ; a**2  |   b   | EMPTY |
        FXCH                        ;   b   | a**2  | EMPTY |
        FMUL    ST(0),ST            ; b**2  | a**2  | EMPTY |
        FADD                        ; a2+b2 | EMPTY |
        FSQRT                       ;  abs  |
                                    ; value |
        CLD
        RET
_ABS_CMPLX      ENDP
```

The C++ interface functions for the complex arithmetic primitives are coded as follows:

```cpp
void AddComplex(double ra, double rbi, double rc, double rdi,\
                double *real, double *imaginary)
// Addition of two complex numbers (a + bi and c + di) by means of
// the formula:
//
//      (a + bi) + (c + di) = (a + c) + i(b + d)
//
// On entry:
//        parameters a, bi, c, and d holds input data
// On exit:
//        parameter real holds real part
//        parameter imaginary holds imaginary part
{
double rPart, iPart;
_asm
{
    FLD     ra
    FLD     rbi
    FLD     rc
```

```
      FLD     rdi
      CALL    ADD_CMPLX
      FSTP    rPart
      FSTP    iPart
}
*real = rPart;
*imaginary = iPart;
return;
}

void SubComplex(double ra, double rbi, double rc, double rdi,\
                double *real, double *imaginary)
// Subtraction of two complex numbers (a + bi and c + di) by means of
// the formula:
//
//          (a + bi) - (c + di) = (a - c) + i(b - d)
//
// On entry:
//          parameters a, b, c, and d holds input data
// On exit:
//          parameter real holds real part
//          parameter imaginary holds imaginary part
{
double rPart, iPart;
_asm
{
      FLD     ra
      FLD     rbi
      FLD     rc
      FLD     rdi
      CALL    SUB_CMPLX
      FSTP    rPart
      FSTP    iPart
}
*real = rPart;
*imaginary = iPart;
return;
}

void MulComplex(double ra, double rbi, double rc, double rdi,\
                double *real, double *imaginary)
// Multiplication of two complex numbers (a + bi and c + di) by means of
// the formula:
//
//          (a + bi)(c + di) = (ac - bd) + i(ad + bc)
//
// On entry:
//          parameters a, b, c, and d holds input data
// On exit:
//          parameter real holds real part
//          parameter imaginary holds imaginary part
{
double rPart, iPart;
_asm
{
      FLD     ra
      FLD     rbi
      FLD     rc
      FLD     rdi
      CALL    MUL_CMPLX
```

```
    FSTP    rPart
    FSTP    iPart
}
*real = rPart;
*imaginary = iPart;
return;
}

void DivComplex(double ra, double rbi, double rc, double rdi,\
                double *real, double *imaginary)
// Division of two complex numbers (a + bi and c + di) by means of
// the formula:
//          (a + bi)                (ac + bd) + i(bc - ad)
//          --------        =       ----------------------
//          (c + di)                      c^2 + d^2
//
// On entry:
//          parameters a, b, c, and d hold input data
// On exit:
//          parameter real holds real part
//          parameter imaginary holds imaginary part
{
double rPart, iPart;
_asm
{
    FLD     ra
    FLD     rbi
    FLD     rc
    FLD     rdi
    CALL    DIV_CMPLX
    FSTP    rPart
    FSTP    iPart
}
*real = rPart;
*imaginary = iPart;
return;
}

double AbsComplex(double ra, double rbi)
// Find the absolute value of a complex number using the formula:
//
//              |a + bi| = SQRT(a^2 + b^2)
//
// On entry:
//          parameters ra and rbi hold input data
//          ST(1) = bi
// On exit:
//          returns |a + bi|
{
double absPart;
_asm
{
    FLD     ra
    FLD     rbi
    CALL    ABS_CMPLX
    FSTP    absPart
}
return absPart;
}
```

SOFTWARE ON-LINE

The functions AddComplex(), SubComplex(), MulComplex, and DivComplex() are contained in the file Calc Primitives.h located in the folder Sample Code\Chapter09\Calc Primitives in the book's on-line software.

9.2.2 Real and Complex Roots of a Quadratic Equation

The procedure _QUAD_FORM_REAL, developed in Section 9.1, fails when the roots of the quadratic equation are complex numbers. Specifically, if the discriminant is negative, then the FSQRT instruction automatically tags ST(0) as indefinite, or NAN, which propagates to the final result. It is possible to design a routine that solves a quadratic equation with complex roots. In order to do this, we must develop different algorithms for obtaining the real and the complex roots of a quadratic equation. In Section 9.2 we applied the quadratic formula to find the real roots. The formula can be broken down as follows:

$$x_1 = \frac{-b + \sqrt{b^2 - 4ac}}{2a}$$

$$x_2 = \frac{-b + \sqrt{b^2 - 4ac}}{2a}$$

Since operations on complex numbers require that the real and the imaginary parts be computed separately, the formulas for the real part x_r and the imaginary part x_i are as follows

$$x_r = \frac{-b}{2a}$$

$$x_i = \frac{\pm\sqrt{b^2 - 4ac}}{2a}$$

The processing routine must inform the caller if the result is real or complex. The calling program then interprets the results accordingly. The following procedure provides a general solution to a quadratic equation with either real or complex roots.

```
            .DATA
; For the quadratic procedures
CONST_A         DT      0       ; Storage for constant a
CONST_B         DT      0       ; Storage for constant b
CONST_C         DT      0       ; Storage for constant c
DISCRIMINANT    DT      0       ; Storage for discriminant

            .CODE
_QUAD_FORMULA   PROC
; Solution of a quadratic equation in the form:
```

```
;
;                         ax^2 + bx + c = 0
;
; in real or complex numbers, by means of the formula
;
;                     -b + SQRT [(b^2 - 4ac)]
;           x(1) = ---------------------
;                             2a
;                     -b - [SQRT (b^2 - 4ac)]
;           x(2) = ---------------------
;                             2a
; where x(1) and x(2) are two possible roots and a, b, and c are
; constants
;
; Solution sets:
;           x(1) and x(2) are the real roots
;           x(r) +/- x(i) are the complex roots
;
; On entry:
;                     ST(0) = c
;                     ST(1) = b
;                     ST(2) = a
; On exit:
;           carry flag clear if roots are real numbers, then
;                     ST(0) = x(1)
;                     ST(1) = x(2)
;           carry flag set if roots are complex, then
;           solution is:
;                     ST(0) = -b/2a = real part = x(r)
;                     ST(1) = i * SQRT (b^2 - 4ac)/2a
;
                                    ;    c    |    b    |    a    |
;*************************|
;      store constants    |
;*************************|
        FSTP      CONST_C     ;    b    |    a    |  EMPTY  |
        FSTP      CONST_B     ;    a    |  EMPTY  |
        FSTP      CONST_A     ;  EMPTY  |
;*************************|
;  evaluate the discriminant |
;     SQRT (b**2 - 4ac)    |
;*************************|
        FLD       CONST_C     ;    c    |  EMPTY  |
        FLD       CONST_A     ;    a    |    c    |  EMPTY  |
        FMULP     ST(1),ST    ;   ac    |  EMPTY  |
        FLD1                  ;    1    |   ac    |  EMPTY  |
        FADD      ST(0),ST    ;    2    |   ac    |  EMPTY  |
        FADD      ST(0),ST    ;    4    |   ac    |  EMPTY  |
        FMULP     ST(1),ST    ;   4ac   |  EMPTY  |
        FLD       CONST_B     ;    b    |   4ac   |  EMPTY  |
        FMUL      ST(0),ST    ;   b^2   |   4ac   |  EMPTY  |
        FXCH                  ;   4ac   |   b^2   |  EMPTY  |
        FSUB                  ; b^2-4ac |  EMPTY - |
;*************************|
;test for real or imaginary |
;*************************|
; If discriminant is positive or zero the roots (or root) are real
; If discriminant is negative the roots are complex
        FTST                  ;   dscr  |  EMPTY  |  EMPTY  |
        FSTSW     STATUS_WW   ; Store status word
```

```
        FWAIT
        MOV     AH,BYTE PTR STATUS_WW + 1
        AND     AH,01000101B    ; Keep C3, C2, and C0
        CMP     AH,00000001B    ; Value for negative and not zero
        JNE     REAL_ROOTS      ; Go if discriminant is not negative
;**********************************************|
;      processing for complex roots           |
;**********************************************|
; At this point the discriminant is negative, therefore the roots
; of the equation are complex
        FCHS                    ; Change discriminant to positive
        FSQRT                   ; dscr   |  EMPTY   |
; Store discriminant
        FSTP    DISCRIMINANT    ; EMPTY - |
;***************************|
;   evaluate the real part  |
;    x(r) of complex root   |
;***************************|
; Calculate -b/2a
        FLD     CONST_A         ;    a    | EMPTY  |
        FLD1                    ;    1    |   a    | EMPTY  |
        FADD    ST(0),ST        ;    2    |   a    | EMPTY  |
        FMULP   ST(1),ST        ;    2a   | EMPTY  |
        FLD     CONST_B         ;    b    |   2a   | EMPTY  |
        FCHS                    ;   -b    |   2a   | EMPTY  |
        FDIV    ST,ST(1)        ;   x(r)  |   2a   | EMPTY  |
;
; ST(0) holds real part of complex root
;***************************|
;   evaluate imaginary part |
;   +/- x(i) of complex root|
;***************************|
; Calculate discriminant/2a
;                               ;   x(r)  |   2a   | EMPTY  |
        FXCH                    ;    2a   |  x(r)  | EMPTY  |
        FLD     DISCRIMINANT    ;   dscr  |   2a   |   x(r) |
        FDIVR                   ;   x(i)  |  x(r)  | EMPTY  |
        FXCH                    ;   x(r)  |  x(i)  | EMPTY  |
        STC                     ; Carry means roots are complex
        RET
;
;**********************************************|
;        processing for real roots            |
;**********************************************|
REAL_ROOTS:
        FSQRT                   ; dscr   |  EMPTY   |
; Store discriminant
        FSTP    DISCRIMINANT    ; EMPTY - |
;***************************|
;   evaluate the root x(1)  |
;***************************|
        FLD     CONST_A         ;    a    | EMPTY  |
        FLD1                    ;    1    |   a    | EMPTY  |
        FADD    ST(0),ST        ;    2    |   a    | EMPTY  |
        FMULP   ST(1),ST        ;    2a   | EMPTY  |
        FLD     CONST_B         ;    b    |   2a   | EMPTY  |
        FCHS                    ;   -b    |   2a   | EMPTY  |
        FLD     DISCRIMINANT    ;   dscr  |  -b    |   2a   |
        FADDP   ST(1),ST        ; -b + dscr|   2a   | EMPTY  |
        FDIVR                   ;   x(1)  | EMPTY  |
```

```
; ST(0) holds root x(1)
;***************************|
;   evaluate the root x(2)  |
;***************************|
          FLD       CONST_A        ;    a     |  x(1)  |
          FLD1                      ;    1     |   a    |  x(1)  |
          FADD      ST(0),ST        ;    2     |   a    |  x(1)  |
          FMULP     ST(1),ST        ;   2a     |  x(1)  |  EMPTY |
          FLD       CONST_B         ;    b     |  2a    |  x(1)  |
          FCHS                      ;   -b     |  2a    |  x(1)  |
          FLD       DISCRIMINANT    ;  dscr    |  -b    |  2a    |
          FSUBP     ST(1),ST        ; -b - dscr|  2a    |  x(1)  |
          FDIVR                     ;  x(2)    |  x(1)  |  EMPTY |
          FXCH                      ;  x(1)    |  x(2)  |  EMPTY |
          CLC                       ; No carry means roots are real
          RET
_QUAD_FORMULA     ENDP
```

Notice that in the QUAD_FORMULA procedure both real roots are returned. However, when the roots are complex, only one is returned. This is possible because complex roots are conjugates of one another. Therefore there is no need to report both imaginary results separately. The calling routine must interpret that the imaginary part of the solution set will have both, a positive and a negative value. Other problems, in trigonometry and calculus, that require a complex solution can be treated in a similar manner. In any case the calling program should be informed that the solution set contains an imaginary or complex component so that this element can be handled accordingly.

The C++ interface function to the _QUAD_FORMULA procedure is as follows:

```
int QuadFormula(double a, double b, double c, double *x1, double *x2)
// Solution of a quadratic equation in the form:
//
//                    ax**2 + bx + c = 0
//
// for its real or imaginary roots, using the formula
//
//                    -b + SQRT[(b^2 - 4ac)]
//         x(1)  =  ————————————————————
//                           2a
//
//                    -b - [SQRT(b^2 - 4ac)]
//         x(2)  =  ————————————————————
//                           2a
//
// where x(1) and x(2) are two possible roots and a, b, and c are
// constants
// Solution sets:
//          x(1) and x(2) are the real roots
//          x(r) +/- x(i) are the complex root
//
// On entry:
//                    ST(0) = c
//                    ST(1) = b
//                    ST(2) = a
// On exit:
//          returns 0 if roots are real numbers, then
```

```
//                      parameter x1 holds first root
//                      parameter x2 holds second root
//          returns 1 if roots are complex, then
//          solution is:
//                      x1= -b/2a = real part = x(r)
//                      x2 = +/- i * SQRT (b^2 - 4ac)/2a
{
double r1, r2;
int rootType = 0;       // 0 if real, 1 if imaginary roots
_asm
{
    FLD     a
    FLD     b
    FLD     c
    CALL    QUAD_FORMULA
    FSTP    r1
    FSTP    r2
    JNC     ROOT_EXIT
    MOV     rootType, 1
ROOT_EXIT:
}
*x1 = r1;
*x2 = r2;
return rootType;
}
```

SOFTWARE ON-LINE

The QuadFormula() function is contained in the file Calc Primitives.h located in the folder Sample Code\Chapter09\Calc Primitives in the book's on-line software.

9.2.3 Polar and Cartesian Coordinates

In traditional mathematics the study of complex numbers leads directly to an alternative plane of trigonometric representation, usually called the polar coordinate system. Conventionally, the polar coordinate system is depicted as based on a point, called the pole, located at the origin of the Cartesian plane, and a ray from this pole, called the polar axis. The polar axis is usually assumed to lie in the positive direction of the x-axis. A point in the polar coordinate system is defined by its directed angle from the polar axis and its directed distance from the pole. Figure 9.4 shows the elements of the polar and Cartesian coordinate systems.

Applications often require routines for converting coordinate pairs between the polar and the Cartesian systems. Several formulas can be used. The following two formulas allow finding the Cartesian coordinates from the polar form:

$$x = r\cos\Theta$$
$$y = r\sin\Theta$$

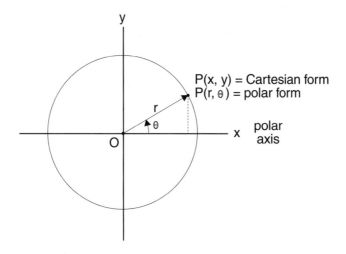

Figure 9.4 *Polar and Cartesian Coordinate Systems*

The following can be used to obtain the polar coordinates from a Cartesian pair:

$$\tan \Theta = \frac{y}{x}$$
$$r = \sqrt{x^2 + y^2}$$

The following procedures convert between the polar and the Cartesian coordinate systems.

```
.CODE
_POLAR_TO_CART     PROC
; Procedure for converting a coordinate pair in polar form to a
; coordinate pair in Cartesian (rectangular) form by means of the
; formulas:
;
;              x = r cos A
;              y = r sin A
;
; where r is the directed distance from the pole, A is the
; directed angle from the polar axis, expressed in degrees,
; and x, y are the Cartesian coordinate pair

; On entry:
;          ST(0)  = A
;          ST(1)  = r
; On exit:
;          ST(0)  = x
;          ST(1)  = y
;
;                                    ;  ST(0)  |  ST(1)  |  ST(2) |
;                                    ;   A     |    r    |  EMPTY |
; Convert degrees to radians
          CALL     _DEG_TO_RAD       ;  A (rads)|    r    |  EMPTY |
```

```
;**************************|
;    solve x = r cos A     |
;**************************|
        FXCH                      ;     r     |     A     |  EMPTY  |
        FLD     ST(1)             ;     A     |     r     |     A   |
        CALL    _COSINE           ;  cos A    |     r     |     A   |
        FMUL    ST,ST(1)          ;     x     |     r     |     A   |
        FXCH    ST(2)             ;     A     |     r     |     x   |
;**************************|
;    solve y = r sin A     |
;**************************|
        CALL    _SINE             ;  sin A    |     r     |     x   |
        FMULP   ST(1),ST          ;     y     |     x     |  EMPTY  |
        FXCH                      ;     x     |     y     |  EMPTY  |
        CLD
        RET
_POLAR_TO_CART    ENDP
;
;
_CART_TO_POLAR    PROC
; Procedure for converting a coordinate pair in Cartesian form to
; a coordinate pair in polar form, by means of the formulas:
;
;                    y
;          tan A = ----
;                    x
;
;          r = SQRT(x**2 + y**2)
;
; where x and y are the Cartesian coordinates, r is the directed
; distance from the pole and A is the directed angle from the
; polar axis, expressed in degrees
;
; On entry:
;          ST(0) = x
;          ST(1) = y
; On exit:
;          ST(0) = A
;          ST(1) = r
;
; Make copies of input values
;                                  | ST(0) | ST(1) | ST(2) | ST(3) |
                                  ;   x   |   y   | EMPTY |
        FLD     ST(0)             ;   x   |   x   |   y   |
        FLD     ST(2)             ;   y   |   x   |   x   |   y   |
        FXCH                      ;   x   |   y   |   x   |   y   |
;**************************|
;    find tan A = y/x      |
;**************************|
;                                  | ST(0) | ST(1) | ST(2) | ST(3) |
;                                  |   x   |   y   |   x   |   y   |
        FDIV                      ;  y/x  |   x   |   y   | EMPTY |
        CALL    _ARC_TANGENT      ;A(rads)|   1   |   x   |   y   |
        FSTP    ST(1)             ;A(rads)|   x   |   y   | EMPTY |
        CALL    _RAD_TO_DEG       ;   A   |   x   |   y   | EMPTY |
;**************************|
; find r = SQRT(x**2 + y**2)|
;**************************|
;                                  | ST(0) | ST(1) | ST(2) | ST(3) |
;                                  |   A   |   x   |   y   | EMPTY |
```

```
        FXCH    ST(2)           ;   y   |   x   |   A   | EMPTY |
        CALL    _PYTHAGORAS     ;   r   |   A   | EMPTY |
        FXCH                    ;   A   |   r   | EMPTY |
        CLD
        RET
_CART_TO_POLAR  ENDP
```

Notice that the procedures listed above assume that the polar angle is expressed in degrees. The procedures can be modified to operate with an angle in radians simply by eliminating the calls to _DEG_TO_RAD and _RAD_TO_DEG. The following C++ interface functions provide access to the polar/Cartesian conversion primitives:

```
void PolarToCart(double angle, double r, double *x, double *y)
// Function for converting a coordinate pair in polar form to a
// coordinate pair in Cartesian (rectangular) form by means of the
// formulas:
//
//              x = r cos A
//              y = r sin A
//
// where r is the directed distance from the pole, A is the
// directed angle from the polar axis, expressed in degrees,
// and x, y are the Cartesian coordinate pair
// On entry:
//          parameters r and angle hold input values
//          ST(1) = r
// On exit:
//          parameter x is x Cartesian coordinate
//          parameter y is y Cartesian coordinate
{
double xCoord, yCoord;
_asm
{
    FLD     r
    FLD     angle
    CALL    POLAR_TO_CART
    FSTP    xCoord
    FSTP    yCoord
}
*x = xCoord;
*y = yCoord;
return;
}


void CartToPolar(double x, double y, double *angle, double *r)
// Function for converting a coordinate pair in Cartesian form to
// a coordinate pair in polar form, by means of the formulas:
//
//                      y
//              tan A = -----
//                      x
//
//              r = SQRT (x^2 + y^2)
//
// where x and y are the Cartesian coordinates, r is the directed
// distance from the pole (radius) and A is the directed angle
// from the polar axis, expressed in degrees
// On entry:
```

```
//          parameters x and y hold the Cartesian coordinates
// On exit:
//          parameter angle is the polar angle
//          parameter r is the polar radius
{
double rValue, aValue;
_asm
{
    FLD     y
    FLD     x
    CALL    CART_TO_POLAR
    FSTP    aValue
    FSTP    rValue
}
*angle = aValue;
*r = rValue;
return;
}
```

SOFTWARE ON-LINE

The functions PolarToCart() and CartToPolar() are contained in the file Calc Primitives.h located in the folder Sample Code\Chapter09\Calc Primitives in the book's on-line software.

Chapter 10

Financial Calculations

Chapter Summary

This chapter is about calculations and formulas used in business, financial management, and accounting. In it we develop a set of fundamental routines that use the math unit to solve common formulas in these fields. The formulas can be applied to calculating interest and for solving annuities and amortization problems. The chapter concludes by exploring the precision limitations of the math unit in financial applications, as well as the possible use of BCD arithmetic to improve the accuracy of calculations.

10.0 Interest Calculations

Interest is the money paid for the use of money. In other words, if borrowing is the renting money, then interest is the rent. The amount borrowed is called the principal and the interest is a percentage of this principal. There are several formulas for calculating interest. In simple interest the calculation is a simple product of the principal, times the interest rate, times the term in years. In compound interest the borrower is forced to pay interest on interest.

10.0.1 Simple Interest

The formula for calculating simple interest is:

$$I = Prt$$

where I is the amount of interest, P is the principal, r is the annual interest rate, and t is the time in years. The following low-level primitive calculates simple interest.

```
_SIMPLE_INTEREST    PROC
; Simple interest calculation according to the formula:
;           I = Prt
; where:
;           P = principal amount (present value)
;           r = Annual interest rate
```

```
;              t = Time in years
;              I = Amount of interest
; On entry:
;              ST(0) = r (interest rate)
;              ST(1) = P (principal)
;              ST(2) = t (time in years)
; On exit:
;              ST(0) = I (interest amount)
;                                   | ST(0) | ST(1) | ST(2) | ST(3) |
;                                   |   r   |   P   |   t   | EMPTY |
        FMULP      ST(1),ST      ;  rP   |   t   | EMPTY |
        FMULP      ST(1),ST      ;   I   | EMPTY |
        CLD
        RET
_SIMPLE_INTEREST   ENDP
```

The C++ interface function is as follows:

```
double SimpleInterest(double term, double i, double p)
{
// Simple interest calculation according to the formula:
//
//           I = Prt
// where:
//           P = principal amount (present value)
//           r = Annual interest rate
//           t = Time in years
//           I = Amount of interest
// On entry:
//           parameter term = time in years
//           parameter i = interest rate (as a decimal fraction)
//           parameter p = principal
// On exit:
//           returns interest amount
double simpInt;
     // Convert interest to percent form
     i /= 100;
_asm
{
    FLD       term     ; Load years
    FLD       p        ; Load principal
    FLD       i        ; Interest
;                               ST0     ST1     ST2
; ASSERT:                     |  i  |  term |  p  |
    CALL      SIMPLE_INTEREST
    FSTP      simpInt  ;|   I   |
}
return simpInt;
}
```

Notice that in the simple interest calculation the term in years is expressed in a floating point variable. This is also true of the term variable in all other financial applications developed in this chapter. Expressing the term as a floating point type allows calculating interest on loans based on fractional terms. For example, in a fifteen month loan the term would be 15/12 or 1.25.

The simple interest formula is quite uncomplicated and the calculations could have been easily done in the interface function. However, we have maintained the same routine structure for consistency and to keep the core calculations independent of the interface.

10.0.2 Compound Interest

In simple interest the charge or fee is based on the principal amount only and is computed once in the term of the loan. Simple interest is often used in management calculations, but in practice most loans or deferred payment agreements are based on a more reasonable model, called compound interest.

Consider a two-year investment of $1000, at a 14 percent interest rate. At simple interest this deposit is held by the borrower for two years, at the end of which the interest rate of 0.14 is multiplied by 2 and applied to the principal. The borrower then returns the principal, plus the accrued interest, to the lender, as follows:

```
Principal:            simple interest accrued:         amount due:
                      (principal * interest * term)
$1000.00                     $280.00                   $1280.00
```

There is an inherent flaw in this model: the borrower is able to use the lender's money for the entire length of the loan not having to pay for it until the loan period has past. In the compound interest model the interest rate is applied to the interest, as well as to the principal. In the following example the same loan is calculated using over the same term, but the interest rate is compounded yearly:

```
    principal:        interest accrued:         amount due:
    first year:
    $1000.00          $140.00                   $1140.00
    second year:
    $1140.00          $159.60                   $1299.60
```

In this case during the second year of the loan the borrower had to pay interest on the principal as well as on the interest accrued on the first year. The difference between simple interest and compound interest calculations in this case is $19.60. Most financial institutions apply compound interest. Semi-annual, quarterly, monthly, and even daily compounding of interest are common.

Two financial terms are often used in relations to compound interest calculations. The *future value* of a loan or investment is the amount of the loan at the end of the loan period. In the case of the previous example, the future value of the loan is $1299.60 if the interest is compounded yearly. The present value of a loan or investment is the principal. In the case of the previous example the present value of the loan is $1000.00.

In financial calculations interest rate and term-of-loan must be expressed in the same units of time. This means that if interest rate is expressed as a yearly rate, then the unit of time for the loan must be one year. For example, suppose that you were calculating monthly payments on a 3-year loan, with a present value of $4000.00, at a yearly interest rate of 14 percent. Since the desired payments are in months, and the interest rate and term are in years, the calculations must perform the following adjustments:

```
                           interest rate = 0.14/12 =   0.01166666..
                           term = 3 * 12 = 36
```

In the code listed in this chapter the low-level core procedures assume that all data is in the same time units. The C++ interface functions perform the time-unit conversions when these are required.

Future Value at Compound Interest

The future value of an investment is the amount of principal plus interest at the end of the loan period. The formula for calculating future value at compound interest is as follows:

$$A = P(i+1)^n$$

where P is the present value or principal, i is the interest rate per compounding period, n is the number of compounding periods, and A is the value of investment at end of the period, or future value. The following low-level procedure performs the calculations.

```
                    .DATA
; Data for all financial procedures discussed in this chapter
PRINCIPAL       DT      0         ; Storage for principal
INTEREST        DT      0         ; Storage for interest rate
TERM            DT      0         ; Storage for term of loan
PAYMENT         DT      0         ; Storage for periodic payment
PRESENT_VAL     DT      0         ; Storage for present value
FUTURE_VAL      DT      0         ; Storage for future value
ANNUITY         DT      0         ; Storage for annuity

                    .CODE
_FUTURE_VALUE   PROC
; Financial procedure to calculate the future value of an investment
; at compound interest:
; On entry:
;           ST(0) = n (term)
;           ST(1) = i (interest rate)
;           ST(2) = P (present value)
; On exit:
;           ST(0) = A (Future value of investment)
;
; Note: interest rate (i) and term (n) must be in the same units
;       of time.
;
;                                 | ST(0) | ST(1) | ST(2) | ST(3) |
;                                 |  n    |   i   |   P   | EMPTY |
        FSTP      TERM            ;   i   |   P   | EMPTY |
        FSTP      INTEREST        ;   P   | EMPTY |
        FSTP      PRESENT_VAL     ; EMPTY |
;***************************|
;   calculate (i + 1)**n    |
;***************************|
        FLD       TERM            ;   n   | EMPTY |
        FLD       INTEREST        ;   i   |   n   | EMPTY |
        FLD1                      ;   1   |   i   |   n   | EMPTY |
        FADD                      ; i + 1 |   n   | EMPTY |
        FXCH                      ;   n   |  i+1  | EMPTY |
```

```
        CALL    _X_TO_Y_MIXED    ;(i+1)^n| EMPTY |
;***************************|
; calculate P*(i + 1)^n     |
;***************************|
        FLD     PRESENT_VAL     ;  P    |(i+1)^n| EMPTY |
        FMULP   ST(1),ST        ;  P    | EMPTY |
        CLD
        RET
_FUTURE_VALUE   ENDP
```

The C++ interface function is as follows:

```
double FutureValue(double term, double i, double p, int ppy)
{
// Financial procedure to calculate the future value of an investment
// at a compound interest, according to the formula:
//
//        A = P(i + 1)^n
//
// where
//        P = present value
//        i = interest rate per compounding period
//        n = number of compounding periods
//        A = Value of investment at end of period
// On entry:
//         parameter term in years
//         parameter i = interest rate
//         parameter p = present value
//         parameter ppy = payment periods per year
// On exit:
//         returns future value of investment
//
// Note: interest rate (i) and term (n) must be in the same units
//        of time.
double futVal;
    // Convert interest to percent form
    i /= 100;
_asm
{
    FLD     p            ; Load present value
    FILD    ppy          ; Time units
    FLD     i            ; Interest
    FDIV    ST,ST(1)     ; t=i/ppy
    FLD     term         ; term
    FMUL    ST,ST(2)     ; term*t
    FSTP    ST(2)
    FXCH
;                    ST0      ST1      ST2
; ASSERT:           |  n  |   i   |  p  |
    CALL    FUTURE_VALUE
    FSTP    futVal
}
return futVal;
}
```

Exponentials in Financial Formulas

Notice that the formula for calculating the future value requires the calculation:

$$(i+1)^n$$

Many financial formulas require calculating powers and roots. Which method is used for obtaining the exponential function affects the accuracy of the result. As discussed in Appendix B, of the exponential primitives in this book, the least accurate one is based on logarithms, while the most accurate ones are based on exponent factoring and binary powers. The advantage of logarithmic methods is that it allows fractional exponents. In financial formulas, the exponent is often the term element. If the formula is to allow fractional terms, then the calculations must be based on the logarithmic method of obtaining the exponential function.

Our solution to this problem is to use the function _X_TO_Y_MIXED developed in Chapter 8 and contained in the Un32_5 module of the MATH32 library. This function uses the binary powering method to calculate the integer portion of the exponential, and the logarithmic method to calculate the fractional part. Consequently, if the term element of a financial expression is an integer, as is more often the case, the exponential is calculated to maximum precision. However, if the term element is fractional, then the fractional part is calculated using logarithms and a valid result is produced, although to less precision.

Present Value at Compound Interest

The present value, or principal, of an investment can be calculated according to the formula:

$$P = A(i+1)^{-n}$$

where i is the interest rate, n is the number of compounding periods, A is the future value, and P is the present value or principal. The following procedure calculates the present value applying the formula.

```
_PRESENT_VALUE      PROC
; Financial procedure to calculate the present value of an investment
; at compound interest.
; On entry:
;           ST(0) = n (term)
;           ST(1) = i (interest rate)
;           ST(2) = A (future value)
; On exit:
;           ST(0) = P (Present value of investment)
;
;                                  | ST(0) | ST(1) | ST(2) | ST(3) |
;                                  |   n   |   i   |   A   | EMPTY |
        FSTP    TERM          ;    i   |   A   | EMPTY |
        FSTP    INTEREST      ;    A   | EMPTY |
        FSTP    FUTURE_VAL    ; EMPTY |
;**************************|
;   calculate (i + 1)**n   |
;**************************|
        FLD     TERM          ;    n   | EMPTY |
        FLD     INTEREST      ;    i   |   n   | EMPTY |
```

```
        FLD1                    ;   1   |   i   |   n   | EMPTY |
        FADD                    ; i + 1 |   n   | EMPTY |
        FXCH                    ;   n   |  i+1  | EMPTY |
        CALL    _X_TO_Y_MIXED   ;(i+1)^n| EMPTY |
; x^-n = 1/x^n
        FLD1                    ;   1   |(i+1)^n| EMPTY |
        FDIVR                   ;(i+1)^-n|
;***************************|
;  calculate A*(i + 1)^n    |
;***************************|
        FLD     FUTURE_VAL      ;   A   |(i+1)^n| EMPTY |
        FMULP   ST(1),ST        ;   P   | EMPTY |
        CLD
        RET
_PRESENT_VALUE  ENDP
```

The C++ interface function is as follows:

```cpp
double PresentValue(double term, double i, double a, int ppy)
{
// Financial procedure to calculate the present value of an investment
// at compound interest, according to the formula:
//
//          P = A(i + 1)^-n
//
// where
//          P = present value
//          i = interest rate per compounding period
//          n = number of compounding periods
//          A = Value of investment at end of period
// On entry:
//          parameter term in years
//          parameter i = interest rate
//          parameter a = future value
//          parameter ppy = payment periods per year
// On exit:
//          returns present value of investment
//
// See note regarding time units in the FutureValue() function.

double presVal;
    // Convert interest to percent form
    i /= 100;
_asm
{
    FLD     a         ; Load future value
    FILD    ppy       ; Time units
    FLD     i         ; Interest
    FDIV    ST,ST(1)  ; i/t
    FLD     term      ; years
    FMUL    ST,ST(2)  ; n*t
    FSTP    ST(2)
    FXCH
    ;                         ST0    ST1    ST2
    ; ASSERT:               |  n  |  i  |  p  |
    CALL    PRESENT_VALUE
    FSTP    presVal
}
return presVal;
}
```

Effective Rate

The effective rate of an investment, also called the effective annual yield, is the simple annual interest rate that would result in an equivalent amount at compound interest. In other words, the effective rate allows determining the simple interest that is equivalent to a compound interest. For example, $1.00 invested at 8 percent simple interest has a future value of $1.08. While $1.00 invested at 8 percent for one year at compound interest, compounded quarterly, has a future value of 1.08243216. In this case the simple interest rate required for a future value of 1.08243216 would be approximately 8.243216%, which is the effective rate of the investment at compound interest. The following formula is used in calculating the effective rate of an investment:

$$Y = \left(\left(\frac{1+i}{n} \right)^{-n} \right) - 1$$

where i is the interest rate, n is the number of compounding periods, and Y is the effective rate of the investment. The following low-level procedure performs the effective rate calculation:

```
_EFFECTIVE_YIELD        PROC
; Financial procedure to calculate the effective annual yield of
; an investment at compound interest
; On entry:
;           ST(0) = i (interest rate)
;           ST(1) = n (term)
; On exit:
;           ST(0) = Y (effective annual yield of investment)
;
;                                     | ST(0) | ST(1) | ST(2) | ST(3) |
;                                     |   i   |   n   | EMPTY | EMPTY |
;*************************|
; calculate (1+i/n)^n      |
;*************************|
        FDIV    ST,ST(1)      ;  i/n  |   n   | EMPTY |
        FLD1                  ;   1   |  i/n  |   n   | EMPTY |
        FADD                  ; 1+i/n |   n   | EMPTY | EMPTY |
        FXCH                  ;   n   | 1+i/n | EMPTY |
        CALL    _X_TO_Y_MIXED ;(1+i/n)^n |
;*************************|
; calculate ((1+i/n)^n) - 1 |
;*************************|
; (1+i/n)^n = w
        FLD1                  ;   1   |   w   | EMPTY |
        FXCH                  ;   w   |   1   | EMPTY |
        FSUB    ST,ST(1)      ;   Y   |   1   | EMPTY |
        FSTP    ST(1)         ;   Y   |
        CLD
        RET
_EFFECTIVE_YIELD  ENDP
```

Following is the C++ interface function.

```
double EffectiveYield(double i, int ppy)
{
// Financial procedure to calculate the effective annual yield
// of an account, according to the formula:
//
//          Y = ((1 + i/n)^n) - 1
//
// where
//          i = interest rate per compounding period
//          n = number of compounding periods
//          Y = effective annual yield of investment
// On entry:
//          parameter i = interest rate
//          parameter ppy = number of compounding periods
// On exit:
//          returns effective annual yield of investment
//
// See note regarding time units in the FutureValue() function.

double yield;
    // Convert interest to percent form
    i /= 100;
_asm
{
    FILD        ppy         ; Periods per year
    FLD         i           ; Interest
                            ;  ST0  |  ST1  |  ST2
                            ;   i   |  ppy  | EMPTY |
; ASSERT:                   |   i   |  ppy  | EMPTY |
    CALL        EFFECTIVE_YIELD
    FSTP        yield
}
```

10.1 Amortization

Loans are often made based on the borrower paying a given amount at regular terms. The payment process is called *amortization* and the individual payments are called *installments*. Several financial calculations relate to the loan amortization, including the calculation of each payment or installment, as well as the annual percentage rate and the add-on interest.

10.1.1 Periodic Payment Calculations

The amount of each periodic payment on a compound interest loan is calculated with the following formula:

$$P = p \times \frac{i}{1 - (i+1)^{-n}}$$

where p is the present value, or principal, i is the interest rate, n is the number of payment periods, or term, and P is the periodic payment required to amortize the loan. The following low-level procedure uses the preceding formula:

```
_PAYMENTS           PROC
; Financial procedure to calculate the amount of each payment on
; a loan.
; On entry:
;           ST(0) = n (term)
;           ST(1) = i (interest rate)
;           ST(2) = p (principal)
; On exit:
;           ST(0) = P (amount of periodic payment)
;
;***************************|
;   store input data        |
;***************************|
;                                    | ST(0) | ST(1) | ST(2) | ST(3) |
;                                    |   n   |   i   |   p   | EMPTY |
        FSTP    TERM         ;    i  |   p   | EMPTY |
        FSTP    INTEREST     ;    p  | EMPTY |
        FSTP    PRINCIPAL    ; EMPTY |
;***************************|
;  calculate 1-(i + 1)^-n   |
;***************************|
        FLD     TERM         ;   n   | EMPTY |
        FLD     INTEREST     ;   i   |   n   | EMPTY |
        FLD1                 ;   1   |   i   |   n   | EMPTY |
        FADD                 ; i + 1 |   n   | EMPTY |
        FXCH                 ;   n   |  i+1  | EMPTY |
        CALL    _X_TO_Y_MIXED ;(i+1)^n| EMPTY |
; x^-n = 1/x^n
        FLD1                 ;   1   |(i+1)^n| EMPTY |
        FDIVR                ;   1/  |
                             ;(i+1)^n| EMPTY |
        FLD1                 ;   1   |1/(i+..| EMPTY |
        FSUBR                ;1-(1/..| EMPTY |
;***************************|
; calculate:                |
;    p * i/1-(i + 1)^-n      |
;***************************|
; den = 1-(i + 1)**-n
;                            ; den   | EMPTY |
        FLD     INTEREST     ;   i   |  den  | EMPTY |
        FDIVR                ; i/den | EMPTY |
        FLD     PRINCIPAL    ;   p   | i/den | EMPTY |
        FMULP   ST(1),ST     ;   P   | EMPTY |
        CLD
        RET
_PAYMENTS           ENDP
```

The following C++ function interfaces with the low-level procedure.

```
double Payments(double term, double i, double p, int ppy)
{
// Financial procedure to calculate the amount of each payment on
// a loan using the formula:
//
//                       i
//      P =  p *  --------------
//                  1-(i + 1)^-n
// where
//      P = amount of periodic payment
//      p = principal (present value)
```

```
//      i = interest rate (in fractional form)
//      n = term of loan (number of payment periods)
//
// On entry:
//          parameter term = years
//          parameter i = interest rate
//          parameter p = principal
//          parameter ppy = payment periods per year
// On exit:
//          returns amount of periodic payment
//
double prin;

    // Convert interest to percent form
    i /= 100;
_asm
{
; Interest rate (i) and term (n) are converted to the same
; same units of time before calling the PAYMENT function
;    For example: to calculate monthly payments on a 3-year
;    loan of $4000.00, at 0.14% yearly
;                   p = 4000
;                   i = 0.14
;                   n = 3
;                 ppy = 12
;             i/ppy = 0.14/12 = 0.0116666 (monthly interest)
;             n*ppy = 3*12   = 36 (payments over loan life)
    FLD     p           ; Load principal
    FILD    ppy         ; Time units
    FLD     i           ; Interest
    FDIV    ST,ST(1)    ; i/ppy
    FLD     term        ; years
    FMUL    ST,ST(2)    ; n*t
    FSTP    ST(2)
    FXCH
    ;                       ST0     ST1     ST2
    ; ASSERT:            |   n   |   i   |   p   |
    CALL    PAYMENTS
    FSTP    prin    ; |   P   |
}
return prin;
}
```

10.1.2 Add-On Interest

Another method for calculating payments on installment loans is based on simple interest. This method is called add-on interest because the interest is added to the present value so that both are paid over the term of the loan. For example, for a loan of $1000.00 at 12 percent interest, the amount to be repaid is $1120.00. If the loan is for a period of 12 months, then the monthly add-on interest payments are $93.34. The formula for calculating add-on interest is simply dividing the simple interest by the loan term, as follows:

$$AOI = \frac{Prt}{n}$$

where n is the number of payments, r is the interest rate, t is the loan time, P is the principal, and AOI is the amount of each payment.

The following low-level procedure performs the calculations:

```
_AOI_PAYMENT      PROC
; Financial procedure to calculate the individual payments
; on a add-on interest loan.
; On entry:
;            ST(0) = r (add-on interest)
;            ST(1) = P (principal)
;            ST(2) = t (number of years)
;            ST(3) = n (number of payments)
; On exit:
;            ST(0) = AOR (AOR payment)
;
;
;                                   | ST(0) | ST(1) | ST(2) | ST(3) |
;                                   |   P   |   r   |   t   |   n   |
;***************************|
;     calculate Prt          |
;***************************|
          FSTP      PRINCIPAL       ; Save principal
          FLD       PRINCIPAL       ; Reload
          CALL      _SIMPLE_INTEREST
                                    ; Prt  |   N   | EMPTY |
          FLD       PRINCIPAL       ;  P   |  Prt  |   N   | EMPTY |
          FADD                      ; P+Prt |   N   | EMPTY |
          FDIV      ST,ST(1)        ; Prt/N |   N   | EMPTY |
          FSTP      ST(1)           ; AOR   | EMPTY |
          CLD
          RET
_AOI_PAYMENT      END
```

The C++ interface function to the _AOI_PAYMENT procedure is as follows:

```
double AoiPayment(double t, double i, double p, int ppy)
{
// Financial procedure to calculate the individual payments
// on a add-on interest loan, according to the formula:
//
//              Pit
//      AOI =   EMPTY—
//               ppy
//
// where
//         ppy = number of payments
//         i = add-on interest rate
//         t = loan time in years
//         P = principal
//      AOI = amount of each add-on interest payment
// On entry:
//         parameter = i (add-on interest rate)
//         parameter = p (principal)
//         parameter ppy = (number of payments per year)
//         parameter t = t (loan time in years)
//         Note:
//              N = t*ppy
// On exit:
```

```
//           Returns amount of each payment
double aoi;
    // Convert interest to percent form
    i /= 100;
_asm
{
; AOI_PAYMENT expects:
;                                STO      ST1      ST2      ST3
;                            |    p   |    i   |    t   |    N   |
; N = ppy * t
;
;                                STO      ST1      ST2      ST3
    FILD      ppy       ;    ppy  | EMPTY |
    FLD       t         ;    t    |  ppy  | EMPTY |
    FMUL      ST,ST(1)  ;    N    |  ppy  | EMPTY |
    FSTP      ST(1)     ;    N    | EMPTY |
    FLD       t         ;    t    |   N   | EMPTY |
    FLD       i         ;    i    |   t   |   N   | EMPTY |
    FLD       p         ;    p    |   i   |   t   |   N   |

; ASSERT:             ;    p    |   i   |   t   |   N   |
    CALL      AOI_PAYMENT
    FSTP      aoi
}
return aoi;
}
```

10.1.3 Annual Percentage Rate

In the example related to add-on interest we considered a one-year loan for $1000.00, at 12 percent interest and calculated that the add-on interest was $1120.00 and the payments were $93.34 per month. In this case we notice that the add-on interest calculation is inherently flawed, since the borrower continues to pay for the full amount of the loan throughout the loan period. In other words, at the sixth month of the loan the borrower has paid back approximately one-half of the future value, yet, the monthly payments continue to be based on the total loan amount. This type of inequity was the cause of a law passed by Congress in 1969 called the *Truth-in-Lending Act*. The law requires that the borrower must be informed of the annual percentage rate of every loan, called the APR. For add-on interests the APR is calculated according to the formula:

$$APR = \frac{2rN}{N+1}$$

where N is the number of payments, r is the add-on interest rate, and APR is the annual percentage rate. For the loan in the preceding paragraph, at an add-on interest of 12 percent, the APR is 22.15. The following low-level procedure calculates the APR:

```
_APR                PROC
; Financial procedure to calculate the annual percentage rate
; APR.
; On entry:
;           ST(0) = r (add-on interest)
;           ST(1) = N (number of payments)
; On exit:
```

```
;              ST(0) = APR (annual percentage rate)
;                                   | ST(0) | ST(1) | ST(2) | ST(3) |
;                                   |   r   |   N   | EMPTY | EMPTY |
;*************************|
;     calculate 2Nr            |
;*************************|
        FMUL    ST,ST(1)            ;   rN  |   N   | EMPTY |
        FADD    ST,ST(0)            ;  2rN  |   N   | EMPTY |
;*************************|
;  calculate 2rN/(N + 1)       |
;*************************|
        FLD1                        ;   1   |  2rN  |   N   | EMPTY |
        FADD    ST,ST(2)            ; 1 + N |  2rN  |   N   | EMPTY |
        FSTP    ST(2)               ; 1 + N |  2rN  | EMPTY |
        FDIV    ST,ST(1)            ;  APR  |  2rN  |
        FSTP    ST(1)               ;  APR  | EMPTY |
        CLD
        RET
_APR    ENDP
```

The following C++ function provides the interface to the _APR procedure.

```
double Apr(double i, int np)
{
// Financial procedure to calculate the annual percentage rate
// APR, according to the formula:
//
//              2rN
//      APR = EMPTY—
//             N + 1
//
// where
//         N = number of payments
//         r = add-on interest rate
//       APR = annual percentage rate
// On entry:
//         parameter i = add-on interest
//         parameter np = number of payments
// On exit:
//         returns APR (annual percentage rate)
double apr;
    // Convert interest to percent form
    i /= 100;
_asm
{
    FILD    np          ; Number of payments
    FLD     i           ; Add-on interest
                        ;  ST0  |  ST1  |  ST2
                        ;   i   |   N   | EMPTY |
    CALL    APR
    FSTP    apr
}
return apr*100;     // Convert to percent
}
```

10.2 Annuities

Each periodic payment made on an interest-bearing account is called an annuity. The notion of an ordinary annuity requires that the amount of each payment is the same, and that the interest rate does not change throughout the loan period.

10.2.1 Annuity Future Value

A common calculation regarding an ordinary annuity is to determine the future value of an account. For example, if you deposit $120.00 per month, on an account bearing 12 percent interest, what is the future value of the investment at the end of the fifth year. The formula for calculating the future value of an annuity is as follows:

$$FV = pm\frac{(i+1)^{n-1}}{i}$$

where pm is the periodic payment amount, i is the interest rate, n is the number of payment periods, and FV is the future value. The following low-level procedure uses the preceding formula to calculate the future value of an ordinary annuity.

```
_ANNUITY_FV      PROC
; Financial procedure to calculate the future value of an
; ordinary annuity.
; On entry:
;           ST(0) = n (number of payment periods)
;           ST(1) = i (interest rate)
;           ST(2) = pm (periodic payment)
; On exit:
;           ST(0) = FV (future value of investment)
;
; Note: interest rate (i) and term (n) must be in the same units
;       of time (see note in the PAYMENTS procedure)
;
;**************************|
;   store input data       |
;**************************|
;                              | ST(0) | ST(1) | ST(2) | ST(3) |
;                              |   n   |   i   |  pm   | EMPTY |
        FSTP     TERM          ;   i   |  pm   | EMPTY |
        FSTP     INTEREST      ;  pm   | EMPTY |
        FSTP     PAYMENT       ; EMPTY |
;**************************|
; calculate (i + 1)^(n - 1) |
;**************************|
        FLD      TERM          ;   n   | EMPTY |
        FLD      INTEREST      ;   i   |   n   | EMPTY |
        FLD1                   ;   1   |   i   |   n   | EMPTY |
        FADD                   ; i + 1 |   n   | EMPTY |
        FXCH                   ;   n   |  i+1  | EMPTY |
        CALL     _X_TO_Y_MIXED ;(i+1)^n| EMPTY |
        FLD1                   ;   1   |(i+1)^n| EMPTY |
        FSUB                   ;(i+1)^n|
                               ;  -1   | EMPTY |
;**************************|
```

```
; calculate:                    |
;            (i + 1)^(n - 1)     |
;    pm *  ----------------      |
;                 i              |
;*************************|
; num = (i + 1)^(n - 1)
;                                    ; num  | EMPTY |
        FLD      INTEREST            ;  i   | num   | EMPTY |
        FDIV                         ; num/i| EMPTY |
        FLD      PAYMENT             ;  pm  | i/num | EMPTY |
        FMULP    ST(1),ST            ;  PV  | EMPTY |
        CLD
        RET
_ANNUITY_FV      ENDP
```

The C++ interface function to the _ANNUITY_FV procedure is as follows:

```
double AnnuityFv(double term, double i, double pm, int ppy)
{
// Financial procedure to calculate the future value of an
// investment using the formula:
//
//                    (i + 1)^(n - 1)
//        FV =   pm *  ---------------
//                           i
// where:
//     FV = future value of an investment
//     pm = periodic payment
//     i = periodic interest rate
//     term = years
//
// On entry:
//         parameter term = years
//         parameter i = yearly interest rate
//         parameter pm  = periodic payment
//         parameter ppy = payment periods per year
// On exit:
//         returns future value of investment
double fValue;
     // Convert interest to percent form
     i /= 100;
_asm
{
; Interest rate (i) and term (n) are converted to the same
; same units of time before calling the ANNUITY_FV function
     FLD      pm       ; Load periodic payment
     FILD     ppy      ; Time units
     FLD      i        ; Interest
     FDIV     ST,ST(1) ; i/t
     FLD      term     ; years
     FMUL     ST,ST(2) ; n*t
     FSTP     ST(2)
     FXCH
     ;                        ST0     ST1     ST2
     ; ASSERT:            |    n   |   i   |   pm  |
     CALL     ANNUITY_FV
     FSTP     fValue
}
return fValue;
}
```

10.2.2 Annuity Present Value

Sometimes we need to know what is the initial sum of money that is equivalent to a certain annuity. In other words, what would we have to invest today, in a lump sum, in order to produce a future value equivalent to the one that would result from the periodic payments of a given annuity. This value, known as the present value of an annuity, is calculated with the following formula:

$$PV = pm \frac{1 - (i+1)^{-n}}{i}$$

where pm is the annuity's periodic payment, i is the rate, n is the term in years, and PV is the present value. The following low-level procedure calculates the present value.

```
_ANNUITY_PV        PROC
; Financial procedure to calculate the present value of an
; annuity.
; On entry:
;          ST(0) = n (term)
;          ST(1) = i (interest rate)
;          ST(2) = pm (periodic payment)
; On exit:
;          ST(0) = PV (present value of investment)
;**************************|
;   store input data       |
;**************************|
;                              | ST(0) | ST(1) | ST(2) | ST(3) |
;                              |   n   |   i   |   pm  | EMPTY |
        FSTP    TERM           ;   i   |   pm  | EMPTY |
        FSTP    INTEREST       ;   pm  | EMPTY |
        FSTP    PAYMENT        ; EMPTY |
;**************************|
; calculate 1-(i + 1)^-n   |
;**************************|
        FLD     TERM           ;   n   | EMPTY |
        FLD     INTEREST       ;   i   |   n   | EMPTY |
        FLD1                   ;   1   |   i   |   n   | EMPTY |
        FADD                   ; i + 1 |   n   | EMPTY |
        FXCH                   ;   n   |  i+1  | EMPTY |
        CALL    _X_TO_Y_MIXED  ;(i+1)^n| EMPTY |
; x^-n = 1/x^n
        FLD1                   ;   1   |(i+1)^n| EMPTY |
        FDIVR                  ;  1/   |
                               ;(i+1)^n| EMPTY |
        FLD1                   ;   1   |1/(i+..| EMPTY |
        FSUBR                  ;1-(1/..| EMPTY |
;**************************|
; calculate:               |
;           1-(i + 1)^-n   |
;   pm *  --------------    |
;               i          |
;**************************|
; num = 1-(i + 1)^-n
;                              ; num   | EMPTY |
        FLD     INTEREST       ;   i   |  num  | EMPTY |
```

```
        FDIV                        ; num/i | EMPTY |
        FLD     PAYMENT             ;  pm   | i/num | EMPTY |
        FMULP   ST(1),ST            ;  PV   | EMPTY |
        CLD
        RET
_ANNUITY_PV     ENDP
```

The C++ function to access the _ANNUITY_PV procedure is as follows:

```
double AnnuityPv(double term, double i, double pm, int ppy)
{
// Financial procedure to calculate the present value of an
// investment using the formula:
//
//                   1-(i + 1)^-n
//      PV =  pm * -------------
//                       i
//
// where:
//      PV = present value of an investment
//      pm = periodic payment
//      i = periodic interest rate
//      n = term in years
//
// On entry:
//          parameter term = years
//          parameter i = interest rate
//          parameter pm = periodic payment
//          parameter ppy = payment periods per year
// On exit:
//          returns present value of investment
double pValue;
    // Convert interest to percent form
    i /= 100;
_asm
{
; Interest rate (i) and term (n) are converted to the same
; same units of time before calling the ANNUITY_PV function
    FLD     pm          ; Load periodic payment
    FILD    ppy         ; Time units
    FLD     i           ; Interest
    FDIV    ST,ST(1)    ; i/t
    FLD     term        ; years
    FMUL    ST,ST(2)    ; n*t
    FSTP    ST(2)
    FXCH
;                               ST0     ST1     ST2
; ASSERT:                    |   n   |   i   |   pm  |
    CALL    ANNUITY_PV
    FSTP    pValue
}
return pValue;
}
```

10.2.3 Annuity Due

In ordinary annuities, such as those discussed previously, the periodic payments are made at the end of each period. An annuity due is one in which the payments are made at the beginning of each period. The result of making the payments at the beginning of each period is that the exponent for the annuity formula must be increased by one. The resulting formula is as follows:

$$A = m\left[\frac{(1+i)^{n+1}-1}{i}-1\right]$$

where pm is the periodic payment, i is the interest rate, n is the number of payment periods, and A is the annuity due at the beginning of each period. The following low-level procedure calculates annuity due.

```
_ANNUITY_DUE      PROC
; Financial procedure to calculate the future value of an
; annuity due.
; On entry:
;             ST(0) = n (term)
;             ST(1) = i (interest rate)
;             ST(2) = pm (periodic payment)
; On exit:
;             ST(0) = A (annuity due)
;
;
;************************|
;    store input data    |
;************************|
;
;                                    | ST(0) | ST(1) | ST(2) | ST(3) |
;                                    |   n   |   i   |  pm   | EMPTY |
          FSTP     TERM              ;   i   |  pm   | EMPTY |
          FSTP     INTEREST          ;  pm   | EMPTY |
          FSTP     PAYMENT           ; EMPTY |
;************************************|
; calculate (i + 1)^(n + 1)  |
;************************************|
          FLD      TERM              ;   n   | EMPTY |
          FLD1                       ;   1   |   n   | EMPTY |
          FADD                       ;  n+1  | EMPTY |
          FLD      INTEREST          ;   i   |  n+1  | EMPTY |
          FLD1                       ;   1   |   i   |  n+1  | EMPTY |
          FADD                       ;  i+1  |  n+1  | EMPTY |
          FXCH                       ;  n+1  |  i+1  | EMPTY |
          CALL     _X_TO_Y_MIXED     ;(i+1) ^
                                     ; (n+1) | EMPTY |
; (i+1)^(n+1) = f1
;************************************|
;    calculate f1 - 1        |
;************************************|
          FLD1                       ;   1   |  f1   | EMPTY |
          FSUBP    ST(1),ST          ; 1-f1  | EMPTY |
;************************************|
; calculate:                 |
;              f1            |
;    pm * ( ----- - 1)       |
;              i            |
;************************************|
;
;                                    ;  f1   | EMPTY |
          FLD      INTEREST          ;   i   |  f1   | EMPTY |
          FDIVP    ST(1),ST          ; f1/i  | EMPTY |
```

```
        FLD1                             ;   1   | f1/i  | EMPTY |
        FSUBP    ST(1),ST                ;1-f1/i | EMPTY |
        FLD      PAYMENT                 ;  pm   |1-f1/i | EMPTY |
        FMULP    ST(1),ST                ;   A   | EMPTY |
        CLD
        RET
_ANNUITY_DUE     ENDP
```

The following code listing is the C++ interface function to the _ANNUITY_DUE procedure.

```
double AnnuityDue(double term, double i, double pm, int ppy)
{
// Financial procedure to calculate the future value of an
// annuity due, using the formula:
//
//                   (i + 1)^(n+1) - 1
//       A =  pm * ------------------- - 1
//                          i
//
//
// where:
//      A = annuity due at the beginning of each period
//      pm = periodic payment
//      i = periodic interest rate (in fractional form)
//      n = number of payment periods
//
// On entry:
//          parameter term is years
//          parameter i is interest rate
//          parameter pm is periodic payment
//          paramenter ppy is payments per year
// On exit:
//          Returns annuity due
//
// Note: interest rate (i) and term (n) must be in the same units
//       of time (see note in the PAYMENTS procedure)

double aDue;
     // Convert interest to percent form
     i /= 100;
_asm
{
; Interest rate (i) and term (n) are converted to the same
; same units of time before calling the ANNUITY_DUE procedure
     FLD      pm            ; periodic payment
     FILD     ppy           ; payments per period
     FLD      i             ; Interest
     FDIV     ST,ST(1)      ; i/t
     FLD      term          ; years
     FMUL     ST,ST(2)      ; n*t
     FSTP     ST(2)
     FXCH
;                      ST0     ST1     ST2
; ASSERT:             |  n  |   i  |  pm  |
     CALL     ANNUITY_DUE
     FSTP     aDue
}
  return aDue;
}
```

10.2.4 Sinking Fund

Businesses or individuals often need to set up an account into which a certain amount is deposited periodically in order to ensure that, at the end of the period, the account will have a given amount of money. The term *sinking fund* refers to the amount of the periodic payment required in order to achieve an annuity of a given future value. The formula for a sinking fund is as follows:

$$pm = \frac{Ai}{(i+1)^{-n} - 1}$$

where i is the interest rate, n is the term of the loan, A is the future value of the annuity, and pm is the periodic payment. The following low-level procedure calculates the sinking fund for an annuity.

```
_SINKING_FUND    PROC
; Financial procedure to calculate the periodic payment
; (sinking fund) required in an annuity.
; On entry:
;            ST(0) = n (term)
;            ST(1) = i (interest rate)
;            ST(2) = A (annuity due)
; On exit:
;            ST(0) = pm (periodic payment required)
;
;***************************|
;    store input data       |
;***************************|
;                             | ST(0) | ST(1) | ST(2) | ST(3) |
;                             |   n   |   i   |   A   | EMPTY |
        FSTP     TERM       ;    i   |   A   | EMPTY |
        FSTP     INTEREST   ;    A   | EMPTY |
        FSTP     ANNUITY    ; EMPTY |
;***************************|
;    calculate A*i          |
;***************************|
        FLD      INTEREST   ;    i   | EMPTY |
        FLD      ANNUITY    ;    A   |   i   | EMPTY |
        FMUL     ST,ST(1)   ;   Ai   |   i   | EMPTY |
        FXCH                ;    i   |   Ai  | EMPTY |
;***************************|
; calculate (i + 1)^n - 1   |
;***************************|
        FLD      TERM       ;    n   |   i   |   Ai  | EMPTY |
        FXCH                ;    i   |   n   |   Ai  |
        FLD1                ;    1   |   i   |   n   | Ai--- |
        FADD                ;  i + 1 |   n   |   Ai  | EMPTY |
        FXCH                ;    n   |  i+1  |   Ai  | EMPTY |
        CALL     _X_TO_Y_MIXED  ;(i+1)^n|  Ai  | EMPTY |
        FLD1                ;    1   |(i+1)^n|  Ai  | EMPTY |
        FSUB                ;(i+1)^n|
                            ;   -1   |   Ai  | EMPTY |
;***************************|
; calculate:                |
;            Ai             |
```

```
;    pm = ---------------   |
;           ((i + 1)^n) - 1 |
;                           |
;***************************|
; ((i + 1)^n) - 1 = f1
;                                        ;  f1  |  Ai  | EMPTY |
        FDIVP     ST(1),ST               ; Ai/f1 | EMPTY |
        CLD
        RET
_SINKING_FUND     ENDP
```

The C++ interface function to the _SINKING_FUND procedure is as follows:

```
double SinkingFund(double term, double i, double a, int ppy)
{
// Financial procedure to calculate the periodic payment
// (sinking fund) required in an annuity, using the formula:
//
//                  Ai
//      pm =  ---------------
//             ((i + 1)^n) - 1
//
// where:
//      pm = periodic payment
//      i = periodic interest rate (in fractional form)
//      n = term of loan (number of payment periods)
//      A = annuity
// On entry:
//          parameter term = n            -> ST(0)
//          parameter i = interest rate -> ST(1)
//          parameter a = A (annuity)    -> ST(2)
// On exit:
//          returns pm (periodic payment required)
//
double pm;

     // Convert interest to percent form
     i /= 100;
_asm
{
; Interest rate (i) and term (n) are converted to the same
; same units of time before calling the SINKING_FUND procedure
     FLD      a           ; Load future value of annuity
     FILD     ppy         ; Time units
     FLD      i           ; Interest
     FDIV     ST,ST(1)    ; i/t
;                         |  ST0  |  ST1  |  ST2  |  ST3  |
;                         ; i/t   |  ppy  |   A   | EMPTY |
     FLD      term        ; n     | i/t   |  ppy  |   A   |
     FMUL     ST,ST(2)    ; n*t   | i/t   |  ppy  |   A   |
     FSTP     ST(2)       ; n*t   | i/t   |   A   | EMPTY |
     FXCH
;                           ST0     ST1     ST2
; ASSERT:                 |  n  |  i  |  A  |
     CALL     SINKING_FUND
     FSTP     pm
}
return pm;
}
```

10.2.5 Number of Compounding Periods

Another calculation often needed in regard to annuities and other investments is to determine the number of compounding periods required to reach a certain future value. The formula in this case is as follows:

$$CP = \frac{\ln\left(\dfrac{fv}{pv}\right)}{\ln\left(1+i\right)}$$

where fv is the future value, pv is the present value, i is the interest rate, ln is the natural log, and CP is the number of compounding periods. The low-level procedure to calculate the number of compounding periods is as follows:

```
_COMP_PERIODS     PROC
; Financial procedure to calculate the number of compounding
; periods necessary for an investment to grow to a predetermined
; future value.
;
;***************************|
;   calculate ln(1 + i)     |
;***************************|
;                            | ST(0) | ST(1) | ST(2) | ST(3) |
;                            |   i   |  fv   |  sv   | EMPTY |
        FLD1                ;   1   |   i   |  fv   |  sv   |
        FADD                ; 1 + i |  fv   |  sv   | EMPTY |
        CALL      _LOG_E    ;ln(1+i)|  fv   |  sv   | EMPTY |
;****************************|
;   calculate:
;       ln(fv/sv)/ln(1 + i)  |
;****************************|
        FXCH      ST(2)     ;  sv   |  fv   |ln(1+i)| EMPTY |
        FDIV                ; fv/sv |ln(1+i)| EMPTY |
        CALL      _LOG_E    ;ln(fv/ |
                            ;  sv)  |ln(1+i)| EMPTY |
        FDIVR               ;  CP   | EMPTY |
        CLD
        RET
_COMP_PERIODS     ENDP
```

Notice that the procedure calls the _LOG_E procedure, developed in Chapter 8, to obtain the natural logs required in the formula. The C++ interface function to the _COMP_PERIODS procedure is as follows:

```
double CompPeriods(double i, double fv, double sv, int cpy)
{
// Financial procedure to calculate the number of compounding
// periods necessary for an investment to grow to a predetermined
// value, by means of the formula:
//
//             ln(fv/sv)
//      CP = ---------
//             ln(1 + i)
//
// where:
```

```
//      CP = number of compounding periods
//      fv = final value
//      sv = starting value
//       i = interest rate
//      ln = log base e (natural logarithm)
// On entry:
//      parameter i = interest rate
//      parameter fv = final value
//      parameter sv = start value
//      parameter cpy = compounding periods per year
// On exit:
//      returns number of compounding period required
//
double cp;

// Convert interest to percent form
    i /= 100;
_asm
{
    FLD     sv          ; Start value
    FLD     fv          ; Final value
    FILD    cpy         ; Time units
    FLD     i           ; Interest
    FDIV    ST,ST(1)    ; i / cpy
    FSTP    ST(1)
;                           ST0     ST1     ST2
; ASSERT:               |  i  |  fv  |  sv  |
    CALL    COMP_PERIODS
    FSTP    cp
}
return cp;
}
```

10.3 Numerical Errors in Financial Calculations

The detection, diagnosing, and management of computational errors is a field on which entire books can, and have, been written. Appendix B is a discussion of computational errors that result in the calculation of the exponential function. Since computational errors are particularly dreadful in financial and accounting applications, we think it is appropriate to mention them at this time. An error of proportionately small magnitude in a financial calculation may not be quantitatively important, however, it could be the cause of an accounting nightmare.

Computer errors in financial calculations are due to several causes. In the first place, the computer is a binary machine and financial problems require decimal numbers. The conversion of decimal-to-binary and binary-to-decimal often introduces errors. This is due to the fact that many decimal numbers cannot be represented exactly in binary. Another type of error originates in the precision limitations of the numeric formats used to represent floating-point. Overflow, underflow, and rounding errors occur when numeric values exceed the precision supported by the adopted format. Furthermore, numerical methods used in computer calculations often produce approximate results. The errors introduced by the use of logarithms in calculating exponentials was discussed in Chapter 8 and is revisited to greater detail in Appendix B. Finally, inexact values can propagate and multiply during calculation, contaminating the final result to create an error of larger magnitude than that of the original one. Or several error conditions can compound either in a single operand, either compensating for each other or augmenting each other.

SOFTWARE ON-LINE

The program Financial Errors, in the CHAPTER10 folder in the book's on-line software, demonstrates some of the errors discussed in the sections that follow. The reader should also refer to Appendix B for a more thorough discussion of error conditions related to the calculation of exponentials.

10.3.1 Conversion Errors

Many decimal numbers do not have an exact binary representation. Computers often surprise us with unexpected numerical values that result from simple errors during data conversion and storage operations. The classical example is a routine with a loop that adds the value 0.10 ten consecutive times, as in the following code fragment:

```
double val = 0.0;
while(val != 1.0)
{
    val = val + 0.10;
}
```

Naively we could expect that the sum reaches the value 1.0 at the tenth iteration of the loop. However, the decimal fraction 0.1 does not have an exact binary representation in the standard floating-point formats. For this reason the code contains an endless loop that never reaches the value 1.0. In Visual C++ 6.0 the consecutive values of the variable val are as follows:

```
0.00000000000000000
0.10000000000000001
0.20000000000000001
0.30000000000000004
0.40000000000000002
0.50000000000000000
0.59999999999999998
0.69999999999999996
0.79999999999999993
0.89999999999999991
0.99999999999999989
1.09999999999999990
```

The project Financial Errors, located in the folder CHAPTER10 folder in the book's CD ROM, contains code to generate and display the conversion error discussed in this session. Decimal/binary conversion errors generate an inexact result.

10.3.2 Representation Errors

Computers cannot store rational numbers in fractional form. For example, the fraction 1/3 must be approximated by the repeating fraction 0.333..33 to however many significant digits are supported in the format. Since there is no exact way of representing the value 1/3 this value is sometimes said to be an unrepresentable number. The error introduced in storing unrepresentable numbers is often compounded by a decimal/binary conversion error. In this manner Visual C++ 6.0 represents the fraction 1/3 as:

```
0.33333333333333331
```

The use of fractional approximations in numerical calculations often causes unexpected results, especially when the representation error is compounded with a conversion or a computational error.

10.3.3 Precision and Computation Errors

Business, financial, and economic problems often deal with very large or very small numbers, for example, the national debt, the gross income of a large corporation, or the per-capita consumption of octopus meat in the central Sahara region.

The floating-point representation of numbers in the math unit follows the formats established in ANSI/IEEE 754. Recall that the double precision format contains 52 explicit significand bits and one implicit bit. The largest decimal significand that can be represented in 53 binary digits is:

```
9007199254740991
```

This value corresponds to a 53-bit significand of all 1 digits. This means that a 53-bit binary significand allows representing decimal numbers to 15 or 16 significant digits.

Independently of the conversion and representation errors mentioned previous, the limit of the number of significant digits that can be encoded in the various floating point formats is often a cause of a errors in calculations. For example, the number

```
988,888,888,888,888,888
```

contains 18 significant digits. In the double format this number is stored in 16 significant digits as follows:

```
988,888,888,888,888,830
```

Similarly, a small number suffers an encoding error due to the number of significant digits that are supported in the format, for example, the value:

```
0.44444444444444444
```

is represented as

```
0.444444444444444
```

In either case the representation suffers a loss of precision. If you were to load the above values into the math unit and perform a multiplication operation, the result would be inexact, as follows:

```
988,888,888,888,888,888 * 0.44444444444444444  = 43950617283906110.0
```

Compare this result with the one obtained by encoding these numbers in BCD20 format and performing BCD multiplication

```
bcd result:       43950617283906172.0049382716049382
math unit result: 43950617283906110.0000000000000000
```

```
        difference:                        62.0049382716049382
```

Cancellation Error

A particularly interesting error, sometimes called a cancellation error, takes place when the operands of an arithmetic operation are a very large and very small number. For example, suppose a C++ program that defines the following double precision numbers:

```
double num1 = 512000000000000;   // 5.12e+14
double num2 = 0.0075;
```

If code now added the variable num1 and num2 the result would be:

```
512000000000000
```

which is the same value as the num1 operand. Notice that in this example both operands are well within the limits of the double precision format. However, the addition operation fails because the sum exceeds the capacity of the double format.

10.4 Financial Software

The low-level financial functions mentioned in this chapter are located in the Un32_8 module of the MATH32 library. The C++ interface functions are in the file Un32_8.h which is found in the Test Un32_8 project, located in the Chapter10 folder. Also in this project is a C++ program that exercises the library and interface routines and demonstrates how to access them from C++. The source and executable files are found in the book's CD ROM.

Chapter 11

Statistical Calculations

Chapter Summary

Statistics has been described as the universal language of science and technology. Descriptive statistics refers to collecting, presenting, and describing data, while inferential statistics allow interpreting descriptive statistics in order to draw conclusions and make decisions. This chapter is about the fundamental calculations and formulas used in descriptive and inferential statistics. In it we develop a minimal set of routines that solve common statistical formulas. The formulas relate to basic data manipulations, measures of central tendency, measures of dispersion, probability, normal distribution, and linear regression analysis.

11.0 About Statistical Data

Computer data used in statistical calculations can originate from many conceivable sources, and be formatted in virtually unlimited ways. Raw data can come directly from sensors and instruments. Primary data can be stored in any type of device and in multiple file formats. Processed data can be in standard or in proprietary formats and data types. For example, the Flexible Image Transport System (FITS) was developed by NASA's Science Office of Standards and Technology (NOST) to provide for the interchange and storage of astronomical data sets. The basic document describing the FITS standard is over 70 pages long and the data can be represented in five different data types. On the other hand, the Jet Propulsion Laboratory of the California Institute of Technology has developed the Planetary Data System (PDS) to further refine astronomical data that specifically relates to planetary science. The documentation for this standard extends to over two hundred pages.

This shows that it would be futile to attempt to anticipate how the data for a statistical calculation is input, manipulated, processed, and formatted. For practical reasons, the functions and routines developed in this chapter assume that data is already stored in arrays. The programmer must be aware that often this is not the case, and that the routines and functions will frequently need to be modified in order to accommodate a particular data source or format.

11.0.1 Data-Type-Flexible Coding

We have attempted to make our routines data-type flexible, within the limitations of the programming language. To a certain degree, type flexibility can be achieved by means of C++ templates. The template mechanism allows creating generic functions that operate with any compatible data type. In C and C++ the same purpose can be achieved with pointers to void, but templates are easier to develop and use. For example, the following template defines a function named Sort which can accept as a parameter an array of any C++ data type.

```
template <class A>
void Sort(A *arrS, int size )
{
...
```

In the function *arrS is a pointer to an array of any data type. The function will effectively sort an array of int, of double, of float, or of any other compatible data type.

11.1 Data Manipulation Primitives

Statistical formulas, as well as the numerical methods presented in Chapter 12, often require arithmetic operations on array data. One of the simplest functions is to obtain the sum of all the elements in an array. In mathematics the Greek capital letter sigma is used to represent summation of a set of addends, for example, the expression

$$\sum_{i=1}^{10} x_i$$

indicates the sum of all values of x_i, from $i = 1$ to $i = 10$. If all possible values of x are included, then the notation is often shortened to

$$\sum x$$

However, the editor programs used in producing computer source code are usually not capable of generating special symbols. For this reason in the code listings we use the word SIG to represent the Greek capital letter sigma, for example:

$$\text{SIG}(x) = \sum x$$

11.1.1 Common Summations

Several standard summations are often required in statistical computations. It is convenient to have available functions to perform these basic functions. Since symbols are not allowed in allowed in C++ function names, we had to come up with some equivalent text designations. Table 11.1 shows the mathematical expression and the function names.

Table 11.1

C++ Function Names for Summation Operations

MATHEMATICAL EXPRESSION	FUNCTION NAME	DESCRIPTION
$\sum x$	= SIGx()	Sum values of x
$\sum x^2$	= SIGxSq()	Square x values, then sum
$\left(\sum x\right)^2$	= SIGxToSq()	Sum x values, then square
$\sum xy$	= SIGxy()	Multiply x * y values, then sum
$n\sum xy$	= SIGxyByn()	Multiply n by SIGxy
$\left(\sum x\right)\left(\sum y\right)$	= SIGxBySIGy()	Multiply SIGx times SIGy
$\sum_{i=a}^{b} x_i$	= SIGxSkipCnt()	Add values of x from i = a to b

The functions listed in Table 11.1 are found in the Un32_9.h file in the Chapter11 folder in the book's CD ROM. The code for the SIGxSkipCnt() function is as follows:

```
template <class A>
double SIGxSkipCnt(A *arr, int skip, int count, int size)
{
// Function to find the sum of values in an array of any
// numeric type, starting at a given offset from the first
// element, for a number of elements
//
// On entry:
//          arr[] is a numeric array
//          skip is the start offset
//          count is the number of elements to use
//          size if the number of elements in the array
//          calculated by the caller, as follows:
//              size = sizeof arr / sizeof arr[0];
//
// On exit:
//          Returns the sum of the array elements
//          Returns 0 if invalid data
// First test for valid input data
//
if((skip + count) > size)
    return 0.0;
double aSum = 0;
for(int x = skip; x < (count + skip); x++)
    aSum = arr[x] + aSum;
return aSum;
}
```

PROGRAMMER'S NOTEBOOK

In C++, when an array is passed as an argument to a function, the number of elements in the array is not available to the function code. If this parameter is required, it must also be passed as an argument by the caller. The conventional way to obtain the number of elements in an array is the expression:

```
sizeof array / sizeof array[0]
```

Here the sizeof operator is first applied to the array name, which returns the number of bytes in the array. This value is then divided by the byte size of each array element, which is obtained by applying the sizeof operator to the array element at offset 0. This problem was resolved in Java, in which a called method can directly determine the size of an array argument.

11.1.2 Sorting

Several statistical functions are easier to implement if the data is first ordered numerically, in other words, if it has been sorted. Donald Knuth defines sorting as the rearrangement of items in an ascending or descending order. Volume III of his series titled *The Art of Computer Programming* (see Bibliography) is titled *Sorting and Searching*. The fact that Knuth devotes much of an entire tome to this topic gives an idea of the complexity and depth of this subject.

Many sorting algorithms have been devised, some of which are quite elaborate. The easiest to understand and implement is probably the infamous bubble sort, which is also one of the least efficient ones. The so-called selection sort is easy to implement and offers acceptable performance for the purpose at hand. The following template function implements the sorting in ascending order of array data.

```
template <class A>

//
void Sort(A *arrS, int size )
{
// Sort an array of any numeric type using the selection
// sort algorithm
    int maxInd;    // Index of largest value in each pass
    int bottom;    // False bottom index for each pass
    int i;         // Local counter
    A temp;        // Temporary storage

    for (bottom = size-1; bottom >= 1; bottom-)
      {
        maxInd = 0;
        for (i = 1; i <= bottom; i++)
            if (arrS[i] > arrS[maxInd])
                maxInd = i;
        temp = arrS[bottom];
        arrS[bottom] = arrS[maxInd];
        arrS[maxInd] = temp;
      }
}
```

11.2 Counting Techniques

A frequent statistical problem consists of determining the number of different ways in which related sets can be counted. For example, assume that there are four candidates for the presidency of a professional association, and three candidates for the secretary position. Also assume that the officers are elected separately and that the same individual cannot hold both offices. The question is, how many combinations of president and secretary can be elected? The graph in Figure 11.1 shows all possible options.

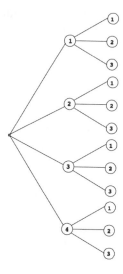

Figure 11.1 *Tree Diagram for an Election Problem*

Since there are three possible secretaries for each of four possible presidents, the total number of combinations of possible pairs are 3 times 4, or 12. This leads to the fundamental counting principle, which states that if any task A can be performed in n different ways, and another independent task B in m different ways, then task A followed by task B can be performed in n * m different ways.

A refinement of the counting principle described in the preceding paragraph relates to the order of the elements in a set. When the order of the elements is important we refer to a permutation, when the order is not important we refer to a combination. For example, any two items out of a set of 3 (labeled a, b, and c) can be associated as follows:

(a,c), (c,a), (a,b), (b,a), (b,c), and (c,b)

which results in six permutations. When the order of the items is not significant, then the following subsets are possible:

{a,c}, {a,b}, and {b,c}

which results in three combinations. Notice that conventional statistical notation uses parenthesis to indicate permutations and curly braces to indicate combinations.

11.2.1 Permutations

In the case of permutation the order in which the elements appear is important. For example, three elements (labeled a, b, and c) can be taken in the following orders:

(a,b,c), (a,c,b), (b,a,c), (b,c,a), (c,a,b), and (c,b,a)

In this case the number of permutations is 6.

The formal definition of a permutation of r elements, from a set S consisting of n elements, is an ordered arrangement of these r elements without repetitions. The conventional notation for the number of permutations is:

$$_nP_r$$

where n is the total number of elements in the set, and r is the number of elements selected from the set. The permutations formula is:

$$_nP_r = \frac{n!}{(n-r)!}$$

In the case of the earlier example consisting of all possible combinations of three elements, the value of n = 3, and the value of r = 3. Therefore:

$$_nP_r = \frac{3!}{(3-3)!} = 6$$

Recall that 0! = 1. The following low-level procedure calculates the number of permutations.

```
.DATA
; This section contains data elements for all the low-level
; procedures in the Un32_9 module of the MATH32 library
NELS        DD          0      ; Storage for n
RELS        DD          0      ; Storage for r
TEMPV       DD          0      ; Temporary storage
; For normal distributions curve
X           DD          0      ; x coordinate of graph
U           DD          0      ; Mean
```

```
SD          DD          0      ; Standard deviation
; Definition of constant e
E           DT          4000ADF85458A2BB4A9BH


            .CODE
_PERMUTATIONS     PROC
; Ordered permutation without repetitions of n elements, in set S,
; taken r at a time, according to the formula:
;
;                 n!
;     nPr = ----------
;               (n - r)!
;
; where:
;           nPr = number of permutations of r elements in a set of
;                 n elements
;           n = number of elements in set S
;           r = number of elements in permuation (r at a time)
;
; Note: code assumes that n and r are integers
;       Uses the _FACTORIAL procedure in the Un32_7 module of
;       the math32 library.
; On entry:
;           ST(0) = n (number of elements)
;           ST(1) = r (taken r at a time)
; On exit:
;           ST(0) = number of possible permutations
;                              | ST(0) | ST(1) | ST(2) | ST(3) |
;                              |   n   |   r   | EMPTY | EMPTY |
; Calculate (n - r)!
            FIST        NELS      ; Save n
            FSUB        ST,ST(1)  ; n-r  |   r   | EMPTY |
            CALL        _FACTORIAL ;(n-r)! |   r   | EMPTY |
            FILD        NELS      ;   n   | (n-r)!| EMPTY |
            CALL        _FACTORIAL ; n!   | (n-r)!| EMPTY |
            FDIV        ST,ST(1)  ; n!/
                                  ;(n-r)! | (n-r)!| EMPTY |
            FSTP        ST(1)
            FRNDINT               ; Round ST to integer
            CLD
            RET
_PERMUTATIONS     ENDP
```

SOFTWARE ON-LINE

The C++ interface function is named Permutations(). The function is located in the Un32_9.h file located in the CHAPTER11 folder in the book's on-line software.

11.2.2 Combinations

We speak of combinations when the order of the items in the resulting sets is not taken into account. In combinations the different orderings of the same items are not counted separately. For example, the number of combinations of two elements, out of a set of three elements (labeled a, b, and c) is as follows:

{a,b}, {a,c}, and {b,a}

The statistical notation for the number of combinations of a subset of r elements, from a set of n elements is:

$$_nC_r$$

The fact that each subset of r elements has $r!$ permutations of its elements, can be expressed as follows:

$$_nP_r = {_nC_r} \times r!$$

This leads to the following formula for the number of combinations:

$$_nC_r = \frac{n!}{r!(n-r)!}$$

The following low-level procedure calculates the number of combinations.

```
_COMBINATIONS          PROC
; Combination of r elements, in set S, that contains
; n elements, according to the formula:
;
;                 n!
;     nCr =    ----------
;              r!(n - r)!
;
; where:
;           nCr = number of combinations of r elements from a set of
;                 n elements
;           n = number of elements in set S
;           r = number of combinations of r elements
;
; Note: code assumes that n and r are integers
;       Uses the _FACTORIAL procedure in the Un32_7 module of
;       the math32 library.
; On entry:
;           ST(0) = n (number of elements)
;           ST(1) = r (taken r at a time)
; On exit:
;           ST(0) = number of possible combinations
;                                 | ST(0) | ST(1) | ST(2) | ST(3) |
;                                 |   n   |   r   | EMPTY | EMPTY |
; Calculate r!(n - r)!
          FISTP        NELS          ; Save n
                                     ;   r   | EMPTY |
          FIST         RELS          ; Save r
;
; Calculate r!
          CALL         _FACTORIAL    ;  r!   |
          FILD         RELS          ;  r    |  r!   | EMPTY |
          FILD         NELS          ;  n    |  r    |  r!   | EMPTY |
; Calculate (n-r)!
          FSUB         ST,ST(1)      ;  n-r  |  r    |  r!   |
```

```
        FSTP        ST(1)          ;  n-r  |    r!  | EMPTY |
        CALL        _FACTORIAL     ;(n-r)! |    r!  | EMPTY |
 ; Calculate r!(n-r)!  =  [d]
        FMUL                       ;  [d]  | EMPTY |
        FILD        NELS           ;   n   |   [d]  | EMPTY |
        CALL        _FACTORIAL     ;   n!  |   [d]  | EMPTY |
        FDIV        ST,ST(1)       ; n!/[d]|   [d]  | EMPTY |
        FSTP        ST(1)
        FRNDINT                    ; Round ST to integer
        CLD
        RET
 _COMBINATIONS     ENDP
```

SOFTWARE ON-LINE

The C++ interface function is named Combinations(). The function is located in the Un32_9.h file located in the CHAPTER11 folder in the book's on-line software.

11.2.3 Binomial Probability

Many problems in statistics deal with two possible outcomes: success or failure, pass or fail, acceptable or defective, boy or girl, and so on. It is possible to calculate the probability of repeated events with two possible outcomes given the following conditions:

1. There must be a fixed number of events.

2. For each event there must be two possible outcomes.

3. Each event must be independent.

4. The probability for each of the two possible outcomes must be the same for each event.

These conditions are sometimes called the Bernoulli conditions since they were first defined by Jacob Bernoulli. In regard to an event that can occur in two equally likely ways, the probability of the same outcome taking place r times out of n trials is expressed by the formula:

$$P(x) = \frac{\dfrac{n!}{r!(n-r)!}}{2^n}$$

where $P(x)$ denotes the probability of obtaining exactly r instances of the same outcome. Notice that the numerator of the right term of the formula is the number of possible combinations of a subset of r elements, from a set of n elements.

The following low-level procedure calculates the binomial probability.

```
_BINOMIAL_PROB      PROC
; In an event that can occur in two equally likely ways,
; calculate the probability of obtaining r instances if the same
```

```
; outcome, according to the formula:
;
;           nCr
;     P = ------
;           2^n
;
; where:
;         nCr = number of combinations of r elements in a set of
;               n elements
;         n = number of trials
;         r = number of repetitions
;         P = binomial probability of event r occurrences out of
;             n trials
;
; Note: code assumes that n and r are integers
; On entry:
;         ST(0) = n (number of elements)
;         ST(1) = r (taken r at a time)
; On exit:
;         ST(0) = binomial probability
;                               | ST(0) | ST(1) | ST(2) | ST(3) |
;                               |   n   |   r   | EMPTY | EMPTY |
        FIST        NELS        ; Save n
; Calculate nCr
        CALL        _COMBINATIONS
;                               | nCr   | EMPTY |
        FISTP       TEMPV       ; Store nCr
; Calculate 2^n
        FLD1                    ;   1   | EMPTY |
        FLD1                    ;   1   |   1   | EMPTY |
        FADD                    ;   2   | EMPTY |
        FILD        NELS        ;   n   |   2   | EMPTY |
        CALL        _X_TO_Y_BYFAC
                                ;  2^n  | EMPTY |
        FILD        TEMPV       ;  nCr  |  2^n  |
        FDIV        ST,ST(1)    ;nCr/2^n|  2^n  |
        FSTP        ST(1)       ;nCr/2^n| EMPTY |
        CLD
        RET
_BINOMIAL_PROB      ENDP
```

SOFTWARE ON-LINE

The C++ interface function is named BinomialProb(). The function is located in the Un32_9.h file located in the CHAPTER11 folder in the book's on-line software.

11.3 Measures of Central Tendency

One of the simplest, and most useful, statistical functions is the determination of the average value in a data set. However, the term average can be interpreted and calculated differently. The different versions of the intuitive notion of average are called the statistical measures of central tendency, or the measures of center.

Four methods of central tendency are defined in elementary statistics: the mean, the median, the midrange, and the mode.

11.3.1 Mean

The mean, or arithmetic mean, is the conventional average. It is calculated by adding all the entries in the data set and dividing by the number of entries. This result is represented by the formula:

$$\bar{x} = \frac{\sum x}{n}$$

where the symbol $\bar{x}$ (pronounced x-bar) is the mean value in the data set. The code for calculating the mean is almost trivial, as shown in the following template function:

```
template <class A>
double Mean(A *arr, int size)
{
// Function to find the mean in an array of any type
// On entry:
//          arr[] is a numeric array
//          size is the number of elements in the array
//          calculated by the caller, as follows:
//              size = sizeof arr / sizeof arr[0];
// On exit:
//          Returns the mean value in the array
double aSum = 0;
for(int x = 0; x < size; x++)
    aSum = arr[x] + aSum;
return aSum/size;
}
```

11.4.2 Median

The median is defined as the value in the middle of the data set. It assumes that the data set is arranged in increasing or decreasing magnitude. That is, it assumes that the data has been previously sorted. One difficulty associated with calculating the median is that a data set with an odd number of entries has an entry located exactly in the middle. However, if there is an even number of entries, then there is not one but two central entries. In the first case (that is, when there are an odd number of entries) the median is the one central element in the set. In the second case (that is, when there is an even number of entries) then the median is the mean of the two central entries. The following C++ template function calculates the median.

```
template <class A>
double Median(A *arr, int size)
{
// Function to find the median in an array of any type
// On entry:
//          arr[] is a numeric array
//          size is the number of elements in the array
//          calculated by the caller, as follows:
//              size = sizeof arr/ sizeof arr[0];
// Terms:
```

```
//          case 1:
//            If arr[] has an odd number of elements,
//            then the median is the value of the middle
//            element
//          case 2:
//            If arr[] has an even number of elements,
//            then the median is the mean of the two middle
//            elements
// On exit:
//          Returns the median value in the array
double median = 0;

int midE = size / 2;
// Sort the copied array

A *arrCopy = new A[size];

// Copy data from original array
for(int j = 0; j < size; j++)
    arrCopy[j] = arr[j];

Sort(arrCopy, size);
if(size % 2 != 0)           // if odd number of elements
   median = (double) arrCopy[midE];
else
   median = (double) ((arrCopy[midE] + arrCopy[midE - 1]) /2.0);

return median;
}
```

Notice that the Median() function creates a copy of the caller's array before sorting. This ensures that the caller's data is not modified.

11.3.3 Midrange

The midrange is defined as the value that is one-half the sum of the lowest and highest values in the data set. The following C++ template function calculates the midrange.

```
template <class A>
double Midrange(A *arr, int size)
{
// Function to find the midrange in an array of any type.
// The midrange is the value midway between the highest
// and the lowest entry.
// On entry:
//          arr[] is a numeric array
//          size is the number of elements in array
//          calculated by the caller, as follows:
//             size = sizeof arr/ sizeof arr[0];
// On exit:
//          Returns the midrange value in the array
double midrange = 0;
// Sort a copy of the array passed as an argument
A *arrCopy = new A[size];

// Copy data from original array
for(int j = 0; j < size; j++)
    arrCopy[j] = arr[j];
```

```
Sort(arrCopy, size);

midrange = (arrCopy[size - 1] + arrCopy[0]) / 2.0;
return midrange;
}
```

11.3.4 Mode

The value that occurs most frequently in a data set is called the mode. If there are no repeated values, then the data set contains no mode. If there is more than one value with the same number of repetitions, then the data set contains more than one mode. In this context the term "multiplier" is used to indicate the number of repetitions of an element in the data set. Thus, the mode is the entry or entries with the highest multiplier.

The mode is the measure of central tendency with the most complex calculation. Here again, many algorithms have been devised for obtaining the mode. The calculations are considerably simplified if the data set is sorted. Once sorted, the repeated data elements are adjacent to each other, and the mode can be obtained with a single pass through the array.

Because a data set can contain no mode, one mode, or multiple modes, the calculating routine must return two values to the caller: one value, which could be called the *mode code*, indicates the number of modes. The second value is the mode itself. This second parameter, which we can call the *mode value*, is significant only if the mode code is not zero. If the routine must report all the modes found in the data set, then the mode value must be a compound data type, such as array. Alternatively we can make the mode value a scalar type and return just one of the identified modes. For example, if the mode routine returns a mode code of 3, and a scalar mode value of 4, then there are three modes in the set, one of which is the value 4. In the case of a scalar mode value, the routine can return either the highest valued or the lowest valued mode.

The flowchart in Figure 11.2, in the following page, depicts the logic used in obtaining the mode or modes in a data set. In this case the mode value is a scalar variable which encodes the mode with the highest index number. The flowchart is based on using two pointers (ePtr and sPtr) to visit all the elements in the sorted data array. Here sPtr is that start offset pointer and ePtr is the end offset pointer. Two variables are used to keep count of the multiplicity of the mode candidates. Another variable (loc) remembers the offset of the last found mode, while the variable numMode keeps track of the number of modes found. The logic proceeds as follows:

1. If the start pointer is at the next-to-the-last array element the entire data set has been visited and the routine ends. In this case the variable numModes holds the number of valid modes, and the array entry located at datArr[loc] is the highest indexed mode found.

2. To avoid comparing the same entries in the data set, the end pointer is now equated to the start pointer plus 1.

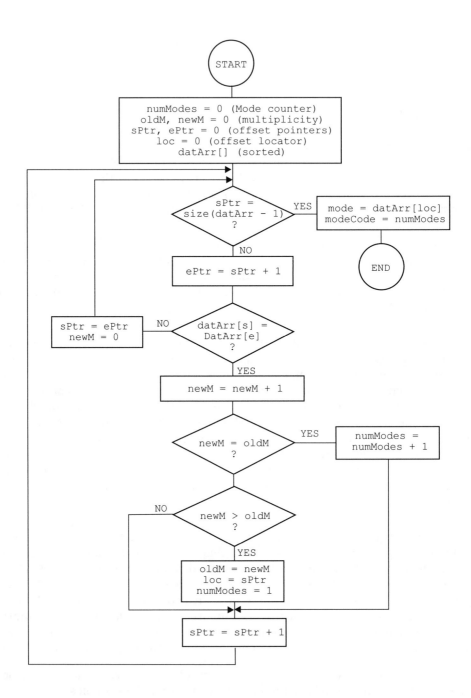

Figure 11.2 *Flowchart for Mode Calculation*

3. When the element located at the start pointer (sPtr) is not equal to the element at the end pointer (ePtr), then the new multiplicity variable (newM) is reset and the start pointer (sPtr) is moved to the position of the end pointer (epTR). Execution resumes at step 1.

4. If the entries located at the two offset pointers are equal, then a mode candidate has been found. In this case the new multiplicity counter (newM) is incremented by one.

5. If the newly found mode candidate has the same multiplicity as the old one, then the number of modes (numModes) is incremented. In this case execution continues at step 7.

6. If a new mode candidate has a higher multiplicity than the old one, the old one is discarded since it is no longer a mode. At that time the location of the mode candidate is stored in a variable (loc) and the modes counter (numModes) is set to 1.

7. The start pointer (sPtr) is incremented by one and execution continues at step 1.

The following C++ function template implements finds the mode in an array of any data type.

```
template <class A>
double Mode(A *arr, int *modeCode, int size)
{
// Function to find the mode, if one exists, in an array
// of any type
// On entry:
//          arr[] is a numeric array
//          modeCode is a user variable to hold a code
//          returned by the function
//          size is the number of elements in the array
//          calculated by the caller, as follows:
//              size = sizeof arr / sizeof arr[0];
// Terms:
//          The mode is the value that occurs most
//          frequently in the array.
//          case 1:
//          If no value occurs more than once, then
//            the array has no mode. Function returns
//            mode = zero, modeCode = 0.
//          case 2:
//            It is possible for an array to have more than
//            one mode. In this case the function returns:
//            mode = mode with highest index
//            modeCode = number of modes.
// On exit:
//          Returns the mode in the array, if there is one
//          Zero if no mode found.
//          Mode is always returned in a variable of type
//          double
//          Modecode holds the number of modes.

double mode = 0;

int numModes = 0;        // Number of modes found
int oldM = 0, newM = 0; // Multiplicity counters
int sPtr = 0, ePtr = 0;      // Offset pointers
int loc = 0;                 // Offset location of mode
```

```
// A match;

// Create a copy of the array passed as an argument
A *arc = new A[size];
// Copy data from original array
for(int j = 0; j < size; j++)
    arc[j] = arr[j];
// Sort the copy
Sort(arc, size);

// Find mode

  while(sPtr < (size - 1))
  {
    ePtr = sPtr + 1;
    if(arc[sPtr] != arc[ePtr]) // No match condition
    {                          // sPtr = ePtr
      sPtr = ePtr;             // Reset multiplier
      newM = 0;                // Continue search
      continue;
    }
    else                       // Match
    {
      newM++;                  // Bump multiplier
        if(newM == oldM)
            numModes++;
        else if(newM > oldM) // New multiplier is larger
        {
            oldM = newM;       // Reset old multiplier
            loc = sPtr;        // Store location of match
            numModes = 1;      // bump modes counter
        }
    }
    sPtr++;
  } // end of loop

//    Case 1: No mode in set
//            numModes = 0;
//    Case 2: One mode in set
//            numModes = 1;
//    Case 3: More than one mode
//            numModes > 1
mode = (double) arc[loc];
*modeCode = numModes;

return mode;
}
```

11.3.5 Weighted Measures of Central Tendency

Sometimes statistical data is furnished in two separate sets. One holds the values and the other one holds the frequency of each data entry. For example:

```
        2   4   5   7   3   1   <= values set

        1   2   1   3   4   2   <= frequency set
```

The weighted mean is expressed by the formula:

$$\bar{x} = \frac{\sum xw}{\sum w}$$

where the x values are the data entries, the w values are the frequency entries, and $\bar{x}$ is the weighted mean. The following C++ template function calculates the weighted mean.

```
template <class A>
double WeightedMean(A *dat, int dist[], int size)
{
// Function to find the weighted mean according to the
// frequency distribution of the data
// On entry:
//          dat[] is a numeric array of the data
//          dist[] is an integer array of the frequency
//          distribution of the data
//          size is the number of elements in the array
//          calculated by the caller, as follows:
//              size = sizeof arr / sizeof arr[0];
// According to the formula:
//
//                SIG xw
//        Wx =    --------
//                SIG w
// where:
//        x is the data elements
//        w is the frequency
//        Wx is the weighted mean
// On exit:
//          Returns the weighted mean value in the data
//          array according to the frequency distribution

double wMean = 0;
A sumwx = 0;
int sumw = 0;

// calculate SIG w * x and SIG w
for(int x = 0; x < size; x++)
{
    sumwx = sumwx + (dat[x] * dist[x]);
    sumw = sumw + dist[x];
}
wMean = (double)sumwx / (double)sumw;

return wMean;
}
```

SOFTWARE ON-LINE

The file Un32_9.h, in the Chapter11 folder in the book's on-line software, also contains C++ template functions for calculating the weighted median and the weighted mode. Also in this file are the functions for calculating the measure of central tendency described earlier in the section.

11.4 Measures of Dispersion

The measures of central tendency do not provide all statistical information about data sets. For example, consider these two data sets:

```
6   7   8   8   11   <= data set 1

0   8   8   9   15   <= data set 2
```

In both data sets the mean, the median, and the mode are 8, however, it is clear that the data in the set 2 is more spread out than the data in set 1. Although the midrange gives us some idea that the data distribution is not uniform, more detail is often necessary. The measures of dispersion provide information about data distribution. They are the *range*, the *variance*, and the *standard deviation*.

11.5.1 Range

The simplest of the measures of dispersion is the range. The range is defined as the difference between the largest and the smallest element in the set. In regards to the data sets in the preceding section, the range is as follows:

```
data set 1, range = 11 - 6 = 5

data set 2, range = 15 - 0 = 15
```

The following C++ template function calculates the range.

```cpp
template <class A>
double Range(A *arr, int size)
{
// Function to find the range of the data set, defined
// as the difference between the largest and the smallest
// element in the set
// On entry:
//         arr[] is a numeric array
//         size is the number of elements in the array
//         calculated by the caller, as follows:
//             size = sizeof arr / sizeof arr[0];
// On exit:
//         Returns the data range

double range = 0;

A largest = 0;    // Largest value
A smallest;

// Obtain the largest value in the array
for(int x = 0; x < size; x++)
{
    if(arr[x] > largest)
        largest = arr[x];
}
smallest = largest;
// Obtain the smallest value in the array
for(x = 0; x < size; x++)
{
    if(arr[x] < smallest)
        smallest = arr[x];
}
range = largest - smallest;
```

```
return range;
}
```

Alternatively, the range can be calculated by first sorting the array, then finding the difference between the last and the first element.

11.4.2 Variance

Since the range uses only the extreme values in the data set, it does not provide any information regarding the rest of the elements. Two sets with the same range can contain totally different data. The variance (as well as the standard deviation, discussed in the following section) takes into account all the values in the data set. The variance is defined as a measure of the spread of the data about the mean. The larger the variance, the greater the data dispersion.

Regarding measures of dispersion, we must make the distinction between functions that apply to a sample set and those that apply to an entire population. The traditional notation uses the letters of the English alphabet when referring to a sample set, and the letters of the Greek alphabet when referring to entire populations. The two formulas for the variance are as follows:

$$s^2 = \frac{\sum (x - \overline{x})^2}{n - 1} \qquad \sigma^2 = \frac{\sum (x - \overline{x})^2}{n}$$

In both formulas $\overline{x}$ is the mean of the data set, the x values are the data values, and n is the number of elements in the data set. The leftmost formula refers to the variance of a sample set, and the rightmost one to the variance of a population. As the denominator gets larger, the difference between formulas becomes smaller. At $n = 30$ the value of x^2 and s^2 is almost the same.

The following C++ template function calculates the variance of a data set.

```
template <class A>
double VarianceS(A *dat, int size)
{
//
// Function to find the variance of a sample set, defined
// by the following formula:
//
//              SIG(x - m)^2
//      vS =  --------------
//              n - 1
// where:
//         x are the sample data
//         m is the mean of the sample data
//         n is the number of data entries
//         vS is the sample variance
//
// On entry:
//         dat[] is a numeric array
//         size is the number of elements in the array
//         calculated by the caller, as follows:
//             size = sizeof arr / sizeof arr[0];
```

```
// On exit:
//           Returns the variance
double varS = 0;
double dMean;
double sumXm = 0;
// Obtain and store mean
dMean = Mean(dat, size);
// Calculate SIG(x - m)^2
for(int x = 0; x < size; x++)
    sumXm = sumXm + (dat[x] - dMean)*(dat[x] - dMean);
varS = sumXm / (size - 1);
return varS;
}
```

SOFTWARE ON-LINE

The C++ template function VarianceP(), in book's on-line software, calculates the variance of a population. Both functions are found in the Un32_9.h file located in the Chapter11 folder in the CD ROM.

11.4.3 Standard Deviation

By far the most used measure of dispersion is the standard deviation. The standard deviation, which is the positive square root of the variance, provides a measure of the data distribution. Like the variance, the standard deviation is defined for a sample or for an entire population. The formulas are as follows:

$$ s = \sqrt{s^2} \qquad \sigma = \sqrt{\sigma^2} $$

Here again, the leftmost formula refers to the standard deviation of a data set, and the rightmost one to the standard deviation of a population.

One of the reasons why standard deviation is often more useful than variance is that standard deviation is expressed in the same units of measure as the data set. For example, if the data set is in years, then the standard deviation is also expressed in years. The variance, on the other hand, would be in years squared, which is not as meaningful.

Traditional textbooks on statistics often list several formulas for calculating the variance, and consequently, the standard deviation. The rational for multiple formulas is to facilitate computations according to the characteristics of the data. For example, one formula for the variance is more suitable when the mean is a decimal fraction, while another one is more suitable when the mean is a whole number. The use of computer algorithms in evaluating statistical functions makes multiple formulas mostly unnecessary.

SOFTWARE ON-LINE

The C++ template functions for calculating the standard deviation are StdDevS() and StdDevP(). The first one refers to a sample and the second one to a population. The functions are found in the file Un32_9.h located in the Chapter11 folder in the book's on-line software.

11.5 Normal Distribution

Standard deviation is defined as measure of the data dispersion. It provides a mechanism for comparing one data set to another one. Several mathematical theorems and rules help understand the meaning of standard deviation.

1. Chebyshev's theorem states that at least 75 percent of the data will be located within two standard deviations on either side of the mean. Also, that at least 89 percent of the data will be within three standard deviations on either side of the mean.

2. The Empirical Rule states that, if data is distributed normally, then approximately 68 percent will be within one standard deviation on either side of the mean, that approximately 95 percent will be within two standard deviations on either side of the mean, and that approximately 99.7 percent of the data will be located within three standard deviations on either side of the mean.

Notice that Chebyshev's theorem refers to any data set, while the empirical rule assumes that the data is distributed normally. The rule is called empirical because it establishes an empirical measure of normality for data distribution.

11.5.1 Normal Curve

The empirical rule is used in statistics to determine if a data set is normally distributed. When data is normally distributed, a Cartesian plot showing data values on the x-axis and the frequencies on the y-axis results in a bell-shaped curve, called the normal frequency or normal distribution curve. Figure 11.3 shows the normal distribution curve or normal curve.

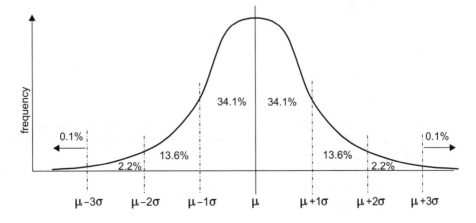

Figure 11.3 *Normal Distribution Curve*

The normal curve in Figure 11.3 extends indefinitely in both directions, never reaching the x-axis. The vertical line at the center of the curve, labeled with the Greek letter m (mu), represents the mean. The curve is symmetrical about the mean. In Figure 11.3, the vertical lines to the left and right of the mean are located at a distance of one standard deviation from the mean. Standard deviation is represented with the Greek letter σ (sigma) in the diagram.

In a normal distribution 34.1 percent of the data is located one standard deviation to the left and to the right of the mean. 13.6 percent of the data is located between one and two standard deviations on either side of the mean. 2.2 percent of the data is located between two and three standard deviations on either side of the mean. Adding these values we see that close to 100 percent of the data is located between three standard deviations to the left and right of the mean. Only 0.1 percent of the data is more than three standard deviations from the mean. The equation of the normal frequency curve is:

$$ y = \frac{e^{-(x-\mu)^2/(2\sigma^2)}}{\sigma\sqrt{2\pi}} $$

where e is the mathematical constant approximately equal to 2.71828183, μ is the mean, and σ is the standard deviation.

PROGRAMMER'S NOTEBOOK

The constant e can be defined to ten bytes of precision in low-level code with the following statement:

```
E         DT          4000ADF85458A2BB4A9BH
```

The procedure named _LOADE located in the Un32_9 module of the MATH32 library returns the constant e to double precision. The value is:

```
2.71828182845904523
```

C++ code can define the constant e as follows:

```
#define E 2.71828182845904523
```

or with the const keyword:

```
const double E = 2.71828182845904523;
```

Since Visual C++ does not support floating-point values to extended precision (10 bytes), 15 bits of significance are lost when the constant is defined in a double precision variable.

Standard Normal Curve

Textbooks on elementary statistics often define a simpler version of the normal curve. The resulting formula is:

$$y = \frac{e^{-x^2/2}}{\sqrt{2\pi}}$$

This simplification of the exponential equation is called the standard normal curve. The standard normal curve makes hand calculations easier and serves didactical purposes. When computers are used to calculate the values of the dependent variable, the equation for the standard normal curve becomes unnecessary.

11.5.2 Calculating f(x)

The core primitive for normal distribution statistics is a routine that returns the value of the dependent variable (y) for any corresponding value of the independent variable (x) according to the first equation in section 11.6.1. Since the equation involves the constant e, as well as an exponential, it is preferable to use low-level code. The following procedure returns y as a function of x using the normal curve equation.

```
_NORMAL_CURVE_Y    PROC
; Calculates y = f(x) for the normal distribution curve based on
; the formula:
;           e^-((x-u)^2/2s^2)
;     Y = -------------------
;             s SQR(2Pi)
; where:
;       x is the Cartesian coordinate in terms of the
;         mean.
;       y is the value of the dependent variable
;       u is the mean
;       s is the standard deviation
; On entry:
;       ST(0) = x
;       ST(1) = u
;       ST(2) = s
;       e is defined as a memory variable
; On exit:
;       ST(0) = Y
;************************|
;    save parameters    |
;************************|
;                           ;  ST0  |  ST1  |  ST2  |
;                           |   x   |   u   |   s   |
        FSTP    X           ;   u   |   s   | EMPTY |
        FSTP    U           ;   s   | EMPTY |
        FSTP    SD          ; EMPTY |
    ; Calculate 2s^2
        FLD1                ;   1   | EMPTY |
        FADD    ST,ST(0)    ;   2   | EMPTY |
        FLD     SD          ;   s   |   2   | EMPTY |
        FMUL    ST,ST(0)    ;  s^2  |   2   | EMPTY |
        FMUL                ;  2s^2 | EMPTY |
    ; Calculate (x-u)^2/2s^2
        FLD     U           ;   u   | 2s^2  | EMPTY |
        FLD     X           ;   x   |   u   | 2s^2  |
        FSUB    ST,ST(1)    ;  x-u  |   u   | 2s^2  |
        FSTP    ST(1)       ;  x-u  | 2s^2  | EMPTY |
        FMUL    ST,ST(0)    ;(x-u)^2| 2s^2  | EMPTY |
```

```
        FDIV        ST,ST(1)      ; (x-u)^2
                                  ; /2s^2 | 2s^2 | EMPTY |
                ; (x-u)^2/2s^2 = w
        FSTP        ST(1)         ;   w   | EMPTY |
    ; calculate e^w
        FLD         E             ;  e    |  w    | EMPTY |
        FXCH                      ;  w    |  e    | EMPTY |
        CALL        _X_TO_Y_MIXED ; In Un32_5
                                  ;  e^w  | EMPTY |
    ;
    ; Calculate e^-w = 1/e^w
        FLD1                      ;   1   | e^w   | EMPTY |
        FDIV        ST,ST(1)      ; 1/e^w | e^w   | EMPTY |
        FSTP        ST(1)         ; e^-w  | EMPTY |
    ;
    ; Calculate denominator: s * SQR 2Pi
        FLDPI                     ;  Pi   | e^-w  | EMPTY |
        FADD        ST,ST(0)      ;  2Pi  | e^-w  | EMPTY |
        FSQRT                     ;SR(2Pi)| e^-w  | EMPTY |
        FLD         SD            ;   s   |SR(2Pi)| e^-w  |
        FMUL                      ;S *
                                  ;SR(2Pi)| e^-w  | EMPTY |
    ; s * SR(2Pi) = v
        FXCH                      ; e^-w  |   v   | EMPTY |
        FDIV        ST,ST(1)      ; f(x)  |   v   | EMPTY |
        FSTP        ST(1)         ; f(x)  | EMPTY |
        CLD
        RET
_NORMAL_CURVE_Y   ENDP
```

The C++ code for normal distribution function is as follows:

```
double NormalDistY(double x, double mean, double sd)
{
// Function to calculate normal distribution given the mean
// and the standard deviation.
//
// On entry:
//        x is the independent variable
//        mean is the mean
//        sd is the standard deviation
// On exit:
//        return f(x) = y for normal distribution
//
double fOfx;
    // Calculate y = f(x)
    _asm
    {
    ; Function expects:          |  ST0  |  ST1  |  ST2  |
                                 ;   x   | mean  |  sd   |
        FLD         sd
        FLD         mean
        FLD         x
        CALL        NORMAL_CURVE_Y
        FSTP        fOfx
    }
return fOfx;
}
```

11.5.3 Probability in Normal Distribution

The normal distribution curve shown in Figure 11.3 can be used to calculate the probability of normal distribution. In this interpretation the area under the curve has a value of 1 and the probability of normal distribution is a part of this total area located between the desired upper- and lower-limits. By looking at Figure 11.3 we can see that the probability of a value lying between the x values of minus one and plus one standard deviations is approximately 0.682. By the same token, the probability of normal distribution between x = mean and x = mean plus one standard deviation is approximately 0.341.

Assume that the life of a light bulb was normally distributed with a mean of 92 hours and a standard deviation of 10 hours. Now suppose that we needed to calculate the normal probability of a light bulb lasting between 97 and 120 hours. The example can be represented using a normal distribution curve, modified to show the mean and standard deviations for this particular case, as in Figure 11.4.

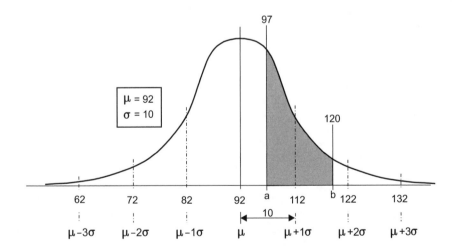

Figure 11.4 *Example of Normal Probability Distribution*

In Figure 11.4 the total probability is represented by the area under the curve that defines y as a function of x. The probability of a light bulb lasting between 97 and 120 hours is then the area under the curve from x = 97, to x = 120. This area is shown in gray in Figure 11.4. The equation that defines the area under the curve from a to b, in Figure 11.4, is the basic integral:

$$Area = \int_{a}^{b} f(x)dx$$

The numerical approximation of the integral is one of the topics of Chapter 12. The following function uses Simpson's rule to calculate the area under the curve.

```
double NormalProbSimp(double a, double b, double mean, double sd)
{
// Calculate normal probability by estimating the area under
// the normal distribution curve, as follows:
//
//      Np = INTEGRAL(from a to b) f(x)dx
//
// where:
//      a is the x coordinate of the first point in the curve
//      b is the x coordinate of the second point
//      f(x) is the normal distribution function:
//                  e^-((x-u)^2/2s^2)
//          f(x) = ------------------
//                     s * SQR 2Pi
//            where:
//          x is the cartesian coordinate
//          f(x) is the dependent variable
//          u is the mean
//          s is the standard deviation
//      Np is the area under the curve from point a to
//      point b.
// Note:
//      Uses the _NORMAL_CURVE_Y procedure to calculate values
//      of the dependent variable
//
// Integration is performed using Simpson's rule, as follows:
//
//    Np =  step/3 * (y1 + 4y2 + 2y3 + 4y4 + 2y5 + 4y6 .. yN+1)
//
// where:
//      N is a constant defining the number of area elements
//          used in the computation
//      y1, y2 .. yN+1 are values of f(x)
//      a is x coordinate of the lower limit of the range
//      b is x coordinate of the upper limit of the range
//      step = (b - a) / N
// The sum consists of 4 elements
//    1. y1
//    2. 4 * (sums of all yj, j even from 2 to N)
//    3. 2 * (sums of all yk, k odd from 3 to N - 1)
//    4. yN+1
//
// Variables
const int N = 20;              // Number of areas
// Local variables
double area = 0;
double xVal;
double fOfx;
double step = (b - a) / N;     // x increment
double tAreas[N + 1];          // Array for f(x)
// Clear data array
```

```
for(int c = 0; c < N + 1; c++)
   tAreas[c] =  0;
// Calculate and store sum element 1
// y1
   _asm
   {
      ; Function expects:            |  ST0  |  ST1  |  ST2  |
                                     ;   x   |  mean |   sd  |
      FLD      sd
      FLD      mean
      FLD      a
      CALL     NORMAL_CURVE_Y
      FSTP     fOfx
   }
   tAreas[0] = fOfx;

// Calculate and store sum element 4
// yN+1
   _asm
   {
      ; Function expects:            |  ST0  |  ST1  |  ST2  |
                                     ;   x   |  mean |   sd  |
      FLD      sd
      FLD      mean
      FLD      b
      CALL     NORMAL_CURVE_Y
      FSTP     fOfx
   }
   tAreas[1] = fOfx;

// Calculate and store sum element 2
// 4 * (sums of all yj, j even from 2 to N)
//
xVal = a + step;            // first j

for(int j = 2; j < (N + 1); j += 2)
{
   _asm
   {
      ; Function expects:            |  ST0  |  ST1  |  ST2  |
                                     ;   x   |  mean |   sd  |
      FLD      sd
      FLD      mean
      FLD      xVal
      CALL     NORMAL_CURVE_Y
      FSTP     fOfx
   }
   // Calculate area for this entry
   tAreas[j] = 4 * fOfx;
   xVal = xVal + (2 * step);
}
// Calculate and store sum element 3
// 2 * (sums of all yk, k odd from 3 to N-1)
//
xVal = a + (2 * step);          // first k
for(int k = 3; k < N; k += 2)
{
   _asm
   {
      ; Function expects:            |  ST0  |  ST1  |  ST2  |
```

```
                              ;   x    |  mean  |   sd   |
        FLD      sd
        FLD      mean
        FLD      xVal
        CALL     NORMAL_CURVE_Y
        FSTP     fOfx
    }
    // Calculate area for this entry
    tAreas[k] = 2 * fOfx;
    xVal = xVal + (2 * step);
}
// At this point there are N + 1 entries in the local array
// Add values to calculate area
for(int z = 0; z <= (N + 1) ; z++)
  area = area + tAreas[z];
return (area * (step / 3));
}
```

The function NormalProb() listed above can be used to calculate the normal probability of the case shown in Figure 11.4. For the calculations we enter 97 as the value for a, 120 for b, 92 for the mean, and 10 for the standard deviation. The result is a probability of approximately 0.306 for a light bulb to last between 97 and 120 hours.

SOFTWARE ON-LINE

The function NormalProbSimp() is found in the Un32_9.h file, located in the Chapter11 folder in the book's on-line sftware.

11.6 Linear Correlation and Regression

Statistical data often indicate relationships between variables. For example, a data set indicating the years of age at which lung cancer was diagnosed can be related to another data set indicating the daily number of cigarettes smoked. Linear correlation analysis is the statistical field that attempts to measure the strength of linear relationships between variables. Figure 11.5 shows scatter diagrams for various types of relationships.

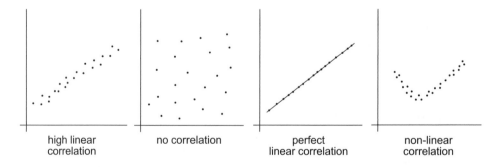

high linear   no correlation   perfect   non-linear
correlation   linear correlation   correlation

Figure 11.5 *Scatter Diagrams and Correlation*

The first scatter plot in Figure 11.5 shows a high linear correlation. The data is located approximately on the path of a straight line. The second scatter plot shows data distributed in a random pattern, with no apparent correlation. In the third diagram all the data is located exactly on the path of a straight line. In this case there is a perfect linear correlation. The fourth diagram shows data that appear to follow a specific path, but this path is not along a straight line. Therefore we can state that there is probably some correlation in the data, but that the relationship is non-linear.

In the sections that follow we concentrate on linear correlation. That is, we assume that the relation between variables is determined by an equation of the form

$$y = f(x)$$
$$= mx + b$$

where m is slope, and b is y-intercept. Nonlinear models in the form of polynomials or quadratic equations are the subject of a field of the Calculus usually called *curve fitting*. Curve fitting techniques are briefly mentioned in Chapter 12. In the present context we exclude nonlinear correlation.

11.6.1 Linear Correlation Coefficient

Establishing data correlation implies that there are at least two data sets. For example, to determine if there is a relationship between the time spent studying and the grade received on an exam we must have one data set for the study time, and another one for the grades. Correlation analysis determines if there is a relation between the data sets, that is, if one variable is somehow related to the other one. A positive correlation indicates y grows as x grows. A negative correlation indicates that y diminishes as x grows.

The quantification of data relationships is a major concern to statisticians. In other words, what is the *significance level* that serves to mark the line where results are attributed to change, or to actual dependencies between variables. Furthermore, the fact that there is a correlation between variables does not necessarily indicate a cause-effect relationship. Suppose a correlation between one data set containing students that listen to loud music, and another one with students with hearing problems. This correlation does not necessarily indicate that listening to loud music results in hearing loss. In fact, it may be due to the fact that students with hearing problems turn the volume higher than those without hearing problems.

The coefficient of linear correlation is based on the notion of covariance. The covariance of two variables (x and y) is defined as the sum of the products of the distances of all values of x and y from a point called the centroid. One limitation of covariance as a measure of linear dependency is that it is not standardized to a unit of measure. If the covariance is standardized by dividing the respective means by the standard deviation the result is a formula called the linear regression coefficient, as follows:

$$r = \frac{n\sum xy - (\sum x)(\sum y)}{\sqrt{n(\sum x^2) - (\sum x)^2}\sqrt{n(\sum y^2) - (\sum y)^2}}$$

The linear correlation coefficient, r, measures the correlation between x and y. The value of r is between –1 and 1. If r is zero, then there is no correlation. A value of r close to 1 indicates a strong positive correlation, that is, y gets larger as x gets larger. A value of r close to –1 indicates a strong negative correlation, that is, y gets smaller as x gets larger. For an interpretation of the various summations in the linear correlation coefficient formula refer to Table 11.1.

The following C++ template function calculates the linear correlation coefficient.

```
template <class A>
double LinCorCoef(A *datX, A *datY, int size)
{
// Function to find the linear correlation coefficient
// between two data sets.
// On entry:
//          datX[] is a numeric data array
//          datY[] is a second numeric data array
//          size is the number of elements in each arrays
//          calculated by the caller, as follows:
//              size = sizeof datX / sizeof datX[0];
// Linear correlation between the data in both arrays is
// calculated according to the formula:
//
//                  n(SIG(xy)) - (SIGx)(SIGy)
//   r = ---------------------------------------------------
//        SQR(n(SIG(x^2))-(SIGx)^2 * SQR(n(SIG(y^2))-(SIGy)^2
//
// where:
//        x are the data elements in the array datX
//        y are the data elements in the array dayY
//        n is the number of elements
//        r is the linear correlation coefficient
// On exit:
//        Returns the linear correlation coefficient

double lcc = 0;
double nume;         // Storage for numerator
double d1, d2;       // Storage for denominator

A sumXY = 0;         // SIGxy
A sumX = 0;          // SIGx
A sumY = 0;          // SIGy
A sumXSq = 0;        // SIGx^2
A sumYSq = 0;        // SIGy^2

// Calcuate sums required in formula
for(int i = 0; i < size; i++)
{
    sumXY = sumXY + (datX[i] * datY[i]);
    sumX = sumX + datX[i];
```

```
    sumY = sumY + datY[i];
    sumXSq = sumXSq + (datX[i] * datX[i]);
    sumYSq = sumYSq + (datY[i] * datY[i]);
}

// Calculate numerator: n(SIGxy) - (SIGx)(SIGy)
nume = (double) size * sumXY - (sumX * sumY);

// Calculate denominator left and right elements
// d1 = SQR(n(SIGx^2)-(SIGx)^2)
// d2 = SQR(n(SIGy^2)-(SIGy)^2)
d1 = (size * sumXSq) - (sumX * sumX);
d2 = (size * sumYSq) - (sumY * sumY);
d1 = sqrt(d1);
d2 = sqrt(d2);
lcc = nume / (d1 * d2);

return lcc;
}
```

Solving the linear regression coefficient formula requires calculating several standard summations. One alternative would have been to call the primitive functions developed earlier in this chapter and listed in Table 11.1. In the LinCorCoef() function we opted for another approach: to calculate all the required summation primitives in a single loop. This makes the function self-contained and easier to follow. The programmer must always be aware of the possibility of doing several calculations in a single loop, thus simplifying the coding.

11.6.2 Linear Regression Analysis

Regression analysis is the field of statistics that attempts to find relationships between variables. When the regression refers to linear relationships it is called *linear regression analysis*. Other types of regression analysis are *curvilinear regression* and *multiple regression*. In curvilinear regression the data distribution follows a non-linear pattern. In multiple regression there are more than two data sets under consideration. In the present context we limit the discussion to linear regression analysis.

We mentioned in section 11.7 that a linear correlation indicates a relationship between variables expressed by the equation:

$$f(x) = mx + b$$

where b is called the y-intercept and m is the slope of the line. Given two linearly related data sets we often need to determine the values of m and b in order to construct the linear equation that defines the relationship line.

The function that is needed is not just any line through the data set, but one that provides the best possible fit. This line, sometimes called the *best-fitting line*, is defined by the equation

$$y' = mx + b$$

where y' is the predicted value, b is the y-intercept, and m is the slope of the line.

In order to ensure that negative and positive values of the data points do not cancel out, we must use the minimum of the sum of the squares of the distances to the proposed line. For this reason the best-fitting line is sometimes called the *least squares line*. Two formulas are required: one to determine the y-intercept b, and another one for the slope m, as follows:

$$m = \frac{n(\sum xy) - (\sum x)(\sum y)}{n(\sum x^2) - (\sum x)^2} \qquad b = \frac{\sum y - m(\sum x)}{n}$$

We have to solve the least squares line using two functions: one returns the y-intercept (b), and the other one the slope (m).

```
template <class A>
double RegrSlope(A *datX, A *datY, int size)
{
// Function to find the slope of a linear regression
// equation related to two data sets.
// On entry:
//          datX[] is a numeric data array
//          datY[] is a second numeric data array
//          size is the number of elements in both arrays
//          calculated by the caller, as follows:
//              size = sizeof datX / sizeof datX[0];
//
// Calculates slope (m) in the formula:
//
//        y' = mx + b
//
//              n(SIGxy) - (SIGx)(SIGy)
//        m =  -------------------------
//              n(SIG(x^2)) - (SIGx)^2
// where:
//          x are the data elements of the array datX
//          y are the data elements of the array dayY
//          n is the number of elements
//          m is the slope of y' = mx + b
// On exit:
//          Returns the slope
double m = 0;
double nume;              // Storage for numerator
double d1, d2;            // Storage for denominator

A sumXY = 0;           // SIGxy
A sumX = 0;            // SIGx
A sumY = 0;            // SIGy
A sumXSq = 0;          // SIGx^2

// Calculate sums required in the formula
for(int i = 0; i < size; i++)
    {
        sumXY = sumXY + (datX[i] * datY[i]);
```

```
    sumX = sumX + datX[i];
    sumY = sumY + datY[i];
    sumXSq = sumXSq + (datX[i] * datX[i]);
}
// Calculate numerator: nSIGxy - (SIGx)(SIGy)
nume = (double) size * sumXY - (sumX * sumY);

// Calculate denominator left and right elements
// d1 = n(SIG(x^2))
// d2 = (SIGx)^2
d1 = size * sumXSq;
d2 = sumX * sumX;
m = nume / (d1 - d2);
return m;
}

template <class A>
double RegrYInt(A *datX, A *datY, int size)
{
// Function to find the y-intercept of a linear regression
// equation between two data sets.
// On entry:
//         datX[] is a numeric data array
//         datY[] is a second numeric data array
//         size is the number of elements in both arrays
//         calculated by the caller, as follows:
//             size = sizeof datX / sizeof datX[0];
//
// Calculates y-intercept (b) in the formula:
//
//       y' = mx + b
//
//             SIGy - m(SIGx)
//       b = ------------------
//                   n
// where:
//       x are the data elements of the array datX
//       y are the data elements of the array dayY
//       n is the number of elements
//       m is the slope of y' = mx + b
// Uses:
//       the slope returned by the RegrYInt()
//       function
// On exit:
//       Returns the y-intercept
double b = 0;
double nume;          // Storage for numerator
double m;             // Storage for y-intercept

A sumX = 0;          // SIGx
A sumY = 0;          // SIGy

// Calculate sums required in the formula
for(int i = 0; i < size; i++)
{
    sumX = sumX + datX[i];
    sumY = sumY + datY[i];
}
// Obtain slope
m = RegrSlope(datX, datY, size);
```

```
// Calculate numerator: SIGy - m(SIGx)
nume = (double) sumY - (m * sumX);
b = nume / size;
return b;
}
```

SOFTWARE ON-LINE

The functions RegrYInt() and RegrSlope () are found in the file Un32_9.h located in the Chapter11 folder in the book's on-line software. The topic of linear least-square interpolation is revisited, with more rigor, in Chapter 12. Additional curve-fitting methods are developed in Section 12.1.3.

Chapter 12

Interpolation, Differentiation, and Integration

Chapter Summary

Numerical analysis is the field of mathematics that studies the methods of approximation used in the solution of mathematical problems. Numerical methods, on the other hand, refers to the computerized solution of the problems of numerical analysis. In this general sense, this book is almost entirely devoted to numerical methods. However, a more restrictive, and perhaps more common definition of numerical methods, focuses on the computerized solution of problems encountered in the calculus. In this chapter we select three common topics of numerical methods: interpolation, differentiation, and integration.

12.0 Interpolation

Interpolation and curve-fitting are numerical techniques by which we estimate the unknown value of a function at a given point using known values at other points. In other words, give a series of values of the abscissa $(x_1, x_2, x_3, ..., x_n)$ and a series of values for the ordinates $(y_1, y_2, y_3, ..., y_n)$ construct a function such that

$$f(x_i) = y_i$$

for each i:

For the problem to have a satisfactory solution we must assume that the values of the ordinates are determined by a smooth mathematical function. Figure 12.1 graphically shows this basic assumption.

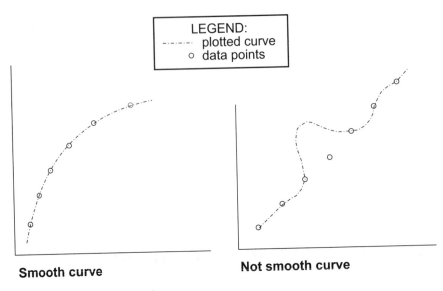

Figure 12.1 *Fundamental Assumption for Interpolation*

In the left-side drawing of Figure 12.1, the y coordinate of the point represented by the black dot is a reasonable approximation, since the function corresponds to a smooth curve. However, in the right-side drawing the point represented by the black dot is outside the curve, since the assumption of smoothness is not met by the function. Also in Figure 12.1, we can see that from the data offered by the other coordinate points it would be impossible to derive the curve on the right-side drawing. In this case interpolation and curve fitting produces invalid results. The expression *well-behaved function* is sometimes used to denote a function that defines a smooth curve.

Two general categories of methods are applied in curve-fitting. The first one is used when the curve passes through all data points. The second one when the curve does not pass through the data points. The nature and origin of the data often determines which case applies. If the data points represent accurate values of a well-behaved function, then we can assume that the curve passes through all points. However, if due to noise or measurement errors the data points are an approximation, then the second type of curve-fitting methods is applicable. Figure 12.2 shows these cases.

Interpolation and curve fitting are core topics of the calculus. The methods and techniques developed in interpolation are also applied to solving problems in differentiation and integration. Here we present a selection of the most used and most common interpolation techniques. For curves that pass through the data points we discuss methods based on linear and Lagrange interpolation. For curves that do not pass through the data points we discuss the least-squares method of curve-fitting.

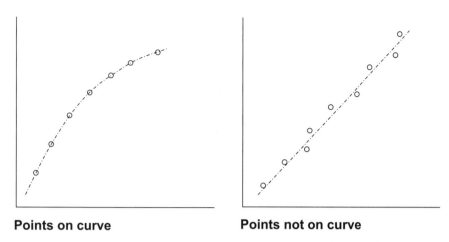

Points on curve   **Points not on curve**

Figure 12.2 *General Types of Interpolation*

12.0.1 Linear Interpolation

The simplest, and perhaps the most used, curve fitting technique is called linear interpolation. The scheme is based on the assumption that unknown values of a function can be reasonably approximated by assuming that two closely located points on the curve are connected by a straight line. Notice that linear interpolation does not assume that the curve is defined by a linear equation, but just that the unknown point is located on a straight line connecting two known points. Figure 12.3 shows the assumption in linear interpolation.

In Figure 12.3 we can determine the coordinates (x_i, y_i) of a point on a curve by assuming that the point lies along a straight line between points a and b, closest on the curve. Notice in Figure 12.3 that the location of the black dot representing the desired point is found by linear interpolation. Therefore it is but an approximation of where the point is actually located, in this case along the dashed-dotted line of the curve.

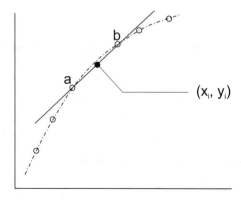

Figure 12.3 *Linear Interpolation*

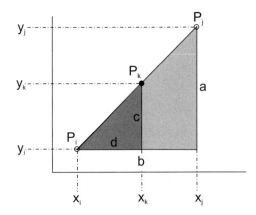

Figure 12.4 *Linear Interpolation by Similar Triangles*

Linear interpolation can be based on similar triangles, as shown in Figure 12.4. The similarity between triangles results in the identities:

$$\frac{a}{b} = \frac{c}{d}$$

$$c = \frac{ad}{b}$$

If the coordinates of point P_i are x_i, y_i, and the coordinates of point P_j are x_j, y_j, then P_k can be calculated as follows:

$$a = y_j - y_i$$
$$b = x_j - x_i$$
$$c = y_k - y_i$$
$$d = x_k - x_i$$

Applying the above result from the similar triangles, we get:

$$c = \frac{(y_j - y_i)(x_k - x_i)}{x_j - x_i}$$

Considering that the y coordinate of point P_k is

$$y_k = c + y_i$$

then, by linear interpolation

$$y_k = \frac{(y_j - y_i)(x_k - x_i)}{x_j - x_i} + y_i$$

The data required for linear interpolation include the coordinates of both adjacent points (P_i and P_j in Figure 12.4) and the x coordinate of the point being interpolated (P_k in Figure 12.4).

We develop two functions that perform linear interpolation. The first one uses the coordinates points of P_j and P_i, as well as the x coordinate of point P_k, to determine the y coordinate of point P_k. This function, called InterpLinPxy() is coded as follows:

```
double InterpLinPky(double pxi,
                    double pyi,
                    double pxj,
                    double pyj,
                    double pxk)
{
// Find the y coordinate of point Pk, which is assumed to be
// along a straight line connecting points Pi and Pj, per
// the following equation:
//
//    Pyk = ((Pyj-Pyi)(Pxk-Pxi)/(Pxj-Pxi))+Pyi
//
// where:
//
//        Pyj is the y coordinate of point Pj
//        Pxj is the x coordinate of point Pj
//        Pyi is the y coordinate of point Pi
//        Pxi is the x coordinate of point Pi
//        Pxk is the x coordinate of the desired point
//        Pyk is the y value found by interpolation
//
// On entry:
//        parameters pxj, pyj, pxi, pyi, pxk are
//        the coordinates of the corresponding points
// Returns:
//        The y coordinate of point Pk found by linear
//        interpolation, as a value of type double.

return (((pyj - pyi)*(pxk - pxi))/(pxj - pxi)) + pyi;

}
```

The second function uses table data containing the x and y coordinates of several points on the curve. The function attempts to find two points adjacent to the desired point. If these exist, the y coordinate of point P_k is calculated by linear interpolation using the function InterpLinPxk(). The function, named InterpLinFromTable() is as follows:

```
double InterpLinFromTable(double *datX,    // x array
                          double *datY,    // y array
```

```
                        double pxk,      // x coordinate
                        int size)        // array size
{
// Find the y coordinate of point Pk, assumed to be
// on a straight line connecting two points, Pi and Pj,
// from the coordinate data contained in two arrays of
// points on the curve. Uses the InterpLinPky() function.
// On entry:
//        array datX[] contains x coordinates of points
//        array datY[] contains y coordinates of points
//        parameter pxk is the x coordinate of point Pk
//        size is the number of elements in both arrays
//        calculated by the caller, as follows:
//                size = sizeof datX / sizeof datX[0];
// Assumptions:
//        Coordinate points in arrays datX[] and datY[]
//        are sorted in increasing order.
// Processing:
//        Routine searches the array of x coordinates
//        until one is found that is larger that the value
//        Pxk. This point is labled Pj. If there is a
//        previous x to point Pj it is labled Pi. The
//        coordinates of Pi and Pj, and the value Pxk
//        are then passed to the InterpLinPky()
//        function.
// Returns:
//        The y coordinate of point Pk found by linear
//        interpolation, as a value of type double.
//        Returns 0 if there are no adjacent points
//        in the data set.
// Find point Pj in array datX[]
double pyj, pyi, pxj, pxi;
double pyk = 0;
int found = 0;

for(int w = 0; w < size; w++)
{
    if(datX[w] > pxk || w != 0)
    {
        pxj = datX[w];
        pxi = datX[w-1];
        pyj = datY[w];
        pyi = datY[w-1];
        found = 1;
        break;
    }
}
if(found == 0)
    return 0;

// Calculate Pyk using InterpLinPky() function
pyk = InterpLinPky(pxi, pyi, pxj, pyj, pxk);
return pyk;
}
```

12.0.2 Lagrange Interpolation

In theory, if there are n data points in a data set, it is possible to find a polynomial expression of degree $n-1$ that defines a curve that passes through every data point. In this sense linear interpolation, as discussed in Section 12.1.1, is a form of polynomial interpolation when the polynomial is of the first degree. The general problem can be expressed as follows: given a set of n points, (x_1, y_1), (x_2, y_2) ... (x_n, y_n), find a unique polynomial of degree $n - 1$ that defines the curve.

The French mathematician Joseph Louis Lagrange defined a now classic solution to the interpolation problem by means of a formula, which is given here without proof:

$$p(x) = \sum_{1 \le j \le n} y_j \prod_{\substack{1 \le i \le n \\ i \ne j}} \frac{x - x_i}{x_j - x_i}$$

Lagrange's formula is based on a set of simple factors. For example, a cubic polynomial (third degree) can be expressed as sum of four terms, as follows:

$$P_3(x) = \frac{(x - x_2)(x - x_3)(x - x_4)}{(x_1 - x_2)(x_1 - x_3)(x_1 - x_4)} y_1 + \frac{(x - x_1)(x - x_3)(x - x_4)}{(x_2 - x_1)(x_2 - x_3)(x_2 - x_4)} y_2 +$$
$$\frac{(x - x_1)(x - x_2)(x - x_4)}{(x_3 - x_1)(x_3 - x_2)(x_3 - x_4)} y_3 + \frac{(x - x_1)(x - x_2)(x - x_3)}{(x_4 - x_1)(x_4 - x_2)(x_4 - x_3)} y_4.$$

where x_i and y_i are values of the independent and dependent variables respectively. Each term of the sum follows a simple pattern: the numerator is a product of linear factors in the form $(x - x_i)$ where one x_i is omitted in each term. For example, in the expansion $P_3(x)$ listed previously, the first term omits x_1, the second term omits x_2, and so on. The omitted x_i is used in constructing the denominator of each term by replacing x with the omitted x_i. Each term is then multiplied by the value of the dependent variable (y_i) corresponding to the x_i omitted in the term.

The computer algorithm for Lagrange interpolation consists of calculating each term of the expansion and accumulating the sum. The outer loop visits every pair in the data set. The inner loop calculates the numerator and the denominator of each term, multiplies by the value of y for each term in the Lagrange expansion, and accumulates the sum. The logic can be expressed in pseudocode as follows:

STEP 1: Define the following variables and controls:

    i (outer loop counter)

    j (inner loop counter)

    pxk (value of x at the interpolation point)

    Ax[] (array of x values)

    Ay[] (array of y values)

    n (number of data pairs in the set)

    Term (variable to hold the Lagrange terms)

Sum = 0 (variable to hold the sum of the Lagrange terms)

STEP 2: If i < n; Term = Ay[i], Step 3

If i = n, Step 6

STEP 3: if j < n, Step 4

if j = n, Step 5

STEP 4: If j i

Term = Term * (pxk – Ax[j]) / (Ax[i] – Ax[j]);

j = j + 1;

Step 3.

STEP 5: Sum = Sum + Term; i = i + 1;

Step 2.

STEP 6: Sum = value of y for pxk.

Done.

The following function calculates the variable of the dependent variable from a data set using Lagrange interpolation. The code implements the algorithm described in the preceding paragraphs.

```
double InterpLagrange(double *datX,   // x array
                      double *datY,   // y array
                      double pxk,     // x coordinate
                      int size)       // array size
{
// Find the y coordinate of point Pk using Lagrange
// interpolation
//
// On entry:
//      array datX[] contains x coordinates of points
//      array datY[] contains y coordinates of points
//      parameter pxk is the x coordinate of point Pk
//      size is the number of elements in both arrays
//      calculated by the caller, as follows:
//              size = sizeof datX / sizeof datX[0];
//
// Returns:
//      The y coordinate of point Pk found by Lagrange
//      interpolation, as a value of type double.
//      Returns 0 if there are no adjacent points
//      in the data set.
double lTerm = 0;    // Each Lagrange term
double pyk = 0;      // Value of the dependent variable
                     // for point pxk
for(int i = 0; i < size; i++)
{
    lTerm =  datY[i];

    for(int j = 0; j < size; j++)
    {
        if(j != i)
            lTerm = lTerm * (pxk - datX[j])/(datX[i]-datX[j]);
```

```
    }
    pyk = pyk + lTerm;
}

return pyk;
}
```

12.0.3 Least-Squares Interpolation

In Section 12.1 we discussed two general types of methods applied in curve-fitting operations. The first type of method assumes that the curve passes through all data points. Linear and Lagrange interpolation, discussed so far, belong to this group. The second type of method does not assume that the curve passes through the data points. Figure 12.2 shows both cases.

In science, engineering, and finance there are many cases in which experimental or sensed data must be considered as an approximation. This implies that we cannot assume that the data exactly defines points on the curve. For these situations we require interpolation methods which provide the best possible fit to the data, on the assumption that the data models a specific, well-behaved function.

In Chapter 11, Section 11.7.2, we discussed regression analysis in the context of statistics and developed methodologies for linear regression. We now revisit the topic of linear regression in more general terms, and generalize the least squares methods of curve fitting to higher-order polynomials.

Linear Models

Linear correlation is based on a relationship between variables expressed by the equation:

$$y = mx + b$$

where b is called the y-intercept and m is the slope of the curve, and y is the value of the dependent variable. In order to construct the linear equation that defines the relationship between the independent and the dependent variables we need to determine the values of m and b. The function desired is the one that provides the best possible fit to the data. In other words, we wish to minimize the vertical distance between the data points and the curve. These distances, sometimes called the *residuals* of the data points, are shown in Figure 12.5.

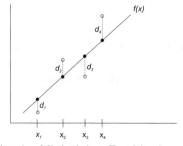

Figure 12.5 *Curve-Fitting by Minimizing Residuals*

In Figure 12.5 the open dots represent the data points, while the distances from the data to the curve, for each x, are the values d1, d2, d3, and d4. To find the best possible fit of the curve, labeled f(x), we must minimize the sum of the distances from the data to the respective points in the curve, since the residuals represent the error in the interpolation. However, if we use the actual values of the distances, the positive and negative values tend to cancel out. This results in an incorrect measurement of the error. A solution is to square the residuals before summing them, thus the name least-squares. After squaring, the value E, called the error sum is expressed by the following formulas:

$$E = \sum_{k=1}^{n} d_k^2$$

In terms of m and b the error sum equation can be expressed as follows:

$$E = \sum_{k=1}^{n} (y_k - mx_k - b)^2$$

We can now apply the differential calculus to obtain the partial derivatives of E with respect to the unknowns, m and b. Since the minimum of m and b occur when the derivatives are set to zero, we can use the equations:

$$\frac{\partial E}{\partial m} = -2\sum_{k=1}^{n} x_k(y_k - mx_k - b) = 0$$

$$\frac{\partial E}{\partial b} = -2\sum_{k=1}^{n} (y_k - mx_k - b) = 0$$

By rearranging the terms and solving for the slope m and the y-intercept b, the equations become:

$$m = \frac{n(\sum xy) - (\sum x)(\sum y)}{n(\sum x^2) - (\sum x)^2} \qquad b = \frac{\sum y - m(\sum x)}{n}$$

The routine to calculate the y coordinate of a point using least squares interpolation first determines the values of m and b in the slope-intercept form of the equation by means of the above formulas. Then the value of the y coordinate for a point P_k can be easily obtained by substituting in the original equation:

$$f(x) = mx + b$$

Calculating the Error Sum

One of the advantages of least-squares interpolation is that we are able to measure the error between the data points and the equation that defines the curve. The error sum is the square of the residuals for each point in the data set. The computation is algorithmically simple since all we have to do is visit each coordinate point in the data set, calculate the square of the value of the residual for the point, and accumulate this value. This error sum can then be used to detect the variation between the data and the curve that models it. A zero value for the error sum indicates that the curve exactly conforms to the data.

Least-Squares Linear Interpolation Function

The following function calculates the y coordinate of a point P_k using least-squares linear interpolation. The function returns the y coordinate in a variable of type double, and the error sum is returned in a parameter passed by reference by the caller.

```
template <class A>
double LinearLSInterp(A *datX,      // Array of x coordinates
                      A *datY,      // Array of y coordinates
                      A pxk,        // x coordinate of point Pk
                      double *eSum, // Caller's variable for
                                    // error sum
                      int size)     // Array size
{
//
// Function to find the y coordinate of point Pk using
// the least-squares methods
// On entry:
// On entry:
//      array datX[] contains x coordinates of points
//      array datY[] contains y coordinates of points
//      parameter pkx is the x coordinate of point Pk
//      size is the number of elements in both arrays
//      calculated by the caller, as follows:
//            size = sizeof datX / sizeof datXY[0];
// STEP 1:
// Calculate slope (m) in the formula:
//
//      y = mx + b
//
//            n(SIGxy) - (SIGx)(SIGy)
//      m = ----------------------------------
//              n(SIGx^2) - (SIGx)^2
// where:
//      x are the data elements of the array datX
//      y are the data elements of the array dayY
//      n is the number of elements
//      m is the y-intercept of y' = mx + b
//
// STEP 2:
// Calculate y-intercept (b) in the formula:
//
//      y = mx + b
//
//            SIGy - m(SIGx)
//      b = ------------------
//                  n
```

```
// where:
//      x are the data elements of the array datX
//      y are the data elements of the array dayY
//      n is the number of elements
//      m is the slope of y = mx + b
//
// STEP 3:
//      Solve for point y applying
//            pyk = mx + b
//

double m = 0;           // slope
double nume;            // Storage for numerator
double d1, d2;          // Storage for denominator
A sumXY = 0;       // SIGxy
A sumX = 0;        // SIGx
A sumY = 0;        // SIGy
A sumXSq = 0;      // SIGx^2
// Variables for calculating y-intercept
double b = 0;           // y-intercept
double pyk;             // Y coordinate of point Pk
// Variables for calculating error sum
double res = 0;         // Residual accumulator
double newY = 0;
double err = 0;         // Accumulator for error sum
// Calculate sums required in formula
for(int i = 0; i < size; i++)
{
    sumXY = sumXY + (datX[i] * datY[i]);
    sumX = sumX + datX[i];
    sumY = sumY + datY[i];
    sumXSq = sumXSq + (datX[i] * datX[i]);
}
// Calculate slope (m)
//
//             n(SIGxy) - (SIGx)(SIGy)
//      m = ----------------------------------
//                 n(SIGx^2) - (SIGx)^2
//

// Start with numerator: nSIGxy - (SIGx)(SIGy)
nume = (double) size * sumXY - (sumX * sumY);

// Calculate denominator left and right elements
// d1 = n(SIGx^2)
// d2 = (SIGx)^2
d1 = size * sumXSq;
d2 = sumX * sumX;
// Now obtain slope
m = nume / (d1 - d2);

// Calculate y-intercept (b)
//
//              SIGy - m(SIGx)
//       b = -----------------
//                   n
//
// Start with numerator: SIGy - m(SIGx)
nume = (double) sumY - (m * sumX);
// Now obtain b
```

```
b = nume / size;

// At this point:
//     variable b holds y-intercept
//     variable m holds slope
//     parameter pxk holds value of x
// Calculate y-coordinate using
//     y' = mx + b
pyk = (m * pxk) + b;

// Calculate error sum using residuals
for(int k = 0; k < size; k++)
{
    newY = (m * datX[k]) + b;
    res = datY[k] - newY;
    err = err + (res * res);
}
// Store error sum in caller's variable
*eSum = err;
// Return y coordinate for point Pk
return pyk;
}
```

SOFTWARE ON-LINE

The interpolation functions listed in the preceding sections are found in the Un32_10.h file, located in the CHAPTER12 folder in the book's on-line software.

Non-Linear Models

If the functional equation is non-linear, then the least-square method described previously in this section cannot be used in curve-fitting. However, certain non-linear equations can be easily transformed into a linear function. Conventional linear regression can then be applied to the transformed function in order to obtain a least-squares straight line fit. A common function that can be easily transformed into a linear form is the general power function:

$$f(x) = bx^m$$

where m and b are constants. Taking the natural logarithm of both sides of this equation results in:

$$\ln f(x) = m \ln x + \ln b$$

The following transformation equations can now be defined:

$$y = f(x)$$
$$b' = \ln b$$
$$x' = \ln x$$
$$y' = \ln y$$

Please note that the primes in the formulas is a convenience in order to allow using the same constants in code listings. It should not be confused with the use of primes to indicate derivatives. The values can be substituted in the slope-intercept form as follows:

$$y' = mx' + b'$$

The slope and y-intercept equations, previously listed, can be modified as follows:

$$m = \frac{n(\sum x'y') - (\sum x')(\sum y')}{n(\sum x'^2) - (\sum x')^2} \qquad b' = \frac{\sum y' - m(\sum x')}{n}$$

The computer algorithm is similar to the one used in the linear model of least-squares interpolation. First the x and y coordinates of the data points are converted to their natural logarithms and stored in local arrays. This data is used to obtain the value of the y-intercept b and the slope m, by substituting in the preceding equations. The natural antilogarithm of b produces the value b, which is substituted in the original formula for the curve to obtain the value of the dependent variable a point P_k. The following procedure performs the processing.

```
double NonLinLSInterp(double *datX,    // Array of x coordinates
                      double *datY,    // Array of y coordinates
                      double  pxk,     // x coordinate of point Pk
                      double *eSum,    // Caller's variable for
                                       // error sum
                      int size)        // Array size
{
// Function to find the y coordinate of point Pk using
// the non-linear least-squares method, assuming that the function
// is of the form:
//
//          f(x) = bx^m
//
// by performing the following transformation:
//
//          ln f(x) = m * ln x + ln b
//
// where:
//          y  = f(x)
//          b' = ln b
//          y' = ln y
//          x' = ln x
//
// On entry:
//          array datX[] contains x coordinates of points
//          array datY[] contains y coordinates of points
//          parameter pkxk is the x coordinate of point Pk
//          size is the number of elements in both arrays
//          calculated by the caller, as follows:
//               size = sizeof datX / sizeof datXY[0];
//
```

```
// PROCESSING:
// STEP 1:
//         Convert the values in arrays datX[] and dat[Y]
//         by taking the natural logs of the values stored there.
//
// STEP 2:
//         Use these values to calculate slope (m)
//
// STEP 3:
//         Calculate y-intercept b' and convert to b.
//
// STEP 4:
//         Solve for point pxyk by applying the original
//         equation:
//
//         y = bx^m
//
// STEP 5:
//         Calculate and store error sum in caller's
//         variable
// Note:
//         Processing uses logarithmic primitives contained
//         in the Un32_5 modules of the MATH32 library.
//         The C++ interface functions to these primitives
//         are included in this module.
//
//
double m = 0;              // slope
double nume;               // Storage for numerator
double d1, d2;             // Storage for denominator
double sumXY = 0;          // SIGxy
double sumX = 0;           // SIGx
double sumY = 0;           // SIGy
double sumXSq = 0;         // SIGx^2
double xTom;               // x to the m power
//
// Variables for calculating y-intercept
double b = 0;              // y-intercept
double lnb = 0;            // ln of y-intercept
double pyk;                // Y coordinate of point Pk
//
// Variables for calculating error sum
double res = 0;            // Residual accumulator
double err = 0;
double newY = 0;
// Arrays for natural logs of x and y
double *lnX = new double[size];
double *lnY = new double[size];
//
// Convert data to ln
for(int l = 0; l < size; l++)
{
    lnX[l] = LogE(datX[l]);
    lnY[l] = LogE(datY[l]);
}

// Calculate sums required in formula
for(int i = 0; i < size; i++)
{
    sumXY = sumXY + (lnX[i] * lnY[i]);
```

```
    sumX = sumX + lnX[i];
    sumY = sumY + lnY[i];
    sumXSq = sumXSq + (lnX[i] * lnX[i]);
}
//
// Calculate slope (m)
//
//              n(SIGxy) - (SIGx)(SIGy)
//       m = -------------------------
//              n(SIGx^2) - (SIGx)^2
//

// Start with numerator: nSIGxy - (SIGx)(SIGy)
nume = (double) size * sumXY - (sumX * sumY);
//
// Calculate denominator left and right elements
// d1 = n(SIGx^2)
// d2 = (SIGx)^2
d1 = size * sumXSq;
d2 = sumX * sumX;
// Now obtain slope
m = nume / (d1 - d2);
//
// Calculate ln of y-intercept (b')
//
//              SIGy - m(SIGx)
//       b' = ----------------
//                   n
//
// Start with numerator: SIGy - m(SIGx)
//
nume = (double) sumY - (m * sumX);
// Now obtain b
lnb = nume / size;
// Solve for b
b = AlogE(lnb);
//
// Calculate y-coordinate using
//    y' = bx^m
//
xTom = XToYMixed(pxk, m);
pyk = b * xTom;
//
// Calculate error sum using equation and residuals
for(int k = 0; k < size; k++)
{
    newY = b * (XToYMixed(datX[k], m));
    res = datY[k] - newY;
    err = err + (res * res);
}
// Store error sum in caller's variable
*eSum = err;
//
return pyk;
}
```

The function NonLinLSInterp() listed previously assumes that all data is located in the first quadrant. This is necessary since logarithmic functions do not exist for negative values.

12.1 Numerical Differentiation

The first derivative of a function at a point P_k can be described graphically as the slope of the tangent at that point, as shown in Figure 12.6.

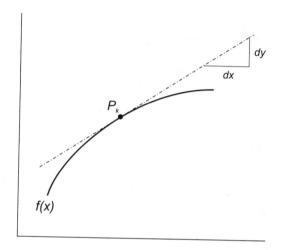

Figure 12.6 *Graphic Interpretation of the Derivative*

In Figure 12.6 the slope of the line tangent to the curve at point P_k (dot-dash line) represents the derivative of $f(x)$ at point P_k. If $f(x)$ is not known, the derivative can be approximated by calculating the slope of a secant line that intersects the curve at two points close to P_k. Figure 12.7 shows this approximation.

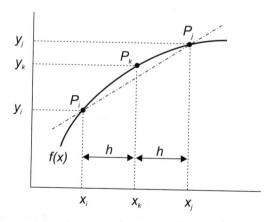

Figure 12.7 *Derivative Approximated by the Slope of a Secant*

In Figure 12.7 the derivative at point P_k can be approximatedly calculated by first drawing a secant line through points P_i and P_j, which are assumed to be close to point P_k. The slope of this secant is our approximation of the derivative at P_k. In this case the equation for the derivative is:

$$y'_k \approx \frac{y_j - y_i}{x_j - x_i}$$

where $h = (x_j - x_i)/2$, as shown in Figure 12.7.

As the data points P_i and P_j get closer together, the derivative increases in accuracy. In other words, as h approaches zero the slope of the secant between points P_i and P_j approximates the slope of the tangent at point P_k.

The preceding equations for the first derivative uses data values at both sides of the point being estimated. They are called expressions of central difference.

Frequently we are forced to estimate the derivative of a point located at the extreme of the data set. In these cases the other reference point available is either to the left or to the right of the point whose derivative is desired. When the reference point is to the left of the point of the desired derivative the formula is said to be a backward difference expression. When the reference point is to the right of the point of the desired derivative then the formula is described as a forward difference expression. Figure 12.8 shows graphically the forward and backward difference cases of the derivative.

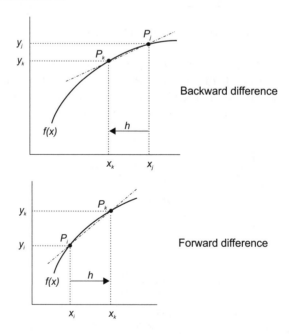

Figure 12.8 *Backward and Forward Difference Cases of the Derivative*

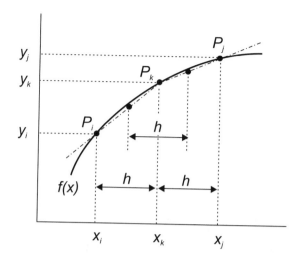

Figure 12.9 *Graphics Interpretation of the Second Derivative*

The forward and backward difference formulas for the derivative are the differences of the y coordinate at both points. Because the derivative is the slope of the tangent at a point on the curve, it expresses the rate of change of the curve at that point. By the same token, the rate of change of the slope, called the second derivative, measures the rate of change of the first derivative, as shown in Figure 12.9.

In Figure 12.9 we draw secant lines between the adjacent points, P_i and P and between points P and P_j. By dividing the difference in the slope of the two lines, by the distance between their midpoints, we obtain the rate of change of the slope, or second derivative. In other words, applying the method of computing the derivative to the formula for the first derivative, and using the two secant lines in Figure 12.9, we are able to obtain the rate of change of the first derivative, which is the second derivative. The second derivative is expressed by the following formula:

$$y'' \approx \frac{\dfrac{y_j - y}{x_j - x} - \dfrac{y_k - y_i}{x - x_i}}{h}$$

Numerical differentiation must be used with caution. Noise or inaccuracies in the data may easily lead to large errors in the calculation of the first derivative. Even larger errors can result in estimating the second derivatives.

Errors in the calculation of the derivative are shown graphically in Figure 12.10. Notice that points P_i and P_j are used in determining the secant. However, since these points are not located on the curve, they do not provide a reasonably good estimate of the slope of the tangent at point P_k.

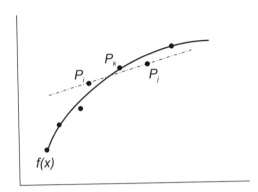

Figure 12.10 *Inaccurate Data Leads to Errors in the Derivative*

When the derivative must be estimated from inaccurate data it is sometimes possible to improve the accuracy by using the least-squares method to obtain a polynomial, then solving for the derivative at points in this curve.

Numerical approximations of the derivative usually return arrays containing the coordinates of points for the first, second, or other derivative. The calculations are based on an unknown function defined by two sets of coordinate points. The derivative of the start and end points in the data set are approximated using the forward difference and backward difference formulas. The intermediate points are estimated using the central difference formula. The following function, named FirstDeriv() performs the calculations for the first derivative.

```
void FirstDeriv(double *datY,    // Array of y coordinates
                double *drv1,    // Array for first derivatives
                int size)        // Size of data arrays

{
// Function to find the first derivative of a set of coordinate
// points using the forward difference, central difference,
// and backward difference formulas:
//
//     Py'k = (Pyk -- Pyi) / h    <= forward difference
//
//     Py'k = (Pyj -- Pyi) / 2h   <= central difference
//
//     Py'k = (Pyj -- Pyk) / h    <= backward difference
//
// where:
//     Py'k is the derivative at point Pk
//     Pyj is the coordinate of a point Pj, to the right of Pk
//     Pyi is the coordinate of a point Pi, to the left of Pk
//     h is the size of the uniform subintervals.
// On entry:
//     array datY[] contains y coordinates of points
//     array drv1[] is used to return the first derivatives
//     size is the number of elements in array datY[]
//     calculated by the caller, as follows:
//             size = sizeof datY / sizeof datY[0];
```

```
// Calculate derivative of first point using forward difference
drv1[0] = (datY[1] - datY[0]) / h;

// Calculate central points using central difference
for(int x = 1; x < (size - 1); x++)
    drv1[x] = (datY[x + 1] - datY[x - 1]) / (2.0 * h);

// Calculate the derivative of the last point using backward
// difference
drv1[size - 1] = (datY[size - 1] - datY[size - 2]) / h;

return;
}
```

Observing Figures 12.7 and 12.8 you can see that the approximation offered by central difference formulas is more accurate than that offered by either forward or backward difference. This error is more noticeable in the calculation of the second derivative. The following function calculates the second derivative using only the central difference.

```
void SecondDeriv(double *datY,    // Array of y coordinates
                 double *drv2,    // Array for second derivatives
                 int size)        // Size of data arrays
{
// Function to find the second derivative of a set of coordinate
// points using the central difference formula:
//
//     Py"k = (Pyj - 2Pyk + Pyi) / h^2
//
// where:
//     Py"k is the second derivative at point Pk
//     Pyj is the coordinate of a point Pj, to the right of Pk
//     Pyi is the coordinate of a point Pi, to the left of Pk
//     h is the size of the uniform subintervals.
// On entry:
//      array datY[] contains y coordinates of points
//      array drv2[] is used to return the second derivatives
//      size is the number of elements in array datY[]
//      calculated by the caller, as follows:
//           size = sizeof datY / sizeof datY[0];
// Note:
//      Since the central difference formula cannot be applied
//      at end points the second derivative at offset [0] and
//      at offset [size - 1] are undefined
//

// Calculate central points using central difference
for(int x = 1; x < (size - 1); x++)
    drv2[x] = (datY[x + 1] - (2 * datY[x]) + datY[x - 1]) / h^2;

return;
}
```

SOFTWARE ON-LINE

The functions FirstDeriv(0 and SecondDeriv() are found in the file Un32_10.h which is located in the Chapter12 folder in the book's on-line software.

12.2 Numerical Integration

The simplest and most intuitive definition of integration is that of an interval of the area under the graph of a function, sometimes called the area under a curve. This concept of the integral is shown in Figure 12.11.

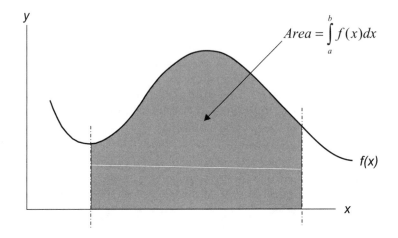

Figure 12.11 *The Integral as the Area under a Curve.*

In Figure 12.11 the area under the curve between $x = a$ and $x = b$ is shaded in gray. The integral of the function $f(x)$ is expressed as:

$$Area = \int_a^b f(x)dx$$

Given the equation of a function, its integral over an interval can often be obtained analytically. However, in science and technology the calculation of the integral must often be based on data points obtained from physical measurements, without the benefit of the curve's equation. In other cases it is difficult, or even impossible, to determine the integral from the equation. Numerical methods allows us to obtain the integral from data points that define the curve, whether or not the curve's equation is available.

Many integration methods have been developed. All of them require values of the dependent variable along the curve. These values can be obtained by applying the curve's equation, or from tabular data. In the examples that follow we assume that the values of the independent variable are furnished as data. Each value of the dependent variable $f(x)$ corresponds to a value of the independent variable x. We start with the simplest, most intuitive, and most useful method of calculating the integral, called the trapezoidal rule.

12.2.1 Integration by the Trapezoidal Rule

Suppose we have a function defined on the interval from a to b, for which values of both the independent and dependent variables are available at points along the abscissa, as shown in Figure 12.12.

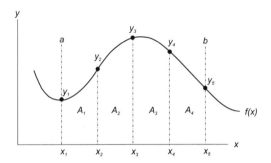

Figure 12.12 *Evaluating the Function at Points Along the x-axis.*

The area under the curve, from point a to point b, can now be represented as four smaller areas from point x_1 to x_2, x_2 to x_3, x_3 to x_4, and x_4 to x_5.

If the points on the curve are now joined with straight lines, we can visualize the areas as a series of trapezoids. The combined areas of these trapezoids approximate the area under the curve, as shown in Figure 12.13. The areas designated $A1$ to $A4$ in Figure 12.12 correspond to the area of the four individual trapezoids. The sum of areas $A1$ to $A4$ is approximately the area under the curve. How close the approximation depends on the number of trapezoids and on the function itself. The more trapezoids, the smaller the error. Also, if in the interval, the function approximates a straight line, then the trapezoidal approximation error is smaller. For example, it can be seen in Figure 12.13 that trapezoid $A4$ is a better approximation of the area for that interval, than trapezoid $A1$ is for its' interval.

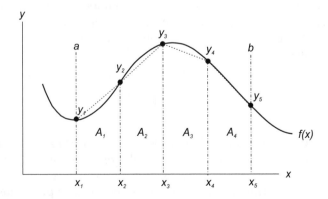

Figure 12.13 *Trapezoidal Approximation of the Integral*

The first trapezoid is represented as in Figure 12.14.

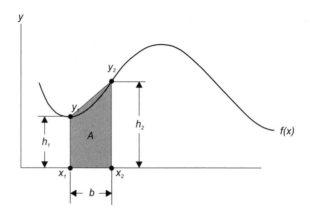

Figure 12.14 *Solving the Trapezoids*

In the case shown in Figure 12.14, the trapezoid labeled A has heights h1 and h2 and base b. The area of this trapezoid is obtained by the formula:

$$Area = \frac{b(h_1 + h_2)}{2}$$

In this case h_1 corresponds with the value y_1, and h_2 with y_2. The base b is obtained by subtracting x_2 minus x_1. Thus, in terms of the coordinate points, the area *A1* of the first trapezoid is calculated by the formula:

$$A_1 = \frac{(x_2 - x_1)(y_1 + y_2)}{2}$$

The remaining trapezoids can be obtained similarly. The general formula is easier to derive if we assume that the coordinate points are equally spaced along the x axis. Therefore the value b is the same for all trapezoids. In this case, the area under a curve defined by N trapezoids, between points a and b, can be approximated by the equation:

$$\int_a^b f(x)dx \approx \frac{b}{2}(y_1 + 2\sum_{k=2}^{N} y_k + y_{N+1})$$

This equation, which is one of the Newton-Cotes formulae, is called the trapezoidal rule. It allows computing the integral from the curve's equation, or from data points obtained experimentally, as long as the x-axis points are equidistant. When the curve's equation is at hand, the calculations can place the data points as close as necessary to approximate the integral to any degree of accuracy. On the other hand, experimentally collected information is limited to the furnished data points.

In cases in which the x-axis points are not equidistant we cannot use the same base for all trapezoids. In this case the formula becomes a sum of the areas of the individual trapezoids:

$$\int_a^b f(x)dx \approx \sum_{k=1}^{N} \frac{(x_k - x_{k+1})(y_k + y_{k+1})}{2}$$

where N is the number of points in the data arrays.

A computer algorithm is easily developed to accommodate the case in which the x-axis coordinates may not be equidistant. This requires calculating the base of each trapezoid individually, which is accomplished by a simple subtraction of the x-coordinate points for the trapezoid, as in the preceding formula. The following function, named TrapRule(), calculates the integral using the trapezoidal rule. During processing the area of each trapezoid is calculated individually. The data is located in two arrays, one defines the x coordinate points and the other one the corresponding y value for each point in the curve.

```
double TrapRule(double *datX,    // Array of x values
                double *datY,    // Array of f(x) values
                int size)        // Size of datY[] array
{
//
// Calculates integral by estimating the area under the curve.
// Integration is performed by adding the areas of individual
// trapezoids, calculated according to the formula:
//
//          (Xk - Xk+1)(Yk + Yk+1)
//     Tk = ----------------------
//                    2
//
// where:
//
//        Tk is the area of the trapezoid
//        Y1,..., Yk, Yk+1 ... Ysize are values of f(x) stored in array
//                  datY[]
//        X1,..., Xk, Xk+1 ... Xsize are values of x stored in array
//                  datX[]
//        size is the number of elements in arrays datX[] and
//        datY[]
//
// The number of trapezoids is one less than the number of
// data points in the arrays datX[] and datY[]
//
// Local variables
double integral = 0;
double base = 0;
double *tAreas = new double[size - 1]; // Array for trapezoid areas
//
// Calculate the area of size-1 trapezoids
for(int i = 0; i < (size - 1); i++)
{
    // Calculate area for this trapezoid
    base = datX[i+1] - datX[i];      // Base for this trapezoid
```

```
    // Store area in local array
    tAreas[i] = (base * (datY[i] + datY[i+1]))/2;
}


//
// At this point there are size - 1 entries in the array of
// trapezoid areas. These values are now added to obtain
// the integral

for(int z = 0; z < (size -1) ; z++)
  integral = integral + tAreas[z];

return integral;
}
```

Other methods of integration have been developed that provide more accurate approximations than the trapezoidal rule. The major advantages of the trapezoidal rule are its ease of implementation, and that it can be used with tabular data in which the x-axis values are not equidistant.

12.2.2 Integration by Simpson's Rule

The trapezoidal rule, described in the preceding section, uses straight line segments to connect data points of the function being integrated. The resulting error of the integral is the area between the function curve and the straight line segment, as shown in Figure 12.15.

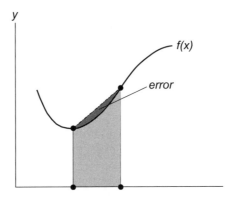

Figure 12.15 *Error in Trapezoidal Integration*

In Figure 12.5 the dark-gray region shows the error between the area of the trapezoid and the area under the function curve. This error can be reduced by finding a curve that more closely matches the function over the interval. Simpson's rule uses a parabola connecting three data points, integrates the area under the parabola, and then sums these areas to approximate the integral of the function. The result is a more accurate approximation than that offered by the trapezoidal rule, but the limitation is that the points along the x-axis must be equidistant. In applying Simpson's rule we estimate the integral by summing areas defined by two subintervals, as shown in Figure 12.16.

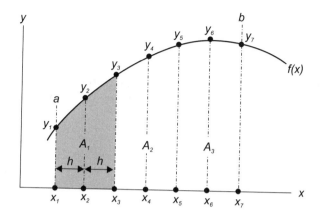

Figure 12.16 *Simpson's Rule Approximation*

In Figure 12.16 the area labeled A_1 is shown in gray. We construct a parabolic arc through the points y_1, y_2, y_3. This arc has the form:

$$y = ax^2 + bx + c$$

This equation can be integrated from x_1 (or $-h$, since x_1 is h units to the left of our midpoint) to x_3 (or h, since x_3 is h units to the right of our midpoint) to find the area A_1:

$$A_1 = \frac{2ah^3}{3} + 2ch$$

Solving for a and c, using the known values of $y_1 = a(-h)^2 + b(-h) + c$, $y_2 = a(0)^2 + b(0) + c$, and $y_3 = a(h)^2 + b(h) + c$, the equation for area A_1 becomes:

$$A_1 = \frac{h}{3}(y_1 + 4y_2 + y_3)$$

Similar manipulations on areas A_2 and A_3 yield:

$$A_2 = \frac{h}{3}(y_3 + 4y_4 + y_5)$$

$$A_3 = \frac{h}{3}(y_5 + 4y_6 + y_7)$$

The integral of the function $f(x)$ from point a to point b is expressed as follows:

$$\int_a^b f(x)dx \approx \frac{h}{3}(y_1 + 4y_2 + 2y_3 + 4y_4 + 2y_5 + 4y_6 + y_7)$$

By dividing the region to be integrated into an even number of N equally spaced intervals and calculating the sum of the terms, we can generalize the preceding formula as follows:

$$\int_{x_1}^{x_{N=1}} f(x)dx \approx \frac{h}{3}(y_1 + 4\sum_{\substack{j \ even}}^{N} y_j + 2\sum_{\substack{k \ odd \\ k \neq 1}}^{N-1} y_k + y_{N+1})$$

In the preceding equation we see that the parenthetical term consists of four elements. The first element is the value of the y coordinate at point a, and the last element is the value of the y coordinate at point b. The second element is the product of 4 times the sum of all values of y_j, where j is an even number in the range 2 to N. The third element is the product of 2 times the sum of all values of y_k, where k is an odd number in the range 3 to $N - 1$. Based on this, the algorithm for computing the integral using Simpson's Rule can be described as follows:

STEP 1: Define N as the number of areas. Define point a as the lower limit of the independent variable and b as the upper limit. Create a local array of N elements to hold sums during computation. Define the size of the step h. Create integer counters j and k.

STEP 2: Calculate and store the value of the dependent variable at point a.

STEP 3: Calculate and store the value of the dependent variable at point b.

STEP 4: Calculate and store the products of 4 times the value of the dependent variable, for all even values of j, from $j = 2$ to $j = N$.

STEP 5: Calculate and store the products of 2 times the value of the dependent variable, for all odd values of k, from $k = 3$ to $k = N - 1$.

STEP 6: Add all values in local array. Multiply sum by $h/3$.

In Section 11.6.3 we used Simpson's rule to approximate the integral in the normal probability distribution function. Since, in that case, the function is known, Simpson's rule is quite suitable. The function NormalProb() can be used as a model to develop others in which the function that defines the dependent variable is known at coding time.

However, if the function is not known, or cannot be used, then Simpson's rule is applicable only if the known data points are equally spaced along the x-axis. Also, since each area requires three data points, there must either be an odd number of data points in the set or we must throw out the first or last data point to get an odd number of points. Within these limitations, Simpson's Rule provides a more accurate approximation of the integral than the trapezoidal rule. The following function, named SimpRule(), calculates the integral from two data sets.

```
double SimpRule(double *datX,   // Array of x values
                double *datY,   // Array of f(x) values
                int size)       // Size of datY[] array
{
// Calculates integral by estimating the area under the curve.
// Integration is performed by means of Simpson's rule,
// as follows:
//    Np  =  step/3 * (y1 + 4y2 + 2y3 + 4y4 + 2y5 + 4y6 .. yN+1)
//           -------   -----------------------------------------
//              |                         |
//              |                         |___ sum term
//              |_____ first term
// where:
//       N is the number of area elements used in the
//       computation. Must be even.
//       y1, y2 .. yN+1 are the observed values of the dependent
//       variable
//       step = difference between x coordinates
//       Np is the integral
// The sum term consists of 4 operations:
//    1. y1
//    2. 4 * (sums of all yj, j even from 2 to N)
//    3. 2 * (sums of all yk, k odd from 3 to N - 1)
//    4. yN+1
// Local variables
double area = 0;
int N = size - 1;        // Number of areas
double *tAreas = new double[N];
double base = 0;
// Calculate base size using the formula:
// b = (b - a)/N
// where a and b are the limit values of the independent
// variable
base = (datX[size - 1] - datX[0]) /N;
// Perform sum term operation 1 and store in local array
tAreas[0] = datY[0];
// Perform sum term operation 4 and store in local array
tAreas[1] = datY[N];
// Perform sum term operation 2 and store sums in local array
// Operation: 4 * (sums of all yj, j even, from 2 to N)
for(int j = 2; j <= N; j += 2)
    tAreas[j] = 4 * datY[j-1];

// Perform sum term operation 3 and store sums in local array
// Operation: 2 * (sums of all yk, k odd, from 3 to N - 1)
for(int k = 3; k <= (N - 1); k += 2)
    tAreas[k] = 2 * datY[k-1];

// At this point there are N number of entries in the
// local array holding sum term. Add values.
 for(int z = 0; z < size ; z++)
  area = area + tAreas[z];

// Apply first term of Simpson's equation
area = (base * area) /3;

return area;
}
```

The function SimpRule(), listed previously, assumes that there are two data sets for the values of the dependent and independent variables. Also that both data sets contain the same number of entries, that the number of entries is odd, and that there are not less than 3 entries.

SOFTWARE ON-LINE

The integral-calculating functions TrapRule() and SimpRule() are found in the file Un32_10.h which is located in the Chapter12 folder in the book's on-line software.

Chapter 13

Linear Systems

Chapter Summary

This chapter includes a review of linear algebra as it relates to the programming problems discussed. Topics include matrices, vectors, scalars, as well as the corresponding numeric operations. The routines developed are those that would be typically found in an array processor package, including scalar and vector operations, matrix addition and multiplication, and the solution of a system of linear equations.

13.0 Linear Equations

A linear equation is one in which the highest power of an unknown is 1. We can represent a linear equation in one variable with the formula:

$$mx + c = 0$$

A representation called the *standard form* facilitates comparing equations of the same type. The standard form for a linear equation is

$$n_1 x_1 + n_2 x_2 + \ldots + n_m x_m = b$$

Notice that, in standard form, each variable (x_1 to x_m) is preceded be a coefficient (n_1 to n_m).

By definition a linear equation is a simple mathematical entity: it cannot contain products of variables, exponentials, trigonometric, or logarithmic functions. The only operation allowed on a single variable is multiplication by a constant, represented by the standard term n_i. Note that this use of the word "linear" relates to the fact that an equation of this type is graphed as a straight line.

Problems that can be represented by a set of linear equations occur frequently in science, engineering, and finance. Numerous examples are found in business, physics, chemistry, economics, computer science, geometry, genetics, ecology, statistics, sociology, and demography. Furthermore, certain branches of mathematics rely heavily on the solution of linear systems. These include modeling, approximation theory, game theory, graph theory, and probability.

13.0.1 Systems of Linear Equations

Consider the following set of linear equations:

$$x + y + 2z = 9$$
$$2x - 4y + 3z = 2$$
$$3x + 6y + 5z = 0$$

The standard form can be generalized by representing a system of m linear equations in n unknowns as follows:

$$a_{11}x_1 + a_{12}x_2 + ... + a_{1n}x_n = b_1$$
$$a_{21}x_1 + a_{22}x_2 + ... + a_{2n}x_n = b_2$$
$$...$$
$$a_{m1}x_1 + a_{m2}x_2 + ... + a_{mn}x_n = b_m$$

In the above expression the coefficients of the unknown terms are written using a double subscript. This encoding is a convenient way for locating the coefficients in a rectangular pattern of rows and columns. For instance, the coefficient a_{34} is located in row number 3, column number 4.

The representation can be further generalized by stating that in the expression

$$a_{ij}$$

i indicates the equation in which the coefficient is located (row), and j the unknown term that it affects (column).

Since the unknown elements of a linear equation cannot have exponents, radicals, or be subject to other mathematical operations, the only variation allowed between terms are signs and coefficients. Based on this fact, the didactical structure can be further simplified by assuming that the unknown terms are ordered in columns (left to right) and by not explicitly representing the plus and the equal sign. The coefficients of the variables and the constants on the right-hand side of the equations may then be extracted from the system of equations. For example:

$$a_{11} \ a_{12} \ ... \ a_{1n} \ b_1$$
$$a_{21} \ a_{22} \ ... \ a_{2n} \ b_2$$
$$. \ . \ . \ .$$
$$. \ . \ . \ .$$
$$a_{m1} \ a_{m2} \ ... \ a_{mn} \ b_m$$

Consider that a solution of a linear equation is a set of numbers that satisfies the equation when these numbers replace the original variables. For example, the equation

$$x + y + 2z = 9$$

has one solution when

$$x = 2, y = 3, \text{and } z = 2$$

because

$$2 + 3 + 2(2) = 9$$

A solution for a linear equation in n variables can be found by assigning values to $(n - 1)$ variables and solving for the remaining unknown. In the previous equation we arbitrarily assigned values $x = 2$ and $y = 3$, and solved for z.

A finite set of linear equations, sometimes called a linear system, has a solution in the set of values that satisfies every equation in the system. For example, the system of linear equations

$$x + y = 5$$
$$2x + 3y = 13$$

has the solution $x = 2$, $y = 3$, because

$$2 + 3 = 5$$
$$2(2) + 3(3) = 13$$

Not all systems of linear equations have a solution. For example, the system

$$x + y = 5$$
$$x + y = 4$$

can have no solution due to intrinsic contradictions. A system of linear equations with no solution is said to be *inconsistent*. If there is a least one solution the system is called *consistent*. Therefore, a system of linear equations can have no solution, one solution, or infinite solutions.

13.0.2 Matrix Representations of Linear Systems

The term *matrix* is used in mathematics to describe an array of numbers stored in a two-dimensional structure of columns and rows. The common notation for matrices uses brackets to enclose the structure. The following is a 4 by 4 matrix of signed, fractional numbers

$$\begin{bmatrix} 1 & 1 & 2 & 9 \\ 2 & -4 & 3 & 2 \\ 3 & 6 & 5 & 0 \end{bmatrix}$$

In linear equations the rectangular array that contains the coefficients of the terms and the values on the right-hand side of the equations is called the *augmented matrix*. For example, the following set of equations in three unknowns

$$x + y + 2z = 9$$
$$2x - 4y + 3z = 2$$
$$3x + 6y + 5z = 0$$

has the augmented matrix

$$\begin{bmatrix} 2 & 3 & -1 & 4.3 \\ 1.2 & 0 & 2 & -3 \\ 1 & 12.7 & -4 & 7 \\ 3 & 0 & -1 & 1.22 \end{bmatrix}$$

Note that the term augmented matrix refers to a matrix that includes the coefficients as well as the constants of the original equations. A matrix that does not include the constant terms is sometimes called the *coefficients matrix*.

Many computer calculations of scientific and technological problems are simplified by operating on data stored in matrix form. One reason why matrices are effective in computer mathematics is that they can be used to store the coefficients of a set of linear equations. The values in the matrix can then be manipulated, following simple rules, in order to obtain a solution to the system (if a solution exists). In this manner matrices become an analytical tool well suited to the iterative processing operations. The solution of a set of linear equations by matrix manipulations is discussed starting at Section 13.3.

The storage of the coefficients of linear equations is just one of many mathematical uses for matrix structures. In computer graphics, matrices are often used to store the coordinates of screen points. In this case, it is possible to transform a graphics image by operating on its matrix of coordinates. For instance, a graphics point, or even an entire figure, can be translated, rotated, or scaled using matrix operations.

13.1 Numeric Data in Matrix Form

A matrix can be visualized as a rectangular pattern of rows and columns containing numeric data. Each *matrix* element is called an *entry*. A matrix is stored in computer memory as a series of ordered numeric items. Each numeric data item (entry) in the matrix takes up memory space according to the storage format. For example, if matrix data is stored as binary floating-point numbers in the FPU formats, each entry takes up the following space:

```
Single precision real ..... 4 bytes,
Double precision real ..... 8 bytes,
Extended precision real .. 10 bytes.
```

Integer matrices will vary from one high-level language to another one and even in different implementations of the same language. Microsoft Visual C++ in 32-bit versions of Windows uses the following data ranges for the integer types:

```
char, unsigned char ....... 1 byte
short, unsigned short ..... 2 bytes
long, unsigned long ....... 4 bytes
```

Various high-level languages store numeric data in matrix-like arrays. If the storage pattern assumes that consecutive entries are elements in the same matrix row, the matrix is said to be in *row-major order*. On the other hand, if consecutive entries are items in the same column, the matrix is said to be in *column-major order*. The following are the storage schemes used by some high-level languages

```
BASIC              Column major order
FORTRAN            Column major order
Pascal             Row major order
C/C++              Row major order
```

13.1.1 Matrices in C++

In C and C++ a matrix can be defined as a multi-dimensional array. Like several other high-level languages, C++ implements multi-dimensional arrays as arrays of arrays. For example:

```
double mat1[4][4] = {
                {2.0, 3.0,  -1.0,  4.3},  // Array 1
                {1.2, 0.0,   2.0, -3.0},  // Array 2
                {1.0, 12.7, -4.0,  7.0},  // Array 3
                {3.0, 0.0,  -1.0,  1.22}, // Array 4
                };
```

In this case mat1[][] is a two-dimensional array, actually consisting of four one-dimensional arrays. When a multi-dimensional array is passed as an argument to a function, the called code must be made aware of its dimensions so that it can access its elements with multiple subscripts. In the case of a two-dimensional array the function must know the number of dimensions and the number of columns in the array argument. The number of rows is not necessary since the array is already allocated in memory. The following demo program shows the definition of a 4-by-4 two-dimensional array and how this array is passed to a function that fills it.

```cpp
#include <iostream.h>
#define ROWS 4
#define COLS 4

int main()
{

// Define a 4-by-4 matrix
int matrx[ROWS][COLS];

// Call function to fill matrix
FillMatrix(matrx);
// Display matrix
    for(i = 0; i < ROWS; i++)
        for(k = 0; k < COLS; k++)
            cout << matrx[i][k] << "\n";
return 0;
}

void FillMatrix(int matx[][COLS])
{
// Fill a matrix of type int
// On entry:
//        matx[][] is caller's matrix
//        Constants ROWS and COLS define the array dimensions
int entry;

    for(int i = 0;i < ROWS; i++)
    {
        cout << "Enter row " << i << "\n";

        for(int j = 0; j < COLS; j++)
        {
            cout << "element " << i << " " << j << ": ";
            cin >> entry;
            matx[i][j] = entry;

        }
    }
}
```

Notice in the preceding program that the function FillMatrix() is made aware that the array passed as an argument is two-dimensional, and that each row consists of 4 columns. This is the information that the code requires for accessing the array with double subscript notation. The major limitation of this approach is that the function FillMatrix() requires the array size to be defined in constants. C++ produces an error if we attempt to define the matrix dimensions using variables. For this reason the function FillMatrix(), as coded in the preceding sample, does not work for filling a two-dimensional 5-by-5 array or, in fact, of any size other than the one for which it was originally coded. Even more complications arise if we attempt to use a template function in relation to multi-dimensional arrays.

An alternative approach for implementing matrices in C++ code is to define the data as a one-dimensional array and let the software handle the partitioning into columns and rows. In this manner we can avoid the drawbacks of passing multi-dimensional arrays as arguments to functions. In addition, with one-dimensional arrays it is easy to use templates in order to create generic functions that operate on arrays of different data types. The following demonstration program implements an matrix fill using one-dimensional arrays and template functions.

```cpp
#include <iostream.h>

int main()
{
  int rows = 4;
  int cols = 4;

  // Matrix is defined as a 2D array
  matrx[16];

  // FillMatrix()
  FillMatrix(matrx, rows, cols);
    // Display matrix

    for(x = 0; x < rows; x++)
        for(y = 0; y < cols; y++)
            cout << mat2[(x * cols) + y] << "\n";
    cout << "\n\n";

  return 0;
}
template <class A>
void FillMatrix(A *matx, int rows, int cols)
{
// Fill a matrix of type int
// On entry:
//        *matx is caller's matrix
//        parameter rows is nomber of rows in matrix
//        parameter cols is number of columns in matrix
A entry;

    for(int i = 0;i < rows; i++)
    {
        cout << "Enter row " << i << "\n";

        for(int j = 0; j < cols; j++)
          {
              cout << "element " << i << " " << j << ": ";
              cin >> entry;
              matx[(i * cols) + j] = entry;
          }
    }
}
```

In the preceding code sample notice that the matrix is defined as a one-dimensional array and that the function FillMatrix() receives the number of rows and columns as parameters. Also that the FillMatrix() function is implemented as a template, which means that it can be used to fill a two-dimensional matrix of any size and data type.

In manipulating matrices the programmer is usually concerned with the following elements:

1. The number of rows in the matrix

2. The number of columns in the matrix

3. The memory space (number of bytes) occupied by each matrix entry

The number of rows and columns determines the dimension of the matrix. It is customary to represent matrix dimensions using the variable M for the number of rows and the variable N for the number of columns. The storage format of the entries determines the memory space occupied by each matrix entry, therefore, the number of bytes that must be skipped in order to index from entry to entry. For this reason the size of each entry is sometimes referred to as the *horizontal skip factor*. By the same token, the number of entries in each matrix row must be used by the program in order to index to successive entries in the same column. This value is sometimes called the *vertical skip factor*. Low-level implementations must use the skip factors to access different matrix entries, as shown later in this chapter. High-level languages (C++ included) access matrix entries using the indices, and usually ignore the byte size of each element.

13.1.2 Locating a Matrix Entry

Each matrix entry is identified by its row and column coordinates. In this context the variable i is often used to designate the entry along a matrix row and the variable j to designate the entry along a matrix column. Thus, any entry in the matrix can be identified by its $_{ij}$ coordinates. The individual matrix is usually designated with an upper case letter. We say that Matrix A is composed of M rows and N columns. The number of entries in the matrix is $M \cdot N$. If each entry takes up s bytes of memory, the matrix memory space can be expressed as $M \cdot N \cdot s$. The following diagram shows a 5-by-4 matrix.

```
                      C O L U M N S
                  0   1   2   3   4
                  |   |   |   |   |
    R     0 ----- X   X   X   X   X
    O     1 ----- X   X   ij  X   X
    W     2 ----- X   X   X   X   X
    S     3 ----- X   X   X   X   X

          M = 5 (total rows)
          N = 4 (total columns)
          i = 1 (row address of entry ij)
          j = 2 (column address of entry ij)
```

Notice that matrix dimensions are stated as the number of rows and columns. Thus, the dimension of the previous matrix is 4-by-5. However, the location within the rows and columns is zero-based. This scheme is consistent with array dimensioning and addressing in C++.

Linear systems software often has to access an individual matrix entry located at the ith row and the jth column. In high-level programming the language itself figures out the horizontal and vertical skip factors. Therefore locating the memory address for a matrix entry is a simple matter of multiplying the row number by the number of columns in the matrix, then adding the offset within the row. If i designates the row, j the column, and cols is the number of columns in each row, then the offset within a matrix implemented in a one-dimensional array is given by the statement:

```
value = (M[(i*cols) + j]);
```

were M is the matrix and value is a variable of the same type as the matrix entries. The following C++ template function returns the matrix element at row i, column j.

```
template <class A>
A Locateij(A *matx, int i, int j, int cols)
{
// Locate and return matrix entry at row i, column j
// On entry:
//       *mat is caller's matrix
//       i = row number
//       j = column number
//       cols = number of matrix columns
 return (matx[(i * cols) + j]);
}
```

13.2 Operations on Matrix Entries

In the terminology of matrix mathematics a vector is a matrix in which one of the elements is of the first order, that is, a matrix that has either a single column or a single row. In this sense we can refer to a matrix as a column vector if its N dimension is 1, and as a row vector if its M dimension is 1. By the same token, a row vector is a matrix consisting of a single row, and a column vector a matrix consisting of a single column. Although, strictly speaking, a vector can be considered as a one-dimensional matrix, the term matrix is more often associated with a rectangular array. Also note that this use of the word *vector* is not related to the geometrical concept of a directed segment in 2-dimensional or 3-dimensional space.

In order to represent individual, undirected quantities, matrix mathematics borrow from analytical geometry the notion of a *scalar*. We say that an individual constant or variable is a scalar quantity, while multi-element structures are either vectors or matrices.

Programs that perform mathematical operations on vectors and matrices are sometimes called *array processors*. The generic designation implied by the word *array* refers to any multi-element structure, whether it be a matrix or a vector. Many array operations require simple arithmetic on the individual entries of the array, for example, adding, subtracting, multiplying or dividing all the entries of an array by a scalar, or finding the square root, powers, logarithmic or trigonometric functions of the individual entries. A second type of array operations refers to arithmetic between two multi-element structures or the manipulation of a single structure, for example, the addition and multiplication of matrices, the calculation of vector products, and matrix inversion. Some matrix arithmetic operations obey rules that differ from those used in scalar operations. Finally, some array operations are oriented towards simplifying and solving systems of linear equations. For example, interchanging rows, multiplying a row by a scalar, and adding a multiple of one row to another row.

13.2.1 Vectors

We have seen that the word vector is often used to refer to the individual rows and columns of a two-dimensional matrix. In this sense vector operations are those that affect the entries in a row or column, and matrix operations are those that affect all the entries in the rectangular array. Vectors constitute one-dimensional arrays of numbers, while matrices constitute a two-dimensional arrays. Hereafter we occasionally refer to the entries in a matrix row as a row vector and the entries in a matrix column as a column vector.

13.2.2 Vector-by-Scalar Operations in C++

Applications are often required to perform discrete operations on the individual elements of matrix rows and columns. According to the terminology presented previously, these can be designated as row and column vector operations. The functions listed in this section perform multiplication, addition, division, and subtraction of a row vector by a scalar and multiplication of a column vector by a scalar. The implementation is based on storing matrix data in one-dimensional arrays, where the rows and columns are handled by code. The functions are coded as templates so that they can be used with any compatible data type.

```
//****************************************************************
//              C++ functions for vector arithmetic
//****************************************************************
template <class A>
void RowMulScalar(A *matx, int i, int cols, A scalar)
{
// Multiply a matrix row times a scalar
// On entry:
//        *matx is caller's matrix
//        i is number of the row
//        cols is number of columns in the matrix
//        scalar is the value to be multiplied by
// On exit:
//        elements in matrix row i are multiplied by scalar
int rowStart = i * cols;
for(int j = 0;j < cols ;j++)
    matx[rowStart + j] *= scalar;
}

template <class A>
void RowPlusScalar(A *matx, int i, int cols, A scalar)
{
// Add a scalar to a matrix row
// On entry:
//        *matx is caller's matrix
//        i is number of the row
```

```
//        cols is number of columns in the matrix
//        scalar is the value to be added
// On exit:
//        Scalar is added to all elements in matrix row i
int rowStart = i * cols;

for(int j = 0;j < cols ;j++)
    matx[rowStart + j] += scalar;
}

template <class A>
void RowMinusScalar(A *matx, int i, int cols, A scalar)
{
// Subtract a scalar from each element in a matrix row
// On entry:
//        *matx is caller's matrix
//        i is number of the row
//        cols is number of columns in the matrix
//        scalar is the value to be subtracted
// On exit:
//        Scalar is subtracted from all elements in matrix row i
int rowStart = i * cols;
for(int j = 0;j < cols ;j++)
    matx[rowStart + j] -= scalar;
}
template <class A>
void RowDivScalar(A *matx, int i, int cols, A scalar)
{
// Divide all elements in a matrix row by a scalar
// On entry:
//        *matx is caller's matrix
//        i is number of the row
//        cols is number of columns in the matrix
//        scalar is the value to be divided by
// On exit:
//        All elements in matrix row i are divided by the
//        scalar
int rowStart = i * cols;

for(int j = 0;j < cols ;j++)
    matx[rowStart + j] /= scalar;
}
template <class A>
void ColMulScalar(A *matx, int j, int rows, int cols, A scalar)
{
// Multiply a matrix column times a scalar
// On entry:
//        *matx is caller's matrix
//        j is column number
//        rows is the number of rows in the matrix
//        cols is number of columns in the matrix
//        scalar is the value to be multiplied by
// On exit:
//        elements in matrix column j are multiplied by scalar

    for(int i = 0;i < rows ;i++)
    {
        matx[(cols * i) + j] *= scalar;
    }
}
```

Since column-level operations are not as common in array processing as row operations, we have provided a single example, which is the ColMulScalar() function. The programmer should be able to use it to develop any other column operations that may be required.

SOFTWARE ON-LINE

The code is found in the file Un32_11.h located in the Chapter13 folder in the book's on-line software.

13.2.3 Low-Level Vector-by-Scalar Operations

Array processors are computationally-intensive applications. Coding these operations in high-level languages is convenient and easy, but sacrifices control, performance, and possibly, precision.

In low level computations the procedure receives the address of the first matrix entry, as well as the row and column parameters required for the operation. For example, to perform a row-level operation the low-level routine must know the address of the matrix, the number of elements in each column, the number of the desired row. In addition, the low-level routine must have available the horizontal skip factor. Using this information code can visit each matrix entry and perform the required operation.

```
;****************************************************************
;              low-level procedures for vector arithmetic
;****************************************************************
       .CODE
_ROW_TIMES_SCALAR   PROC    USES esi edi ebx ebp
; Procedure to multiply a matrix row vector by a scalar
; On entry:
;         ST(0) = scalar multiplier
;         ESI -> matrix containing the row vector
;         EAX = number of row vector (0 based)
;         ECX = number of columns in matrix
;         EDX = horizontal skip factor
; On exit:
;         entries of specified row vector multiplied by ST(0)

; Formula for offset of start of vector is
;              offset = [ ((i-1) * N * s) ]
; where i represents the matrix row and s the column
; AL holds 0-based number of the desired row vector
; CL holds the number of entries per row (N)
; DL holds skip factor
       MOV     AH,0            ; Clear high-order byte
       MUL     CL              ; AX = AL * CL
; Second multiplication assumes that product will be less than
; 65535. This assumption is reasonable since the matrix space
; assigned is 400 s
       PUSH    DX              ; Save before multiply
       MOV     DH,0            ; Clear high-order byte
       MUL     DX              ; AX = AX * DL
       POP     DX              ; Restore DX
```

```
            ADD     ESI,EAX         ; Add offset to pointer
            MOV     DH,0            ; Clear high-order byte
; At this point:
;    ESI -> first entry in the matrix row
;    ST(0) holds scalar multiplier
;    ECX = number of entries in row
;    EDX = byte length of each matrix entry
ENTRIES:
            CALL    FETCH_ENTRY
            FMUL    ST,ST(1)        ; Multiply by ST(1)
            CALL    STORE_ENTRY
            ADD     ESI,EDX         ; Index to next entry
            LOOP    ENTRIES
            RET
_ROW_TIMES_SCALAR       ENDP
;*****************************************************************

_ROW_PLUS_SCALAR    PROC    USES esi edi ebx ebp
; Procedure to add a scalar to a matrix row
; On entry:
;          ST(0) = scalar multiplier
;          ESI -> matrix containing the row vector
;          EAX = number of row vector (0 based)
;          ECX = number of columns in matrix
;          EDX = horizontal skip factor
; On exit:
;          entries of row vector multiplied by ST(0)
;
; Formula for offset of start of vector is
;              offset = [ ((i-1) * N * s) ]
; AL holds 0-based number of the desired row vector
; CL holds the number of entries per row (N)
; DL holds skip factor (8 for double precision)
            MOV     AH,0            ; Clear high-order byte
            MUL     CL              ; AX = AL * CL
; Second multiplication assumes that product will be less than
; 65535. This assumption is reasonable since the matrix space
; assigned is 400 s
            PUSH    DX              ; Save before multiply
            MOV     DH,0            ; Clear high-order byte
            MUL     DX              ; AX = AX * DL
            POP     DX              ; Restore DX
            ADD     ESI,EAX         ; Add offset to pointer
            MOV     DH,0            ; Clear high-order byte
; At this point:
;    ESI -> first entry in the matrix row
;    ST(0) holds scalar multiplier
;    ECX = number of entries in row
;    EDX = byte length of each matrix entry
ENTRIES_A:
            CALL    FETCH_ENTRY
            FADD    ST,ST(1)        ; Add scalar
            CALL    STORE_ENTRY
            ADD     ESI,EDX         ; Index to next entry
            LOOP    ENTRIES_A
            RET
_ROW_PLUS_SCALAR        ENDP
;*****************************************************************
;
_ROW_DIV_SCALAR             PROC
```

```
; Procedure to divide a matrix row vector by a scalar
; On entry:
;            ST(0) = scalar divisor
;            ESI -> matrix containing the row vector
;            EAX = number of row vector (0 based)
;            ECX = number of columns in matrix
;            EDX = horizontal skip factor
; On exit:
;            Entries of row vector divided by ST(0)
;            ST(0) is preserved
; Algorithm:
;             Division is performed by obtaining the reciprocal of
;             the divisor and using the multiplication routine
;                             |   ST(0)   |   ST(1)   |  ST(2)
;                          ;  divisor  |     ?     |     ?
        FLD     ST(0)     ;  divisor  |  divisor  |     ?
        FLD1              ;     1     |  divisor  |  divisor
        FDIV    ST,ST(1); 1/divisor  |     1     |  divisor
        FSTP    ST(1)     ; 1/divisor |  divisor  |     ?
        CALL    _ROW_TIMES_SCALAR
        FSTP    ST(0)     ;  divisor  |     ?     |     ?
        CLD
        RET
_ROW_DIV_SCALAR            ENDP
;****************************************************************
_ROW_MINUS_SCALAR         PROC
; Procedure to subtract a scalar from the entries in a matrix
; row
;
; On entry:
;          ST(0) = scalar to subtract
;          ESI -> matrix containing the row vector
;          EAX = number of row vector (0 based)
;          ECX = number of columns in matrix
;          EDX = horizontal skip factor
;
; On exit:
;          Scalar subtracted from entries of the row vector
; Algorithm:
;          Subtraction is performed by changing the sign of the
;          subtrahend and using the addition routine
;                           |   ST(0)   |   ST(1)   |  ST(2)
;                           |     #     |     ?
        FCHS              ;     -#     |     ?
        CALL    _ROW_PLUS_SCALAR
        FCHS              ;     #      |     ?
        CLD
        RET
_ROW_MINUS_SCALAR         ENDP
```

PROGRAMMER'S NOTEBOOK

In the preceding routines scalar subtraction is performed by changing the sign of the scalar addend, while division is accomplished by multiplying by the reciprocal of the divisor. Also notice that sign inversion of a row vector can be obtained by using − 1 as a scalar multiplier.

Notice that the row operations procedures listed previously receive the horizontal skip factor in the EDX register. The core procedures _ROW_TIMES_SCALAR and _ROW_PLUS_SCALAR then call the auxiliary procedures FETCH_ENTRY and STORE_ENTRY to access and store the matrix entries. FETCH_ENTRY and STORE_ENTRY determine the type of data access required according to the value in the EDX register. If the value in EDX is 4, then the data is encoded in single precision format. If the value is 8 then the data is in double precision. If the value is 10, then the data is in extended precision. This mechanism allows creating low-level code that can be used with any of the three floating point types in ANSI/IEEE 754. The C++ interface routines, which are coded as template functions, use the sizeof operator on a matrix entry to determine the data type passed by the caller.

Visual C++ Version 6, in Win32 operating systems, defines the size of int, long, unsigned long, and float data types as 4 bytes. Therefore it is not possible to use the size of a data variable to determine if an argument is of integer or float type. For this reason the interface routines listed in this section can only be used with float-type arguments. Attempting to pass integer matrices or scalars will result in undetected computational errors. The C++ interface functions to the low-level row-operation procedures are as follows:

```
//*****************************************************************
//            C++ interface functions to Un32_13 module
//*****************************************************************
template <class A>
void RowTimesScalarLL(A *matx, int i, int cols, A scalar)
{
// Multiply a matrix row times a scalar using low-level code
// in the Un32_13 module
// On entry:
// On entry:
//      *matx is caller's matrix (floating point type)
//      i is number of the row
//      cols is number of columns in the matrix
//      scalar is the value to add (floating point type)
// Routine expects:
//      ST(0) holds scalar
//      ESI -> matrix
//      EAX = row vector number
//      ECX = number of columns in matrix
//      EDX = horizontal skip factor
// On exit:
//      elements in matrix row i are multiplied by scalar
int eSize = sizeof(matx[0]);

_asm
{
        MOV     ECX,cols            // Columns to ECX
        MOV     EAX,i               // Row number to EAX
        MOV     ESI,matx            // Address to ESI
        FLD     scalar              // Scalar to ST(0)
        MOV     EDX,eSize           // Horizontal skip
        CALL    ROW_TIMES_SCALAR
}
return;
}
```

```
template <class A>
void RowPlusScalarLL(A *matx, int i, int cols, A scalar)
{
// Add a scalar to each entry in a matrix row
// using low-level code
// On entry:
//        *matx is caller's matrix (floating point type)
//        i is number of the row
//        cols is number of columns in the matrix
//        scalar is the value to add (floating point type)
// Routine expects:
//        ST(0) holds scalar
//        ESI -> matrix
//        EAX = row vector number
//        ECX = number of columns in matrix
//        EDX = horizontal skip factor
// On exit:
//        scalar is added to elements in matrix row I

int eSize = sizeof(matx[0]);

    _asm
{

        MOV     ECX,cols            // Columns to ECX
        MOV     EAX,i               // Row number to EAX
        MOV     ESI,matx            // Address to ESI
        FLD     scalar              // Scalar to ST(0)
        MOV     EDX,eSize           // Horizontal skip
        CALL    ROW_PLUS_SCALAR
}
return;
}

template <class A>
void RowDivScalarLL(A *matx, int i, int cols, A scalar)
{
// Divide a matrix row by a scalar using low-level code
// in the Un32_13 module
// On entry:
//        *matx is caller's matrix (float type)
//        i is number of the row
//        cols is number of columns in the matrix
//        scalar is the value to divide by (float type)
// Routine expects:
//        ST(0) holds scalar
//        ESI -> matrix
//        EAX = row vector number
//        ECX = number of columns in matrix
//        EDX = horizontal skip factor
// On exit:
//        elements in matrix row i are divided by scalar
int eSize = sizeof(matx[0]);
_asm
{

        MOV     ECX,cols            // Columns to ECX
        MOV     EAX,i               // Row number to EAX
        MOV     ESI,matx            // Address to ESI
        FLD     scalar              // Scalar to ST(0)
```

```
        MOV     EDX,eSize           // Horizontal skip
        CALL    ROW_DIV_SCALAR
}
return;
}


template <class A>
void RowMinusScalarLL(A *matx, int i, int cols, A scalar)
{
// Subtract a scalar from a matrix row using low-level code
// in the Un32_13 module
//
// On entry:
//        *matx is caller's matrix (float type)
//        i is number of the row
//        cols is number of columns in the matrix
//        scalar is the value to subtract (float type)
//
// Routine expects:
//        ST(0) holds scalar
//        ESI -> matrix
//        EAX = row vector number
//        ECX = number of columns in matrix
//        EDX = horizontal skip factor
// On exit:
//        Scalar is subtracted from elements in matrix row I

int eSize = sizeof(matx[0]);
_asm
{
        MOV     ECX,cols            // Columns to ECX
        MOV     EAX,i               // Row number to EAX
        MOV     ESI,matx            // Address to ESI
        FLD     scalar              // Scalar to ST(0)
        MOV     EDX,eSize           // Horizontal skip
        CALL    ROW_MINUS_SCALAR
}
return;
}
```

SOFTWARE ON-LINE

The C++ interface routines are found in the file Un32_12.h located in the Chapter13\Test Un32_12.h folder in the book's on-line software.

13.2.4 Matrix-by-Scalar Operations

Often we need to perform scalar operations on all entries in a matrix, with the more useful operations being scalar multiplication, division, addition, and subtraction. In this section we present the routines to perform these matrix-by-scalar calculations. Because matrix-by-scalar manipulations are computationally intensive, we develop the routines in low-level code and provide C++ interface functions to the assembly language procedures. Here we list the low-level code and the interface routines for performing matrix-by-scalar multiplication. The low-level code is as follows:

```
        .CODE
_MAT_TIMES_SCALAR    PROC    USES esi edi ebx ebp
; Procedure to multiply a matrix by a scalar
; On entry:
;           ST(0) = scalar multiplier
;           ESI -> matrix containing the row vector
;           EAX = number of rows
;           ECX = number of columns
;           EDX = horizontal skip factor
; On exit:
;           entries of matrix multiplied by ST(0)
; Total number of entries is M * N
        MOV      AH,0               ; Clear high-order byte
        MUL      CL                 ; AX = AL * CL
        MOV         ECX,EAX         ; Make counter in CX
; At this point:
;    ESI -> first entry in the matrix
;    ST(0) holds scalar multiplier
;    ECX = number of entries in matrix
;    EDX = byte length of each matrix entry (4, 8, or 10 bytes)
MAT_MUL:
        CALL     FETCH_ENTRY
        FMUL     ST,ST(1)           ; Multiply by ST(1)
        CALL     STORE_ENTRY
        ADD      ESI,EDX            ; Index to next entry
        LOOP     MAT_MUL
        CLD
        RET
_MAT_TIMES_SCALAR    ENDP
```

The C++ interface function is named MatTimesScalarLL(). The code is as follows:

```
template <class A>
void MatTimesScalarLL(A *matx, int rows, int cols, A scalar)
{
// Multiply a matrix times a scalar using low-level code
// in the Un32_13 module
// On entry:
//      *matx is caller's matrix (type double)
//      rows is number of the rows in matrix
//      cols is number of columns in the matrix
//      scalar is the value to multiply by (floating point type)
// Routine expects:
//      ST(0) holds scalar
//      ESI -> matrix
//      EAX = row vector number
//      ECX = number of columns in matrix
//      EDX = horizontal skip factor
// On exit:
//      elements in matrix are multiplied by scalar
int eSize = sizeof(matx[0]);
_asm
{

        MOV      ECX,cols           // Columns to ECX
        MOV      EAX,rows           // Rows to EAX
        MOV      ESI,matx           // Address to ESI
        FLD      scalar             // Scalar to ST(0)
```

```
        MOV     EDX,eSize          // Horizontal skip
        CALL    MAT_TIMES_SCALAR
}
return;
}
```

SOFTWARE ON-LINE

The routines for matrix-by-scalar addition, multiplication, division, and subtraction are found in the UN32_13.ASM module of the MATH32 library. The C++ interface routines are in the file Un32_12.h located in the Chapter13/Test Un32_12 folder in the book's on-line software.

13.2.5 Matrix-by-Matrix Operations

Two of the matrix-by-matrix operations defined in linear algebra are matrix addition and multiplication. Matrix addition is the process of adding the corresponding entries of two matrices. This implies that the operation is valid only if the matrices are of the same size. The addition process in the case C = A + B consists of locating each corresponding entry in matrices A and B and storing their sum in the same location in matrix C.

The matrix multiplication of C = A * B is rather counter-intuitive. Instead of multiplying the corresponding elements of two matrices, matrix multiplication consists of multiplying each of the entries in a row of matrix A, by each of the corresponding entries in a column of matrix B, and adding these products to obtain an entry of matrix C. For example

$$A = \begin{bmatrix} A_{11} & A_{12} & A_{13} \\ A_{21} & A_{22} & A_{23} \end{bmatrix} \quad B = \begin{bmatrix} B_{11} & B_{12} & B_{13} & B_{14} \\ B_{21} & B_{22} & B_{23} & B_{24} \\ B_{31} & B_{32} & B_{33} & B_{34} \end{bmatrix} \quad C = A \times B = \begin{bmatrix} C_{11} & C_{12} & C_{13} & C_{14} \\ C_{21} & C_{22} & C_{23} & C_{24} \end{bmatrix}$$

The entries in the product matrix C are obtained as follows:

First row of matrix:

$$
\begin{aligned}
C_{11} &= (A_{11}*B_{11}) + (A_{12}*B_{21}) + (A_{13}*B_{31}) \\
C_{12} &= (A_{11}*B_{12}) + (A_{12}*B_{22}) + (A_{13}*B_{32}) \\
C_{13} &= (A_{11}*B_{13}) + (A_{12}*B_{23}) + (A_{13}*B_{33}) \\
C_{14} &= (A_{11}*B_{14}) + (A_{12}*B_{24}) + (A_{13}*B_{34})
\end{aligned}
$$

Second row of matrix C:

$$
\begin{aligned}
C_{21} &= (A_{21}*B_{11}) + (A_{22}*B_{21}) + (A_{23}*B_{31}) \\
C_{22} &= (A_{21}*B_{12}) + (A_{22}*B_{22}) + (A_{23}*B_{32}) \\
C_{23} &= (A_{21}*B_{13}) + (A_{22}*B_{23}) + (A_{23}*B_{33}) \\
C_{24} &= (A_{21}*B_{14}) + (A_{22}*B_{24}) + (A_{23}*B_{34})
\end{aligned}
$$

Matrix multiplication requires a series of products, which are obtained using as factors the entries in the rows of the first matrix and the entries in the columns of the second matrix. Therefore, matrix multiplication is defined only if the number of columns of the first matrix is equal to the number of rows in the second matrix. This relationship can be visualized as follows:

```
        matrix A                    matrix B
        R   C                       r    c
            |_____|
```

where R, C represent the rows and columns of the first matrix, and r, c represent the rows and columns of the second matrix. By the same token, the product matrix (C) will have as many rows as the first matrix (A) and as many columns as the second matrix (B). In the previous example, since matrix A is a 2-by-3 matrix, and matrix B is a 3-by-4 matrix, matrix C will be a 2-by-4 matrix.

Here again, since matrix addition and multiplication are computationally intensive we implement the operations in low-level code. The C++ interface routines provide access to the low-level primitives.

Matrix Addition

The following low-level procedure performs matrix addition. The procedure requires that both matrices be of the same dimension, that is, that they have the same number of columns and rows. Notice that the data element is used in both addition and multiplication routines.

```
            .486
            .MODEL flat
            .DATA
;***************************************************|
;   Data for this matrix addition and multiplication |
;***************************************************|
ELEMENT_CNT     DW      0       ; Storage for total number of
                                ; entries in matrix C
;
MAT_A_ROWS      DB      0       ; Rows in matrix A
MAT_A_COLS      DB      0       ; Columns in matrix A
MAT_B_ROWS      DB      0       ; Rows in matrix B
MAT_B_COLS      DB      0       ; Columns in matrix B
MAT_C_ROWS      DB      0       ; Rows in matrix C
MAT_C_COLS      DB      0       ; Columns in matrix C
SKIP_FACTOR     DD      0       ; Element size
;
; Control variables for matrix multiplication
PROD_COUNT      DB      0       ; Number of products in each
                                ; multiplication iteration
WORK_PRODS      DB      0       ; Working count for number of
                                ; products
WORK_ROWS       DB      0       ; Number of rows in matrices A
                                ; and C
WORK_COLS       DB      0       ; Number of columns in matrices B
                                ; and C
            .CODE
;********************************
;       matrix addition
;********************************
```

```
_ADD_MATRICES     PROC      USES esi edi ebx ebp
; Procedure to add all the corresponding entries of two matrices
; of the same size, as follows:
;
;    A=                   B=                  C=(A+B)
;    A11   A12   A13      B11   B12   B13     A11+B11   A12+B12   A13+B13
;    A21   A22   A23      B21   B22   B23     ....
;    A31   A32   A33      B31   B32   B33                         A33+B33
;
; On entry:
;      ESI -> first matrix (A)
;      EDI -> second matrix (B)
;      EBX -> storage area for addition matrix (C)
;              Code assumes that matrix C is correctly
;              dimensioned
;      EAX = number of rows in matrix
;      ECX = number of columns in matrix
;      EDX = horizontal skip factor
;
; On exit:
;      AX = 0 if matrices are the same size, then matrix C
;      contains sum of A + B
;
;      AX = 1 if matrices are of different size and the matrix
;      sum is undefined
;
; Note: matrix addition is defined only regarding two matrices of
;       the same size. Matrices must be of type float and of the
;       same format
;***********************|
;   test for equal size    |
;***********************|
        CMP     AX,CX              ; Test for matrices of equal size
        JE      GOOD_SIZE          ; Go if same size
;***********************|
;       DATA ERROR         |
;***********************|
; At this point matrices cannot be added
        MOV     AX,1               ; Error code
         CLD
        RET
;***********************|
;   store matrix parameters |
;***********************|
; Calculate number of entries by multiplying matrix rows times
; matrix columns
GOOD_SIZE:
        PUSH    EDX                        ; Save register
        MUL     CX                         ; Rows times columns
        MOV     ELEMENT_CNT,AX             ; Store number of entries
         POP    EDX
; At this point:
;      ESI -> first matrix (A)
;      EDI -> second matrix (B)
;      EBX -> storage area for addition matrix (A+B)
;***********************|
;   perform matrix addition  |
;***********************|
A_PLUS_B:
; ESI -> matrix entry in matrix A
```

```
; EDX = entry size (4, 8, or 10 bytes)
        CALL    FETCH_ENTRY      ; ST(0) now holds entry of A
; Fetch entry in matrix B
        XCHG    ESI,EDI          ; ESI -> matrix B entry
        CALL    FETCH_ENTRY      ; ST(0) = matrix B entry
                                 ; ST(1) = matrix A entry
        XCHG    ESI,EDI          ; Reset pointer
; Add entries
        FADD                     ; ST(0)  |  ST(1)  |  ST(2)
                                 ; eA + eB | EMPTY  |
        XCHG    EBX,ESI          ; ESI -> matrix C entry
; Store sum
        CALL    STORE_ENTRY      ; Store sum in matrix C and pop
                                 ; stack
        XCHG    EBX,ESI          ; Restore pointers
; Update entries counter
        DEC     ELEMENT_CNT      ; Counter for matrix entries
        JNZ     NEXT_MAT_ELE     ; Continue if not end of matrix
;***************************|
;    end of matrix addition |
;***************************|
        MOV     AX,0             ; No error flag
        CLD
        RET
;***************************|
;   index matrix pointers   |
;***************************|
; Add entry size to each matrix pointer
NEXT_MAT_ELE:
        ADD     ESI,EDX          ; Add size to pointer
        ADD     EDI,EDX
        ADD     EBX,EDX
        JMP     A_PLUS_B
;
_ADD_MATRICES   ENDP
```

The C++ interface function to the _ADD_MATRICES procedure is as follows:

```
template <class A>
void AddMatrices(A *matA, A *matB, A *matC, int rows, int cols)
{
// Perform matrix addition: C = A + B using low-level code in the
// Un32_13 module
// On entry:
//       *matA and *matB are matrices to be added
//       *matC is matrix for sums
//       rows is number of the rows in matrices
//       cols is number of columns in the matrices
// Requires:
//       All three matrices must be of the same dimensions
//       All three matrices must be of the same floating
//       point data type
// Routine expects:
//       ESI -> first matrix (A)
//       EDI -> second matrix (B)
//       EBX -> storage area for addition matrix (C)
//       EAX = number of rows in matrices
//       ECX = number of columns in matrices
//       EDX = horizontal skip factor
// On exit:
```

```
//      returns matC[] = matA[] + matB[]

int eSize = sizeof(matA[0]);

_asm
{
        MOV     ECX,cols            // Columns to ECX
        MOV     EAX,rows            // Rows to EAX
        MOV     ESI,matA            // Address to ESI
        MOV     EDI,matB
        MOV     EBX,matC
        MOV     EDX,eSize           // Horizontal skip
        CALL    ADD_MATRICES
}
return;
}
```

Matrix Multiplication

The following low-level procedure performs matrix multiplication. The procedure requires that the number of columns in the first matrix be the same as the number of rows in the second matrix. The matrix for results must be capable of storing a number of elements equal to the product of the number of rows of the first matrix by the number of columns of the second matrix. The following low-level procedure performs matrix multiplication. The data variables for the _MUL_MATRICES procedure were defined in the _ADD_MATRICES procedure, listed previously.

```
            .CODE
;******************************************************************
;                   matrix multiplication
;******************************************************************
_MUL_MATRICES    PROC     USES esi edi ebx ebp
; Procedure to multiply two matrices (A and B) for which a matrix
; product (A * B) is defined. Matrix multiplication requires that
; the number of columns in matrix A be equal to the number of
; rows in matrix B, as follows:
;              A                      B
;           R    C                 r     c
;                |_____ = _____|
;
; Example:
;
;    A=(2 by 3)             B=(3 by 4)
;    A11   A12   A13        B11   B12   B13   B14
;    A21   A22   A23        B21   B22   B23   B24
;                          B31   B32   B33   B34
;
; The product matrix (C) will have 2 rows and 4 columns
;    C=(2 by 4)
;    C11   C12   C13   C14
;    C21   C22   C23   C24
;
; In this case the product matrix is obtained as follows:
;    C11 = (A11*B11)+(A12*B21)+(A13*B31)
;    C12 = (A11*B12)+(A12*B22)+(A13*B32)
;    C13 = (A11*B13)+(A12*B23)+(A13*B33)
;    C14 = (A11*B14)+(A12*B24)+(A13*B34)
```

```
;
;    C21 = (A21*B11)+(A22*B21)+(A23*B31)
;    C22 = (A21*B12)+(A22*B22)+(A23*B32)
;    C23 = (A21*B13)+(A22*B23)+(A23*B33)
;    C24 = (A21*B14)+(A22*B24)+(A23*B34)
;
; On entry:
;      ESI -> first matrix (A)
;      EDI -> second matrix (B)
;      EBX -> storage area for products matrix (C)
;      AH = rows in matrix A
;      AL = columns in matrix A
;      CH = rows in matrix B
;      CL = columns in matrix B
;      EDX = number of bytes per entry
; Assumes:
;      Matrix C is dimensioned as follows:
;      Columns of C = columns of B
;      Rows of C = rows of A
; On exit:
;      Matrix C is the products matrix

; Note: the entries of matrices A, B, and C must be of type float
;       and of the same data format
;
; Store number of product in each multiplication iteration
         MOV     PROD_COUNT,AL
; At this point:
;      AH = rows in matrix A
;      AL = columns in matrix A
;      CH = rows in matrix B
;      CL = columns in matrix B
; Store matrix dimensions
         MOV     MAT_A_ROWS,AH
         MOV     MAT_A_COLS,AL
         MOV     MAT_B_ROWS,CH
         MOV     MAT_B_COLS,CL
; Store skip factor
         MOV     SKIP_FACTOR,EDX
; Calculate total entries in matrix C
; Columns in C = columns in B
;    Rows in C = rows in A
         MOV     MAT_C_COLS,CL
         MOV     MAT_C_ROWS,AH
; Calculate number of products
         MOV     AH,0              ; Clear high byte of product
         MUL     CL                ; Rows times columns
         MOV     ELEMENT_CNT,AX    ; Store  count
; At this point:
;      ESI -> first matrix (A)
;      EDI -> second matrix (B)
;      EBX -> storage area for products matrix (A*B)
         MOV     START_BMAT,EDI            ; Storage for pointer
;*************************|
; initialize row and column |
;         counters          |
;*************************|
; Set up work counter for number of rows in matrix C
; This counter will be used in determining the end of the
; matrix multiplication operation
```

```
        MOV     AL,MAT_C_ROWS               ; Rows in matrix C
        MOV     WORK_ROWS,AL                ; To working counter
; Reset counter for number of columns in matrix C
; This counter will be used in resetting the matrix pointers at
; the end of each row in the products matrix
        MOV     AL,MAT_C_COLS               ; Columns in matrix C
        MOV     WORK_COLS,AL                ; To working counter
;***************************|
;   perform multiplication  |
;***************************|
NEW_PRODUCT:
; Save pointers to matrices A and B
        PUSH    ESI             ; Pointer to A
        PUSH    EDI             ; Pointer to B
; Load 0 as first entry in sum of products
        FLDZ
                                ; ST(0)  |  ST(1)  |  ST(2)
;                               |   0    |   ?     |   ?    |
; Store number of products in work counter
        MOV     AL,PROD_COUNT           ; Get count
        MOV     WORK_PRODS,AL           ; Store in work counter
A_TIMES_B:
; Fetch entry in current row of matrix A
        MOV     EDX,SKIP_FACTOR         ; size to DL
; ESI -> matrix entry in current row of matrix A
        CALL    FETCH_ENTRY     ; ST(0) now holds entry of A
        XCHG    ESI,EDI         ; ESI -> matrix B
        CALL    FETCH_ENTRY     ; ST(0) = matrix B
                                ; ST(1) = matrix A
        XCHG    ESI,EDI         ; Reset pointer
; Multiply s
                                ; ST(0)   |  ST(1)  |  ST(2)
        FMULP   ST(1),ST        ; eA * eB |previous |  EMPTY |
                                ;         |  sum    |  EMPTY |
        FADD                    ; p sum   | EMPTY-  |
; Test for last entry in product column
        DEC     WORK_PRODS      ; Is this last product
        JZ      NEXT_PRODUCT    ; Go if at end of products column
;***************************|
;     next product          |
;***************************|
; Index to next column of matrix A
        ADD     ESI,SKIP_FACTOR  ; Add size to pointer
; Index to next row in the same column in matrix B
        MOV     EAX,EDX                 ; Horizontal skip factor to AL
        MUL     MAT_B_COLS              ; Times number of columns
        ADD     EDI,EAX         ; Add to pointer
        JMP     A_TIMES_B       ; Continue in same product column
;***************************|
;     store product         |
;***************************|
NEXT_PRODUCT:
; Restore pointers to start of current A row and B column
        POP     EDI             ; B matrix pointer
        POP     ESI             ; A matrix pointer
; At this point ST(0) has sum of products
; Store this sum as entry in products matrix (by DS:BX)
        XCHG    EBX,ESI         ; ESI -> matrix C
; Store sum
        MOV     EDX,SKIP_FACTOR  ; size to DL
```

```
        CALL    STORE_ENTRY     ; Store sum in matrix C and pop
                                ; stack
        XCHG    EBX,ESI         ; Restore pointers
; Index to next entry in matrix C
        ADD     EBX,SKIP_FACTOR     ; Add size to pointer
;**************************|
;  test for last column in |
;       matrix C           |
;**************************|
; WORK_COLS keeps count of current column in matrix C
        DEC     WORK_COLS       ; Is this the last column in C
        JE      NEW_C_ROW       ; Go if last row
; Index to next column in matrix B
        ADD     EDI,SKIP_FACTOR     ; Add size to pointer
        JMP     NEW_PRODUCT
;**************************|
;     index to new row     |
;**************************|
; First test for end of processing
NEW_C_ROW:
        DEC     WORK_ROWS       ; Row counter in matrix C
        JNE     NEXT_C_ROW      ; Go if not last row of C
;**************************|
;       end of matrix      |
;       multiplication     |
;**************************|
        JMP     MULT_M_EXIT


;**************************|
;  next row of matrix C    |
;**************************|
; At the start of every new row in the products matrix, the
; matrix B pointer must be reset to the start of matrix B
; and the matrix A pointer to the start entry of the next
; row of matrix A
NEXT_C_ROW:
        MOV     EDI,START_BMAT  ; EDI -> start of B
        MOV     AH,0            ; Clear high byte of adder
; Pointer for matrix A
        MOV     EAX,SKIP_FACTOR     ; Entry size of A
        MUL     MAT_A_COLS          ; Size times columns
        ADD     ESI,EAX             ; ESI -> next row of A
; Reset counter for number of columns in matrix C
        MOV     AL,MAT_C_COLS       ; Columns in matrix C
        MOV     WORK_COLS,AL        ; To working counter
        JMP     NEW_PRODUCT         ; Continue processing
;*********************|
;        EXIT         |
;*********************|
MULT_M_EXIT:
        CLD
        RET
_MUL_MATRICES    ENDP
```

The C++ interface function to the _MUL_MATRICES procedure is as follows:

```
template <class A>
bool MulMatrices(A *matA, A *matB, A *matC,
                int rowsA, int colsA,
                int rowsB, int colsB)
```

```
{
// Perform matrix multiplication: C = A * B using low-level code in the
// Un32_13 module
// On entry:
//       *matA and *matB are matrices to be multiplied
//       *matC is matrix for product
//       rowsA is number of the rows in matrix A
//       colsA is number of columns in the matrix A
//       rowsB is number of the rows in matrix B
//       colsB is number of columns in the matrix B
// Requires:
//       All three matrices must be of the same dimensions
//       All three matrices must be of the same float
//       data type
// Asumes:
//        The number of rows in matrix C is the product of the
//        columns of matrix B times the rows or matrix A
// Routine expects:
//       ESI -> first matrix (A)
//       EDI -> second matrix (B)
//       EBX -> storage area for addition matrix (C)
//       AH = number of rows in matrix A
//       AL = number of columns in matrix A
//       CH = number of rows in matrix B
//       CL = number of columns in matrix B
//       EDX = horizontal skip factor
// On exit:
//       returns true if matC[] = matA[] * matB[]
//       returns false if columns of matA[] not = rows
//       of matB[]. If so, matC[] is undefined

int eSize = sizeof(matA[0]);
// Test for valid matrix sizes:
//     columns of matA[] = rows of matB[]
if(colsA != rowsB)
   return false;
_asm
{
        MOV     AH,BYTE PTR rowsA
        MOV     AL,BYTE PTR colsA
        MOV     CH,BYTE PTR rowsB
        MOV     CL,BYTE PTR colsB
        MOV     ESI,matA          // Address to registers
        MOV     EDI,matB
        MOV     EBX,matC
        MOV     EDX,eSize         // Horizontal skip
        CALL    MUL_MATRICES
}
return true;
}
```

SOFTWARE ON-LINE

The routines for matrix addition and multiplication are found in the
UN32_13.ASM module of the MATH32 library. The C++ interface routines are in
the file Un32_12.h located in the Chapter13/Test Un32_12 project folder in the
book's on-line software.

13.3 The Solution of a Linear System

A method for solving a system of linear equations is based on performing systematic, elementary manipulations that simplify the system. The computer algorithms are based on the methodical elimination of unknowns. A simplified system is obtained by performing one or more of the following operations:

1. Multiplying an equation by a non-zero constant

2. Interchanging two equations

3. Adding a multiple of one equation to another equation

In Section 13.0.2 we used an augmented matrix to hold the coefficients and the constant terms of a set of linear equations. In this case each horizontal row in the augmented matrix corresponds to one of the equations in the original system. In searching for a solution of a linear system we can operate on the matrix entries in the same manner as we operate on the terms of a linear equation.

Two general strategies have been used in developing methods for the solution of a system of linear equations: the direct and the iterative approach. In both cases the condition for a valid solution is a non-singular system with a square matrix of coefficients. The *direct approach* is based on the systematic elimination of variables. The method known as *Gaussian elimination*, described later in this chapter, belongs to this category. The *iterative approach*, based on obtaining a series of gradual approximations, requires that the matrix of coefficients have a diagonal of non-zero entries. Probably the best known iterative methods are those of Jacobi and Gauss-Siedel. Iterative methods are more useful with large matrices that contain many zero entries.

13.3.1 Gauss-Jordan Elimination

One of the best-known algorithms for solving a system of linear equations is due to Carl Friedrich Gauss (1777–1855). It is interesting to note that a mathematician who died almost a century before the advent of computers developed a method that is so well suited to digital machines. This is not a unique case, numeric algorithms by Newton, Fourier, and others, are widely used in the computer solutions of mathematical problems that occur in the physical and social sciences.

An important variation of Gaussian elimination is credited to the French mathematician Camille Jordan (1838–1922). A detailed description of this method, known as Gauss-Jordan elimination, is found in textbooks on linear algebra (such as the one by Howard Anton, listed in the Bibliography). The Gauss-Jordan method is based on the systematic reduction of the augmented matrix representing a linear system, to a structure so simple that the solution can be obtained by inspection. Two unique matrix forms are part of the Gauss-Jordan method. The first one is called the *row-echelon form*. The following characteristics identify a matrix in row-echelon form:

1. The first non-zero entry in a row is a 1 (except if the row consists entirely of zeros). The first 1-entry in a row is called the leading 1.

2. The leading 1 occurs farther to the right in each successive row.

3. Rows consisting entirely of zeros are grouped toward the bottom of the matrix.

The following matrix is in row-echelon form.

$$\begin{bmatrix} 1 & 0 & 7 & 4 \\ 0 & 1 & 4 & 2 \\ 0 & 0 & 1 & 3 \end{bmatrix}$$

Note that in the row-echelon form there is a diagonal of 1-valued entries and that all entries below this diagonal are zeros. Another way of stating this property of the row-echelon form is to say that all entries below a leading 1 must be zero, and that, in each row, the leading one is found further to the right than in the preceding row.

The second matrix form encountered in Gauss-Jordan elimination is named the *reduced row-echelon form*. A matrix is said to be in reduced row-echelon form if it has zero entries above and below each leading 1. The following matrix is in reduced row-echelon form.

$$\begin{bmatrix} 1 & 0 & 0 & 1 \\ 0 & 1 & 0 & 2 \\ 0 & 0 & 1 & 3 \end{bmatrix}$$

Suppose a system of linear equations whose coefficients are contained in a square, augmented matrix. After this matrix is transformed to the reduced row-echelon form, the column that originally held the constant terms (rightmost column) will now contain the solution set for the linear system. Gauss-Jordan elimination consists of a series of steps that can be followed in order to systematically reduce the matrix, first to the row-echelon form, and then to the reduced row-echelon form.

For example, the following system of linear equations

$$3x + 2y - z = 1$$
$$6x + 6y + 2z = 12$$
$$3x - 2y + z = 11$$

can be represented in the matrix

$$\begin{bmatrix} 3 & 2 & -1 & 1 \\ 6 & 6 & 2 & 12 \\ 3 & -2 & 1 & 11 \end{bmatrix}$$

The analytical solution of the equations yields the results:

$$x = 2$$
$$y = -1$$
$$z = 3$$

The solution matrix to this system is:

$$\begin{bmatrix} 1 & 0 & 0 & 2 \\ 0 & 1 & 0 & -1 \\ 0 & 0 & 1 & 3 \end{bmatrix}$$

Notice that since the matrix is in reduced row-echelon form, the fourth column contains the system's solution.

13.3.2 Errors in Gaussian Elimination

One of the first numerical methods analyzed for an inherent sensitivity to errors was Gaussian elimination. Note that the method itself is exact and that errors in Gaussian elimination are introduced during rounding operations. Since floating-point representations allow a limited precision, rounding errors are unavoidable in digital computers.

One effort at reducing rounding errors during Gaussian elimination is a variation called *pivotal condensation*. The Gauss-Jordan algorithm presented in Section 13.3.3 uses pivotal condensation. Two implementations of pivoting have been used in practice: *partial pivoting* and *complete pivoting*. It has been shown that Gaussian methods with pivoting, are stable in relation to rounding errors, which is not the case with methods that do not use pivots.

Software can further reduce rounding errors by performing all Gaussian calculations in the extended format. This approach is valid if, as recommended, the input matrix entries are encoded in one of the lesser precision formats. However, current implementations of Visual C++ on the PC do not allow the use of extended precision. This is one of the reasons why we preferred to develop the Gauss-Jordan routine in assembly language, and provide a C++ interface function. The procedure listed in Section 13.5 uses Gauss-Jordan elimination with pivotal condensation. The calculations take place in the extended precision format. At the conclusion of the processing, the results are converted to the caller's double precision array. A more detailed error analysis of the code could further improve its precision by manipulating the FPU rounding control modes or by introducing additional rounding operations during processing.

13.4 A Gauss-Jordan Algorithm

The validity of all Gaussian methods of elimination is based on the assumption that a matrix of a non-singular system can be converted to the reduced row echelon form by systematically performing elementary row operations. Once the matrix is in this form, the solution set is found at the right-most column.

The following computer algorithm is based on a method named Gauss-Jordan elimination with pivoting.

PHASE 1. Transform the matrix to row-echelon form

1. Designate the top row of the matrix as the current top row. Designate the first column of the matrix as the pivot column.

2. Find the element in the pivot column (searching down from the current top row) which has the largest absolute value. This element will be referred to as the *pivot*. The row containing the pivot will be referred to as the *pivot row*. If no non-zero value is found, then go to PHASE 2.

3. If the pivot row is not the currently designated top row of the matrix, exchange the pivot row with the current top row.

4. Divide entries in the row currently designated as the top row by the value of the pivot. Continue at PHASE 2 if the designated top row is the last row of the matrix.

5. Search down the pivot column below the current pivot position to find the first non-zero entry. Designate this value as the *factor*. Designate the row where the factor was found as the *temporary top row*. Exit LOOP A if no non-zero value was found during the Search.

6. Change the sign of the factor. Multiply the each entry of the current top row by the factor.

7. Add the row obtained in step 6 to the temporary top row.

8. Continue at step 5 until all entries in the pivot column below the current top row are zero.

9. Designate the next row below the current top row as the current top row. Designate the column to the right of the current pivot column as the new pivot column. Continue at step 2.

PHASE 2. Transform the matrix to reduced row-echelon form

10. Designate the bottom row of the matrix as the current bottom row.

11. Find the first leading 1 in the bottom row. Designate the column containing this leading 1 as the one's column. If no leading 1 is found, go to step 15.

12. Search up the one's column for the first non-zero entry. Designate this value as the factor. Designate the row in which the factor was found as the temporary top row. Continue at step 15 if no valid factor is found.

13. Change sign of factor. Multiply the currently designated bottom row by the factor.

14. Add the row obtained in step 13 to the row currently designated as the temporary top row. Continue at step 12.

15. Designate the row above the current bottom row as the bottom row. End of calculation if this is the first row of the matrix, if not, continue at STEP 11.

13.5 Solution of a Linear System

The procedure _GAUSS_SOLVE, found in the Un32_13 module of the MATH32 library, is based on the algorithm described in Section 13.4. The processing assumes a non-singular augmented matrix containing the coefficients and constant terms in a set of linear equations. The C++ interface function to the _GAUSS_SOLVE procedure is coded as follows:

```
void GaussSolve(double *matS,
                double *matR,
                int rowsS, int colsS)
{
// Use Gauss-Jordan elimination with pivoting to solve a system
// of linear equations stored in an augmented matrix.
// Uses GAUSS_SOLVE procedure in Un32_13.asm
// On entry:
//        *matS is the source matrix
//        *matR is the results matrix
//        rowsS is number of the rows in matrix S
//        colsS is number of columns in the matrix S
// Requires:
//        Both matrices must be of the same dimensions
//        Both matrices must be of double precision type
// Routine expects:
//        ESI -> first matrix (S)
//        EDI -> second matrix (R)
//        AH = number of rows in matrix S
//        AL = number of columns in matrix S
// On exit:
//        returns the solution to the linear system in a
//        matrix in row-echelon form

_asm
{
        MOV     AH,BYTE PTR rowsS
        MOV     AL,BYTE PTR colsS
        MOV     ESI,matS          // Address to ESI
        MOV     EDI,matR
        CALL    GAUSS_SOLVE
}
return;
}
```

The GaussSolve() function, listed previously, receives two identical matrices as parameters. One contains the linear system and the second one is used to store the results of the calculations. However, on occasions you may wish to store the results in the same matrix that originally contained the linear system. If so, you can pass the same argument as both parameters. This is possible because the low-level procedure _GAUSS_SOLVE copies the caller's data to its own extended precision internal buffers for processing.

SOFTWARE ON-LINE

The procedure _GAUSS_SOLVE is found in the Un32_13 module of the MATH32 library. The C++ interface function is in the file Un32_12.h, located in the Chapter13\Test Un32_12 folder in the book's on-line software.

Chapter 14

Solving and Parsing Equations

Chapter Summary

This chapter presents the manipulations and processing methods required for evaluating algebraic equations, usually called equation parsing. Numerical parsing operations are useful in the development of many types of mathematical routines often found in graphics, scientific, financial, and engineering applications, and in the processing performed by language compilers and interpreters. Parsing numerical expressions is considered among the common routines that are most difficult to implement in code.

14.0 Function Mapping

In solving a numerical expression in string form, code must analyze the text string that contains the equation. Typically this equation defines a functional relation in the form:

$$y = f(x)$$

where $f(x)$ represents a functional expression, for example:

$$y = x(TANx + 1)$$

To solve the equation, code must apply the fundamental rules of algebra in order to determine the order in which its lexical elements must be evaluated, and then proceed to calculate one variable in terms of the other one. In designing a routine that performs these operations the following elements must be taken into account:

1. The equation must be expressed in a way that leaves no uncertainties regarding the constants, variables, and operations represented in the string. At the same time, the equation must consist of standard ASCII symbols available in a computer keyboard. Equation grammar refers to the rules and conventions that ensure that the equation

string is singularly interpreted. For example, a particular version of equation grammar may required that the mathematical expression:

$$y = 4x^2$$

be represented in text form as

$$y = 4*x^2$$

In this case the $*$ symbol explicitly represents the multiplication, while the $^\wedge$ symbol allows the representation of powers without a superscript. In defining equation grammar any conventional symbolism can be adopted, as long as the resulting expression is algebraically meaningful and the rules and symbols are consistent.

2. Computer calculations usually require that numerical values in the equation string be converted to binary encodings and stored in data structures that can be conveniently accessed by the code. In the case of a compiler, this operation includes looking up the current value of identifiers that represent variables and constants. In any case, the code must be able to replace the representation of the independent variable (x) with its numerical value.

3. The processing must follow the rules of algebra in performing the various operations indicated in the equation. This means that parenthetical elements must be resolved according to their level of nesting, and that the operations must read left-to-right and performed in the conventional order, that is: exponentiation first, then multiplication and division, with additions and subtractions being performed last.

4. As the processing performs the sequence of steps required in evaluating the numerical expression, the representation of the expression is updated accordingly. This means that, when a calculation is performed, its expression in the equation string is replaced by the partial result. The process continues until a final, singular value is reached.

In addition to the more-or-less standard operations listed above, a sophisticated equation-solving routine may be capable of other functions. For example: checking the equation string for errors of syntax, making sure that the constants and variables have a valid lifetime and scope, and confirming that parenthetical terms are correctly paired.

Many approaches have been used successfully in parsing equations and mathematical expressions. The literature refers to the conventional arithmetic representation in which the operators are placed between the operands as infix notation. For example, the expression

$$y = x + 3$$

is said to be in infix form. In 1951 Jan Lukasiewicz, a Polish logician, devised a notation that did not require the use of parenthesis to show precedence. This representation scheme, know as Polish notation, was later extended to algebraic expressions.

Polish notation eliminates the need for parenthesis by placing the operators before the operands. For example, the above previous expression can be written in Polish form as

$$y = +x3$$

Because in conventional Polish notation the operator precedes the operands, it is sometimes called a prefix form. A variation on Polish notation, called reverse Polish notation, or RPN, places the operator after the operand, as follows:

$$y = x3+$$

RPN is sometimes called the postfix (or suffix) form. In the United States reverse Polish notation was popularized by its use in some Hewlett-Packard calculators.

If an expression in the prefix form of Polish notation is evaluated right-to-left, the operators can be processed as they are encountered in the text string. The same applies to the left-to-right evaluation of an expression in reverse Polish (RPN).

Polish notation methodology is particularly suited for stack-based implementations since the numerical elements can be retrieved from the stack during evaluation, in the order in which they are required. For this reason many language processors and equation solving applications begin by re-formatting the algebraic expression entered by the user into Polish or reverse Polish form.

In this chapter we present an equation-solving routine that uses the FPU in performing all numerical calculations. Since the FPU cannot conveniently access an external stack structure, and since its internal stack is limited to eight elements, we adopted an array-based approach. The algorithm is based on the longhand algebraic method for solving equations, with or without parenthetical elements. This straight-forward algebraic approach is suited to FPU implementations.

14.1 Developing a Parser

One of the most difficult programming tasks in developing a language compiler or interpreter is coding the parser. A parser can be considered an expert system on the grammar and syntax of the particular language. Mak (1991) states that "The brain of the compiler is the parser." In the dynamics of language compilation and interpretation, the parser receives data from a software element called the *scanner*. It is the scanner that breaks the source statements into its component parts, namely: tokens, identifiers, constants, and operators. The parser uses its knowledge of the semantics of the language to perform the operations indicated in the scanned expression. Figure 14.1, on the following page, shows the steps performed in evaluating a simple mathematical expression.

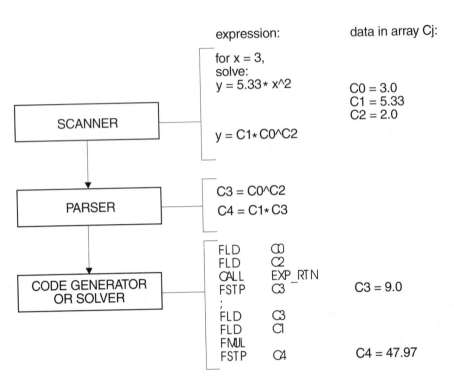

Figure 14.1 *Evaluating the Expression y = 5.33 * x^2*

In Figure 14.1 the first step performed by the scanner consists of identifying the numeric values in the equation string. Once the values are identified, they are converted from ASCII to binary and stored in the array named C. In this example C0 is the value of the independent variable (x), while C1 and C2 are the constant terms in the original equation string. At this point the numeric values in the original expression can be replaced by references to their locations in array C, since the binary values are stored in the array C. The parser operates on this simplified version of the mathematical expression. In the case of a language compiler, the parser typically calls the services of a code generator, which produces the assembly language source text. In an equation solving application, or an interpreted language, the parser uses an "executer" or *solver* routine which directly calculates the value of the function or expression.

The reader should consider that the example in Figure 14.1 is an extremely simple one. In a real world case, the expression to be solved may contain keywords of the specific language, identifiers that represent variables and constants, operators, white space symbols such as blanks and tabulation codes, hard-coded constants, as well as parenthesis or other precedence-indicating symbols, possibly to several levels of nesting. In addition to interpreting the keywords and arithmetic operators of the specific language, the compiler or interpreter must be capable of performing logical functions on the operands, and of implementing decision and iterative structures according to the syntax.

Since this book is not about language compilers or interpreters, our discussion of parsers is limited to the operations required for evaluating a function expressed in an algebraic equation. Nevertheless, an equation evaluating routine must perform, to a lesser degree of complication, all the functions of a fully implemented language parser. The processing logic contained in the routines listed in the sections that follow can be used as a simplified model for the development of a full-fledged parser as required by a compiler or interpreter program.

14.2 Evaluating User Equations

The approach to developing a parsing routine, as required by an equation-solving or function-mapping application, can be based on the processing requirements listed in Section 14.0. In the following sections we consider each of these requirements separately.

14.2.1 Equation Grammar

The first requirement is that the equation be expressed in a way that leaves no uncertainty regarding the encoded operations. A possible starting point is to identify all of the recognized arithmetic operators. Two types of arithmetic operators can be identified. Unary operators take a single operand, for example, when the + and – symbols are used to indicate the sign of a value. Trigonometric functions are unary. Conventional operators, called binary operators, require two operands. This is the case for operators that indicate exponentiation, multiplication, division, addition, and subtraction. The following lists the operators used by the processing routines developed later in this chapter:

```
                    operation           symbol or keyword
Trigonometry:
                    sine                SIN
                    cosine              COS
                    tangent             TAN
                    arcsine             ASIN
                    arcosine            ACOS
                    arctangent          ATAN
Arithmetic:
                    exponentiation      ^
                    multiplication      *
                    division            /
                    addition            +
                    subtraction         –
Special symbols:
                    assignment          =
                    forced precedence   (, )
                    variable x          x, X
                    variable y          y, Y
```

Notice that the exponentiation function can be used to calculate roots, for example, the square root of x would be represented as x^0.5. In spite of this fact, most languages and applications will profit from a separate square root function. Also notice that the arithmetic operators + and – serve a double purpose: they can be used to express an operation, as well as the sign of a numeric quantity. The processing has to take special steps to distinguish between the unary and binary actions of the + and – signs.

Other symbols may include the = sign, which serves to mark the start of the solvable term, and the letters x and y to represent the variables. Parenthesis are used to group operations with forced precedence. The processing logic first executes the operations enclosed in parenthesis, as is consistent with the rules of algebra. Since parenthetical terms can be nested, the processing logic must determine the level of nesting in order to solve the expression from the innermost to the outermost component.

14.2.2 Equation Syntax

The expression of the equation must be meaningful and consistent. This is determined by the adopted syntax rules. For example, equation syntax may require that binary arithmetic operators must have a numeric expression directly to their left and right. Therefore the expression

$$y = 3.0 * x$$

is meaningful. In this case the constant 3.0 and the variable x represent numeric values. On the other hand, the expression

$$y = / * 4.2$$

is not consistent with the rule, since the / operator to the left of the * operator, does not resolve to a numeric value. By the same token, syntax rules often require that unary arithmetic operators and trigonometric operators must have a numeric value directly to their right, for example:

$$y = SINx$$

is a valid expression but

$$y = SIN * x$$

is not. In this case the * operator, to the right of the SIN operator, does not resolve to a numeric value.

Regarding parenthesis, equation syntax usually requires that they be matched and that they enclose solvable expressions, for example:

$$y = 2 * (x^2 - 5)$$

is a legal expression but

$$y = 2 * ()$$

is meaningless. Nevertheless, the processing logic of the equation evaluating routine may be able to correctly resolve some expressions with imperfect syntax. For example, the parentheses are superfluous in the equation

$$y = 2 * (x^2)$$

However, the evaluating routine may be able to correctly evaluate the expression in spite of its syntax flaw.

The resolving routine can adopt several options. The most desirable processing method would be to check that the expression meets all adopted grammar and syntax requirements. This implies searching the equation string for invalid or incorrectly used symbols, expressions, or keywords, checking numerical terms associated with the operators, certifying the pairing and nesting of parenthetical expressions, and examining numerical operands for range and for invalid symbols. On the other hand, the code may assume that the equation syntax is correct and proceed to solve it with little or no pre-checking.

This second approach is the one used by the solving and parsing routines developed later in this chapter. We leave it to the programmer to provide grammar and syntax checking operations that ensure the correctness and consistency of the expression.

14.2.3 Symbol Table and Numeric Data

Typically a full-fledged parser, as used in a language compiler or interpreter, creates and manages a data structure that collects and stores all necessary information regarding the identifiers present in the program. This structure is usually called the symbol table. The symbol table is used in locating the data type and actual numerical value of identifiers. For example, in solving the expression:

```
y = x * const_1
```

the processing routine looks up the identifiers x and const_1 in the symbol table. A typical symbol table contains the data types of x and const_1 as well as a reference that allows accessing the value of the identifier from its respective memory storage. Creating and managing the symbol table is one of the principal functions of the compilation engine.

A simple equation evaluating routine, such as the one developed in this chapter, need not use identifiers, except for those that represent the independent and dependent variables. Therefore, the symbol table is reduced to a rudimentary structure. On the other hand, the processing routine must find ways for locating, encoding, and storing the numeric values present in the equation string. For example, in the equation string

```
y = 3.33 * x / 1.5
```

code must be able to identify the variable x and determine and store its value. Also, the constants 3.33 and 1.5 must be converted from ASCII into binary floating-point format, and the resulting values stored for future reference. These manipulations are usually performed by the scanner module, as shown in Figure 14.1.

The equation evaluating routine listed later in this chapter performs all of these operations. First, the value of the independent variable (x) is stored at offset 0 of an array of floating-point values in extended precision format. Then the values for the hard-coded constants that appear in the equation string are converted to binary floating-point (in extended precision format) and stored in consecutive entries of the array. In the equation string, the variable x and the numeric ASCII strings are replaced with a reference to the array element that holds the binary encoding, as in Figure 14.1.

In the processing routines developed later in this chapter, the array reference is encoded in a single byte scheme which is designated as the Cj format. The single-byte encoding is convenient since it allows replacing a single-digit constant or variable without expanding the equation string. The j element in this designation is the entry number, or index, in the array. The Cj value is calculated by adding 80H to the offset of the entry. Therefore offset number 3 in the array is represented by the value 83H. The excess 80H format for array entry references makes possible for the software to identify these references in the equation string. This is possible because all other text characters in the equation string are in the range 0H to 7FH.

14.3 An Equation-Solving Algorithm

The parser contained in the procedure named _CALCULATE_Y, listed later in this chapter, receives from the caller an ASCII string that represents an equation where the dependent variable (y) is expressed as a function of the independent variable (x). The syntax requires that the element on the left of the = sign contain no expression other than the dependent variable. Therefore the expression

```
2 * y = x^3
```

is not legal. Instead, the equation must be entered in the form

```
y = x^3/2 .
```

This is compatible with the constraints of most high-level languages. For example, C language requires that the element to the left of the = sign (called an lvalue) must be a variable name or a specific element of an array.

14.3.1 The _EVALUATE Procedure

The equation string can contain any of the operators listed in Section 14.2.1, according to the rules of syntax mentioned in Section 14.2.2. The core of our equation solving engine is a low-level routine named _EVALUATE which is capable of applying the rules of algebra to numerically evaluate an expression that contains no parenthetical elements. The _EVALUATE procedure is based on the following logic:

1. It receives from the caller a text string containing a mathematical expression that does not include parenthetical elements. This requirement makes it possible that the expression be resolved left-to-right following the basic rules of algebra.

2. Code assumes that the FPU Stack Top register contains the value of the independent variable.

3. The elements that can be found in the expression string are the legal keywords and operators, constants in decimal or exponential format, the independent variable represented by the letters x or X, as well as single-byte references (in Cj format) to numeric data stored in the working array.

4. The _EVALUATE procedures attempts to find a numeric solution to the mathematical expression received from the caller by applying the rules of elementary algebra.

Processing assumes that the expression syntax is consistent and that it meets the grammatical and syntax requirements. The routine makes practically no checks for consistency or correctness in the mathematical expression. A commercial application of this routine will certainly require a preliminary check of the expression string for correct grammar and syntax.

As _EVALUATE calculates the value of the caller's expression, it also simplifies it. The simplification consists of replacing the portion of the text string that represents an operation with the partial results obtained during processing, much as it is done in the longhand method of equation solving. For example, suppose that the expression

```
x^2 + 7 * x - TANx + 4.33
```

is to be evaluated for a value of x = 10. The first processing step of the scanning stage of the _EVALUATE procedure is to store the value of the independent variable at offset 0 of the internal array, and replace the letter "x" in the original expression with a reference in Cj format. Since the Cj reference for x is always 80H, which corresponds with index C0, the expression becomes

```
C0^2 + 7 * C0 - TANC0 + 4.33.
```

Notice that, in order to simplify this illustration we have not used the H postfix for references in Cj format. The reader should understand that the value C0 is in the above expression is actually offset C0H in the working array.

The second step is to replace the keywords in the expression string with single-byte tokens. The valid keywords and tokens are as follows:

keyword:	token:
SIN	01H
COS	02H
TAN	03H
ASIN	04H
ACOS	05H
ATAN	06H

Tokenization of keywords is a convenient simplification since the parser can more easily work with one-byte tokens than multi-character keywords. After the keyword TAN is tokenized with 03H the expression string appears as follows:

```
C0^2 + 7 * C0 - 03C0 + 4.33.
```

The next step is to locate and store hard-coded constants in the expression string. The scanning proceeds left-to-right from the start of the expression. The constants encountered are 2, 7, and 4.33. These are converted to floating-point format and stored in the working array. Then the equation is edited so that the constants are replaced by their array references in Cj format, as follows:

```
C0^C1 + C2 * C0 - 03C0 + C3
```

At this time the working array appears as follows:

Cj reference	data
C0	10.0 (value of x)
C1	2.0
C2	7.0

```
C3                    4.33
```

The _EVALUATE procedure now performs the operations indicated in the expression string, left to right, according to the following rules: trigonometric operations are performed first, then exponentiation, then multiplication and division, finally addition and subtraction. As each operation is performed, the partial result is stored in the working array and the text that represents the resolved element in the expression string is replaced with the corresponding array reference. Therefore the expression

```
C0^C1 + C2 * C0 - 03C0 + C3
```

is first simplified by calculating the tangent, indicated by token 03, as follows:

```
C0^C1 + C2 * C0 - C4 + C3
```

C4 now holds TAN 10. The next simplification consists of performing the exponentiation operation indicated by the ^ operator. The equation now becomes:

```
C5 + C2 * C0 - C4 + C3
```

C5 holds 100, which is the value C0^C1, or 10^2. Next, the multiplication operation indicated by the * operator is performed. Now the equation becomes:

```
C5 + C6 - C4 + C3
```

At this point the variables have the following values:

```
C5 = 100
C6 = 7 * 10 = 70
C4 = TAN 10 = 0.176326
C3 = 4.33
```

Next, the addition and subtraction operations are performed, left to right, as follows:

```
100 + 70 = 170
170 - 0.176326 = 169.8236
169.8236 + 4.33 = 174.1536
```

This value is stored at offset 9 of the working array, which is represented by C9H in Cj format.

SOFTWARE ON-LINE

The _EVALUATE procedure, found in the module Un32_14.asm of the MATH32 library, implements the logic previously described. The file is located in the book's on-line software source.

The C++ interface function to the _EVALUATE procedure is as follows:

```
double Evaluate(char *equation,
                double varX)
{
// Function to evaluate an equation, or part of an equation,
// that contains no parenthetical components, according to
// the following rules of equation grammar:
```

```
//
//              SIN (sine)         01H
//              COS (cosine)       02H
//              TAN (tangent)      03H
//              ASIN (arcsine)     04H
//              ACOS (arcosine)    05H
//              ATAN (arctangent)  06H
//
//
// Recognized symbols:
//                      ^ = exponentiation
//                      * = multiplication
//                      / = division
//                      + = addition (or unary +)
//                      - = subtraction (or unary -)
//
//
// Other elements of equation grammar:
//              1. ASCII constants, for example:
//                      1, 2.33, or -1.2E+233
//              2. The independent variable, represented by the
//                 letter x or X.
//              3. Binary constants, represented by a single digit
//                 entry in excess 80H format
//
//
// Rules of precedence:
//              1. Equation is evaluated left-to-right
//              2. Trigonometric functions are evaluated first
//              3. then exponentiation operations
//              4. then multiplication and division
//              5. addition and subtraction are performed last
//
// On entry:
//              equation[] is an equation string terminated in 0H
//              Equal sign and elements to its left must be removed
//              by the calling routine
//              Variable varX is the independent variable
// Uses:
//              The _EVALUATE procedure in the Un32_14 module of
//              the MATH32 library
//
// On exit:
//              returns the solution of equation or mathematical
//              expression passed by the caller in a type double
//

double result;
_asm
{

        MOV     ESI,equation        // Address to ESI
        FLD     varX
        CALL    EVALUATE
        FSTP    result
}
return result;
}
```

SOFTWARE ON-LINE

The C++ Evaluate() function is in the Un32_13.h file located in the Chapter14/Test Un33_13 project folder in the book's CD on-line software.

14.3.2 _CALCULATE_Y Procedure

Once equipped with a routine that evaluates the dependent variable in a simple mathematical expression, it is possible to tackle the more complex task of evaluating an expression that can include parenthetical terms. The simplest approach is to isolate parenthetical elements that can be treated as simple expressions, and proceed to use _EVALUATE to find the numerical solution of these elements.

The _CALCULATE_Y procedure is based on the following program elements:

1 It receives from the caller a text string containing an equation in the form

$$y = w$$

where w is a mathematical expression that can include parenthetical elements.

2. The routine assumes that the FPU Stack Top register contains the value of the independent variable.

3. The elements that can be found in the equation string are the legal keywords and operators, constants in decimal or exponential format, the independent variable represented by the letters x or X, the dependent variable (y), and the = sign.

4. The _CALCULATE_Y procedures attempts to find a numeric solution to the equation received from the caller by applying the rules of elementary algebra. In this case the processing assumes that the expression's syntax is consistent and that it meets the grammatical requirements of the application or language.

The processing logic for the CALCULATE_Y procedure consists of preparatory and processing steps. The preparatory steps (called preliminary steps in the code listing) consist of copying the right-hand term of the caller's equation from the caller's buffer into a working buffer. It then resets the routine's variables and pointers and clears the working buffers.

The processing phase consists of solving the equation by following these steps:

STEP 1. The equation string is searched for a "(" symbol. If no "(" symbol is found then the equation contains no parenthetical terms and can be passed entirely to the EVALUATE. In this case execution continues at STEP 6.

STEP 2. If a "(" symbol is found, its offset in the equation string is stored in a working variable (named LEFT_PARS).

STEP 3. The equation string is then searched for a "(" or ")" symbol to the right of the "(" symbol previously found.

STEP 4. If a "(" symbol is found then the equation contains a nested parenthetical expression. Execution continues at STEP 2.

STEP 5. If a ")" symbol is found in the equation string then an elementary parenthetical expression is located between the offset stored at LEFT_PARS and the current ")" symbol. The expression between parenthesis is copied to a working buffer and passed to the EVALUATE procedure. When execution returns, the parenthetical expression is erased from the original equation string and replaced with the numerical result returned by the _EVALUATE procedure. This result is stored in the working array buffer in excess C0H form. Execution then continues at STEP 1.

STEP 6. At this point either the original equation contained no parenthetical elements or all parenthetical elements have been resolved. The equation is now passed to the _EVALUATE procedure, which returns the final solution.

The exit logic of the _CALCULATE_Y procedure is based on the assumption that the _EVALUATE procedure returns $Cj > 1$ if the last iteration of the procedure performed calculations. In this case the result is found in $ST(0)$. If $Cj = 1$, then the final iteration of _EVALUATE performed no calculations. In this case the result is the x variable if no values are stored in the high range of the working array, or the highest numbered entry in the high range of the working array. This is due to the fact that values in the high range of the working array are entered by the _CALCULATE_Y procedure.

The following is the C++ interface function to the _CALCULATE_Y procedure.

```
double CalculateY(char *equation,
                  double varX)
{
// Function to calculate the dependent variable in a mathematical
// expression that may include parenthetical terms, possibly nested.
// This function uses the _EVALUATE procedure to solve the
// expressions enclosed in parenthesis, or to resolve an expression
// without parenthesis.
//
// On entry:
//         equation[] is an array of char containing the equation
//         text in the format:
//                 y = w
//                 where w is an expression that may include
//                 parenthetical terms, the independent
//                 variable, numeric constants, and the keywords
//                 and operators listed in the header of the
//                 _EVALUATE procedure
//                 The equation must terminate in 0H or 0DH
//         ST(0) = value of the independent variable (x)
//
// Uses:
//          The _EVALUATE procedure in the Un32_14 module of
//          the MATH32 library
// On exit:
//          returns the solution of equation or mathematical
//          expression passed by the caller, in a type double
//

double result;
_asm
{
```

```
        MOV     ESI,equation            // Address to ESI
        FLD     varX
        CALL    CALCULATE_Y
        FSTP    result
}
return result;
}
```

SOFTWARE ON-LINE

The _CALCULATE_Y procedure is found in the module Un32_14.asm of the
MATH32 library. The CalculateY() function is in the Un32_13.h file located in
the Chapter14/Test Un33_13 project folder in the book's on-line software.

Chapter 15

Neural Networks

Chapter Summary

This chapter is about the use of neural networks in artificial intelligence. In it we cover the fundamental topics, at entry-level complexity. The chapter starts with an overview of the biological neuron and its artificial representation in computers. It proceeds with topics of increasing complexity, starting with the simple logical networks, then the perceptron and the Adaline networks. The chapter concludes with backpropagation, which is perhaps the most used and useful mechanism in AI technology.

15.0 Reverse-Engineering the Brain

One approach to creating intelligent machines is based on emulating the brain. This thinking was originally based on the assumption that a reasoning machine that operates in a brain-like manner will show some degree of intelligence. In the beginning of the twentieth century the work of Santiago Ramon y Cajal, Colgi, Hebb, and others, provided a better understanding of the brain's component cells and of its architecture and operation. The neuron, the building block of the nervous system, was identified and described. Over the years, scientists gained a better understanding of the neuron's complex electrical and biochemical operation. Figure 15.1 (on the following page) shows the neuron's basic components.

15.0.1 The Biological Neuron

The neuron is an information processing unit that receives input signals from complex, and extremely large, tree-like elements called dendrites. The dendrites connect with the main body of the cell, called the soma. Located in the soma is the nucleus. Inputs from the dendrites, which can be excitatory or inhibitory, are collected in the soma. When the excitatory element of the input reaches a certain threshold level, the neuron fires and transmits an electrical signal along the axon. At the end of the axon there are the axonic endings, which are usually connected to the dendrites of other neurons. The space between the axon of one neuron and the dendrite of another one is

389

called the synapse. A single neuron has from 1000 to 10,000 synapses. The excitement of the axonic ending of a neuron releases a chemical substance, called a neurotransmitter, that establishes the communication of information between neurons.

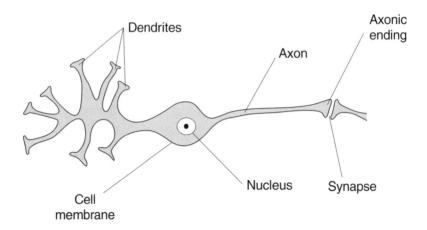

Figure 15.1 *Representation of a Neuron*

The electrical activity of a neuron is triggered by many different mechanisms, including light, pressure, and the action of chemical substances. When the applied potential is more positive than negative, the signal is said to be excitatory. Otherwise, the signal is inhibitory. In one case the neuron fires, and in the other one it is prevented from firing. The axon serves as a transmission line for the signals that end up in the axonic endings. Neurons meet at the synapse. The separation between the axonic ending and the dendrite, called the synaptive cleft, is about 200 millionth of a millimeter.

Current research establishes that the synapse serves as a biocomputer. However, the way information is acquired and transmitted in the physical neuron is quite different from the methods used in machine computation. For example, experiments in biological pattern recognition show that patterns are matched in a single step, rather than bit-by-bit, as is the case in a computing machine. Also, that there are several types of synapses, in addition to the conventional axon-dendrite connection shown in Figure 15.1. For example, the axonic ending of one neuron can contact the trunk of a dendrite tree, the soma, or even the axon of another neuron. This leads to a complicated structure of enabling, modulating, and inactive synapses. Furthermore, some neurons feed back onto themselves directly, or via other neurons. All of these mechanisms appear to control the firing action of a neuron in ways that are not clearly understood.

When we reverse-engineer the brain we must add the staggering dimensions of the brain network itself to the inherent complexity of the neuron. It is estimated that the human cerebral cortex contains approximately 100 billion cells and each neuron may have over 1000 dendrites. This results in over 100,000 billion synapses. Since each neuron can fire about 100 times per second, the human brain can execute over 10,000 trillion synaptical activities per second.

In our efforts to duplicate the brain's thought processes we have encountered three major difficulties:

1. Our imperfect understanding of the brain's physiological, electrical, and chemical operation and their interaction.

2. Our inability to synthetically match the brain's complexity and connectivity.

3. The unsuitability of the von Neumann computer paradigm to simulate the non-discrete, massively parallel data processing system that is the brain.

In spite of these limitations, artificial neural networks are one of the most successful and promising fields of artificial intelligence research.

15.0.2 The Artificial Neuron

Early research on the physiology of the brain cell led to mathematical representations of the neuron. In the 1940s a neurophysiologist, Warren S. McCulloch, and a mathematician, Walter Pitts, developed the first mathematical model of a nerve cell. Their model assumed that each neuron contained some biological threshold that determined whether the sum of the excitatory and inhibitory stimuli received at the dendrites would make the cell fire and that the nerve cell contained internal elements that served to enhance or diminish each of the input signals. These elements are called synaptic weights.

Mathematically, the McCulloch-Pitts neuron can be described as performing a sum-of-products operation of the stimuli (or inputs) times the synaptic weights, as follows:

$$S = \sum_{j=1}^{n} w_j i_j$$

where S is the weighted sum, w represents the synaptic weights, i is the input, and j is the instance of the input and the weight. The threshold condition (T) that determines the firing of a neuron is expressed in the equation:

$$T = \sum_{j=1}^{n} w_j i_j \geq \Theta$$

where T is a predefined constant. If T equals zero, then any positive value of S causes the neuron to fire. The conditions that determine the firing or not firing of a neuron are called its activation function.

For computational convenience the conventional model of a neuron also contains an additional input value, called the bias. The bias provides an easy way to make adjustments in the weighted sum. In most models the bias consists of a dedicated input line and its corresponding weight, usually assigned the subscript value of zero.

If the input value for the bias line is set to 1, and the bias weight to zero, then the bias has no effect on the weighted sum. In order to include this bias term, the neuron equation is modified so that the range of the bias instance starts at zero, as follows:

Figure 15.2 is a conventional graphic depiction of an artificial neuron.

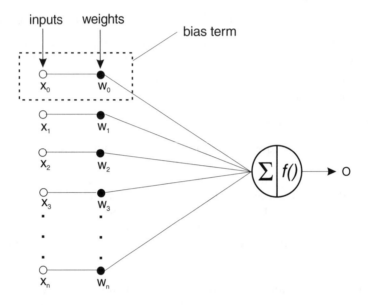

Figure 15.2 *Simple Artificial Neuron*

In Figure 15.2 the various inputs (x) and their corresponding weights (w) generate a weighted sum, which is used by the activation function to produce the neuron output (O). When the activation function is nonlinear, the output is not continuous and the neuron's response is bounded. One common nonlinear activation function, called a hard limiter, produces a binary output. In this case the neuron's output is 1 if the neuron fires, and 0 otherwise. Figure 15.3 graphically shows the output of a hard limited activation function.

In other types of artificial neurons the activation function is not hard limited, that is, the output is not binary. The sigmoid (s-shaped) and ramp-type activation functions are commonly used in more advanced neurons. Later in this chapter you will see the use of neurons with sigmoid activation functions. Figure 15.4 shows the curves of a hard limited, a ramp type, and a sigmoid-shaped activation function.

The coding for a neuron is simple. The following C++ function performs the action of a neuron with a hard limited, binary activation function.

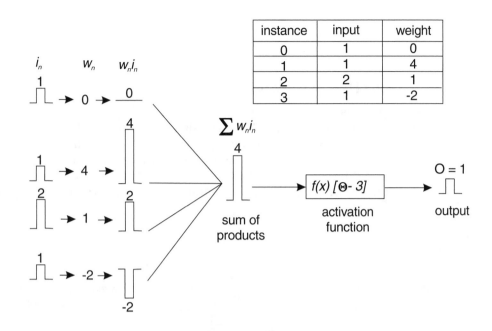

Figure 15.3 *Hard Limited Activation Function*

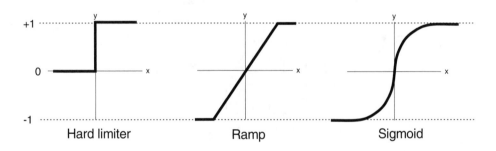

Figure 15.4 *Neuron Output Curves*

```
int NeuronB(int numValues,        // number of data elements
            int weights[],        // array of weights
            int values[])         // array of inputs
{
// Computes sum-of-products of = values[x] * weight[x]
// for numValues number of array elements
// then applies the threshold function:
//        if(sum < 0)
//            output = 0
//        else
//            output = 1
int result = 0;
for(int j = 0; j < numValues; x++)
    result += weights[j] * values[j];
if(result < 0)
```

```
      return 0;
else
      return 1;
}
```

15.0.3 Artificial Neural Networks

The neuron is a processing unit that usually forms part of a more complex structure called an artificial neural network, or ANN. The simplest possible network consists of a single neuron whose weights, and perhaps one or more input lines, have been preset to perform a specific task. Such a network can be described as a specialized neural unit. By selecting suitable weights, a binary output neuron can be used in a simple network to perform the logic AND operation. This unit is shown in Figure 15.5.

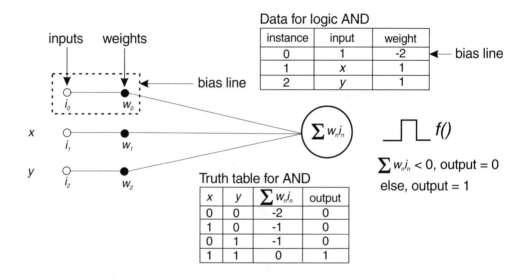

Figure 15.5 *Logic AND Specialized Neural Unit*

The neural unit of Figure 15.5 assumes that the input for the x and y variables are binary values. The bias line is preset with an input value of 1 and a weight of −2. The weights for instances 1 and 2 are also preset at 1, as shown in the table at the top of the illustration. The truth table at the bottom of the figure demonstrates that the unit produces the correct result for a binary AND operation for all possible input values. In network diagrams individual neurons are sometimes called nodes. We use the term "unit" in the sense of a dedicated, or specialized, node.

By manipulating the preset values for the weights, and the input for the bias line, we can create neural units for performing logical OR and NOT operations. Figure 15.6 shows the data and truth tables for OR and NOT neurons.

Data for Logic OR

instance	input	weight
0	1	-1
1	x	1
2	y	1

Data for Logic NOT

instance	input	weight
0	1	0
1	x	-2
2	0	1

Truth table for OR

x	y	$\sum w_n i_n$	output
0	0	-1	0
1	0	0	1
0	1	0	1
1	1	1	1

Truth table for NOT

x	$\sum w_n i_n$	output
0	-1	1
1	0	0

Figure 15.6 *Data and Truth Tables for OR and NOT Neurons*

In C++, a conventional software model of a neuron can be a function that performs the sum-of-products calculation, then applies an activation function. The following function can be considered as a neural unit for performing the logic AND operation.

```
int LogicAnd(int x, int y)
{

// Pre conditions:
// 1. Receives values of x and y in integer variables. Caller
//     is responsible for ensuring binary values only (0 or 1)
// 2. Function sets weights and bias line for performing logical
//     AND operation, using a neuron that follows the
//     McCulloch-Pitts model
// Post conditions:
//     Returns logical 0 or 1

  int weights[3];          // Array for weights
  int inputs[3];           // Array for inputs
  // Set weights
  weights[0] = -2;         // Bias line weight
  weights[1] = 1;
  weights[2] = 1;

  inputs[0] = 1;           // Bias line input
  inputs[1] = x;
  inputs[2] = y;
  // Call Neuron()
  return (Neuron(3, weights, inputs));
}
```

Note that the LogicAnd() function calls the NeuronB() function previously discussed; also, that the neural unit function contains the weight table required for the AND operation and sets the input value for the bias line accordingly.

SOFTWARE ON-LINE

The project named Neural Units, contained in the Chapter 15 folder in the book's on-line software, contains exercises of a McCulloch-Pitts neuron and several neural units.

15.1 The Network as a Classifier

One of the common applications of a simple network is in classification. Classification consists of assigning an object to one of several predefined groups or classes. Neural networks that perform classification have been found useful in character recognition, evaluation of loan applications, in analyzing sonar data, and other applications. The neural units developed in the preceding section can be considered as linear classifiers. Figure 15.7 shows how the logical AND, OR, and NOT operations can be interpreted as a mapping function.

In all three images of Figure 15.7 a Cartesian plane is divided into two areas. In one area are located all the instances of the logic operation in which the result is 1 (white dots), and in the other area all instances in which the result is 0 (black dots). Thus, the neural network can be interpreted as a classifier that processes the inputs and places the result in the corresponding area.

15.1.1 Multiple-Node Networks

A classification problem is said to be linearly separable if there exists a straight line in the Cartesian plane that separates all possible outputs. This is the case in the three examples in Figure 15.7. However, there are simple classification problems that cannot be solved by a single node network. Figure 15.8 shows the possible outputs of the logic XOR operation.

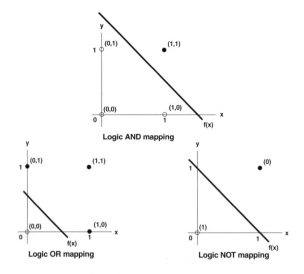

Figure 15.7 *2D Mapping of Logic Operations*

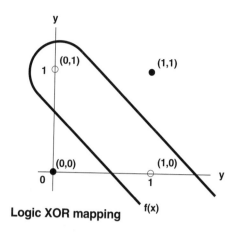

Figure 15.8 *Mapping of the XOR Operation*

In the case shown in Figure 15.8 it is not possible to construct a straight line that separates the black dots from the white ones. Therefore, the classification function has to be a nonlinear, as shown in the illustration. In other words, the XOR function is not linearly separable. For this reason it cannot be implemented in a single neural unit. However, the XOR function can be obtained by combining primitive logic operations, that is, by performing several AND, OR, and NOT operations. In this context we refer to a mechanism that requires several neural units as a network. Figure 15.9 shows a network for performing the XOR function.

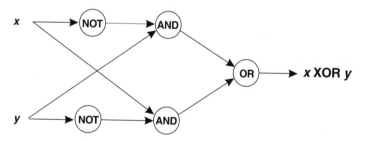

Figure 15.9 *XOR Network*

The following function implements the logic XOR operation. The function assumes that there are neural units available to perform logic AND, OR, and NOT.

```
int LogicXor(int weights[], int values[])
{
// This function performs a logic XOR using a network based on the
// following relationship:
//
//     x XOR y = (NOT(x) AND y) OR (NOT(y) AND x)
//
// Stage 1:
//     calculate outputA = NOT(x) AND y
//               outputB = NOT(y) AND x
```

```
// Stage 2:
//    calculate result = outputA OR outputB

    int temp[] = {0, 0};
    int result;
    int outputA, outputB;

    // Save data in temp[] array
    temp[0] = values[0];
    temp[1] = values[1];
    // Call LogicNot() and LogicAnd()operations, as follow:
    //    outputA = NOT(x) AND y
    //    outputB = NOT(y) AND x
    values[0] = LogicNot(weights, values); // NOT x
    outputA = LogicAnd(weights, values);   // x AND y
    // Invert x and y
    values[0] = temp[1];         // Variable y
    values[1] = temp[0];
    // ASSERT:
    //       values[0] = y
    //       values[1] = x
    values[0] = LogicNot(weights, values); // NOT y
    outputB = LogicAnd(weights, values);   // x AND y
    // ASSERT:
    //       Stage 1 concluded
    // State 2:
    //       perform outputA OR output B
    values[0] = outputA;            // Update values[] array
      values[1] = outputB;
    result = LogicOr(weights, values);    // Perform OR
    return result;

}
```

15.1.2 Software Model for Neural Nets

Biological information-processing systems, such as the brain, are massively parallel. In biological systems interconnected neurons appear to process information simultaneously, or almost simultaneously. Many efforts have been made at imitating this architecture in computer hardware. However, the conventional software model of a neuron lacks parallelism. In a von Neumann machine processing operations take place sequentially, instead of simultaneously. In this sense the graphics in Figure 15.9 suggest that the NOT and AND operations take place in parallel, since the participating data elements are not interrelated. Only the OR operation requires data from both processing branches. However, the processing that takes place in the LogicXor() function, previously listed, is necessarily sequential. In the sections that follow we often depict networks graphically, and then implement these networks in conventional code. The sequential limitations of software-based networks are often the cause of their inefficiency.

15.2 The Perceptron

Perhaps the most interesting and unique properties of biological neural networks is their ability to learn and adapt. The McCulloch-Pitts neuron can be configured to perform a specific task by manipulating the weights, as in the neural units described previously, but they cannot learn. The second generation model for artificial neurons, called the perceptron, was originally defined by Frank Rosenblatt in 1958. The perceptron is a more advanced neuron unit because it is capable of supervised learning.

The original perceptron was a pattern classifier based on optical inputs. It was an optical and mechanical device in which weights were controlled by means of potentiometers. The notion of the perceptron was later generalized to that of a classifier that tolerates input noise, and uses a supervised learning algorithm. One implementation of the perceptron is in a network based on the McCulloch-Pitts neuron. However, the hard delimiter in the original McCulloch-Pitts neuron produced a binary output. In order to use the McCulloch-Pitts neuron model in a perceptron it is convenient to modify its output so that it is +1 or –1. The following function conforms to these requirements.

```
int MPNeuron(double x, double y, double wn[])
{

// 1. The function MPneuron() models the McCulloch-Pitts neuron.
//    The function receives the following parameters:
//        x is the value of the independent variable
//        y is the value of the dependent variable
//        wn[] is an array of three weights
//    mpNeuron() computes:
//    f(x) = f((w[0] * x1) + (w[1] * x2) + (w[2] x bias))
//    where bias is a constant with the value 1.0
//    Returns:
//    if f(x) is positive, lpNeuron() returns 1
//    if f(x) is negative, lpNeuron() returns -1
//

double result = 0;
// Routine logic, per McCulloch-Pitts model
result = (wn[0] * x1) + (wn[1] * x2) + (wn[2] * bias);
if(result < 0)
    return -1;
else
    return 1;
}
```

15.2.1 Perceptron as a Classifier

The classification problem consists of processing features that identify data into a specific pattern space. Figure 15.7 shows the logic functions AND, OR, and NOT as a classification problem in 2D space. Figure 15.10 shows a more general case of the classification problem.

In Figure 15.10, on the following page, there are ten objects, represented by square and circular shapes according to their location in the Cartesian plane. The dotted line is a linear function that separates the two groups. In this case the two classes are linearly separable. The classification problem consists of finding the equation of a line that separates the two groups of objects.

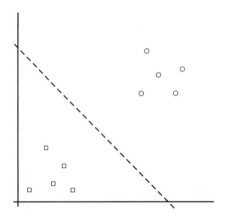

Figure 15.10 *The Classification Problem*

Consider a straight line in the form

$$ax + by + c = 0$$

and recall that in the McCulloch-Pitts neuron the function consists of the weighted sum of three inputs:

$$f(x) = f(w_0 x + w_1 y + w_2)$$

where w0, w1, and w2 are the three weights. Assume that the input value for the bias line is 1.

Comparing the McCulloch-Pitts function with the conventional equation we see that the coefficient of x corresponds to the first weight, the coefficient of y to the second weight, and the constant c to the bias. In terms of a neural network we can restate the classification problem as finding the weights that satisfy the activation function for each pair in the data set. In the case of the McCulloch-Pitts neuron the activation function can be expressed as follows:

$$if(w_0 x + w_1 y + w_2) < 0, output = -1$$

$$if(w_0 x + w_1 y + w_2) \geq 0, output = 1$$

The value of +1 or –1 returned by the activation function is used as the classifier.

15.2.2 Perceptron Learning

Since the synaptic weights control the output, the perceptron's response can be changed by manipulating the weights. Therefore, the perceptron learning process consists of changing its weights whenever the actual response does not correspond to the expected one. This mechanism is called supervised learning. In the case of the perceptron, supervised learning is based on the following assumptions:

1. The perceptron is able to distinguish between an erroneous response and a correct response. This implies that the training data set must contain information to allow determining a correct response.

2. Whenever an incorrect response is detected, weights are changed in order to make the error less likely to occur if the same inputs are presented again.

3. The perceptron learns only by mistakes. If the response is correct, no change in the weights takes place. In other words, no effort is made to improve a correct response.

The last perceptron rule requires some explanation. Figure 15.10 shows a classification problem in which the dotted line serves as a classifier function that separates data into two fields. In Figure 15.10 the dotted line is approximately half-way between the clusters formed by the two groups. When an unclassified input is presented to the perceptron it is reasonable to expect that it will be correctly classified, assuming that the new input conforms to the same clustering as the previous ones. However, the function represented by the dotted line in Figure 15.10 is not the only classification function that correctly maps the data. Figure 15.11 shows an alternative mapping function for the same data set.

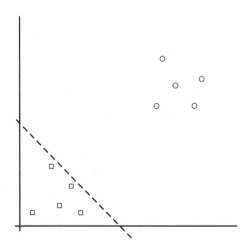

Figure 15.11 *Alternative Mapping Function*

The function represented by the dotted line in Figure 15.11 correctly classifies all points in the data set. However, it is less likely that the function in Figure 15.11 will continue to correctly map new inputs if the clustering continues to follow the same statistical pattern. Statistically speaking, we can say that the mapping function in Figure 15.10 is "more correct" than the one in Figure 15.11. Since the perceptron makes no weight adjustments in case of correct responses, it is possible that it will return a mapping function such as the one in Figure 15.11.

A minimum distance classification function is one that evaluates class membership according to some minimum distance from some central point in the region. The perceptron does not ensure this minimum distance, but merely that all data points are correctly classified.

15.2.3 Training the Perceptron

In perceptron training a set of patterns is presented in random order. The process of visiting all the data in a single pattern is called an iteration. Visiting all the data sets in the pattern is called a learning epoch. If the perceptron's mapping function produces no errors in an epoch, then learning has concluded. If not, the procedure is repeated until the perceptron produces all correct responses, or until an alternative terminating condition is reached. Table 15.1 contains 10 data entries representing the coordinate pairs in Figures 15.10 and 15.11.

Table 15.1

Data Set for Classification in Figure 15.10

ENTRY	X	Y	GROUP
1	1.0	1.0	1
2	9.4	6.4	−1
3	2.5	2.0	1
4	4.0	3.0	1
5	13.5	9.0	−1
6	10.5	9.0	−1
7	2.5	4.5	1
8	4.6	1.0	1
9	12.0	10.5	−1
10	11.0	13.4	−1

The perceptron uses each of the data entries in Table 15.1 and proceeds to calculate the output using an initial set of weights. The initial weights can be set to zero or chosen at random.

Suppose we use the following initial weights: 0.75, 0.5, and − 0.6. Starting with the first data set in Table 15.1 we calculate the net function as follows:

$$f(net)^1 = f[(0.75 \times 1) + (0.5 \times 1) + (0.6 \times 1)]$$
$$f(net)^1 = f(0.65) = 1$$

In this case the perceptron's output (+1) coincides with the expected output in Table 15.1. Therefore, no weight adjustment is necessary for this first iteration. The net function for the second data pair is then calculated as follows:

$$f(net)^2 = [(0.75 \times 9.4) + (0.5 \times 6.4) + (0.6 \times 1)]$$
$$f(net)^2 = f(9.65) = 1$$

However, in this case the output generated by the net function (+1) does not coincide with the expected one from Table 15.1, (−1). Accordingly, the perceptron must then adjust the weights following the formula previously mentioned. If we fix the value of ? at 0.2, then the calculation of the new weights for iteration number 3 becomes:

$$w_3 = w_2 + 0.2(-1-1)x_2$$

$$w_3 = \begin{bmatrix} 0.75 \\ 0.50 \\ -0.60 \end{bmatrix} - 0.4 \begin{bmatrix} 9.4 \\ 6.4 \\ 1.0 \end{bmatrix} = \begin{bmatrix} -3.01 \\ -2.06 \\ -1.00 \end{bmatrix}$$

The next data pair in Table 15.1 (entry number 3) is now evaluated with these new weights. Here again, if the result coincides with the expected result, no action is taken. If the value returned by the net function is different from the one in the table, then the data has been incorrectly mapped and a weight adjustment takes place. This process continues until no weight change has taken place during one epoch, or until another terminating condition is reached.

Rosenblatt developed the following formula for changing the synaptic weights in a perceptron:

$$w_i(t+1) = w_i(t) + \eta[d(t) - y(t)]x_i(t)$$

where

$$[d(t) - y(t)]$$

is the difference between the actual and the target output,

$$x_i(t)$$

is the input at the current iteration, and h represents a positive learning constant.

15.2.4 A Perceptron Function

The following function simulates a perceptron in software. The function uses the MPNeuron() function previously listed.

```
int Perceptron(double x[],       // array of x values
               double y[],       // array of y values
               int expected[],   // array of expected classification
               int dataSize,     // number of elements in arrays
               double weights[]) // array of 3 weights
{
// The function Perceptron() is a single layer classifier that
```

```
// uses the MPneuron() function.
// Terms:
// Array data is defined for dataSize entries of x[] and y[].
// Array expected[] contains either +1 or -1 according to the
// expected mapping of the corresponding values of x and y.
// There is a limit of 500 tests in each training session
// Perceptron() returns 1 if succesful, 0 otherwise
//
// The training constant eta is 0.2, therefore the learning
// increment is either +2*eta or -2*eta (2eta = 0.4)
// The function can be modified so that the learning increment
// is passed as a parameter
//
// Terminology:
//     the variables are labeled x[k], y[k], expected[k] for each
//     kth iteration.
//
// Training algorithm for Perceptron():
// The perceptron is trained according to the following logic:
// 1. x[k], y[k], and w[] are fed into the MPneuron() for each
//     iteration in the data set, from k = 0 to k = dataSize.
//
// 2. At each iteration k, the result returned by MPneuron() is
//     compared with the expected results (as stored in the expected[k]
//     array).
//     If the values match, execution proceeds at the next iteration
//     using the same weights. If not, then the following actions
//     take place:
//     a. If the difference between the expected and the actual
//        output of MPneuron() is negative, then the learning
//        increment (li) = -0.4, otherwise li = 0.4
//     b. A new set of weights is created by adding or subtracting,
//        from the current weight, the product of the input[k]
//        times the learning increment, as follows:
//        if li < 0 then
//           new w[0] = old w[0] -(li * x[k])
//           new w[1] = old w[1] -(li * y[k])
//           new w[2] = old w[2] -(li * expected[k])
//        if li >= 0
//           new w[0] = old w[0] +(li * x[k])
//           new w[1] = old w[1] +(li * y[k])
//           new w[2] = old w[2] +(li * expected[k])
// 3. Execution continues until a valid function is found, that is,
//     one that correctly maps all the values in the data set, or until
//     number of iterations defined in the constant PLIMIT is reached.

// Local variables
double li = 0.4;          // Learning increment is +/-0.4
int k = 0;                // Iteration counter
int thisTest = 0;         // Test counter
int thisResult = 0;       // Value returned by lbNeuron()
int thisDif = 0;          // expected[k] - thisResult
int wc = 0;               // Weight change flag

// Clear global variables
changeCnt = 0;
testCnt = 0;

while(thisTest < PLIMIT)
  {
```

```
for (k = 0; k < dataSize; k++)
{
    if(k == 0)
      wc = 0;
      thisResult = MPNeuron(x[k], y[k], weights);
      thisDif = expected[k] - thisResult;
      // Test for expected result, if not change weights
      if(thisResult != expected[k])
      {
          // ASSERT: weights need changing
          if(thisDif < 0)
          {
          weights[0] = weights[0] - (li * x[k]);
          weights[1] = weights[1] - (li * y[k]);
          weights[2] = weights[2] - (li * bias);
      }
      else
      {
          weights[0] = weights[0] + (li * x[k]);
          weights[1] = weights[1] + (li * y[k]);
          weights[2] = weights[2] + (li * bias);
      }
      // Set weight change flag
      wc = 1;
      // Count weight changes
      changeCnt++;
      }
} // End of for loop
// At this point one full epoch has taken place
// Test for new weight change during this epoch
if(wc == 0)
  return 1;          // Return success!
// if not
testCnt++;                  // Increment global test counter
} // end of while loop
// At this point the limit in the number of iterations has
// been reached without finding a valid equation
return 0;
}
```

SOFTWARE ON-LINE

The Perceptron() function is exercised in the project named Perceptron, located in the Chapter15 folder in the book's on-line software.

15.3 The Adaline

In 1960 Bernard Widrow and Marcian Hoff, of Stanford University, developed an adaptive system specialized in signal processing. The system was named Adaline, for Adaptive Linear Neuron. The Adaline was very similar to the perceptron, except that the learning rule used by Adeline was more robust.

15.3.1 Widrow-Hoff Learning

The rules developed by Widrow and Hoff for the Adaline have since been called Widrow-Hoff learning, the delta rule, or the least mean squares (LMS) algorithm. Like the perceptron, the Adaline changes the weights only in case of an incorrect response. In the Adaline a small quantity ?, represented by the Greek letter delta, is added to the weights of the current iteration. This quantity is represented as follows:

$$\Delta w_j$$

where w is the weight at the jth input. The new weight is calculated in this manner:

$$w_n = w_o + \Delta w_o$$

where w_n is the new weight, w_o is the old weight, and w_o is the small increment (delta quantity) added to the old weight. The formula for the increment is

$$\Delta w_j = \eta [d(t) - y(t)] x_j(t)$$

15.3.2 A Neuron for Adaline

Recall Rosenblatt's formula for changing the synaptic weights in the perceptron:

$$w_i(t+1) = w_i(t) + \eta [d(t) - y(t)] x_i(t)$$

Note that, in the perceptron, the values of d and y are limited to +1 and –1. Therefore, the bracketed term can be equal only to +2 or –2. The correction made to the ith weight is

$$\pm 2\eta \times x_i(t)$$

The Adaline, on the other hand, makes no such assumption. In the case of the Adaline the value returned by the activation function is not binary, but continuous. This requires a neuron with a different activation function. A neuron-like function with a continuous output can be coded as follows:

```
double CNeuron(double x, double y, double wn[])
{
// 1. CNeuron()receives the following paramenters:
//       x is the value of the independent variable
//       y is the value of the dependent variable
//       w[] is an array of three weights
//       bias is a constant defined as 1.0
//    CNeuron() computes:
//       f(x) = f((w[0] * x) + (w[1] * y) + (w[2] x bias))
//    Returns:
//       f(x) in a variable of type double
//
double result = 0;
```

```
// Routine logic
result = (wn[0] * x) + (wn[1] * y) + (wn[2] * bias);
return result;
}
15.4.3 Adaline() Function
The following function simulates an Adaline network in software.

int Adaline(double x[],        // array of x values
            double y[],        // array of y values
            int expected[],    // array of expected classification
            int dataSize,      // number of elements in arrays
            double wt[])       // array of 3 weights
{
// The function Adaline() is a single layer classifier that
// uses the MPNeuron() function. Adeline() net uses the LMS
// training algorithm.
// Terms:
// Array data is defined for dataSize entries of x[] and y[].
// Array expected[] contains 1 or -1 according to the mapping
// of the corresponding values of x and y.
// There is a limit of 500 epochs in each training session
// Adaline() returns 1 if succesful, 0 otherwise
// Terminology:
//    the variables are labeled x[k], y[k], expected[k] for each
//    kth epoch.
//
// Training algorithm for Adaline():
// The Adaline is trained according to the following logic:
// 1. x[k], y[k], and w[] are fed into the MPneuron() for each
//    iteration in the data set, from k = 0 to k = dataSize.
//
// 2. At each jth iteration the result returned by MPNeuron() is
//    stored in the local int variable netVal according to the
//    following rule:
//      if netj is positive, netVal = 1
//      if netj is negative, netVal = -1
//    netVal is then compared with the expected results (as stored in
//    the expected[k] array).
//    If the values match, execution proceeds at the next iteration
//    using the same weights.
//    If not, then the new weight is calculated according to the
//    formula:
//    wn = wo + Deltaw
//    where wn is the new weight
//    wo is the old weight,
//    Deltaw is an increment applied to the old weight:
//
//    Deltaw = n(dj - netj)ij
//    n is the learning rate (0.1)
//    dj is the desired output value (1 or -1) from expected[j]
//    netj is the value returned by the MPNeuron() for test case j
//    ij is the input vector
//
// 3. Execution continues until a valid function is found, that is,
//    one that correctly maps all the values in the data set, or until
//    number of iterations defined in the constant ALIMIT is reached.

// Local variables

int ALIMIT = 500;
```

```
double n = 0.1;           // learning increment
int k = 0;                // Iteration counter
int thisTest = 0;         // Test counter
double thisNet = 0.0;     // Value returned by MPNeuron()
int netVal = 1;           // Binary result of net
int wc = 0;               // Weight change flag
double deltaW = 0.0;      // Increment for weight change

// Clear global variables
changeCnt = 0;
testCnt = 0;

while(thisTest < ALIMIT)
  {

  // Start of epoch
  for (k = 0; k < dataSize; k++)
   {
      if(k == 0)
         wc = 0;
      thisNet = CNeuron(x[k], y[k], wt);
      // Convert to binary result
      if(thisNet < 0)
        netVal = -1;
      else
        netVal = 1;
      // Test for expected result, if not change weights
      if(netVal != expected[k])
      {
         // ASSERT: weights need changing
         // Calculate deltaw = n(dj - netj)ij
         // For the first input vector
         deltaW = n * x[k] * (expected[k] - thisNet);
         wt[0] = wt[0] + deltaW;
         // For the second input vector
         deltaW = n * y[k] * (expected[k] - thisNet);
         wt[1] = wt[1] + deltaW;
         // For the bias vector
         deltaW = n * (expected[k] - thisNet);
         wt[2] = wt[2] + deltaW;
         // Set weight change flag
         wc = 1;
         // Count weight changes
         changeCnt++;
      }

   } // End of epoch
      // At this point one epoch has taken place
      // Test for no weight change during epoch. This indicates a
      // valid function.
       if(wc == 0)
         return 1;         // Return success!
      // if not
      testCnt++;                  // Increment global test counter
      thisTest++;                 // Iteration counter
  } // end of while loop
  // At this point the limit in the number of iterations has
  // been reached without finding a valid equation
  return 0;
}
```

SOFTWARE ON-LINE

The function Adaline() is exercised in the project named Perceptron, located in the Chapter15 folder in the book's on-line software.

15.4 Improving the Classification Function

When either the perceptron or the Adaline is used as a classifier, the results returned are the first ones found that satisfy the mapping of the data set. If we are seeking the fastest possible solution to the classification problem, then the fact that the perceptron and the Adaline stop at the first function that satisfies the data set may be a desirable feature. However, if we are seeking not the first, but statistically the best possible mapping function, the results returned by the perceptron or the Adaline may not be ideal.

15.4.1 Calculating the Error Sum

In improving the mapping function we need some way of measuring the difference between the returned function and the ideal one. In Chapter 12 we looked at some linear and non-linear methods for calculating the error sum. One simple approach, called least-squares interpolation, consists of minimizing the differences between the curve and the data points (see Figure 12.5). The residuals are first added, then squared in order to eliminate possible sign compensations. Consider the equation of a straight line in the form

$$ax + bx + c = 0$$

where a, b, and c are the three weights. Since the error sum requires an equation in slope-intercept form ( ), we must calculate m and b from the weights, as follows:

$$m = \frac{-a}{b}$$

$$b = \frac{-a}{c}$$

The following function can be used to calculate the error sum given a data set that contains the weights and the input values.

```
double GetErrorSum(double wt[],     // Array of weights
                   double x[],      // Array of x values
                   double y[],      // Array of y values
                   int dataSize)    // Number of data elements
{
// This function returns the error sum given the x and y values
// and the weights.
// In the equation form ax + by + c = 0:
// w[0] = a
// w[1] = b
```

```
// w[2] = c
// Then, the slope-intercept form (y = mx + b) is determined from
// the weights, as follows:
//        y = -a/bx - a/c
// which makes:
//        m = -a/b
//        b = -a/c
// Given the equation for the line, the error sum is determined by
// calculating each value of the dependent variable and subtracting
// this value from the corresponding one in the data arrays.
// These differences, called the residuals, are added and then
// squared to eliminate sign compensations. One-half of this result
// is returned as the error sum.

double m = 0;           // Slope
double b = 0;           // Y-intercept
int j = 0;
double newY = 0;        // Temporary storage for y
double res = 0;         // Accumulator for residuals
double errSum = 0;      // Error sum

// First calculate m and b
m = -wt[0]/wt[1];
b = -(wt[2]/wt[1]);
// Use m and b to calculate error sum
    for(j = 0; j < dataSize; j++)
    {
        newY = (m * x[j]) + b;
        res = y[j] - newY;
        errSum = errSum + (res * res);
    }
    return errSum/2;
}
```

15.4.2 Improving Perceptron Results

The GetErrorSum() function, in the preceding section, allows measuring the difference between the actual output and the ideal output according to the data. It provides a measure of the error that can be used to improve the results returned by the perceptron or Adaline networks. One possible approach is to use the perceptron to calculate several different mapping functions, estimate the error sum of each one, then select the best one. Since the mapping function returned by the perceptron depends on the initial weights, it makes sense to find some mechanism for changing the initial weights for each attempt. For example, one could select random initial weights, call the perceptron function a number of times, compare the error sum in each case, and return the function with the smallest error. This logic, which can be described as a brute-force, random weights trainer, is implemented in the following function.

```
void BFRTrainer(double x[],     // array of x values
                double y[],     // array of y values
                int expected[], // array of expected classification
                int dataSize)   // number of elements in arrays

{
// The function BFRTrainer() trains the Perceptron() by means of a
// brute-force algorithm in which the weights are selected at
```

```
// random during each iteration. For each iteration that produces
// a valid equation the error sum is calculated and stored.
// Calculations are repeated 400 times and the weights with the
// smallest error sum are returned.
// Function logic:
// 1. Routine calculates an initial random set of weights in
//    the range 20 to -20.
//    Perceptron() is called with these weights. If Perceptron()
//    returns a valid equation, the error sum for this
//    equation is remembered.
// 2. A new set of random weights is calculated and Perceptron()
//    is again called with these new weights. If Perceptron
//    returns a valid equation, the error sum for this
//    new equation is compared to the error sum of the previous
//    one. The weights for the lower error sum are preserved and
//    the other ones are discarded.
// 3. This is repeated 400 iterations after which the weights
//    with the smallest error sum are stored in the global
//    array named bestWeights[]. The lowest error sum is
//    calculated and returned in a global variable
int limit = 400;                    // Number of iterations
int counter = 0 ;
// Set initial weights
double patWeights[] = {0.0, 0.0, 0.0};
double weights[] = {0.0, 0.0, 0.0};
oldESum = 100000;                   // Start with large error sum
// initialize random number'
InitRandom();
// Set initial pattern weights at random
patWeights[0] = Random();
patWeights[1] = Random();
patWeights[2] = Random();
while(counter < limit)
{
    // Set weights to local weight pattern
    weights[0] = patWeights[0];
    weights[1] = patWeights[1];
    weights[2] = patWeights[2];
    if(Perceptron(x, y, expected, 10, weights))
    {
        thisESum = GetErrorSum(weights, x, y, dataSize);
        // Store best weights
        if(thisESum < oldESum)
        {
            bestWeights[0] = weights[0];
            bestWeights[1] = weights[1];
            bestWeights[2] = weights[2];
            oldESum = thisESum;
        }
    }
    // Reset pattern weights using random numbers
    patWeights[0] = Random();
    patWeights[1] = Random();
    patWeights[2] = Random();
      counter++;
}
  // Calculate error sum and store in global variable
  bestESum = GetErrorSum(bestWeights, xData, yData, dataSize);
  return;
}
```

SOFTWARE ON-LINE

The BFRTrainer() function, the GetErrorSum() function, and a support function to generate random weights, are demonstrated in the Perceptron project and contained in the Chapter15 folder in the book's on-line software.

15.5 Backpropagation Networks

In the late 1960s the severe limitations of single-layer networks, such as the perceptron and Adaline, were presented in a seminal book by Minsky and Papert. Interest in neural networks declined during the 1970s and early 1980s as a result. It was the discovery and dissemination of multilayer neural networks that rekindled interest in the field.

A multilayer neural network consists of three independent layers: an input layer that contains nodes that receive the input signal, a hidden layer of computational nodes, and an output layer that provides the system's results. These systems have been called multilayer perceptrons, feed-forward neural networks, back-error propagation networks, and most commonly, backpropagation networks. The last two names refer to the learning algorithm used in these models. Backpropagation networks have been the most successful and widely accepted of the neural network paradigms.

The simplest architecture for a backpropagation network consists of the three layers previously mentioned, although more than three layers are also common. Figure 15.12 shows a multilayer network.

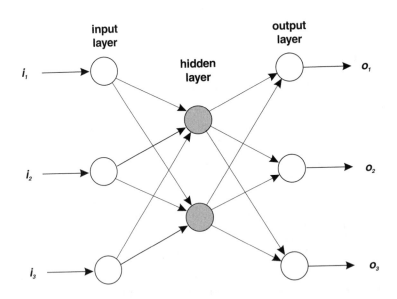

Figure 15.12 *Multilayer Network for Backpropagation*

Figure 15.12 represents a fully-connected, feed-forward network. It is fully connected because all the nodes receive inputs from every node in the previous layer. It
is a feed-forward network because nodes connect to the adjacent layers only. For
example, if one or more nodes in the output layer were directly connected to nodes
in the input layer, then the network would lose its feed-forward characteristic. By
the same token, if one or more nodes in the output layer did not receive input from
one or more nodes in the hidden layer, then the network would not be fully connected. Most learning schemes used in backpropagation are based on
fully-connected, feed-forward networks.

15.5.1 Nonlinear Neurons

Two of the neurons so far implemented in code use a nonlinear activation function,
that is, the output is limited to two possible values. In the neuron function named
NeuronB() (used in the Neural Units project) the output is either 0 or 1. In the neuron
modeled by the MPneuron() function (used in the Perceptron project) the output is +1
or –1. On the other hand, the Adaline() function requires a continuous output from the
neuron. For the Adaline we developed a third neuron, named CNeuron(), that returns
a continuous function in a variable of type double. This function is linear.

But backpropagation requires that we measure the rate-of-change of the output
function in order to make more accurate weight corrections. For this reason, the activation function for backpropagation must meet other requirements, namely, the
function must be continuous, monotonically increasing, invertible, differentiable at
all points, and must approach its saturation value as the net function approaches infinity. Several sigmoid functions meet these requirements, one of the most often
used is called the logistic function, represented in the following equation:

$$f(x) = \frac{1}{1 + e^{(-x)}}$$

where x is the network output.

One of the advantages of the logistic function is that its derivative is very easy to
compute. The derivative of the logistic is expressed as follows:

$$f'(x) = f(x)[1 - f(x)]$$

Figure 15.13, on the following page, shows a graph of the logistic and its derivative.

In Figure 15.13 the logistic function is represented by the solid line and its derivative by the dashed line. The following function, named NeuronS(), returns the logistic function of an array of values and weights.

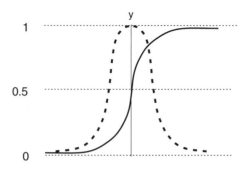

Figure 15.13 *The Logistic Function and Its Derivative*

```
double NeuronS(double xIn[], double wts[], int size)
{
// 1. NeuronS()receives the following parameters:
//        xIn[] is and array of values
//        wts is an array of weights
//        size is the number of elements in data arrays
//
//      NeuronS() computes:
//      f(x) = the sum of products of xIn[] by wts[]
//      Returns the logistic function:
//      f(x) = 1/1+e^-f(x)
//
double sum = 0;          // f(x)
double result = 0;       // 1/1-e^-f(x)
int i;

// Routine logic
for(i = 0; i < size; i++)
   sum += wts[i] * xIn[i];

result = 1/(1+exp(-sum));

return result;
}
```

15.5.2 Backpropagation Algorithm

Backpropagation is a generalized application of the Widrow-Hoff learning algorithm described earlier. Since backpropagation is a form of supervised learning, it requires that the correct outputs be furnished in the data set. Once the network is trained, the weights are used to computer new values from new inputs. The network uses a feed-forward mechanism whereby inputs are first presented to the input layer. These values are passed directly onto the first hidden layer, where the hidden layer neurons compute the weighted sum of the inputs. The results from the first hidden layer are then passed to other hidden layers, or to the output layer, according to the specific system network architecture. The output layer, in turn, performs the weighted sum of the inputs to produce output values.

Assuming a 3-layer network with a single input layer, a single hidden, and a single output layer, backpropagation can be summarized in the following steps:

1. The input data, and the corresponding weights, are presented to the hidden layer neurons.

2. The hidden layer performs the weighted sum of the inputs and presents the results to the neurons in the output layer, which calculate the network outputs.

3. A supervisor or trainer element calculates the error for each output cell, as follows:

$$e = t - \hat{t}$$

where e is the error, is the desired output from the training data set, and (usually called t-hat) is the actual output from the last layer. The supervisor also calculates the derivative of the logistic function:

$$f'(x) = f(x)[1 - f(x)]$$

and the total error signal:

$$\Delta O = f'(x)e$$

4. The error signal of each output unit is calculated by multiplying the total error signal by the corresponding output unit weight.

5. The error signal of the hidden unit neurons is calculated using the derivatives of the hidden unit responses.

6. The error factor for correcting weights is calculated as follows:

   where h is the learning constant and x is the input data.

7. Using the results from steps 3 to 6, the weights of the hidden layer and the output layer are updated.

15.5.3 Software Model for Backpropagation

The following function, named Backprop323() performs backpropagation in a network with 3 input layers, 2 hidden layers, and 3 output layers following the algorithm described in Section 15.6.2.

```
void BackProp232(double inD[],   // Input array of 3 elements
                 double outD[],  // 3 elements with desired output
                 double wHL1[],  // 2, 3-element hidden layer weigths
                 double wHL2[],
                 double wOL1[],  // 3, 2-element output layer weights
                 double wOL2[],
                 double wOL3[],
                 double eDif[])  // Storage for 3 error differences
{
double eta = 1.0;    // Learning rate
double hidN[2];      // Storage for hidden neuron
double outN[3];      // Storage for output layer results
```

```
double e[3];           // Difference between actual and expected respose
double fDx[3];         // Derivative of output unit responses
double deltaO[3];      // error signals
double outError[2];    // hidden unit responsability for output error
double fDhu[2];        // derivatives of hidden unit responses
double eSignal[2];     // error signal to use in backpropagation
double deltaZ1[3];     // factors for output unit weight corrections
double deltaZ2[3];
int j;

// Step 1:
// Presents the input data, stored in the global array named inD[]
// to the hidden layer neurons using weights tables wHL1[] and wHL2[].
hidN[0] = NeuronS(inD, wHL1, 3);
hidN[1] = NeuronS(inD, wHL2, 3);

// Step 2:
// Outputs from the hidden layer (in hidN[]) are presented to
// the 3 output layers using the weights tables wOL1[], wOL2[], and
// wOL3[]
outN[0] = NeuronS(hidN, wOL1, 2);
outN[1] = NeuronS(hidN, wOL2, 2);
outN[2] = NeuronS(hidN, wOL3, 2);

// Step 3:
// First calculate e = t - t^ for each output cell
// Then calculate the derivative of the output unit responses:
//     Logistic function:      f(x) = 1/1+e^-f(x)
//     Derivative:             fDx = f(x)[1-f(x)]
// Finally, calculate the error signal:
// deltaE = f'(x) * e
for(j = 0; j < 3; j++)
{
    e[j] = outD[j]-outN[j];
    fDx[j] = outN[j] * (1-outN[j]);
    deltaO[j] = fDx[j] * e[j];
}

// Step 4:
// Calculate error signal of output units:
// Multiply error signal of each output unit (in deltaO[]) by the
// output unit weights (in wO1[], wO2[], and wO3[])
outError[0] = wOL1[0] * deltaO[0] +\
              wOL2[0] * deltaO[1] +\
              wOL3[0] * deltaO[2];
outError[1] = wOL1[1] * deltaO[0] +\
              wOL2[1] * deltaO[1] +\
              wOL3[1] * deltaO[2];

// Step 5:
// Calculate error signal of hidden units
// Start with the derivative of the hidden unit responses:
fDhu[0] = hidN[0] * (1-hidN[0]);
fDhu[1] = hidN[1] * (1-hidN[1]);
eSignal[0] = fDhu[0] * outError[0];
eSignal[1] = fDhu[1] * outError[1];

// Step 6:
// Correct the output weights (in wO1[], wO2[], and wO3[])
// As follows:
```

```
//          Z = Z + deltaZ
//          deltaZ = eta * inD[] * eSignal[]
deltaZ1[0] = eta * (inD[0] * eSignal[0]);
deltaZ1[1] = eta * (inD[1] * eSignal[0]);
deltaZ1[2] = eta * (inD[2] * eSignal[0]);
//
deltaZ2[0] = eta * (inD[0] * eSignal[1]);
deltaZ2[1] = eta * (inD[1] * eSignal[1]);
deltaZ2[2] = eta * (inD[2] * eSignal[1]);

// Update weights in hidden layer unit
wHL1[0] = wHL1[0] + deltaZ1[0];
wHL1[1] = wHL1[1] + deltaZ1[1];
wHL1[2] = wHL1[2] + deltaZ1[2];

wHL2[0] = wHL2[0] + deltaZ2[0];
wHL2[1] = wHL2[1] + deltaZ2[1];
wHL2[2] = wHL2[2] + deltaZ2[2];

// Step 7:
// Update weights in output unit
wOL1[0] = wOL1[0] + (eta * (hidN[0] * deltaO[0]));
wOL1[1] = wOL1[1] + (eta * (hidN[1] * deltaO[0]));

wOL2[0] = wOL2[0] + (eta * (hidN[0] * deltaO[1]));
wOL2[1] = wOL2[1] + (eta * (hidN[1] * deltaO[1]));

wOL3[0] = wOL3[0] + (eta * (hidN[0] * deltaO[2]));
wOL3[1] = wOL3[1] + (eta * (hidN[1] * deltaO[2]));

// Return error differences
for(j = 0; j < 3; j++)
   eDif[j] = e[j];

return;
}
```

Note that the backpropagation function previously listed assumes a network with a specific architecture. Although it is possible to code backpropagation functions that allow changing the number of layers, or the number of nodes in each layer, the coding would be more complicated. Since networks are usually designed to solve a specific problem that requires a given number of nodes and layers, such a generic function may not be very useful.

15.5.4 Executing the Network

The backpropagation function in the preceding section serves to calculate the weights during a single iteration of the data set. In order to use a specific network, once trained, it is necessary to develop a function that iterates the data set through the network. This requires presenting the data from the input layers to the hidden layer neurons, then the output from the hidden layer neurons to the output layer neurons. Here again, developing a function that conforms to a specific network architecture simplifies the programming. The following function named Network232() iterates the data set through a fully-connected, feed-forward network with a 3-node input layer, a 2-node hidden layer, and a 3-node output layer.

```
void Network232(double inD[],     // Input array of 3 elements
          double outD[],     // Array of 3 elements for results
          double wHL1[],     // 2, 3-element hidden layer weigths sets
          double wHL2[],
          double wOL1[],     // 3, 2-element output layer weights sets
          double wOL2[],
          double wOL3[])
{
// Local variables
double outHL1[2];

// Function to execute a network with 3 neurons in the input layer,
// 2 in the hidden layer, and 3 in the output layer
// Step 1:
// Presents the input data, stored in the array named inD[]
// to the hidden layer neurons using weights tables wHL1[] and wHL2[].
outHL1[0] = NeuronS(inD, wHL1, 3);
outHL1[1] = NeuronS(inD, wHL2, 3);

// Step 2:
// Outputs from the hidden layer (in outHL1[]) are presented
// to the 3 output layers using the weights in wOL1[], wOL2[],
// and wO3[]
outD[0] = NeuronS(outHL1, wOL1, 2);
outD[1] = NeuronS(outHL1, wOL2, 2);
outD[2] = NeuronS(outHL1, wOL3, 2);

// Done

return;
}
```

15.5.5 A Backpropagation Trainer

Training the backpropagation network consists of repeatedly calling a backpropagation function until a certain terminating condition is reached. This terminating condition can be in the form of a maximum number of epochs, or a predefined error limit. The following function, named BPTrain323() performs backpropagation training of a network with the 323 architecture.

```
int BPTrain323(double inD[],     // Input array of 3 elements
          double outD[],     // Array of 3 elements for results
          double wHL1[],     // 2, 3-element hidden layer weigths sets
          double wHL2[],
          double wOL1[],     // 3, 2-element output layer weights sets
          double wOL2[],
          double wOL3[],
          double *eTotal)    // Error total returned to caller
{
// This function uses BackProp323() and Network323() to train the
// neural network until a LIMIT number of training epochs is reached
// or when a predefined error difference (ELIMIT) is achieved
int LIMIT = 5000;
int count = 0;
double ELIMIT = 0.0001;

int k;
double eData[3];        // Storage for error data returned by
                        // BackProp323()
```

```
double eSum;              // Error sum

while(LIMIT > count)
{
      BackProp232(inD, outD, wHL1, wHL2, wOL1, wOL2, wOL3, eData);
      // Init error sum
      eSum = 0;
      // Check for error limit
      // First add error data
      for(k = 0; k < 3; k++)
          eSum += (eData[k] * eData[k])/2;
      // Test for limit
      if(eSum < ELIMIT)
      {
          *eTotal = eSum;
          return 1;
      }
      // Bump epoch counter
      count++;
}
// Epoch limit was reached
*eTotal = eSum;
return 0;
}
```

SOFTWARE ON-LINE

The backpropagation functions are located in the Backprop project found in the Chapter15 folder in the book's on-line software.

Part II

Application Development

Chapter 16

The C++ Language on the PC

Chapter Summary

This chapter introduces the C++ programming language and describes its various PC implementations. Also describes the various elements and use of flowcharts in program design. The chapter concludes with a walkthrough of a C++ Console Application using Microsoft Developers Studio and Visual C++.

16.0 Introducing C++

C++ is a general-purpose programming language that evolved from a previous language named C. C++ is defined as a statically typed, free-form language that supports procedural programming, data abstraction, and object-orientation. C++ is one of the most used and popular programming languages for the PC. The name stems from C's "++" operator which is used to increment the value of a variable.

C, the direct ancestor of C++, is a computer programming language originally designed and implemented by Dennis Ritchie in 1972 while working at the Bell Laboratories. The first version of C language ran under the UNIX operating system on a DEC PDP 11 machine. The predecessor of C is a language called BCPL (Basic Combined Programming Language) developed in 1969 by Martin Richards of Cambridge University. The name C language originated in the fact that Bell Laboratories' version of BCPL, which was developed by Ken Thompson, is named B.

It is generally accepted that C, and its descendant C++, are members of the ALGOL family of programming languages. Therefore they are closely related to the algebraic languages, such as ALGOL and Pascal, and not as similar to BASIC or FORTRAN.

16.0.1 Evolution of C++

Bjarne Stroustrup developed C++ while working at Bell Labs in 1983. The original concept was that C++ would be an enhancement to the C language that would add object oriented facilities. The enhancement ended up including classes, virtual functions, a single line comment style, operator overloading, improved type-checking, multiple inheritance, template functions, and some degree of exception handling. The C++ programming was standardized in 1998 as ISO/IEC 14882:1998. The current version dates from 2003.

During its evolution many libraries have been added to the C++ language. One of the first ones was (named iostream) provided much needed stream-level input and output functions. The new operators named cin() and cout() replaced the more awkward C functions named printf() and scanf().

16.0.2 Advantages of the C++ Language

The following are the most often cited advantages of C language:

1. C++ is not a specialized programming language, therefore, it is suitable for developing a wide range of applications, from major system program to minor utilities.

2. Although C++ is a relatively small language it contains all the necessary operators, data types, and control structures to make it generally useful.

3. The language includes an abundant collection of library functions for dealing with input/output, data and storage manipulations, system interface, and other primitives not directly implemented in the language.

4. C++ data types and operators closely match the characteristics of the computer hardware. This makes C++ programs efficient as well as easy to interface with assembly language programs.

5. The C++ language is not tied to any particular environment or operating system. The language is available on machines that range from microcomputers to mainframes. For this reason, C++ programs are portable, that is, they are relatively easy to adapt to other computer systems.

6. Object orientation is optional in C++. Applications that do not require object orientation can turn-off this feature.

16.0.3 Disadvantages of the C++ Language

The following are the most often noted disadvantages of C++:

1. Because C++ is not a language of very high level, it is not the easiest to learn. Beginners find that some constructions in C++ are complicated and difficult to grasp.

2. The rules of C++ language are not very strict and the compiler often permits considerable variations in the coding. This allows some laxity in style which often leads to incorrect or inelegant programming habits.

3. Instead of being line C, a small and compact language, C++ has become large and complex.

4. C++ library functions are devised to operate on a specific machine. This sometimes complicates the conversion of C++ software to other implementations or systems.

5. The insecurities of C were not resolved in C++.

16.1 PC Implementations of C and C++

Several software companies have developed C and C++ compilers for the PC. Some of these products have gained and lost popular favor as other new versions or implementations were introduced to the market. Historically some of the better known implementations of C for the PC were the Microsoft C and Quick C compilers, IBM C 2 compiler (which is a version of Microsoft C licensed to IBM), Intel iC 86 and iC 286 compilers, Borland Turbo C, Turbo C Professional, Lattice C, Aztec C, Zortech C, and Metaware High C.

Currently C++ development systems have almost completely replaced the old C compilers. The most popular ones are the Microsoft versions furnished as part of Developer's Studio usually referred to as Visual C++. The last version of Visual C++ that did not include .NET support was version 6. After version 6, the .NET suffix was added to C++ compiler versions, thus Visual C++ .NET in Developer's Studio Version 7, Visual C++ .NET 2003 and Visual C++ .NET 2005. However, in the 2005 version the .NET designation was removed and there is a Visual C++ 2005 version. Microsoft's explanation for the name change is that the .NET designation had led user's to believe that the 7.0 and 7.1 compilers were intended exclusively for .NET development, which is not the case.

Borland Corporation had developed and marketed several C++ systems for the PC over the years. The current version is Borland C++ Builder 2006 for Microsoft Windows. Two variations are available in the Borland product: the Professional and the Enterprise editions.

16.2 Flowcharts and Software Design

A set of logical instructions designed to perform a specific task is called a program. The document containing a set of instructions for starting a computer system could be described as a power up program for a human operator. By the same token, a computer program is a set of logical instructions that will make the computer perform a certain and specific task. Note that in both cases the concept of a program requires a set of instructions that follow a logical pattern and a predictable result. A set of haphazard instructions that lead to no predictable end can hardly be considered a program.

Programmers have devised logical aids that help them make certain that computer programs follow an invariable sequence of options and their associated actions. One of the most useful of these aids is called a flowchart. A flowchart is a graphical representation of the options and actions that form a program. Flowcharts use graphic symbols to enclose different types of program operations. The most common ones are:

1. Rectangle. Used to represent processing operations such as arithmetic calculations and data manipulations.

2. Parallelogram. Represents input and output functions, such as keyboard input, display, and printing operations.

3. Diamond. Indicates a decision or logical comparison whose result is a yes or no.

4. Circle. Used to represent a program termination point such as the end of execution or an error exit.

5. Flow lines. Are used to connect other flowchart symbols indicating the direction of program flow.

The use of a flowchart is best illustrated with an example. Figure 16.1 is a flowchart of the program for turning on a computer system in which the components can be connected to the power line in three possible ways:

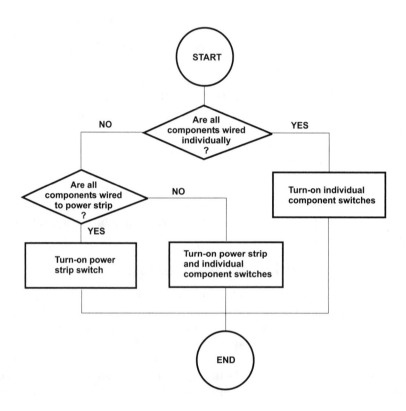

Figure 16.1 *A Machine Start-Up Flowchart*

1. All computer components are directly connected to individual wall outlets.

2. All computer components are connected to a power strip and the power strip is connected to the wall outlet.

3. Some components are connected to a power strip and some are connected directly to the wall outlets.

Note in the flowchart of Figure 16.1 that the diamond flowchart symbols are used to represent program decisions. These decisions correspond to the principles of Aristotelian logic, therefore, there must be two and not more than two answers to the question. These possible answers are usually labeled YES and NO in the flowchart. Decisions are the crucial points in the program's logic. A program that requires no decisions or comparisons consists of such simple logic that a flowchart would be trivial and unnecessary. For instance, a flowchart that consists of three processing steps: start, solve problem, and end, is logically meaningless.

Computing machines of the present generation are not equipped with humanlike intelligence. Therefore some assumptions that are obvious when dealing with human beings are invalid regarding computers. Computer programs can leave no loose ends and make no assumptions of reasonable behavior. You cannot tell a computer "well, you know what I mean." The programmer uses flowcharts to make certain that each processing step is clearly specified.

Regarding the flowchart graphics it should be noted that arrowheads are optional in flowcharts if execution proceeds from the top down, as in Figure 16.1. However, if execution flows up, to a higher flowchart level, then the connecting lines should have arrowheads indicating the direction of program flow.

16.3 The C++ Console Application

Microsoft Visual C++ provides a text-based program type that is useful in developing C++ programs that do not require a graphical user interface. For example, an engineer requires a small program to perform a set of numerical calculations. If the application is not to be distributed to other users and does not require graphics output, then the console application provides a simple way of developing a C++ program.

To create a console application you must first start Visual C++ in Developer's Studio. Once the development environment is displayed, select the New command in the File menu. This command displays the selection box shown in Figure 16.2.

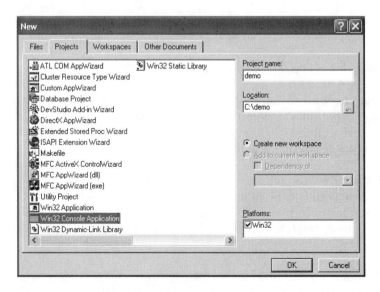

Figure 16.2 *Visual C++ New File Command*

Once the New File box is displayed you must make sure that the Project tab is active and select the Win32 Console Application option towards the bottom of the Project Screen. Next you enter the name of your project and its location in the machine's file system. In this example we have named our project demo1. The button to the right of the input box labeled "location:" allows browsing the machine's file system.

Once you click the OK button, the following screen contains several radio buttons which allow selecting between an empty project, a simple application, a "Hello, World" application, or an application that support MFC (Microsoft Foundation Classes). In this example we will select "A Simple Application" as shown in Figure 16.3.

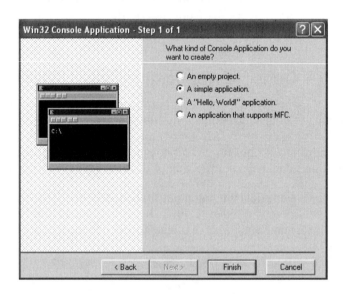

Figure 16.3 *Selecting "A Simple Application" Option*

Clicking the button labeled "Finish" produces a notification screen that informs the user that a skeleton project has been created, its name, and the presence of two precompiled headers named "Stdafx.h" and "StdAfx.cpp." Figure 16.4 is a screen capture of this notification.

When you press the OK button execution returns to the main Visual Studio screen. This screen is divided into three panes. The one on the right is the editor. The one at the bottom shows development operations and is called the Build pane. The pane at the left, called the Workspace pane shows the program components.

The Workspace pane now shows two buttons on its bottom margin. One, labeled Class View, is currently selected. This mode of the Workspace pane is used in object-oriented development. At this time you should click the button labeled File View and then the plus sign (+) control to the left of the project files entry.

This action will display the source, header, and resource files currently in the project. Also a ReamMe.txt file that contains information about the components of the console application that you have just created.

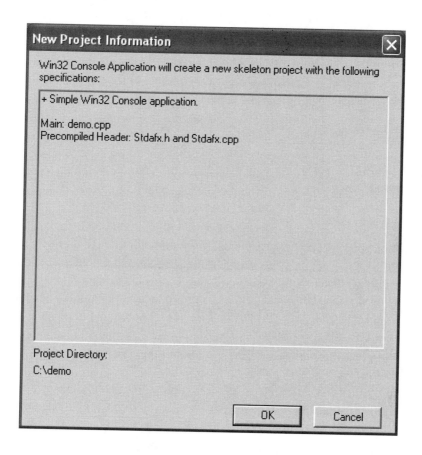

Figure 16.4 *Visual C++ New Project Information Screen*

If you now expand the Source Files entry by clicking on its plus sign button a list of two text files will be displayed. The first one, named demo.cpp (assuming that you named your program "demo") will be the source file created by Visual Studio. You can use this file as a start point for your program. The file named StdAfc.cpp contains a single statement to include stdafx.h with your source. You can view stdafx.h by opening the Header Files group and double clicking the filename. Figure 16.5, on the following page, shows the Visual Studio screen at this point.

You should now edit the source file to suit your own programming style and start coding the console application. The Build menu contains command to build the program (named Rebuild All) and to execute the code within the development environment. For example, the following code listing shows a modification of the simple application file into a Hello, World program.

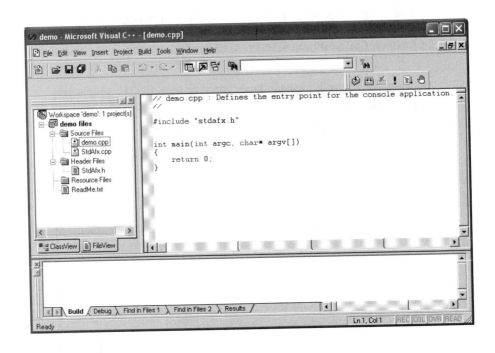

Figure 16.5 *Visual C++ Screen for a New Project*

```
// demo.cpp
// First demonstration program for a console application
// Coded by: Julio Sanchez
// Date: March 21, 2007 //
#include "stdafx.h"
#Include <iostream.h>

int main(){
     cout < "\nHello World\n\n";
     return 0;
}
```

Notice that we have eliminated the parameters passed to the main() function since our program does not use them. Also that we have added an include line for the iostream library. This will allow us to use the powerful input and output functions in this library (cin() and cout()). When this program executes, a Hello, World screen is displayed.

Chapter 17

Event-Driven Programming

Chapter Summary

This chapter deals with the principles of Windows programming. In it we explain the differences between a DOS and a Windows program and provide a model, called event-driven programming, that is suitable for Windows applications. We also describe the fundamental components of a Windows program, its file structure, and visual features.

17.0 Graphical Operating Systems

The operating systems used in the first generation of digital computers required that the user toggle a series of binary switches, each one with an attached light, in order to enter data into the machine. In these systems, the bootstrap sequence consisted of a predefined sequence of switch-toggling and button-pressing actions which the operator memorized and executed in a ritual-like fashion. To handle one of these machines you had to belong to a select class of binary-speaking, hexadecimal-minded experts who devoted most of their life to deciphering the arcane mysteries of hardware and software.

In the years that followed, several inventions and adaptations simplified computer input and output. IBM Corporation, that had developed punched cards for its business machines, adapted this technology to its first line of computers. Teletype machines (called TTYs) had been invented for use in the communications industry but were soon adapted for use as computer devices. With the TTY, the typewriter keyboard found its way into computing. Next came the use of Cathode Ray Tubes, from commercial television, as a way of displaying text without having to print it on paper. An added bonus was that the CRT could also be used to display pictures.

Other devices followed suit. Some have become standard elements of the technology while others have disappeared. They include the lightpen, the touch screen, the graphic tablet, the joystick, and the mouse. The common element in all of these devices is that they allow the user to visually interact with the machine. The idea for an interactive input device came from the work of Allan Kay at the Xerox Palo Alto Research Center in the early 1970s. Dr. Kay was attempting to design a computer that could be used by preschool children, who were too young to read or to type commands in text form. A possible approach was to use small screen objects, called icons, that represented a familiar thing. A mechanical device (the mouse) allowed moving these graphics objects on the screen in order to interact with the system. Interactive graphics and the graphical user interface were the result.

Steve Wozniak, one of the founders of Apple computers, relates that he visited Xerox PARC and saw the mouse and the graphical user interface. He immediately concluded that the interactive way of controlling a computer was the way of the future. Back at Apple, Wozniak started the development of an operating system that supported graphical, mouse-controlled, icon-based, user interaction. After one or two unsuccessful tries, the Macintosh computer was released. Soon thereafter, Microsoft delivered a graphical operating system for the PC, called Windows.

17.1 Enter Windows

Many realized the advantages of a more reasonable and physical interaction with a computing machine. However, it took many years for the actual implementation of this idea in an effective operating system that would be preferred by the majority of users. Douglas Engelbart demonstrated a viable mouse interface in 1968, but it took fifteen years for the mouse to become a standard computer component. In order to implement a graphical user interface (GUI) it was necessary to have not only a pointing device that worked but also a graphics-capable video terminal. In addition, the software would have to provide a set of graphics services in a device independent manner.

In the PC world, the evolution of hardware and software components into a graphical operating system took approximately one decade. The first versions were rather crude and achieved little popular acceptance. It was not until Windows 95 that Microsoft's graphical operating system for the PC became the standard.

Originally, the development of a graphical operating system for the PC was a joint effort between Microsoft and IBM. But soon, the two companies had major strategic and tactical differences. IBM's efforts resulted in a product called Operating System/2, or OS/2. It seemed that Windows and OS/2 would share the PC operating system market. However, for reasons unrelated to its technical merits, OS/2 quickly lost ground to Windows. OS/2 Warp version 4.0 was on the market until 2005 but is no longer sold by IBM.

17.1.1 Text-Based and Graphical Programs

To a programmer, DOS and Windows are in different worlds. DOS programs have complete and unrestricted access to the machine's hardware. Once the code gains control it can do whatever it pleases. Its only limitations are the hardware capabilities and the programmer's skills. Although the designers of DOS-like operating systems often list rules that "well behaved programs" should follow, there is no way of enforcing these rules. Intentionally or by error a DOS program can raise havoc with the system by deleting files from storage and even attempting to physically destroy a hardware device. Each DOS application has total control over all system resources. It can allocate all memory to itself, set whatever video mode is convenient, control the printer and the communications lines, and manage the mouse and the keyboard. Resources need not be shared since the operating system is dormant while an application is in the foreground. In this environment, more than one application rarely execute simultaneously.

Windows applications must share resources between themselves and with the operating system. Memory, CPU, display hardware, communications lines and devices, mouse, keyboard, and disk storage are all shared. Each program operates in its private address space and has limited access to other memory areas. Code cannot access the hardware devices directly but must to do so through operating system services called the Application Program Interface (API). This mechanism ensures that all programs are well behaved.

Windows and DOS graphics programming are quite different. Because all resource are shared, Windows must control access to all devices and resources, including memory, the video system, communications devices, and input and output components. Figure 17.1 shows how DOS and Windows programs access system resources.

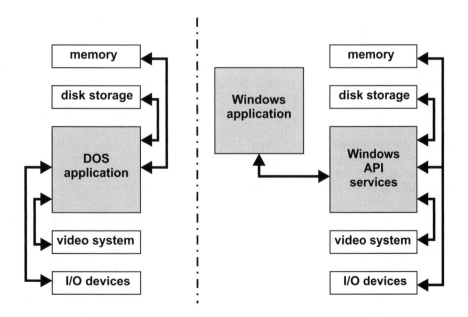

Figure 17.1 *DOS and Windows Applications Access to System Resources*

17.1.2 Graphics Services

Windows is a graphical operating system; therefore, a Windows application is a graphics program. Since Windows applications use a graphical user interface, and considering that an application's access to resources must be controlled, then Windows is forced to provide a host of graphics services to applications. In a protected mode environment, applications cannot access memory or devices directly. If an application attempts to do so, the CPU notifies the operating system (Windows in this case) which proceeds first to halt and eventually to destroy the offending program. The general rule under Windows is behave or you will be destroyed. The Windows programmer has no direct control over the machine resources but the operating system provides services which allow supervised access.

17.2 Programming Models

The nature of the operating system environment determines the programming models. In DOS, a program is "the god of the machine." The running code has unrestricted access and control. It executes on its own strengths, implements its own functionality, and calls the operating system only to request a specific service. It shares the machine with no other program. Therefore, the programming model for a DOS program is a set of sequential instructions and program constructs. Control does not return to the operating system until the application terminates. This form of interaction between the application and the operating system gives rise to the sequential programming model.

The conceptual model for a Windows program is quite different. In this case, the code has no direct access to devices and resources and must share the machine with other applications and with the operating system itself. The current versions of Windows implement preemptive multitasking. This means that the operating system can switch the foreground (CPU access) from one application to another. If an application misbehaves, Windows can simply turn it off. Therefore, it is the operating system that is "the god of the machine," not the running program.

17.2.1 Event-Driven Programs

A new programming model is necessary to accommodate this mode of interaction between an application and the system code. The model is sometimes called event-driven programming. In event-driven programming, synchronization between the operating system and the application is in the form of program events. For example, when the user changes the size of an application's window (a user event), the operating system takes the appropriate action and then notifies the application code (a system event). The application, in turn, may decide to take its own action to update its display area, or it may decide not to take any action at all. In either case, it returns control to the operating system. The event-driven model, although simple and effective, can appear odd to a programmer used to working in DOS. In the case of a Windows program, the application's code is no longer a sequential set of instructions, but a series of blocks of code which execute when the corresponding message is received.

The event-driven model is implemented by means of messages passed among the participants. In this example, we can say that in response to the user's window resizing event, the mouse device sent a message to the operating system, which in turn, sent a message to the application. However, these messages are not like the interrupts so often used in DOS programming. DOS interrupts can run concurrently; that is, a new interrupt can take place before the previous one has concluded. On the other hand, Windows messages are queued and must wait in line until their turn comes for processing.

DOS programmers have to change their mindset when working in Windows. Figure 17.2 shows two models of programming: the sequential model in a DOS program and the event-driven model in a Windows programs. In DOS, the application receives control at load time and retains it during its entire life span. In a Windows program, a user input event sends a message to the operating system. The operating system takes the appropriate action and sends a message to the corresponding application. The application looks at the message and decides if an action is required or if the message should be ignored. In either case, it returns control to Windows. At termination time, the application sends a message to Windows requesting to be ended.

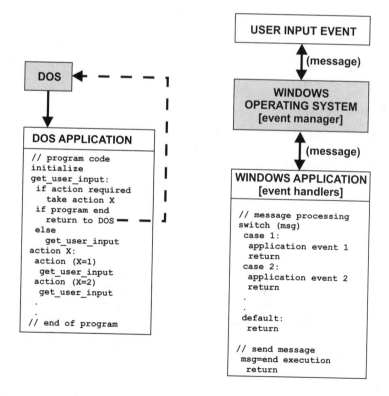

Figure 17.2 *Sequential and Event-Driven Programming Models*

The Event Manager

The principal difference between sequential and event-driven programs is that one is active and the other one is passive. Event-driven code need not provide loops to monitor user input since this function is performed by Windows. The application stays dormant until it is notified by the operating system that an event has taken place. It is the operating system code which must monitor the hardware devices in order to detect actions which must be handled by the code. This function is called event management, and Windows is the event manager.

It is the event manager's responsibility to detect events and to make sure that applications that must react to the events are promptly notified. This is usually called an event dispatch operation. The event manager monitors hardware devices that can generate events, such as the keyboard and the mouse. The software routine that checks for action on the event generating devices is sometimes called the event loop.

The Event Handler

In the event-driven model, the application is the event handler. Its function is to wait for an event to occur and then perform the corresponding action. The event handler does not monitor event-generating devices directly, and often does not provide the first level response to an event. It simply performs the action or actions that are within its responsibilities and returns control to the event manager. In Windows all of the complications of the device interface are performed at the system level. In this sense, event-driven programs are easier to code since the "dirty work" is done by the operating system.

17.2.2 Event Types

Events can be grouped into several types, although the classification is not very rigorous. Often one event can be placed into one or another group according to our own definitions. However, three general groups can be delimited without much overlap:

- System events
- Control events
- Program events

One event often triggers another one of the same or a different type. For example, a system-generated event can generate a control event. Or a program event can be the cause of a system event, which, in turn, generates another system event, and so forth. This interaction between events creates an event chain, and it is this interconnection between related events that sometimes makes it difficult to pinpoint a particular event type.

System Events

In this group are those events that originate in the operating system software. For example, if the user presses the left mouse button while the cursor is inside the client area, Windows sends a WM_LBOTTONDOWN message to the application. This message indicates to the application code that a certain action has taken place (or is about to take place) to which, perhaps, the program should respond. Many other system-generated events are possible, including a user typing on the keyboard, selecting a menu item, moving the mouse cursor, dragging an object with the mouse, resizing a window, or operating a scroll bar control.

Control Events

Control events relate to graphics control objects that are so abundant in the Windows environment. Among them are buttons, list boxes, combo boxes, scroll bars, up-down controls, and many others. A control event takes place when user interaction with a control requires a reaction from the operating system or the application. In this case, the control sent a message to the operating system. If necessary, the operating system can then generate an event to notify the application of the user's action.

Much of the programming required to implement a user interface consists of re-sponding to control events. The reason is that Windows programs have available many standard controls. Programmers usually find it more convenient to use one of these pre-canned components than to create a customized one. Windows controls are discussed in greater detail later in this chapter.

Program Events

In Figure 17.1 we can see that the termination of a Windows application requires that the program code send the corresponding message to the operating system. In this case, the application originates an event in the form of a request to terminate execution. This type of event is called a programmer-created or program event.

17.2.3 Event Modeling

A program event is generated when the application code sends a message to the oper-ating system requiring a service or action. However, program events are rarely at the origin of the event chain. For example, the program event that requests program termi-nation often originates in a control event where the user indicated a desire to end the program. Figure 17.3, on the following page, shows the event chain that takes place when a user clicks on the close button on a Window's title bar.

In Figure 17.3 the message sent by the operating system to the application is la-beled WM_DESTROY. This is one of many standard messages that are used by Win-dows. The WM_DESTROY message indicates to an application that the operating system has already destroyed the window. Therefore, the WM_DESTROY message is received after the fact. Another Windows message named WM_CLOSE is sent to an application to notify it that it is time to clean up and terminate. WM_CLOSE is a sort of warning. This means that Figure 17.3 is actually a rough sketch of a process which actually entails several other steps.

The Windows messaging system that results from event generation and handling mechanisms can get quite complicated. The programmer or program designer often needs to create a model of an event chain. Several modeling tools have been devel-oped for graphically depicting processes and data. Data flow and entity-relationship diagrams are well-known tools used by program designers. However, these conventional tools are not well suited to modeling event-driven systems. Ward and Mellor and also Hatley and Pribhai have introduced variations of the data flow diagrams which are better suited for representing systems in which events take place in real time. In Figure 17.3 we have used symbols from conventional data flow diagrams and from the Ward and Mellor extensions in order to model an event chain. More elaborate and detailed representations could include symbols for data stores, control stores, and for processes.

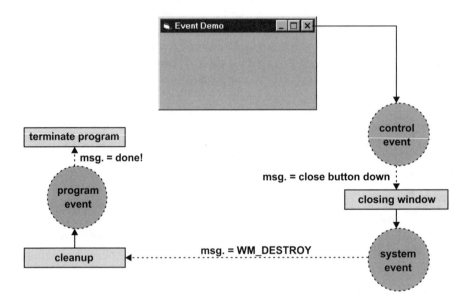

Figure 17.3 *An Event Chain*

17.3 File Structure of a Windows Program

In addition to adopting an event-driven programming model in replacement of the conventional sequential model, the Windows programmer also needs to become familiar with the file structure of a Windows application. Many DOS programs have simple file structures. In many cases, the only files external to the source are the various include files listed in the program header. Internally, the DOS compiler sets the necessary linker options, references the corresponding libraries, and creates the executable almost transparently to the programmer. Windows compilers are also designed to automatically manage much of the complications of program generation; however, Windows programs usually require more files than their DOS counterparts. Several of these files have no equivalent in DOS systems.

17.3.1 Source Files

The source files in Windows are of the same type as in DOS programming. In either case, they contain the language statements and constructs of a C or C++ program. Specific filename extensions are associated with different source files type, these are: .C, .CPP, and .H. The .C extension depicts a program in C. The presence of this extension usually indicates that the program is coded in straight C; therefore, it does not include any keywords and constructs that are part of C++. The .CPP extension indicates a source file that can contain C++-specific keywords and constructs. The .H extension corresponds to a header file, usually referenced at the beginning of a program by means of the #include statement.

17.3.2 Library Files

C and C++ are small languages: they do not provide functions for performing input and output operations. For this reason, few useful programs in C or C++ contain no other functions than those explicitly coded in the source. Most DOS programs, and virtually all Windows programs, require the support of library functions.

Library files differ from source files. All the program files containing source code are merged at compile time. Library functions are incorporated at link time, although the internal mechanisms of C and C++ development systems sometimes prevent us from noticing the difference. What often happens is that an #include statement in a source file references a header file, which in turn, contains references to other sources or to object code in a library. These references are usually made by means of the extern declarator which serves to indicate that a variable or function has external linkage. Figure 17.4 shows the different source, object, and library files that can be part of a C or C++ program.

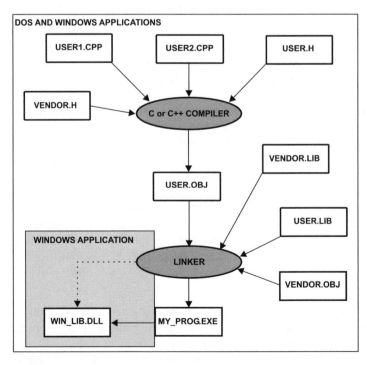

Figure 17.4 *File Structure of a Windows Program*

17.3.3 Resource Files

Windows programs have access to data items stored within the executable file, or in a library file, called resources. Although resources are data, and though they are often associated with the program's .EXE file, they are unique. In the first place, resources cannot be modified at run time. They are actually read-only files and are not directly accessible to program code. Furthermore, resources do not reside in the program's data area. At load time program resources are usually left on a disk file until they are needed. Typical resources are icons, cursors, dialog boxes, menus, bitmaps, and character strings.The current generation of software tools for developing Windows applications, such as Microsoft Visual C++ and Borland C++, contain special tools for working with resources. In the Microsoft system there are several specialized resource editors.

Resources, which have no equivalent in DOS programming, provide several advantages to Windows applications:

• User interface components are isolated from the rest of the code.

• Applications can be modified without changing the resources.

• Data can be shared among applications by reusing the same resources.

• Graphical elements in the user interface can be created easily.

• Graphical elements in the user interface can be modified without changing the application code.

Several special file types are used by Windows to manage resources. The resource script file (extension .RC) is an ASCII text file from which the binary resources are created. Resource script files can be created manually by the programmer or they can be automatically generated by the resource facilities in the development environment. The Resource Compiler program, which is part of the development system, produces binary resource files (extension .RES) from the ASCII script files (extension .RC). In Microsoft Visual C++, the command line version of the Resource Compiler is named RC.EXE; in Borland C++ it is named BRCC.EXE or BRCC32.EXE.

Before Windows 95, the Resource Compiler also had the task of adding the binary resources to the executable file created by the linker. The Microsoft linker for Windows 95 and later versions can directly manipulate files in both .OBJ and .RES formats.

17.3.4 Make Files

During the evolution of software development systems, the component programs became increasingly more powerful. Consequently, the number switches, options, and modes that could be selected in compilers, linkers and other tools also grew rapidly. It became progressively more difficult for the programmer to remember and enter all of these options every time that a project had to be compiled or updated. Various forms of so-called make programs and utilities were developed to remedy this situation. By means of a make program the programmer would record all the switches and options in a text file and the make utility would automatically apply them in a particular project.

In DOS programming, the simplest implementation of a make utility is a batch file. An additional refinement is a program that reads the switches and options selected in the development environment and applies them to the assembler, compiler, linker, or other software tools. In most environments, the use of the make program is optional; however, in the Windows platform the options and controls are so intricate that most programmers would not consider entering them manually.

All major Windows C and C++ development systems for the PC have a make utility. In Microsoft's Visual C++, it is named NMAKE.EXE. In the Borland system, it is MAKE.EXE. The text file for either system has the extension .MAK. Make files consist of instructions in ASCII text format. They can be created manually or generated automatically by the development environment. The make programs can be entered from the command line or they can be activated from inside the development environment, usually by selecting a specific menu option.

17.3.5 Object Files

The compiler program generates a file called an object file, which has the extension .OBJ. The Microsoft Visual C++ development system uses the Common Object File Format (COFF) specification which originated in the UNIX operating system. The Microsoft version of COFF adds additional header data in order to make the format compatible with DOS and with 16-bit versions of Windows.

The compilers in Windows development systems have many different switches, options, and modes of operation selectable by the user. In Visual C++, the compiler options are selected through the Settings command of the Build menu. The Project Settings dialog box contains a C/C++ tab which activates the compiler switches and controls, as shown in Figure 17.5, on the following page.

Once the C/C++ tab is active in the Project Settings, a drop-down list box labeled Category is displayed. When the Category box is expanded (see Figure 17.5), eight options appear. Selecting any one of these options activates a dialog box with a new set of selectable choices. The Warning Level and Optimizations drop down list boxes are also shown in Figure 17.5. The compiler warning level determines the severity of warnings for which the compiler generates a message. Optimizations consist of four predefined modes: default, disable, maximize speed, and minimize size. In addition, an entry labeled Customize (not shown in Figure 17.5) is also in the list box. Selecting this option activates another list box of customized optimizations when the Optimizations options in selected in the Category list box.

Mastering all the compiler options and switches requires considerable experience with the development environment. Programmers often use the default settings and make modifications to these settings when necessary.

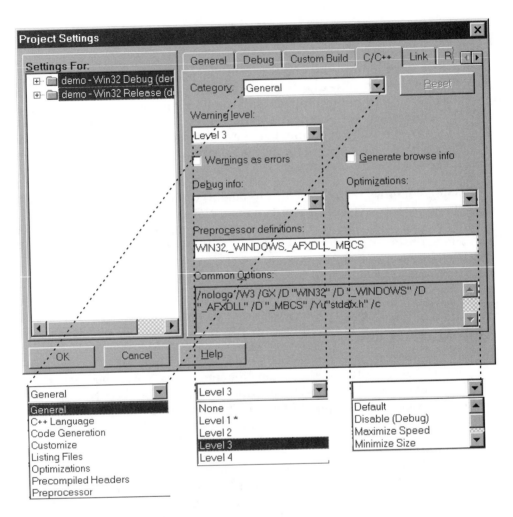

Figure 17.5 *Visual C++ the Compiler Options*

17.3.6 Executable Files

In both DOS and Windows, the executable file is generated by a linker program. The link command can be entered manually from the command line, through a batch file or make program, or automatically by the development environment. The link operation takes an object file generated by the compiler or assembler (extension .OBJ), the library files generated by a library manager (extension .LIB), files in Common Object File Format (extension .COFF), and the binary resource files created by the resource compiler (extension .RES), and creates an executable file with the extension .EXE. In addition, the Microsoft linker automatically converts 32-bit Object Module Format (.OMF) files into COFF. The output of the linker can also be a dynamic link library (extension .DLL).

The linker is the tool that generates the executable file; however, the Windows development environments can activate the linker functions indirectly. In

Microsoft Visual C++, the CL program can be used to compile source files, or to compile and then link object files into executables. Make files can also reference the linker operation. As in the case of the compiler, there are considerable numbers of switches and options that can be selected at link time. In Visual C++, the linker options are activated through the Settings command of the Build menu. The Project Settings dialog box contains a Link tab, as shown in Figure 17.6.

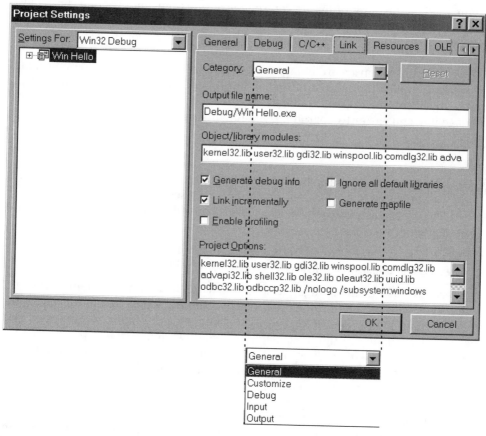

Figure 17.6 *Visual C++ Linker Options*

When the Link tab is active in the Project Settings, a drop-down list box labeled Category is displayed. Five options appear when the Category box is expanded (see Figure 17.6). Selecting any one of these options activates a dialog box with a new set of selectable choices.

Learning all the linker switches and options requires a mastery of the development system and as well as an intimate knowledge of the Windows operating system. Here again, programmers often change the default linker options only when necessary, but this attitude often results in sacrifices in functionality and performance.

17.3.7 Dynamic Linking

One of the unique characteristics of Windows is dynamic linking. In DOS the linker operates statically: it takes one or more object and library files and merges them into an executable which physically includes all the code in both the sources and the library modules. Windows programs use conventional libraries and source files at link time, but also a special type of file called a dynamic-link library or DLL. In dynamic linking, the library files are referenced at link time, but the code is not physically incorporated into the executable. When the program runs, the required run-time libraries are loaded into memory and the references are resolved.

Dynamic link libraries are binary files that have functions which can be shared by several applications. In addition to sharing code between applications, DLLs can be used to break up an application into separate components that are more manageable and easier to upgrade. Dynamic linking has several advantages:

- Several processes can share the same code thus saving memory space and reducing access time.

- DLLs can be modified and upgraded independently of the applications that use them. Programs that use static libraries must be recompiled when the libraries are modified. This mechanism can be effectively used in providing after-sale support for a program.

- DLLs can be accessed from different programming languages or environments as long as the applications follows the calling conventions.

There are also disadvantages associated with DLLs. One of them is that the code is not self-contained since it depends on the DLLs files being present on the target system. If a process that uses load-time dynamic linking references a DLL that is not available on the host machine, Windows immediately terminates the program. In the case of applications that depend on run-time dynamic linking, the program is not terminated if the DLL is not loaded, but the functionality associated with the missing DLL are not available to the code.

Dynamic link libraries are stored in different formats and associated with various filename extensions. The standard extension is .DLL but files with the extension .EXE, .DRV, .FON, and others can also perform as dynamic-link libraries. The system-level DLL files, such as KERNEL32.DLL, USER32.DLL, and GDI32.DLL, are found in the WINxx\SYSTEM directory. Files with the extension .DLL are loaded automatically by Windows. Others must be explicitly loaded by the program module using the LoadLibary() or LoadLibaryEx() API functions.

In addition to DLLs, two other types of libraries and their associated files are used in Windows programming:

- Object libraries (extension .LIB) contain object code that is added to the application at link time. Object libraries are used in static linking. The standard C libraries named LIBC.LIB and LIBCMT.LIB are in this group.

- Import libraries (extension .LIB) are a special form of an object library which contains no code. Import libraries provide information to the linker regarding dynamic-link libraries. Access to the dynamic link library GDI32.DLL is by means of the import library named GDI32.LIB.

Chapter 18

The Window Program Components

Chapter Summary

Windows applications have unique characteristics. In this chapter we discuss some of the more original features: Windows naming conventions, numeric constants, and Windows handles. Windows programs are also visually different since a typical Windows application has a graphics window and interactive controls. The chapter introduces the use of these graphical components as well as elements of programming style, such as program comments and assertions notation. Programming templates are described as a means of simplifying the coding effort.

18.0 "Hello, World"

Experienced and talented DOS programmers are often intimidated by their first encounter with Windows programs. Windows programming has been described as weird, convoluted, and awkward. In reality Windows programming is not more difficult than DOS programming, although Windows programs, at first, appear more complicated.

In the first place, Windows text-only programs can be coded using the Console Application facilities described in Chapter 16. Regarding programs that require graphics or a Graphical User Interface, then Windows programs are definitively easier to design and code than their DOS counterparts. Because Windows is a graphics environment it contains a multitude of services, canned routines, and coding aids that are not available in DOS. Anyone who has ever implemented a graphical user interface for a DOS application will greatly appreciate the graphics facilities in Windows.

Kernighan and Ritchie, in their book *The C Programming Language*, (published by Prentice-Hall in 1978) listed the code for a short program that displays the message "hello, world!" on the screen. This program has since been known as the Hello World program, and many other authors of programming books have followed suit. The C++ version of the Hello World program is as follows:

```
#include <iostream.h>
void main() {
  cout < "\nHello, World!";
}
```

In a text-based operating system, such as DOS or UNIX, the Hello World program is quite simple. The message is displayed at the current position of the text cursor, with whatever attributes are active. It assumes that the video system is a "glass teletype" that scrolls automatically when the text reaches the end of the screen. In this text-based paradigm there are no provisions for graphical user interface or for multitasking.

A Windows program that performs these same functions must deal with many other levels of complexity. In the first place, it must coexist and share resources with the operating system and with other applications. In Chapter 17 we discussed the model of an event-driven program that is necessary in this case. In addition, a Windows application is a graphics-based not a text-based program. Finally, the program executes as a window which can be overlapped, overlaid, minimized, maximized, to which keyboard and mouse input can be given or withdrawn, and many other functions. As can be expected, the code is more lengthy, although it is relatively simple considering the functionality achieved. The following is a listing for a Windows version of the Hello World program:

```
/*************************************************************
    WIN_HELLO.CPP   Displays the message "Hello World from
    Windows!" in the client area.
*************************************************************/
#include <windows.h>
LRESULT CALLBACK WndProc (HWND, UINT, WPARAM, LPARAM) ;
// WinMain()
int WINAPI WinMain (HINSTANCE hInstance, HINSTANCE
                    hPrevInstance, PSTR szCmdLine,
                    int iCmdShow)
{
  static char szAppName[] = "WinHello" ;
  HWND        hwnd ;
  MSG         msg ;
  // Defining a structure of type WNDCLASSEX
  WNDCLASSEX wndclass ;
  wndclass.cbSize        = sizeof (wndclass) ;
  wndclass.style         = CS_HREDRAW | CS_VREDRAW ;
  wndclass.lpfnWndProc   = WndProc ;
  wndclass.cbClsExtra    = 0 ;
  wndclass.cbWndExtra    = 0 ;
  wndclass.hInstance     = hInstance ;
  wndclass.hIcon         = LoadIcon (NULL, IDI_APPLICATION) ;
  wndclass.hCursor       = LoadCursor (NULL, IDC_ARROW) ;
  wndclass.hbrBackground = (HBRUSH) GetStockObject
                             (WHITE_BRUSH) ;
  wndclass.lpszMenuName  = NULL ;
  wndclass.lpszClassName = szAppName ;
  wndclass.hIconSm       = LoadIcon (NULL, IDI_APPLICATION) ;
  // Registering the structure wmdclass
  RegisterClassEx (&wndclass) ;
  // CreateWindow()
  hwnd = CreateWindow (szAppName,
```

```
        "WIN_HELLO Program",      // window caption
        WS_OVERLAPPEDWINDOW,      // window style
        CW_USEDEFAULT,            // initial x position
        CW_USEDEFAULT,            // initial y position
        500,                      // initial x size
        300,                      // initial y size
        NULL,                     // parent window handle
        NULL,                     // window menu handle
        hInstance,                // program instance handle
        NULL) ;                      // creation parameters
  ShowWindow (hwnd, iCmdShow) ;
  UpdateWindow (hwnd) ;
  // Message loop
     while (GetMessage (&msg, NULL, 0, 0))
        {
        TranslateMessage (&msg) ;
        DispatchMessage (&msg) ;
        }
     return msg.wParam ;
}
// Window procedure
LRESULT CALLBACK WndProc (HWND hwnd, UINT iMsg, WPARAM wParam,
                     LPARAM lParam)
{
    HDC         hdc ;
    PAINTSTRUCT ps ;
    RECT        rect ;
    switch (iMsg)
        {
        case WM_PAINT :
            hdc = BeginPaint (hwnd, &ps) ;
            GetClientRect (hwnd, &rect) ;
            DrawText (hdc, "Hello World from Windows!",
                    1, &rect, DT_SINGLELINE | DT_CENTER |
                    DT_VCENTER) ;
            EndPaint (hwnd, &ps) ;
            return 0 ;
        case WM_DESTROY :
            PostQuitMessage (0) ;
            return 0 ;
        }
    return DefWindowProc (hwnd, iMsg, wParam, lParam) ;
}
```

Figure 18.1, on the following page, is a screen capture of the WIN_HELLO program.

18.1 Naming Conventions

A programmer used to working in a non-Windows environment will immediately notice that a Windows program has certain unfamiliar elements. One of them is the unusual names assigned to variables. For example, in the WIN_HELLO program, we find the following line:

```
    static char szAppName[] = "WinHello" ;
```

If you are a C or C++ programmer the statement itself is quite familiar: it is the declaration and initialization of a static array of type char. What could strike you as unusual is the array name szAppName. The identifier "szAppName" appears strange because it follows the conventions recommended by Microsoft for variable names. Reputedly, this naming style originated with a Microsoft programmer of Hungarian descent named Charles Simonyi, and is, in his honor, called Hungarian notation.

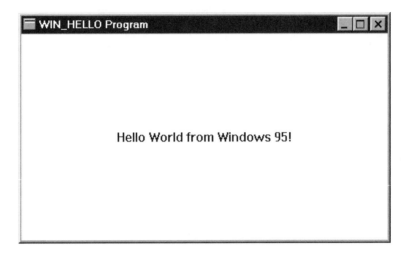

Figure 18.1 *The WIN_HELLO Program Screen*

In Hungarian notation the variable name should provide a way for identifying the data type. In szAppName the prefix sz stands for string terminated in zero. The word or words that follow the prefix describe the variable contents, in this case the application name. Table 18.1 lists the standard variable name prefixes recommended by Microsoft.

The merits of Hungarian notation have been passionately debated. Those who support it claim that it helps avoid data type mismatch errors. The detractors mention that linking the name and the data type forces the programmer to alter the name whenever the data type is changed, which can lead to unnecessary program editing. Also that Microsoft itself does not follow Hungarian notation consistently, there being many cases in which this style leads to confusion.

Perhaps the most noted source of confusion caused by Hungarian notation is the wParam passed to the Windows procedure (see WinProc in the WIN_HELLO program listing). In this case, wParam indicates a word according to the naming convention, but when the Win32 API was developed, wParam was changed to a 32-bit unsigned integer. However, for consistency with older code, the Windows documentation still refers to it as wParam.

In principle, we agree that Hungarian notation creates more problems that it solves. However, deviating from it only makes things worse since Microsoft has officially adopted it in Windows and in the Visual C++ documentation. In this book, we use Hungarian notation in the sample programs and the code listing, specially when referring to Windows data types and structures. In our own variables we use it rather loosely, and only for the sake of simplicity and uniformity.

Table 18.1

Standard Prefixes Recommended by Microsoft

PREFIX	DATA TYPE	INTERPRETATION
b/f	BOOL	zero = false, non-zero = true (f stands for flag)
by	BYTE	8-bit unsigned integer
ch	CHAR	single ASCII character
cx/cy	INT	count of x values and count of y values
d	DOUBLE	IEEE double precision real
dw	DWORD	32-bit integer
h	HANDLE	32-bit unsigned integer handle
hwnd	HWND	handle to a Window
i	INT	16-bit integer value
l	LONG	32-bit unsigned integer value
Lp/np	FAR*	32-bit far pointer
n	SHORT	16-bit unsigned integer value
p	pointer	any pointer, usually far
pt	POINT	32-bit coordinates, labeled x and y, packed in a structure
rgb	RGB	red, green, and blue values packed in 32-bits as follows: 00BB GGRR
x/y	INT	x-coordinate and y-coordinate
s	array of char	string
sz	array of char	string terminated in zero
w	WORD	16-bit unsigned integer

Regarding function names, there is less reason for debate. Microsoft suggests that programmers use a verb-noun model that describes what the function does and what it operates on: for example, the Windows API functions BeginPaint(), GetClientRect(), DrawText(), and EndPaint() referenced in the WIN_HELLO program listed earlier in this chapter. The function name begins with a capital letter and so does every other word in the function name. The underscore is not used as a separator. Since this style is familiar to most C and C++ programmers there is no reason to adopt a different one for our own functions.

18.2 Constants and Handles

The Windows header files define many numeric constants which can be a source of bewilderment to the uninitiated. For example, in the WIN_HELLO program listed previously, we find the upper-case identifiers CS_HREDRAW, CS_VREDRAW, WS_OVERLAPPEDWINDOW, and CW_USEDEFAULT, among others. These, and many other identifiers, are used as placeholders for numeric constants. The rationale is that symbolic names are more significant than numeric values and that they can be changed transparently to the code and the documentation. Table 18.2 lists the meaning of some common prefixes used for numeric constants. Many others are also used by Windows functions.

Table 18.2

Common Windows Prefixes for Numeric Constants

PREFIX	TYPE
BS	button style
CBS	combo-box style
CS	class style
CW	create window related
DT	draw text related
DS	dialog box style
ES	edit class style
IDI	ID number of an icon
IDC	ID number of a cursor
LBS	list-box style
SBS	scroll-bar style
SS	static class style
WS	window style
WM	window message

18.2.1 Windows Handles

In the context of Windows programming, a handle is a token that represents an item, object, or resource. For example, the following line in the WIN_HELLO program

```
hwnd = CreateWindw (szAppName, ...
```

is a call to the CreateWindows() function which returns a value assigned to the variable hwnd. In turn, hwnd is defined as a variable of the type HWND earlier in the code listing, as follows:

```
HWND        hwnd ;
```

In this case, the value returned by the CreateWindow() call is a handle to a window of the type HWND as defined in the windows.h header file. After a handle has been obtained it can be used to reference the object or data item that it represents. For example, later in the listing of the WIN_HELLO program we see that the handle to the program's window is used to obtain system information. The call

```
GetClientRect (hwnd, &rect) ;
```

passes the handle to the Windows API function in order to retrieve the coordinates of the windows client area, which are stored by the function in the structure variable named rect. In addition, Windows uses this same handle to identify a particular window to your program. If your application has several windows open simultaneously, it can tell to which of them the message refers, by examining the handle.

DOS programmers first learn about handles in relation to disk files. In DOS, there is little use for handles outside of disk files and standard devices. Windows handles, on the other hand, refer to many types of objects and resources in the environment, including handles to windows, to strings, to icons, to menus, to cursors, to instances of a program, to graphical objects (including pens, brushes, fonts, and bitmaps), to dynamically allocated memory, to regions, and to color palettes. Table 18.3 lists some of the more frequently used handles in Windows programming.

Table 18.3

Common Windows Handle Types

HANDLE IDENTIFIER	REFERS TO
HACCEL	an accelerator
HBITMAP	a bitmap
HBRUSH	a brush
HCURSOR	a cursor
HDC	a device context
HFONT	a font
HICON	an icon
HINSTANCE	a program instance
HKEY	a registry key
HMENU	a menu
HPALETTE	a GDI palette
HPEN	a GDI pen
HRGN	a GDI region
HWND	a Window

18.3 Visual Elements

Visually, a Windows application has a characteristic appearance that makes it different from a typical text-based program, and even from other programs in graphics, multitasking environments. Most of the elements that make a Windows program different are the components of the main window and the input/output controls used in implementing the graphical user interface.

18.3.1 The Main Window

Although it is possible for a Windows program to execute without output to the display device, most Windows applications will have a main window. The main window is a program's principal means of input and output and its only access to the screen. Figure 18.2, on the following page, shows the basic components of a program's main window.

Some of the elements of the window in Figure 18.2 are present in all program windows, others must be present but can be configured in different ways, and a third group is optional. The following are the fundamental building blocks of a program window:

- The main window display has a title bar, although the caption can be left blank.

- The control buttons to the right of the title bar are used to minimize, maximize, and close the program window. The programmer can select which of these control buttons are displayed.

- The icon on the left side of the title bar activates the system menu. Programs can use the Windows default icons or one of their own.

- The menu bar is optional. A typical menu bar contains one or more drop-down menus. Each drop-down menu consists of commands which are activated by a mouse click or by using the Alt key and the underlined letter code. Menu commands that expand into submenus are usually indicated by trailing ellipses.

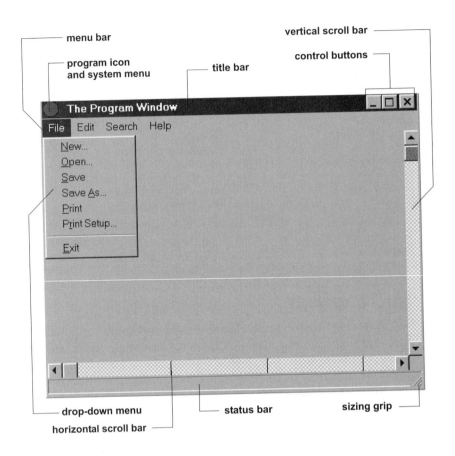

Figure 18.2 *Elements of the Program Window*

- The program main window, as well as many input/output controls, can have vertical or horizontal scroll bars. Windows notifies the application of user's action on the scroll bars, but the application must provide the required processing.

- Windows programs can have a status bar at the bottom of the screen which can contain a single text display area (as the one in Figure 18.2) or be divided into multiple areas. The status bar, as its name indicates, is used to display messages that inform the user about the program's status. A status bar includes sizing grips on the right-hand side which are used to change the window's size.

- The screen zone assigned to each program window is called the client area. The dimensions and graphics attributes of the client area can be obtained from the operating system using the window's handle.

18.3.2 Controls

The window in Figure 18.2 contains several controls: the buttons on the title bar, the menu items and command, the sizing grip, the scroll bars, and the button that activates the system menu. Buttons, scroll bars, menu commands, and sizing grips are just a few of the many control components that are available in Windows. These graphical components that are used in implementing input/output operations are generically called controls. Some controls have been around since the original Windows operating system; others were added with the various versions and revisions; still others are furnished as add-on or after-market products.

Controls are classified in two groups: predefined controls and common controls. The differences between them are mostly related to the message type used to communicate with the application and to historical considerations. Programmers can create their own customized controls and provide processing routines to operate them. In this section we refer to standard controls that are part of Windows; these include: buttons, check boxes, edit boxes, list boxes, combo boxes, scroll bars, dialog boxes, toolbars, ToolTips, status bars, tree views and tree lists, progress bars, up-down controls, trackbars, and many others. Figure 18.3 shows some of the controls frequently used by Windows applications.

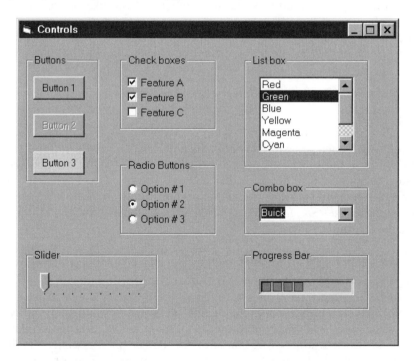

Figure 18.3 *Windows Program Controls*

18.3.3 Other Visual Components

In addition to the input/output controls such as those shown in Figure 18.3, Windows applications have available a rich selection of additional program elements. Every Windows user is familiar with dialog boxes, list and tree views of files, tab controls, toolbars and ToolTips, and property sheets. Wizards are used to assist the user in executing complex tasks. Much of Windows programming is learning how to implement these canned program components in your own code and adapt them to perform the tasks that your program requires.

18.4 Programming Style

Books have been written on styles of coding, most of them not very useful to the working programmer. Furthermore, to discuss style may seem like a contradiction in itself. If style is the distinctive originality that identifies an individual, then we should let it be the result of each one's personality and not try to influence it by our opinions and recommendations. In part this is true, however, in the context of Windows programming a bad coding style easily leads to programs that are undecipherable and to the proverbial "spaghetti code."

As we mentioned earlier in this chapter, Windows programming is not difficult or profound, but many Windows programs end up being extremely complicated. Well-organized and commented sources let you quickly find the fundamental routines and grasp the basics of the processing operations. A good style promotes the organization of the code into execution blocks and the use of comments to describe the mechanism by which the program accomplishes its basic purposes. On the other hand, program comments should not be used to explain the fundamentals of a programming language nor trivial or obvious manipulations. Sometimes we see a line of code that contains several flaws, for example:

```
int var1, var2 = 22;    // Creating and initializing
                        // Variables
```

In this line, we can spot several style problems. In the first place, modern programming techniques advise that each variable be declared separately. Multiple declarations and initializations are possible in C and C++, but the code is clearer if you devote a single text line to each variable. Another flaw in the sample line is that the variable names are not indicative of their function in the program. Finally, the comment is useless since any C programmer knows that the statement is a variable declaration.

18.4.1 Commented Headers

One way to make sure that the necessary information is included in the source is to use a pre-designed comment block. Software companies often furnish their programmers with standard headers that are to be attached to all source files. These headers serve to remind the programmer of all the items of information that should be included with each source file. The following can be used as a checklist for information that may be included in the program's commented header block:

- Developers copyright notice
- Program name
- Programmer or programmer's name and other personal information
- Name of the source file
- List of related sources, program modules, and support files, including libraries, DLLs, and required add-ons
- Description of the development system and resources
- Program hardware and software requirements

- Program update chronology listing dates and modifications

- Program test history and bug fixes

The following is the commented header used by the author's software company:

```
//*******************************************************
//*******************************************************
//                        PROGNAME.CPP
//        Copyright (c) 20?? by Skipanon Software Associates
//                      ALL RIGHTS RESERVED
//*******************************************************
//*******************************************************
// Date:                          Coded by:
// Filename:                      Module name:
//                                Source file:
// Description:
//
//*******************************************************
// Libraries and software support:
//
//*******************************************************
// Development environment:
//
//*******************************************************
// System requirements:
//
//*******************************************************
// Start date:
// Update history:
//            DATE              MODIFICATION
//
//
//*******************************************************
// Test history:
//   TEST PROTOCOL            DATE         TEST RESULTS
//
//*******************************************************
// Programmer comments:
//
//
//*******************************************************
//*******************************************************
```

18.4.2 Assertions Notation

During the past few years, there has been a movement in the programming community that favors more formal methods of program specifications. It is claimed that programs that are specified mathematically can be easily tested for consistency, thus avoiding many design problems and defects. Furthermore, that formal specifications will eventually lead to automated coding, that is, to machines that can write programs without the intervention of a human programmer.

The jury is still out regarding the practical viability of formal methods of program specification, but most programmers agree that it is a good idea to reduce the ambiguity and inconsistency in the coding. A semiformal method of specifications called assertions notation has been progressively gaining favor. It consists of a specific style of comments that are inserted into the code so as to define the state of computation. The following types of assertion are used as comment headers:

ASSERT

The ASSERT comment header describes the state of computation at a particular point in the code. The idea is to state what is certainly known about variables and other program objects so that this information can be used in designing and testing the code that follows.

Special symbols and terms are associated with the ASSERT header. The term Assigned is used to represent variables and constants initialized to a value. The symbol && is used to represent logical AND, while the symbol || represents logical OR regarding the terms of the assertion. The symbol == is used to document a specific value and the symbol ?? indicates that the value is unknown or undeterminable at the time. Finally, the symbol —>, or the word Implies, is used to represent logical implication. The programmer is also free to use other symbols and terms from mathematics, logic, or from the project's specific context, as long as they are clear and consistent.

INV

The INV comment header represents a loop invariant. It is used to describe a state of computation at a certain point within a loop. The intention of the loop invariant is to document the fundamental tasks to be performed by the loop. Since the invariant is inserted inside the loop body, its assertion must be true before the body executes, during every loop iteration, and immediately after the loop body has executed for the last time. Loop invariants are also used in code optimization techniques.

PRE and POST

The PRE and POST comment headers are associated with functions. They establish the conditions that are expected at the function's entry and exit points. The PRE: assertion describes the elements (usually variables and constants) that must be supplied by the caller. The POST: assertion describes the state of computation at the moment the function concludes its execution. Taken together, they represent the terms of a contract between caller and function. If the caller ensures that the PRE: assertion is true at call time, then the function guarantees that the POST: assertion is satisfied at return time.

FCTVAL

The FCTVAL assertion represents the value returned by a function. In C, this value is associated with the function name. Therefore, it should not be used to represent variables modified by the function.

18.4.3 Programming Templates

One objection often made to API-level programming for Windows is coding complications. In the program named WIN_HELLO, listed previously in this chapter, you saw that the code for a simple hello, world program in Windows can be quite elaborate.

One way in which to reduce the difficulties of Windows programming is by means of code templates that provide a standard framework of operations. A further simplification is by means of "wrappers" that allow accessing API services indirectly, thus avoiding some of the coding difficulties. Development environments such as the MFC, Visual Basic, Power Builder, and Delphi are based on these two simplifications, although implemented differently in each language.

Our approach precludes the use of "wrappers" for accessing API services; however, it is possible to simplify direct API-level programming by means of source code templates. A template is a program, or a part of a program, from which all the "specifics" have been removed. The template method is reminiscent of how a sculptor uses a wood-and-wire frame to build a statue by adding and manipulating the modeling clay. In the case of sculpturing, the frame has a body, a head, arms, and legs; however it has no sex, or age, or clothing attire. The same frame is used to build a statue of a naked cave man and that of a fully-dressed lady in the court of Louis XIV.

The programming version of the sculptor's wood-and-wire frame is a coding template. If we take the idea of a standard header block and add to it the generic portions of the WIN_HELLO program, we come up with the following source code template for a Windows program:

```
//**************************************************************
//**************************************************************
//  TEMPL01.CPP
//  Copyright (c) 20?? by
//  ALL RIGHTS RESERVED
//**************************************************************
//**************************************************************
// Date:                      Coded by:
// Filename:                  Module name:
//                            Source file:
// Program description:
//
//**************************************************************
// Libraries and software support:
//
//**************************************************************
// Development environment:
//
//**************************************************************
// System requirements:
//
//**************************************************************
// Start date:
// Update history:
//          DATE              MODIFICATION
//
//**************************************************************
// Test history:
//  TEST PROTOCOL       DATE        TEST RESULTS
//
//**************************************************************
// Programmer comments:
```

```
//
//
//****************************************************************
//****************************************************************
#include <windows.h>
//
// Predeclaration of the Windows Procedure
LRESULT CALLBACK WndProc (HWND, UINT, WPARAM, LPARAM) ;
//
//*****************************
//        WinMain
//*****************************
int WINAPI WinMain (HINSTANCE hInstance, HINSTANCE hPrevInstance,
                    PSTR szCmdLine, int iCmdShow)
{
     static char szAppName[] = "AppName" ; // Application name
     HWND         hwnd ;
     MSG          msg ;
     // Defining a structure of type WNDCLASSEX
     WNDCLASSEX  wndclass ;
     wndclass.cbSize        = sizeof (wndclass) ;
     wndclass.style         = CS_HREDRAW | CS_VREDRAW ;
     wndclass.lpfnWndProc   = WndProc ;
     wndclass.cbClsExtra    = 0 ;
     wndclass.cbWndExtra    = 0 ;
     wndclass.hInstance     = hInstance ;
     wndclass.hIcon         = LoadIcon (NULL,IDI_APPLICATION);
     wndclass.hCursor       = LoadCursor (NULL, IDC_ARROW) ;
     wndclass.hbrBackground = (HBRUSH) GetStockObject
                                  (WHITE_BRUSH) ;
     wndclass.lpszMenuName  = NULL ;
     wndclass.lpszClassName = szAppName ;
     wndclass.hIconSm       = LoadIcon (NULL,IDI_APPLICATION);

     // Registering the structure wmdclass
     RegisterClassEx (&wndclass) ;
     // CreateWindow()
     hwnd = CreateWindow (szAppName,
           "Window Caption",           // window caption
           WS_OVERLAPPEDWINDOW,        // window style
           CW_USEDEFAULT,              // initial x position
           CW_USEDEFAULT,              // initial y position
           CW_USEDEFAULT,              // initial x size
           CW_USEDEFAULT,              // initial y size
           NULL,                       // parent window handle
           NULL,                       // window menu handle
           hInstance,                  // program instance
                                       // handle
           NULL) ;                     // creation parameters
     ShowWindow (hwnd, iCmdShow) ;
     UpdateWindow (hwnd) ;
     // Message loop
     while (GetMessage (&msg, NULL, 0, 0))
           {
           TranslateMessage (&msg) ;
           DispatchMessage (&msg) ;
           }
     return msg.wParam ;
}
//*****************************
```

```
//    Windows Procedure
//***************************
LRESULT CALLBACK WndProc (HWND hwnd, UINT iMsg, WPARAM wParam, LPARAM
lParam)
{
    HDC         hdc ;
    PAINTSTRUCT ps ;
    RECT        rect ;
    switch (iMsg)
        {
    // Windows message processing
        case WM_PAINT :
            hdc = BeginPaint (hwnd, &ps) ;
            GetClientRect (hwnd, &rect) ;
        // Initial display operations here
            EndPaint (hwnd, &ps) ;
            return 0 ;
        // End of program execution
        case WM_DESTROY :
            PostQuitMessage (0) ;
            return 0 ;
        }
    return DefWindowProc (hwnd, iMsg, wParam, lParam) ;
}
```

You can use this template as a frame on which to build a Windows application. It is important to remember that a programming template is not a sample program, inasmuch as a sculptor's frame is not a statue. Some templates may compile correctly while others may not. Each template contains instructions regarding its use. Typically, the programmer removes these notes when they are no longer necessary.

SOFTWARE ON-LINE

The directory Chapter_18\TEMPLATES in the book's on-line software contains the electronic source for all the templates listed in the text.

Chapter 19

A First Windows Program

Chapter Summary

This chapter introduces API programming in Windows. It cannot be denied that Windows programming is complicated: no matter where we start, or how we approach it, the discussion soon gets into the details of many options, modes, controls, switches, and alternatives. There are well over one thousand API services, many of which take a dozen or more parameters. The event-driven mechanism that is at the heart of a Windows program often becomes complex: system and application code may have to exchange eight or ten messages to accomplish a simple task. In this chapter we cover the fundamental components of a Windows program and follow in the process of creating a Windows program from a template.

19.0 Preliminary Steps

Our approach to Windows programming is to avoid class libraries or other wrappers, such as The Microsoft Foundation Classes (MFC). At an initial level of Windows programming the use pre-canned interfaces may have some attraction, however, in high-performance graphics these packages are, at best, a nuisance and more often a major hindrance. On the other hand, we do take advantage of the editing and code generating facilities provided by Developer Studio and use the program-generating wizards, since there is no control or performance price to be paid in this case.

Before we can create a major graphics application we must be able to construct the Windows code framework that supports it. Fabricating a program requires not only knowledge of the programming language, but also skills in using the development environment. For example, to create an icon for your program's title bar you need to know about the API services that are used in defining and loading the icon, but you also need to have skills in using the icon editor that is part of Developer Studio. Even after the icon has been created and stored in a file, you need to follow a series of steps that make this resource available to the program.

19.1 The Program Project

We assume that you have already installed one of the supported software development products. The text is compatible with Microsoft Visual C++ Version 5.0 and later. We used Visual C++ Version 6.0 in creating the sample programs for this book. The following section describes the steps in creating a new project in Microsoft Developer Studio, inserting a source code template into the project, modifying and saving the template with a new name, and compiling the resulting file into a Windows executable.

19.1.1 Creating a Project

You start Developer Studio by double-clicking on the program icon on the desktop, or selecting it from the Microsoft Visual C++ program group. The initial screen varies with the program version, the Windows configuration, the options selected when Developer Studio was last executed, and the project under development. Version 5.0 introduced the notion of a project workspace, also called a workspace, as a container for several related projects. In version 5 the extension .mdp, used previously for project files, was changed to .dsw, which now refers to a workspace. The dialog boxes for creating workspaces, projects, and files were also changed. The workspace/project structure and the basic interface are also used in Visual C++ Version 6.0.

We start by creating a project from a template file. The walkthrough is intended to familiarize the reader with the Developer Studio environment. Later in this chapter you will learn about the different parts of a Windows program and develop a sample application. We call this first project Program Zero Demo, for the lack of a better name. The project files are found in the Program Zero project folder in the book's on-line software package.

A project is located in a workspace, which can include several projects. Project and workspace can be located in the same folder or subfolder or in different ones, and can have the same or different names. In the examples and demonstration programs used in this book we use the same folder for the project and the workspace. The result of this approach is that the workspace disappears as a separate entity, simplifying the creation process.

A new project is started by selecting the New command from the Developer Studio File menu. Once the New dialog box is displayed, click on the Project tab option and select a project type from the displayed list. In this case our project is Win32 Application. Make sure that the project location entry corresponds to the desired drive and folder. If not, click the button to the right of the location text box and select another one. Next, enter a project name in the corresponding text box at the upper right of the form. The name of the project is the same one used by Development Studio to create a project folder. In this example we create a project named Program Zero Demo which is located in a folder named 3DB_PROJECTS. You can use these same names or create ones of your liking. Note that as you type the project name it is added to the path shown in the location text box. At this point the New dialog box appears as in Figure 19.1.

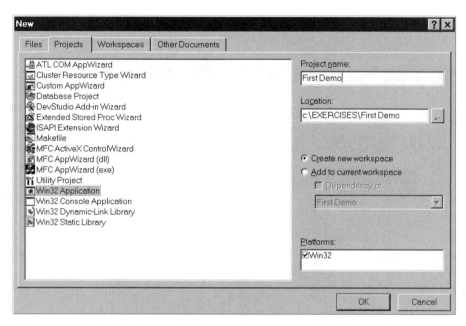

Figure 19.1 *Using the New Command in Developer Studio File Menu*

Make sure that the radio button labeled Create new workspace is selected so that clicking the OK button on the dialog box creates both the project and the workspace. At this point, you have created a project, as well as a workspace of the same name, but there are no program files in it yet. How you proceed from here depends on whether you are using another source file as a base or template or starting from scratch.

If you wish to start a source file from scratch, click on Developer Studio Project menu and select Add To Project and New commands. This action displays the same dialog box as when creating a project, but now the Files tab is open. In the case of a source file, select the C++ Source File option from the displayed list and type a file name in the corresponding text box. The dialog appears as shown in Figure 19.2, on the following page.

The development method we use in this book is based on using source code templates. To use a template as a base, or another source file, you have to follow a different series of steps. Assuming the you have created a project, the next step is to select and load the program template or source file. We use the template named Templ01.cpp. If you have installed the book's software in your system, the template file is in the path 3DB/Templates.

To load the source file into your current project, open Developer Studio Project menu and select Add To Project item and then the Files commands. This action displays an Insert Files into Project dialog box. Use the buttons to the right of the Look in text box to navigate into the desired drive and folder until the desired file is selected. Figure 16-3 shows the file Templ01.cpp highlighted and ready for inserting into the project.

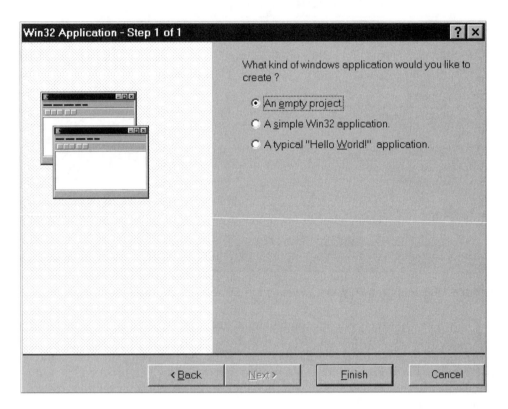

Figure 19.2 *Creating a New Source File in Developer Studio*

When using a template file to start a new project you must be careful not to destroy or change the original source. The template file is usually renamed once it is inserted into the project. It is possible to insert a template file in a project, rename it, delete it from the project, and then reinsert the renamed file. However, it is easier to rename a copy of the template file before it is inserted into the project. The following sequences of operations are used:

1. Click the File menu and select the Open command. Navigate through the directory structure to locate the file to be used as a template. In this case the file Templ01.cpp is located in 3DB/Templates folder.

2. With the cursor still in Developer Studio editor pane, open the File menu and click on the Save As command. Navigate through the directory structure again until you reach the PROJECTS\Program Zero Demo folder. Save the file using the name Prog_zero.cpp.

3. Click on the Project menu and select the commands Add to Project and Files. Locate the file named Prog_Zero.cpp in the Insert Files into Project dialog box, select it, and click the OK button.

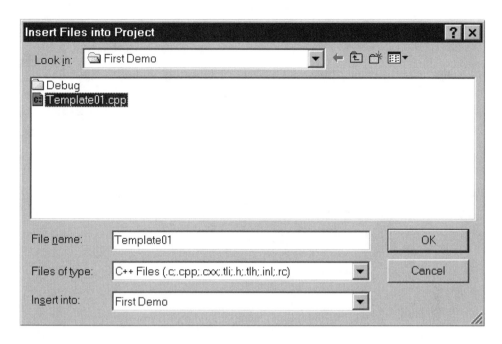

Figure 19.3 *Inserting an Existing Source File Into a Project*

The file Prog_zero.cpp now appears in the Program Zero Demo file list in Developer Studio workspace pane. It is also displayed in the Editor window.

The Developer Studio main screen is configurable by the user. Furthermore, the size of its display areas is determined by the system resolution. For this reason, it is impossible to depict a Developer Studio screen display that matches the one that every user will see. In the illustrations and screen snapshots throughout this book we have used a resolution of 1152-by-854 pixels in 16-bit color with large fonts. However, our screen format may not exactly match yours. Figure 19.4, on the following page, shows a full screen display of Developer Studio with the file Prog_zero.cpp loaded in the Editor area.

The Project Workspace pane of Developer Studio was introduced in Version 4.0. It has four possible views: Class View, File View, Info View, and Resource View. The Resource View is not visible in Figure 19.4. In order to display the source file in the editor pane, you must first select File View tab and double-click on the Prog_zero.cpp filename.

At this point, you can proceed to develop the new project using the renamed template file as the main source. The first step is to make sure that the development software is working correctly. To do this open the Developer Studio Build menu and click the Rebuild All command. Developer Studio compiles and builds your program, which is at this stage nothing more than the renamed template file. The results are shown in the Output area. If compilation and linking took place without error, reopen the Build menu and select the Execute Prog_zero.exe command button. If everything is in order, a do-nothing program executes in your system.

Editor pane

Project Worskpace pane

Editor window controls

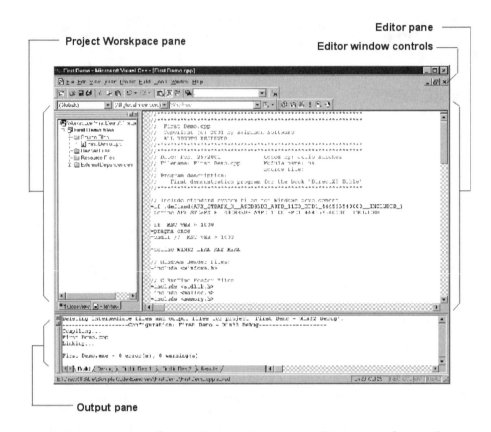

Output pane

Figure 19.4 *Developer Studio Project Workspace, Editor, and Output Panes*

Now click the Save command on the File menu to make sure that all project files are saved on your hard drive.

19.2 Elements of a Windows Program

The template file Templ01.cpp, which we used and renamed in the previous example, is a bare bones windows program with no functionality except to display a window on the screen. Before proceeding to edit this template into a useful program, you should become acquainted with its fundamental elements. In this section, we take apart the template file Templ01.cpp for a detailed look into each of its components. The program contains two fundamental components: WinMain() and the Windows procedure.

19.2.1 WinMain()

All Windows GUI applications must have a WinMain() function. WinMain() is to a Windows GUI program what main() is to a DOS application. It is usually said that WinMain() is the program's entry point, but this is not exactly true. C/C++ compilers generate a startup code that calls WinMain(), so it is the startup code and not WinMain() that is actually called by Windows. The WinMain() header line is as follows:

```
|------------------- Return type
|    |------------- One of the standard calling conventions
|    |                 defined in windows.h
|    |     |------- Function name
|    |     |
|    |     |     [            parameter list ....
--- ------ ------- -----------------------------------------------
int WINAPI WinMain (HINSTANCE hInstance, HINSTANCE hPrevInstance,
                    PSTR szCmdLine, int iCmdShow) {
```

WINAPI is a macro defined in the windows.h header file which translates the function call to the appropriate calling convention. Recall that calling conventions refer to how the function arguments are placed in the stack at call time, and if the caller or the called routine is responsible for restoring stack integrity after the call. Microsoft Basic, FORTRAN, and Pascal push the parameters onto the stack in the same order in which they are declared. In these languages the stack must be restored by the caller. In C and C++, the parameters are pushed in reverse order, and the stack is restored automatically after the call returns. For historical reasons (and to take advantage of hardware features of the Intel processors) Windows requires the Pascal calling convention. In previous versions of Windows the calling convention for WinMain() was PASCAL or FAR PASCAL. You can still replace WINAPI for FAR PASCAL and the program will compile and link correctly, but the use of the WINAPI macro makes your program more portable.

Parameters

Most often parameters are passed to WinMain() by Windows, but some can be passed by whatever program executes your application. Your code can inspect these parameters to obtain information about the conditions in which the program executes. Four parameters are passed to WinMain():

- HINSTANCE is a handle-type identifier. The variable hInstance is an integer that identifies the instance of the program. Consider that in a multitasking environment there can be several copies (instances) of the same program running simultaneously. Windows sets this value and passes it to your code. Your program needs to access this parameter to enter it in the WNDCLASSEX structure; also when calling the CreateWindow() function. Because the handle to the instance is required outside of WinMain() by many functions of the Windows API, the template file stores it in a public variable, named pInstance. In general, the use of public variables is undesirable in Windows programming, but this case is one of the valid exceptions to the rule.

- The variable hPrevInstance is also of type HINSTANCE. This parameter is included in the call for compatibility with previous versions of Windows, which used a single copy of the code to run more than one program instance. In 16-bit Windows the first instance had a special role in the management of resources. Therefore, an application needed to know if it was the first instance. hPrevInstance held the handle of the previous instance. In Windows 95/NT and later this parameter is unused and its value is set to NULL.

- PSTR szCmdLine. This is a pointer to a string that contains the command tail entered by the user when the program is executed. It works only when the program name is entered from the DOS command line or from the Run dialog box. For this reason, it is rarely used by code.

- int iCmdShow. This parameter determines how the window is to be initially displayed. The program that executes your application (normally Windows) assigns a value to this parameter, as shown in Table 19.1.

Table 19.1

WinMain() Display Mode Parameters

VALUE	MEANING
SW_HIDE	Hides the window and activates another window
SW_MINIMIZE	Minimizes the specified window and activates the top-level window in the system's list
SW_RESTORE	Activates and displays a window. If the window is minimized or maximized, Windows restores it to its original size and position (same as SW_SHOWNORMAL)
SW_SHOW	Activates a window and displays it in its current size and position
SW_SHOWMAXIMIZED	Activates a window and displays it as a maximized window
SW_SHOWMINIMIZED	Activates a window and displays it as an icon
SW_SHOWMINNOACTIVE	Displays a window as an icon. The active window remains active
SW_SHOWNA	Displays a window in its current state. The active window remains active
SW_SHOWNOACTIVATE	Displays a window in its most recent size and position. The active window remains active
SW_SHOWNORMAL	Activates and displays a window. If the window is minimized or maximized, Windows restores it to its original size and position (same as SW_RESTORE)

19.2.2 Data Variables

The program file Templ01.cpp defines several variables. One of them, the handle to the program's main window, is defined globally. The other ones are local to WinMain() or the windows procedure. The variable defined globally is:

```
HWND        hwnd;
```

HWND is a 16-bit unsigned integer which serves as a handle to a window. The variable hwnd refers to the actual program window. The variable is initialized when we make the call to CreateWindow() service, described later in this section.

The variables defined in WinMain() are as follows:

```
    static char szClassName[] = "MainClass" ; // Class name
    MSG         msg ;
```

The first one is and array of char that shows the application's class name. In the template it is given the name MainClass, which you can replace for a more meaningful one. The application class name must be the same one used in the WNDCLASSEX structure. MSG is a message-type structure of which msg is a variable. The MSG structure is defined in the Windows header files as follows:

```
typedef struct tagMSG {        // msg
    HWND   hwnd;       // Handle to window receiving message
    UINT   message;    // message number
    WPARAM wParam;     // Context-dependent additional information
    LPARAM lParam;     // about the message
    DWORD  time;       // Time at which message was posted
    POINT  pt;         // Cursor position when message was posted
} MSG;
```

The comments to the structure members show that the variable holds information that is important to the executing code. The values of the message variable are reloaded every time a new message is received.

19.2.3 WNDCLASSEX Structure

This structure is defined in the windows header files, as follows:

```
typedef struct tagWNDCLASSEX {
UINT        cbSize;
UINT        style;
WNDPROC     lpfnWndProc;
int         cbClsExtra;
int         cbWndExtra;
HINSTANCE   hInstance;
HICON       hIcon;
HCURSOR     hCursor;
HBRUSH      hbrBackground;
LPCSTR      lpszMenuName;
LPCSTR      lpszClassName;
HICON       hIconSm;
} WNDCLASSEX;
```

The WNDCLASSEX structure contains window class information. It is used with the RegisterClassEx() and GetClassInfoEx() functions. The structure is similar to the WNDCLASS structure used in 16-bit Windows. The differences between the two structures is that WNDCLASSEX has a cbSize member, which specifies the size of the structure, and the hIconSm member, which contains a handle to a small icon associated with the window class. In the template file Templ01.cpp the structure is declared and the variable initialized as follows:

```
// Creating a WNDCLASSEX structure
WNDCLASSEX  wndclass ;
wndclass.cbSize        = sizeof (WNDCLASSEX) ;
wndclass.style         = CS_HREDRAW | CS_VREDRAW ;
wndclass.lpfnWndProc   = WndProc ;
wndclass.cbClsExtra    = 0 ;
wndclass.cbWndExtra    = 0 ;
wndclass.hInstance     = hInstance ;
wndclass.hIcon         = LoadIcon (NULL, IDI_APPLICATION) ;
wndclass.hCursor       = LoadCursor (NULL, IDC_ARROW) ;
wndclass.hbrBackground = (HBRUSH) GetStockObject
```

```
wndclass.lpszMenuName  = NULL ;
wndclass.lpszClassName = szClassName ;
wndclass.hIconSm       = LoadIcon (NULL, IDI_APPLICATION) ;
```

The window class is a template that defines the characteristics of a particular window, such as the type of cursor and the background color. The class also specifies the address of the windows procedure that carries out the work for the window. The structures variables define the window class, as follows:

cbSize specifies the size, in bytes, of the structure. The member is set using the sizeof operator in the statement:

```
sizeof(WNDCLASSEX);
```

style specifies the class style or styles. Two or more styles can be combined by means of the C bitwise OR (|) operator. This member can be any combination of the values in Table 19.2.

Table 19.2

Summary of Window Class Styles

SYMBOLIC CONSTANT	ACTION
CS_BYTEALIGNCLIENT	Aligns the window's client area on the byte boundary (in the x direction) to enhance performance during drawing operations. This style affects the width of the window and its horizontal position on the display.
CS_BYTEALIGNWINDOW	Aligns a window on a byte boundary (in the direction) to enhance performance during operations that involve moving or sizing the window. This style affects the width of the window and its horizontal position on the display.
CS_CLASSDC	Allocates one device context to be shared by all windows in the class. Window classes are process specific; therefore, different threads can create windows of the same class.
CS_DBLCLKS	Sends double-click messages to the window procedure when the user double-clicks the mouse while the cursor is within a window belonging to the class.
CS_GLOBALCLASS	Allows an application to create a window of the class regardless of the value of the hInstance parameter passed to the CreateWindowEx() function. If you do not specify this style, the hInstance parameter passed to CreateWindowEx() function must be the same as the one passed to the RegisterClass() function.
CS_HREDRAW	Redraws the entire window if a movement or size adjustment changes the width of the client area.
CS_NOCLOSE	Disables the Close command on the System menu.

(continues)

Table 19.2

Summary of Window Class Styles (continued)

SYMBOLIC CONSTANT	ACTION
CS_OWNDC	Allocates a unique device context for each window in the class.
CS_PARENTDC	Specifies that child windows inherit their parent window's device context. Specifying CS_PARENTDC enhances an application's performance.
CS_SAVEBITS	Saves, as a bitmap, the portion of the screen image obscured by a window. Windows uses the saved bitmap to recreate the screen image when the window is removed. This style is useful for small windows (such as menus or dialog boxes) that are displayed briefly and then removed before other screen activity takes place.
CS_VREDRAW	Redraws the entire window if a movement or size adjustment changes the height of the client area.

Of these, the styles CS_HREDRAW and CS_VREDRAW are the ones most commonly used. They can be ORed to produce a window that is automatically redrawn if it is resized vertically or horizontally, as implemented in the Templ01.cpp code.

lpfnWndProc is a pointer to the window procedure, described later in this chapter. In the template Templ01.cpp it is initialized to the name of the Windows procedure, as follows:

```
wndclass.lpfnWndProc    = WndProc;
```

cbClsExtra is a count of the number of extra bytes to be allocated following the window-class structure. The operating system initializes the bytes to zero. In the template this member is set to zero.

cbWndExtra is a count of the number of extra bytes to allocate following the window instance. The operating system initializes the bytes to zero. In the template this member is set to zero.

hInstance is a handle to the instance of the window procedure.

hIcon is a handle to the class icon. If this member is NULL, an application must draw an icon whenever the user minimizes the application's window. In the template this member is initialized by calling the LoadIcon() function.

hCursor is a handle to the class cursor. If this member is NULL, an application must explicitly set the cursor shape whenever the mouse moves into the application's window. In the template this member is initialized by calling the LoadCursor() function.

hbrBackground is a background brush. This member can be a handle to the physical brush to be used for painting the background, or it can be a color value. If it is a color value, then it must be one of the standard system colors listed in Table 19.3.

Table 19.3

Common Windows Standard System Colors

SYMBOLIC CONSTANT	MEANING
COLOR_ACTIVEBORDER	Border color of the active window
COLOR_ACTIVECAPTION	Caption color of the active window
COLOR_APPWORKSPACE	Window background of MDI clients
COLOR_BACKGROUND	Desktop color
COLOR_BTNFACE	Face color for buttons
COLOR_BTNSHADOW	Shadow color for buttons
COLOR_BTNTEXT	Text color on buttons
COLOR_CAPTIONTEXT	Text color for captions, size boxes, and scroll boxes
COLOR_GRAYTEXT	Color for dissabled text
COLOR_HIGHLIGHT	Color of a selected item
COLOR_HIGHLIGHTTEXT	Text color of a selected item
COLOR_INACTIVEBORDER	Border color of inactive window
COLOR_INACTIVECAPTION	Caption color of an inactive window
COLOR_MENU	Background color of a menu
COLOR_MENUTEXT	Text color of a menu
COLOR_Scroll bar	Color of a scroll bar's gray area
COLOR_WINDOW	Background color of a window
COLOR_WINDOWFRAME	Frame color of a window
COLOR_WINDOWTEXT	Text color of a window

When this member is NULL, an application must paint its own background whenever it is required to paint its client area. In the template this member is initialized by calling the GetStockObject() function.

lpszMenuName is a pointer to a null-terminated character string that specifies the resource name of the class menu, as it appears in the resource file. If you use an integer to identify the menu, then you must use the MAKEINTRESOURCE macro. If this member is NULL, the windows belonging to this class have no default menu, as is the case in the template file.

lpszClassName is a pointer to a null-terminated string or it is an atom. If this parameter is an atom, it must be a global atom created by a previous call to the GlobalAddAtom() function. The atom, a 16-bit value, must be in the low-order word of lpszClassName; the high-order word must be zero. If lpszClassName is a string, it specifies the window class name. In Templ01.cpp this member is set to the szClassName[] array.

In Windows 95/NT and later hIconSm is a handle to a small icon that is associated with the window class. This is the icon shown in dialog boxes that list filenames and by Windows Explorer. A Windows 95/98 application can use a predefined icon in this case, using the LoadIcon function with the same parameters as for the hIcon member. In Windows NT this member is not used and should be set to NULL. Windows 95/NT and later applications that set the small icon to NULL still have the default small icon displayed on the task bar.

In most cases it is better to create both the large and the small icon than to let Windows create the small one from the large bitmap. Later in this chapter we describe how to create both icons as a program resource and how to make these resources available to the application.

Contrary to what has sometimes been stated, the LoadIcon() function cannot be used to load both large and small icons from the same resource. For example, if the icon resource is named IDI_ICON1, and we proceed as follows:

```
wndclass.hicon       = LoadIcon (hInstance,
                         MAKEINTRESOURCE(IDI_ICON1);
.
.
.
wndclass.hiconSm  = LoadIcon (hInstance,
                         MAKEINTRESOURCE(IDI_ICON1);
```

the result is that the large icon is loaded from the resource file, but not the small one. This happens even if the resource file contains both images. Instead, you must use the LoadImage() function, as follows:

```
wndclass.hIcon         = (HICON)LoadImage(hInstance,
                           MAKEINTRESOURCE(IDI_ICON1),
                           IMAGE_ICON,        // Type
                           32, 32,            // Pixel size
                           LR_DEFAULTCOLOR) ;
.
.
.
wndclass.hIconSm       = (HICON)LoadImage(hInstance,
                           MAKEINTRESOURCE(IDI_ICON1),
                           IMAGE_ICON,        // Type
                           16, 16,            // Pixel size
                           LR_DEFAULTCOLOR) ;
```

Now both the large and the small icon resources are loaded correctly and are used as required. Also notice that the value returned by LoadImage() is typecast into HICON. This manipulation became necessary starting with version 6 of Microsoft Visual C++ due to changes made to the compiler in order to improve compatibility with the ANSI C++ standard.

19.2.4 Registering the Windows Class

Once your code has declared the WNDCLASSEX structure and initialized its member variables, it has defined a window class that encompasses all the structure attributes. The most important ones are the window style (wndclass.style), the pointer to the Windows procedure (wndclass.lpfnWndProc), and the window class name (wndclass. lpszClassName). The RegisterClassEx() function is used to notify Windows of the existence of a particular window class, as defined in the WNDCLASSEX structure variable. The address-of operator is used to reference the location of the specific structure variable, as in the following statement:

```
RegisterClassEx (&wndclass) ;
```

The RegisterClassEx() function returns an atom (16-bit integer). This value is non-zero if the class is successfully registered. Code should check for a successful registration since you cannot create a windows otherwise. The following construct ensures that execution does not proceed if the function fails.

```
if(!RegisterClassEx (&wndclass))
  return(0);
```

This coding style is the one used in the template Templ01.cpp.

19.2.5 Creating the Window

A window class is a general classification. Other data must be provided at the time the actual windows is created. The CreateWindowEx() function receives the additional information as parameters. CreateWindowEx() is a Windows 95 version of the CreateWindow() function. The only difference between them is that the new version supports an extended window style passed as its first parameter.

The CreateWindowEx() function is very rich in arguments, many of which apply only to special windows styles. For example, buttons, combo boxes, list boxes, edit boxes, and static controls can all be created with a CreateWindowEx() call. At this time, we refer only to the most important function parameters that relate to the a program's main window.

```
In the file Templ01.cpp the call to CreateWindowEx is coded as follows:
  hwnd = CreateWindowEx (
        WS_EX_LEFT,              // left aligned (default)
        szClassName,             // pointer to class name
        "Window Caption",        // window caption (title bar)
        WS_OVERLAPPEDWINDOW,     // window style
        CW_USEDEFAULT,           // initial x position
        CW_USEDEFAULT,           // initial y position
        CW_USEDEFAULT,           // initial x size
        CW_USEDEFAULT,           // initial y size
        NULL,                    // parent window handle
        NULL,                    // window menu handle
        hInstance,               // program instance handle
        NULL) ;                  // creation parameters
```

The first parameter passed to the CreateWindowEx() function is the extended window style introduced in the Win32 API. The one used in the file Templ01.cpp, WS_EX_LEFT, acts as a placeholder for others that you may want to select, since it is actually the default value. Table 19.4 lists some of the most common extended styles. The second parameter passed to the CreateWindowEx() function call is either a pointer to a string with the name of the window type, a string enclosed in double quotation marks, or a predefined name for a control class.

In the template file, szClassName is a pointer to the string defined at the start of WinMain(), with the text "MainClass." You can edit this string in your own applications so that the class name is more meaningful. For example, if you were coding an editor program you may rename the application class as "TextEdClass." However, this is merely a name used by Windows to associate a window with its class; it is not displayed as a caption or used otherwise.

Table 19.4

Common Windows Extended Styles

SYMBOLIC CONSTANT	MEANING
WS_EX_ACCEPTFILES	The window created with this style accepts drag drop files.
WS_EX_APPWINDOW	A top level window is forced onto the application taskbar when the window is minimized.
WS_EX_CLIENTEDGE	Window has a border with a sunken edge.
WS_EX_CONTEXTHELP	The title bar includes a question mark. When the user clicks the mark, the cursor changes to a question mark with a pointer. If the user then clicks a child window, it receives a WM_HELP message.
WS_EX_CONTROLPARENT	Allows the user to navigate among the child windows of the window by using the TAB key.
WS_EX_DLGMODALFRAME	Window that has a double border. Optionally the window can be created with a title bar by specifying the WS_CAPTION style in the dwStyle parameter.
WS_EX_LEFT	Window has generic "left aligned" properties. This is the default.
WS_EX_MDICHILD	Creates an MDI child window.
WS_EX_NOPARENTNOTIFY	Specifies that a child window created with this style does not send the WM_PARENTNOTIFY message to its parent window when it is created or destroyed.
WS_EX_OVERLAPPEDWINDOW	Combines the WS_EX_CLIENTEDGE and WS_EX_WINDOWEDGE styles.
WS_EX_PALETTEWINDOW	Combines the WS_EX_WINDOWEDGE, WS_EX_TOOLWINDOW, and WS_EX_TOPMOST styles.
WS_EX_RIGHTSCROLLBAR	Specifies that the vertical scroll bar (if present) is to the right of the client area. This is the default.
WS_EX_STATICEDGE	Creates a window with a three dimensional border style intended to be used for items that do not accept user input.
WS_EX_TOOLWINDOW	Creates a tool window. This type of window is intended to be used as a floating toolbar.
WS_EX_TOPMOST	Creates a window which is placed, that is A window created with this style should be placed above all non topmost windows and should stay stays above them, even when the window is deactivated.
WS_EX_TRANSPARENT	Creates a windows that is A window created with this style is to be transparent. That is, any windows that are beneath it are not obscured by it.
WS_EX_WINDOWEDGE	Specifies a window with a border with a raised edge.

Control classes can also be used as a window class name. These classes are the symbolic constants BUTTON, Combo box, EDIT, List box, MDICLIENT, Scroll bar, and STATIC.

The third parameter can be a pointer to a string or a string enclosed in double quotation marks entered directly as a parameter. In either case, this string is used as the caption to the program window and is displayed in the program's title bar. Often this caption coincides with the name of the program. You should edit this string to suit your own program.

The fourth parameter is the window style. Over 25 styles are defined as symbolic constants. The most used ones are listed in Table 19.5.

Table 19.5

Window Styles

SYMBOLIC CONSTANT	MEANING
WS_BORDER	Window that has a thin-line border.
WS_CAPTION	Window that has a title bar (includes the WS_BORDER style).
WS_CHILD	Child window. This style cannot be used with the WS_POPUP style.
WS_CLIPCHILDREN	Excludes the area occupied by child windows when drawing occurs within the parent window.
WS_CLIPSIBLINGS	Clips child windows relative to each other. When a particular child window receives a WM_PAINT message, this style clips all other overlapping child windows out of the region of the child window to be updated. If WS_CLIPSIBLINGS is not specified and child windows overlap, it is possible to draw within the client area of a neighboring child window.
WS_DISABLED	Window is initially disabled. A disabled window cannot receive input from the user.
WS_DLGFRAME	Window has a border of a style typically used with dialog boxes. The window does not have a title bar.
WS_HSCROLL	Window that has a horizontal scroll bar.
WS_ICONIC	Window is initially minimized. Same as the WS_MINIMIZE style.
WS_MAXIMIZE	Window is initially maximized.
WS_MAXIMIZEBOX	Window that has a Maximize button. Cannot be combined with the WS_EX_CONTEXTHELP style.
WS_MINIMIZE	Window is initially minimized. Same as the WS_ICONIC style.
WS_MINIMIZEBOX	Window has a Minimize button. Cannot be combined with the WS_EX_CONTEXTHELP style.
WS_OVERLAPPED	Overlapped window. Has a title bar and a border.

(continues)

Table 19.5

Window Styles (continued)

SYMBOLIC CONSTANT	MEANING
WS_OVERLAPPEDWINDOW	Overlapped window with the WS_OVERLAPPED, WS_CAPTION, WS_SYSMENU, WS_THICKFRAME, WS_MINIMIZEBOX, and WS_MAXIMIZEBOX styles. Same as the WS_TILEDWINDOW style.
WS_POPUP	Pop-up window. Cannot be used with the WS_CHILD style.
WS_POPUPWINDOW	Pop-up window with WS_BORDER, WS_POPUP, and WS_SYSMENU styles. The WS_CAPTION and WS_POPUPWINDOW styles must be combined to make the System menu visible.
WS_SIZEBOX	Window that has a sizing border. Same as the WS_THICKFRAME style.
WS_SYSMENU	Window that has a System-menu box in its title bar. The WS_CAPTION style must also be specified.
WS_TILED	Overlapped window. Has a title bar and a border. Same as the WS_OVERLAPPED style.
WS_TILEDWINDOW	Overlapped window with the WS_OVERLAPPED, WS_CAPTION, WS_SYSMENU, WS_THICKFRAME, WS_MINIMIZEBOX, and WS_MAXIMIZEBOX styles. Same as the WS_OVERLAPPEDWINDOW style
WS_VISIBLE	Window is initially visible.
WS_VSCROLL	Window that has a vertical scroll bar.

The style defined in the template file Templ01.ccp is WS_OVER-LAPPEDWINDOW. This style creates a window that has the styles WS_OVER-LAPPED, WS_CAPTION, WS_SYSMENU, WS_THICKFRAME, WS_MINIMIZEBOX, and WS_MAXIMIZEBOX. It is the most common style of windows.

The fifth parameter to the CreateWindowEx() service defines the initial horizontal position of the window. The value CS_USERDEFAULT (0x80000000) determines the use of the default position. The template file uses the same CS_USERDEFAULT symbolic constant for the y position, and the windows x and y size.

Parameters nine and ten are set to NULL since this window has no parent and no default menu.

The eleventh parameter, hInstance, is a the handle to the instance that was passed to WinMain() by Windows.

The last entry, called the creation parameters, can be used to pass data to a program. A CREATESTRUCT-type structure is used to store the initialization parameters passed to the windows procedure of an application. The data can include an instance handle, a new menu, the window's size and location, the style, the window's name and class name, and the extended style. Since no creation parameters are passed, the field is set to NULL.

The CreateWindowEx() function returns a handle to the window of type HWND. The template file Templ01.cpp stores this handle in a global variable named hwnd. The reason for this is that many functions in the Windows API require this handle. By storing it in a global variable we make it visible throughout the code.

If CreateWindowsEx() fails, it returns NULL. Code in WinMain() can test for this error condition with the statement:

```
if(!hwnd)
   return(0);
```

We do not use this test in the template file Templ01.cpp because it is usually not necessary. If WinMain() fails, you may use the debugger to inspect the value of hwnd after CreateWindowEx() in order to make sure that a valid handle was returned.

19.2.6 Displaying the Window

CreateWindowEx() creates the window internally but does not display it. To display the window your code must call two other functions: ShowWindow() and UpdateWindow(). ShowWindow() sets the window's show state and UpdateWindow() updates the window's client area. In the case of the program's main window, ShowWindow() must be called once, using as a parameter the iCmdShow value passed by Windows to WinMain(). In the template file the call is coded as follows:

```
ShowWindow (hwnd, iCmdShow) ;
```

The first parameter to ShowWindow() is the handle to the window returned by CreateWindowEx(). The second parameter is the window's display mode parameter, which determines how the window must be initially displayed. The display mode parameters are listed in Table 19.1, but in this first call to ShowWindow() you must use the value received by WinMain().

UpdateWindow() actually instructs the window to paint itself by sending a WM_PAINT message to the windows procedure. The processing of the WM_PAINT message is described later in this chapter. The actual code in the template file is as follows:

```
UpdateWindow (hwnd) ;
```

If all has gone well, at this point your program is displayed on the screen. It is now time to implement the message passing mechanisms that are at the heart of event-driven programming.

19.2.7 The Message Loop

In an event-driven environment there can be no guarantee that messages are processed faster than they originate. For this reason Windows maintains two message queues. The first type of queue, called the system queue, is used to store messages that originate in hardware devices, such as the keyboard and the mouse. In addition, every thread of execution has its own message queue. The message handling mechanism can be described with a simplified example: when a keyboard event occurs, the device driver software places a message in the system queue. Windows uses information about the input focus to decide which thread should handle the message. It then moves the message from the system queue into the corresponding thread queue.

A simple block of code, called the message loop, removes a messages from the thread queue and dispatches it to the function or routine which must handle it. When a special message is received, the message loop terminates, and so does the thread. The message loop in Templ01.cpp is coded as follows:

```
while (GetMessage (&msg, NULL, 0, 0))
    {
    TranslateMessage (&msg) ;
    DispatchMessage (&msg) ;
    }
  return msg.wParam ;
```

The while statement calls the function GetMessage(). The first parameter to GetMessage() is a variable of the structure type MSG, described in Appendix A. The structure variable is filled with information about the message in the queue, if there is one. If no message is waiting in the queue, Windows suspends the application and assigns its time slice to other threads of execution. In an event-driven environment, programs act only in response to events. No event, no message, no action.

The second parameter to GetMessage() is the handle to a window for which to retrieve a message. Most applications set this parameter to NULL, which signals that all messages for windows that belong to the application making the call should be retrieved. The third and the fourth parameter to GetMessage() are the lowest and the highest message numbers to be retrieved. Threads that only retrieve messages within a particular range can use these parameters as a filter. When the special value 0 is assigned to both of these parameters (as is the case in our message loop) then no filtering is performed and all messages are passed to the application.

There are two functions inside the message loop. TranslateMessage() is a keyboard processing function that converts keystrokes into characters. The characters are then posted to the message queue. If the message is not a keystroke that needs translation, then no special action is taken. The DispatchMessage() function sends the message to the windows procedure, where it is further processed and either acted upon, or ignored. The windows procedure is discussed in the following section. GetMessage() returns 0 when a message labeled WM_QUIT is received. This signals the end of the message loop; at this point execution returns from WinMain(), and the application terminates.

19.3 The Window Procedure

At this moment in a program's execution the window class has been registered, the window has been created and displayed, and all messages are being routed to your code. The windows procedure, sometimes called the window function, is where you write code to handle the messages received from the message loop. It is in the windows procedure where you respond to the events that pertain to your program.

Every window must have a window procedure. Although the name WinProc() is commonly used, you can use any other name for the windows procedure provided that it appears in the procedure header, the prototype, in the corresponding entry of the WNDCLASSEX structure, and that it does not conflict with another name in your application. Also, a Windows program can have more than one windows procedure. The program's main window is usually registered in WinMain() but others can be registered elsewhere in an application. Here again, each windows procedure corresponds to a window class, has its own WNDCLASSEX structure, as well as a unique name.

```
In the template, the windows procedure is coded as follows:
    |----------------------- Return type, equivalent to a long type
    |         |--------------- Same as FAR PASCAL calling convention.
    |         |                 Used in windows and dialog procedures.
    |         |       |------- Procedure name
    |         |       |     [ parameter list ...                    ]
------- -------- ------- ----------------------------------------
LRESULT CALLBACK WndProc (HWND hwnd, UINT iMsg, WPARAM wParam,
                          LPARAM lParam) {
```

The windows procedure is of callback type. The CALLBACK symbol was first introduced in Windows 3.1 and is equivalent to FAR PASCAL, and also to WINAPI, since all of them currently correspond to the __stdcall calling convention. Although it is possible to substitute __stdcall for CALLBACK in the function header, it is not advisable, since this could compromise the application's portability to other platforms or to future versions of the operating system.

The return value of a windows procedure is of type LRESULT, which is a 32-bit integer. The actual value depends on the message, but it is rarely used by application code. However, there are a few messages for which the windows procedure is expected to return a specific value. It is a good idea to check the Windows documentation when in doubt.

19.3.1 Windows Procedure Parameters

The four parameters to the windows procedure are the first four fields in the MSG structure. The MSG structure is discussed earlier in this chapter. Since the windows procedure is called by Windows, the parameters are provided by the operating system at call time, as follows:

- hwnd is the handle to the window receiving the message. This is the same handle returned by CreateWindow().

- iMsg is a 32-bit unsigned integer (UINT) that identifies each particular message. The constants for the various messages are defined in the windows header files. They all start with the letters WM_, which stand for window message.

- wParam and lParam are called the message parameters. They provide additional information about the message. Both values are specific to each message.

The last two members of the message structure, which correspond to the message's time of posting and cursor position, are not passed to the windows procedure. However, application code can use the functions GetMessageTime() and GetMessagePos() to retrieve these values.

19.3.2 Windows Procedure Variables

The implementation of the windows procedure in Templ01.cpp starts by declaring a scalar of type HDC and two structure variables of type HWND and MSG respectively. The variables are as follows:

- hdc is a handle to the device context. A device context is a data structure maintained by Windows which is used in defining the graphics objects and their attributes, as well as their associated graphics modes. Devices such as the video display, printers, and plotters, must be accessed through a handle to their device contexts, which is obtained from Windows.

- ps is a PAINTSTRUCT variable. The structure is defined by Windows as follows:

```
typedef struct tagPAINTSTRUCT {
    HDC     hdc;              // identifies display device
    BOOL    fErase;           // not-zero if background must be erased
    RECT    rcPaint;          // Rectangle structure in which painting is
                              // requested
    BOOL    fRestore;         // RESERVED
    BOOL    fIncUpdate;       // RESERVED
    BYTE    rgbReserved[32];  // RESERVED
} PAINTSTRUCT;
```

The structure contains information that is used by the application to paint its own client area.

- rect is a RECT structure variable. The RECT structure is also defined by Windows:

```
typdef struct _RECT {
    LONG    left;     // x coordinate of upper-left corner
    LONG    top;      // y of upper-left corner
    LONG    right;    // x coordinate of bottom-right corner
    LONG    bottom;   // y of bottom-right
} RECT;
```

The RECT structure is used to define the corners of a rectangle, in this case of the application's display area, which is also called the client area.

19.3.3 Message Processing

The windows procedure receives and processes messages. The message can originate as follows:

- Some messages are dispatched by WinMain(). In this group are the messages placed in the thread's message queue by the DispatchMessage() function in the message loop. Messages handled in this manner are referred to as queued messages. Queued messages originate in keystrokes, mouse movements, mouse button clicks, the system timer, and in orders to redraw the window.

- All other messages come directly from Windows. These are called nonqueued messages.

The windows procedure examines each message, queue or nonqueued, and either takes action or passes the message back for default processing. In the template file Templ01.cpp the message processing skeleton is coded as follows:

```
switch (iMsg)
    {
// Windows message processing
    // Preliminary operations
    case WM_CREATE:
        return (0);

    // Redraw window
    case WM_PAINT :
        hdc = BeginPaint (hwnd, &ps) ;
        GetClientRect (hwnd, &rect) ;
    // Initial display operations here
        EndPaint (hwnd, &ps) ;
        return 0 ;

    // End of program execution
    case WM_DESTROY :
        PostQuitMessage (0) ;
        return 0 ;
    }
    return DefWindowProc (hwnd, iMsg, wParam, lParam) ;
```

Messages are identified by uppercase symbolic constants that start with the characters WM_ (window message). Over two hundred message constants are defined in Windows. Three messages are processed in the template file: WM_CREATE, WM_PAINT and WM_DESTROY.

When the Windows procedure processes a message it must return 0. If it does not process a particular message, then the function DefWindowsProc() is called to provide a default action.

WM_CREATE Message Processing

The WM_CREATE message is sent to an application as a result of the CreateWindowEx() function in WinMain(). This message gives the application a chance to perform preliminary initialization, such as displaying a greeting screen, or playing a sound file. In the template, the WM_CREATE processing routine does nothing. It serves as a placeholder where the programmer can insert the appropriate code.

WM_PAINT Message Processing

The WM_PAINT message informs the program that all or part of the client window must be repainted. This happens when the user minimizes, overlaps, or resizes the client window area. Recall that the style of the program's main window is defined in the template with the statement:

```
    wndclass.style    = CS_HREDRAW | CS_VREDRAW ;
```

This style determines that the screen is redrawn if it is resized vertically or horizontally.

In WM_PAINT, processing begins with the BeginPaint() function. BeginPaint() serves to prepare the window for a paint operation by filling a variable of type PAINTSTRUCT, previously discussed. The call to BeginPaint() requires the hwnd variable, which is the handle to the window that is to be painted. Also a variable ps, of a structure of type PAINTSTRUCT, which is filled by the call. During BeginPaint() Windows erases the background using the currently defined brush.

The call to GetClientRect() requires two parameters. The first one is the handle to the window (hwnd), which is passed to the windows procedure as a parameter. In the template file this value is also stored in a public variable. The second parameter is the address of a structure variable of type RECT, where Windows places the coordinates of the rectangle that defines the client area. The left and top values are always set to zero.

Processing ends with EndPaint(). EndPaint() notifies Windows that the paint operation has concluded. The parameters passed to EndPaint() are the same ones passed to BeginPaint(): the handle to the window and the address of the structure variable of type PAINTSTRUCT.

WM_DESTROY Message Processing

The WM_DESTROY message is received by the windows procedure when the user takes an action to destroy the window, usually clicking the Close button or selecting the Close or Exit commands from the File or the System menus. The standard processing performed in WM_DESTROY is:

```
PostQuitMessage (0) ;
```

The PostQuitMessage() function inserts a WM_QUIT message in the message queue, thus terminating the GetMessage loop and ending the program.

19.3.4 The Default Windows Procedure

The code in the template file contains a return statement for each of the messages that it handles. For example:

```
case WM_PAINT :
    hdc = BeginPaint (hwnd, &ps) ;
    GetClientRect (hwnd, &rect) ;
// Initial display operations here
    EndPaint (hwnd, &ps) ;
    return 0 ;
```

The last statement in this routine returns a value of zero to Windows. The Windows documentation states that zero must be returned when an application processes the WM_PAINT message. Some Windows messages, not many, require a return value other than zero.

Many of the messages received from Windows, or retrieved from the message queue, are of no interest to your application. In this case, code must provide a default action for those messages that it does not handle. Windows contains a function, named DefWindowsProc(), that ensures this default action. DefWindowsProc() provides specific processing for those messages that require it, thus implementing a default behavior. For those messages that can be ignored, DefWindowsProc() returns zero. Your application uses the return value of DefWindowsProc() as its own return value from the Windows procedure. This action is coded as follows in the template file:

```
return DefWindowProc (hwnd, iMsg, wParam, lParam) ;
```

The parameters passed to DefWindowsProc() are the same message parameters received by your windows procedure from the operating system.

19.4 The WinHello Program

In the first walkthrough, at the beginning of this chapter, we used the template file Templ01.cpp to create a new project, which we named Program Zero Demo. Program Zero Demo resulted in a do-nothing program since no modifications were made to the template file at that time. In the present walkthrough we proceed to make modifications to the template file in order to create a Windows program different from the template. This project, which we named Hello Windows, is a Windows version of the classic "Hello World" program.

We first create a new project and use the template file Templ01.cpp as the source code base for it. In order to do this we must follow all the steps in the first walkthrough, except that the project name is now Hello Windows and the name template file Templ01.cpp is copied and renamed WinHello.cpp. After you have finished all the steps in the walkthrough you will have a project named Hello Windows and the source file named WinHello.cpp listed in the Project Workspace and displayed in the Editor Window. After the source file is renamed, you should edit the header block to reflect the file's new name and the program's purpose. Figure 19.5 shows the Developer Studio screen at this point.

The project Hello Windows, which we are about to code, has the following features:

- The caption displayed on the program title bar is changed to "Hello Windows."

- When the program executes it displays a greeting message on the center of its client area.

- The program now contains a customized icon. A small version of the icon is displayed in the title bar and a larger one is used when the program's executable is represented by a shortcut on the Windows desktop.

Once you have created the project named Hello Windows and included in it the source file WinHello.cpp, you are ready to start making modifications to the source and inserting new elements into the project.

19.4.1 Modifying the Program Caption

The first modification that we make to the source is to change the caption that is displayed on the title bar when the program executes. This requires editing the third parameter passed to the CreateWindowsEx() function in WinMain(). The parameter now reads "Hello Windows." Throughout this book we use the project's name, or a variation of it, as the title bar caption. Our reason for this is to make it easy to find the project files from a screen snapshot of the executable.

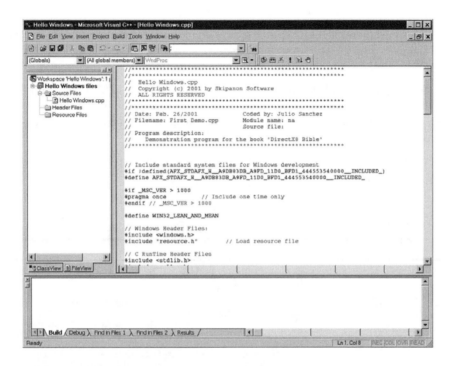

Figure 19.5 *The Hello Windows Project and Source File*

19.4.2 Displaying Text in the Client Area

The second modification requires entering a call to the DrawText() API function in the case WM_PAINT processing routine. The routine now is:

```
case WM_PAINT :
    hdc = BeginPaint (hwnd, &ps) ;
    GetClientRect (hwnd, &rect) ;

    // Display message in the client area
    DrawText (hdc,
              "Hello World from Windows",
              -1,
              &rect,
              DT_SINGLELINE | DT_CENTER | DT_VCENTER);

    EndPaint (hwnd, &ps) ;
    return 0 ;
```

The call to DrawText() requires five parameters. When calls require several parameters, we can improve the readability of the source by devoting a separate text line to each parameter, or to several associated parameters, as in the previous listing.

- The first parameter to DrawText() is the handle to the device context. This value was returned by the call to BeginPaint(), described previously in this chapter.

- The second parameter to DrawText() points to the string to be displayed. The string can also be enclosed in double quotation marks, as in the previous listing.

- The third parameter is – 1 if the string defined in the second parameter terminates in NULL. If not, then the third parameter is the count of the number of characters in the string.

- The fourth parameter is the address of a structure of type RECT which contains the logical coordinates of the area in which the string is to be displayed. The call to GetClientRect(), made in the WM_PAINT message intercept, filled the members of the rect structure variable.

- The fifth parameter is the text formatting options.

Table 19.6 lists the most used of these controls.

Table 19.6

Symbolic Constant in DrawText() Function

SYMBOLIC CONSTANT	MEANING
DT_BOTTOM	Bottom-justifies text. Must be combined with DT_SINGLELINE.
DT_CALCRECT	This constant is used to determine the width and height of the rectangle. If there are multiple lines of text, DrawText uses the width of the rectangle in the RECT Structure variable supplied in the call and extends the Base of the rectangle to bound the last line of text. If there is only one line of text, DrawText modifies the right side of the rectangle so that text is not drawn.
DT_CENTER	Centers text horizontally.
DT_EXPANDTABS	Expands tab characters. The default number of characters per tab is eight.
DT_EXTERNALLEADING	Includes the font external leading in line height. Normally, it is not included.
DT_LEFT	Aligns text to the left.
DT_NOCLIP	Draws without clipping. The function executes somewhat faster when DT_NOCLIP is used.
DT_NOPREFIX	DrawText interprets the control character & as a command to underscore the character that follows. The control characters && prints a single &. By specifying DT_NOPREFIX, this processing is turned off.
DT_RIGHT	Aligns text to the right.
DT_SINGLELINE	Displays text on a single line only. Carriage returns and linefeeds are ignored.
DT_TOP	Top-justifies text (single line only).
DT_VCENTER	Centers text vertically (single line only).
DT_WORDBREAK	Breaks words. Lines are automatically broken between words if a word extends past the edge of the rectangle specified by the lpRect parameter. A carriage return-linefeed sequence also breaks the line.

19.4.3 Creating a Program Resource

The last customization that you have to perform on the template file is to create two cus-tomized icons, which are associated with the program window. The icons correspond to the hIcon and hIconSm members of the WNDCLASSEX structure described previously and listed in Appendix F. hIcon is the window's standard icon. Its default size is 32-by-32 pixels, although Windows automatically resizes this icon as required. The standard icon is used on the Windows desktop when a shortcut is created and in some file listing modes of utilities like Windows Explorer. The small icon is 16-by-16 pixels, which makes it one-fourth the size of the large one. This is the icon shown in dialog boxes that list filenames, by Windows Explorer, and in the program's title bar. Windows NT uses a scaled version of the standard icon when a smaller one is required.

An icon is a resource. Resources are stored in a read-only, binary data files, which the application can access by means of a handle. We introduce icons at this time be-cause other program resources such as cursors, menus, dialog boxes, bitmaps, and fonts are handled similarly. The icons that we create in this walkthrough are consid-ered an application-defined resource.

The most convenient way of creating and using resources is to take advantage of the facilities in the development environment. Visual C++ provides several resource editors, and Developer Studio facilitates the creation and manipulation of the support files required for using resources. Graphics programmers often want to retain the highest possible control over their code; however, the use of these facilities in creat-ing and managing resources does not compromise this principle. The files created by the development environment are both visible and editable. As you gain confidence and knowledge about them you can progressively take over some or all of the opera-tions performed by the development software. In this book we sometimes let the de-velopment environment generate one or more of the program files and then proceed to edit them so that it better suits our purpose.

The convenience of using the automated functions of the development environ-ment is made evident by the fact that a simple resource often requires several soft-ware elements. For example, a program icon requires the following components:

- A bitmap that graphically encodes the icon. If the operating system and the application supports the small icon, then two bitmaps are required.

- A script file (also called a resource definition file) which lists all the resources in the ap-plication and may describe some of them in detail. The resource script can also reference other files and may include comments and preprocessor directives. The resource com-piler (RC.EXE) compiles the script file into a binary file with the extension .RES. This bi-nary file is referenced at link time. The resource file has the extension .RC.

- The script file uses a resource header file, with the default filename "resource.h", which contains preprocessor directives related to the resources used by the application. The application must reference this file with an #include statement.

19.4.4 Creating the Icon Bitmap

Developer Studio provides support for the following resources: dialog boxes, menus, cursors, icons, bitmaps, toolbars, accelerators, string tables, and version controls. Each resource has either a graphics editor or a wizard that helps create the resource. In this discussion we refer to either one of them as a resource editor.

Resource editors can be activated by clicking on the Resource command in the Insert menu. At this time Developer Studio displays a dialog box with an entry for each type of resource. Alternatively, you can access the resource editors faster by displaying the Resource toolbar. In Visual C++ 4 and later this is accomplished by clicking on the Toolbars command in the View menu, and then selecting the checkbox for the Resource option. In Versions 5 and 6 select the Customize command in the Tools menu, open the Toolbars tab in the Customize dialog box and select the checkbox for the Resource option. The Graphics and Colors boxes should also be checked to display the normal controls in the resource editors. The resulting toolbar is identical in both cases. Once the Resource toolbar is displayed, you can drag it into the toolbar area or to any other convenient screen location. The Insert Resource dialog screen and the resource toolbar are shown in Figure 19.6.

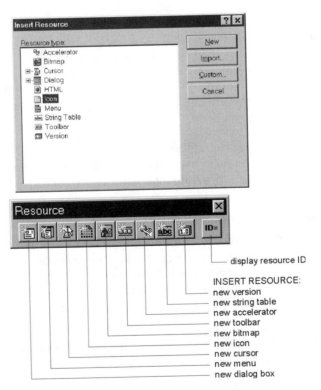

Figure 19.6 *Developer Studio Insert Resource Dialog Screen and Toolbar*

You can activate the icon editor either by selecting the icon option in the Resource dialog box or by clicking the appropriate button on the toolbar. The icon editor is simple to use and serves well in most cases. It allows creating the bitmap for several sizes of icons. Although the interface to the icon editor is simple, it is also powerful and flexible. You should experiment with the icon editor, as well as with the other resource editors, until you have mastered all their options and modes. Figure 19.7 shows the icon editor in Developer Studio.

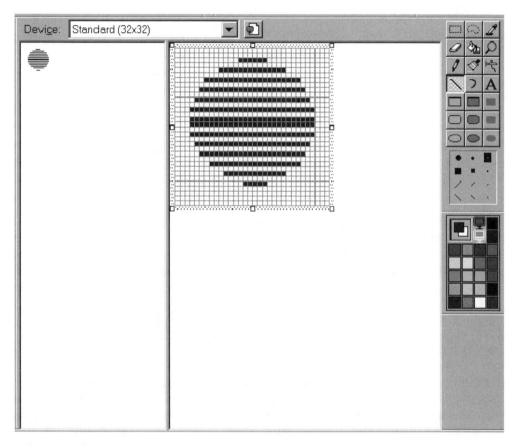

Figure 19.7 *Creating An Icon Resource with Developer Studio Icon Editor*

The toolbar on the right of the icon editor is similar to the one used in the Windows Paint utility and in other popular graphics programs. There are several tools that allow drawing lines, curves, and geometrical figures in outline or filled form. Also, there is a palette box from which colors for foreground and background can be selected.

Developer Studio makes possible the creation of a large and a small icon in the same resource. To request the small icon, click on the New Device Image button and then select the 16-by-16 icon. The two icons, 32-by-32 pixels and 16-by-16 pixels, can be developed alternatively by selecting one of them in the Open Device Image scroll box in the icon editor. Windows automatically uses the large and the small icon as required.

In the WinHello program the WNDCLASSEX structure is edited to support user-created large and small icons, as follows:

// The program icon is loaded in the hIcon and hIconSm

```
//    structure members
WNDCLASSEX  wndclass ;
wndclass.hIcon          = (HICON) LoadImage(hInstance,
                          MAKEINTRESOURCE(IDI_ICON1),
                          IMAGE_ICON,          // Type
                          32, 32,              // Pixel size
                          LR_DEFAULTCOLOR) ;
    .
    .
    .
wndclass.hIconSm        = (HICON) LoadImage(hInstance,
                          MAKEINTRESOURCE(IDI_ICON1),
                          IMAGE_ICON,          // Type
                          16, 16,              // Pixel size
                          LR_DEFAULTCOLOR) ;
```

The MAKEINTRESOURCE macro is used to convert an integer value into a resource. Although resources can also be referenced by their string names, Microsoft recommends the use of the integer value. The name of the icon resource, IDI_ICON1, can be obtained from the resource script file. However, an easier way of finding the resource name is to click the Resource Symbols button on the Resource toolbar (labeled ID=) or select the Resource Symbols command in the View menu. Either the symbolic name or the numerical value for the icon resource that is shown on the Resource Symbols screen can also be used in the MAKEINTRESOURCE macro.

In the process of creating an icon bitmap, Developer Studio also creates a new script file, or adds the information to an existing one. However, when working outside of the MFC, you must manually insert the script file into the project. This is done by selecting the Add to Project command in the Project menu and then clicking on the Files option. In the Insert Files into Project dialog box, select the script file, which in this case is the one named Script1.rc, and then press the OK button. The script file now appears on the Source Files list in the Files View window of the Project Workspace.

In addition to the script file, Developer Studio also creates a header file for resources. The default name of this file is resource.h. In order for resources to be available to the code you must enter an #include statement in the main source file, as follows:

```
#include "resource.h"
```

Notice that the double quotation marks surrounding the filename indicates that it is in the current folder.

Figure 19.8 *Screen Snapshot of the WinHello Program*

At this point, all that is left to do is to compile the resources, the source files, and link the program into an executable. This is done by selecting the Rebuild All command in the Build menu. Figure 19.8 shows the screen display of the WinHello program.

19.5 WinHello Program Listing

The following is a listing of the WinHello cpp source file that is part of the Hello Windows project .

```
//*************************************************************************
//   PROJECT: Hello Windows
//   Source: WinHello.cpp
//   Chapter reference: 19
//*************************************************************************
//
//   Description:
//     A Hello Windows demonstration program
//   Topics:
//        1. Create a program icon
//        2. Display a text message in the client area
//*************************************************************************
```

```
//
//
#include <windows.h>          // Standard Windows header
#include "resource.h"          // Load resource file for icon

// Predeclaration of the window procedure
LRESULT CALLBACK WndProc (HWND, UINT, WPARAM, LPARAM) ;

//***********************************************************************
//                              WinMain
//***********************************************************************


int WINAPI WinMain (HINSTANCE hInstance, HINSTANCE hPrevInstance,
                    PSTR szCmdLine, int iCmdShow)
{
    static char szAppName[] = "Demo" ; // Class name
    HWND         hwnd ;
    MSG          msg ;

    // Defining a structure of type WNDCLASSEX
    //   The program icon is loaded in the hIcon and hIconSm
    //   structure members
    WNDCLASSEX   wndclass ;
    wndclass.cbSize       = sizeof (wndclass) ;
    wndclass.style        = CS_HREDRAW | CS_VREDRAW ;
    wndclass.lpfnWndProc  = WndProc ;
    wndclass.cbClsExtra   = 0 ;
    wndclass.cbWndExtra   = 0 ;
    wndclass.hInstance    = hInstance ;
    wndclass.hIcon        = (HICON)LoadImage(hInstance,
                                    MAKEINTRESOURCE(IDI_ICON1),
                                    IMAGE_ICON,
                                    32, 32,
                                    LR_DEFAULTCOLOR) ;
    wndclass.hCursor      = LoadCursor (NULL, IDC_ARROW) ;
    wndclass.hbrBackground = (HBRUSH) GetStockObject
                              (WHITE_BRUSH) ;
    wndclass.lpszMenuName  = NULL ;
    wndclass.lpszClassName = szAppName ;
    wndclass.hIconSm      = (HICON)LoadImage(hInstance,
                                    MAKEINTRESOURCE(IDI_ICON1),
                                    IMAGE_ICON,
                                    16, 16,
                                    LR_DEFAULTCOLOR) ;
    // Registering the structure wmdclass

    RegisterClassEx (&wndclass) ;

    // CreateWindow()
    hwnd = CreateWindowEx (
            WS_EX_LEFT,               // Left aligned (default)
            szAppName,                 // pointer to class name
            "Hello Windows",          // window caption
            WS_OVERLAPPEDWINDOW,      // window style
            CW_USEDEFAULT,            // initial x position
            CW_USEDEFAULT,            // initial y position
            CW_USEDEFAULT,            // initial x size
            CW_USEDEFAULT,            // initial y size
            NULL,                     // parent window handle
```

```
            NULL,                      // window menu handle
            hInstance,                 // program instance handle
            NULL) ;                    // creation parameters

    ShowWindow (hwnd, iCmdShow) ;
    UpdateWindow (hwnd) ;

    // Message loop
    while (GetMessage (&msg, NULL, 0, 0))
            {
            TranslateMessage (&msg) ;
            DispatchMessage (&msg) ;
            }
    return msg.wParam ;

}

//****************************
//    Windows Procedure
//****************************

LRESULT CALLBACK WndProc (HWND hwnd, UINT iMsg, WPARAM wParam,
LPARAM lParam)
{

    PAINTSTRUCT ps ;
    RECT        rect ;
    HDC         hdc;

    switch (iMsg)
        {
    // Windows message processing
        case WM_CREATE:
            return 0;

        case WM_PAINT :
            hdc = BeginPaint (hwnd, &ps) ;
            GetClientRect (hwnd, &rect) ;

            // Display message in the client area
            DrawText (hdc,
                    "Hello World from Windows",
                    -1,
                    &rect,
                    DT_SINGLELINE | DT_CENTER | DT_VCENTER);

            EndPaint (hwnd, &ps) ;
            return 0 ;

        // End of program execution
        case WM_DESTROY :
            PostQuitMessage (0) ;
            return 0 ;
        }
    return DefWindowProc (hwnd, iMsg, wParam, lParam) ;

}
```

SOFTWARE ON-LINE

The source file for the Hello World program is found in the Chapter 19 folder in the book's software on-line.

Chapter 20

Text Display

Chapter Summary

In this chapter we discuss the part of Windows programming that refers to text display and rendering operations. The discussion on text programming at this point serves as an introduction into Windows application development. Because understanding text programming requires knowledge of the fundamental concepts of Windows programming. These are the client area, the Windows coordinate system, the display context, and the mapping mode.

20.0 Text in Windows

Computer systems, including the PC, have historically differentiated between text and graphics. The original notion was that programs could either execute in textual form, by displaying messages composed of alphabetical and numeric characters, or they could use pictures and images to convey information. When the VGA (Video Graphics Array) video standard was released in 1987, it defined both text and graphics modes, with entirely different features and programming. Even in Windows, which is a graphics environment by design, there is a distinction between console-based applications and graphics-based applications. In console-based applications, Windows refers to a Console User Interface, or CUI, and in graphics-based applications, to a Graphics User Interface, or GUI. When you select the New command in the Developer Studio File menu, the Projects tab contains an option for creating a Win32 Console Application.

In fact, in the Windows environment, the distinction between text and graphics programs is not clear. The text-related functions in the API, which are more than twenty, are actually part of the GDI (Graphics Device Interface). In Windows text is a another graphics resource.

Here we consider Windows text operations as related to GUI programming. Console-based applications are not discussed in this chapter. In addition, text manipulations and programming provide an introduction to topics related to client area access and control, which are at the core of Windows programming.

20.1 The Client Area

The part of the window on which a program can draw is called the client area. The client area does not include the title bar, the sizing border, nor any of the optional elements such as the menu, toolbar, status bar, and scroll bars. The client area is the part of the program window that you access to convey information to the user and on which your application displays child windows and program controls.

DOS programmers own the device, whether working on graphics or on text modes. Once a DOS text program has set a video mode, it knows how many characters can be displayed in each text line, and how many text lines fit on the screen. By the same token, a DOS graphics program knows how many pixel rows and columns are in its domain. Some Windows programs use a fixed-size window, but in most cases, a Windows application cannot make assumptions regarding the size of its client area. Normally, the user is free to resize the screen vertically, horizontally, or in both directions simultaneously. There is practically no limit to how small it can be made, and it can be as large as the entire Windows application area. Writing code that can reasonably accommodate the material displayed to any size of the client area is one of the challenges of Windows programming.

20.2 Device and Display Contexts

The notion of a device context is that of a Windows data structure that stores information about a particular display device, such as the video display or a printer. All Windows functions that access the GDI require a handle to the device context as a parameter. The device context is the link between your application, the GDI, and the device-dependent driver that executes the graphics command on the installed hardware. Figure 20.1 is a schematic diagram of this relationship.

In Figure 20.1 we see that the Windows application uses one of several available operations to obtain a device context. The call to BeginPaint(), used in TEMPL01.CPP and in the WinHello program listed in Chapter 16, returns the handle to the device context. BeginPaint() is the conventional way of obtaining the handle to the device context in a WM_PAINT handler. The GetDC() function is often used to obtain the handle to the device context outside of WM_PAINT. In either case, from now on, a particular device context data structure is associated with the application.

Once a device context has been obtained, GDI calls examine the device context attributes to determine how to perform a drawing operation. In Figure 20.1 we see some of the DC attributes: the background color, the brush, and the current position of the drawing pen. There are many attributes associated with a common display context. For example, the default stock pen is defined as BLACK_PEN in the device context. If this stock pen is not changed, the GDI uses it in all drawing operations. The application can, however, change the stock pen in the device context to NULL_PEN or WHITE_PEN by calling SelectPen().

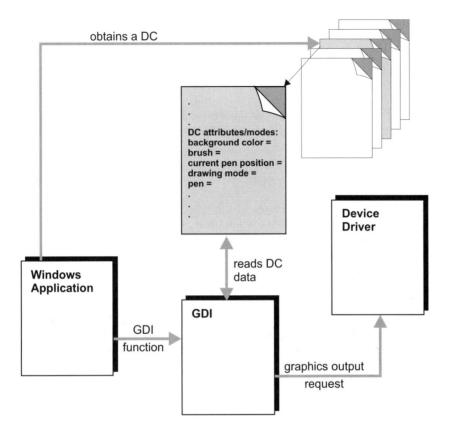

Figure 20.1 *The Device Context, Application, GDI, and Device Driver*

20.2.1 The Display Context

The video display is a device that requires most careful handling in a multitasking environment. Several applications, as well as the system itself, usually share the display device. The notions of child and parent windows, client and non-client areas, desktop windows, and of applications area, all relate to this topic. The display context is a special device context for a display device.

The principal difference between a device context and the display context is that a device context allows access to the entire device, while the display context limits access to the output area of its associated window. A display context usually refers to one of the following areas:

- The window's client area

- The window's entire surface, including the non-client area

- The entire desktop surface

Application output is usually limited to the client area, therefore, this is the default display context.

Since the display context is a specialization of the term device context, it is correct to refer to the display context as a device context. The reverse, however, is not always true. For example, the printer device context is not a display context. Previously we referred to the display context as a device context, which is acceptable. Windows documentation does not always use these terms rigorously. This has been the cause of some misunderstanding. The fact that Windows documentation sometimes uses the term display device context as equivalent to display context has added to the confusion.

20.2.2 Display Context Types

According to the application's needs, there are four possible classes of display contexts: common DC, single DC, private DC, and parent DC. The type of display context for a window is defined in the WNDCLASSEX structure. During the call to RegisterClassEx() we establish the type of display context for the windows class. This is determined by the value entered in the wndclass.style member of WNDCLASSEX.

In Table 19.2 there are three constants that refer to the display context types: CS_OWNDC, CS_CLASSDC, and CS_PARENTDC. When no display type constant is entered in the wndclass.style, then the display context type is common, which is the default. In the case of a common display context, Windows resets all attributes to the default values each time the handle is retrieved. This requires the application to reset each attribute that is different from the default settings.

The class display context is enabled with the CS_CLASSDC constant at the time of registering the window class. In this case, Windows initializes the display context attributes once, for all windows of the class. When the display context is retrieved, Windows resets the device origin and the clipping region, but not the other attributes. All windows of this class obtain the same attributes with the handle to the display context. One disadvantage of a class display context is that if one window makes changes to the display context, these changes remain in effect for all subsequent windows that use it.

The parent display context is enabled by entering the CS_PARENTDC constant in the WNDCLASSEX structure. In this case, Windows creates a common display context and sets its clipping region to the same as that of the parent. The result is that a child window can draw to its parent's client area. The most common use of a parent display context is in drawing controls inside dialog boxes. Round-off errors that result from calculating the bounding box for dialog boxes sometimes cause controls that are clipped at display time. Using a parent display context solves this problem.

The private display context is associated with a window when the CS_OWNDC constant is used in the wndclass.style member of WNDCLASSEX. At registration time, each window created from the class is given a private display context. Because each window has its own display context permanently associated, it need be retrieved only once. All attributes assigned to a private display context are retained until they are explicitly changed. In some types of applications the use of a private display context minimizes coding and improves performance.

Applications that often make changes to the client area, as is the case with many graphics programs, can often profit from a private display context. In order to accomplish this, several changes have to be made to the TEMPL01.CPP program file. In the first place, an OR operation must be performed between the CS_OWNDC constant and the other values in the wndclass.style member of WNDCLASSEX, as follows:

```
// Defining a structure of type WNDCLASSEX
  WNDCLASSEX   wndclass ;
  wndclass.cbSize        = sizeof (wndclass) ;
  wndclass.style         = CS_HREDRAW | CS_VREDRAW | CS_OWNDC;
    .
    .
    .
```

The remaining changes take place in the Windows procedure. In the first place, you must declare a variable of type HDC. This variable must have static scope so that its value is preserved between reentries of the windows procedure. The display context can be obtained during WM_CREATE processing, which executes at the time the window is created. This is possible because the display context is private. In this case, you can use the GetDC() function to obtain the handle to the display context, as in the following code fragment:

```
LRESULT CALLBACK WndProc (HWND hwnd, UINT iMsg, WPARAM wParam,
                          LPARAM lParam) {
     // Local variables
     PAINTSTRUCT        ps ;
     RECT               rect ;
     static HDC         hdc;        // Handle to private DC

     switch (iMsg)
        {
     // Windows message processing
        case WM_CREATE:
             hdc = GetDC(hwnd);     // Obtain handle to
                                    // private DC
             return 0;
    .
    .
    .
```

The private display context is available during WM_PAINT message intercept, and need not be retrieved during each iteration. Therefore, the return value from BeginPaint() can be discarded and the EndPaint() function becomes unnecessary, as in the following code fragment:

```
case WM_PAINT :
     BeginPaint (hwnd, &ps) ;
     GetClientRect (hwnd, &rect) ;
     // Display message in the client area
     DrawText (hdc,
               "Demo program using a private DC",
               -1,
```

```
                    &rect,
                        DT_SINGLELINE | DT_CENTER | DT_VCENTER);
            return 0 ;

                .

                .

                .
```

The project named Private DC Demo, in the book's on-line software package, contains the full source for a private DC demonstration. You can use the source file TEMPL02.CPP as a template for creating applications that use a private display context.

20.2.3 Window Display Context

Applications sometimes wish to draw not only on the client area, but elsewhere in the window. Normally, areas such as the title bar, menus, status bar, and scroll bars are inaccessible to code that uses one of the display context types previously mentioned. You can, however, retrieve a window-level display context. In this case, the display context's origin is not at the top-left corner of the client area, but at the top-left corner of the window. The GetWindowDC() function is used to obtain the handle to the window-level display context and the ReleaseDC() function to release it. In general, drawing outside of the client area should be avoided, since it can create problems to the application and to Windows.

20.3 Mapping Modes

One of the most important attributes of the display context is the mapping mode, since it affects practically all drawing operations. The mapping mode is actually the algorithm that defines how logical units of measurement are translated into physical units. To understand mapping modes we must start with logical and device coordinates.

The programmer specifies GDI operations in terms of logical coordinates, or logical units. The GDI sends commands to the device driver in physical units, also called device coordinates. The mapping mode defines the logical units and establishes the methods for translating them into device coordinates. This translation can be described as a mapping operation. In regards to the display device, as well as in most printers, device coordinates are expressed in pixels. Logical coordinates depend on the selected mapping mode. Windows defines six fixed-size mapping modes, as shown in Table 20.1.

Two other mapping modes, not listed in Table 20.1, are MM_ANISOTROPIC and MM_ISOTROPIC. These modes can be used for shrinking and expanding graphics by manipulating the coordinate system. These two scalable mapping modes are useful in performing powerful graphics manipulation.

Table 20.1

Windows Fixed-Size Mapping Modes

MAPPING MODE	LOGICAL UNITS	X-AXIS	Y-AXIS
MM_TEXT	pixel	right	down
MM_LOWMETRIC	0.1 mm	right	up
MM_HIGHMETRIC	0.01 mm	right	up
MM_LOENGLISH	0.01 inch	right	up
MM_HIENGLISH	0.001 inch	right	up
MM_TWIPS	1/1440 inch	right	up

The default mapping mode, MM_TEXT, is also the most used one. In MM_TEXT, the logical coordinates coincide with the device coordinates. Programmers who learned graphics in the DOS environment usually feel very comfortable with this mapping mode. Note that the name MM_TEXT refers to how we normally read text in the Western languages: from left-to-right and top-to-bottom. The name is unrelated to text display.

The selection of a mapping mode depends on the needs and purpose of the application. Two of the mapping modes, MM_LOMETRIC and MM_HIMETRIC, are based on the metric system (millimeters). MM_LOENGLISH and MM_HIENGLISH are based on the English system of measurement (inches). MM_TWIPS is based on a unit of measurement used in typography called the twip, which is equivalent to 1/20th of a point, or 1/1440 inch. An application that deals with architectural or technical drawings, in which dimensions are usually in inches or millimeters, can use one of the mapping modes based on metric or English units of measurement. A graphics design program, or a desktop publishing application, would probably use the MM_TWIPS mapping mode.

The SetMapMode() function is used to change the mapping mode in the device context. One of the parameters in the call is the handle to the device context; the other parameter is one of the predefined mapping mode constants. For example, to change the mapping mode to LO_METRIC, you would code:

```
static int      oldMapMode;
   .
   .
   .
oldMapMode = SetMapMode (hdc, LO_METRIC);
```

The function returns the previous mapping mode, which can be stored in an integer variable. Later on, the original mapping mode can be restored as follows:

```
SetMapMode (hdc, oldMapMode);
```

SetMapMode() returns zero if the function call fails.

20.3.1 Screen and Client Area

Windows uses several coordinate systems. The basic unit of measurement is the pixel, also called a device unit. Horizontal values increase from left to right and vertical values from top to bottom. The origin of the coordinate system is the top-left corner of the drawing surface. Three different extents are used in relation to the device area: screen, client area, and window coordinate systems.

The screen coordinate system refers to the entire display area. This coordinate system is used when location and size information refer to the entire video display. The call to CreateWindowEx(), in the program WINHELLO.CPP and most of the template files, uses the symbolic constant CW_USEDEFAULT. This constant lets Windows select a position and size for the program's window. Alternatively, we could have specified the window's location and size in device units. For example, the following call to CreateWindowEx() locates the window at 20-by-20 pixels from the screen's upper-left corner and forces a dimension of 400-by-500 pixels:

```
// CreateWindow()
hwnd = CreateWindowEx (
        WS_EX_LEFT,                 // Left aligned (default)
        szClassName,                // pointer to class name
        "WinHello Program",         // window caption
        WS_OVERLAPPEDWINDOW,        // window style
        20,                         // initial x position
        20,                         // initial y position
        400,                        // initial x size
        500,                        // initial y size
        NULL,                       // parent window handle
        NULL,                       // window menu handle
        hInstance,                  // program instance handle
        NULL) ;                     // creation parameters
```

Other Windows functions, such as those that return the mouse cursor position, the location of a message box, or the location and size of the windows rectangle, also use screen coordinates.

Client area coordinates are relative to the upper-left corner of the client area, not to the video display. The default unit of measurement is the pixel. The function ClientToScreen() can be used to obtain the screen coordinates of a point in the client area. ScreenToClient() obtains the client area coordinates of a point defined by its screen coordinates. In either function, the x and y coordinates are passed and returned in a structure of type POINT.

Window coordinates refer to the top-left corner of the window itself, not to the client area. Applications that use the window display context, mentioned earlier in this chapter, use windows coordinates.

20.3.2 Viewport and Window

The terms viewport and window, when used in relation to logical and device coordinates, can be the source of some confusion. In the first place, Windows documentation uses the term viewport in a way that does not coincide with its most accepted meaning. In graphics terminology, a viewport is a specific screen area set aside for a particular graphics function. In this sense, the notion of a viewport implies a region within the application's window.

In Windows, the viewport is often equated with the client area, the screen area, or the application area, according to the bounds of the device context. The one characteristic element of the viewport is that it is expressed in device units, which are pixels. The window, on the other hand, is expressed in terms of logical coordinates. Therefore, the unit of measurement of a window can be inches, millimeters, twips, or pixels in the six fixed-sized mapping modes, or one defined by the application in the two scalable mapping modes.

In regards to viewports and windows, there are two specific boundaries that must be considered: the origin and the extent. The origin refers to the location of the window or viewport, and the extent to its width and height. The origin of a window and a viewport can be set to different values in any of the mapping modes. Function calls to set the window and the viewport extent are ignored when any one of the six fixed-sized mapping modes is selected in the device context. However, in the two scalable mapping modes, MM_ISOTROPIC and MM_ANISOTROPIC, both the origin and the extent of the viewport and the window can be set separately.

A source of confusion is that both the viewport and the window coincide in the default mapping mode (MM_TEXT). In the fixed-size mapping modes, the extent of the viewport and the window cannot be changed, as mentioned in the preceding paragraph. This should not be interpreted to mean that they have the same value. Actually, the measurement in units of length of the viewport and the window extent is meaningless. It is the ratio between the extent that is useful. For example, if the viewport extent is 20 units and the window extent is 10 units, then the ratio of viewport to window extent is of 20/10, or 2. This value is used as a multiplier when converting between window and device coordinates. Other factors that must be taken into account in these conversions are the location of the point, the origin of the viewport, and the origin of the window. Figure 20.2, on the following page, is a simplified, schematic representation of the concepts of viewport and window.

In Figure 20.2, the dimension of the logical units is twice that of the device units, in both axes. Therefore, the ratio between the window extension and the device extension (xVPExt / xWExt and yVPExt / yWExt) equals 2. The point located at xW, yW is at window coordinates xW = 8, yW = 9, as shown in the illustration. To convert to device coordinates, we apply the corresponding formulas. In calculating the x-axis viewport coordinate of the point xW, yW, we proceed as follows:

```
xVP = (xW - xWOrg) x (xVPExt / xWExt) + xVPOrg
xVP = (8 -(- 16)) x 2 + 0
xVP = 48
```

This means that in the example in Figure 20.2, the point at window coordinates x = 8, y = 9, located in a window whose origin has been displaced 16 logical units on the x-axis, and 5.5 logical units in the y-axis, is mapped to viewport coordinates xVP = 48, yVP = 25. Note that the sample calculations do not include the y-coordinate.

20.4 Programming Text Operations

Text operations in console-based applications are usually a simple task. The text characters are displayed using whatever font is selected at the system level, and at the screen line and column where the cursor is currently positioned. In analogy with the old Teletype machines, this form of text output programming is said to be based on the model of a "glass TTY." But even when the program takes control of the display area, the matter of text output is no more complicated than selecting a screen line and a column position.

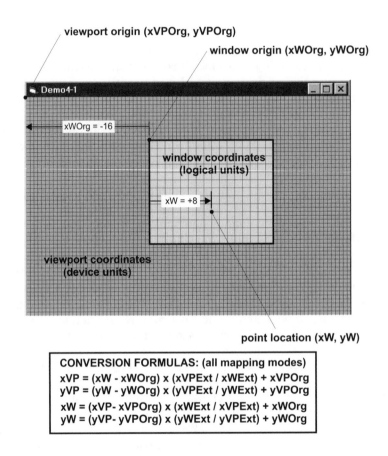

Figure 20.2 *Viewport and Window Coordinates*

In graphics programming, and particularly in Windows graphics, the coding of text operations often becomes a major task, to the point that Windows text programming is considered a specialty field. In this sense, it is possible to speak of bitmapped graphics, of vector graphics, and of text graphics. Developing a GDI-based text-processing application, such as a Windows word processing or desktop publishing program, involves a great amount of technical complexity. In addition to programming skills, it requires extensive knowledge of typography, digital composition, and graphics arts. At present, we are concerned with text graphics in a non-specialized context. That is, text display is one of the functionality that is normally necessary in implementing a Windows application. But even in this more general sense, text programming in Windows is not without some complications.

20.4.1 Typefaces and Fonts

A collection of characters of the same design is called a typeface. Courier, Times Roman, and Helvetica are typefaces. Courier is a monospaced typeface that originated in typewriter technology. The characters in the Courier typeface all have the same width. Times Roman is a typeface developed in the nineteenth century by an English newspaper with the purpose of making small type readable when printed on newspaper stock. Times Roman uses short, horizontal lines of a different thickness. To some, these elements resemble hooks; for which the typeface is called serif (hook, in French). On the other hand, the characters in the Helvetica typeface have the same thickness; therefore, it is called a sans-serif typeface (without hooks).

Times Roman and Helvetica are proportionally spaced fonts; that is, each character is designed to an ideal width. In a proportionally spaced font, the letter "w" is wider than the letter "i." In Windows, proportionally spaced fonts are sometimes called variable pitch fonts. They are more pleasant and easier to read than monospaced fonts, but digits displayed in proportionally spaced fonts do not align in columns. Figure 20.3 shows text in Courier, Times Roman, and Helvetica typefaces.

```
These lines are in Courier typeface.
All characters have the same width.
```

These lines are in *Times Roman* typeface.
Times Roman is a serif typeface of
great readability.

These lines are in *Helvetica* typeface.
Helvetica is a sans-serif typeface
often used for display type.

Figure 20.3 *Courier, Times Roman, and Helvetica Typefaces*

A group of related typefaces is called a typeface family; for example, Helvetica Bold and Helvetica Oblique are typeface families. A font is a collection of characters of the same typeface and size. In this sense you can speak of the Times Roman 12-point font. Type style is a term used somewhat loosely in reference to specific attributes applied to characters in a font. Boldface (dark), roman (straight up), and italics (slanted towards the right) are common type styles.

Historically, Windows fonts have been of three different types: raster, vector, and TrueType. Raster fonts are stored as bitmaps. Vector fonts, sometimes called stroke fonts, consist of a set of drawing orders required to produce each letter. TrueType fonts, introduced in Windows 95, are similar to PostScript fonts. They are defined as lines and curves, can be scaled to any size, and rotated at will. TrueType fonts are more versatile and have the same appearance on the screen as when printed. TrueType fonts also assure portability between applications. Programmers working in Windows 95/NT and later deal mostly with TrueType fonts.

For reasons related to copyright and trademark laws, some Windows fonts have names that differ from the traditional typefaces. For example, Times New Roman is the Windows equivalent of Times Roman, and the Helvetica typeface is closely approximated by the Windows versions called Arial, Swiss, and Switzerland.

The default Windows font is named the system font. In current versions of Windows, the system font is a proportionally spaced font. It is also a raster font, therefore, the characters are defined as individual bitmaps. Figure 20.4 is a screen snapshot of a Windows program that demonstrates the screen appearance of the various non-TrueType fonts.

Figure 20.4 *Windows Non-TrueType Fonts*

20.4.2 Text Formatting

In order to display text in a graphics, multitasking environment (one in which the screen can be resized at any time), code must be able to obtain character sizes at run time. For example, in order to display several lines of text you must know the height of the characters so that the lines are shown at a reasonable vertical distance from each other. By the same token, you also need to know the width of each character, as well as the width of the client area, in order to handle the end of each text line.

The GetTextMetrics() function provides information about the font currently selected in the display context. GetTextMetrics() requires two parameters: the handle to the device context and the address of a structure variable of type TEXTMETRICS. Table 20.2 lists the members of the TEXTMETRIC structure:

Table 20.2

TEXTMETRIC structure

TYPE	MEMBER	CONTENTS
LONG	tmHeight	Character height (ascent + descent)
LONG	tmAscent	Height above the baseline
LONG	tmDescent	Height below the baseline
LONG	tmInternalLeading	Internal leading
LONG	tmExternalLeading	External leading
LONG	tmAveCharWidth	Width of the lowercase letter "x"
LONG	tmMaxCharWidth	Width of widest letter in font
LONG	tmWeight	Font weight
LONG	tmOverhang	Extra width per string added to some synthesized fonts
LONG	tmDigitizedAspectX	Device horizontal aspect
LONG	tmDigitizedAspectY	Device vertical aspect. The ratio of tmDigitizedAspectX / tmDigitizedAspecty members is the aspect ratio of the device for which the font was designed.
BCHAR	tmFirstChar	First character in the font
BCHAR	tmLastChar	Last character in the font
BCHAR	tmDefaultChar	Character used as a substitute for Those not implemented in the font
BCHAR	tmBreakChar	Character used as a word break in Text justification
BYTE	tmItalic	Nonzero if font is italic
BYTE	tmUnderlined	Nonzero if font is underlined
BYTE	tmStruckOut	Nonzero if font is strikeout
BYTE	tmPitchAndFamily	Contains information about the font family in the four low-order bits of the following constants:

CONSTANT	BIT	MEANING
TMPF_FIXED_PITCH	0	fixed pitch font
TMPF_VECTOR	1	vector font
TMPF_TRUETYPE	2	True Type font
TMPF_DEVICE	3	device font

BYTE	tmCharSet	Specifies the font's character set

Notice that in printing and display technology, the baseline is an imaginary horizontal line that aligns the base of the characters, excluding descenders. The term leading (pronounced "led-ing") refers to the space between lines of type, usually measured from the baseline of one line to the baseline of the next one. Figure 20.5 shows the vertical character dimensions represented by the corresponding members of the TEXTMETRIC structure.

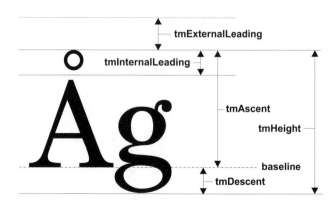

Figure 20.5 *Vertical Character Dimensions in the TEXTMETRIC Structure*

Text metric values are determined by the font installed in the device context. For this reason, where and how an application obtains data about text dimensions depend on the font and on how the device context is handled. An application that uses the system font, and no other, need only obtain text metric values once in each session. Since the system font does not change during a Windows session, these values are valid throughout the program's lifetime. However, if an application changes device contexts or fonts during execution, then the text metric values may also change.

In the simplest case, a text processing application can obtain text metric data while processing the WM_CREATE message. Usually, the minimal data required for basic text manipulations is the character height and width. The height is calculated by adding the values in the tmHeight and tmExternalLeading members of the TEXTMETRIC structure for the current display context (see Figure 20.5). The width of the lowercase characters can be obtained from the tmAveCharWidth member.

The calculation of the average width of uppercase characters is somewhat more complicated. If the font currently selected in the display context is monospaced (fixed pitch, in Windows terminology), then the width of the uppercase characters is the same as the lowercase ones. However, if the current font is proportionally spaced (sometimes called a variable pitch font in Windows), then you can obtain an approximation of the width of the uppercase characters by calculating 150 percent of the width of the lowercase ones. We have seen that the tmPitchAndFamily member of TEXTMETRIC has the low-order bit set if the font is monospaced. We can logically AND this value with a binary 1 in order to test if the font is monospaced. Assuming that a TEXTMETRIC structure variable is named tm, and that the width of the lowercase characters is stored in an integer variable named cxChar, the code would be as follows:

```
int cxCaps;      // Storage for width of uppercase characters
 if(tm.tmPitchAndFamily & 0x1)
    cxCaps = (3 * cxChar) / 2;    // 150 percent
else
    cxCaps = cxChar;              // 100 percent
```

More compact coding results from using the ? operator, as follows:

```
cxCaps = (tm.tmPitchAndFamily & 1 ? 3 : 2) * cxChar / 2;
```

The values can be stored in static variables for future use. The following code fragment shows the usual processing in this case:

```
static int cxChar;      // Storage for lowercase character width
static int cxCaps;      // Storage for uppercase character width
static int cyChar;      // Storage for character height plus
                        // leading
   .
   .

   .
 case WM_CREATE :
   hdc = GetDC (hwnd) ;
   GetTextMetrics (hdc, &tm) ;
```

```
cxChar  = tm.tmAveCharWidth ;
cxCaps  = (tm.tmPitchAndFamily & 1 ? 3 : 2) * cxChar / 2;
cyChar  = tm.tmHeight + tm.tmExternalLeading ;

ReleaseDC (hwnd, hdc) ;
return 0 ;
```

In addition to information about text dimensions, text processing applications also need to know the size of the client area. The problem in this case is that in most applications, the size of the client area can change at any time. If the window was created with the style WM_HREDRAW and WM_VREDRAW, a WM_SIZE message is sent to the Windows procedure whenever the client area size changes vertically or horizontally. A WM_PAINT message automatically follows. The application can intercept the WM_SIZE message and store, in a static variable, the vertical and horizontal dimensions of the client area. The size of the client area can be retrieved from these variables whenever you need to redraw to the window. Traditionally, the variables named cxClient and cyClient are used to store these values. The low word of the lParam value, passed to the Windows procedure during WM_SIZE, contains the width of the client area, and the high word contains the height. The code can be as follows:

```
static int cxClient;     // client area width
   static int cyClient;     // client area height
   .
   .
   .
   case WM_SIZE:
     cxClient = LOWORD (lParam);
     cyClient = HIWORD (lParam);
     return 0;
```

20.4.3 Paragraph Formatting

The logic needed for text formatting at the paragraph level is as follows: First, we determine the character dimensions by calling the GetTextMetric() and then reading the corresponding members of a TEXTMETRIC structure. Next, we obtain the size of the client area during WM_SIZE processing by means of the high- and low-word of the lParam argument. This information is sufficient for performing exact calculation on a monospaced font. In the case of a proportionally spaced font, we are forced to deal in approximations, since what we have obtained is the average width of lower-case characters and an estimate of the width of the upper-case ones.

GetTextExtentPoint32(), a function that has suffered several transformations in the various versions of Windows, computes the exact width and height of a character string. The function takes as a parameter the handle to the device context, since the string size calculated is based on the currently installed font. Other parameters are the address of the string, its length in characters, and the address of a structure of type SIZE where information is returned to the caller. The SIZE structure contains only two members: one for the x dimension and another one for the y dimension. The value returned by GetTextExtentPoint32() is in logical units.

Putting it all together: suppose you have a rather long string, one that requires more than one screen line, stored in a static or public array, and you want to display this string breaking the screen lines only at the end of words. Since in Windows the length of each line in the client area can be changed at any time by the user, the code would have to dynamically adjust for this fact. Placing the processing in a WM_PAINT message handler ensures that the display is updated when the client area changes in size. This also requires that we intercept the WM_SIZE message to recalculate the size of the client area, as discussed previously. The processing logic in WM_PAINT could be as follows:

1. Step through the string, pausing at each space, and calculate the string length using GetTextExtentPoint32(). Keep count of the number of characters to the previous space, or the beginning of the text string in the case of the first word.

2. If the length of the string is larger than could fit in the client area, then backtrack to the previous space and display the string to that point. Reset the string pointer so that the new string starts at the last character displayed. Continue at step 1.

3. If the end of the string has been reached, display the string starting at the last space and exit the routine.

The actual implementation requires a few other processing details. For example, you may want to leave a margin of a couple of characters on the left and right sides of the display area. In addition, the code would need to manipulate pointers and counters to keep track of the string positions and the number of characters to the previous space. One possible algorithm is reminiscent of the classic case of a circular buffer with two pointers: one to the buffer head and another one to the tail. Figure 20.6 graphically shows the code elements in one of many possible implementations.

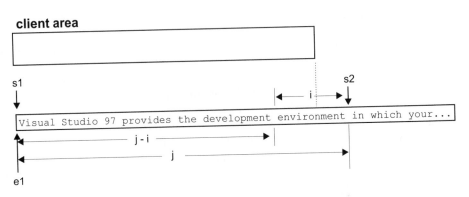

Figure 20.6 *Processing Operations for Multiple Text Lines*

In Figure 20.6, the pointer that signals the start of the string is designated with the letter s and the one for the end of the string with the letter e. s1 and e1 is the start position for both pointers. The variable i is a counter that holds the number of characters since the preceding space, and j holds the number of characters in the current substring. The code steps along the string looking for spaces. At each space, it measures the length of the string and compares it to the horizontal dimension of the client area. When the s pointer reaches location s2, the substring is longer than the display space available in the client area. The variable i is then used to reset the pointer to the preceding space and to decrement the j counter. The substring is displayed starting at e1, for a character count of j. Pointers and counters are then reset to the new sub-string and processing continues until the end of the string is found.

A demonstration program named TEX1_DEMO is furnished in the book's software on-line. The message to be displayed is stored in a public string, as follows:

```
// Public string for text display demonstration
char TextMsg[] = {"Visual Studio 97 provides the development "
"environment in which your programming and Web site "
"development packages run. This integrated set of tools runs"
  .
  .
  .
"spreadsheet programs." };
```

The processing operations, located in the Windows procedure, are coded as follows:

```
LRESULT CALLBACK WndProc (HWND hwnd, UINT iMsg, WPARAM wParam,
                          LPARAM lParam) {
 static int  cxChar, cxCaps, cyChar ;// Character dimensions
 static int  cxClient, cyClient; // Client area parameters
    HDC         hdc ;             // handle to device context
    int j;                        // Offset into string
    int i;                        // characters since last
                                  // space
    char *startptr, *endptr;      // String pointers
    int cyScreen;                 // Screen row holder

 // Structures
    PAINTSTRUCT ps;
    TEXTMETRIC  tm;
    SIZE        textsize; // Test string size

 switch (iMsg)
        {
    case WM_CREATE :
        hdc = GetDC (hwnd) ;
        GetTextMetrics (hdc, &tm) ;

        // Calculate and store character dimensions
        cxChar = tm.tmAveCharWidth ;
        cxCaps = ((tm.tmPitchAndFamily & 1) ? 3 : 2) *\
                cxChar / 2 ;
        cyChar = tm.tmHeight + tm.tmExternalLeading ;

        ReleaseDC (hwnd, hdc) ;
        return 0 ;

    case WM_SIZE:
        // Determine and store size of client area
        cxClient = LOWORD(lParam);
        cyClient = HIWORD(lParam);
        return 0;
```

```
case WM_PAINT :
    hdc = BeginPaint (hwnd, &ps) ;
    // Initialize variables
    cyScreen = cyChar;          // screen row counter
    startptr = TextMsg;         // start position pointer
    endptr = TextMsg;           // end position pointer
    j = 0;                      // length of string
    i = 0;                      // characters since last
                                // space
    // Text line display loop
    // INVARIANT:
    //          i = characters since last space
    //          j = length of current string
    // startptr = pointer to substring start
    //   endptr = pointer to substring end

    while(*startptr) {
       if(*startptr == 0x20){    // if character is
                                 // space
          GetTextExtentPoint32 (hdc, endptr, j,\
                                &textsize);
    // ASSERT:
    //        textsize.cx is the current length of the
    //        string
    //        cxClient is the abscissa of the client area
    //        (both in logical units)
    // Test for line overflow condition. If so, adjust
    // substring to preceding space and display
       if(cxClient - (2 * cxChar) < textsize.cx) {
          j = j - i;
          startptr = startptr - i;
          TextOut (hdc, cxChar, cyScreen, endptr, j);
          cyScreen = cyScreen + cyChar;
          endptr = startptr;
          j = 0;
          }
    // End of space character processing.
    // Reset chars-to-previous-space counter, whether
    // or not string was displayed
          i = 0;
          }
    // End of processing for any text character
    // Update substring pointer and counters
       startptr++;
       j++;
       i++;
       }
    // End of while loop
    // Display last text substring
       j = j - i;
       TextOut (hdc, cxChar, cyScreen, endptr, j);
       EndPaint (hwnd, &ps);
       return 0 ;
case WM_DESTROY :
    PostQuitMessage (0) ;
    return 0 ;
}
return DefWindowProc (hwnd, iMsg, wParam, lParam) ;
}
```

In Figure 20.7 there are two screen snapshots of the TEX1_DEMO program in the Text Demo No 1 project folder. The first one shows the text line as originally displayed in our system. The second one shows them after the client area has been resized.

Figure 20.7 *Two Screen Snapshots of the TEX1_DEMO Program*

Notice that the TEX1_DEMO program uses a variable (j) to store the total size of the substring (see Figure 20.6). In C++ it is valid to subtract two pointers in order to determine the number of elements between them. The code in the TEX1_DEMO program could have calculated the number of elements in the substring by performing pointer subtraction.

20.4.4 The DrawText() Function

Another useful text display function in the Windows API is DrawText(). This function is of a higher level than TextOut() and, in many cases, text display operations are easier to implement with DrawText(). DrawText() uses a rectangular screen area that defines where the text is to be displayed. In addition, it recognizes some control characters embedded in the text string as well as a rather extensive collection of format controls, which are represented by predefined constants. The following are the general forms for TextOut() and DrawText()

```
TextOut (hdc, nXStart, nYStart, lpString, cbString);
DrawText (hdc, lpString, nCount, &rect, uFormat);
```

In both cases, hdc is the handle to the device context and lpString is a pointer to the string to be displayed. In TextOut() the second and third parameters (xXstart and nYStart) are the logical coordinates of the start point in the client area, and the last parameter is the string length. In DrawText() the third parameter (nCount) is the string length in characters. If this parameter is set to –1 then Windows assumes that the string is zero-terminated. The positioning of the string in DrawText() is by means of a rectangle structure (type RECT). This structure contains four members; two for the rectangle's top-left coordinates, and two for its bottom-right coordinates. The values are in logical units. The last parameter (uFormat) is any combination of nineteen format strings defined by the constants listed in Table 20.3.

Table 20.3

String Formatting Constants in DrawText()

SYMBOLIC CONSTANT	MEANING
DT_BOTTOM	Specifies bottom-justified text. Must be combined with DT_SINGLELINE.
DT_CALCRECT	Returns width and height of the rectangle. In the case of multiple text lines, DrawText() uses the width of the rectangle pointed to by lpRect and extends its base to enclose the last line of text. In the case of a single text line, then DrawText() modifies the right side of the rectangle so that it encloses the last character. In either case, DrawText() returns the height of the formatted text, but does not draw the text.
DT_CENTER	Text is centered horizontally.
DT_EXPANDTABS	Expands tab characters. The default number of characters per tab is eight.
DT_EXTERNALLEADING	Includes the font's external leading in the line height. Normally, external leading is not included in the height of a line of text.
DT_LEFT	Specifies text that is aligned flush-left.
DT_NOCLIP	Draws without clipping. This improves performance.
DT_NOPREFIX	Turns off processing of prefix characters. Normally, DrawText() interprets the ampersand (&) mnemonic-prefix character as an order to underscore the character that follows. The double ampersands (&&) is an order to print a single ampersand symbol. This function is turned off by DT_NOPREFIX.
DT_RIGHT	Specifies text that is aligned flush-right.
DT_SINGLELINE	Specifies single line only. Carriage returns and linefeed are ignored.
DT_TABSTOP	Sets tab stops. The high-order byte of nFormat is the number of characters for each tab. The default number of characters per tab is eight.
DT_TOP	Specifies top-justified text (single line only).
DT_VCENTER	Specifies vertically centered text (single line only).
DT_WORDBREAK	Enables word-breaking. Lines are automatically broken between words if a word would extend past the edge of the rectangle specified by lpRect. A carriage return (\n) or linefeed code (\r) also breaks the line.

The program TEX2_DEMO, located in the Text Demo No 2 project folder on the book's on-line software package, is a demonstration of text display using the DrawText() function. Following are the excerpts from the program code:

```
LRESULT CALLBACK WndProc (HWND hwnd, UINT iMsg, WPARAM wParam,
                          LPARAM lParam) {

   static int  cxChar, cyChar ;       // Character dimensions
   static int  cxClient, cyClient;    // Client area parameters
   HDC       hdc ;                     // handle to device context

   // Structures
   PAINTSTRUCT ps;
   TEXTMETRIC  tm;
   RECT        textRect;

   switch (iMsg) {
      case WM_CREATE :
           hdc = GetDC (hwnd) ;
           GetTextMetrics (hdc, &tm) ;

           // Calculate and store character dimensions
           cxChar = tm.tmAveCharWidth ;
           cyChar = tm.tmHeight + tm.tmExternalLeading ;

           ReleaseDC (hwnd, hdc) ;
           return 0 ;

      case WM_SIZE:
           // Determine and store size of client area
           cxClient = LOWORD(lParam);
           cyClient = HIWORD(lParam);
           return 0;

      case WM_PAINT :
           hdc = BeginPaint (hwnd, &ps) ;

           // Initialize variables
             SetRect (&textRect,        // address of structure
                     2 * cxChar,              // x for start
                     cyChar,                  // y for start
                     cxClient -(2 * cxChar), // x for end
                     cyClient);               // y for end

           // Call display function using left-aligned and
           //wordbreak controls
           DrawText( hdc, TextStr, -1, &textRect,
                   DT_LEFT | DT_WORDBREAK);

           EndPaint (hwnd, &ps);
           return 0 ;

      case WM_DESTROY :
           PostQuitMessage (0) ;
           return 0 ;
       }

   return DefWindowProc (hwnd, iMsg, wParam, lParam) ;
}
```

20.5 Text Graphics

Comparing the listed processing operations with those used in the TEX1_DEMO program (previously in this chapter) you can see that the processing required to achieve the same functionality is simpler using DrawText() than TextOut(). This observation, however, should not mislead you into thinking that DrawText() should always be preferred. The interpretation of the reference point at which the text string is displayed when using TextOut() depends on the text-alignment mode set in the device context. The GetTextAlign() and SetTextAlign() functions can be used to retrieve and change the eleven text alignment flags. This feature of TextOut() (and its newer version TextOutExt()) allow the programmer to change the alignment of the text-bounding rectangle and even to change the reading order to conform to that of the Hebrew and Arabic languages.

Windows 95/NT GDI and later supports the notion of paths. Paths are discussed in detail in Chapter 21. For the moment, we define a path, rather imprecisely, as the outline produced by drawing a set of graphical objects. One powerful feature of TextOut(), which is not available with DrawText(), is that when it is used with a TrueType font, the system generates a path for each character and its bounding box. This can be used to display text transparently inside other graphics objects, to display character outlines (called stroked text), and to fill the text characters with other graphics objects. The resulting effects are often powerful.

20.5.1 Selecting a Font

The one limitation of text display on paths is that the font must be TrueType. Therefore, before getting into fancy text graphics, you must be able to select a TrueType font into the device context. Font manipulations in Windows are based on the notion of a logical font. A logical font is a description of a font by means of its characteristics. Windows uses this description to select the best matching font among those available.

Two API functions allow the creation of a logical font. CreateFont() requires a long series of parameters that describe the font characteristics. CreateFontIndirect() uses a structure in which the font characteristics are stored. Applications that use a single font are probably better off using CreateFont(), while programs that change fonts during execution usually prefer CreateFontIndirect(). Note that the item list used in the description of a logical font is the same in both functions. Therefore, storing font data in structure variables is an advantage only if the structure can be reused. The description that follows refers to the parameters used in the call to CreateFont(), which are identical to the ones used in the structure passed by CreateFontIndirect().

The CreateFont() function has one of the longest parameter lists in the Windows API: fourteen in all. Its general form is as follows:

```
HFONT CreateFont( nHeight, nWidth, nEscapement, int nOrientation,
                  fnWeight, fdwItalic, fdwUnderline, fdwStrikeOut,
                  fdwCharSet, fdwOutputPrecision, fdwClipPrecision,
                  fdwQuality, fdwPitchAndFamily,
```

```
LPCTSTR lpszFace);
```

Following are brief descriptions of the function parameters.

- nHeight (int) specifies the character height in logical units. The value does not include the internal leading, so it is not equal to the tmHeight value in the TEXTMETRIC structure. Also note that the character height does not correspond to the point size of a font. If the MM_TEXT mapping mode is selected in the device context, it is possible to convert the font's point size into device units by means of the following formula:

- hHeight = (point_size * pixels_per_inch) / 72

- The pixels per inch can be obtained by reading the LOGPIXELSY index in the device context, which can be obtained by the call to GetDeviceCaps(). For example, to obtain the height in logical units of a 50-point font we can use the following expression:

```
50 * GetDeviceCaps (hdc, LOGPIXELSY) / 72
```

- nWidth (int) specifies the logical width of the font characters. If set to zero, the Windows font mapper uses the width that best matches the font height.

- nEscapement (int) specifies the angle between an escapement vector, defined to be parallel to the baseline of the text line, and the drawn characters. A value of 900 (90 degrees) specifies characters that go upward from the baseline. Usually this parameter is set to zero.

- nOrientation (int) defines the angle, in tenths of a degree, between the character's base line and the x-axis of the device. In Windows NT the value of the character's escapement and orientation angles can be different. In Windows 95 they must be the same.

- fnWeight (int) specifies the font weight. The constants listed in Table 20.4 are defined for convenience:

Table 20.4

Character Weight Constants

WEIGHT	CONSTANT
FW_DONTCARE	= 0
FW_THIN	= 100
FW_EXTRALIGHT	= 200
FW_ULTRALIGHT	= 200
FW_LIGHT	= 300
FW_NORMAL	= 400
FW_REGULAR	= 400
FW_MEDIUM	= 500
FW_SEMIBOLD	= 600
FW_DEMIBOLD	= 600
FW_BOLD	= 700
FW_EXTRABOLD	= 800
FW_ULTRABOLD	= 800
FW_HEAVY	= 900
FW_BLACK	= 900

- fdwItalic (DWORD) is set to 1 if font is italic.

- fdwUnderline (DWORD) is set to 1 if font is underlined.

- fdwStrikeOut (DWORD) is set to 1 if font is strikeout.

- fdwCharSet (DWORD) defines the font's character set. The following are predefined character set constants:

```
ANSI_CHARSET
DEFAULT_CHARSET
SYMBOL_CHARSET
SHIFTJIS_CHARSET
GB2312_CHARSET
HANGEUL_CHARSET
CHINESEBIG5_CHARSET
OEM_CHARSET
```

Windows 95 and later:

```
JOHAB_CHARSET
HEBREW_CHARSET
ARABIC_CHARSET
GREEK_CHARSET
TURKISH_CHARSET
THAI_CHARSET
EASTEUROPE_CHARSET
RUSSIAN_CHARSET
MAC_CHARSET
BALTIC_CHARSET
```

The DEFAULT_CHARSET constant allows the name and size of a font to fully describe it. If the font does not exist, another character set can be substituted. For this reason, this field should be used carefully. A specific character set should always be defined to ensure consistent results.

- fdwOutputPrecision (DWORD) determines how closely the font must match the values entered in the fields that define its height, width, escapement, orientation, pitch, and font type. Table 20.5 lists the constants associated with this parameter.

Table 20.5

Predefined Constants for Output Precision

PREDEFINED CONSTANT	MEANING
OUT_CHARACTER_PRECIS	Not used.
OUT_DEFAULT_PRECIS	Specifies the default font mapper behavior.
OUT_DEVICE_PRECIS	Instructs the font mapper to choose a Device font when the system contains multiple fonts with the same name.
OUT_OUTLINE_PRECIS	Windows NT: This value instructs the font mapper to choose from TrueType and other outline-based fonts. Not used in Windows 95 and later versions.
OUT_RASTER_PRECIS	Instructs the font mapper to choose a raster font when the system contains multiple fonts with the same name.

(continues)

Table 20.5

Predefined Constants for Output Precision (continued)

PREDEFINED CONSTANT	MEANING
OUT_STRING_PRECIS	This value is not used by the font mapper, but it is returned when raster fonts are enumerated.
OUT_STROKE_PRECIS	Windows NT: This value is not used by the font mapper, but it is returned when TrueType, other outline-based fonts, and vector fonts are enumerated. Windows 95 and later: This value is used to map Vector fonts, and is returned when TrueType or Vector fonts are enumerated.
OUT_TT_ONLY_PRECIS	Instructs the font mapper to choose from only TrueType fonts. If there are no TrueType fonts installed in the system, the font mapper returns to default behavior.
OUT_TT_PRECIS	Instructs the font mapper to choose a TrueType font when the system contains multiple fonts with the same name.

If there is more than one font with a specified name, you can use the OUT_DEVICE_PRECIS, OUT_RASTER_PRECIS, and OUT_TT_PRECIS constants to control which one is chosen by the font mapper. For example, if there is a font named Symbol in raster and TrueType form, specifying OUT_TT_PRECIS forces the font mapper to choose the TrueType version. OUT_TT_ONLY_PRECIS forces the font mapper to choose a TrueType font, even if it must substitute one of another name.

- fdwClipPrecision (DWORD) specifies the clipping precision. This refers to how to clip characters that are partially outside the clipping region. The constants in Table 20.6 are recognized by the call.

Table 20.6

Predefined Constants for Clipping Precision

PREDEFINED CONSTANT	MEANING
CLIP_DEFAULT_PRECIS	Default clipping behavior.
CLIP_CHARACTER_PRECIS	Not used.
CLIP_STROKE_PRECIS	Not used by the font mapper, but is returned when raster, vector, or TrueType fonts are enumerated. Windows NT: For compatibility, this value is always returned when enumerating fonts.
CLIP_MASK	Not used.
CLIP_EMBEDDED	Specify this flag to use an embedded read-only font.
CLIP_LH_ANGLES	The rotation for all fonts depends on whether the orientation of the coordinate system is left- or right-handed. If not used, device fonts always rotate counterclockwise.
CLIP_TT_ALWAYS	Not used.

- fdwQuality (DWORD) specifies the output quality. This value defines how carefully GDI must attempt to match the logical font attributes to those of an actual physical font. The constants in Table 20.7 are recognized by CreateFont().

Table 20.7

Predefined Constants for Output Precision

PREDEFINED CONSTANT	MEANING
DEFAULT_QUALITY	Appearance of the font does not matter.
DRAFT_QUALITY	Appearance of the font is less important than when the PROOF_QUALITY value is used.
PROOF_QUALITY	Character quality of the font is more important than exact matching of the logical-font attributes. When PROOF_QUALITY is used, the quality of the font is high and there is no distortion of appearance.

- fdwPitchAndFamily (DWORD) defines the pitch and the family of the font. The two low-order bits specify the pitch, and the four high-order bits specify the family. Usually, the two bit fields use a logical OR for this parameter. Table 20.8 lists the symbolic constants recognized by CreateFont() for the font pitch and the family values.

Table 20.8

Pitch and Family Predefined Constants

TYPE	VALUE	MEANING
PITCH:		
	DEFAULT_PITCH	
	FIXED_PITCH	
	VARIABLE_PITCH	
FAMILY:		
	FF_DECORATIVE	Novelty fonts (such as Old English)
	FF_DONTCARE	Don't care or don't know.
	FF_MODERN	Fonts with constant stroke width, with or without serifs, such as Pica, Elite, and Courier New.
	FF_ROMAN	Fonts with variable stroke width and with Serifs. Such as MS Serif.
	FF_SCRIPT	Fonts designed to look like handwriting, such as Script and Cursive.
	FF_SWISS	Fonts with variable stroke width and without serifs,such as MS Sans Serif.

- lpszFace (LPCTSTR) points to a null-terminated string that contains the name of the font's typeface. Alternatively, the typeface name can be entered directly inside double quotation marks. If the requested typeface is not available in the system, the font mapper substitutes with an approximate one. If NULL is entered in this field, a default typeface is used. Example typefaces are Palatino, Times New Roman, and Arial. The following code fragment shows a call to the CreateFont() API for a 50-point, normal weight, high quality, italic font using the Times New Roman typeface.

```
HFONT       hFont;        // handle to a font
// Create a logical font
hFont = CreateFont (
    50 * GetDeviceCaps (hdc, LOGPIXELSY) / 72, //height
    0,                      // width
```

```
        0,                          // escapement angle
        0,                          // orientation angle
        FW_NORMAL,                  // weight
        1,                          // italics
        0,                          // not underlined
        0,                          // not strikeout
        DEFAULT_CHARSET,            // character set
        OUT_DEFAULT_PRECIS,         // precision
        CLIP_DEFAULT_PRECIS,        // clipping precision
        PROOF_QUALITY,              // quality
        DEFAULT_PITCH | FF_DONTCARE, // pitch and family
        "Times New Roman");          // typeface name

   // Select font into the display context
   SelectObject (hdc, hFont);
```

20.5.2 Drawing with Text

Once a TrueType font is selected in the display context, you can execute several manipulations that treat text characters as graphics objects. One of them is related to the notion of a path, introduced in Windows NT and also supported by Windows 95 and later. A path is the outline generated by one or more graphics objects drawn between the BeginPath() and EndPath() functions. Paths are related to regions and to clipping, topics covered in detail in Chapter 21.

The TextOut() function has a unique property among the text display functions: it generates a path. For this to work, a TrueType font must first be selected into the display context. Path drawing operations are not immediately displayed on the screen but are stored internally. Windows provides no handles to paths, and there is only one path for each display context. Three functions are available to display graphics in a path: StrokePath() shows the path outline, FillPath() fills and displays the path's interior, and StrokeAndFillPath() performs both functions. You may question the need for a FillAndStrokePath() function since it seems that you could use StrokePath() and FillPath() consecutively to obtain the same effect. This is not the case. All three path-drawing APIs automatically destroy the path. Therefore, if two of these functions are called consecutively, the second one has no effect.

The path itself has a background mix mode, which is delimited by the rectangle that contains the graphics functions in the path. The background mix mode is a display context attribute that affects the display of text, as well as the output of hatched brushes and nonsolid pens. Code can set the background mix mode to transparent by means of the SetBkMode() function. This isolates the text from the background. The program TEX3_DEMO, located in the Text Demo No 3 folder in the book's on-line software package, is a demonstration of text display inside paths. One of the text lines is stroked and the other one is stroked and filled. The program first creates a logical font and then selects it into the display context. Processing is as follows:

```
case WM_PAINT :
   hdc = BeginPaint (hwnd, &ps) ;
   .
   .
   .
```

```
// Start a path for stroked text
// Set background mix to TRANSPARENT mode
BeginPath (hdc);
SetBkMode(hdc, TRANSPARENT);          // background mix
TextOut(hdc, 20, 20, "This Text is STROKED", 20);
EndPath(hdc);
// Create a custom black pen, 2 pixels wide
aPen = CreatePen(PS_SOLID, 2, 0);
SelectObject(hdc, aPen);              // select it into DC
StrokePath (hdc);                     // Stroke the path

// Second path for stroked and filled text
BeginPath (hdc);
SetBkMode(hdc, TRANSPARENT);
TextOut(hdc, 20, 110, "Stroked and Filled", 18);
EndPath(hdc);
// Get and select a stock pen and brush
aPen = GetStockObject(BLACK_PEN);
aBrush = GetStockObject(LTGRAY_BRUSH);
SelectObject(hdc, aPen);
SelectObject(hdc, aBrush);
StrokeAndFillPath (hdc);              // Stroke and fill path

// Clean-up and end WM_PAINT processing
DeleteObject(hFont);
EndPaint (hwnd, &ps);
```

Figure 20.8 is a screen snapshot of the TEXTDEM3 program.

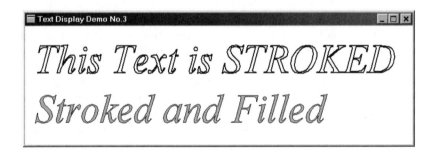

Figure 20.8 *Screen Snapshot of the TEXTDEM3 Program*

Chapter 21

Keyboard and Mouse Programming

Chapter Summary

Most applications require user input and control operations. The most common input devices are the keyboard and the mouse. In this chapter we discuss keyboard and mouse programming in Windows.

21.0 Keyboard Input

Since the first days of computing, typing on a typewriter-like keyboard has been an effective way of interacting with the system. Although typical Windows programs rely heavily on the mouse device, the keyboard is the most common way to enter text characters into an application.

The mechanisms by which Windows monitors and handles keyboard input are based on its message-driven architecture. When the user presses or releases a key, the low-level driver generates an interrupt to inform Windows of this action. Windows then retrieves the keystroke from a hardware register, stores it in the system message queue, and proceeds to examine it. The action taken by the operating system depends on the type of keystroke, and on which application currently holds the keyboard foreground, called the input focus. The keystroke is dispatched to the corresponding application by means of a message to its Windows procedure.

The particular way by which Windows handles keystrokes is determined by its multitasking nature. At any given time, several programs can be executing simultaneously, and any one of these programs can have more than one thread of execution. One of the possible results of a keystroke (or a keystroke sequence) is to change the thread that holds the input focus, perhaps to a different application. This is the reason why Windows cannot directly send keyboard input to any specific thread.

It is the message loop in the WinMain() function of an application that retrieves keyboard messages from the system queue. In fact, all system messages are posted to the application's message queue. The process makes the following assumptions: first, that the thread's queue is empty; second, that the thread holds the input focus; and third, that a keystroke is available at the system level. In other words, it is the application that asks Windows for keystrokes; Windows does not send unsolicited keystroke data.

The abundance of keyboard functions and keyboard messages makes it appear that Windows keyboard programming is difficult or complicated. The fact is that applications do not need to process all keyboard messages, and hardly ever do so. Two messages, WM_CHAR and WM_KEYDOWN, usually provide code with all the necessary data regarding user keyboard input. Many keystrokes can be ignored, since Windows generates other messages that are more easily handled. For example, applications can usually disregard the fact that the user selected a menu item by means of a keystroke, since Windows sends a message to the application as if the menu item had been selected by a mouse click. If the application code contains processing for menu selection by mouse clicks, then the equivalent keyboard action is handled automatically.

21.1 Input Focus

The application that holds the input focus is the one that gets notified of the user's keystrokes. A user can visually tell which window has the input focus since it is the one whose title bar is highlighted. This applies to the parent window as well as to child windows, such as an input or dialog box. The application can tell if a window has the input focus by calling the GetFocus() function, which returns the handle to the window with the input focus.

The Windows message WM_SETFOCUS is sent to the window at the time that it receives the input focus, and WM_KILLFOCUS at the time it loses it. Applications can intercept these messages to take notice of any change in the input focus. However, these messages are mere notifications; application code cannot intercept these messages to prevent losing the input focus.

Keyboard data is available to code holding the input focus at two levels. The lower level, usually called keystroke data, contains raw information about the key being pressed. Keystroke data allows code to determine whether the keystroke message was generated by the user pressing a key or by releasing it, and whether the keystroke resulted from a normal press-and-release action or from the key being held down (called typematic action). Higher-level keyboard data relates to the character code associated with the key. An application can intercept low-level or character-level keystroke messages generated by Windows.

21.1.1 Keystroke Processing

Four Windows messages inform application code of keystroke data: WM_KEYDOWN, WM_SYSKEYDOWN, WM_KEYUP, and WM_SYSKEYUP. The keydown-type messages are generated when a key is pressed, sometimes called the make action. The keyup-type messages are generated when a key is released, called the break action. Applications usually ignore the keyup-type message. The "sys-type" messages, WM_SYSKEYDOWN and WM_SYSKEYUP, relate to system keys. A system keystroke is one generated while the Alt key is held down.

When any one of these four messages takes place, Windows puts the keystroke data in the lParam and wParam passed to the window procedure. The lParam contains bit-coded information about the keystroke, as shown in Table 21.1.

Table 21.1

Bit and Bit Fields in the lParam of a Keystroke Message

BITS	MEANING
0-15	Repeat count field. The value is the number of times the keystroke is repeated as a result of the user holding down the key (typematic action).
16-23	OEM scan code. The value depends on the original equipment manufacturer.
24	Extended key. Bit is set when the key pressed is one duplicated in the IBM Enhanced 101- and 102-key keyboards, such as the right-hand ALT and CTRL keys, the / and Enter keys on the numeric keypad, or the Insert, Delete, Home, PageUp, PageDown, and End keys.
25-28	Reserved.
29	Context code. Bit is set if the Alt key is down while the key is pressed. Bit is clear if the WM_SYSKEYDOWN message is posted to the active window because no window has the keyboard focus.
30	Previous key state. Key is set if the key is down before the message is sent. Bit is clear if the key is up. This key allows code to determine if the keystroke resulted from a make or break action.
31	Transition state. Always 0 for a WM_SYSKEYDOWN Message.

The wParam contains the virtual-key code, which is a hardware-independent value that identifies each key. Windows uses the virtual-key codes instead of the device-dependent scan code. Typically, the virtual-key codes are processed when the application needs to recognize keys that have no associated ASCII value, such as the control keys. Table 21.2, on the following page, lists some of the most used virtual-key codes.

Notice that originally, the "w" in wParam stood for "word," since in 16-bit Windows the wParam was a word-size value. The Win32 API expanded the wParam from 16 to 32 bits. However, in the case of the virtual-key character codes, the wParam is defined as an int type. Code can typecast the wParam as follows:

Table 21.2

Virtual-Key Codes

SYMBOLIC NAME	HEX VALUE	KEY
VK_CANCEL	0x01	Ctrl + Break
VK_BACK	0x08	Backspace
VK_TAB	0x09	Tab
VK_RETURN	0x0D	Enter
VK_SHIFT	0x10	Shift
VK_CONTROL	0x11	Ctrl
VK_MENU	0x12	Alt
VK_PAUSE	0x13	Pause
VK_CAPITAL	0x14	Caps Lock
VK_ESCAPE	0x1B	Esc
VK_SPACE	0x20	Spacebar
VK_PRIOR	0x21	Page Up
VK_NEXT	0x22	Page Down
VK_END	0x23	End
VK_HOME	0x24	Home
VK_LEFT	0x25	Left arrow
VK_UP	0x26	Up arrow
VK_RIGHT	0x27	Right arrow
VK_DOWN	0x28	Down arrow
VK_SNAPSHOT	0x2C	Print Screen
VK_INSERT	0x2D	Insert
VK_DELETE	0x2E	Delete
VK_MULTIPLY	0x6A	Numeric keypad *
VK_ADD	0x6B	Numeric keypad +
VK_SUBTRACT	0x6D	Numeric keypad -
VK_DIVIDE	0x6F	Numeric keypad /
VK_F1..VK_F12	0x70..0x7B	F1 .. F12

```
int       aKeystroke;
char      aCharacter;
  .
  .
aKeystroke = (int) wParam;
aCharacter = (char) wParam;
```

Simple keystroke processing can be implemented by intercepting
WM_KEYDOWN. Occasionally, an application needs to know when a system-level
message is generated. In this case, code can intercept WM_SYSKEYDOWN. The
first operation performed in a typical WM_KEYDOWN or WM_SYSKEYDOWN

handler is to store in local variables the lParam, the wParam, or both. In the case of the wParam code can cast the 32-bit value into an int or a char type as necessary (see the preceding Tech Note).

Processing the keystroke usually consists of performing bitwise operations in order to isolate the required bits or bit fields. For example, to determine if the extended key flag is set, code can logically AND with a mask in which bit 24 is set and then test for a non-zero result, as in the following code fragment:

```
unsigned long     keycode;
  .
  .

WM_KEYDOWN:
    keycode = lParam;            // store lParam
    if(keycode & 0x01000000) {   // test bit 24
    // ASSERT:
    //   key pressed is extended key
```

Processing the virtual-key code, which is passed to your intercept routine in the lParam, consists of comparing its value with the key or keys that you wish to detect. For example, to know if the key pressed was the Backspace, you can proceed as in the following code fragment:

```
int     virtkey;
  .
  .

WM_KEYDOWN:
    virtkey = (int) lParam;      // cast and store lParam
    if(virtkey == VK_BACK) {     // test for Backspace
    // ASSERT:
    //    Backspace key pressed
```

21.1.2 Determining the Key State

An application can determine the state of any virtual-key by means of the GetKeyState() service. The function's general form is as follows:

```
SHORT GetKeyState(nVirtKey);
```

GetKeyState() returns a SHORT integer with the high-order bit set if the key is down and the low-order bit set if it is toggled. Toggle keys are those which have a keyboard LED to indicate their state: Num Lock, Caps Lock, and Scroll Lock. The LED for the corresponding key is lit when it is toggled and unlit otherwise. Some virtual-key constants can be used as the nVirtKey parameter of GetKeyState(). Table 21.3, on the following page, lists the virtual-keys.

Take note that in testing for the high-bit set condition returned by GetKeyState() you may be tempted to bitwise AND with a binary mask, as follows:

```
    if(0x8000 & (GetKeyState(VK_SHIFT))) {
```

Table 21.3

Virtual-Keys used in GetKeyState()

PREDEFINED SYMBOL	KEY	RETURNS
VK_SHIFT	Shift	State of left or right Shift keys
VK_CONTROL	Ctrl	State of left or right Ctrl keys
VK_MENU	Alt	State of left or right Alt keys
VK_LSHIFT	Shift	State of left Shift key
VK_RSHIFT	Shift	State of right Shift key
VK_LCONTROL	Ctrl	State of left Ctrl key
VK_RCONTROL	Ctrl	State of right Ctrl key
VK_LMENU	Alt	State of left Alt key
VK_RMENU	Alt	State of right Alt key

The following statement is a test for the left Shift key pressed.

```
if(GetKeyState(VK_LSHIFT) < 0) {
// ASSERT:
//          Left shift key is pressed
```

Although, in many cases, such operations produce the expected results, its success depends on the size of a data type, which compromises portability. In other words, if GetKeyState() returns a 16-bit integer, then the mask 0x8000 effectively tests the high-order bit. If the value returned is stored in 32 bits, however, then the mask must be the value 0x80000000. Since any signed integer with the high-bit set represents a negative number, it is possible to test the bit condition as follows:

```
if(GetKeyState(VK_SHIFT) < 0) {
```

This test does not depend on the operand's bit size.

21.1.3 Character Code Processing

Applications often deal with keyboard input as character codes. It is possible to obtain the character code from the virtual-key code since it is encoded in the wParam of the WM_KEYDOWN, WM_SYSKEYDOWN, WM_KEYUP, and WM_SYSKEYUP messages. The codes for the alphanumeric keys are not listed in Table 21.1, however, there is also a virtual-key code for each one. The virtual-key codes for the numeric keys 0 to 9 are VK_0 to VK_9, and the ones for the alphabetic characters A through Z are VK_A through VK_Z.

This type of processing is not without complications. For example, the virtual-key code for the alphabetic characters does not specify if the character is in upper- or lower-case. Therefore, the application would have to call GetKeyState() in order to determine if the <Shift> key was down or the Caps Lock key toggled when the character key was pressed. Furthermore, the virtual-key codes for some of the character keys, such as ;, =, +, <, are not defined in the windows header files. Applications must use the numeric values assigned to these keys or define their own symbolic constants.

Fortunately, character code processing in Windows is much easier. The TranslateMessage() function converts the virtual-key code for each character into its ANSI (or Unicode) equivalent and posts it in the thread's message queue. TranslateMessage() is usually included in the program's message loop. After TranslateMessage(), the message is retrieved from the queue, typically by GetMessage() or PeekMessage(). The final result is that an application can intercept WM_CHAR, WM_DEADCHAR, WM_SYSCHAR, and WM_SYSDEADCHAR in order to obtain the ANSI character codes that correspond to the virtual-key of a WM_KEYDOWN message.

Dead-type character messages refer to the diacritical characters used in some foreign language keyboards. These are marks added to characters to distinguish them from other ones, such as the acute accent (á) or the circumflex (â). In English language processing, WM_DEADCHAR and WM_SYSDEADCHAR are usually ignored.

The WM_SYSCHAR message corresponds to the virtual-key that results from WM_SYSKEYDOWN. WM_SYSCHAR is posted when a character key is pressed while the Alt key is held down. Since Windows also sends the message that corresponds to a mouse click on the system item, applications often ignore WM_SYSCHAR.

This leaves us with WM_CHAR for general purpose character processing. When the WM_CHAR message is sent to your Windows procedure, the lParam is the same as for WM_KEYDOWN. However, the wParam contains the ANSI code for the character, instead of the virtual-key code. This ANSI code, which is approximately equivalent to the ASCII code, can be directly handled and displayed without additional manipulations. Processing is as follows:

```
char        aChar;              // storage for character
  .
  .
case WM_CHAR:
   aChar = (char) wParam;
// ASSERT:
//       aChar holds ANSI character code
```

21.1.4 Keyboard Demonstration Program

The program KBR_DEMO.CCP, located in the Keyboard Demo folder on the book's on-line software package, is a demonstration of the keyboard processing routines described previously. The program uses a private device context; therefore, the font is selected once, during WM_CREATE processing. KBR_DEMO uses a typewriter-like, TrueType font, named Courier. Courier is a monospaced font (all characters are the same width). This makes possible the use of standard character symbols to produce a graph of the bitmaps. Figure 21.1, on the following page, is a screen snapshot of the KBD_DEMO program.

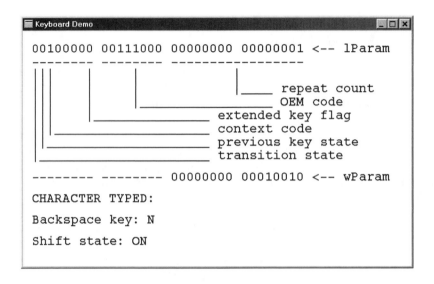

Figure 21.1 *KBR_DEMO Program Screen*

Figure 21.1 shows the case in which the user has typed the Alt key. Note that the wParam value 00010010 is equivalent to 0x12, which is the virtual-key code for the Alt key (see Table 21.1). The critical processing in the KBD_DEMO program is as follows:

```
LRESULT CALLBACK WndProc (HWND hwnd, UINT iMsg, WPARAM wParam,
                          LPARAM lParam) {

    static int    cxChar, cyChar ;     // Character dimensions
    static int    cxClient, cyClient;  // Client area parameters
    static HDC        hdc ;            // handle to private DC
    unsigned long    keycode;         // storage for keystroke
    unsigned long    keymask;         // bit mask
    unsigned int     virtkey;         // virtual-key
    int              i, j;            // counters
    char             aChar;           // character code

    // Structures
    PAINTSTRUCT ps;
    TEXTMETRIC  tm;
    RECT        textRect;             // RECT-type
    HFONT       hFont;
    .
    .
    .
    case WM_PAINT :
        // Processing consists of displaying the text messages
        BeginPaint (hwnd, &ps) ;
        // Initialize rectangle structure
        SetRect (&textRect,           // address of structure
            2 * cxChar,               // x for start
            cyChar,                   // y for start
            cxClient -(2 * cxChar),   // x for end
            cyClient);                // y for end
        // Display multi-line text string
```

```
        DrawText( hdc, TextStr0, -1, &textRect,
          DT_LEFT | DT_WORDBREAK);

        // Display second text string
        SetRect (&textRect,              // address of structure
          2 * cxChar,                    // x for start
          13 * cyChar,                   // y for start
          cxClient -(2 * cxChar),        // x for end
          cyClient);                     // y for end
        // Display text string
          DrawText( hdc, TextStr1, -1, &textRect,
          DT_LEFT | DT_WORDBREAK);
          .
          .
          .
          EndPaint (hwnd, &ps);
          return 0 ;
// Character code processing
case WM_CHAR:
      aChar = (char) wParam;
      // Test for control codes and replace with space
      if (aChar < 0x30)
        aChar = 0x20;

      // Test for shift key pressed
      if(GetKeyState (VK_SHIFT) < 0) {
        i = 0;          // counter
        j = 13;         // string offset
        for(i = 0; i < 3; i++){
          TextStr4[j] = StrON[i];
          j++;
        }
      }
      else {
        i = 0;          // counter
        j = 13;         // string offset
        for(i = 0; i < 3; i++){
          TextStr4[j] = StrOFF[i];
          j++;
        }
      }

    TextStr2[17] = aChar;
    return 0;

// Scan code and keystroke data processing
// Display space if a system key
case WM_SYSKEYDOWN:
    TextStr2[17] = 0x20;

case WM_KEYDOWN:
// Store bits for lParam in TextStr0[]
   keycode = lParam;      // get 32-bit keycode value
   i = 0;                 // counter for keystroke bits
   j = 0;                 // offset into string
   keymask = 0x80000000;// bitmask

   for (i = 0; i < 32; i++) {
     // Test for separators and skip
```

```
        if(i == 8 || i == 16 || i == 24) {
          TextStr0[j] = 0x20;
          j++;
        }
      // Test for 1 and 0 bits and display digits
      if(keycode & keymask)
          TextStr0[j] = '1';
      else
          TextStr0[j] = '0';
      keymask = keymask >> 1;
      j++;
      }

// Store bits for wParam in TextStr1[]
   keycode = wParam;      // get 32-bit keycode value
   i = 0;                 // counter for keystroke bits
   j = 18;                // initial offset into string
   keymask = 0x8000;      // bitmask

   // 16-bit loop
     for (i = 0; i < 16; i++) {
       // Test for separators and skip
       if(i == 8) {
          TextStr1[j] = 0x20;
          j++;
       }
     // Test for 1 and 0 bits and display digits
       if(keycode & keymask)
          TextStr1[j] = '1';
       else
          TextStr1[j] = '0';
          keymask = keymask >> 1;
          j++;
       }

// Test for Backspace key pressed
virtkey = (unsigned int) wParam;
     if (virtkey == VK_BACK)
        TextStr3[15] = 'Y';
     else
        TextStr3[15] = 'N';

// Force WM_PAINT message
InvalidateRect(NULL, NULL, TRUE);
return 0;
   .
   .
   .
```

21.2 The Caret

In the MS DOS environment, the graphic character used to mark the screen position at which typed characters are displayed is called the cursor. The standard DOS cursor is a small, horizontal bar that flashes on the screen to call the user's attention to the point of text insertion. In Windows, the word cursor is used for an icon that marks the screen position associated with mouse-like pointing. Windows applications signal the location where keyboard input is to take place by means of a flashing, vertical bar called the caret.

In order to avoid confusion and ambiguity, Windows displays a single caret. The system caret, which is a shared resource, is a bitmap that can be customized by the application. The window with the input focus can request the caret to be displayed in its client area, or in a child window.

21.2.1 Caret Processing

Code can intercept the WM_SETFOCUS message to display the caret. WM_KILLFOCUS notifies the application that it has lost focus and that it should therefore destroy the caret. Caret display and processing in WM_SETFOCUS usually starts by calling CreateCaret(). The function's general form is as follows:

```
BOOL CreateCaret(hwnd, hBitmap, nWidth, nHeight);
```

The first parameter is the handle to the window that owns the caret. The second one is an optional handle to a bitmap. If this parameter is NULL then a solid caret is displayed. If it is (HBITMAP) 1, then the caret is gray. If it is a handle to a bitmap, the other parameters are ignored and the caret takes the form of the bitmap. The last two parameters define the caret's width and height, in logical units. Applications often determine the width and height of the caret in terms of character dimensions.

CreateCaret() defines the caret shape and size but does not set its screen position, nor does it display it. To set the caret's screen position you use the SetCaretPos() function, which takes two parameters, the first one for the caret's x-coordinate and the second one for the y-coordinate. The caret is displayed on the screen using ShowCaret(), whose only argument is the handle to the window.

Applications that use the caret usually intercept WM_KILLFOCUS. This ensures that they are notified when the window loses the keyboard focus, at which time the caret must be hidden and destroyed. The HideCaret() function takes care of the first action. Its only parameter is the handle to the window that owns the caret. DestroyCaret(), which takes no parameters, destroys the caret, erases it from the screen, and breaks the association between the caret and the window. Applications that use the caret to signal the point of input often display the characters typed by the user. But since the caret is a graphics object, it must be erased from the screen before the character is displayed. Otherwise, the caret symbol itself, or parts of it, may pollute the screen. A program that processes the WM_CHAR message to handle user input usually starts by hiding the caret, then the code processes the input character, and finally, resets the caret position and redisplays it.

21.2.2 Caret Demonstration Program

The CAR_DEMO program, located in the Caret Demo folder on the book's software on-line, is a demonstration of caret processing during text input. The program displays an entry form and uses the caret to signal the current input position. When the code detects the Enter key, it moves to the next line in the entry form. The Backspace key can be used to edit the input. When Backspace is pressed, the previous character is erased and the caret position is updated. Program logic keeps track of the start location of each input line so that the user cannot backspace past this point. The Esc key erases the caret and ends input. Note that since user input is not stored by the program, the text is lost if the screen is resized or if the application looses the input focus. Figure 21.2 is a screen snapshot of the CAR_DEMO program.

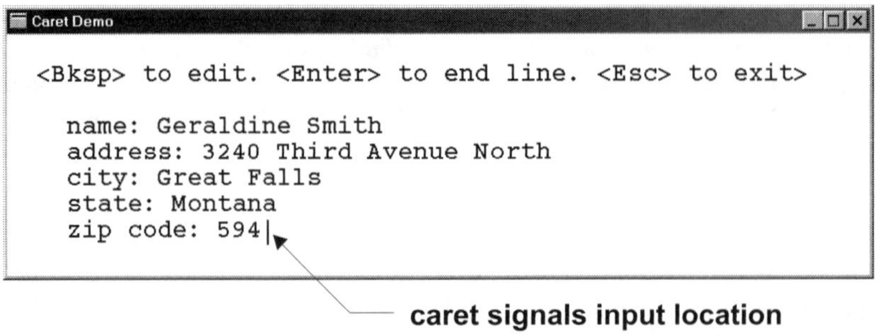

caret signals input location

Figure 21.2 *CAR_DEMO Program Screen*

Figure 21.2 shows execution of the CAR_DEMO program. The following are excerpts of the program's processing:

```
LRESULT CALLBACK WndProc (HWND hwnd, UINT iMsg, WPARAM wParam,
                          LPARAM lParam) {
static int   cxChar, cyChar ;      // character dimensions
static int   cxClient, cyClient;   // client area parameters
static int   xCaret, yCaret;       // caret position
static int   xLimit;               // left limit of line
static int   formEnd = 0;          // 1 if Esc key pressed
static int   lineNum = 1;          // input line
static HDC   hdc ;                 // handle to private DC
char         aChar;                // storage for character code

// Structures
PAINTSTRUCT ps;
TEXTMETRIC  tm;
RECT        textRect;
HFONT       hFont;

switch (iMsg) {

    case WM_CREATE :
        .
        .
        .
        // Calculate and store character dimensions
        cxChar = tm.tmAveCharWidth ;
        cyChar = tm.tmHeight + tm.tmExternalLeading ;
        // Store size of client area
        cxClient = LOWORD(lParam);
        cyClient = HIWORD(lParam);
        // Store initial caret position
        xCaret = xLimit = 10;
        yCaret = 3;
        return 0 ;
        .
        .
        .
    case WM_PAINT :
        BeginPaint (hwnd, &ps) ;
        // Initialize rectangle structure
        SetRect (&textRect,                // address of structure
                 2 * cxChar,               // x for start
```

```
                cyChar,                        // y for start
                cxClient -(2 * cxChar),    // x for end
                cyClient);                     // y for end
   // Display multi-line text string
   DrawText( hdc, TextStr1, -1, &textRect,
            DT_LEFT | DT_WORDBREAK);

   EndPaint (hwnd, &ps);
   return 0 ;


// Character input processing
case WM_CHAR:

HideCaret(hwnd);
aChar = (char) wParam;

switch (wParam) {    // wParam holds virtual-key code
case '\r':               // Enter key pressed
     yCaret++;
     aChar = 0x20;
// cascaded tests set x caret location in new line
     if(yCaret == 4)     // in address: line
        xCaret = xLimit = 13;

     if(yCaret == 5)     // in city: line
        xCaret = xLimit = 10;

     if(yCaret == 6)     // in state: line
        xCaret = xLimit = 11;

     if(yCaret == 7)     // in zip code: line
        xCaret = xLimit = 14;

     if(yCaret > 7) {     // Enter key ignored on
                          // last line
        yCaret--;
   }
  break;

case '\b':         // Backspace key pressed
   if (xCaret > xLimit) {
       aChar = 0x20;          // Replace with space
       xCaret--;
       // Display the blank character
       TextOut (hdc, xCaret * cxChar, yCaret * cyChar,
            &aChar, 1);
       }
     break;

case 0x1b:       // Esc key processing
     formEnd = 1;
     // Destroy the caret
     HideCaret(hwnd);
     DestroyCaret();
     break;

default:
   // Display the character if Esc not pressed
   if(formEnd == 0) {
```

```
             TextOut (hdc, xCaret * cxChar, yCaret * cyChar,
             &aChar, 1);
                xCaret++;
             }
      break;
   }
   if(formEnd == 0) {
       SetCaretPos(xCaret * cxChar, yCaret * cyChar);
       ShowCaret(hwnd);
   }
   return 0;

case WM_SETFOCUS:
    if(formEnd == 0) {
        CreateCaret (hwnd, NULL, cxChar / 4, cyChar);
        SetCaretPos(xCaret * cxChar, yCaret * cyChar);
        ShowCaret(hwnd);
    }
    return 0;
case WM_KILLFOCUS:
    // Destroy the caret
    HideCaret(hwnd);
    DestroyCaret();
    return 0;
.
.
.
```

21.3 Mouse Programming

The use of a mouse as an input device dates back to the work at Xerox PARC, which pioneered the ideas of a graphical user interface. Since mouse and GUI have been interrelated since their original conception, one would assume that a graphical operating system, such as Windows, would require the presence of a mouse device. This is not the case. Windows documentation still considers the mouse an option and recommends that applications provide alternate keyboard controls for all mouse-driven operations.

During program development, you can make sure that a mouse is available and operational by means of the GetSystemMetrics() function, as follows:

```
    assert (GetSystemMetrics(SM_MOUSEPRESENT));
```

In this case, the assert macro displays a message box if a mouse is not present or not operational. The developer can then choose to ignore the message, debug the code, or abort execution. In the release version of a program that requires a mouse you can use the abort macro to break execution. For example:

```
    if (!GetSystemMetrics(SM_MOUSEPRESENT))
        abort();
```

Alternatively, an application can call PostQuitMessage(). This indicates to Windows that a thread has made a termination request and it posts a WM_QUIT message. PostQuitMessage() has an exit code parameter that is returned to Windows, but current versions of the operating system make no use of this value. The objection to using PostQuitMessage() for abnormal terminations is that execution ends abruptly, without notification of cause or reason. In this case the program should display a message box informing the user of the cause of program termination.

Windows supports other devices such as pens, touch screens, joysticks, and drawing tablets, which are all considered mouse input. The mouse itself can have up to three buttons, labeled left, middle, and right buttons. A one-button mouse is an anachronism and the three-button version is usually associated with specialized systems. The most common one is the two-button mouse, where the left button is used for clicking, double-clicking, and dragging operations and the right button activates context-sensitive program options.

An application can tell how many buttons are installed in the mouse by testing the SM_CMOUSEBUTTONS with the GetSystemMetrics() function. If the application requires a certain number of buttons, then the assert or abort macros can be used, as previously shown. For example, a program that requires a three-button mouse could test for this condition as follows:

```
assert (GetSystemMetrics(SM_CMOUSEBOUTTONS) == 3);
```

If the three-button mouse is required in the release version of the program, then the code could be as follows:

```
if(GetSystemMetrics(SM_CMOUSEBUTTONS) != 3))
    abort();
```

Notice that the assert macro is intended to be used in debugging. If the condition is false, the macro shows information about the error and displays a message box with three options: abort, debug, and ignore. Assert has no effect on the release version of the program; it is as if the statement containing assert had been commented out of the code. For this reason conditions that must be evaluated during execution of the release version of a program should not be part of an assert statement.

The abort macro can be used to stop execution in either version. Abort provides no information about the cause of program termination.

Programs that use the assert macro must include the file assert.h. VERIFY and other debugging macros are available when coding with the Foundation Class Library, but they are not implemented in ANSI C.

21.3.1 Mouse Messages

There are 22 mouse messages currently implemented in the Windows API. Ten of these messages refer to mouse action on the client area, and ten to mouse action in the nonclient area. Of the remaining two messages WM_NCHITTEST takes place when the mouse is moved either over the client or the nonclient area. It is this message that generates all the other ones. WM_MOUSEACTIVATE takes place when a mouse button is pressed over an inactive window, an event that is usually ignored by applications.

The abundance of Windows messages should not lead you to think that mouse processing is difficult. Most applications do all their mouse processing by intercept-

ing two or three of these messages. Table 21.4 lists the mouse messages most frequency handled by applications.

Table 21.4

Frequently Used Client Area Mouse Messages

MOUSE MESSAGE	DESCRIPTION
WM_LBUTTONDOWN	Left button pressed
WM_LBUTTONUP	Left button released
WM_RBUTTONDOWN	Right button pressed
WM_RBUTTONUP	Right button released
WM_RBUTTONDBLCLK	Right button double-clicked
WM_LBUTTONDBLCLK	Left button double-clicked
WM_MOUSEMOVE	Mouse moved into client area

Table 21.4 lists only client area mouse messages; nonclient area messages are
usually handled by the default windows procedure.

Mouse processing is similar to keyboard processing, although mouse messages
do not require that the window have the input focus. Once your application gains
control in a mouse message handler, it can proceed to implement whatever action
is required. However, there are some differences between keyboard messages and
mouse messages. To Windows, keyboard input is always given maximum attention. The operating system tries to assure that keyboard input is always preserved. Mouse messages, on the other hand, are expendable. For example, the
WM_MOUSEMOVE message, which signals that the mouse cursor is over the application's client area, is not sent while the mouse is over every single pixel of the
client area. The actual rate depends on the mouse hardware and on the processing
speed. Therefore, it is possible, given a small enough client area and a slow
enough message rate, that code may not be notified of a mouse movement action
over its domain. Mouse programming must take this possibility into account.

In client area mouse messages, the wParam indicates which, if any, keyboard
or mouse key was held down while the mouse action took place. Windows defines
five symbolic constants to represent the three mouse keys and the keyboard Ctrl
and Shift keys. These constants are listed in Table 21.5.

Table 21.5

Virtual Key Constants for Client Area Mouse Messages

CONSTANT	ORIGINATING CONDITION
MK_CONTROL	Ctrl key is down.
MK_LBUTTON	Left mouse button is down.
MK_MBUTTON	Middle mouse button is down.
MK_RBUTTON	Right mouse button is down.
MK_SHIFT	Shift key is down.

Code can determine if one of the keys was held down by ANDing with the corresponding constant. For example, the following fragment can be used to determine if the Ctrl key was held down at the time that the left mouse button was clicked in the client area:

```
case WM_LBUTTONDOWN:
    if(wParam & MK_CONTROL) {
    // ASSERT:
    //   Left mouse button clicked and <Ctrl> key down
```

The predefined constants represent individual bits in the operand; therefore, you must be careful not attempt to equate the wParam with any one of the constants. For example, the MK_LBUTTON constant is always true in the WM_LBUTTONDOWN intercept, for this reason the following test always fails:

```
case WM_LBUTTONDOWN:
    if(wParam == MK_CONTROL) {
```

On the other hand, you can determine if two or more keys were held down by performing a bitwise OR of the predefined constants before ANDing with the wParam. For example, the following expression can be used to tell if either the Ctrl keys or the Shift keys were held down while the left mouse button was clicked:

```
if(wParam & (MK_CONTROL | MK_SHIFT)) {
    // ASSERT:
    //    Either the <Ctrl> or the <Shift> key was held down
    //    when the mouse action occurred
```

To test if both the <Ctrl> and the <Shift> keys were down when the mouse action occurred, you can code as follows:

```
if((wParam & MK_CONTROL) && (wParam & MKSHIFT)) {
    // ASSERT:
    //    The <Ctrl> and <Shift> key were both down when the
    //    mouse action occurred
```

21.3.2 Cursor Location

Applications often need to know the screen position of the mouse. In the case of the client area messages, the lParam encodes the horizontal and vertical position of the mouse cursor when the action takes place. The high-order word of the lParam contains the vertical mouse position and the low-order word the horizontal position. Code can use the LOWORD and HIWORD macros to obtain the value in logical units. For example:

```
int     cursorX, cursorY;      // Storage for coordinates
  .
  .
  .
case WM_MOUSEMOVE:
    cursorX = LOWORD(lParam)
    cursorY = HIWORD(lParam);
    // ASSERT:
    //    Variables now hold x and y cursor coordinates
```

21.3.3 Double-Click Processing

Handling mouse double-clicks requires additional processing as well as some fore-thought. In the first place, mouse double-click messages are sent only to windows that were created with the CS_DBLCLKS style. The CS_DBLCLKS style is described in Table 21.2. The structure of type WNDCLASSES for a windows that it to receive mouse double-clicks can be defined as follows:

```
// Defining a structure of type WNDCLASSEX
   WNDCLASSEX  wndclass ;
   wndclass.cbSize    = sizeof (WNDCLASSEX) ;
   wndclass.style     = CS_HREDRAW | CS_VREDRAW |
                        CS_DBLCLKS;
   .
   .
   .
```

Three client area mouse messages are related to the double-click action, one for each mouse button. If the window class includes the CS-DBLCLKS type, then client area double-click messages take place. WM_LBUTTONDBLCLK intercepts double-clicks for the left mouse button, WM_RBUTTONDBLCLK for the right mouse button, and WM_MBUTTONDBLCLK for the center button.

The double-click notification occurs when a mouse button is clicked twice within a predefined time interval. The double-click speed is set by selecting the Mouse Properties option in the Windows Control Panel. The SetDoubleClickTime() function can also be used to change the double-click interval from within an application, although it is not a good idea to do this without user participation. The default double-click time is 500 msec (one-half second). In addition, the two actions of a double-click must occur within a rectangular area defined by Windows, according to the display resolution. If the mouse has moved outside of this rectangle between the first and the second clicks, then the action is not reported as a double-click. The parameters for the double-click rectangle can be retrieved with the GetSystemMetrics() function, using the predefined constant SM_CXDOUBLECLK for the x-coordinate, and SM_CYDOUBLECLK for the y-coordinate.

A double-click intercept receives control on the second click, because at the time of the first click it is impossible to know if a second one is to follow. Therefore, if the code intercepts normal mouse clicks, it also receives notification on the first click of a double-click action. For this reason, programs are usually designed so that the action taken as a result of a double-click is a continuation of the one taken on a single click. For example, selecting an application file in Windows Explorer by means of a single mouse click has the effect of highlighting the file-name. If the user double-clicks, the file is executed. In this case the double-click action complements the single-click one. Although it is possible to implement double-click processing without this constraint, the programming is more complicated and the user interface becomes sluggish.

21.3.4 Capturing the Mouse

The mouse programming logic so far discussed covers most of the conventional programming required for handling mouse action inside the active window. By intercepting the client area messages, not the nonclient area ones, we avoid being notified of actions that usually, do not concern our code. However, there are common mouse operations that cannot be implemented by processing client area messages only. For example, a Windows user installs a program icon on the desktop by right-clicking on the icon and then dragging it outside of the program group window. When the right mouse button is released, Windows displays a menu box that includes options to move or copy the program item, to create a shortcut, or to cancel the operation. In this case, the action requires crossing the boundary of the active window. Therefore, client area messages cease as soon as this boundary is reached.

Another case is a drawing program that uses a mouse dragging operation to display a rectangular outline. The rectangle starts at the point where the button is clicked, and ends at the point where the button is released. But what happens if the user crosses over the client area boundary before releasing the mouse button? In this case the application is not notified of the button release action since it occurs outside the client area. Furthermore, if the drawing action is performed during the WM_MOUSEMOVE intercept, the messages also stop being sent to the applications windows procedure as soon as the client area boundary is crossed. It would be a dangerous assumption to implement this function assuming that the user never crosses the boundary of the program's client area.

Problems such as these are solved by capturing the mouse, which is done by the SetCapture() function. The only parameter to SetCapture() is the handle of the capturing window. Once the mouse is captured, all mouse actions are assumed to take place in the client area, and the corresponding message intercepts in the application code are notified. The most obvious result of a mouse capture is that the client area message handlers are active for mouse actions that take place outside the client area. Only one window can capture the mouse, and it must be the active one, also called the foreground window. While the mouse is captured all system keyboard functions are disabled. The mouse capture ends with the call to ReleaseCapture(). GetCapture() returns the handle to the window that has captured the mouse, or NULL if the mouse capture fails.

Applications should capture the mouse whenever there is a possibility, even a remote one, of the user crossing the boundary of the client area during mouse processing. Implementing a simple drag-and-drop operation usually requires capturing the mouse. Mouse operations that take place between windows, whether they be child windows or not, also require capturing the mouse. Multitasking operations are limited during mouse capture. Therefore, it is important that the capture is released as soon as it is no longer necessary.

21.3.5 The Cursor

The screen image that corresponds to the mouse device is called the cursor. Windows provides thirteen built-in cursors from which an application can select. In addition, you can create your own customized cursor and use it instead of a standard one. There are over twenty Windows functions that relate to cursor operations; however, even programs that manipulate cursor images hardly ever use more than a couple of them. Figure 21.3 shows the Windows built-in cursors and their corresponding symbolic names.

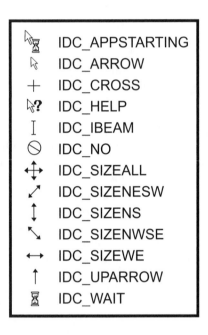

Figure 21.3 *Windows Built-In Cursors*

Code that manipulates cursor images must be aware of Windows cursor-handling operations. A mouse-related message not yet discussed is WM_SETCURSOR. This message is sent to your window procedure, and to the default window procedure, whenever a noncaptured mouse moves over the client area, or when its buttons are pressed or released. In the WM_SETCURSOR message, the wParam holds the handle to the window receiving the message. The low-order word of lParam is a code that allows determining where the action takes place, usually called the hit code. The high-order word of the lParam holds the identifier of the mouse message that triggered WM_SETCURSOR.

One of the reasons for WM_SETCURSOR is to give applications a chance to change the cursor; also for a parent window to manipulate the cursor of a child window. The problem is that Windows has a mind of its own regarding the cursor. If your application ignores the WM_SETCURSOR message, the default window procedure receives the message anyway. If Windows determines (from the hit code) that the cursor has moved over the client area of a window, then the default window procedure sets the cursor to the class cursor defined in the hCursor member of the WNDCLASSEX structure in WinMain(). If the cursor is in a nonclient area, then Windows sets it to the standard arrow shape.

What all of this means to your application code is that if you ignore the WM_SETCURSOR message, and don't take other special provisions, Windows continuously changes the cursor according to its own purposes, probably interfering with your own manipulations. The simplest solution is to intercept WM_SETCURSOR and return a nonzero value. In this case the window procedure halts all further cursor processing. You could also use the WM_SETCURSOR intercept to install your own cursor or cursors, however, the disadvantage of this approach is that WM_SETCURSOR does not provide information about the cursor's screen location.

An alternate method is to perform cursor manipulations at one of the mouse message intercepts, or any other message handler for that matter. For example, code can implement cursor changes at WM_MOUSEMOVE. In this case the lParam contains the cursor's horizontal and vertical position. Child windows can use this intercept to display their own cursors. In this case the hCursor field of the WNDCLASSEX structure is usually set to NULL, and the application takes on full responsibility for handling the cursor.

Applications that manipulate the cursor often start by setting a new program cursor during WM_CREATE processing. In cursor processing there are several ways of achieving the same purpose. The methods described are those that the authors have found more reliable. To create and display one of the built-in cursors you need a variable to store the handle to the cursor. The LoadCursor() and SetCursor() functions can then be used to load and display the cursor. To load and display the IDC_APPSTARTING cursor code can be as follows:

```
HCURSOR    aCursor;
.
.
.
aCursor = LoadCursor(NULL, IDC_APPSTARTING);
SetCursor (aCursor);
```

The first parameter of the LoadCursor() function is the handle to the program instance. This parameter is set to NULL to load one of the built-in cursors. Any of the symbolic names in Figure 21.3 can be used. The cursor is not displayed until the SetCursor() function is called, using the cursor handle as a parameter.

Graphics applications sometimes need one or more special cursors to suit their own needs. In the Visual C++ development environment, creating a custom cursor is made easy by the image editor. The process described for creating a program icon in previously in the section titled "Creating a Program Resource," is almost identical to the one for creating a custom cursor. Briefly reviewing:

1. In the Insert menu select the Resource command and then the Cursor resource type.

2. Use the editor to create a cursor. Note that all cursors are defined in terms of a 32-by-32 bit monochrome bitmap.

3. A button on the top bar of the editor allows positioning the cursor's hot spot. The default position for the hot spot is the upper left corner.

4. In the process of creating a cursor, Developer Studio also creates a new script file, or adds the information to an existing one. You must manually insert the script file into the project by selecting the Add to Project command from the Project menu and then selecting the Files option. In the "Insert Files into Project" dialog box select the script file and then click the OK button. The script file now appears on the Source Files list in the Files View window of the Project Workspace.

5. In addition to the script file, Developer Studio also creates a header file for resources. The default name of this file is resource.h. In order for resources to be available to the code you must enter an #include statement for the resource.h file in your source.

In order to use the custom cursor in your code you must know the symbolic name assigned to this resource, or its numeric value. The information can be obtained by selecting the Resource Symbols command from the View menu, or clicking the corresponding button on the toolbar.

The LoadCursor() function parameters are different for a custom cursor than for a built-in one. In the case of a custom cursor, you must enter the handle to the instance as the first parameter, and use the MAKEINTRESOURCE macro to convert the numeric or symbolic value into a compatible resource type. For example, if the symbolic name of the custom cursor is IDC_CURSOR1, and the handle to the instance is stored in the variable pInstance (as is the case in the template files furnished in this book) you can proceed as follows:

```
HCURSOR        aCursor;       // handle to a cursor
  .
  .
  .
aCursor = LoadCursor(pInstance,
                 MAKEINTRESOURCE(IDC_CURSOR1));
SetCursor(aCursor);
```

21.4 Mouse and Cursor Demonstration Program

The program named MOU_DEMO, located in the Mouse Demo project folder of the book's software on-line, is a demonstration of some of the mouse handling operations previously described. At this point in the book we have not yet covered the graphics services, or the implementation of user interface functions. For these reasons, it is difficult to find a meaningful demonstration for mouse operations.

MOU_DEMO monitors the left and the right mouse buttons. Clicking the left button changes to one of the built-in cursors. The cursors are displayed are the same ones as in Figure 21.3. Clicking the right mouse button displays a customized cursor in the form of the letter "A." The hot spot of the custom cursor is the vertex of the "A." When the mouse is moved in the client area, its position is displayed on the screen. Figure 21.4 is a screen snapshot of the MOU_DEMO program.

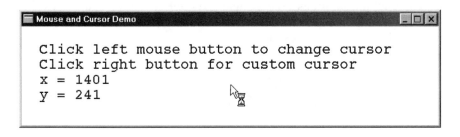

Figure 21.4 *MOU_DEMO Program Screen*

The program's first interesting feature is that no class cursor is defined in the WNDCLASSEX structure. Instead, the hCursor variable is initialized as follows:

```
wndclass.hCursor    = NULL;
```

Since the program has no class cursor, one is defined during WM_CREATE processing, with the following statements:

```
// Select and display a cursor
aCursor = LoadCursor(NULL, IDC_UPARROW);
SetCursor(aCursor);
```

In this code, the variable aCursor, of type HCURSOR, is declared in the windows procedure. Toggling the built-in cursors is performed in the WM_LBUTTONDOWN message intercept. The coding is as follows:

```
case WM_LBUTTONDOWN:
curNum++;        // bump to next cursor

switch (curNum) {
case 1:
   aCursor = LoadCursor(NULL, IDC_WAIT);
   SetCursor(aCursor);
   break;
case 2:
   aCursor = LoadCursor(NULL, IDC_APPSTARTING);
   SetCursor(aCursor);
   break;
case 3:
   aCursor = LoadCursor(NULL, IDC_CROSS);
   SetCursor(aCursor);
   break;
   .
   .
   .
case 12:
   aCursor = LoadCursor(NULL, IDC_UPARROW);
   SetCursor(aCursor);
   curNum = 0;
   break;
}
```

Note that the static variable curNum, defined in the window procedure, is used to keep track of the cursor being displayed and to index through all thirteen cursor images. The custom cursor is created using the cursor editor that is part of Visual Studio. The display of the custom cursor is implemented during WM_RBUTTONDOWN processing:

```
case WM_RBUTTONDOWN:
    aCursor = LoadCursor(pInstance,
            MAKEINTRESOURCE(IDC_CURSOR1));
    SetCursor(aCursor);
    return 0;
```

The movement of the mouse in the client area is detected by intercepting the WM_MOUSEMOVE message. The processing consists of obtaining the cursor coordinates from the low-order and high-order words of lParam, and converting the numeric values into ASCII strings for display. The code uses _itoa() for this purpose. The ASCII values are placed on the corresponding string arrays. The processing is as follows:

```
case WM_MOUSEMOVE:
    cursorX = LOWORD(lParam);
    cursorY = HIWORD(lParam);
    // Convert integer to ASCII string
    _itoa(cursorX, CurXStr + 4, 10);
    _itoa(cursorY, CurYStr + 4, 10);
    // Display x coordinate of mouse cursor
    // First initialize rectangle structure
    SetRect (&textRect,         // address of structure
        2 * cxChar,        // x for start
        3 * cyChar,             // y for start
        cxClient -(2 * cxChar),  // x for end
        cyClient);          // y for end
    // Erase the old string
    DrawText( hdc, CurXBlk, -1, &textRect,
        DT_LEFT | DT_WORDBREAK);
    // Display new string
    DrawText( hdc, CurXStr, -1, &textRect,
        DT_LEFT | DT_WORDBREAK);
 // Display y coordinate of mouse cursor
    .
    .
    .
    return 0;
```

In order to avoid having Windows change the cursor as it moves into the client area, the code intercepts the WM_SETCURSOR message, as follows:

```
case WM_SETCURSOR:
    return 1;
```

When running the MOU_DEMO program notice that if the cursor is moved at a rather fast rate out of the client area, toward the left side or the top of the screen, the last value displayed for the diminishing coordinate may not be zero. This is due to the fact, mentioned earlier in this section, that WM_MOUSEMOVE messages are not sent to the window for every pixel of screen travel. Mouse programming must also take this into account and use greater-than and smaller-than comparisons to determine screen areas of cursor travel.

Chapter 22

Graphical User Interface Elements

Chapter Summary

This chapter is about programming the Windows graphical user interface (GUI). The Windows GUI consists of child windows and built-in controls, such as status bars, toolbars, ToolTips, trackbars, up-down controls, and many others. The discussion also includes general purpose controls such as message boxes, text boxes, combo boxes, as well as the most used of the common controls. All of these components are required to build a modern Windows program; it is difficult to imagine a graphics application that does not contain most of these elements.

22.0 Window Styles

One of the members of the WNDCLASSEX structure is the windows style. Previously in the book we briefly discussed windows styles and listed the constants that can be used to define this member. Since the eleven style constants can be ORed with each other, many more windows styles can result. Furthermore, when you create a window using the CreateWindow() function, there are twenty-seven window style identifiers (see Table 19.5). In addition, the CreateWindowEx() function provides twenty-one style extensions (see Table 19.4). Although the number of possible combinations of all these elements is very large, in practice, about twenty window styles, with unique properties, are clearly identified, all of which are occasionally used. This list can be further simplified into three general classes (overlapped, pop-up, and child windows) and three variations (owned, unowned, and child), which give rise to five major styles.

In the sections that follow we discuss four specific window styles:

- Unclassed child windows. These are windows that are related to a parent window but that do not belong to one of the predefined classes.

- Basic controls. These are child windows that belong to one of the standard control classes: BUTTON, Combo box, EDIT, LISTBOX, MDICLIENT, SCROLLBAR, and STATIC.

- Dialog boxes. A special type of pop-up window, that usually includes several child window controls, typically used to obtain and process user input.

- Common controls. A type of controls introduced in Windows 3.1, which include status bars, toolbars, progress bars, animation controls, list and tree view controls, tabs, property sheets, wizards, rich edit controls, and a new set of dialog boxes.

Several important topics related to child windows and window types are not discussed, among them are OLE control extensions, ActiveX controls, and multiple document interface (MDI). OCX controls relate to OLE automation and ActiveX controls are used mostly in the context of Web programming.

22.1 Child Windows

The simplest of all child windows is one that has a parent but does not belong to any of the predefined classes. Sometimes these are called "unclassed" child windows. However, if we refer to the "classed" child windows as controls, then the "unclassed" windows can be simply called "child windows." These are the designations used in the rest of the book: we refer to unclassed child windows simply as child windows and the classed variety as controls.

A child window must have a parent, but it cannot be an owned or an unowned window. The child window can have the appearance of a main window, that is, it can have a sizing border, a title bar, a caption, one or more control buttons, an icon, a system menu, a status bar, and scroll bars. The one element it cannot have is a menu, since an application can have a single menu and it must be on the main window. On the other hand, a child window can be defined just as an area of the parent window. Moreover, a child window can be transparent; therefore, invisible on the screen. The conclusion is that it is often impossible to identify a child window by its appearance.

A child window with a caption bar can be moved inside its parent client area; however, it will be automatically clipped if moved outside of the parent. The child window overlays a portion of its parent client area. When the cursor is over the child, Windows sends messages to the child, not to the parent. By the same token, mouse action on the child window's controls, or its system menu, is sent to the child. A child window can have its own window procedure and perform input processing operations independently of the parent. When the child window is created or destroyed, or when there is a mouse-button-down action on the child, a WM_PARENTNOTIFY message is sent to the parent window. One exception to parent notification is if the child is created with the WS_EX_NOPARENTNOTIFY style.

A child window is created in a similar manner as the parent window, although there are some important variations. Creating a child window involves the same steps as in creating the main window. You must first initialize the members of the WNDCLASSEX structure. Then the window class must be registered. Finally, the window is actually created and displayed when a call is made to CreateWindow() or CreateWindowEx() function.

There are not many rules regarding when and where an application creates a child window. The child window can be defined and registered in WinMain() and displayed at the same time as the main window. Or the child window can be created as the result of user input or program action. We have already mentioned the great number of windows styles and style combinations that can be used to define a child window. Some of these styles are incompatible, and others are ineffective when combined.

The styles used in creating the child window determine how it must be handled by the code. For example, if a child window is created with the WS_VISIBLE style, then it is displayed as it is created. If the WS_VISIBLE style is not used, then to display the child window you have to call ShowWindow() with the handle to the child window as the first parameter, and SW_SHOW, SW_SHOWNORMAL, or one of the other predefined constants, as the second parameter.

In operation, the child window provides many features that facilitate program design. For instance, a child window has its own window procedure, which can do its own message processing. This procedure receives the same parameters as the main window procedure and is notified of all the windows messages that refer to the child. The child window can have its own attributes, such as icons, cursors, and background brush. If the main window is defined with an arrow cursor and the child window with a cross cursor, the cursor changes automatically to a cross as it travels over the child, and back to an arrow as it leaves the child's client area. The fact that each window does is own message processing considerably simplifies the coding. Screen environments with multiple areas, such as the ones in Visual Studio, Windows Explorer, and many other applications, are implemented by means of child windows.

Parent and child windows can share the same display context or have different ones. In fact, each window can have any of the display contexts described previously in the text. If the child window is declared with the class style CS_PARENTDC, then it uses the parent's display context. This means that output performed by the child takes place in the parent's client area, and the child has no addressable client area of its own. On the other hand, parent and child can have separate device contexts. If both windows are declared with the class style CS_OWNDC, discussed previously, then each has its own display context with a unique set of attributes. If there is more than one child window, they can be declared with the class style CS_CLASSDC, and the children share a single device context, which can be different from the one of the parent window.

Each child window is given its own integer identifier at the time it is created. Since child windows can have no menus, the HMENU parameter passed to CreateWindows() or CreateWindowsEx() is used for this purpose. The child window uses this identifier in messages sent to its parent, which enables the parent to tell to which child window the message belongs, if more than one is enabled. If multiple child windows are given the same numeric identification then it may be impossible for the parent to tell them apart.

22.1.1 Child Windows Demonstration Program

The program named CHI_DEMO, located in the Child Window Demo project folder on the book's software on-line, is a demonstration of a program with a child window. The program displays an overlapped child window inside the parent window. When the left mouse button is clicked inside the child window, a text message is displayed in its client area. The same happens when the left mouse button is clicked in the parent's client area. At the same time, the old messages in the parent or the child windows are erased. Figure 22.1 is a screen snapshot of the CHI_DEMO program.

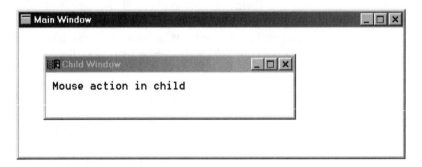

Figure 22.1 *CHI_DEMO Program Screen*

The program uses a child window, which is defined using the WS_OVERLAPPEDWINDOW style. This style, which is the same one used in the parent window, gives both the parent and the child a title bar with caption, a system menu, a border, and a set of control buttons to close, minimize and restore. The child window is created during WM_CREATE message processing of the parent window, as follows:

```
LRESULT CALLBACK WndProc(HWND hwnd, UINT iMsg, WPARAM wParam,
                         LPARAM lParam) {
  PAINTSTRUCT ps ;
  WNDCLASSEX  chiclass ;
  switch (iMsg) {
    case WM_CREATE:
      hdc = GetDC (hwnd) ;
      // The system monospaced font is selected
      SelectObject (hdc, GetStockObject (SYSTEM_FIXED_FONT)) ;
      // Create a child window
      chiclass.cbSize        = sizeof (chiclass) ;
      chiclass.style         = CS_HREDRAW | CS_VREDRAW
                               | CS_OWNDC;
      chiclass.lpfnWndProc   = ChildWndProc ;
      chiclass.cbClsExtra    = 0 ;
      chiclass.cbWndExtra    = 0 ;
      chiclass.hInstance     = pInstance ;
      chiclass.hIcon         = NULL;
      chiclass.hCursor       = LoadCursor (NULL, IDC_CROSS) ;
      chiclass.hbrBackground = (HBRUSH) GetStockObject
                               (WHITE_BRUSH);
      chiclass.lpszMenuName  = NULL;
      chiclass.lpszClassName = "ChildWindow" ;
      chiclass.hIconSm       = NULL;
```

```
        RegisterClassEx (&chiclass) ;

        hChild = CreateWindow ("ChildWindow",
                "A Child Window",  // caption
                WS_CHILD | WS_VISIBLE |
                WS_OVERLAPPEDWINDOW ,
                40, 40,         // x and y of window location
                400, 100,       // x and y of window size
                hwnd,           // handle to the parent window
                (HMENU) 1001,   // child window designation
                pInstance,      // program instance
                NULL) ;
        // Make sure child window is valid
        assert(hChild != NULL);
        return 0 ;
        .
        .
        .
```

Note that the child is defined with the styles WS_CHILD, WS_VISIBLE, and WS_OVERLAPPEDWINDOW. The WS_VISIBLE class style ensures that the child becomes visible as soon as CreateWindows() is executed. The child window is assigned the arbitrary value 1001 in the HMENU parameter to CreateWindow(). The child has a private DC, the same as the parent, but the DCs are different. The assert statement ensures, during program development, that the child window is a valid one.

During the parent's WM_PAINT message processing a call is made to UpdateWindow() with the handle of the child window as a parameter. The result of this call is that the child's window procedure receives a WM_PAINT message.

The window procedure for the child, named ChildWndProc() in the demo program, is prototyped in the conventional manner and its name defined in the lpfnWndProc member of the child's WNDCLASSEX structure. The child's window procedure is coded as follows:

```
LRESULT CALLBACK ChildWndProc (HWND hChild, UINT iMsg, WPARAM wParam,
                               LPARAM lParam) {
 switch (iMsg) {
  case WM_CREATE:
      childDc = GetDC(hChild);
      SelectObject (childDc, GetStockObject
                   (SYSTEM_FIXED_FONT)) ;
      return 0;
    case  WM_LBUTTONDOWN:
    // Display message in child and erase text in parent
    TextOut(childDc, 10, 10, "Mouse action in child ", 22);
    TextOut(hdc, 10, 10, "                         ", 22);
return 0;

  case WM_DESTROY:
      return 0;
  }
 return DefWindowProc (hChild, iMsg, wParam, lParam) ;
}
```

During the WM_CREATE processing of the child's windows procedure, the code obtains a handle to the child's DC. Also, the system fixed font is selected into the DC at this time.

In the CHI_DEMO program we have declared several public variables: the handles to the windows of the parent and the child and the handles to their display context. This stretches one of the fundamental rules of Windows programming: to keep public data at a minimum. In this case, however, we achieve a substantial simplification in the coding, since now the parent can have access to the child's device context, and vice versa. Therefore, when the user clicks the left mouse button in the child's client area, a text message is displayed in the child window and the one in the parent window is simultaneously erased. Similar processing takes place when the left mouse button is clicked in the parent's client area.

22.2 Basic Controls

These are the traditional controls that have been around since the Win16 APIs. They are predefined child windows that belong to one of the standard window classes. Table 22.1 lists the predefined classes used for controls.

Table 22.1

Predefined Control Classes

CLASS NAME	MEANING
BUTTON	A small rectangular child window representing a button. The user clicks a button to turn it on or off. Button controls can be used alone or in groups, and they can be labeled or not. Button controls typically change appearance when clicked.
COMBOBOX	Consists of a list box and a selection field similar to an edit control (see description). Depending on its style, you can or cannot edit the contents of the selection field. If the list box is visible, typing characters into the selection field highlights the first list box entry that matches the characters typed. By the same token, selecting an item in the list box displays the selected text in the selection field.
EDIT	A rectangular child window into which you type text. You select the edit box and give it the keyboard focus by clicking it or moving to it by pressing the Tab key. You can enter text into an Edit control if it displays a flashing caret. You use the mouse to move the cursor inside the box, to select characters to be replaced, or to position the cursor for inserting new characters. The backspace key deletes characters. Edit controls use a variable-pitch system font and display characters from the ANSI character set. The WM_SETFONT message can be used to change the default font. During input, tab characters are expanded into ss many spaces as are required to move the care to the next tab stop. Tab stops are preset eight spaces apart.

(continues)

Table 22.1

Predefined Control Classes (continued)

CLASS NAME	MEANING
LISTBOX	A list of character strings. It is used to present a list of names, such as filenames, from which you can select. Selection is made by clicking an item in the list box. The selected string is highlighted, and a notification message is sent to the parent window. When the item list is too long for the window, you can use a vertical or horizontal scroll bar. If the scroll bar is not needed, it is automatically hidden.
SCROLLBAR	A rectangular control with a scroll box and direction arrows at both ends. The scroll bar sends a notification message to its parent window whenever the user clicks it. The parent window is responsible for updating the position of the scroll box when necessary. Scroll bar controls have the same appearance and function as scroll bars used in ordinary windows. Unlike scroll bars, however, scroll bar controls can be positioned anywhere in a window and for any purpose. The scroll bar class also includes size box controls, which is a small rectangle that you can expand to change the size of the window.
STATIC	A simple text field, box, or rectangle, used to label, group, or separate other controls. Static controls take no input and provide no output.

Figure 22.2 shows buttons of several types, a list box, a combo box, and a scroll bar control.

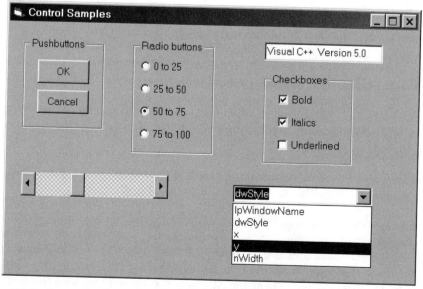

Figure 22.2 *Buttons, List Box, Combo Box, and Scroll Bar Controls*

In conventional Windows programming basic controls are not frequently used in the client area of the main window. Most often you see them in message boxes or input boxes, described later in this chapter. For this reason, Developer Studio does not provide a resource editor for inserting controls in the client area, although it does contain a powerful editor for dialog boxes. In spite of this, the use of basic controls in child windows adds considerable power to a programmer's toolkit. The result is a completely customizable message, dialog box, toolbar, or other child window, in which you are free from all the restrictions of the built-in versions of these components. The price for this power and control is that you must implement all the functionality in your own code.

The CreateWindow() or CreateWindowEx() functions are used to build any one of the controls in Table 22.1. If the control is created using the WS_VISIBLE window style, then it is displayed immediately on the window whose handle is passed as a parameter to the call. If not, then the ShowWindow() function has to be called in order to display it. The call returns a handle to the created control, or NULL if the operation fails. The following code fragment shows creating a button control.

```
static HWND     hwndRadio1;      // Handle to control
    . . .
    hwndRadio1 = CreateWindow (
      "BUTTON",             // Control class name
      "Radio 1",                  // Button name text
      WS_CHILD | WS_VISIBLE | BS_RADIOBUTTON /| WS_SIZEBOX,
      20,                       // x coordinate of location
      60,                       // y coordinate
      100, 30,                  // button size
      hChild,                   // Handle to parent window
      (HMENU) 201,              // control id number
      pInstance,                // Instance handle
      NULL) ;                   // Pointer to additional data
```

Because controls belong to predefined classes, they need not be registered as a window class. Therefore, the WNDCLASSEX structure and the call to RegisterClass() or RegisterClassEx() are not required in this case. In the case of a main window, the eighth parameter of CreateWindow() is the handle to its menu. Since controls cannot have a menu, this parameter is for the control's numeric designation, the same as with a child window. Thereafter, this numeric value, which can be also a predefined constant, identifies the control. If the control is to be addressable, this identification number should be unique.

In addition to the general window style, each of the predefined control classes has its own set of attributes. The prefixes are shown in Table 22.2.

The class-specific styles are ORed with the window style constants passed in the third parameter to CreateWindow(). Note that in the previous code fragment the BS_RADIOBUTTON constant is included in the field. There are several variations of the button class. The buttons in the first group of Figure 22.2, labeled Pushbuttons, are plain pushbuttons. They appear raised when not pushed and sunken after being pushed by the user. Pushbuttons operate independently. These buttons are usually created with the BS_PUSHBUTTON and BS_DEFPUSHBUTTON styles.

Table 22.2

Prefix for Predefined Window Classes

PREFIX	CONTROL TYPE
BS	button
CBS	combo box
ES	edit box
LBS	list box
SBS	scroll bar
SS	static

Radio buttons are reminiscent of the buttons in the radios of old-style automobiles: pushing one button automatically pops out all the others. The styles BS_RADIOBUTTON and BS_AUTORADIOBUTTONS are used for creating this type of button. Radio buttons contain a circular area with a central dot that indicates the button's state.

Another variation is the checkbox. A checkbox can have two or three states. A two-state checkbox can be checked or unchecked, while the three-state style can also be grayed. Checkboxes, like regular buttons, operate independently of each other. Two-state checkboxes are created with the BS_CHECKBOX style. The three-state version requires ORing the BS_3STATE constant with BS_CHECKBOX.

A unique style of button is the groupbox, which is enabled with the button style BS_GROUPBOX. A groupbox is used to enclose several buttons or controls in a labeled frame. It is unique in the sense that it is defined as a button, but a groupbox does not respond to user input, nor does it send messages to the parent window. Figure 22.2 shows three group boxes, one for each type of button.

Three types of controls are designed for manipulating text: the edit box, the combo box, and the list box. You select an edit box control for input by clicking it or tabbing until it has the input focus. When a caret is displayed, you can enter text until the rectangle is filled. If the edit box control is created with the ES_AUTOSCROLL style, then you can enter more characters than fit in the box since the text automatically scrolls to the left, although this practice is not recommended since part of the input disappears from the screen. If the edit box is defined with the ES_MULTILINE style then you can enter more than one text line. However, this style can create conflicts if the active window contains a default pushbutton that also responds to the Enter key. The built-in solution to this problem is that the default style of edit box requires the Ctrl+Enter key combination to end an input line. However, if the edit box is created with the style ES_WANTRETURN, then the Enter key alone serves as a line terminator.

The list box control displays a list of text items from which the user can select one or more. Code can add or remove strings from the list box. Scroll bars can be requested for a list box. If the list box is created with the LBS_NOTIFY style then the parent window receives a message whenever the user clicks or double-clicks an item. The LBS_SORT style makes the list box sort items alphabetically.

The combo box is a combination of a textbox and a list box. The user can enter text on the top portion of the combo box, or drop down the list box and select an item from it. Alternatively, the edit function of the combo box can be disabled. Figure 22.2 shows a combo box.

Scroll bar controls can be vertical or horizontal and be aligned at the bottom, top, left, or right of a rectangle defined at call time. It is important to distinguish between window and control scroll bars. Any window can have scroll bars if it is defined with the WS_VSCROLL or WS_HSCROLL styles. Scroll bar controls are individual scroll bars which can be positioned anywhere on the parent's client area. Both, windows and control scroll bars send messages to the parent window whenever a user action takes place. Scroll bar controls are of little use by themselves but provide a powerful and convenient way of obtaining user input; for example, a scroll bar control which allows the user to move up or down a numeric range without typing values. In this case the scroll bar is usually combined with another control that displays the selected value. The CON_DEMO program, in this chapter, has an example of this use of a scroll bar control.

Static controls do not interact with the user since they cannot receive mouse or keyboard input. The principal use of static controls is to display rectangular frames of several colors and borders, and to provide feedback from another control. The CON_DEMO program, described later in this chapter, which is found in the Controls Demo project folder in the book's software pckage, has a child window with a static control that displays the position of a scroll bar.

22.2.1 Communicating with Controls

Controls are child windows and child windows can communicate with their parents. As is the case in all Windows functions, controls communicate with their parent window by means of a message passing mechanism. The messages passed to the parent window depend on the type of control. This communication works both ways: a control sends a message to its parent window informing it that a certain user action has taken place, or the parent window sends a message to a control requesting that it take a certain action or report some item of information stored internally. For example, when the user clicks on a pushbutton control, a WM_COMMAND message is sent to the parent window. When a parent window needs to know if a radio button is checked or unchecked it sends a BM_GETCHECK message to the radio button control.

WM_COMMAND is used to inform the parent window of action on a menu, on a control, or of an accelerator keystroke. The high-order word of the wParam is zero if the message originates in a menu, and one if it originates in an accelerator keystroke. If the message originated in a control, then the high-word of the wParam is a control-specific notification code. Table 22.3 lists the notification codes for the button controls.

Table 22.3

Notification Codes for Buttons

NOTIFICATION CODE	ACTION
BN_CLICKED	Button was clicked
BN_DBLCLK	Button was double-clicked
BN_SETFOCUS	Button has gained keyboard focus
BN_KILLFOCUS	Button has lost keyboard focus

In the case of a control, the low-order word of the wParam contains the control identifier. This identifier is the number assigned to the control in the hMenu parameter of CreateWindows() or CreateWindowsEx(). Usually, an application defines a symbolic constant for each control, since this is a mnemonic aid and helps to make sure that no two controls are assigned the same value. One or more #define statements can be used as follows:

```
#define    WARMBUTTON      101
#define    HOTBUTTON       102
#define    COLDBUTTON      103
```

A switch statement on the low word of wParam can later be used to tell which button been pressed by the user, for example:

```
int buttonID, buttonNotify;
.
.
case WM_COMMAND:
    buttonID =       LOWORD(wParam);
    buttonNotify =  HIWORD(wParam);
    //eliminate non-control actions
    if(buttonNotify <= 1)
      return 0;
    switch (buttonID):
      case WARMBUTTON:
        if(buttonNotify == BN_CLICKED)
     // ASSERT:
     //    Tested button was clicked
     .
     .
     .
```

Some controls store information about their state or other data. For example, a three-state checkbox can be in a checked, unchecked, or indeterminate state. Table 22.4 lists the checkbox constants that define the three settings. These are used with three-state checkboxes and radio buttons.

Table 22.4

Notification Codes for Three-State Controls

NOTIFICATION CODE	ACTION
BST_CHECKED	Control is checked
BST_INDETERMINATE	Control is checked and grayed
BST_UNCHECKED	Control is unchecked

If you send a BM_GETCHECK message to a three-state checkbox or radio button it responds with one of these values. Suppose a three-state checkbox, with identification code CHKBOX1, and handle hwndChkBox1, which you wished to change from the checked to indeterminate state; it can be coded as follows:

```
LRESULT butMsg;
int     buttonID, buttonNotify;
.
.
.
case WM_COMMAND:
    buttonID =      LOWORD(wParam);
    buttonNotify =  HIWORD(wParam);

    //eliminate non-control actions
    if(buttonNotify <= 1)
      return 0;
    switch (buttonID):
      case CHKBOX1:
        butMsg = SendMessage(hwndChkBox1,   // handle
                             BM_GETCHECK,   // message
                             0, 0L);        // must be zero
      if(butMsg == BST_CHECKED)
      // ASSERT:
      //      checkbox is in checked state
      SendMessage(hwndChkBox1,
            BM_SETCHECK,          // order to set new state
            BST_INDETERMINATE,    // change to this state
            0, 01);
      .
      .
      .
```

Note, in the previous code fragment, that we used the SendMessage() function to communicate with the control. SendMessage() is used to send a message to a window or windows bypassing the message queue. In contrast, the PostMessage() function places the message in the thread's message queue. In communicating with a control, the first parameter to SendMessage() is the control's handle and the second one is the message to be sent. The third parameter is zero when we wish to obtain information from a control, and it contains a value or state when we wish to change the data stored. The BM_GETCHECK message returns a value, of type LRESULT, which is one of the notification codes in Table 22.4. The BM_SETCHECK message is used to change the button's state.

Scroll bar controls have a unique way of communicating with the parent window. Like main windows scroll bars, scroll bar controls send the WM_VSCROLL and WM_HSCROLL messages, the first one in the case of a vertical scroll bar action and the second one in the case of a horizontal scroll bar. The lParam is set to zero in windows scroll bars and to the scroll bar handle in the case of a scroll bar control. The high-order word of the wParam contains the position of the scroll box and the low-order word the scroll box value, which is one of the SB prefix constants listed in Table 22.5.

Table 22.5

User Scroll Request Constants

VALUE	MEANING
SB_BOTTOM	Scroll to the lower right
SB_ENDSCROLL	End scrolling
SB_LINELEFT	Scroll left by one unit
SB_LINERIGHT	Scroll right by one unit
SB_PAGELEFT	Scroll left by the width of the window
SB_PAGERIGHT	Scroll right by the width of the window
SB_THUMBPOSITION	Scrolls to the absolute position. The current position is specified by the nPos parameter
SB_THUMBTRACK	Drags scroll box to the specified position. The current position is specified by the NPos parameter
SB_TOP	Scroll to the upper left

In processing scroll bar controls the first step is to make sure that the message originates in the control being monitored. When the scroll action does not originate in windows scroll bars, or on those of another control, the processing usually consists in determining the new position for the scroll box. Two functions in the Windows API, SetScrollInfo() and GetScrollInfo(), provide all necessary functionality for scroll bar operation. SetScrollInfo() is used to set the minimum and maximum positions for the scroll box, to define the page size, and to set the scroll box to a specific location. GetScrollInfo() retrieves the information regarding these parameters. Four other functions, SetScrollPos(), SetScrollRange(), GetScrollPos(), and GetScrollRange() are furnished. In theory, these last four functions are furnished for backward compatibility, although they are often easier to implement in code that the new versions.

A program that implements a horizontal scroll bar usually starts by creating a scroll bar control. You can use the SBS_HORZ scroll bar style and determine its vertical and horizontal size in the sixth and seventh parameters to CreateWindows(), as follows:

```
#define  SCROLLBAR    401          // scroll bar id code
static   HWND         hwndSB;      // handle for the scroll bar
    .
// create a scroll bar class child window
    hwndSB = CreateWindow ("SCROLLBAR",  // Control class name
           "",                     // Button name text
           WS_CHILD | WS_VISIBLE | SBS_HORZ ,
           20,                     // x coordinate of location
           140,                    // y coordinate
           150, 25,                // dimensions
           hChild,                 // handle to parent window
           (HMENU) SCROLLBAR,      // child window id.
           pInstance,              // instance handle
           NULL) ;
```

Once the scroll bar is created, you must determine its range, set the initial position of the scroll box, and define its page size, if page operations are implemented. This last value determines how much the scroll box moves when the bar itself is clicked. All of this can be done with a single call to SetScrollInfo(), in which case the parameters are stored in a SCROLLINFO-type structure, as follows:

```
// Store parameters in SCROLLINFO structure members
   scinfo.cbSize = sizeof(SCROLLINFO);    // structure size
   scinfo.fMask  = SIF_POS | SIF_RANGE | SIF_PAGE; mask
   scinfo.nMin   = 0;                 // minimum value
   scinfo.nMax   = 99;                // maximum value
   scinfo.nPage  = 0;                 // page size
   scinfo.nPos   = 50;                // initial position
// Store scroll bar information
   SetScrollInfo(hwndSB, SB_CTL, &scinfo, TRUE);
   //              |       |        |           |___ redraw
   //              |       |        |____ address of SCROLLINFO
   //              |       |_____ refers to a scroll bar
   //              |                control
   //              |_____ handle to the scroll bar control
```

Manipulating the scroll bar requires intercepting the corresponding scroll bar messages. The current position of the scroll box is usually stored in a local variable, in this case the variable is named sbPos. Since this is a horizontal scroll bar, you can intercept the WM_HSCROLL message and then make sure that it refers to the scroll bar you are monitoring.

```
static   int        sbPos;         // position of scroll box
   .
   .
   .
case WM_HSCROLL:
     // Make sure action refers to local scroll bar
     // not the Windows scroll bars
     if(hwndSB == (HWND) lParam) {
         switch (LOWORD (wParam))  // Scroll code
           {
         case SB_LINELEFT:     // Scroll left one unit
            if(sbPos > 0)
            sbPos--;
            break;
         case SB_LINERIGHT:    // Scroll right one unit
            if(sbPos < 99)
            sbPos++;
            break;
         // Processing for user dragging the scroll box
          case SB_THUMBTRACK:
          case SB_THUMBPOSITION:
             sbPos = HIWORD (wParam);
             break;
         }
         // Display scroll box at new position
         SetScrollPos(hwndSB,
                      SB_CTL,
                      sbPos,
                      TRUE);
```

```
}
return 0;
```

Finally, there is the static class of controls that are often used for text fields, for labeling boxes, and for drawing frames and rectangles. Although static controls are frequently limited to labeling and simple drawing operations, they can be made to receive mouse input by means of the SS_NOTIFY style. Furthermore, the text in a static control can be changed at run time. The CON_DEMO program, described in the following section, located in the Controls Demo project folder on the book's software on-line, has two static controls. One is used to display the position of the scroll bar, and the other one is a black frame that surrounds the scroll bar buttons.

22.2.2 Controls Demonstration Program

The program named CON_DEMO, in the book's software package, is a demonstration of some of the basic controls described in previous sections and of the programming required to operate them. The controls are contained in a child window, much like the one created in the CHI_DEMO program already described. Figure 22.3 is a labeled screen snapshot of the CON_DEMO program.

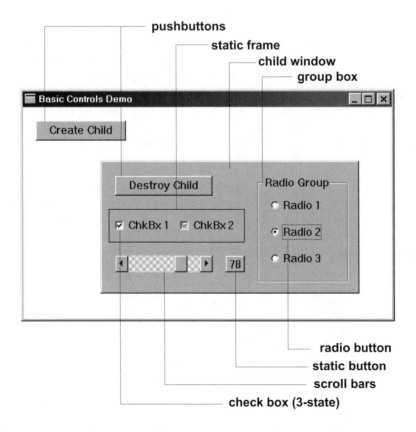

Figure 22.3 *CON_DEMO Program Screen*

The program's main screen contains a pushbutton that displays the child window. In the remainder of this section we have selected some excerpts from the program code to demonstrate the processing.

At the start of the code, the child windows and controls are defined as symbolic names. This is a useful simplification in applications that manipulate several resources or program elements that are identified by numeric values. The advantage is that the information is centralized for easy access and that it ensures that there are no repeated values.

```
// Constants for child windows and controls
#define   CHILD1        1001
#define   CREATEWIN     102
#define   DESTROYWIN    103
#define   RADGROUP      104
#define   RADIO1        201
#define   RADIO2        202
#define   RADIO3        203
#define   CHKBOX1       301
#define   CHKBOX2       302
#define   SCROLLBAR     401
#define   SCRBARWIN     501
#define   FRAME         502
```

The numeric values assigned to individual controls are arbitrary; it is a good idea, however, to follow a pattern for numbering resources and controls, since this avoids chaos in large programs. For example, child windows can be assigned a four-digit number, controls a three-digit number, and so forth. It is also recommended practice to use a dense set of integers for representing related controls, since there are Windows functions that operate on this assumption. Following this rule, the radio buttons in the CON_DEMO program are numbered 201, 202, and 203, and the checkboxes have numbers 301 and 302.

The creation of the child window in the CON_DEMO program is almost identical to the one in CHI_DEMO, previously described. The individual controls are created in the child window using the CreateWindow() function with the parameter set required in each case. The handles for the individual controls are defined as static variables in the child windows procedure, as follows:

```
LRESULT CALLBACK ChildWndProc (HWND hChild, UINT iMsg, WPARAM
                              wParam, LPARAM lParam) {
   static HWND   hwndChildBut1;     // Handle to child's button
   static HWND   hwndRadio1, hwndRadio2, hwndRadio3;
   static HWND   hwndChkBx1, hwndChkBx2;
   static HWND   hwndSB, hwndVal;
   static HWND   hwndGrpBox1;
   static HWND   hwndFrame;
   .
   .
   .
```

The code in the child window intercepts the WM_CREATE message. During message processing it installs the system's fixed font in the display context and then proceeds to create the individual controls. A bool-type variable, named childStatus, is used to store the state of the child window. This variable is TRUE if the child window is displayed. This avoids creating more than one copy of the child. The first control created in the child window is the pushbutton that destroys it and returns execution to the parent. Before that, the system's fixed font is selected into the display context. Coding is as follows:

```
switch (iMsg) {
   case WM_CREATE:
     // Test that child window is not already displayed
     if(childStatus)
       return 0;
     // ASSERT:
     //       child window is not displayed
     childStatus = TRUE;     // child window is displayed
     childDc = GetDC(hChild);   // handle to private DC
         SelectObject (childDc,
                       GetStockObject (SYSTEM_FIXED_FONT));
     // Place destroy button on child window
     hwndChildBut1 = CreateWindow (
        "BUTTON",                 // Control class name
        "Destroy Child",          // Button name text
        WS_CHILD | WS_VISIBLE | BS_PUSHBUTTON,
        20, 20,                   // x and y location
        150,                      // Window width
        30,                       // Window height
        hChild,                   // Handle to parent window
        (HMENU) DESTROYWIN,       // Child window id.
        pInstance,                // Instance handle
        NULL) ;
         .
         .
         .
```

The user interaction with the controls is monitored and processed in the WM_COMMAND message intercept of the child window. First, the notification code and the button identifier are stored in local variables. A switch statement on the button identification code allows directing the processing to the routines for each of the buttons. The code examines the notification code to make sure that the intercept is due to action on a button control, and not on an accelerator key or a menu item.

```
   case WM_COMMAND:
     buttonID = LOWORD (wParam);
     buttonNotCode = HIWORD (wParam);
     switch (buttonID) {
       if(buttonNotCode <= 1)
         return 0;

       case DESTROYWIN:
         if(buttonNotCode == BN_CLICKED) {
           childStatus = FALSE;
           DestroyWindow(hChild);
           UpdateWindow(hwnd);
           }
           break;
       // Radio button # 1 action
       case RADIO1:
```

```
// Set radio button ON
   SendMessage(hwndRadio1, BM_SETCHECK, 1, 0L);
   SendMessage(hwndRadio2, BM_SETCHECK, 0, 0L);
   SendMessage(hwndRadio3, BM_SETCHECK, 0, 0L);
   break;
   .
   .
   .
   .
```

22.3 Menus

The menu is one of the most important elements of the Windows user interface. It occupies the line below the title bar. Often, only the program's main window has a top-level menu. There has been considerable uncertainty regarding the names of the various elements in a menu. The following designations are based on Microsoft's The Windows Interface Guidelines for Software Design, listed in the Bibliography.

- The menu bar is a screen line directly below the title bar, which contains entries called the menu titles, or just the menus.

- Each menu title (menu) activates a drop-down box, which contains one or more menu items. Menu items are usually arranged in a single column, although Windows supports multiple column menus.

- Menu items can be of three types: menu commands, child menus, and separators. A menu command is a menu item that executes a program function directly. A child menu, also called a cascading or hierarchical menu, is a submenu, which can in turn contain menu commands, child menus, and separators. Items that activate a child menu are usually marked by a triangular arrow to the right of its name. A separator is a screen line that is used to group related menu items.

- Pop-up menus are activated by clicking the second mouse button. They are usually unrelated to the program's menu bar.

- Access keys are keystrokes that can be used instead of mouse button action to access menu items. Access keys are underscored in the menu title and in menu items. To activate a menu title by means of the access key you must hold down the Alt key. Once a drop down menu is displayed, access to the contained items is by pressing the corresponding access key. The Alt key is not required in this case.

- Shortcut keys are keystroke combinations that allow accessing a menu item directly. Shortcut keys are usually a Ctrl + key combination or a function key. Windows documentation sometimes calls these shortcut keys accelerators, but The Windows Interface Guidelines for Software Design prefers the former name.

   There are also some style considerations regarding the design and implementation of menus. Although the design of the user interface is a topic outside the scope of this book, there are several general principles worth mentioning.

- A menu title should be a single word that represents the items that it contains. Each menu title should have an access key, which activates the menu when used in conjunction with the Alt key. Access keys are underlined in the menu bar. No two menu titles should have the same access key.

- Cascading menus should be used sparingly since they add complexity to the interface. Their purpose is to reduce the number of entries in the main menu and to logically organize hierarchical entries. The user should never have to navigate through more than two levels of cascading menus to reach a command.

- Menu items that are not active or are currently unavailable should be disabled and displayed in gray characters. Alternatively, a permanently inactive item can be removed from a menu.

- If a menu command requires additional data to execute, it should be followed by an ellipsis (...). The ellipsis serves as a visual key that information for executing a command is incomplete. Typically, commands with ellipses display a dialog box where the additional data is supplied. However, commands that obviously generate other informational actions should not be followed by ellipses; for example, a Properties command is expected to display information, therefore it should not have ellipsis.

- Check boxes are used in menus to indicate the status of a menu item. A checked item signals that it is functional. Code should check and uncheck items during processing to update their status.

- All menu items should have access keys, but items on the same drop down menu cannot have the same access key. The first choice for an access keys is the first character in the menu title or entry. If the first character is already used as an access key, then the next one in the item name that is not used as an access key should be selected.

- Shortcut keys that activate menu commands are best implemented with the Crtl key followed by a mnemonic letter associated with the entry. Function keys can also be used. For example, Ctrl + S can be used for a Save command and Ctrl + P for a Print command. The most used commands should be assigned a shortcut.

Figure 22.4 shows some of the most common elements in a menu.

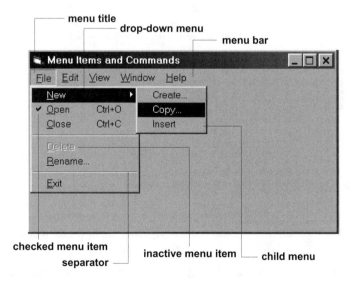

Figure 22.4 *Common Menu Elements*

22.3.1 Creating a Menu

There are several ways to create a menu. Before the Visual Studio and other development environments came into existence, menus were created using API functions. CreateMenu() creates an empty menu and returns its handle. InsertMenuItem() can be used to populate the menu with components. AppendMenu() adds a component to an existing menu. Other functions, such as DeleteMenu(), DestroyMenu(), DrawMenuBar(), ModifyMenu() and RemoveMenu() are also available. Finally, the LoadMenuIndirect() function can be used to load a menu from a memory resident menu template.

1. From the Developer Studio Insert menu, select the Resource command. Select the Menu resource type in the dialog box, and click New.

2. Create the main menu entries in your program (the menu titles) as well as the menu items in each of the drop-down menus. At this time you can assign an identification code to each menu item, define child menus (called pop-up in the input form), determine if the item is initially grayed, checked, or inactive, assign shortcut keys, and other menu attributes. Details on how to use the menu editor are available in Developer Studio online Help. Figure 22.5 shows the Developer Studio menu editor screen.

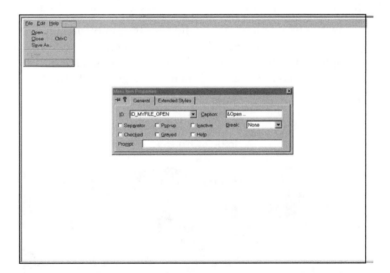

Figure 22.5 *Developer Studio Menu Editor*

3. Once you have finished creating the menu, click on the close button of the menu editor window. If the application already has a script file, the new menu is added to it. If not, Developer Studio prompts you to save the new script file.

4. Skip this step if a script file has already been inserted into the project. If not, open the Project menu, select Add to Project, and then Files. In the Insert Files into Project

dialog box, select the script file and then click OK. The script file now appears in the project workspace window.

5. Select the Resource View button in the project workspace pane and click + on Script Resources. Click + on Menu. Note the identifier name for the menu resource, which is IDR_MENU1 if this is the first menu created.

6. Enter the menu identifier in the wndclass structure defined in WinMain(), as follows:

```
wndclass.lpszMenuName  = MAKEINTRESOURCE(IDR_MENU1);
```

7. Developer Studio creates a header file named resource.h which assigns numeric values to the program resources. The file is saved under the name "resource.h" and stored in the project's main directory. The main source file must reference this header file in an include statement, such as:

```
#include "resource.h"
```

8. To recompile the program with the new menu, select Rebuild All from the Build menu.

9. To edit the menu, double-click on the corresponding IDR_MENU1 icon.

If you receive a redefinition of symbol error at build time there are two possible solutions: one is to comment-out the redefined symbol in the file named afxres.h located in Msdev\Mfc\Include directory. The other one is to edit the resource, in this case the menu, and change the name in the ID: field. Changing the afxres.h file is a permanent way of avoiding this error, but the development system cannot be used for MFC applications if afxres.h has been altered.

22.3.2 Menu Item Processing

There are several intercept messages related to application menu processing. WM_MENUSELECT is sent when the mouse cursor moves among the menu items, and WM_INITITEM when the user selects an item from a menu. However, most applications do all their menu processing in the WM_COMMAND message intercept. In the case of a menu, the lParam is 0 and the wParam contains the menu ID code, which is the identification number and its corresponding string constant found in the resource.h file. System menus notify the application through the WM_SYSCOMMAND message. The following code fragment shows the intercept routine for the item named Open in the File menu:

```
case WM_COMMAND:
   switch (LOWORD (wParam)) {
   case ID_MYFILE_OPEN:
      // ASSERT:
      //   Menu item resource named ID_MY-FILE_OPEN
      //   was activated by user
 . . .
```

An important fringe benefit from using the menu editor in Developer Studio is that access keys are automatically detected and vectored to the corresponding handler. Suppose that in the preceding code fragment the Open command was defined so that the letter O is preceded by the & symbol in the editor screen. In this case, when the user presses the "O" key while the File menu is open, a WM_COMMAND message with the key code ID_MYFILE_OPEN is sent to the handler.

22.3.3 Shortcut Keys

Shortcut keys require a special treatment so that the keystrokes are vectored to the desired handler. It is recommended that shortcut keys be listed in the same line as the menu item. In order to do this you must insert the text for the control keystroke, preceded by \t in the caption window of the Menu Item Properties editor screen. In this case \t indicates a Tab code which displays the following text on the next tab field. Figure 22.6 shows the insertion of a shortcut key designation in Developer Studio menu editor.

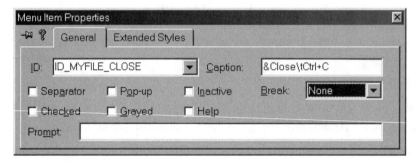

Figure 22.6 *Developer Studio Insertion of a Shortcut Key Code*

But the shortcut key label is only a caption and has no effect on the processing. In order to associate a shortcut key with a menu item you must create an accelerator table. The following steps can be followed:

1. Select Resource from the Developer Studio Insert menu.. Select the Accelerator resource type in the dialog box and click New.

2. Create an accelerator table. The table includes an identification field that contains the resource ID, a key field for the keystroke that activates the shortcut, and a type field that specifies the properties of the key. Figure 22.7 shows the Accel Properties dialog box in the accelerator editor.

3. Once created, the accelerator table becomes a program resource whose name can be found in the Resource tab of Developer Studio project workspace pane, or by clicking the Resource Symbols command in the View menu or its corresponding toolbar button. Developer Studio assigns the name IDR_ACCELERATOR1 to the first accelerator table; normally, there is one per application.

4. The accelerator table must now be loaded into the application and processed so that the corresponding messages are sent to the windows procedure. This requires using the LoadAccelerator() function. Its parameters are the handle to the program's instance and an identifier of the accelerator table. LoadAccelerator() returns a handle to the accelerator, of type HACCEL. Processing of accelerator keys is by means of the TranslateAccelerator() function, which takes as parameters the handle to the window whose messages are to be translated, the handle to the accelerator table returned by LoadAccelerator(), and a pointer to a message structure. Both functions are usually included in WinMain(), as in the following code fragment:

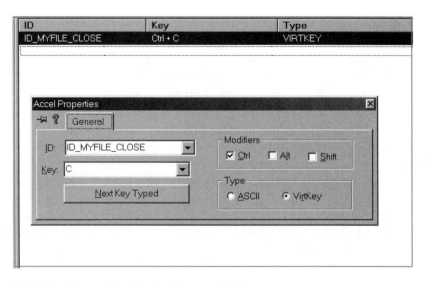

Figure 22.7 *Developer Studio Accelerator Editor*

```
LRESULT CALLBACK WinMain (HINSTANCE hInstance, HINSTANCE
                          hPrevInstance, PSTR szCmdLine,
                          int iCmdShow) {
    static char szAppName[] = "Demo" ;
    HWND        hwnd ;
    MSG         msg ;
    HACCEL      hAccel;               // Handle to accelerator
    .
    .
    .
    ShowWindow (hwnd, iCmdShow) ;
    UpdateWindow (hwnd) ;
    // Load accelerators
       hAccel = LoadAccelerators (hInstance,
                 MAKEINTRESOURCE (IDR_ACCELERATOR1));
    // Message loop
    while (GetMessage (&msg, NULL, 0, 0)) {
        if (!TranslateAccelerator (hwnd, hAccel, &msg)) {
            TranslateMessage (&msg) ;
            DispatchMessage (&msg) ;
        }
    }
    return msg.wParam ;
}
```

22.3.4 Pop-Up Menus

A pop-up menu is a context-sensitive submenu that is activated by clicking the right mouse button. The pop-up menu is unrelated to the application's main menu and implemented differently. The items in a pop-up menu should be related to the context in which the right mouse button is pressed. Therefore, in a full-featured application, the processing usually requires calculating the screen coordinates where the mouse action takes place, or the object currently selected, in order to determine which, among several pop-up menus, is to be activated.

As with the program's main menu, there are several methods for creating a pop-up menu. You can use the menu editor to create a pop-up menu; however, a little trickery is required since pop-up menus have no title and the menu editor does not allow creating menu items without first entering the title. The following steps can be used to create and install a simple pop-up menu:

1. Use the menu editor to create the pop-up menu. In order to create a drop down menu you have to enter a temporary menu title. Since this title is used by Developer Studio name mangler to create the item id, it may be a good idea to used the menu title "popup1."

2. Under the temporary menu title (popup1 is the suggested one), enter the menu items as you would for a program menu. You can use all the attributes available and there can be child menus in the pop-up. Once you have finished creating the menu, double-click on the temporary menu title (popup1) and erase all the characters in the caption field. This creates a drop down menu with no menu title. To see the drop down menu you have to click on the left corner of the menu editor's title bar. This can be a little deceptive, since at times it may seem that the drop down menu has disappeared.

3. When you close the menu editor, a new menu resource appears in the Resource tab of the Program window. If this is your second menu it is named IDR_MENU2. The new menu is now included in your script resource file.

4. You need to load the pop-up menu and obtain its handle. This can be done in the WM_CREATE message intercept of the window that contains it. It requires the use of the LoadMenu() function, which returns a handle to the menu resource. The GetSubMenu() function converts this handle into a submenu handle, which can then be used by the code. Processing is usually as follows:

```
static   HMENU    pMenu;    // Handle to pop-up menu
    .

    .
    .
case WM_CREATE:
    hdc = GetDC(hwnd);
    // Get handle to pop-up menu
    pMenu = LoadMenu(pInstance,
                (MAKEINTRESOURCE(IDR_MENU2)));
    pMenu = GetSubMenu(pMenu, 0);
    return 0;
```

5. Once you have its handle, the pop-up menu can be displayed. The TrackPopupMenu() function is used to define the screen location where the pop-up menu is shown, its position relative to the mouse cursor, and to define which mouse button actions, if any, are tracked when an item is selected. If the pop-up menu is activated by the right mouse button, as is usually the case, then the menu display code can be placed at the corresponding message intercept, as in the following code fragment.

```
case WM_RBUTTONDOWN:
    // Get mouse coordinates
    aPoint.x = LOWORD(lParam);
```

```
    aPoint.y = HIWORD(lParam);
    ClientToScreen(hwnd, &aPoint);
    TrackPopupMenu(pMenu,
                  TPM_LEFTALIGN | TPM_TOPALIGN |\
                  TPM_LEFTBUTTON,
                  aPoint.x, aPoint.y,
                  0,
                  hwnd,
                  NULL);
    return 0;
```

In the preceding code sample we start by obtaining the mouse coordinates from the lParam. One problem is that TrackPopupMenu() requires the horizontal and vertical coordinates in screen units, and the WM_RBUTTONDOWN message intercept reports the mouse position in client area units. For this reason, the ClientToScreen() function is necessary to convert client area into screen coordinates.

The TrackPopupMenu() function displays the pop-up menu. Its first parameter is the handle to the menu obtained during WM_CREATE processing. The second parameter is one or more bitwise constants. In this case we have established that the display position is relative to the upper left corner of the menu box, and that the left mouse button is the one tracked for menu selections. The display points are entered as the third and fourth parameters to the call. The fifth one is reserved (must be zero), the sixth one is the handle to the window that owns the pop-up menu, and the last one defines a RECT-type structure in which the user can click without erasing the pop-up menu. If this value is NULL then the shortcut menu disappears if the user clicks outside of its area.

6. Intercepting action on the pop-up menu is at WM_COMMAND message processing. For example, if the id of the first item in the pop-up menu is ID_POPUP1_UNDO, then the case statement at the intercept point has this label, as follows:

```
case WM_COMMAND:
    switch (LOWORD (wParam)) {
    .
    .
    .
    case ID_POPUP1_UNDO:
    // Assert:
    //    User clicked "undo" item on pop-up menu
```

22.3.5 The Menu Demonstration Program

The program named MEN_DEMO, contained in the Menu Demo project folder on the book's software on-line package, is a trivial demonstration of an application with a main menu, a shortcut key (accelerator) to access one of the menu items, and a pop-up menu that is displayed when the user right-clicks on the client area. Processing consists of a message box that lists the menu item selected by the user.

22.4 Dialog Boxes

Dialog boxes are a programming aid; they provide no new functionality. Everything that can be done in a dialog box can also be done in a child window, as described earlier in this chapter.

What dialog boxes do for the programmer is to prepackage a series of functions that are frequently needed. Also, dialog boxes perform much of the processing and housekeeping operations for you. They handle the keyboard focus, passing keyboard input from one control to another one, they monitor mouse movements, and they provide a special procedure for tracking action on the controls contained in the dialog box. When used in conjunction with the dialog box editor in Developer Studio, dialog boxes are easy to create and implement in code. Windows 3.1 introduced an extension to the concept of dialog boxes, usually called the common dialog boxes. The common dialog boxes are a set of prepackaged services for operations that are usually required in many applications. These include opening and saving files, selecting a font, selecting or changing color attributes, searching and replacing text strings, and controlling the printer. The common dialog boxes are discussed later in this section.

22.4.1 Modal and Modeless

There are two general types of dialog boxes: those that suspend the application until the user interacts with the dialog box, and those that do not. The first type, which are the most common ones, are called modal dialog boxes. The second type, which are often seen in floating toolbars, are called modeless dialog boxes. Modal dialog boxes do not prevent the user from switching to another application. Although, upon return to the original thread, it is the modal dialog box that retains the foreground. *The Windows Interface Guidelines for Software Design* (see Bibliography) recommends that modal dialog boxes should have an OK button, to accept and process input, and a Cancel button to abort execution and discard the users action with the dialog box.

22.4.2 The Message Box

The simplest of all dialog boxes is used to display a message on the screen, which the user acknowledges having read by pressing a button. A special function in the Windows API allows creating message boxes directly, without having to use the dialog box editor or manipulate a program resource. The message box contains a title, a message, any one of several predefined icons, and one or more pushbuttons. The general form of the function call is as follows:

```
int MessageBox(hwnd, lpText, lpCaption, uType);
```

where hwnd is the handle to the window that owns the message box, lpText is a pointer to the text message to be displayed (or the message string itself), lpCaption is a pointer to the caption (or the caption string itself), and uType is one of several bit flags that control the behavior of the message box. Table 22.6 lists the most useful bit flags used in the MessageBox() function.

Table 22.6
Often Used Message Box Bit Flags

SYMBOLIC CONSTANT	MEANING
MB_ABORTRETRYIGNORE	Contains three push buttons: Abort, Retry, and Ignore.
MB_OK	Contains one push button: OK. This is the default.
MB_OKCANCEL	Contains two push buttons: OK and Cancel.
MB_RETRYCANCEL	The message box has two buttons: Retry and Cancel.
MB_YESNO	Contains two push buttons: Yes and No.
MB_YESNOCANCEL	Contains three push buttons: Yes, No, and Cancel.

(continues)

Table 22.6

Often Used Message Box Bit Flags (continued)

SYMBOLIC CONSTANT	MEANING
Icon Flags:	
MB_ICONEXCLAMATION	Exclamation-point icon.
MB_ICONWARNING	Exclamation-point icon.
MB_ICONINFORMATION	Question mark icon.
MB_ICONASTERISK	Lowercase letter i icon in a circle.
MB_ICONQUESTION	Question-mark icon.
MB_ICONSTOP	Stop-sign icon.
MB_ICONERROR	Hand icon.
MB_ICONHAND	Hand icon.
Default Button Flags:	
MB_DEFBUTTON1	The first button is the default button.
MB_DEFBUTTON2	The second button is the default button.
MB_DEFBUTTON3	The third button is the default button.
MB_DEFBUTTON4	The fourth button is the default button.
Modality Flags:	
MB_APPLMODAL	User must respond to the message box before continuing work in the window. However, the user can move to the window of another application and work in those windows.
MB_SYSTEMMODAL	Same as MB_APPLMODAL except that the message box has the WS_EX_TOPMOST style. Use system-modal message boxes to notify the user of serious errors that require immediate attention.
MB_TASKMODAL	Same as MB_APPLMODAL except that all the top-level windows belonging to the current task are disabled if the hwnd parameter is NULL.
Other Flags:	
MB_HELP	Adds a Help button to the message box. Choosing the Help button or pressing F1 generates a Help event.
MB_RIGHT	The text is right-justified.
MB_SETFOREGROUND	The message box becomes the foreground window. Internally, Windows calls the SetForegroundWindow function for the message box.
MB_TOPMOST	Message box is created with the WS_EX_TOPMOST window style.

For example, the following statement creates a message box labeled "Menu Action," with the text string "File Close Requested," which contains an exclamation sign icon, and a button labeled OK:

```
MessageBox (hwnd,
          "File Close Requested",
          "Menu Action",
           MB_ICONEXCLAMATION | MB_OK);
```

Figure 22.8 shows the resulting message box.

Figure 22.8 *Simple Message Box*

22.4.3 Creating a Modal Dialog Box

Developer Studio provides a dialog box editor, which is a tool for creating dialog boxes. Once the dialog box has been created, it becomes another program resource that can be referenced in the code. The dialog box editor can be used to create simple message boxes; however, in this case it is easier to use the MessageBox() function described in the previous section. Dialog boxes are useful when they are used to obtain user input.

A unique feature of dialog boxes is that they contain their own processing. In a sense, the dialog box procedure is like your window procedure.

You create a modal dialog box by means of the DialogBox() function, with the following standard form:

```
int DialogBox (hInstance, lpTemplate, hwndParent, lpDiaProc);
```

where hInstance is the handle to the program instance that contains the dialog box, lpTemplate identifies the dialog box template or resource, hwndParent is the handle to the owner window, and lpDiaProc is the name of the dialog box procedure. It is this procedure that receives control when the dialog box is created. The following code fragment shows the creation of a dialog box at the time that a menu command with the id ID_DIALOG_ABOUT is intercepted:

```
ID_DIALOG_ABOUT:
  DialogBox (pInstance,
             MAKEINTRESOURCE (IDD_DIALOG1),
             hwnd,
             (DLGPROC) AboutDlgProc);
```

In this case the dialog box resource is named IDD_DIALOG1, and the dialog box procedure that receives control is AboutDlgProc(). The dialog box procedure's general form is as follow:

```
BOOL DialogProc (hwndDlg, uMsg, wParam, lParam);
```

where DialogProc is the name of the procedure defined in the lpDiaProc field of the DialogBox() function. The first parameter passed to the dialog procedure (hwndDlg) is the handle to the dialog box. The second one is the Windows message. The wParam and lParam values contain message-specific information, as is the case in the window procedure.

As soon as the dialog box is created, and before it is displayed, Windows send the WM_INITDIALOG message to the dialog box procedure. Typically, the dialog box procedure intercepts the message to initialize controls and perform other housekeeping functions. In WM_INITDIALOG the wParam contains the handle to the control that has focus, which is the first visible and not disabled control in the box. The application returns TRUE to accept this default focus. Alternatively, the application can set the focus to another control, in which case it returns FALSE.

The dialog box procedure receives messages for the controls in the dialog box. These messages can be intercepted in the same manner as those sent to the window procedure. The following code fragment is a dialog box procedure for a dialog box that contains a single button:

```
BOOL CALLBACK AboutDlgProc (HWND hDlg, UINT iMsg, WPARAM wParam,
                           LPARAM lParam) {
 switch (iMsg) {
    case WM_INITDIALOG :
      return TRUE ;
    // Dialog box controls message intercepts
    case WM_COMMAND :
      switch (LOWORD (wParam)) { // Get control id
         case IDOK :
           EndDialog (hDlg, 0) ;
           return TRUE ;
         }
    break ;
    }
 return FALSE ;
 }
```

Notice that, unlike a window function, AboutDlgProc() does not return control via the default window procedure. In general, a dialog box procedure returns FALSE to indicate that default processing is to be provided by Windows and TRUE when no further processing is required. The exception is the WM_INITDIALOG message in which the return value refers to the acceptance or rejection of the default focus, as discussed previously.

Notice that dialog procedures, like all window procedures, have to be of type CALLBACK. Failing to declare a window procedure, or a callback procedure, with this type, can be the source of unpredictable errors, such as the General Protection Fault.

You can create a dialog box by means of the following steps:

1. Select Resource from the Developer Studio Insert menu. Select the Dialog resource type in the dialog box and click New.

2. The dialog box editor executes by displaying a blank form and a floating toolbox containing controls that can be inserted in the dialog box. If the toolbar is not visible, you can show it on the editor screen by opening the Tools menu, then selecting Customize, and checking the Controls box in the Toolbars tab. The controls include all those already mentioned and some others. To add a control to the dialog box you drag it onto the form and then use the handles to size it. Double-clicking on the form, or on one of the controls, displays a Dialog Properties window which allows defining the attributes of that particular element. Figure 22.9, on the following page, shows the dialog box editor with the Dialog Properties windows for the form and the Controls toolbox.

3. Once you have finished creating the dialog box, click the Close button of the menu editor window. If the application already has a script file, the dialog box is added to it. If not, Developer Studio prompts you to save the new script file.

4. Skip this step if a script file has already been inserted into the project. If not, open the Project menu, select Add to Project, and then select Files. In the Insert Files into Project dialog box, select the script file and then click OK. The script file now appears in Developer Studio project workspace pane.

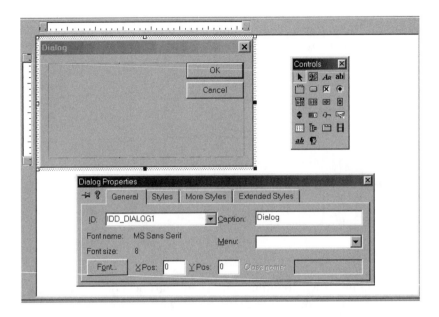

Figure 22.9 *Developer Studio Dialog Editor*

5. Select Resource View button in project workspace pane and click + on Script Resources. Click + on Dialog. Note the identifier name for the dialog box resource, (usually IDD_DIALOG1) if this is the first dialog box created.

6. Your code must now create the dialog box, usually by intercepting the corresponding menu command and calling DialogBox(). Also, the dialog box procedure has to intercept the WM_INITDIALOG message and provide handlers for the controls contained in the box, as previously described.

22.4.4 Common Dialog Boxes

Windows 3.1 introduced the common dialog boxes as a set of prepackaged services for performing routine operations required in many applications. The idea behind them is to standardize frequent input functions so that they appear the same in different program functions and even in different applications. For example, the common dialog box used to select a filename and to browse through the disk storage system is the same if you are opening or saving a file. Furthermore, two applications that manipulate files can use the same common dialog box, giving the user a familiar interface. The following operations can be performed by means of common dialog boxes: opening and saving files, selecting fonts, selecting or changing color attributes, searching and replacing text strings, and controlling the printer. Common dialog boxes have a modal behavior, that is, the program is suspended until the user closes the dialog box.

Common dialog boxes are processed internally by Windows; therefore, they do not have a dialog box procedure.

The identification for the menu item or other resource that activates the common dialog box usually serves as the message intercept. The processing is done directly in the intercept routine. Each common dialog box is associated with a structure that is used to pass information to it and to receive the results of the user's action. Programs that use common dialog boxes should include the commdlg.h header file.

For example, the menu item ID_DIALOG_COLORSELECTOR can intercept user action in WM_COMMAND message processing and then proceed to fill a variable of the structure type CHOOSECOLOR (see Appendix F) as in the following code fragment:

```
LRESULT CALLBACK WndProc (HWND hwnd, UINT iMsg, WPARAM wParam,
                          LPARAM lParam) {
   HDC           hdc ;
   TEXTMETRIC    tm ;
   HBRUSH        hBrush;        // Handle to brush

   // Variables for color and font common dialog
   static CHOOSECOLOR cc ;    // Structure
   static COLORREF    custColors[16] ; // Array for custom
                                       // colors
   int    i;                 // counter for custom color display
   switch (iMsg) {
      ...
      case WM_COMMAND:
         switch (LOWORD (wParam)) {
         // Color Selector common dialog
            case ID_DIALOG_COLORSELECTOR:
               cc.lStructSize    = sizeof (CHOOSECOLOR) ;
               cc.hwndOwner      = hwnd ;
               cc.hInstance      = NULL ;
               cc.rgbResult      = RGB (0x80, 0x80, 0x80) ;
               cc.lpCustColors   = custColors ;
               cc.Flags          = CC_RGBINIT | CC_FULLOPEN ;
               cc.lCustData      = 0L ;
               cc.lpfnHook       = NULL ;
               cc.lpTemplateName = NULL ;
               ...
```

Once the structure variable is filled with the necessary data, the application can call the ChooseColor() function to display the common dialog box. ChooseColor() requires a single parameter: the address of the previously mentioned structure. Most of the structure members are obvious. The lpCustColors member is an array of 16 COLORREF-type values that holds the RGB values for the custom colors in the dialog box. The Flags members are bit flags that determine the operation of the dialog box. In the previous example we set CC_RGBINIT bit so that the rgbREsult member holds the initial color selection. The values 0x80 for each of the red, green, and blue components produce a middle gray color. The last three members of the structure are used for customizing the dialog box. The constant CC_FULLOPEN causes the dialog box to open in the full display mode, that is, with the controls necessary for the user to create custom colors.

ChooseColor() returns TRUE if the user clicks the OK button on the dialog box. Therefore, the coding continues as follows:

```
if (ChooseColor (&cc) == TRUE) {
// ASSERT:
//    structure members have color data selected by user
// Clear the client window
hdc = GetDC(hwnd);
InvalidateRect (hwnd, NULL, TRUE) ;
UpdateWindow (hwnd) ;
```

The colors selected by the user are stored in two members of the CHOOSECOLOR structure: rbgResult holds the solid color box, and the array variable custColor holds the 16 custom colors. The code now creates a solid brush, using the color stored in the rgbResult member, and displays a rectangle filled with this color. Then a loop displays the first eight of the 16 custom colors:

```
hBrush = CreateSolidBrush(cc.rgbResult);
// Select the brush in the DC
SelectObject (hdc, hBrush) ;
// Draw a rectangle using the brush
Rectangle (hdc, 20, 20, 100, 100) ;
// Display first eight custom colors using the
// color triplets stored in the custColors array
    for (i = 0; i < 8; i++) {
        hBrush = CreateSolidBrush(custColors[i]);
            SelectObject (hdc, hBrush) ;
            Rectangle (hdc, 20+(20 * i), 120,
                40+(20 * i), 140) ;
// Clear and exit
    DeleteObject (SelectObject (hdc, hBrush)) ;
    ReleaseDC (hwnd, hdc);
}
return 0 ;
  ...
```

Figure 22.10 shows the color dialog box as displayed by this code.

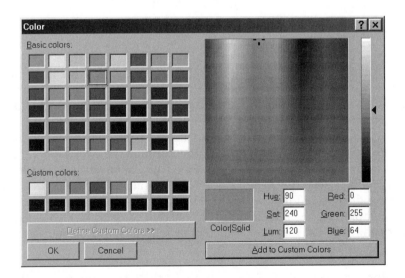

Figure 22.10 *Color Selection Common Dialog Box*

22.4.5 The Dialog Box Demonstration Program

The program named DIA_DEMO, contained in the Dialog Box Demo project folder on the book's on-line software package, is a trivial demonstration of several dialog boxes. The Dialog menu contains commands for creating a modeless dialog box, three different modal dialog boxes (one of them with a bitmap), and for the color and font common dialog boxes. The code demonstrates how information obtained by modal and common dialog boxes is passed to the application.

22.5 Common Controls

Windows 95 introduced a new set of controls that supplement the ones that existed previously. They are also available in Windows NT version 3.51 and later. These controls, sometimes referred to as the new common controls, allow the implementation of status bars, toolbars, trackbars, progress bars, animation controls, image lists, list view controls, tree view controls, property sheets, tabs, wizards, and rich edit controls. It is evident from this list that one could devote an entire volume to their discussion. Table 22.7 is a list of the Windows common controls first implemented in Windows 95.

Table 22.7

Original Set of Common Controls

CONTROL	DESCRIPTION
Frame Window Controls:	
toolbar	Displays a window with command-generating buttons.
ToolTip	Small pop-up window that describes purpose of a toolbar button or other tool.
status bar	Displays status information at the bottom screen line.
Explorer-type Controls:	
list view	Displays a list of text with icons.
tree view	Displays a hierarchical list of items.
Miscellaneous Controls:	
animation	Displays successive frames of an AVI video clip.
header	Appears above a column of text. Controls width of text displayed.
hotkey	Enables user to perform an action quickly.
image list	A collection of images used to manage large sets of icons or bitmaps. It isn't really a control, but supports lists used by other controls.
progress bar	Indicates progress of a long operation.
rich edit	Allows the user to edit with character and paragraph formatting.
slider	Displays a slider control with optional tick marks.
spin button	Displays a pair of arrow buttons user can click to increment or decrement a value.
tab	Displays divider-like elements used in tabbed dialog. boxes or property sheets.

Before we can implement the new common controls, some preliminary steps are required. The reason is that the common controls library is not automatically referenced at link time, nor is it initialized for operation. The following operations are necessary:

1. The common controls library, named Comctl32.lib, must be included in the list of libraries referenced by the linker program. This is accomplished by opening the Project

menu and selecting the Settings command. In the Project Settings dialog box, open the Link tab. The "Object/library modules" edit box contains a list of all the referenced libraries, separated from one another by a space. Position the caret between two library entries and type "Comctl32.lib." Click the OK button.

2. The program code must include the common controls header file. This is accomplished with the statement:

```
#include <commctrl.h>
```

3. The InitCommonControls() function must be called before the common controls are used. This function takes no parameters and returns nothing. The initialization can be placed in WinMain(), as follows:

```
InitCommonControls();
```

4. Rich edit controls reside in their own library, named Riched32.dll and have their own header file, named richedit.h. To use library controls your program must include the statement:

```
LoadLibrary ("RICHED32.DLL");
```

At this point the application can implement common controls. In this section we sample some of the common controls that are more frequently found in graphics applications, namely toolbars and ToolTip controls. These, together with the status bar controls, are sometimes called the frame window controls. Some of the common controls are available in the toolbar of Developer Studio dialog box editor. The resource editor contains a specific toolbar editor for creating this type of common control. Most common controls can also be created by means of the CreateWindow() or CreateWindowEx() functions. Others have a dedicated function, such as CreateToolbarEx().

22.5.1 Common Controls Message Processing

Most common controls send WM_NOTIFY messages. One notable exception is the toolbar controls, which send WM_COMMAND. In processing common controls messages we follow similar methods as in processing menu selections.

The WM_NOTIFY message contains the ID of the control in wParam and a pointer to a structure in lParam. The structure is either an NMHDR structure, or more frequently, a larger structure that has an NMHDR structure as its first member. The common notifications (whose names start with NM_) and the ToolTip control's TTN_SHOW and TTN_POP notifications, are the only cases in which the NMHDR structure is actually used by itself. The format of the NMHDR structure is as follows:

```
typedef struct tagNMHDR {
    HWND hwndFrom;
    UINT idFrom;
    UINT code;
} NMHDR;
```

where hwndFrom is the handle to the controls sending the message, idFrom is the control identifier, and code is one of values in Table 22.8.

Table 22.8

Common Control Notification Codes

CODE	ACTION IN CONTROL OR RESULTS
NM_CLICK	User clicked left mouse button.
NM_DBLCLK	User double-clicked left mouse button.
NM_RCLICK	User clicked right mouse button.
NM_RDBLCLK	User double-clicked right mouse button.
NM_RETURN	User pressed the Enter key.
NM_SETFOCUS	Control has been given input focus.
NM_KILLFOCUS	Control has lost input focus.
NM_OUTOFMEMORY	Control could not complete an operation because there was not enough memory available.

Most often notifications pass a pointer to a larger structure that contains an NMHDR structure as its first member. For example, the list view control uses the LVN_KEYDOWN notification message, which is sent when a key is pressed. In this case the pointer is to an LV_KEYDOWN structure, defined as follows:

```
typedef struct tagLV_KEYDOWN {
    NMHDR hdr;
    WORD wVKey;
    UINT flags;
} LV_KEYDOWN;
```

Since the NMHDR member is the first one in this structure, the pointer in the notification message can be cast to either a pointer to an NMHDR or a pointer to an LV_KEYDOWN.

22.5.2 Toolbars and ToolTips

A toolbar is a window containing graphics buttons or other controls. It is usually located between the client area and the menu bar. Although Windows applications have been using toolbars for a long time, there was no system support for toolbars until the release of the WIN-32 API. The most common use of toolbars is to provide fast access to menu commands. Toolbars often include separators, which are spaces in the toolbar that allow grouping associated buttons. A ToolTip is a small pop-up window that is displayed when the mouse is left on a toolbar button for more than one-half second. ToolTips usually consist of a short text message that explains the function of the toolbar button or control. Figure 22.11 shows a program containing a toolbar with nine buttons.

Figure 22.11 *Toolbar*

Note in Figure 22.11 that separators are used to group the toolbar buttons. In this case, the first group of buttons correspond with functions in the File menu, the second group with functions in the Edit menu, and so forth. Normally, not all menu commands have a toolbar button, but only the ones most often used.

22.5.3 Creating a Toolbar

There are several ways to create a toolbar. You can define the toolbar in code, using standard buttons furnished in Visual C++. You can use a pre-made bitmap of toolbar buttons, which can be converted into a toolbar resource and then edited. You can create custom buttons using the toolbar editor. Or you can use a combination of these methods. In this section we follow the simplest method, but even then, you must be careful to perform the steps in the same order in which we list them. The toolbar creation tools in Developer Studio were designed to be used in MFC programming; therefore, the system makes assumptions regarding the order in which the steps are performed. If you are careless in this respect, you may end up having to do some manual editing of the resource files.

We must accept that there are complications in creating a toolbar outside of the MFC, however, much suffering can be avoided if the toolbar is not created until the program menu has been defined. The idea is to use the same identification codes for the toolbar as for the corresponding menu items, such that message processing takes place at the same intercept routine. For example, if the first toolbar button in Figure 22.11 corresponds to the New command in the File menu, then both the button and the menu item could be named ID_FILE_NEW. The same applies to the other buttons in the toolbar. In the following description about the creation of a toolbar we assume that the identification strings have been defined for the corresponding menu entries:

```
File menu:
        ID_FILE_NEW
        ID_FILE_OPEN'
        ID_FILE_SAVE
Edit menu:
        ID_EDIT_CUT
        ID_EDIT_COPY
        ID_EDIT_PASTE
Print menu:
        ID_PRT_PRINT
Help menu:
        ID_HLP_ABOUT
        ID_HLP_HELP
```

All of the toolbar buttons in Figure 22.11 correspond to standard buttons contained in Developer Studio. These buttons can be loaded into the toolbar by referencing their system names, or by loading a bitmap that contains them. In the current example we use the bitmap approach.

Toolbars require that each of the buttons be defined in a structure of type TBBUTTON (see Appendix F). Your program, usually in the window procedure, creates an array of structures, with one entry for each button in the toolbar. The button separators must be included. In the case of the screen in Figure 22.11, the array of TBBUTTON structure is as follows:

```
// Array for attributes for toolbar buttons
 TBBUTTON tbb[] = {
0, ID_FILE_NEW,     TBSTATE_ENABLED, TBSTYLE_BUTTON, 0, 0, 0, 0,
1, ID_FILE_OPEN,    TBSTATE_ENABLED, TBSTYLE_BUTTON, 0, 0, 0, 0,
2, ID_FILE_SAVE,    TBSTATE_ENABLED, TBSTYLE_BUTTON, 0, 0, 0, 0,
0, 0,               TBSTATE_ENABLED, TBSTYLE_SEP,    0, 0, 0, 0,
3, ID_EDIT_CUT,     TBSTATE_ENABLED, TBSTYLE_BUTTON, 0, 0, 0, 0,
4, ID_EDIT_COPY,    TBSTATE_ENABLED, TBSTYLE_BUTTON, 0, 0, 0, 0,
5, ID_EDIT_PASTE,   TBSTATE_ENABLED, TBSTYLE_BUTTON, 0, 0, 0, 0,
0, 0,               TBSTATE_ENABLED, TBSTYLE_SEP,    0, 0, 0, 0,
6, ID_PRT_PRINT,    TBSTATE_ENABLED, TBSTYLE_BUTTON, 0, 0, 0, 0,
0, 0,               TBSTATE_ENABLED, TBSTYLE_SEP,    0, 0, 0, 0,
7, ID_HLP_ABOUT,    TBSTATE_ENABLED, TBSTYLE_BUTTON, 0, 0, 0, 0,
8, ID_HLP_HELP,     TBSTATE_ENABLED, TBSTYLE_BUTTON, 0, 0, 0, 0, }; /*
 |-----------|  |-------------|  |-----------|| see note |
 |           |                |              |    below
 |           |                |              |--- One or more
 |           |                |                   button styles
 |           |                |----------- One or more state flags
 |           |------------------------- Command ID mapped to
 |                                       button
 |------------------- Zero-based index to button image in
 |                    bitmap (excluding separators)
Note:
      0, 0, 0, 0
      |  |  |  |----------- index of button string
      |  |--|-------------- application defined value
      |-------------------- Reserved
*/
```

Table 22.9 lists the style flags used with toolbars.

Table 22.9

Toolbar and Toolbar Button Style Flags

STYLE	DESCRIPTION
Toolbar Styles:	
BSTYLE_ALTDRAG	Allows the user to change the position of a toolbar button by dragging it while holding down the Alt key. If this style is not specified, the user must hold down the Shift key while dragging a Button. Note that the CCS_ADJUSTABLE style must be specified to enable toolbar buttons to be dragged.
TBSTYLE_TOOLTIPS	Creates a ToolTip control that an application can use to display descriptive text for the buttons in the toolbar.
TBSTYLE_WRAPABLE	Creates a toolbar that can have multiple lines of buttons. Toolbar buttons can "wrap" to the next line when the toolbar becomes too narrow to include all buttons on the same line. Wrapping occurs on separation and non-group boundaries.
Toolbutton Styles:	
TBSTYLE_BUTTON	Creates a standard push button.
TBSTYLE_CHECK	Button toggles between the pressed and not pressed states each time the user clicks it. The button has a different background color when it is in the pressed state.
TBSTYLE_CHECKGROUP	Creates a check button that stays pressed until another button in the group is pressed.

(continues)

Table 22.9

Toolbar and Toolbar Button Style Flags (continued)

STYLE	DESCRIPTION
Toolbutton Styles:	
TBSTYLE_GROUP	Creates a button that stays pressed until another button in the group is pressed.
TBSTYLE_SEP	Creates a separator. A button that has this style does not receive user input and is not assigned a button number.

Table 22.10 lists the toolbar states

Table 22.10

Toolbar States

TOOLBAR STATE	DESCRIPTION
TBSTATE_CHECKED	The button has the TBSTYLE_CHECKED style and is being pressed.
TBSTATE_ENABLED	The button accepts user input. A button not having this state does not accept user input and is grayed.
TBSTATE_HIDDEN	The button is not visible and cannot receive user input.
TBSTATE_INDETERMINATE	The button is grayed.
TBSTATE_PRESSED	The button is being pressed.
TBSTATE_WRAP	A line break follows the button. The button must also have the TBSTATE_ENABLED state.

The bitmap for the toolbar in Figure 22.11 is furnished with Developer Studio. We have made a copy of this bitmap and you can find it in the Resource directory on the book's on-line software package. The name of the bitmap is toolbar.bmp. The process of creating a toolbar from a toolbar bitmap requires that you follow a certain sequence. The price to pay for changing the order of operations is that you may end up with incorrect resource files that must be manually edited. The following operations result in the toolbar resource:

1. Select Resource from the Insert menu. Select the Bitmap resource type and click the Import button.

2. In the Import Resource dialog editor, edit the filename field for that of a bitmap file. This is accomplished by entering "*.bmp". Now you can search though the file system until you find the toolbar bitmap. In this case the desired bitmap has the name "toolbar.bmp." Select the bitmap and click on the button labeled Import.

3. The toolbar bitmap is now loaded into the bitmap editor. The toolbar bitmap is shown in Figure 22.12. The buttons are labeled according to the identifications assigned in the TBBUTTON structure members listed previously.

4. Now you must convert the bitmap into a toolbar resource. This is accomplished by opening the Image menu and clicking on the Toolbar editor command. The New Toolbar Resource dialog box with the pixel size of normal toolbar buttons is displayed, which is 16 pixels wide and 15 pixels high. Click OK and the toolbar editor appears with the bitmap converted into a toolbar.

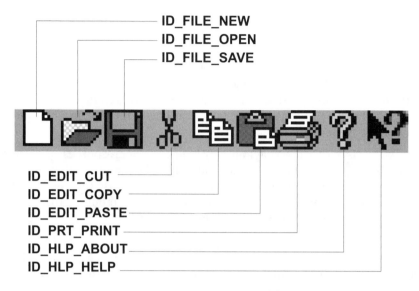

ID_FILE_NEW
ID_FILE_OPEN
ID_FILE_SAVE

ID_EDIT_CUT
ID_EDIT_COPY
ID_EDIT_PASTE
ID_PRT_PRINT
ID_HLP_ABOUT
ID_HLP_HELP

Figure 22.12 *"Toolbar.bmp" Button Identification Codes*

5. You can now proceed to edit the toolbar and assign identification codes to each of the buttons. Note that there is a blank button at the end of the toolbar, which is used for creating custom buttons. You can click on the blank button and use the editor to create a new button image. To delete a button, click on it and drag it off the toolbar. To reposition a button, click on it and drag it to it new location. To create a space in the toolbar drag the button so that it overlaps half the width of its neighbor button. To assign an identification code to a toolbar button, double-click on the button and enter the new identification in the ID: edit box of the Toolbar Button Properties dialog box. At this time you may enter the corresponding identification codes for all the buttons in the toolbar. Figure 22.13 shows the toolbar editor once the separators have been inserted.

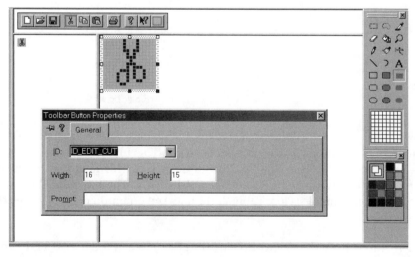

Figure 22.13 *Developer Studio Toolbar Editor*

Figure 22.13 also shows the Toolbar Button Properties dialog box open and the new identification code in the ID: edit box.

6. Once the identification codes have been assigned to all the buttons, click the button labeled X to close the editor. Also close the next screen and save the resource file under the default name, or assign it a new one. Some programmers like to give the resource file the same base name as the application's main module. The extension for the resource file must be .RC.

7. The next step is one that you have already done for other resources: open the Project menu, click the Add To Project command, select Files, and add the resource file to the project. The toolbar is now in the project. You can use the Resource Symbols command in the View menu, or the corresponding toolbar button, to make sure that the identification codes are correct and coincide with those in the BBUTTON structure, and the menu items.

8. Displaying the toolbar requires a call to the CreateToolbarEx() function. The call returns a handle to the toolbar, which is of type HWND since the toolbar is a window. In this example, the call is as follows:

```
#define     ID_TOOLBAR     400     // Toolbar id number
      .
      .
      .
HWND        tbHandle;       // Handle to the toolbar
      .
      .
      .
case WM_CREATE:
  // Create toolbar
  tbHandle = CreateToolbarEx (hwnd,  // Handle to window
           WS_CHILD | WS_VISIBLE |
           WS_CLIPSIBLINGS |
           CCS_TOP | TBSTYLE_ToolTipS, // Window styles
           ID_TOOLBAR,              // Toolbar identifier
           9,                       // Number of button images
                                    // in toolbar bitmap
           hInst,                   // Module instance
           IDB_BITMAP1,             // Bitmap ID
           tbb,                     // TBBUTON structure
           12,                      // Number of buttons
                                    // plus separators
           0, 0, 0, 0,
           sizeof (TBBUTTON));
```

The second parameter in the call refers to controls bits that define the style, position, and type of toolbar. The window style WS_CHILD is always required and most toolbars use WS_VISIBLE and WS_CHILDREN. The bits with the CCS_ prefix are common control styles. Table 22.11 lists the common control styles that refer to toolbars.

In the current call to CreateToolbarEx() we used CCS_TOP and the TBSTYLE_TOOLTIPS in order to create a toolbar displayed above the application's client area, and to provide ToolTip support.

Table 22.11

Toolbar Common Control Styles

STYLE	DESCRIPTION
CCS_ADJUSTABLE	Allows toolbars to be customized by the user. If this style is used, the toolbar's owner window must handle the customization notification messages sent by the toolbar.
CCS_BOTTOM	Causes the toolbar to position itself at the bottom of the parent window's client area and sets the width to be the same as the parent window's width.
CCS_NODIVIDER	Prevents a two-pixel highlight from being drawn at the top of the control.
CCS_NOHILITE	Prevents a one-pixel highlight from being drawn at the top of the control.
CCS_NOMOVEY	Causes the toolbar to resize and move itself horizontally, but not vertically, in response to a WM_SIZE message. If the CCS_NORESIZE style is used, this style does not apply.
CCS_NOPARENTALIGN	Prevents the toolbar from automatically moving to the top or bottom of the parent window. Instead, it keeps its position within the parent window despite changes to the size of the parent window.
CCS_NORESIZE	Prevents the toolbar from using the default width and height when setting its initial size or a new size. Instead, the control uses the width and height specified in the request for creation or sizing.
CCS_TOP	Causes the control to position itself at the top of the parent window's client. The width is set to the size of the parent window. This is the default style.

22.5.4 Standard Toolbar Buttons

The common controls library contains bitmaps for standard toolbar buttons that can be referenced by name and used by application code. In this case no toolbar bitmap is required; therefore, the button images cannot be edited in Developer Studio. There are a total of fifteen button images in two sizes, 24-by-24 pixels and 16-by-16 pixels. When using the standard toolbar buttons, the TBBUTTON structure must be filled differently than when using a toolbar bitmap resource. The parameters of CreateToolbarEx() are also different. The following code fragment shows the TBBUTTON structure for loading all fifteen standard toolbar buttons:

```
TBBUTTON tbb[] = {
// File group
STD_FILENEW,   ID_FILE_NEW,   TBSTATE_ENABLED, TBSTYLE_BUTTON,
               0, 0, 0, 0,
STD_FILEOPEN,  ID_FILE_OPEN,  TBSTATE_ENABLED, TBSTYLE_BUTTON,
               0, 0, 0, 0,
STD_FILESAVE,  ID_FILE_SAVE,  TBSTATE_ENABLED, TBSTYLE_BUTTON,
               0, 0, 0, 0,
0,             0,             TBSTATE_ENABLED, TBSTYLE_SEP,
               0, 0, 0, 0,
// Edit group
STD_COPY,      ID_EDIT_COPY,  TBSTATE_ENABLED, TBSTYLE_BUTTON,
               0, 0, 0, 0,
STD_CUT,       ID_EDIT_CUT,   TBSTATE_ENABLED, TBSTYLE_BUTTON,
               0, 0, 0, 0,
```

```
STD_PASTE,     ID_EDIT_PASTE,  TBSTATE_ENABLED, TBSTYLE_BUTTON,
                               0, 0, 0, 0,
STD_FIND,      ID_EDIT_FIND,   TBSTATE_ENABLED, TBSTYLE_BUTTON,
                               0, 0, 0, 0,
STD_REPLACE,   ID_EDIT_REPLACE,  TBSTATE_ENABLED,
                               TBSTYLE_BUTTON,
                               0, 0, 0, 0,
STD_UNDO,      ID_EDIT_UNDO,   TBSTATE_ENABLED, TBSTYLE_BUTTON,
                               0, 0, 0, 0,
STD_REDOW,     ID_EDIT_REDO,   TBSTATE_ENABLED, TBSTYLE_BUTTON,
                               0, 0, 0, 0,
STD_DELETE,    ID_EDIT_DELETE, TBSTATE_ENABLED,
                               TBSTYLE_BUTTON,
                               0, 0, 0, 0,
0,             0,              TBSTATE_ENABLED, TBSTYLE_SEP,
                               0, 0, 0, 0,
// Print group
STD_PRINTPRE,  ID_PRINT_PREVIEW,  TBSTATE_ENABLED,
                               TBSTYLE_BUTTON,
                               0, 0, 0, 0,
STD_PRINT,     ID_PRINT_PRINT, TBSTATE_ENABLED,
                               TBSTYLE_BUTTON,
                               0, 0, 0, 0,
0,             0,              TBSTATE_ENABLED, TBSTYLE_SEP,
                               0, 0, 0, 0,
// Help and properties group
STD_PROPERTIES,ID_PROPS,       TBSTATE_ENABLED, TBSTYLE_BUTTON,
                               0, 0, 0, 0,
STD_HELP,      ID_HELP,        TBSTATE_ENABLED, TBSTYLE_BUTTON,
                               0, 0, 0, 0, } ;
```

The call to CreateToolbarEx() is also different. The fourth parameter, which indicates the number of button images in the toolbar bitmap is set to zero in the case of standard buttons. The fifth parameter, which in the case of a toolbar bitmap is set to the application instance, is now the constant HINST_COMMCTRL defined in the common controls library. The sixth parameter is the constant IDB_STD_SMALL_COLOR. The resulting call to CreateToolbarEx() is as follows:

```
tbHandle = CreateToolbarEx (hwnd,
          WS_CHILD | WS_VISIBLE | CCS_TOP |
          TBSTYLE_WRAPABLE,
          ID_TOOLBAR,          // Toolbar ID number
          0,                   // Number of bitmaps (none)
          (HINSTANCE)HINST_COMMCTRL, // Special resource instance for
                                     //   standard buttons
          IDB_STD_SMALL_COLOR, // Bitmap resource ID
          tbb,                 // TBBUTTON variable
          18,                  // Count of buttons plus separators
          0, 0, 0, 0,          // Not required for standard
                               //   buttons
          sizeof (TBBUTTON));
```

The program named TB1_DEMO, located in the Toolbar Demo No 1 project folder in the book's on-line software package, is a demonstration of using the standard toolbar buttons. When you click on any of the toolbar buttons, a message box is displayed which contains the button's name. Figure 22.14 is a screen snapshot of the TN1_DEMO program.

Figure 22.14 *TB1_DEMO Program Screen*

22.5.5 Combo Box in a Toolbar

Windows programs, including Developer Studio, often contain a combo box as part of the toolbar. This application of the combo box is a powerful one. For example, the combo box that is part of Developer Studio standard menu bar is used to remember search strings that have been entered by the user. At any time, you can inspect the combo box and select one of the stored strings for a new search operation. Not only does it save you the effort of retyping the string, it is also a record of past searches.

The position of the combo box in the toolbar is an important consideration. If the combo box is to the right of the last button in the toolbar, then it is a matter of calculating the length of the toolbar in order to position the combo box. However, if, as is often the case, the combo box is located between buttons in the toolbar, or at its start, then code must make space in the toolbar. The method suggested by Nancy Cluts in her book *Programming the Windows 95 User Interface* (see Bibliography) is based on adding separators to make space for the combo box. Since each separator is eight pixels wide, we can calculate that for a 130-pixels wide combo box we would need at least 17 separators. In many cases a little experimentation may be necessary to find the number of separators.

The creation of the combo box requires calling CreateWindow() with "COMBOBOX" as the fist parameter. If the combo box is to have a series of string items, as is usually the case, then it is created with the style CBS_HASSTRINGS. If the combo box is to have an edit box feature, then the CBS_DROPDOWN style is used. If it is to have a list of selectable items but no editing possibilities, then the CBS_DROPDOWNLIST style is used. The following code fragment shows the creation of a combo box in a toolbar:

```
static HWND   cbHandle;       // Handle to combo box
static HWND   tbHandle;       // Handle to toolbar
 .
 .
 .
cbHandle = CreateWindow ("COMBOBOX",
          NULL,                       // No class name
          WS_CHILD | WS_VISIBLE | WS_BORDER |
          CBS_HASSTRINGS |CBS_DROPDOWNLIST,
          0,                  // x origin
          0,                  // y origin
          130,                // width
          144,                // height
```

```
       tbHandle,                // Parent window handle
       (HMENU) IDR_MENU1,       // Menu resource ID
       pInstance,               // Application instance
       NULL);
```

Once the combo box is created, we need to add the text strings with which it is originally furnished. This is accomplished by a series of calls to SendMessage() with the message code CB_INSERTSTRING. Typical coding is as follows:

```
char *szStrings[] = { "Visual C++",
                      "Borland C",
                      "Pascal",
                      "Fortran 80",
                      "Visual Basic" };
  .
  .
  .
//Add strings to combo box
  for (i=0; i < 5; i++)
      SendMessage(cbHandle,
                  CB_INSERTSTRING,
                  (WPARAM)-1,
                  (LPARAM)szStrings[i]);
```

The program TB2_DEMO, located in the Toolbar Demo No 2 project folder on the book'son-line software package, demonstrates the creation of a toolbar that includes a combo box.

22.5.6 ToolTip

A ToolTip is a small window that contains a brief descriptive message. Although ToolTips can be activated in relation to any screen object, we are presently concerned with ToolTips associated with a toolbar. For a toolbar to support ToolTips, it must have been created with the TBSTYLE_TOOLTIPS, listed in Table 22.9.

Providing ToolTip support for toolbar buttons is straightforward and simple. However, when you need to furnish ToolTips for other elements in the toolbar, such as the combo box previously mentioned, then ToolTip processing may get more complicated. The first step in creating ToolTips is retrieving a handle for the ToolTip window. This is usually performed in the WM_CREATE message intercept. It consists of calling SendMessage() with the first parameter set to the toolbar handle and the second parameter set to the TB_GETTOOLTIPS message identifier. The following code fragment shows the creation of a three-button toolbar and its corresponding ToolTip window:

```
LRESULT CALLBACK WndProc (HWND hwnd, UINT iMsg, WPARAM wParam,
                          LPARAM lParam) {
  static HWND     tbHandle;     // Handle to toolbar
  static HWND     hWndTT;       // Handle to ToolTip
     .
     .
     .
   switch (iMsg)
         {
   case WM_CREATE:
      // Create a toolbar
```

```
tbHandle = CreateToolbarEx (hwnd,   // Handle to window
            WS_CHILD | WS_VISIBLE |
            WS_CLIPSIBLINGS |
            CCS_TOP | TBSTYLE_TOOLTIPSS, // Window styles
            0,                  // Toolbar identifier
            3,                  // Number of button images
                                // in toolbar bitmap
            pInstance,          // Module instance
            IDB_BITMAP1,        // Bitmap ID
            tbb,                // TBBUTON structure
            3,                  // Number of buttons
                                // plus separators
            0, 0, 0, 0,         // Not required
            sizeof (TBBUTTON));
// Get the handle to the ToolTip window.
   hWndTT = (HWND)SendMessage(tbHandle,
                  TB_GETToolTipS, 0, 0);
   .
   .
   .
```

Once you create the ToolTip window and obtain its handle, the next step is to create and initialize a structure of type TOOLINFO. The coding proceeds as follows:

```
if (hWndTT) {
  // Fill in the TOOLINFO structure.
  lpToolInfo.cbSize = sizeof(lpToolInfo);
  lpToolInfo.uFlags = TTF_IDISHWND | TTF_CENTERTIP;
  lpToolInfo.hwnd = hwnd;
  lpToolInfo.uId = (UINT)tbHandle;
  lpToolInfo.hinst = pInstance;
  lpToolInfo.lpszText = LPSTR_TEXTCALLBACK;
}
```

The first flag, TTF_IDISHWND, indicates that the fourth structure member (uId) is a handle to a window, in this case, the toolbar. The flag TTF_CENTERTIP determines that the ToolTip is displayed below the window specified in the uId member, here again, the toolbar. Finally, the lpszText member is set to the constant LPSTR_TEXTCALLBACK, which makes the control send the TTN_NEEDTEXT notification message to the owner window. The values entered in the other structure members are self-explanatory.

Processing of ToolTip messages, as is the case with most controls, takes place at the WM_NOTIFY message intercept. At the time the message handler receives control, the lParam is a pointer to a structure of type HMHDR (see Appendix F), or to a larger structure that has NMHDR as its first member. The third member of the HMHDR structure contains the control-specific notification code. This parameter is TTN_NEEDTEXT when text is required for a ToolTip. Therefore, code can switch on this structure member and provide processing in a case statement, as shown in the following code fragment:

```
LPNMHDR        pnmh;          // Pointer to HMHDR structure
TOOLTIPINFO    lpToolTipInfo;
LPTOOLTIPTEXT  lpToolTipText;
static char    szBuf[128]; // Buffer for ToolTip text
   .
```

```
    .
    .
  case WM_NOTIFY:
     pnmh = (LPNMHDR) lParam;
     switch (pnmh->code) {
         case TTN_NEEDTEXT:
         // Display ToolTip text.
             lpToolTipText = (LPTOOLTIPTEXT)lParam;
             LoadString (pInstance,
                         lpToolTipText->hdr.idFrom,
                         szBuf,
                         sizeof(szBuf));
                         lpToolTipText->lpszText = szBuf;
         break;
       default:
           return TRUE;
           break;
     }
   return 0;
 break;
```

Note that the TTN_NEEDTEXT message intercept contains a pointer to a structure of type TOOLTIPTEXT in the lParam (see Appendix F). The first member of TOOLlTIPTEXT (hdr) is a structure of type NMHDR, and the idFrom member of HMHDR is the identifier of the control sending the message. Code uses this information and the LoadString() function to move the text into the buffer named szBuf. The text moved into szBuf comes from the lpszText member of a structure variable of type TOOLTIPTEXT. This second member is a pointer to a text string defined as a string resource in the application's executable.

The string resource that contains the messages that are displayed with each ToolTip is the last missing element of ToolTip implementation. You create the string resource by opening the Insert menu and selecting the Resource command. In the Insert Resource dialog box select String Table and then click on the New button. An example of a resource table is seen in Figure 22.15.

ID	Value	Caption
ID_FILE_NEW	40001	Create a new file
ID_FILE_OPEN	40002	Open current file
ID_FILE_SAVE	40003	Save current file

Figure 22.15 *Developer Studio Resource Table Editor*

The resource table consists of three entries: the id, the value, and the caption fields. You fill the id field so that it contains the same identification code as the button for which you are providing a ToolTip. In the caption field, you enter the text that is to be displayed at the ToolTip. Developer Studio automatically fills the value field for the one assigned to the corresponding toolbar button. Double-clicking on the entry displays a dialog box where these values can be input.

SOFTWARE ON-LINE

The program named TT_DEMO, located in the ToolTip Demo project folder on the book's on-line software package, is a demonstration of the processing required for the implementation of ToolTip controls.

Chapter 23

Drawing Lines and Curves

Chapter Summary

This chapter is on using some of the graphic services in the Windows Graphics Device Interface. It discusses the simpler of these services, which are used for reading and setting individual pixels and for drawing lines and curves in a two-dimensional space. The described graphics functions are among the most often used in conventional Windows graphics.

The chapter includes a discussion on the architecture of a Windows graphics application, the GDI itself, and a more extensive look at the Windows Device Context. It is in the Device Context where system-level graphics information is stored. Applications must often read these attributes. It also covers Windows graphics objects and their attributes, that is, pens, brushes, bitmaps, palettes, fonts, paths, and regions. Also some of the attributes of the Device Context: color, mix mode, background mode, pen position, and arc direction. These attributes determine how graphics output takes place.

23.0 Drawing in a Window

Windows programs are event-driven; applications share resources with all other running programs and with the operating system. This determines that a graphics program cannot make exclusive use of the display, or of other system resources, since these are part of a pool that is accessible to all code in a multitasking environment. The following implications result from this architecture:

- A typical Windows application must obtain information about the system and the display device before performing output operations. The application must know the structure and dimensions of the output surface, as well as its capabilities, in order to manage the display function.

- In Windows, output to devices is performed by means of a logical link between the application, the device driver, and the hardware components. This link is called a device context. A display context is a special device context for the display device. Applications that draw to a window using conventional Windows functions must first obtain the display context. The handle to this display context is passed as a parameter to all API drawing functions.

- Unlike a DOS program, a Windows applications cannot draw to the screen and assume that the resulting image remains undisturbed for unlimited time. On the contrary, a Windows program must take into account that the video display is a shared resource. Windows notifies the application that its client area needs to be painted or repainted by posting a WM_PAINT message to the program's message queue. A well-designed Windows program must be able to redraw its client area upon receiving this message.

The first two of these topics, that is, obtaining the device context handle and the display device attributes, are discussed in a separate section later in this chapter. Here we are concerned with the mechanisms used by Windows applications for accessing the display device in a way that is consistent with the multitasking nature of the environment.

23.1 The Redraw Responsibility

Windows applications are burdened with the responsibility redrawing their client area at any time. This is an obligation to be taken seriously since it implies that code must have ways for reconstructing the display on demand. What data structures and other controls are necessary to redraw the screen, and how code handles this responsibility, depends on the application itself. In some programs the screen redraw burden is met simply by keeping tabs of which of several possible displays is active. In other applications the screen redraw obligation can entail such elaborate processing that it becomes a major consideration in program design.

The operating system, or your own code, sends the WM_PAINT message whenever the client area, or a portion thereof, needs to be redrawn. Application code responds to its screen redraw responsibility during the WM_PAINT message intercept. The following events cause the operating system to send WM_PAINT:

- The user has brought into view a previously hidden area of the application window. This happens when the window has been moved or uncovered.

- The user has resized the window.

- The user has scrolled the window contents.

The WM_PAINT message is not produced when a window is merely moved to another position on the desktop, since in this case, the client area has not been changed by the translation. Therefore, the operating system is able to maintain the screen contents because no new graphics elements were introduced or removed, and the screen size remains the same. However, the operating system cannot anticipate how an application handles a screen resizing operation. There are several possible processing options: are the screen contents scaled to the new dimension of the client area, or is their original size maintained? Are the positions of the graphics elements changed as a consequence of the resize operation, or do they remain in the same place? Not knowing how these alternatives are to be handled, the operating system responds by sending WM_PAINT to the application and letting it take whatever redraw action it considers appropriate. The same logic applies when the client area is scrolled or when portions of the window are uncovered.

There are other times during which Windows attempts to restore the application's screen, but may occasionally post the WM_PAINT message if it fails in this effort. These occasions are when a message or dialog box is displayed, when a menu is pulled down, or when a tooltip is enabled. Finally, there are cases in which Windows always saves and restores the screen automatically, for example, when the mouse cursor or a program icon is dragged or moved across the client area.

23.1.1 The Invalid Rectangle

In an effort to minimize the processing, Windows keeps tabs of which portion of the application's client area needs to be redrawn. This notion is based on the following logic: it is wasteful for application code to repaint the entire screen when only a small portion of the program's client area needs to be redrawn. In practice, for simpler programs, it is often easier to assume that the entire client area needs redrawing than to get into the complications of repainting parts of the screen. However, in more complex applications, particularly those that use multiple child windows, it may save considerable time and effort if code can determine which of these elements need redrawing and which can be left unchanged.

The screen area that needs to be redrawn is called the update region. The smallest rectangle that binds the update region is called the invalid rectangle. When the WM_PAINT message is placed in the message queue, Windows attaches to it a structure of type RECT that contains the dimensions and location of the invalid rectangle. If another screen area becomes invalid before WM_PAINT is posted, Windows then makes the necessary correction in the invalid rectangle. This scheme saves posting more than one WM_PAINT message on the queue. Applications can call GetUpdateRect() to obtain the coordinates of the top-left and bottom-right corner of the update region.

An application can force Windows to send a WM_PAINT message to its own window procedure. This is accomplished by means of the InvalidateRect() or InvalidateRgn() functions. InvalidateRect() has the effect of adding a rectangle to a window's update region. The function has the following general form:

```
BOOL InvalidateRect(
HWND hWnd,          // 1
CONST RECT* lpRect, // 2
BOOL bErase         // 3
        );
```

The first parameter identifies the window whose update region has changed. The second parameter is a pointer to a structure variable of type RECT that contains the coordinates of the rectangle to be added to the update region. If this parameter is NULL, then the entire client area is added to the update region. The third parameter is a flag that indicates if Windows should erase or not erase the background. If this parameter is TRUE, then the background is erased when the BeginPaint() function is called by the application. If it is FALSE, the background remains unchanged.

23.1.2 Screen Updates On-Demand

The standard reply of an application that has received a WM_PAINT message is to redraw its own client area. This implies that the application has been designed so that a screen update takes place every time a WM_PAINT message is received. In this case the application design has to take into account the message-driven characteristic of a Windows program.

Consider a program that contains three menu commands: one to display a circle, another one to display a rectangle, and a third one to display a triangle. When the user clicks on any one of the three menu items, a WM_COMMAND message is posted to the application's message queue. The low word of the WPARAM encodes which one of the menu items was selected. Application code usually switches on this value in order to field all possible commands. However, the screen should not be updated during WM_COMMAND processing. What code can do at this point is set a switch that indicates the selected command. In this example a static variable of type int, named drawMode, could be set to 1 to indicate a circle drawing request, to 2 to indicate a rectangle, and to 3 to indicate a triangle. After this switch is set according to the menu command entered by the user, code calls InvalidateRect() so that Windows posts WM_PAINT to the application's own message queue. The application then processes WM_PAINT inspecting the value of the drawMode variable. If the value is 1 it draws a circle, if its is 2 it draws a rectangle, and if it is 3 it draws a triangle.

To a non-Windows programmer this may appear to be quite a round-about way of doing things. Why not draw the geometrical figures at the time that the menu commands are received? The problem with drawing as the commands are received is that if the window is resized or covered there is no mechanism in place to restore its screen image. The result would be either a partially or a totally blank client area. However, if the screen updates take place during WM_PAINT message processing, then when Windows sends WM_PAINT to the application because of a screen contents change, the application redraws itself and the client area is correctly restored.

On the other hand, not all screen drawing operations can take place during WM_PAINT message processing. Applications sometimes have to perform display functions that are directly linked to a user action; for example, a rubber-band image that is drawn in direct and immediate response to a mouse movement. In this case code cannot postpone the drawing until WM_PAINT is received.

23.1.3 Intercepting WM_PAINT

The WM_PAINT message is generated only for windows that were created with the styles CS_HREDRAW or CS_VREDRAW. Receiving WM_PAINT indicates to application code that all or part of the client area must be repainted. The message can originate in Windows, typically because the user has minimized, overlapped, or resized the client area. Or also because the application itself has produced the message by calling InvalidateRect() or InvalidateRgn(), as previously discussed.

Typically, WM_PAINT processing begins with the BeginPaint() function. BeginPaint() prepares the window for a paint operation. In the first place it fills a variable of type PAINTSTRUCT, which is defined as follows:

```
typedef struct tagPAINTSTRUCT {
    HDC     hdc;                // Identifies display device
    BOOL    fErase;            // TRUE if background must be
                               // erased
    RECT    rcPaint;           // Rectangle structure specifying
                               // the update region
    BOOL    fRestore;          // RESERVED
    BOOL    fIncUpdate;        // RESERVED
    BYTE    rgbReserved[32];   // RESERVED
} PAINTSTRUCT;
```

If the screen erasing flag is set, BeginPaint() uses the window's background brush to erase the background. In this case, when execution returns from BeginPaint() code can assume that the update region has been erased. At this point the application can call GetClientRect() to obtain the coordinates of the update region, or proceed on the assumption that the entire client area must be redrawn.

Processing ends with EndPaint(). EndPaint() notifies Windows that the paint operation has concluded. The parameters passed to EndPaint() are the same ones passed to BeginPaint(): the handle to the window and the address of the structure variable of type PAINTSTRUCT. One important consequence of the EndPaint() function is that the invalid region is validated. Drawing operations by themselves have no validating effect. Placing the drawing operations between BeginPaint() and EndPaint() functions automatically validates the invalid region so that other WM_PAINT messages are not produced. In fact, placing the BeginPaint() EndPaint() functions in the WM_PAINT intercept, with no other processing operation, has the effect of validating the update region. The DefWindowProc() function operates in this manner.

The project Pixel and Line Demo in the book's software on-line demonstrates image display and update in response to WM_PAINT messages. The processing uses a static variable to store the state of the display. A switch construct in the WM_PAINT routine performs the screen updates, as in the following code fragment:

```
// Drawing command selector
static int    drawMode = 0;
//
//                      0 = no menu command active
//                          Active menu command:
//                      1 = Set Pixel
//                      2 = LineTo
//                      3 = Polyline
//                      4 = PolylineTo
//                      5 = PolyPolyline
//                      6 = Arc
//                      7 = AngleArc
//                      8 = PolyBezier
//                      9 = PolyDraw
        . . .
//*****************************
//      menu command processing
```

```
//*******************************
case WM_COMMAND:
switch (LOWORD (wParam)) {
        //****************************
//     SetPixel command
        //****************************
case ID_DRAWOP_PIXELDRAW:
drawMode = 1;    // Command to draw line
InvalidateRect(hwnd, NULL, TRUE);
break;
//***************************
//    LineTo command
//***************************
case ID_DRAWOP_LINE_LINETO:
drawMode = 2;    // Command to draw line
InvalidateRect(hwnd, NULL, TRUE);
break;
    . . .
//*******************************
//      WM_PAINT processing
//*******************************
case WM_PAINT :
BeginPaint (hwnd, &ps) ;
switch(drawMode)
        {
// 1 = SetPixel command
case 1:
pixColor = RGB(0xff, 0x0, 0x0); // Red
for (i = 0; i < 1000; i++) {
x = i * cxClient / 1000;
y = (int) (cyClient / 2 *
(1- sin (pix2 * i / 1000)));
SetPixelV (hdc, x, y, pixColor);
                }
break;
// 2 = LineTo command
case 2:
// Create a solid blue pen, 4 pixels wide
SelectObject(hdc, bluePen4);
MoveToEx (hdc, 140, 140, NULL);
LineTo (hdc, 300, 140);
LineTo (hdc, 300, 300);
LineTo (hdc, 140, 300);
LineTo (hdc, 140, 140);
break;
    . . .
```

23.2 The Graphics Device Interface

The Graphics Device Interface (GDI) consists of a series of functions and related data structures that applications can use to generate graphics output. The GDI can output to any compatible device, but most frequently the device is either the video display, a graphics hard copy device (such as a printer or plotter), or a metafile in memory. By means of GDI functions you draw lines, curves, closed figures, paths, bitmapped images, and text. The objects are drawn according to the style selected for drawing objects, such as pens, brushes, and fonts. The pen object determines how lines and curves are drawn; the brush object determines how the interior of closed figures is filled. Fonts determine the attributes of text.

Output can be directed to physical devices, such as the video display or a printer, or to a logical device, such as a metafile. A metafile is a memory object that stores output instructions so that they can later be used to produce graphics on a physical device. It works much like a tape recording that can be played back at any time, any number of times.

The GDI is a layer between the application and the graphics hardware. It ensures device-independence and frees the programmer from having to deal with hardware details of individual devices. The device context, mentioned previously, is one of the fundamental mechanisms used by the GDI to implement device-independent graphics. The GDI is a two-dimensional interface, which contains no 3D graphics primitives or transformations. It is also a static system, with very little support for animation. Therefore, the GDI is not capable of doing everything that a graphics programmer may desire, but within these limitations, it provides an easy and convenient toolkit of fundamental functions.

GDI functions can be classified into three very general categories:

- Functions that relate to the device context. These are used to create and release the DC, to get information about it, and to get and set its attributes.

- Drawing primitives. These are used to draw lines and curves, fill areas, and display bitmaps and text.

- Functions that operate on GDI objects. These perform manipulation of graphics objects such as pens, brushes, and bitmaps, which are not part of the device context.

23.2.1 Device Context Attributes

The GDI can output to any compatible device, including hard copy graphics devices and memory. For this reason, when referring to the GDI functions, we always use the term device context, instead of the more restrictive display context. Previously we discussed the fundamentals of the device context and developed a template file TEMPL02.CPP, found in the Templates directory on the book's on-line software package; it creates a program that uses a private device context. A private device context has the advantage that it need be retrieved only once and that attributes assigned to it are retained until they are explicitly changed. In the following examples and demonstration programs, we continue to use a private device context to take advantage of these simplifications.

The mapping modes are among the most important attributes of the device context. Two scalable mapping modes, named MM_ANISOTROPIC and MM_ISOTROPIC, are used in shrinking and expanding graphics by manipulating the coordinate system. They provide a powerful image manipulation mechanism and are discussed in Chapter 21. For now, we continue to use the default mapping mode, MM_TEXT, in the demonstrations and examples.

Device context operations belong to two types: those that obtain information and those that set attributes. For example, the GDI function GetTextColor() retrieves the current text color from the device context, while the function SetTextColor() is used to change the text color attribute. Although these functions are sometimes referred to as get- and set-types, the function names do not always start with these words. For example, the SelectObject() function is used to both get and set the attributes of pens, brushes, fonts, and bitmaps.

Graphics applications often need to obtain information regarding the device context. For example, a program may need to know the screen resolution or the number of display colors. One of the most useful functions for obtaining information regarding the capabilities of a device context is GetDeviceCaps(). The call to GetDeviceCaps() requires two parameters: the first one is the handle to the device context, and the second one is an index value that identifies the capability being queried. Table 23.1 lists some of the most useful information returned by this function.

Table 23.1

Information Returned by GetDeviceCaps()

INDEX	MEANING
DRIVERVERSION	Version number of device driver.
TECHNOLOGY	Any one of the following:

Value	Meaning
DT_PLOTTER	Vector plotter
DT_RASDISPLAY	Raster display
DT_RASPRINTER	Raster printer
DT_RASCAMERA	Raster camera
DT_CHARSTREAM	Character stream
DT_METAFILE	Metafile
DT_DISPFILE	Display file

INDEX	MEANING
HORZSIZE	Width of the physical screen (millimeters).
VERTSIZE	Height of the physical screen (millimeters).
HORZRES	Width of the screen (pixels).
VERTRES	Height of the screen (raster lines).
LOGPIXELSX	Number of pixels per logical inch along the screen width.
LOGPIXELSY	Number of pixels per logical inch along the screen height.
BITSPIXEL	Number of color bits per pixel.
PLANES	Number of color planes.
NUMBRUSHES	Number of device-specific brushes.
NUMPENS	Number of device-specific pens.
NUMFONTS	Number of device-specific fonts.
NUMCOLORS	Number of entries in the color table, if the device has a color depth of no more than 8 bits per pixel. Otherwise, −1 is returned.
ASPECTX	Relative width of a device pixel used for line drawing.
ASPECTY	Relative height of a device pixel used for line drawing.
ASPECTXY	Diagonal width of the device pixel.

(continues)

Table 23.1

Information Returned by GetDeviceCaps() (continued)

INDEX	MEANING
CLIPCAPS	Flag indicating clipping capabilities of the device. Value is 1 if the device can clip to a rectangle. Otherwise, it is 0.
SIZEPALETTE	Number of entries in the system palette.
NUMRESERVED	Number of reserved entries in the system palette.
COLORRES	Actual color resolution of the device, in bits per pixel.
PHYSICALWIDTH	For printing devices: the width of the physical page, in device units.
PHYSICALHEIGHT	For printing devices: the height of the physical page, in device units.
PHYSICALOFFSETX	For printing devices: the distance from the left edge of the physical page to the left edge of the printable area, in device units.
PHYSICALOFFSETY	For printing devices: the distance from the top edge of the physical page to the top edge of the printable area, in device units.
RASTERCAPS	Value that indicates the raster capabilities of the device, as follows:

Capability	Meaning
RC_BANDING	Requires banding support.
RC_BITBLT	Capable of transferring bitmaps.
RC_BITMAP64	Supports bitmaps larger than 64K.
RC_DI_BITMAP	Supports SetDIBits() and GetDIBits functions.
RC_DIBTODEV	Capable of supporting the SetDIBitsToDevice function.
RC_FLOODFILL	Capable of performing flood fills.
RC_PALETTE	Palette-based device.
RC_SCALING	Capable of scaling.
RC_STRETCHBLT	Capable of performing the StretchBlt function.
RC_STRETCHDIB	Capable of performing the StretchDIBits function.

INDEX	MEANING
CURVECAPS	Indicates the curve capabilities of the device, as follows:

Value	Meaning
CC_NONE	Does not support curves.
CC_CIRCLES	Device can draw circles.
CC_PIE	Device can draw pie wedges.
CC_CHORD	Device can draw chord arcs.
CC_ELLIPSES	Device can draw ellipses.
CC_WIDE	Device can draw wide borders.
CC_STYLED	Device can draw styled borders.
CC_WIDESTYLED	Device can draw wide and styled borders.
CC_INTERIORS	Device can draw interiors.
CC_ROUNDRECT	Device can draw rounded rectangles.

(continues)

Table 23.1

Information Returned by GetDeviceCaps() (continued)

INDEX	MEANING
LINECAPS	Indicates the line capabilities of the device, as follows:

Value	Meaning
LC_NONE	Does not support lines.
LC_POLYLINE	Device can draw a polyline.
LC_MARKER	Device can draw a marker.
LC_POLYMARKER	Device can draw multiple markers.
LC_WIDE	Device can draw wide lines.
LC_STYLED	Device can draw styled lines.
LC_WIDESTYLED	Device can draw lines that are wide and styled.
LC_INTERIORS	Device can draw interiors.

INDEX	MEANING
POLYGONALCAPS	Indicates the polygon capabilities of the device, as follows:

Value	Meaning
PC_NONE	Does not support polygons.
PC_POLYGON	Device can draw alternate-fill polygons.
PC_RECTANGLE	Device can draw rectangles.
PC_WINDPOLYGON	Device can draw winding-fill polygons.
PC_SCANLINE	Device can draw a single scanline.
PC_WIDE	Device can draw wide borders.
PC_STYLED	Device can draw styled borders.
PC_WIDESTYLED	Device can draw borders that Are wide and styled.
PC_INTERIORS	Device can draw interiors.

INDEX	MEANING
TEXTCAPS	Indicates the text capabilities of the device, as follows:

Value	Meaning
TC_OP_CHARACTER	Device is capable of character output precision.
TC_OP_STROKE	Device is capable of stroke output precision.
TC_CP_STROKE	Device is capable of stroke clip precision.
TC_CR_90	Device is capable of 90-degree character rotation.
TC_CR_ANY	Device is capable of any character rotation.
TC_SF_X_YINDEP	Device can scale independently in the x- and y-directions.
TC_SA_DOUBLE	Device is capable of doubled character for scaling.
TC_SA_INTEGER	Device uses integer multiples only for character scaling.
TC_SA_CONTIN	Device uses any multiples for exact character scaling.

(continues)

Table 23.1

Information Returned by GetDeviceCaps() (continued)

VALUE	MEANING
TC_EA_DOUBLE	Device can draw double-weight characters.
TC_IA_ABLE	Device can italicize.
TC_UA_ABLE	Device can underline.
TC_SO_ABLE	Device can draw strikeouts.
TC_RA_ABLE	Device can draw raster fonts.
TC_VA_ABLE	Device can draw vector fonts.
TC_SCROLLBLT	Device cannot scroll using a bit-block transfer.

23.2.2 DC Info Demonstration Program

The program named DCI_DEMO, located in the DC Info Demo project folder on the book's on-line software package, shows how to obtain device context information. The menu labeled "DC Info" contains commands for displaying the most used general device context capabilities, the device driver version, as well as the specific line and curve drawing capabilities. Figure 23.1 shows the various menu commands in the DCI_DEMO program.

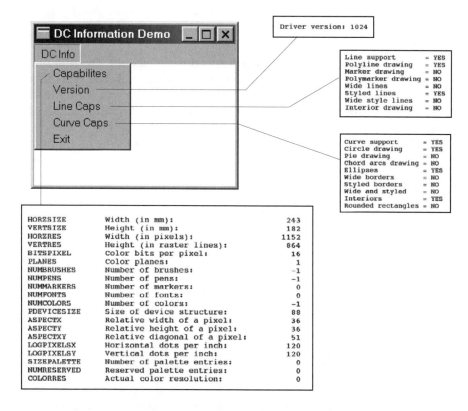

Figure 23.1 *Screen Snapshots of the DC Info Program*

The Capabilities command in the DC Info menu displays the device context values for some of the most used elements returned by the GetDeviceCaps() function. To simplify the programming, the data required during processing is stored in a header file named DC_Caps.h, which can be found in the project directory. The header file is formatted as follows:

```
// Header file for DC Info Demo project
// Contains array of structures

#define LINES ((int) (sizeof DCcaps / sizeof DCcaps [0]))
struct
       {
int   iIndex ;
char *szLabel ;
char *szDesc ;
       }
DCcaps [] =
         {
         HORZSIZE,         "HORZSIZE",        "Width (in mm):",
         VERTSIZE,         "VERTSIZE",        "Height (in mm):",
         HORZRES,          "HORZRES",         "Width (in pixels):",

           .

           .

           .

         NUMRESERVED,      "NUMRESERVED",     "Reserved palette entries:",
         COLORRES,         "COLORRES",        "Actual color resolution:"
         } ;
```

Each entry in the array of structures contains three elements. The first one (int iIndex) is the index name required in the GetDeviceCaps() call. The two other elements are strings used at display time. Processing takes place in a loop in which the number of iterations is determined by the constant LINES, which is calculated by dividing the number of entries in the structure by the number of elements in each entry. This coding allows us to change the number of entries in the array without having to change the loop.

```
// Obtain and display DC capabilities
for (i = 0 ; i < LINES ; i++) {
TextOut (hdc, cxChar, cyChar * (1 + i),
DCcaps[i].szLabel,
strlen (DCcaps[i].szLabel)) ;
TextOut (hdc, cxChar + 16 * cxCaps, cyChar * (1 + i),
DCcaps[i].szDesc,
strlen (DCcaps[i].szDesc)) ;
SetTextAlign (hdc, TA_RIGHT | TA_TOP) ;
TextOut (hdc, cxChar + 16 * cxCaps + 40 * cxChar,
cyChar * (1 + i), szBuffer,
wsprintf (szBuffer, "%5d",
GetDeviceCaps (hdc, DCcaps[i].iIndex))) ;
SetTextAlign (hdc, TA_LEFT | TA_TOP) ;
      }
break;
```

In the previous code fragment, the first TextOut() call displays the szLabel variable in the DCcaps structure. The second call to TextOut() displays the szDesc string. The value in the device context is obtained with the GetDeviceCaps() function that is part of the third call to TextOut(). In this case the iIndex element in the array is used as the second parameter to the call. The wsprintf() function takes care of converting and formatting the integer value returned by GetDeviceCaps() into a displayable string.

Obtaining and displaying the driver version is much simpler. The coding is as follows:

```
// Get driver version
_itoa(GetDeviceCaps(hdc, DRIVERVERSION),
szVersion + 16, 10);
// Initialize rectangle structure
    SetRect (&textRect,                    // address of structure
             2 * cxChar,                   // x for start
             cyChar,                       // y for start
             cxClient,                     // x for end
             cyClient);                    // y for end
DrawText( hdc, szVersion, -1, &textRect,
DT_LEFT | DT_WORDBREAK);
break;
```

In this case we use the _itoa() function to convert the value returned by GetDeviceCaps() into a string. SetRect() and DrawText() are then used to format and display the string.

Obtaining and displaying the curve drawing and line drawing capabilities of the device context requires different processing. These values (see Table 23.1) are returned as bit flags associated with an index variable. For example, we make the call to GetDeviceCaps() using the index constant CURVECAPS as the second parameter. The integer returned by the call contains all the bit flags that start with the prefix CC (CurveCaps) in Figure 23.1. Code can then use a bitwise AND to test for one or more of curve drawing capabilities. The following code fragment shows one possible approach for obtaining curve-drawing capabilities:

```
// Get curve drawing capabilities
curvecaps = GetDeviceCaps (hdc, CURVECAPS);
// Test individual bit flags and change default
// string if necessary
if (curvecaps & CC_NONE)
strncpy(szCurvCaps + 21, strNo, 3);
if (curvecaps & CC_CIRCLES)
strncpy(szCurvCaps + (26 + 21), strYes, 3);
    .
    .
    .
if (curvecaps & CC_ROUNDRECT)
strncpy(szCurvCaps + (9 * 26 + 21), strYes, 3);
// Initialize rectangle structure
        SetRect (&textRect,                // address of
                                           // structure
                 2 * cxChar,               // x for start
                 cyChar,                   // y for start
                 cxClient,                 // x for end
                 cyClient);                // y for end

DrawText( hdc, szCurvCaps, -1, &textRect,
```

```
DT_LEFT | DT_WORDBREAK);
break;
```

Each of the if statements in the processing routine tests one of the bit flags returned by GetDeviceCaps(). If the bit is set, then a text string containing the words YES or NO is moved into the display string. When all the bits have been examined, the message string named szCurvCaps is displayed in the conventional manner.

23.2.3 Color in the Device Context

Monochrome displays are a thing of the past. Virtually all Windows machines have a color display and most of them can go up to 16.7 million displayable colors. In graphics programming you will often have to investigate the color capabilities of a device as well as select and manipulate colors.

In Windows programming, colors are defined by the relative intensity of the red, green, and blue primary components. Each color value is encoded in 8 bits, therefore, all three primary components require 24 bits. Since no C++ data type is exactly 24 bits, however, the color value in Windows is stored in a type called COLORREF, which contains 32 bits. The resulting encoding is said to be in RGB format, where the letters stand for the red, green, and blue components, respectively. Figure 23.2 shows the bit structure of the COLORREF type.

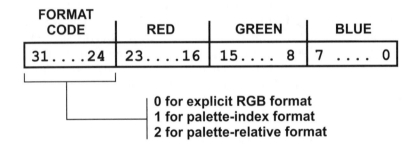

Figure 23.2 *COLORREF Bitmap*

Windows provides a macro named RGB, defined in the windows.h header file; it simplifies entering the color values into a data variable of type COLORREF. The macro takes care of inserting the zeros in bits 24 to 31, and in positioning each color in its corresponding field. As the name RGB indicates, the first value corresponds to the red primary, the second one to the green, and the third one to the blue. For example, to enter a middle-gray value, in which each of the primary colors is set to 128, proceed as follows:

```
COLORREF    midGray;  // Variable of type COLORREF
midGray = RGB(128, 128, 128);
```

The COLORREF data type is also used to encode palette colors. Windows uses the high-order 8 bits to determine if a color value is in explicit RGB, palette-index, or palette-relative format. If the high-order byte is zero, then the color is an explicit RGB value; if it is 1 then it is a palette-index value; if it is 2 then the color is a palette-relative value. Using the RGB macro when creating explicit-RGB values ensures that the high-order byte is set correctly.

Obtaining color information from the device context requires careful consideration. Note in Table 23.1 that the index constant NUMCOLORS is valid only if the color depth is no more than 8 bits per pixel. The device queried in Figure 23.1 has 16 bits per pixel; therefore, the NUMCOLORS value is set to −1. By the same token, the COLORRES index constant is valid only if the device sets the RC_PALETTE bit. In Figure 23.1 the value of this field is 0. The two most useful constants for obtaining general color depth information are PLANES and BITPIXEL. PLANES returns the number of color planes and BITPIXEL returns the number of bits used in encoding each plane.

23.3 Graphic Objects and GDI Attributes

We should first mention that Windows graphics objects are not objects in the object-oriented sense. Windows graphics objects are pens, brushes, bitmaps, palettes, fonts, paths, and regions. Of these, pens and brushes are the objects most directly related to pixel and line drawing operations.

23.3.1 Pens

The pen graphics object determines a line's color, width, and style. Windows uses the pen currently selected in the device context with any of the pen-based drawing functions. Three stock pens are defined: BLACK_PEN, WHITE_PEN, and NULL_PEN. The default pen is BLACK_PEN, which draws solid black lines. Applications refer to a pen by means of its handle, which is stored in a variable of type HPEN. The GetStockObject() function is used to obtain a handle to one of the stock pens. The pen must be selected into the device context before it is used, as follows:

```
HPEN      aPen;        // handle to pen
  .
  .
  .
aPen = GetStockObject (WHITE_PEN);
SelectObject (hdc, aPen);
```

The two functions can be combined in a single statement, as follows:

```
SelectObject (hdc, GetStockObject (WHITE_PEN));
```

In this case, no pen handle variable is required. SelectObject() returns the handle to the pen previously installed in the device context. This can be used to save the original pen so that it can be restored later.

Drawing applications sometimes require one or more custom pens, which have a particular style, width, and color. Custom pens can be created with the functions CreatePen(), CreatePenIndirect(), and ExtCreatePen(). In the CreatePen() function the pen's style, width, and color are passed as parameters. CreatePenIndirect() uses a structure of type LOGPEN to hold the pen's style, width, and color. ExtCreatePen(), introduced in Windows 95, is the more powerful of the three. The iStyle parameter is a combination of pen type, styles, end cap style, and line join attributes. The constants used in defining this parameter are listed in Table 23.2.

Table 23.2

Values Defined for the ExtCreatePen() iStyle Parameter

PEN TYPE	DESCRIPTION
PS_GEOMETRIC	Pen is geometric.
PS_COSMETIC	Pen is cosmetic. Same as those created with CreatePen() and CreatePenIndirect(). Width must be 1 pixel.
Pen Style	
PS_ALTERNATE	Windows NT: Pen sets every other pixel. (cosmetic pens only.) Windows 95: Not supported.
PS_SOLID	Pen is solid.
PS_DASH	Pen is dashed.
PS_DOT	Pen is dotted.
PS_DASHDOT	Pen has alternating dashes and dots.
PS_DASHDOTDOT	Pen has alternating dashes and double dots.
PS_NULL	Pen is invisible.
PS_USERSTYLE	Windows NT: Pen uses a styling array supplied by the user. Windows 95: Not supported.
PS_INSIDEFRAME	Pen is solid. Any drawing function that takes a bounding rectangle, the dimensions of the figure are shrunk so that it fits entirely in the bounding rectangle. Geometric pens only.
End Cap Style (only in stroked paths)	
PS_ENDCAP_ROUND	End caps are round.
PS_ENDCAP_SQUARE	End caps are square.
PS_ENDCAP_FLAT	End caps are flat.
Join Style (only in stroked paths)	
PS_JOIN_BEVEL	Joins are beveled.
PS_JOIN_MITER	Joins are mitered when they are within the current limit set by the SetMiterLimit() function. If it exceeds this limit, the join is beveled. SetMiterLimit() is discussed in Chapter 21.
PS_JOIN_ROUND	Joins are round.

The standard form of the ExtCreatePen() function is as follows:

```
HPEN ExtCreatePen (iStyle,      // pen style
                   iWidth,      // pen width
                   &aBrush,     // pointer to a LOGBRUSH
                                // structure (next section)
dwStyleCount,// length of next parameter
lpStyle);    // dot-dash pattern array
```

The second parameter to ExtCreatePen() defines the pen's width. If the pen is a geometric pen, then its width is specified in logical units. If it is a cosmetic pen then the width must be set to 1.

A geometric pen created with ExtCreatePen() has brush-like attributes. The third parameter is a pointer to LOGBRUSH. The LOGBRUSH structure, described in the following section, is defined as follows:

```
struct tagLOGBRUSH {
UINT      lbStyle;
COLORREF  lbColor;
LONG      lbHatch;
} LOGBRUSH
```

If the pen is a cosmetic pen, then the lbStyle member must be BS_SOLID and the lbColor member defines the pen's color. In this case the lbHatch member, which sets a brush's hatch pattern, is ignored. If the pen is geometric, then all three structure members are meaningful and must be used to specify the corresponding attributes.

The fourth parameter, dwStyleCount, determines the length of the fifth parameter. The fifth parameter, lpStyle, is a pointer to an array of doubleword values. The first value in the array is the length of the first dash of a user-defined pen style, the second one is the length of the first space, and so on. If the pen style does not contain the PS_USERSTYLE constant, then the fourth parameter must be zero, and the fifth parameter must be NULL. Note that PS_USERSTYLE is supported in Windows NT but not in Windows 95 or later versions.

The end cap styles determine the appearance of the line ends. Three constants are defined for round, square, and flat line ends. The end join style determines the appearance of the connecting point of two lines. Both styles are available only for geometric pens. Figure 23.3, on the following page, shows the pen styles and the effects of the different end caps and joins.

Note in Figure 23.3 that the difference between square and flat caps is that the square style extends the line by one-half its width. The white lines in the end cap style insert are drawn with the white stock pen, to better show the style's effect. The NULL_PEN style creates a pen that draws with transparent ink, therefore it leaves no mark as it moves on the drawing surface. This style is occasionally used in creating figures that are filled with a particular brush style but have no border.

23.3.2 Brushes

The brush object determines the attributes used in filling a solid figure. The outline of these figures is determined by the brush selected in the device context. A brush has a style, color, and hatch pattern. There are several stock brushes: WHITE_BRUSH, LTGRAY_BRUSH, GRAY_BRUSH, DKGRAY_BRUSH, BLACK_BRUSH, and NULL_BRUSH. All stock brushes are solid, that is, they fill the entire enclosed area of the figure. The NULL_BRUSH is used to draw figures without filling the interior. If a solid figure is drawn with the NULL_PEN, then it is filled but has no outline.

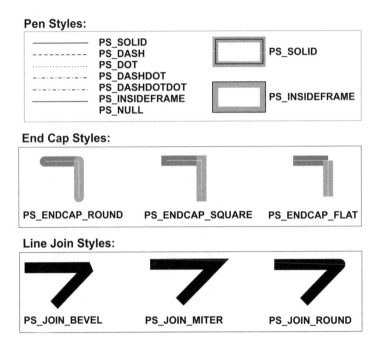

Figure 23.3 *Pen Syles, End Caps, and Joins*

Applications refer to a brush by its handle, which is stored in a variable of type HBRUSH. The GetStockObject() function is used to obtain a handle to one of the stock brushes. The brush must be selected into the device context before use, as follows:

```
HBRUSH     aBrush;       // handle to brush
   .
   .
   .
aBrush = GetStockObject (WHITE_BRUSH);
SelectObject (hdc, aBrush);
```

As in the case of a pen, the two functions can be combined in a single statement, as follows:

```
SelectObject (hdc, GetStockObject (WHITE_BRUSH));
```

In this case, no brush handle variable is required. SelectObject() returns the handle to the brush previously installed in the device context. This can be used to save the original brush so that it can later be restored.

A custom brush is created by means of the CreateBrushIndirect() function. The call returns a handle to the brush, of type HBRUSH. The only parameter is a pointer to a structure of type LOGBRUSH which holds the brush style, color, and hatch pattern. The LOGBRUSH structure is also used by the ExtCreatePen() previously described. Table 23.3 lists the predefined constants used for members of the LOGBRUSH structure.

The foreground mix mode attribute of the device context, also called the drawing mode, determines how Windows combines the pen or brush color with the display surface when performing drawing operations. The mixing is a raster operation based on a boolean function of two variables: the pen and the background. For this reason it is described as a binary raster operation, or ROP2. All four boolean primitives are used in setting the mix mode: AND, OR, NOT, and XOR. The function for setting the foreground mix mode is SetROP2(). GetROP2() returns the current mix mode in the device context. The general form of the SetROP2() function is as follows:

Table 23.3

Constants in the LOGBRUSH Structure Members

BRUSH STYLE	DESCRIPTION
BS_DIBPATTERN	A pattern brush defined by a device-independent bitmap. If lbStyle is BS_DIBPATTERN, the lbHatch member contains a handle to a packed DIB. Note: DIB stands for Device Independent Bitmap. DIBs were discussed previously.
BS_DIBPATTERNPT	Same as BS_DIBPATTERN but the lbHatch member contains a pointer to a packed DIB.
BS_HATCHED	Hatched brush.
BS_HOLLOW	Hollow brush.
BS_NULL	Same as BS_HOLLOW.
BS_PATTERN	Pattern brush defined by a memory bitmap.
BS_SOLID	Solid brush.

BRUSH COLOR	DESCRIPTION
DIB_PAL_COLORS	The color table consists of an array of 16-bit indices into the currently realized logical palette.
DIB_RGB_COLORS	The color table contains literal RGB values.

HATCH STYLE	DESCRIPTION
HS_BDIAGONAL	A 45-degree upward, left-to-right hatch.
HS_CROSS	Horizontal and vertical cross-hatch.
HS_DIAGCROSS	45-degree crosshatch.
HS_FDIAGONAL	A 45-degree downward, left-to-right hatch.
HS_HORIZONTAL	Horizontal hatch.
HS_VERTICAL	Vertical hatch.

```
int SetROP2(
        HDC hdc,          // 1
        int fnDrawMode    // 2
            );
```

23.3.3 Foreground Mix Mode

The first parameter is the handle to the device context and the second parameter is one of sixteen mix modes defined by Windows. The function returns the previous mix mode, which can be used to restore the original condition. Table 23.4 lists the ROP2 mix modes. The center column shows how the pen (P) and the screen (S) pixels are logically combined at draw time. The boolean operators correspond to the symbols used in C. Figure 23.4, on the following page, shows the brush hatch patterns.

Table 23.4

Mix Modes in SetROP2()

CONSTANT	BOOLEAN OPERATION	DESCRIPTION
R2_BLACK	0	Pixel is always 0.
R2_COPYPEN	P	Pixel is the pen color. This is the default mix mode.
R2_MASKNOTPEN	~P&S	Pixel is a combination of the colors common to both the screen and the inverse of the pen.
R2_MASKPEN	P&S	Pixel is a combination of the colors common to both the pen and the screen.
R2_MASKPENNOT	P&~S	Pixel is a combination of the colors common to both the pen and the inverse of the screen.
R2_MERGENOTPEN	~P|S	Pixel is a combination of the screen color and the inverse of the pen color.
R2_MERGEPEN	P|S	Pixel is a combination of the pen color and the screen color.
R2_MERGEPENNOT	P|~S	Pixel is a combination of the pen color and the inverse of the screen color.
R2_NOP	S	Pixel remains unchanged.
R2_NOT	~S	Pixel is the inverse of the screen color.
R2_NOTCOPYPEN	~P	Pixel is the inverse of the pen color.
R2_NOTMASKPEN	~(P&S)	Pixel is the inverse of the color.
R2_MASKPEN		Pixel is color of mask.
R2_NOTMERGEPEN	~(P|S)	Pixel is the inverse of the _MERGEPEN color.
R2_NOTXORPEN	~(P^S)	Pixel is the inverse of the R2_XORPEN color.
R2_WHITE	1	Pixel is always 1.
R2_XORPEN	P^S	Pixel is a combination of the colors in the pen and in the screen, but not in Both.

Legend:
    ~ = boolean NOT | = boolean OR
    & = boolean AND   ^ = boolean XOR

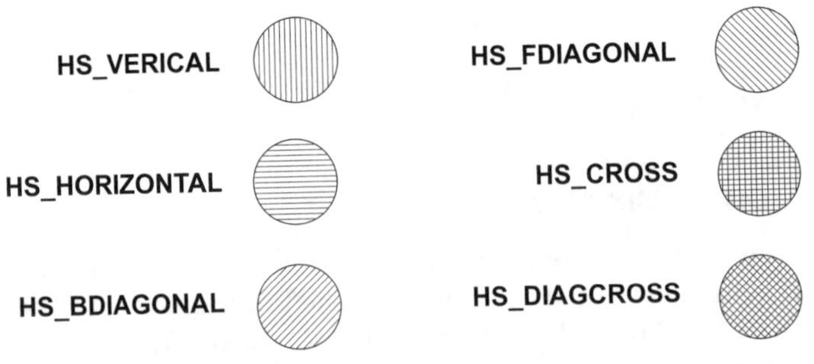

HS_VERICAL   **HS_FDIAGONAL**

HS_HORIZONTAL   **HS_CROSS**

HS_BDIAGONAL   **HS_DIAGCROSS**

Figure 23.4 *Brush Hatch Patterns*

23.3.4 Background Modes

Windows recognizes two background modes that determine how the gaps between dots and dashes are filled when drawing discontinuous lines, as well as with text and hatched brushes. The background modes, named OPAQUE and TRANSPARENT, are set in the device context by means of the SetBkMode() function. The function's general form is as follows:

```
int SetBkMode(
          HDC hdc,       // 1
          int iBkMode    // 2
               );
```

The first parameter is the handle to the device context, and the second one the constants OPAQUE or TRANSPARENT. If the opaque mode is selected, the background is filled with the current screen background color. If the mode is TRANSPARENT, then the background is left unchanged.

The background mode affects lines that result from a pen created with CreatePen() or CreatePenIndirect(), but not by those created with ExtCreatePen().

23.3.5 Current Pen Position

Many GDI drawing functions start at a screen location known as the current pen position, or the current position. The pen position is an attribute of the device context. The initial position of the pen is at logical coordinates (0, 0). Two functions relate directly to the current pen position: MoveToEx() and GetCurrent Position(). Some drawing functions change the pen position as they execute. The MoveToEx() function is used to set the current pen position. Its general form is as follows:

```
BOOL MoveToEx(
          HDC hdc,            // 1
          int X,              // 2
          int Y,              // 3
          LPPOINT lpPoint     // 4
               );
```

The first parameter is the handle to the device context. The second and third parameters are the x- and y-coordinates of the new pen position, in logical units. The fourth parameter is a pointer to a structure of type POINT that holds the x- and y-coordinates of the previous current pen position. If this parameter is set to NULL the old pen position is not returned. The function returns a boolean that is TRUE if the function succeeds and FALSE if it fails.

The GetCurrentPositionEx() function can be used to obtain the current pen position. Its general form is as follows:

```
BOOL MoveToEx(
          HDC hdc,            // 1
          int X,              // 2
          int Y,              // 3
          LPPOINT lpPoint     // 4
               );
```

The second parameter is a pointer to a structure variable of type POINT that receives the coordinates of the current pen position. The function returns TRUE if it succeeds and FALSE if it fails.

Drawing functions whose names contain the word "To" use and change the current pen position, these are: LineTo(), PolylineTo(), and PolyBezierTo(). Windows is not always consistent in this use of the word "To", since the functions AngleArc() and PolyDraw() also use and update the current pen position.

23.3.6 Arc Direction

One start-point and one end-point on the circumference of a circle define two different arcs: one drawn clockwise and one drawn counterclockwise. The exception is when the start and end points coincide. Figure 23.5 shows this possible ambiguity.

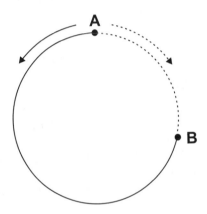

Figure 23.5 *The Arc Drawing Direction*

In Figure 23.5 the solid line arc is drawn counterclockwise from point A to point B, while the dotted line arc is drawn clockwise between these same points. The SetArcDirection() function is used to resolve this problem. The function's general form is as follows:

```
int SetArcDirection(
            HDC hdc,             // 1
            int ArcDirection     // 2
               );
```

The second parameter is either the constant AD_CLOCKWISE, or the constant AD_COUNTERCLOCKWISE. The function returns the previous arc drawing direction.

23.4 Pixels, Lines, and Curves

The lowest-level graphics primitives are to set a screen pixel to a particular attribute and to read the attributes of a screen pixel. In theory, with functions to set and read a pixel, all the other graphics operations can be developed in software. For example, a line can be drawn by setting a series of adjacent pixels, a closed figure can be filled by setting all the pixels within its boundaries, and so on. However, in actual programming practice these simple primitives are not sufficient. In the first place, high-level language code requires considerable overhead in performing the pixel set and read operations. To draw lines and figures by successively calling these functions would be prohibitively time-consuming. On the other hand, there are cases in which the programmer must resort to pixel-by-pixel drawing since other higher-level functions are not available.

There are eleven functions in the Windows API that can be used to draw lines. For one of them, StrokePath(), we postpone the discussion until later, since we must first discuss paths in greater detail. Table 23.5 lists the remaining ten line-drawing functions.

Table 23.5

Line-Drawing Functions

FUNCTION	DRAWING OPERATION
LineTo()	A straight line from current position up to a point. Pen position is updated to line's end point.
PolylineTo()	One or more straight lines between the current position and points in an array. Pen position is used for the first line and updated to end point of last line.
Polyline()	A series of straight line segments between points defined in an array.
PolyPolyLine()	Multiple polylines.
ArcTo()	An elliptical arc updating current pen position.
Arc()	An elliptical arc without updating current pen position.
AngleArc()	A segment of arc starting at current pen position.
PolyBezier()	One or more Bezier curves without updating the current pen position.
PolyBezierTo()	One or more Bezier curves updating the current pen position.
PolyDraw()	A set of lines and Bezier curves.
StrokePath()	See Chapter 21.

23.4.1 Pixel Operations

Two Windows functions operate on single pixels: SetPixel() and GetPixel().SetPixel() is used to set a pixel at any screen location to a particular color attribute. GetPixel() reads the color attribute of a pixel at a given screen location. The general form of SetPixel() is as follows:

```
COLORREF SetPixel(
        HDC hdc,            // 1
        int X,              // 2
        int Y,              // 3
        COLORREF crColor    // 4
            );
```

The first parameter is the handle to the device context. The second and third parameters are the x- and y-coordinates of the pixel to set, in logical units. The fourth parameter contains the pixel color in a COLORREF type structure. The function returns the RGB color to which the pixel was set, which may not coincide with the one requested in the call because of limitations of the video hardware. A faster version of this function is SetPixelV(). It takes the same parameters but returns a boolean value that is TRUE if the operation succeeded and FALSE if it failed. In most cases SetPixelV() is preferred over SetPixel() because of its better performance. The following code fragment shows how to draw a box of 100-by-100 pixels using the SetPixelV() function:

```
int x, y, i, j;          // control variables
COLORREF    pixColor;
      .
      .
      .
x = 120;           // start x
y = 120;           // start y
pixColor = RGB(0xff, 0x0, 0x0); // Red
// Draw a 100-by-100 pixel box
for (i = 0; i < 100; i++) {
for (j = 0; j < 100; j++) {
SetPixelV (hdc, x, y, pixColor);
x++;
            }
x = 120;
y++;
      }
```

23.4.2 Drawing with LineTo()

The simplest of all line-drawing functions is LineTo(). The function requires three parameters: the handle to the device context, and the coordinates of the end points of the line. The line is drawn with the currently selected pen. The start point is the current pen position, for this reason LineTo() is often preceded by MoveToEx() or another drawing function that sets the current pen position. LineTo() returns TRUE if the function succeeds and FALSE if it fails, but most often the return value is not used by code. If the LineTo() function succeeds, the current pen position is reset to the line's end point; therefore, the function can be used to draw a series of connected line segments.

The following code fragment draws a rectangle using four lines:

```
HPEN           bluePen4;     // handle for a pen
int            x, y, i, j;   // local variables
    .
    .
    .
// Create and select pen
bluePen4 = CreatePen (PS_SOLID, 4, RGB (0x00, 0x00, 0xff);
SelectObject (hdc, bluePen4);

// Set current pen position for start point
MoveToEx (hdc, 140, 140, NULL);
  LineTo (hdc, 300, 140);       // draw first segment
  LineTo (hdc, 300, 200);       // second segment
  LineTo (hdc, 140, 300);       // third segment
```

```
LineTo (hdc, 140, 140);        // last segment
```

23.4.3 Drawing with PolylineTo()

The PolylineTo() function draws one or more straight lines between points contained in an array of type POINT. The current pen position is used as a start point and is reset to the location of the last point in the array. PolylineTo() provides an easier way of drawing several connected line segments, or an unfilled closed figure. The function uses the current pen. Its general form is as follows:

```
BOOL PolylineTo(
        HDC hdc,               // 1
        CONST POINT *lppt,     // 2
        DWORD cCount           // 3
            );
```

The second parameter is the address of an array of points that contains the x- and y-coordinate pairs. The third parameter is the count of the number of points in the array. The function returns TRUE if it succeeds and FALSE otherwise. The following code fragment shows the drawing of a rectangle using the PolylineTo() function:

```
HPEN        redPen2;
POINT       pointsArray[4];  // array of four points
.
.
.
// Create a solid red pen, 2 pixels wide
redPen2 = CreatePen (PS_SOLID, 2, RGB(0xff, 0x00, 0x00));
SelectObject (hdc, redPen2);
// Fill array of points
pointsArray[0].x = 300; pointsArray[0].y = 160;
pointsArray[1].x = 300; pointsArray[1].y = 300;
pointsArray[2].x = 160; pointsArray[2].y = 300;
pointsArray[3].x = 160; pointsArray[3].y = 160;
// Set start point for first segment
MoveToEx (hdc, 160, 160, NULL);
// Draw polyline
PolylineTo (hdc, pointsArray, 4);
```

23.4.4 Drawing with Polyline()

The Polyline() function is similar to PolylineTo() except that it does not use or change the current pen position. Therefore, you need one more entry in the array of points to draw a figure with Polyline() since the initial position of the drawing pen cannot be used as the starting point for the first line segment. The following code fragment shows drawing a rectangle using the Polyline() function.

```
HPEN        blackPen;
POINT       pointsArray[4];  // array of four points
.
.
.
// Create a solid red pen, 2 pixels wide
blackPen = CreatePen (PS_DASH, 1, 0);
SelectObject (hdc, blackPen);
// Fill array of points
pointsArray[0].x = 160; pointsArray[0].y = 160;
```

```
pointsArray[1].x = 300; pointsArray[1].y = 160;
pointsArray[2].x = 300; pointsArray[2].y = 300;
pointsArray[3].x = 160; pointsArray[3].y = 300;
pointsArray[4].x = 160; pointsArray[4].y = 160;
// Draw polyline
Polyline (hdc, pointsArray, 5);
```

23.4.5 Drawing with PolyPolyline()

As the function name implies, PolyPolyline() is used to draw several groups of lines
or "polylines." Since the points array contains sets of points for more than one
polyline, the function requires an array of values that holds the number of points for
each polyline. PolyPolyline(), like Polyline(), does not use or change the current
pen position. The function's general form is as follows:

```
BOOL PolyPolyline(
        HDC hdc,                          // 1
        CONST POINT *lppt,                // 2
        CONST DWORD *lpdwPolyPoints,      // 3
        DWORD cCount                      // 4
            );
```

The second parameter is an array containing vertices of the various polylines.
The third parameter is an array that contains the number of vertices in each of the
polylines. The fourth parameter is the count of the number of elements in the
third parameter, which is the number of polylines to be drawn. The function returns TRUE if it succeeds and FALSE otherwise. The following code fragment
shows the drawing of two polylines, each with five vertices, using the
PolyPolyline() function.

```
 POINT        pointsArray[10];      // array of points
 DWORD        vertexArray[2];        // vertices per polyline
// Fill array of points for first polyline
pointsArray[0].x = 160; pointsArray[0].y = 160;
pointsArray[1].x = 300; pointsArray[1].y = 160;
pointsArray[2].x = 300; pointsArray[2].y = 300;
pointsArray[3].x = 160; pointsArray[3].y = 300;
pointsArray[4].x = 160; pointsArray[4].y = 160;

// Fill array of points for second polyline
pointsArray[5].x = 160; pointsArray[5].y = 230;
pointsArray[6].x = 230; pointsArray[6].y = 160;
pointsArray[7].x = 300; pointsArray[7].y = 230;
pointsArray[8].x = 230; pointsArray[8].y = 300;
pointsArray[9].x = 160; pointsArray[9].y = 230;

// Fill number of vertices in array
vertexArray[0] = 5;
vertexArray[1] = 5;
// Draw two polylines
PolyPolyline (hdc, pointsArray, vertexArray, 2);
```

Figure 23.6 shows the figures that result from executing the previous code sample. The second polyline is shown in dashed lines to visually distinguish it from
the first one. However, in an actual drawing there is no way of changing pens inside a call to PolyPolyline().

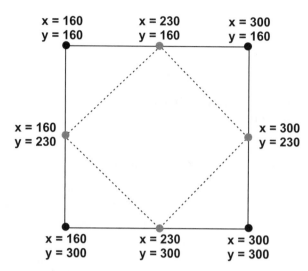

Figure 23.6 *Coordinates of Two Polylines in the Sample Code*

23.4.6 Drawing with Arc()

The Arc() function draws an elliptical arc. It is also used to draw circles, since the circle is a special case of the ellipse. The function's general form is as follows:

```
BOOL Arc(
          HDC hdc,           // 1
          int nLeftRect,     // 2
          int nTopRect,      // 3
          int nRightRect,    // 4
          int nBottomRect,   // 5
          int nXStartArc,    // 6
          int nYStartArc,    // 7
          int nXEndArc,      // 8
          int nYEndArc       // 9
              );
```

The second and third parameters are the x- and y-coordinates of the upper-left corner of a rectangle that contains the ellipse, while the fourth and fifth parameters are the coordinates of its lower-right corner. By using a bounding rectangle to define the ellipse, the Windows API avoids dealing with elliptical semi-axes. However, whenever necessary, the bounding rectangle can be calculated from the semi-axes. The sixth and seventh parameters define the coordinates of a point that sets the start point of the elliptical arc. The last two parameters set the end points of the elliptical arc. The elliptical arc is always drawn in the counterclockwise direction. The SetArcDirection() function has no effect in this case.

The coordinates of the start and end points of the elliptical arc need not coincide with the arc itself. Windows draws an imaginary line from the center of the ellipse to the start and end points. The point at which this line (or its prolongation) intersects the elliptical arc is used as the start or end point. If the start and end points are the same, then a complete ellipse is drawn. The following code fragment draws an elliptical arc:

```
Arc (hdc,
        150, 150,             // upper-left of rectangle
        350, 250,             // lower-right
        250, 260,             // start point
        200, 140;             // end point
```

Figure 23.7 shows the location of each of the points in the preceding call to the Arc() function and the resulting ellipse.

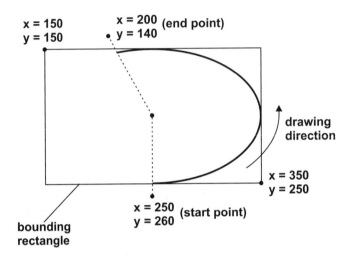

Figure 23.7 *Coordinates of an Elliptical Arc in Sample Code*

23.4.7 Drawing with ArcTo()

ArcTo() is a version of the Arc() function that updates the current pen position to the end point of the elliptical arc. This function requires Windows NT Version 3.1 or later. It is not available in Windows 95 or later. The function parameters are identical to those of the Arc() function.

23.4.8 Drawing with AngleArc()

The AngleArc() function draws a straight line segment and an arc of a circle. The straight line segment is from the current pen position to the arc's starting point. The arc is defined by the circle's radius and two angles: the starting position, in degrees, relative to the x-axis, and the angle sweep, also in degrees, relative to the starting position. The arc is drawn in a counterclockwise direction. The function's general form is as follows:

```
BOOL AngleArc(
        HDC hdc,              // 1
        int X,                // 2
        int Y,                // 3
        DWORD dwRadius,       // 4
        FLOAT eStartAngle,    // 5
        FLOAT eSweepAngle     // 6
              );
```

The second and third parameters are the coordinates of the center of the circle that defines the arc, in logical units. The fourth parameter is the radius of the circle, also in logical units. The fifth parameter is the start angle in degrees, relative to the x-axis. The last parameter is the sweep angle, also in degrees, relative to the angle's starting position. Figure 23.8 shows the various elements in the AngleArc() function.

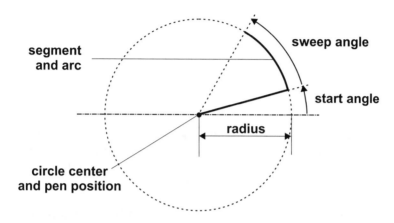

Figure 23.8 *AngleArc() Function Elements*

The AngleArc() function is not available in Windows 95 or later; however, it can be emulated in code. Microsoft Developers Network contains the following listing which allows implementing the AngleArc() function in software:

```
BOOL AngleArc2(HDC hdc, int X, int Y, DWORD dwRadius,
float fStartDegrees, float fSweepDegrees) {
int iXStart, iYStart;   // End point of starting radial line
  int iXEnd, iYEnd;       // End point of ending radial line
  float fStartRadians;    // Start angle in radians
  float fEndRadians;      // End angle in radians
  BOOL bResult;           // Function result
float fTwoPi = 2.0f * 3.141592f;
/* Get the starting and ending angle in radians */
if (fSweepDegrees > 0.0f) {
fStartRadians = ((fStartDegrees / 360.0f) * fTwoPi);
fEndRadians = (((fStartDegrees + fSweepDegrees) / 360.0f) *
fTwoPi);
} else {
fStartRadians = (((fStartDegrees + fSweepDegrees) / 360.0f) *
fTwoPi);
fEndRadians =  ((fStartDegrees / 360.0f) * fTwoPi);
  }

/* Calculate a point on the starting radial line via */
/* polar -> cartesian conversion */
iXStart = X + (int)((float)dwRadius * (float)cos(fStartRadians));
iYStart = Y - (int)((float)dwRadius * (float)sin(fStartRadians));

/* Calculate a point on the ending radial line via */
/* polar -> cartesian conversion */
iXEnd = X + (int)((float)dwRadius * (float)cos(fEndRadians));
iYEnd = Y - (int)((float)dwRadius * (float)sin(fEndRadians));
```

```
/* Draw a line to the starting point */
LineTo(hdc, iXStart, iYStart);

/* Draw the arc */
bResult = Arc(hdc, X - dwRadius, Y - dwRadius,
X + dwRadius, Y + dwRadius,
iXStart, iYStart,
iXEnd, iYEnd);

// Move to the ending point - Arc() wont do this and ArcTo()
// wont work on Win32s or Win16 */
MoveToEx(hdc, iXEnd, iYEnd, NULL);

return bResult;
}
```

Notice that the one documented difference between the preceding listing of AngleArc2(), and the GDI AngleArc() function, is that if the value entered in the sixth parameter exceeds 360 degrees, the software version will not sweep the angle multiple times. In most cases this is not a problem.

The program named PXL_DEMO, in the Pixel and Line Demo project folder on the book's on-line software package, uses the AngleArc2() function to display a curve similar to the one in Figure 23.8.

23.4.9 Drawing with PolyBezier()

In mechanical drafting, a spline is a flexible edge that is used to connect several points on an irregular curve. Two French engineers, Pierre Bezier and Paul de Casteljau, almost simultaneously discovered a mathematical expression for a spline curve that can be easily adapted to computer representations. This curve is known as the Bezier spline or curve, since it was Bezier who first published his findings. The Bezier curve is defined by its end points, called the nodes, and by one or more control points. The control points serve as magnets or attractors that "pull" the curve in their direction, but never enough for the curve to intersect the control point. Figure 23.9 shows the elements of a simple Bezier curve.

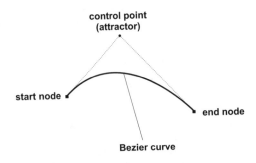

Figure 23.9 *The Bezier Spline*

The Bezier curve in Figure 23.9 can be generated by a geometrical method that consists of creating a series of progressively smaller line segments. The process, sometimes called the divide and conquer method, starts by joining the half-way points between the nodes and the attractor, thus creating a new set of nodes and a new attractor. The process continues until a sufficiently accurate approximation of the spline is reached. Figure 23.10 shows the progressive steps in creating a Bezier spline by this method.

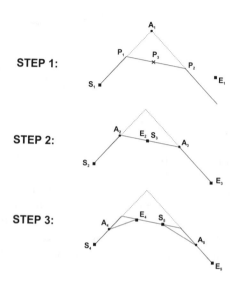

Figure 23.10 *Divide-and-Conquer Method of Creating a Bezier Curve*

In Step 1of Figure 23.10, we see the start node S1, the end node E1, and the attractor A1. We first find a point midway between S1 and A1 and label it P1. Another point midway between A1 and E1 is labeled P2. Points P1 and P2 are joined by a line segment, whose midpoint is labeled P3. In Step 2 we can see two new figures. The first one has nodes at S2 and E2, and the attractor at A2. The second figure has nodes at S3 and E3, and the attractor at A3. In Step 3 we have joined the midpoints between the nodes and the attractors with a line segment, thus continuing the process. The two new figures have their new respective nodes and attractors, so the process can be again repeated. In Step 3 we can see how the resulting line segments begin to approximate the Bezier curve in Figure 23.9.

The divide and conquer process makes evident the fundamental assumption of the Bezier spline: the curve is in the same direction and tangent to straight lines from the nodes to the attractors. A second assumption is that the curve never intersects the attractors. The Bezier formulas are based on these assumptions.

The Bezier curve generated by the divide and conquer method is known as a quadratic Bezier. In computer graphics the most useful Bezier is the cubic form. In the cubic form the Bezier curve is defined by two nodes and two attractors. The development of the cubic Bezier is almost identical to that of the quadratic. Figure 23.11, on the following page, shows the elements of a cubic Bezier curve.

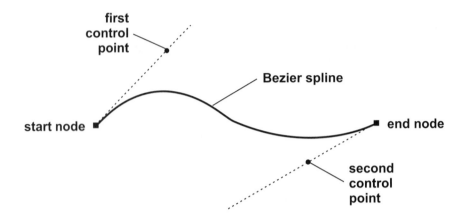

Figure 23.11 *Elements of the Cubic Bezier*

The PolyBezier() function, introduced in Windows 95, draws one or more cubic Bezier curves, each one defined by its nodes and two attractors. The function can be called to draw multiple Bezier curves. In this case the first curve requires four parameters, and all the other curves require three parameters. This is because the end node of the preceding Bezier curve serves as the start node for the next one. PolyBezier() does not change the current pen position. The Bezier curve is drawn using the pen selected in the device context. The function's general form is as follows:

```
BOOL PolyBezier(
        HDC hdc,                // 1
        CONST POINT *lppt,      // 2
        DWORD cPoints           // 3
            );
```

The first parameter is the handle to the device context. The second parameter is the address of an array of points that contains the x- and y-coordinate pairs for the nodes and control points. The third parameter is the count of the number of points in the array. This value must be one more than three times the number of curves to be drawn. For example, if the PolyBezier() function is called to draw four curves, there must be 13 coordinate pairs in the array $(1 + (3 * 4))$. The function returns TRUE if it succeeds and FALSE otherwise.

The Bezier data is stored in the array of points in a specific order. In the first Bezier curve, the first and fourth entries are the nodes and the second and third are attractors. Note that in the array the first and fourth entries are at offset 0 and 3 respectively, and the second and third entries are at offset 1 and 2. If there are other Bezier curves in the array, the first node is not explicit in the data, since it coincides with the end node of the preceding curve. Therefore, after the first curve, the following two entries are attractors, and the third entry is the end node. Table 23.6 shows the sequence of nodes and control points for an array with multiple Bezier curves.

Table 23.6

Nodes and Control Points for the PolyBezier() Function

NUMBER	OFFSET	TYPE
1	0	Start node of curve 1
2	1	First attractor of curve 1
3	2	Second attractor of curve 1
4	3	End node of curve 1
5	4	First attractor of curve 2
6	5	Second attractor of curve 2
7	6	End node of curve 2
8	7	First attractor of curve 3
9	8	Second attractor of curve 3
10	9	End node of curve 3

The following code fragment shows the drawing of a Bezier curve using the PolyBezier() function:

```
POINTS        pointsArray[4];  // Array of x/y coordinates
   ...
// Fill array of points for Bezier spline
// Entries 0 and 3 are nodes
// Entries 1 and 2 are attractors
pointsArray[0].x = 150; pointsArray[0].y = 150;
pointsArray[1].x = 200; pointsArray[1].y = 75;
pointsArray[2].x = 280; pointsArray[2].y = 190;
pointsArray[3].x = 350; pointsArray[3].y = 150;
// Draw a Bezier spline
PolyBezier (hdc, pointsArray, 4);
```

The resulting Bezier curve is similar to the one in Figure 23.9.

23.4.10 Drawing with PolyBezierTo()

The PolyBezierTo() function is very similar to PolyBezier() except that the start node for the first curve is the current pen position, and the current pen position is updated to the end node of the last curve. The return value and parameters are the same for both functions. In the case of PolyBezierTo() each curve is defined by three points: two control points and the end node. Table 23.7, on the following page, shows the sequence of points stored in the points array for PolyBezierTo().

23.4.11 Drawing with PolyDraw()

PolyDraw() is the most complex of the Windows line-drawing functions. It creates the possibility of drawing a series of line segments and Bezier curves, which can be joint or disjoint. PolyDraw() can be used in place of several calls to MoveTo(), LineTo(), and PolyBezierTo() functions. All the figures are drawn with the pen currently selected in the device context. The function's general form is as follows:

```
BOOL PolyDraw(
        HDC hdc,               // 1
        CONST POINT *lppt,     // 2
        CONST BYTE *lpbTypes,  // 3
        int cCount             // 4
            );
```

Table 23.7

Nodes and Control Points for the PolyBezierTo() Function

NUMBER	OFFSET	TYPE
1	0	First attractor of curve 1
2	1	Second attractor of curve 1
3	2	End node of curve 1
4	3	First attractor of curve 2
5	4	Second attractor of curve 2
6	5	End node of curve 2
7	6	First attractor of curve 3
8	7	Second attractor of curve 3
9	7	End node of curve 3

The second parameter is the address of an array of points that contains x- and y-coordinate pairs. The third parameter is an array of type BYTE that contains identifiers that define the purpose of each of the points in the array. The fourth parameter is the count of the number of points in the array of points. The function returns TRUE if it succeeds and FALSE otherwise. Table 23.8 lists the constants used to represent the identifiers entered in the function's third parameter.

Table 23.8

Constants for PolyDraw() Point Specifiers

TYPE	MEANING
PT_MOVETO	This point starts a disjoint figure. The point becomes the new current pen position.
PT_LINETO	A line is to be drawn from the current position to this point, which then becomes the new current pen position.
PT_BEZIERTO	This is a control point or end node for a Bezier curve. This constant always occurs in sets of three. The current position defines the start node for the Bezier curve. The other two coordinates are control points. The third entry is the end node.
PT_CLOSEFIGURE	The figure is automatically closed after the PT_LINETO or PT_BEZIERTO type for this point is executed. A line is drawn from the end point to the most recent PT_MOVETO or MoveTo() point. The PT_CLOSEFIGURE constant is combined by means of a bitwise OR operator with a PT_LINETO or PT_BEZIERTO constant. This indicates that the corresponding point is the last one in a figure and that the figure is to be closed.

The PolyDraw() function is not available in Windows 95 or later. Microsoft has published the following code for implementing the function in software:

```
//****************************
// Win95 version of PolyDraw()
// as published by Microsoft)
//****************************
BOOL PolyDraw95(HDC   hdc,            // handle of a device context
                CONST LPPOINT lppt,   // array of points
```

```
                CONST LPBYTE lpbTypes,     // line and curve identifiers
                int   cCount)             // count of points
{
int i;
for (i=0; i<cCount; i++)
switch (lpbTypes[i]) {
case PT_MOVETO :
MoveToEx(hdc, lppt[i].x, lppt[i].y, NULL);
break;

case PT_LINETO | PT_CLOSEFIGURE:
case PT_LINETO :
LineTo(hdc, lppt[i].x, lppt[i].y);
break;

case PT_BEZIERTO | PT_CLOSEFIGURE:
case PT_BEZIERTO :
PolyBezierTo(hdc, &lppt[i], 3);
i+=2;
break;
    }

return TRUE;
}
```

Notice that in the function PolyDraw95() the processing for closed and open figures takes place in the same intercepts. Therefore, there is no closing action implemented. When using this software implementation, including the PT_CLOSEFIGURE constant has no effect on the drawing. We have coded the following modification, named PolyDraw95A(), which closes open figures:

```
//*****************************
// Win95 version of PolyDraw()
//         improved!
//*****************************
BOOL PolyDraw95A (HDC  hdc,           // handle to device context
CONST LPPOINT lppt,  // array of points
CONST LPBYTE lpbTypes, // array of identifiers
int  cCount)          // count of points
{
int i;
  static long lastPenx, lastPeny;   // Storage for last pen position
  POINT       currentPoints[1];

// Store initial position of drawing pen
GetCurrentPositionEx (hdc, currentPoints);
lastPenx = currentPoints[0].x;
lastPeny = currentPoints[0].y;

for (i=0; i<cCount; i++)
switch (lpbTypes[i]) {
case PT_MOVETO :
MoveToEx(hdc, lppt[i].x, lppt[i].y, NULL);
// Store position for closed figures
lastPenx = lppt[i].x;
lastPeny = lppt[i].y;
break;

case PT_LINETO | PT_CLOSEFIGURE:
```

```
LineTo(hdc, lppt[i].x, lppt[i].y);
LineTo(hdc, lastPenx, lastPeny);
break;

case PT_LINETO :
LineTo(hdc, lppt[i].x, lppt[i].y);
break;

case PT_BEZIERTO | PT_CLOSEFIGURE:
// Store start points of Bezier for closing
GetCurrentPositionEx (hdc, currentPoints);
lastPenx = currentPoints[0].x;
lastPeny = currentPoints[0].y;
// Draw curve
PolyBezierTo(hdc, &lppt[i], 3);
i+=2;
// Close with line
LineTo(hdc, lastPenx, lastPeny);
break;

case PT_BEZIERTO :
// Draw Bezier
PolyBezierTo(hdc, &lppt[i], 3);
i+=2;
break;
    }

return TRUE;
}
```

The following code fragment displays several open and close figures using the PolyDraw() function or its software version Polydraw95A():

```
POINT        pointsArray[16];      // array of points
BYTE         controlArray[16];
 .
 .
 .
// In this example, pen is moved to start position externally
MoveToEx (hdc, 150, 50, NULL);

// Filling array of points for three lines
  // offset:        purpose:
  //    0           end point of line 1
  //    1           start of line 2
  //    2           end of line 2
  //    3           start of line 3
  //    4           end of line 3
pointsArray[0].x = 250; pointsArray[0].y = 50;
pointsArray[1].x = 150; pointsArray[1].y = 70;
pointsArray[2].x = 250; pointsArray[2].y = 70;
pointsArray[3].x = 150; pointsArray[3].y = 90;
pointsArray[4].x = 250; pointsArray[4].y = 90;

// Move to start node of Bezier curve
pointsArray[5].x = 150; pointsArray[5].y = 150;

// Filling array of points for first Bezier spline
pointsArray[6].x = 200; pointsArray[6].y = 75;
pointsArray[7].x = 280; pointsArray[7].y = 190;
```

```
pointsArray[8].x = 350; pointsArray[8].y = 150;

// Filling array for closed figure
pointsArray[9].x = 200; pointsArray[9].y = 200;
pointsArray[10].x = 300; pointsArray[10].y = 200;
pointsArray[11].x = 300; pointsArray[11].y = 300;
pointsArray[12].x = 200; pointsArray[12].y = 300;

// Filling array for second Bezier spline
pointsArray[13].x = 300; pointsArray[13].y = 90;
pointsArray[14].x = 350; pointsArray[14].y = 40;
pointsArray[15].x = 350; pointsArray[15].y = 40;
pointsArray[16].x = 400; pointsArray[16].y = 90;

// Filling control array
controlArray[0]  = PT_LINETO;
controlArray[1]  = PT_MOVETO;
controlArray[2]  = PT_LINETO;
controlArray[3]  = PT_MOVETO;
controlArray[4]  = PT_LINETO;
controlArray[5]  = PT_MOVETO;
controlArray[6]  = PT_BEZIERTO;
controlArray[7]  = PT_BEZIERTO;
controlArray[8]  = PT_BEZIERTO;
controlArray[9]  = PT_MOVETO;
controlArray[10] = PT_LINETO;
controlArray[11] = PT_LINETO;
controlArray[12] = PT_LINETO | PT_CLOSEFIGURE;
controlArray[13] = PT_MOVETO;
controlArray[14] = PT_BEZIERTO | PT_CLOSEFIGURE;
controlArray[15] = PT_BEZIERTO;
controlArray[16] = PT_BEZIERTO;
// Drawing lines and Bezier curves
PolyDraw95A (hdc, pointsArray, controlArray, 17);
```

Figure 23.12 is an approximation of the figures that result from the previous code sample.

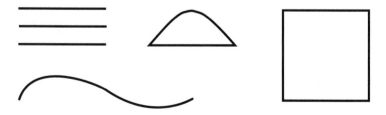

Figure 23.12 *Approximate Result of the PolyDraw() Code Sample*

23.4.12 Pixel and Line Demonstration Program

The program named PXL_DEMO, located in the Pixel and Line Demo project folder of the book's software on-line, is a demonstration of the drawing functions discussed in this chapter. Pixel-level functions are used to display the point plot of a sine curve. Also, the program contains a function named DrawDot(), which uses the SetPixelV() function to draw a black screen dot by setting five adjacent pixels. The demo program displays a pop-up menu, named Line Functions, which has menu commands for exercising LineTo(), PolyLineTo(), PolyLine(), PolyPolyLine(), Arc(), AngleArc(), PolyBezier(), and PolyDraw(). Code for implementing PolyDraw() and AngleArc() in software is also included in the demo program.

Chapter 24

Drawing Solid Figures

Chapter Summary

In this chapter we continue exploring the graphics functions in the Windows GDI, concentrating on geometrical elements that contain an interior region, in addition to a perimeter or outline. These are called solid or closed figures. The interior area allows them to be filled with a given color, hatch pattern, or bitmap image. At the same time, the perimeter of a closed figure can be rendered differently than the filled area. For example, the circumference of a circle can be outlined with a 2-pixel wide black pen, and the circle's interior filled with 1-pixel wide red lines, slanted at 45 degrees, and separated from each other by 10 pixels.

24.0 Closed Figures

Closed figures allow several graphics manipulations. For instance, a solid figure can be used to define the output area to which Windows can perform drawing operations. This area, called the clipping region, allows you to produce unique and interesting graphics effects, such as filling geometrical figures with text or pictures.

Some closed figures are geometrically simple: a rectangle, an ellipse, or a symmetrical polygon. More complex figures are created by combining simpler ones. A region is an area composed of one or more rectangles, polygons, or ellipses. Regions are used to define irregular areas that can be filled, to clip output, or to establish an area for mouse input.

Paths are relatively new graphics objects, since they were introduced with Windows NT, and are also supported in Windows 95 and later. A path is the route the drawing instrument follows in creating a figure, or set of figures. It is used to define the outline of a graphics object. After a path is created, you can draw its outline (called stroking the path), fill its interior, or both. A path can also be used in clipping, or converted into a region. Paths and regions add a powerful dimension to Windows graphics.

24.1 Closed Figure Elements

A closed figure has both a perimeter and an interior. The perimeter of a closed figure is drawn using the current pen and the GDI line-related attributes discussed in Chapter 20. The interior is filled using the current brush, also partly discussed in Chapter 20. There are several closed figures that can be drawn with the Windows GDI; among them are ellipses, polygons, chords, pies, and rectangles. Later in this chapter we see that the Windows names for some of these figures are not geometrically correct. Areas bound by complex lines, such as irregular polygons, Bezier curves, and text characters, can also be filled.

Like lines and curves, closed figures have attributes that determine their characteristics. Most of the attributes that relate to closed figures are described in Chapter 23. These include the mix mode, the background mode, the arc direction, the brush pattern, the pen styles, as well as the brush, pen, and background colors. Two attributes that are specific to closed figures are the brush origin and the polygon filling mode.

24.1.1 Brush Origin

Figure 23.4, in the preceding chapter, shows the various hatch patterns that can be used with a brush. Windows locates the hatch pattern in reference to coordinates (0,0). It is important to know that this origin is in device units, not in logical units. The hatch pattern is a bitmap. In Windows 95 and later, the bitmap is 8-by-8 pixels. In Windows NT, it can have any size. The painting process consists of repeating the bitmap horizontally and vertically until the area is filled.

In some cases the default origin of the bitmap produces undesirable results. This usually happens when the alignment of a filled figure does not coincide with that of the brush hatch pattern. Figure 24.1 shows two rectangles, one filled with an unaligned hatch pattern and the other one filled with an aligned hatch pattern.

The SetBrushOrgEx() function can be used to reposition the hatch bitmap in relation to the origin of the client area. The function's general form is as follows:

```
BOOL SetBrushOrgEx(
        HDC hdc,         // 1
        int nXOrg,       // 2
        int nYOrg,       // 3
        LPPOINT lppt     // 4
            );
```

The second parameter specifies the x-coordinate of the new brush origin. In Windows 95 and later, the range is 0 to 7. In Windows NT, the range cannot be greater than the width of the bitmap. In either case, if the entered value exceeds the width of the bitmap, it is adjusted by performing the modulus operation:

```
xOrg = xOrg % bitmap width
```

The third parameter is the y-coordinate of the new brush origin. Its range and adjustments are the same as for the second parameter. The fourth parameter is a pointer to a POINT structure that stores the origin of the brush previously selected in the device context. If this information is not required, NULL can be entered in this parameter. The function returns TRUE if the operation succeeds and FALSE otherwise.

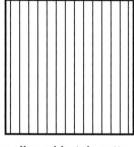 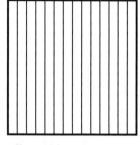

unaligned hatch pattern   **aligned hatch pattern**

Figure 24.1 *Brush Hatch Patterns*

A call to SetBrushOrgEx() sets the origin of the next brush that an application selects into the device context. Note that the first parameter of the SetBrushOrgEx() function is the handle to the device context, and that the brush variable is nowhere to be found in the parameter list. Therefore, the brush origin is associated with the device context, not with a particular brush. The origin in the device context is assigned to the next brush created.

The following code fragment shows the display of two rectangles. The brush origin is changed for the second one. The Rectangle() function is described later in this chapter.

```
static HBRUSH      vertBrush1, vertBrush2;
LOGBRUSH           brush1;
.
.
.
// Create a brush
brush1.lbStyle = BS_HATCHED;
brush1.lbColor = RGB(0x0, 0xff, 0x0);
brush1.lbHatch = HS_VERTICAL;
vertBrush1 = CreateBrushIndirect (&brush1);
SelectObject (hdc, (HGDIOBJ)(HBRUSH) vertBrush1);
// Draw a rectangle with this brush
Rectangle (hdc, 150, 150, 302, 300);

// Create a new hatched brush with offset origin
brush1.lbStyle = BS_HATCHED;
brush1.lbColor = RGB(0x0, 0x0, 0x0);
brush1.lbHatch = HS_VERTICAL;
// Offset the new brush 6 pixels
SetBrushOrgEx (hdc, 5, 0, NULL);
vertBrush2 = CreateBrushIndirect (&brush1);
SelectObject (hdc, (HGDIOBJ)(HBRUSH) vertBrush2);
// Draw a rectangle with the new brush
Rectangle (hdc, 350, 150, 502, 300);
```

The results of executing this code are similar to the rectangles in Figure 24.1. The GetBrushOrg() function can be used to retrieve the origin of the current brush.

Notice that Windows documentation recommends that to avoid brush misalignment an application should call SetStretchBltMode() with the stretching mode set to HALFTONE before calling SetBrushOrgEx().

24.1.2 Object Selection Macros

The Windows header file windowsx.h contains four macros that can be used in selecting a pen, brush, font, or bitmap. The advantage of using these macros is that the objects are automatically typecast correctly. The macros are named SelectPen(), SelectBrush(), SelectFont() and SelectBitmap(). They are all defined similarly. The SelectBrush() macro is as follows:

```
#define SelectBrush (hdc, hbr) \
        ((HBRUSH) SelectObject ((hdc, (HGDIOBJ)(HBRUSH)(hbr)))
```

You can use these macros to easily produce code that is correct and more portable. Programs that use the object selection macros must contain the statement:

```
#include <windowsx.h>
```

24.1.3 Polygon Fill Mode

The polygon fill mode attribute determines how overlapping areas of complex polygons and regions are filled. The polygon fill mode is set with the SetPolyFillMode() function. The function's general form is as follows:

```
int SetPolyFillMode(
        HDC hdc,               // 1
        int iPolyFillMode      // 2
            );
```

The second parameter is one of two constants: ALTERNATE and WINDING. ALTERNATE defines a mode that fills between odd-numbered and even-numbered polygon sides, or in other words, those areas that can be reached from the outside of the polygon by crossing an odd number of lines, excluding the vertices. This fill algorithm is based on what is called the parity rule.

The WINDING mode is based on the nonzero winding rule. In the WINDING mode, the direction in which the figure is drawn determines whether an area is to be filled. A polygon line segment is drawn either in a clockwise or a counterclockwise direction. The term winding relates to the clockwise and counterclockwise drawing of the polygon segments. An imaginary line, called a ray, is extended from an enclosed area in the figure, to a point distant from the figure and outside of it. The ray must be on a positive x-direction. Every time the ray crosses a clockwise winding, a counter is incremented. The same counter is decremented whenever the line crosses a counterclockwise winding. The winding counter is examined when the ray reaches the outside of the figure. If the winding counter is nonzero, the area is filled.

In figures that have a single interior region the fill mode is not important. This is not the case in figures that have enclosed areas. A typical case is a polygon in the shape of a five-pointed star with an enclosed pentagon. In this case, the ALTERNATE mode does not fill the interior pentagon, while the WINDING fill mode does. In more complex figures the same rules apply, although they may not be immediately evident. Figure 24.2 shows the results of the polygon fill modes in two different figures. Recall that in the WINDING fill mode, the direction in which the line segments are drawn is significant.

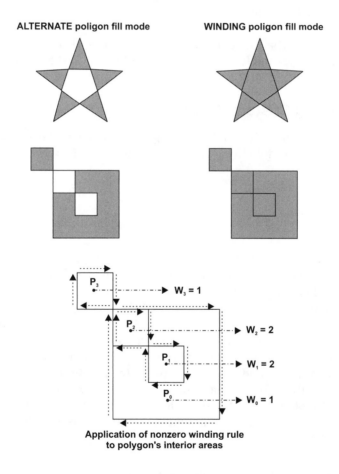

Figure 24.2 *Effects of the Polygon Fill Modes*

In Figure 24.2 you can see the application of the nonzero winding rule to the interior areas of a complex polygon. For example, the ray from point P1 to the exterior of the figure crosses two clockwise segments (windings). Therefore, it has a winding value of 2. Since the winding is nonzero, the area is filled. The same rule can be applied to other points in the figure's interior, as shown in Figure 24.2.

Notice that some Windows documentation states that in the WINDING mode all interior areas of a figure are filled. This oversimplification is not correct. If the interior segments of the polygon in Figure 24.2 were drawn in the opposite direction, some areas would have zero winding and would not be filled. The program named FIL_DEMO, located in the Filled Figure Demo project folder, in the book's on-line software package, contains the menu command Polygon (2), on the Draw Figures pop-up menu, which displays a complex polygon that has an unfilled interior in the WINDING mode.

You can retrieve the current polygon fill mode with the GetPolyFillMode() function. The only parameter to the call is the handle to the device context. The value returned is either ALTERNATE or WINDING.

24.1.4 Creating Custom Brushes

In Chapter 23 we mentioned that a logical brush can be created with the CreateBrushIndirect() function. CreateBrushIndirect() has the following general form:

```
HBRUSH CreateBrushIndirect (
CONST LOGBRUSH* lplb     // 1
);
```

The only parameter of the function is the address of a structure of type LOGBRUSH. The structure members are divided into three groups: brush style, brush color, and hatch style. The function returns the handle to the created brush. Table 24.1 lists the members of the LOGBRUSH structure.

Once the brush is created, it can be selected into the device context by either calling the SelectObject() function or the SelectBrush() macro discussed previously in this chapter. In addition, you can create specific types of brushes easier by using the functions CreateSolidBrush(), CreateHatchBrush(), or CreatePatternBrush(). CreateSolidBrush() is used to create a brush of a specific color and no hatch pattern. The function's general form is as follows:

```
HBRUSH CreateSolidBrush (COLORREF colorref);
```

The only parameter is a color value in the form of a COLORREF type structure. The color value can be entered directly using the RGB macro. For example, the following code fragment creates a solid blue brush:

```
static HBRUSH        solidBlueBrush;
.
.
.
solidBlueBrush = CreateSolidBrush ( RGB (0x0, 0x0, 0xff));
```

Table 24.1

LOGBRUSH Structure Members

BRUSH STYLE	DESCRIPTION
BS_DIBPATTERN	A pattern brush defined by a device-independent bitmap. If lbStyle is BS_DIBPATTERN, the lbHatch member contains a handle to a packed DIB.
BS_DIBPATTERNPT	Same a BS_DIBPATTERN but the lbHatch member contains a pointer to a packed DIB.
BS_HATCHED	Hatched brush.
BS_HOLLOW	Hollow brush.
BS_NULL	Same as BS_HOLLOW.
BS_PATTERN	Pattern brush defined by a memory bitmap.
BS_SOLID	Solid brush.

BRUSH COLOR	DESCRIPTION
DIB_PAL_COLORS	The color table consists of an array of 16-bit indices into the currently realized logical palette.
DIB_RGB_COLORS	The color table contains literal RGB values.

HATCH STYLE	DESCRIPTION
HS_BDIAGONAL	A 45-degree upward, left-to-right hatch.
HS_CROSS	Horizontal and vertical cross-hatch.
HS_DIAGCROSS	45-degree crosshatch.
HS_FDIAGONAL	A 45-degree downward, left-to-right hatch.
HS_HORIZONTAL	Horizontal hatch.
HS_VERTICAL	Vertical hatch.

CreatehatchBrush() creates a logical brush with a hatch pattern and color. The function's general form is as follows:

```
HBRUSH CreateHatchBrush(
        int fnStyle,       // 1
        COLORREF clrref    // 2
            );
```

The first parameter is one of the hatch style identifiers listed in Figure 24.2. The second parameter is a color value of COLORREF type. The function returns the handle to the logical brush.

If an application requires a brush with a hatch pattern different from the ones predefined in Windows, it can create a custom brush with its own bitmap. In Windows 95 and later, the size of the bitmap cannot exceed 8-by-8 pixels, but there is no size restriction in Windows NT. The function's general form is as follows:

```
HBRUSH CreatePatternBrush (
HBITMAP hbmp        // 1
);
```

The function's only parameter is a handle to the bitmap that defines the brush. The bitmap can be created with CreateBitmap(), CreateBitmapIndirect() or CreateCompatibleBitmap() functions. These functions are described in Chapter 25.

24.2 Drawing Closed Figures

There are seven Windows functions that draw closed figures, shown in Table 24.2.

Table 24.2
Windows Functions for Drawing Closed Figures

FUNCTION	FIGURE
Rectangle()	Rectangle with sharp corners
RoundRect()	Rectangle with rounded corners
Ellipse()	Ellipse or circle
Chord()	Solid figure created by an arc on the circumference of an ellipse connected by a chord
Pie()	Pie-shaped wedge created by joining the end points of an arc on the perimeter of an ellipse with the center of the arc
Polygon()	Closed polygon
PolyPolygon()	Series of closed polygons, possibly overlapping

All functions that draw closed figures use the pen currently selected in the device context for the figure outline, and the current brush for filling the interior. All of the line attributes discussed in Chapter 20 apply to the perimeter of solid figures. The programmer has control of the width of the perimeter, its line style, and its color. By selecting NULL_PEN you can draw a figure with no perimeter. The fill is determined by the current brush. Windows approximates the color of the brush according to the device capabilities.

This often requires manipulating dot sizes by a process called dithering. Dithering is a technique that creates the illusion of colors or shades of gray by treating the targeted areas as a dot pattern. The process takes advantage of the fact that the human eye tends to blur small spots of different color by averaging them into a single color or shade. For example, a pink color effect can be produced by mixing red and white dots.

The brush can be any one of the stock brushes: WHITE_BRUSH, LTGRAY_BRUSH, GRAY_BRUSH, DKGRAY_BRUSH, BLACK_BRUSH, and NULL_BRUSH. All stock brushes are solid. NULL_BRUSH is used to draw figures without filling the interior. The GetStockObject() function is used to obtain a handle to one of the stock brushes. Since stock brushes need not be stored locally, the most common case is that the stock brush is retrieved and installed in the device context at the time it is needed. SelectBrush() and GetStockObject() can be combined as follows:

```
SelectBrush (hdc, GetStockObject (WHITE_BRUSH));
```

The creation and installation of custom brushes was discussed previously in this chapter.

24.2.1 Drawing with Rectangle()

The simplest solid-figure drawing function is Rectangle(). This function draws a rectangle using the current pen for the outline and fills it with the current brush. The function's general form is as follows:

```
BOOL Rectangle(
        HDC hdc,            // 1
        int nLeftRect,      // 2
        int nTopRect,       // 3
        int nRightRect,     // 4
        int nBottomRect     // 5
        );
```

The second and third parameters are the coordinates of the upper-left corner of the rectangle. The fourth and fifth parameters are the coordinates of the lower-right corner. The function returns TRUE if it succeeds and FALSE if it fails. Figure 24.3 shows a rectangle drawn using this function.

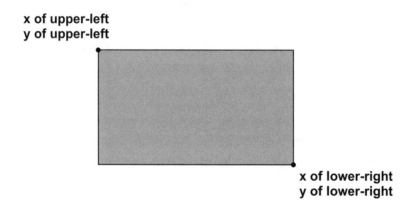

Figure 24.3 *Figure Definition in the Rectangle() Function*

24.2.2 Drawing with RoundRect()

The RoundRect() function draws a rectangle with rounded corners. Like all the solid figure drawing functions, it uses the current pen for the outline and fills the figure with the current brush. The function's parameter list is as follows:

```
BOOL RoundRect(
        HDC hdc,            // 1
        int nLeftRect,      // 2
        int nTopRect,       // 3
        int nRightRect,     // 4
        int nBottomRect,    // 5
        int nWidth,         // 6
        int nHeight         // 7
        );
```

The second and third parameters are the coordinates of the upper-left corner of the bounding rectangle. The fourth and fifth parameters are the coordinates of the lower-right corner. The sixth parameter is the width of the ellipse that is used for drawing the rounded corner arc. The seventh parameter is the height of this ellipse. The function returns TRUE if it succeeds and FALSE if it fails. Figure 24.4 shows the values that define a rounded-corner rectangle drawn using this function.

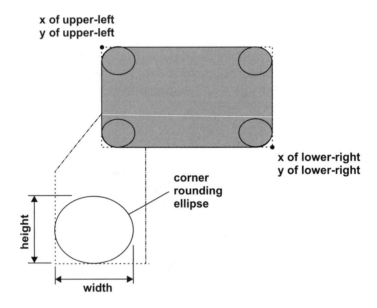

Figure 24.4 *Definition Parameters for the RoundRect() Function*

24.2.3 Drawing with Ellipse()

The Ellipse() function draws a solid ellipse. Ellipse() uses the current pen for the outline and fills the figure with the current brush. The function's general form is as follows:

```
BOOL Ellipse(
        HDC hdc,         // 1
        int nLeftRect,   // 2
        int nTopRect,    // 3
        int nRightRect,  // 4
        int nBottomRect  // 5
        );
```

The second and third parameters are the coordinates of the upper-left corner of a rectangle that binds the ellipse. The fourth and fifth parameters are the coordinates of the lower-right corner of this rectangle. The function returns TRUE if it succeeds and FALSE if it fails. Figure 24.5 shows an ellipse drawn using this function.

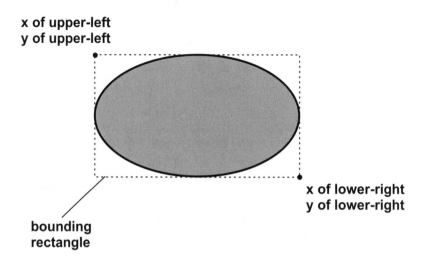

Figure 24.5 *Figure Definition in the Ellipse() Function*

24.2.4 Drawing with Chord()

Chord() draws a solid figure composed of an arc of an ellipse whose ends are connected to each other by a straight line, called a secant. The Chord() function is related to the Arc() function described in Chapter 20. The parameters that define the elliptical arc are the same for the Arc() as for the Chord() function. The function's general form is as follows:

```
BOOL Chord(
        HDC hdc,          // 1
        int nLeftRect,    // 2
        int nTopRect,     // 3
        int nRightRect,   // 4
        int nBottomRect,  // 5
        int nXRadial1,    // 6
        int nYRadial1,    // 7
        int nXRadial2,    // 8
        int nYRadial2     // 9
        );
```

The second and third parameters are the x- and y-coordinates of the upper-left corner of a rectangle that contains the ellipse, while the fourth and fifth parameters are the coordinates of its lower-right corner. The sixth and seventh parameters define the coordinates of a point that sets the start point of the secant. The last two parameters set the end points of the secant. The elliptical arc is always drawn in the counterclockwise direction. The SetArcDirection() function has no effect in this case.

The coordinates of the start and end points of the secant need not coincide with the elliptical arc, since Windows prolongs the secant until it intersects the elliptical arc. Figure 24.6, on the following page, shows the elements that define the figure.

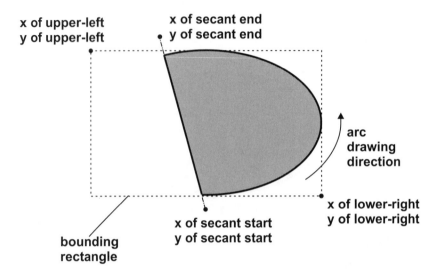

Figure 24.6 *Figure Definition in the Chord() Function*

Notice that the name of the Chord() function does not coincide with its mathematical connotation. Geometrically, a chord is the portion of a secant line that joins two points on a curve, not a solid figure.

24.2.5 Drawing with Pie()

Pie() draws a solid figure composed of the arc of an ellipse whose ends are connected to the center by straight lines. In Windows terminology the two straight lines are called radials. The Pie() function is related to the Arc() function described in Chapter 20. The parameters that define the elliptical arc are the same for the Arc() as for the Pie() functions. It is also similar to the Chord() function previously described. The difference between Chord() and Pie() is that in Chord() the line points are connected to each other and in Pie() they are connected to the center of the ellipse. The function's general form is as follows:

```
BOOL Pie(
        HDC hdc,          // 1
        int nLeftRect,    // 2
        int nTopRect,     // 3
        int nRightRect,   // 4
        int nBottomRect,  // 5
        int nXRadial1,    // 6
        int nYRadial1,    // 7
        int nXRadial2,    // 8
        int nYRadial2     // 9
        );
```

The second and third parameters are the x- and y-coordinates of the upper-left corner of a rectangle that contains the ellipse, while the fourth and fifth parameters are the coordinates of its lower-right corner. The sixth and seventh parameters define the coordinates of the end point of the start radial line. The last two parameters set the coordinates of the end points of the end radial line. The elliptical arc is always drawn in the counterclockwise direction. The SetArcDirection() function has no effect in this case.

The coordinates of the start and end points of the radials need not coincide with the elliptical arc, since Windows prolongs these lines until they intersect the elliptical arc. Figure 24.7 shows the elements that define the figure.

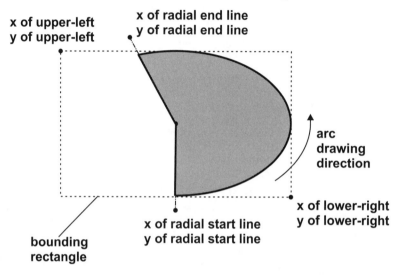

Figure 24.7 *Figure Definition in the Arc() Function*

24.2.6 Drawing with Polygon()

The Polygon() function is similar to the Polyline() function described in Chapter 20. The main difference between a polygon and a polyline is that the polygon is closed automatically by drawing a straight line from the last vertex to the first one. The polygon is drawn with the current pen and filled with the current brush. The inside of the polygon is filled according to the current polygon fill mode, which can be ALTERNATE or WINDING. Polygon fill modes were discussed in detail earlier in this chapter. The function's general form is as follows:

```
BOOL Polygon(
        HDC hdc,                // 1
        CONST POINT *lpPoints,  // 2
        int nCount              // 3
        );
```

The second parameter is the address of an array of points that contains the x- and y-coordinate pairs of the polygon vertices. The third parameter is the count of the number of vertices in the array. The function returns TRUE if it succeeds and FALSE otherwise.

When drawing the lines that define a polygon you can repeat the same segment. It is not necessary to avoid going over an existing line. When the WINDING fill mode is selected, however, the direction of each edge determines the fill action. The following code fragment shows the drawing of a complex polygon that is defined in an array of structures of type POINT.

```
// Arrays of POINT structures for polygon vertices
POINT  polyPoints1[] = {
       { 100, 100 }, // 1
       { 150, 100 }, // 2
       { 150, 150 }, // 3
       { 300, 150 }, // 4
       { 300, 300 }, // 5
       { 150, 300 }, // 6
       { 150, 150 }, // 7
       { 200, 150 }, // 8
       { 200, 200 }, // 9
       { 250, 200 }, // 10
       { 250, 250 }, // 11
       { 200, 250 }, // 12
       { 200, 200 }, // 13
       { 150, 200 }, // 14
       { 150, 150 }, // 15
       { 100, 150 }  // 16
} ;
     .   .   .
// Draw the polygon using array data
SetPolyFillMode (hdc, ALTERNATE);
Polygon (hdc, polyPoints1, 17);
```

Figure 24.8 shows the figure that results from this code when the ALTERNATE fill mode is active. The polygon vertices are numbered and labeled.

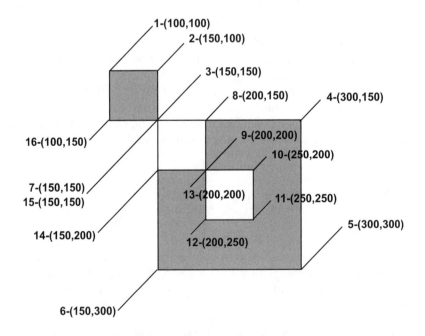

Figure 24.8 *Figure Produced by the Polygon Program*

24.2.7 Drawing with PolyPolygon()

As the function name implies, PolyPolygon() is used to draw several closed polygons. The outlines of all the polygons are drawn with the current pen and the interiors are filled with the current brush and according to the selected fill mode. The polygons can overlap. Unlike the Polygon() function, the figures drawn with PolyPolygon() are not automatically closed. The PolyPolygon() function is similar to the PolyPolyline() function described in Chapter 20. Like the PolyPolyline() function, PolyPolygon() requires an array of values that holds the number of points for each polygon. The function's general form is as follows:

```
BOOL PolyPolygon(
        HDC hdc,                    // 1
        CONST POINT *lpPoints,      // 2
        CONST INT *lpPolyCounts,    // 3
        int nCount                  // 4
        );
```

The second parameter is an array containing the vertices of the various polygons. The third parameter is an array that contains the number of vertices in each of the polygons. The fourth parameter is the count of the number of elements in the third parameter, which is also the number of polygons to be drawn. The function returns TRUE if it succeeds and FALSE otherwise. The following code fragment shows the drawing of four polygons, each one with four vertices, using the PolyPolygon() function.

```
// Arrays of POINT structures for holding the vertices
// of all four polygons
POINT   polyPoly1[] = {
  { 150, 150 }, // 1   |
  { 300, 150 }, // 2   |
  { 300, 300 }, // 3   |-- first polygon
  { 150, 300 }, // 4   |
  { 150, 150 }, // 5   |

  { 200, 200 }, // 6   |
  { 250, 200 }, // 7   |
  { 250, 250 }, // 8   |-- second polygon
  { 200, 250 }, // 9   |
  { 200, 200 }, // 10  |

  { 150, 150 }, // 11  |
  { 200, 150 }, // 12  |
  { 200, 200 }, // 13  |-- third polygon
  { 150, 200 }, // 14  |
  { 150, 150 }, // 15  |

  { 100, 100 }, // 11  |
  { 150, 100 }, // 12  |
  { 150, 150 }, // 13  |-- fourth polygon
  { 100, 150 }, // 14  |
  { 100, 100 }, // 15  |
  } ;

// Array holding the number of segments in each
// polygon
```

```
int  vertexArray[] = {
    { 5 },
    { 5 },
    { 5 },
    { 5 }
};
    .
    .
    .
// Draw the polygon using array data
SetPolyFillMode (hdc, ALTERNATE);
PolyPolygon (hdc, polyPoly1, vertexArray, 4);
```

The resulting polygon is identical to the one in Figure 24.8.

24.3 Operations on Rectangles

Rectangular areas are often used in Windows programming. Child windows are usually in the form of a rectangle, as are message and input boxes as well as many other graphics components. For this reason, the Windows API includes several functions that operate on rectangles. These are listed in Table 24.3.

Table 24.3

Windows Functions Related to Rectangular Areas

FUNCTION	FIGURE
FillRect()	Fills the interior of a rectangle using a brush defined by its handle
FrameRect()	Draws a frame around a rectangle
InvertRect()	Inverts the pixels in a rectangular area
DrawFocusRect()	Draws rectangle with special dotted pen to indicate that the object has focus

One common characteristic of all the rectangular functions is that the rectangle coordinates are stored in a structure of type RECT. The use of a RECT structure is a more convenient way of defining a rectangular area than by passing coordinates as function parameters. It allows the application to easily change the location of a rectangle, and to define a rectangular area without hard-coding the values, thus making the code more flexible. The RECT structure is as follows:

```
typdef struct _RECT {
    LONG    left;      // x coordinate of upper-left corner
    LONG    top;       // y of upper-left corner
    LONG    right;     // x coordinate of bottom-right corner
    LONG    bottom;    // y of bottom-right
} RECT;
```

24.3.1 Drawing with FillRect()

The FillRect() function fills the interior of a rectangular area, whose coordinates are defined in a RECT structure. The function uses a brush specified by its handle. The filled area includes the upper-left corner of the rectangle but excludes the bottom-right corner. The function's general form is as follows:

```
int FillRect(
        HDC hDC,              // 1
        CONST RECT *lprc,     // 2
        HBRUSH hbr            // 3
        );
```

The second parameter is a pointer to a structure of type RECT that contains the rectangle's coordinates. The third parameter is the handle to a brush or a system color. If a handle to a brush, it must have been obtained with CreateSolidBrush(), CreatePatternBrush(), or CreateHatchBrush() functions described previously. Additionally, you may use a stock brush and obtain its handle by means of GetStockObject(). The function returns TRUE if it succeeds and FALSE if it fails. Table 24.4 lists the constants that are used to identify the system colors in Windows.

Table 24.4

Windows System Colors

VALUE	MEANING
COLOR_3DDKSHADOW	Dark shadow display elements
COLOR_3DFACE,	
COLOR_BTNFACE	Face color for display elements
COLOR_3DHILIGHT,	
COLOR_3DHIGHLIGHT,	
COLOR_BTNHILIGHT,	
COLOR_BTNHIGHLIGHT	Highlight color for edges facing the light source
COLOR_3DLIGHT	Light color for edges facing the light source
COLOR_3DSHADOW,	
COLOR_BTNSHADOW	Shadow color for edges facing away from the light source
COLOR_ACTIVEBORDER	Active window border
COLOR_ACTIVECAPTION	Active window caption
COLOR_APPWORKSPACE	Background color of multiple document interface. (MDI) applications
COLOR_BACKGROUND,	
COLOR_DESKTOP	Desktop color
COLOR_BTNTEXT	Text on push buttons
COLOR_CAPTIONTEXT	Text in caption, size box, and scroll bar arrow box
COLOR_GRAYTEXT	Grayed (disabled) text Set to 0 if the current display driver does not support it
COLOR_HIGHLIGHT	Item(s) selected in a control
COLOR_HIGHLIGHTTEXT	Text of item(s) selected in a control
COLOR_INACTIVEBORDER	Inactive window border
COLOR_INACTIVECAPTION	Inactive window caption
COLOR_INACTIVECAPTIONTEXT	Color of text in an inactive caption
COLOR_INFOBK	Background color for ToolTip controls
COLOR_INFOTEXT	Text color for ToolTip controls
COLOR_MENU	Menu background
COLOR_MENUTEXT	Text in menus
COLOR_SCROLLBAR	Scroll bar gray area
COLOR_WINDOW	Window background
COLOR_WINDOWFRAME	Window frame
COLOR_WINDOWTEXT	Text in windows

24.3.2 Drawing with FrameRect()

The FrameRect() function draws border around a rectangular area, whose coordinates are defined in a RECT structure. The width and height of this border are one logical unit. The border is drawn with a brush, not with a pen. The brush is specified by its handle. The function's general form is as follows:

```
int FrameRect(
        HDC hDC,                // 1
        CONST RECT *lprc,       // 2
        HBRUSH hbr              // 3
        );
```

The second parameter is a pointer to a structure of type RECT that contains the coordinates. The third parameter is the handle to a brush, which must have been obtained with CreateSolidBrush(), CreatePatternBrush(), or CreateHatchBrush() functions described previously. Additionally, you may use a stock brush and obtain its handle by means of GetStockObject(). The function returns TRUE if it succeeds and FALSE if it fails.

Because the borders of the rectangle are drawn with a brush, rather than with a pen, the function is used to produce figures that cannot be obtained by other means. For example, if you select a brush with the vertical hatch pattern HS_VERTICAL, the resulting rectangle has dotted lines for the upper and lower segments since this is the brush pattern. The vertical segments of the rectangle are displayed as solid lines only when the rectangle's side coincides with the brush's bitmap pattern. Another characteristic of the FrameRect() function is that dithered colors can be used to draw the rectangle's border.

24.3.3 Drawing with DrawFocusRect()

The DrawFocusRect() function draws a rectangle of dotted lines. The rectangle's interior is not filled. The function's name relates to its intention, not to its operation, since the drawn rectangle is not given the keyboard focus automatically. The DrawFocusRect() function uses neither a pen nor a brush to draw the perimeter. The dotted lines used for the rectangle are one pixel wide, one pixel high, and are separated by one pixel. The function's general form is as follows:

```
BOOL DrawFocusRect(
        HDC hDC,                // 1
        CONST RECT *lprc        // 2
        );
```

The second parameter is a pointer to a structure of type RECT that contains the coordinates. The function returns TRUE if it succeeds and FALSE if it fails. Figure 24.9 shows a rectangle drawn with the DrawFocusRect() function.

There are several unique features of the DrawFocusRect() function. The most important feature is that the rectangle is displayed by means of an XOR operation on the background pixels. This ensures that it is visible on most backgrounds. Also, that the rectangle can be erased by calling the function a second time with the same parameters. This is a powerful feature of this function since an application can call

DrawFocusRect() to draw a rectangle around an object or background, and then erase the rectangle and restore the display without having to preserve the overdrawn area.

Figure 24.9 *Rectangle Drawn with DrawFocusRect()*

The area that contains a rectangle drawn with DrawFocusRect() cannot be scrolled. In order to scroll this area you can call DrawFocusRect() a second time to erase the rectangle, scroll the display, then call the function again to redraw the focus rectangle.

24.3.4 Auxiliary Operations on Rectangles

Windows provides several auxiliary functions designed to facilitate manipulating structures of type RECT. Although these functions have no unique functionality, they do simplify the coding. Table 24.5 lists these auxiliary functions.

Table 24.5

Rectangle-Related Functions

FUNCTION	FIGURE
SetRect()	Fills a RECT structure variable with coordinates
CopyRect()	Copies the data in a RECT structure variable to another one
SetEmptyRect()	Fills a RECT structure variable with zeros thus creating an empty rectangle
OffsetRect()	Translates a rectangle along the x- and y-axis
InflateRect()	Increases or decreases the width and height of a rectangle
IntersectRect()	Creates a rectangle that is the intersection of two other rectangles
UnionRect()	Creates a rectangle that is the union of two other rectangles
SubratctRect()	Creates a rectangle that is the difference between two other rectangles
IsRectEmpty()	Determines if a rectangle is empty
PtInRect()	Determines if a point is located within the perimeter of a rectangle
EqualRect()	Determines if two rectangles are equal

The function SetRect() is used to set the coordinates in a RECT structure. It is equivalent to entering these values into the structure member variables. The function's general form is as follows:

```
BOOL SetRect(
        LPRECT lprc,   // 1
        int xLeft,     // 2
        int yTop,      // 3
        int xRight,    // 4
        int yBottom    // 5
        );
```

The first parameter is a pointer to the structure variable that references the rectangle to be set. The second and third parameters are the x and y-coordinates of the upper-left corner. The fourth and fifth parameters are the coordinates of the lower-right corner.

The CopyRect() function is used to copy the parameters from one rectangle structure variable to another one. The function's parameters are the addresses of the destination and source structures. Its general form is as follows:

```
BOOL CopyRect(
        LPRECT lprcDst,       // 1
        CONST RECT *lprcSrc   // 2
        );
```

The first parameter is a pointer to a structure of type RECT that receives the copied coordinates. The second parameter is a pointer to the structure that holds the source coordinates.

The function SetRectEmpty() takes as a parameter the address of a structure variable of type RECT and sets all its values to zero. The result is an empty rectangle that does not show on the screen. Its general form is as follows:

```
BOOL SetEmptyRect (LPRECT rect);
```

The function's only parameter is the address of the RECT structure that is to be cleared.

Notice that there is a difference between an empty rectangle and a NULL rectangle. An empty rectangle is one with no area, that is, one in which the coordinate of the right side is less than or equal to that of the left side, or the coordinate of the bottom side is less than or equal to that of the top side. A NULL rectangle is one in which all the coordinates are zero. The Foundation Class Library contains different member functions for detecting an empty and a NULL rectangle. The Windows API, however, has no function for detecting a NULL rectangle.

OffsetRect() translates a rectangle along both axes. The function's general form is as follows:

```
BOOL OffsetRect(
        LPRECT lprc,   // 1
        int dx,        // 2
        int dy         // 3
        );
```

The first parameter is a pointer to a structure variable of type RECT that contains the parameters of the rectangle to be moved. The second parameter is the amount to move the rectangle along the x-axis. The third parameter is the amount to move the rectangle along the y-axis. Positive values indicate movement to the right or down. Negative values indicate movement to the left or up.

In reality, the OffsetRect() function does not move the rectangle, but simply changes the values in the RECT structure variable referenced in the call. Another call to a rectangle display function is necessary in order to show the translated rectangle on the screen. Figure 24.10 shows the effect of OffsetRect().

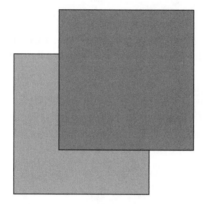

Figure 24.10 *Effect of the OffsetRect() Function*

In Figure 24.10, the light-gray rectangle shows the original figure. The OffsetRect() function was applied to the data in the figure's RECT structure variable, adding 50 pixels along the x-axis and subtracting 50 pixels along the y-axis. The resulting rectangle is shown with a dark-gray fill.

InflateRect() serves to increase or decrease the size of a rectangle. The function's general form is as follows:

```
BOOL InflateRect(
        LPRECT lprc,    // 1
        int dx,         // 2
        int dy          // 3
        );
```

The first parameter is a pointer to a structure variable of type RECT that contains the rectangle to be resized. The second parameter is the amount to add or subtract from the rectangle's width. The third parameter is the amount to add or subtract from the rectangle's height. In both cases, positive values indicate an increase of the dimension and negative values a decrease. The InflateRect() function does not change the displayed rectangle, but modifies the values in the RECT structure variable referenced in the call. Another call to a rectangle display function is necessary in order to show the modified rectangle on the screen. Figure 24.11, on the following page, shows the effect of the InflateRect() function.

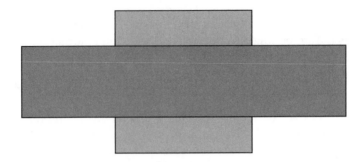

Figure 24.11 *Effect of the InflateRect() Function*

In Figure 24.11, the light-gray rectangle shows the original 150-by-150 pixels figure. The InflateRect() function was applied to increase the width by 100 pixels and decrease the height by 75 pixels. The results are shown in the dark-gray rectangle.

The IntersectRect() function applies a logical AND operation on two rectangles to create a new rectangle that represents the intersection of the two figures. If there are no common points in the source rectangles, then an empty rectangle is produced. The function's general from is as follows:

```
BOOL IntersectRect (LPRECT, CONST LPRECT, CONST LPRECT);
                    ------  ------------  ------------
                       |         |             |
                       1         2             3
```

The first parameter is the address of a RECT structure variable where the intersection coordinates are placed. The second parameter is a pointer to a RECT structure variable that holds the coordinates of the first rectangle. The third parameter is a pointer to a RECT structure variable with the coordinates of the second rectangle. Figure 24.12 shows the effect of the IntersectRect() function.

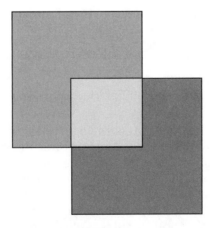

Figure 24.12 *Effect of the IntersectRect() Function*

The UnionRect() function applies a logical OR operation on two rectangles to create a new rectangle that represents the union of the two figures. If there are no common points in the source rectangles, then an empty rectangle is produced. The resulting image is the smallest rectangle that contains both sources. The function's general form is as follows:

```
BOOL UnionRect(
         LPRECT lprcDst,        // 1
         CONST RECT *lprcSrc1,  // 2
         CONST RECT *lprcSrc2   // 3
         );
```

The first parameter is the address of a RECT structure variable where the union rectangle coordinates are placed. The second parameter is a pointer to a RECT structure variable that holds the coordinates of the first rectangle, and the third parameter is a pointer to a RECT structure variable with the coordinates of the second rectangle. Figure 24.13 shows the effect of the UnionRect() function.

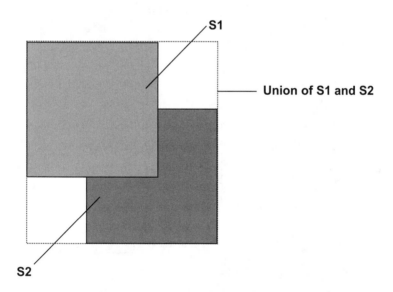

Figure 24.13 *Effect of the UnionRect() Function*

The SubtractRect() function creates a new rectangle by subtracting the coordinates of two source rectangles. The function's general form is as follows:

```
BOOL SubtractRect(
         LPRECT lprcDst,        // 1
         CONST RECT *lprcSrc1,  // 2
         CONST RECT *lprcSrc2   // 3
         );
```

The first parameter is the address of a RECT structure variable where the resulting coordinates are placed. The second parameter is a pointer to a RECT structure variable that holds the coordinates of the first source rectangle. It is from this rectangle that the coordinates of the second source rectangle are subtracted. The third parameter is a pointer to a RECT structure variable with the coordinates of the second source rectangle. The coordinates of this rectangle are subtracted from the ones of the first source rectangle.

The result of the operation must be a rectangle, not a polygon or any other non-rectangular surface. This imposes the restriction that the rectangles must completely overlap in either the vertical or the horizontal direction. If not, the co-ordinates of the resulting rectangle are the same as those of the first source rectangle. Figure 24.14 shows three possible cases of rectangle subtraction.

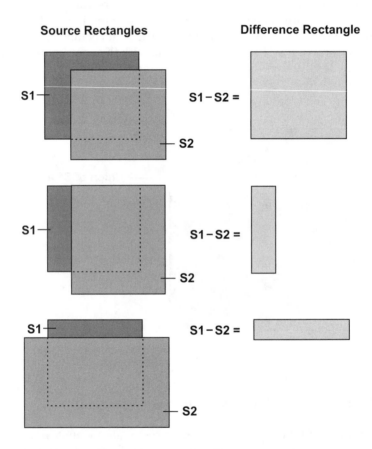

Figure 24.14 *Cases in the SubtractRect() Function*

The IsRectEmpty() function determines whether a rectangle is empty. An empty rectangle is one with no area, that is, one in which the width and/or the height are zero or negative. The function's general form is as follows:

```
BOOL IsRectEmpty(
         CONST RECT *lprc    // 1
         );
```

The only parameter is the address of the RECT structure variable that contains the rectangle's parameters. The function returns TRUE if the rectangle is empty and FALSE otherwise.

The PtInRect() function determines whether a point lies within a rectangle. A point that lies on the rectangle's top or left side is considered to be within the rectangle, but a point within the right or bottom side is not. The function's general form is as follows:

```
BOOL PtInRect(
        CONST RECT *lprc,    // 1
        POINT pt             // 2
        );
```

The first parameter is the address of a RECT structure variable that contains the rectangle's dimensions. The second parameter is a structure of type POINT which holds the coordinates of the point being tested. The function returns TRUE if the point is within the rectangle and FALSE otherwise.

The EqualRect() function determines whether two rectangles are equal. For two rectangles to be equal all their coordinates must be identical. The function's general form is as follows:

```
BOOL EqualRect(
        CONST RECT *lprc1,   // 1
        CONST RECT *lprc2    // 2
        );
```

The first parameter points to a RECT structure variable that contains the parameters of one rectangle. The second parameter points to a RECT structure variable with the parameters of the second rectangle. The function returns TRUE if both rectangles are equal and FALSE otherwise.

24.3.5 Updating the Rectangle() Function

All of the rectangle operations described in the preceding section receive the coordinates in a structure of type RECT. The basic rectangle-drawing function, Rectangle(), receives the figure coordinates as parameters to the call. This difference in data formats, which is due to the evolution of Windows, makes it difficult to transfer the results of a rectangle operation into the Rectangle() function. To solve this problem we have coded a rectangle-drawing function, called DrawRect(), which takes the figure coordinates from a RECT structure. The function is as follows:

```
BOOL DrawRect (HDC hdc, LPRECT aRect) {
  return (Rectangle (hdc, aRect->left,
        aRect->top,
                        aRect->right,
                        aRect->bottom));
}
```

Since DrawRect() uses Rectangle() to draw the figure, the return values are the same for both functions.

24.4 Regions

A region is an area composed of one or more polygons or ellipses. Since a rectangle is a polygon, a region can also be (or contain) a rectangle, or even a rounded rectangle. In Windows programming, regions are used for three main purposes:

- To fill or frame an irregular area

- To clip output to an irregular area

- To test for mouse input in an irregular area

From these uses we can conclude that the main role of a region is to serve as a boundary. Regions can be combined logically, copied, subtracted, and translated to another location. In Windows 95 and later versions a new region can be produced by performing a rotation, scaling, reflection, or shearing transformation on another region.ere. Presently we deal with the simpler operations on regions. There is a rich set of functions that relate to regions and region operations. These are listed in Table 24.6.

Table 24.6

Region-Related GDI Functions

FUNCTION	ACTION
CREATING REGIONS:	
CreateRectRgn()	Creates a rectangular-shaped region, given the four coordinates of the rectangle
CreateRectRgnIndirect()	Creates a rectangular-shaped region, given a RECT structure with the coordinates of the rectangle
CreateRoundRectRgn()	Creates a region shaped like a rounded-corner rectangle, given the coordinates of the rectangle and the dimensions of the corner ellipse
CreateEllipticRgn()	Creates an elliptically-shaped region from a bounding rectangle
CreateEllipticRegionIndirect()	Creates an elliptically-shaped region from the parameters of a bounding rectangle in a RECT structure
CreatePolygonRgn()	Creates a polygon-shaped region from an array of points that define the polygon
CreatePolyPolygonRgn()	Creates one or more polygon-shaped regions from an array of points that define the polygons
PathToRegion()	Converts the current path into a region
ExtCreateRgn()	Creates a region based on a transformation performed on another regions.
COMBINING REGIONS:	
CombineRgn()	Combines two regions into one by performing a logica subtraction, or copy operation
FILLING AND PAINTING REGIONS:	
FillRgn()	Fills a region using a brush

(continues)

Table 24.6

Region-Related GDI Functions (continued)

FUNCTION	ACTION
GetPolyFillMode()	Gets fill mode used by FillRgn()
SetPolyFillMode()	Sets the fill mode for FillRgn()
FrameRgn()	Frames a region using a brush
PaintRgn()	Paints the interior of a region with the brush currently selected in the device context
InvertRgn()	Inverts the colors in a region
REGION STATUS AND CONTROL:	
SetWindowRgn()	Sets the window regions. The window region is the area where the operating system allows drawing operations to take place
GetWindowRgn()	Retrieves the window region established by SetWindowRgn()
OffsetRgn()	Moves a region along the x- or y-axis
SelectClipRgn()	Makes a region the current clipping region
ExtSelectClipRgn()	Combines a region with the current clipping region
GetClipRgn()	Gets handle of the current clipping region
ValidateRgn()	Validates the client area removing the area in the region from the current update region
InvalidateRgn()	Forces a WM_PAINT message by invalidating a screen area defined by a region
OBTAIN REGION DATA:	
PtInRegion()	Tests if a point is located within a region
RectInRegion()	Tests if a given rectangle overlaps any part of a region
EqualRgn()	Tests if two regions are equal
GetRgnBox()	Retrieves a region's bounding box
GetRegionData()	Retrieves internal structure information about a region

In the sections that follow we discuss some of the region-related functions. Other region operations are discussed, in context, later in the book.

24.4.1 Creating Regions

A region is a GDI object, hence, it must be explicitly created. The functions that create a region return a handle of type HRGN (handle to a region). With this handle you can perform many region-based operations, such as filling the region, drawing its outline, and combining it with another region. You often create two or more simple regions by calling their primitive functions, and then combine them into a more complex region, usually by means of the CombineRgn() function.

CreateRectRgn() is used to create a rectangular region. The function's general form is as follows:

```
HRGN CreateRectRgn(
        int nLeftRect,    // 1
        int nTopRect,     // 2
```

```
        int nRightRect,    // 3
        int nBottomRect    // 4
        );
```

The first and second parameters are the coordinates of the upper-left corner of the rectangle. The third and fourth parameters are the coordinates of the lower-right corner.

CreateRectRgnIndirect() creates a rectangular-shaped region, identical to the one produced by CreateRectRgn(); the only difference is that CreateRectRgnIndirect() receives the coordinates in a RECT structure variable. The function's general form is as follows:

```
HRGN CreateRectRgnIndirect(
        CONST RECT *lprc    // 1
            );
```

CreateRoundRectRgn() creates a region shaped like a rounded-rectangle. Its general form is as follows:

```
HRGN CreateRoundRectRgn(
        int nLeftRect,      // 1
        int nTopRect,       // 2
        int nRightRect,     // 3
        int nBottomRect,    // 4
        int nWidthEllipse,  // 5
        int nHeightEllipse  // 6
        );
```

The first and second parameters are the coordinates of the upper-left corner of the bounding rectangle. The third and fourth parameters are the coordinates of the lower-right corner. The fifth parameter is the width of the ellipse that is used for drawing the rounded corner arc. The sixth parameter is the height of this ellipse. The shape of the resulting region is the same as that of the rectangle in Figure 24.4.

CreateEllipticRgn() creates an elliptically-shaped region. The function's general form is as follows:

```
HRGN CreateEllipticRgn(
        int nLeftRect,     // 1
        int nTopRect,      // 2
        int nRightRect,    // 3
        int nBottomRect    // 4
);
```

The first and second parameters are the coordinates of the upper-left corner of a rectangle that bounds the ellipse. The third and fourth parameters are the coordinates of the lower-right corner of this bounding rectangle. The shape of the resulting region is similar to the one in Figure 24.5.

CreateEllipticRegionIndirect() creates an elliptically-shaped region identical to the one produced by CreateEllipticRgn() except that in this case the parameters are read from a RECT structure variable. The function's general form is as follows:

```
HRGN CreateEllipticRgnIndirect(
        CONST RECT *lprc    // 1
);
```

CreatePolygonRgn() creates a polygon-shaped region. The call assumes that the polygon is closed; no automatic closing is provided. The function's general form is as follows:

```
HRGN CreatePolygonRgn(
        CONST POINT *lppt,    // 1
        int cPoints,          // 2
        int fnPolyFillMode    // 3
);
```

The first parameter is the address of an array of points that contains the x- and y-coordinate pairs of the polygon vertices. The second parameter is the count of the number of vertices in the array. The third parameter specifies the polygon fill mode, which can be ALTERNATE or WINDING. ALTERNATE defines a mode that fills between odd-numbered and even-numbered polygon sides, that is, those areas that can be reached from the outside of the polygon by crossing an odd number of lines. WINDING mode fills all internal regions of the polygon. These are the same constants as used in the SetPolyFillMode() function described earlier in this chapter. In the CreatePolygonRgn() function call the fill mode determines which points are included in the region.

CreatePolyPolygonRgn() creates one or more polygon-shaped regions. The call assumes that the polygons are closed figures. No automatic closing is provided. CreatePolyPolygonRgn() is similar to PolyPolygon(). The function's general form is as follows:

```
HRGN CreatePolyPolygonRgn(
        CONST POINT *lppt,         // 1
        CONST INT *lpPolyCounts,   // 2
        int nCount,                // 3
        int fnPolyFillMode         // 4
);
```

The first parameter is a pointer to an array containing vertices of the various polygons. The second parameter is a pointer to an array that contains the number of vertices in each of the polygons. The third parameter is the count of the number of elements in the second parameter, which is the same as the number of polygons to be drawn. The fourth parameter specifies the polygon fill mode, which can be ALTERNATE or WINDING. These two constants have the same effect as described in the CreatePolygonRgn() function.

All the region-creation functions discussed so far return the handle to the region if the call succeeds, and NULL if it fails.

A region can be created from a path by means of the PathToRegion() function. Paths are discussed later in this chapter. The ExtCreateRgn() function allows creating a new region by performing a transformation on another region. Transformations are not discussed in this book.

24.4.2 Combining Regions

Sometimes a region consists of a simple, primitive area such as a rectangle, and ellipse, or a polygon. On other occasions a region is a complex figure, composed of two or more simple figures of the same or different types, which can overlap, be adjacent, or disjoint. The CombineRgn() function is used to create a complex region from two simpler ones. The function's general form is as follows:

```
int CombineRgn(
        HRGN hrgnDest,      // 1
        HRGN hrgnSrc1,      // 2
        HRGN hrgnSrc2,      // 3
        int fnCombineMode   // 4
);
```

The first parameter is the handle to the resulting combined region. The second parameter is the handle to the first source region to be combined. The third parameter is the handle to the second source region to be combined. The fourth parameter is one of five possible combination modes, listed in Table 24.7.

Table 24.7

Region Combination Modes

MODE	EFFECT
RGN_AND	The intersection of the two combined regions
RGN_COPY	A copy of the first source region
RGN_DIFF	Combines the parts of the first source region that are not in the second source region
RGN_OR	The union of two combined regions
RGN_XOR	The union of two combined regions except for any overlapping area

CombineRgn() returns one of four integer values, as shown in Table 24.8.

Table 24.8

Region Type Return Values

VALUE	MEANING
NULLREGION	The region is empty
SIMPLEREGION	The region is a single rectangle
COMPLEXREGION	The region is more complex than a single rectangle
ERROR	No region was created

One property of CombineRgn() is that the destination region, expressed in the first parameter, must exist as a region prior to the call. Creating a memory variable to hold the handle to this region is not sufficient. The region must have been first created by means of one of the region-creation functions, otherwise CombineRgn() returns ERROR. The following code fragment shows the required processing for creating two simple regions and then combining them into a complex region using the RGN_AND combination mode:

```
HRGN        rectRgn, ellipRgn, resultRgn;
   .   .   .
// Create a rectangular region
rectRgn = CreateRectRgn (100, 100, 300, 200);
// Create an elliptical region
ellipRgn = CreateEllipticRgn (200, 100, 400, 200);
// Create a dummy region for results. Skipping this
// step results in an ERROR from the CombineRgn() call
resultRgn = CreateRectRgn (0, 0, 0, 0);
// Combine regions and fill
CombineRgn (resultRgn, rectRgn, ellipRgn, RGN_AND);
FillRgn (hdc, resultRgn, redSolBrush);
```

Figure 24.15 shows the results of applying the various region combination modes on two simple, overlapping regions.

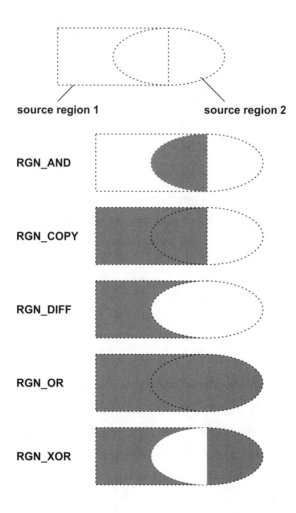

Figure 24.15 *Regions Resulting from CombineRgn() Modes*

In addition to the CombineRgn() function, the windowsx.h header files defines several macros that facilitate region combinations. These macros implement the five combination modes that are entered as the last parameter of the CombineRgn() call. They are as follows:

```
CopyRgn       (hrgnDest, hrgnScr1);
IntersectRgn  (hrgnDest, hrgnScr1, hrgnSrc2);
SubtractRgn   (hrgnDest, hrgnScr1, hrgnScr2);
UnionRgn      (hrgnDest, hrgnSrc1, hrgnSrc2);
XorRgn        (hrgnDest, hrgnSrc1, hrgnScr2);
```

In all of the macros, hrgnDest is the handle to the destination region, while hrgnScr1 and hrgnSrc2 are the handles to the source regions.

24.4.3 Filling and Painting Regions

Several functions relate to filling, painting, and framing regions. The difference between filling and painting is that fill operations require a handle to a brush, while paint operations use the brush currently selected in the device context.

The FillRgn() function fills a region using a brush defined by its handle. The function's general form is as follows:

```
BOOL FillRgn(
         HDC hdc,       // 1
         HRGN hrgn,     // 2
         HBRUSH hbr     // 3
         );
```

The first parameter is the handle to the device context. The second one is the handle to the region to be filled. The third parameter is the handle to the brush used in filling the region. The function returns TRUE if it succeeds and FALSE if it fails.

PaintRgn() paints the interior of a region with the brush currently selected in the device context. The function's general form is as follows:

```
BOOL PaintRgn(
         HDC hdc,       // 1
         HRGN hrgn      // 2
         );
```

The second parameter is the handle to the region to be filled. The function returns TRUE if it succeeds and FALSE if it fails.

FrameRgn() draws the perimeter of a region using a brush defined by its handle. The function's general form is as follows:

```
BOOL FrameRgn(
         HDC hdc,       // 1
         HRGN hrgn,     // 2
         HBRUSH hbr,    // 3
         int nWidth,    // 4
         int nHeight    // 5
         );
```

The second parameter is the handle to the region to be filled. The third one is the handle to the brush used in filling the region. The fourth parameter specifies the width of the brush, in logical units. The fifth parameter specifies the height of the brush, also in logical units. The function returns TRUE if it succeeds and FALSE if it fails. If the width and height of the brush are different, then oblique portions of the image are assigned an intermediate thickness. The result is similar to using a calligraphy pen. Figure 24.16 shows a region drawn with the FrameRgn() function.

Figure 24.16 *Region Border Drawn with FrameRgn()*

The InvertRgn() function inverts the colors in a region. In a monochrome screen, inversion consists of turning white pixels to black and black pixels to white. In a color screen, inversion depends on the display technology. In general terms, inverting a color produces its complement. Therefore, inverting blue produces yellow, inverting red produces cyan, and inverting green produces magenta. The function's general form is as follows:

```
BOOL InvertRgn(
        HDC hdc,      // 1
        HRGN hrgn     // 2
        );
```

The second parameter is the handle to the region to be inverted. The function returns TRUE if it succeeds and FALSE if it fails.

24.4.4 Region Manipulations

Several functions allow the manipulation of regions. These manipulations include moving a region, using a region to define the program's output area, setting the clipping region, obtaining the clipping region handle, and validating or invalidating a screen area defined by a region. The region manipulations related to clipping are discussed in the following section.

A powerful, but rarely used function in the Windows API is SetWindowRgn(). It allows you to redefine the window area of a window, thus redefining the area where drawing operations take place. In a sense, SetWindowRgn() is a form of clipping that includes not only the client area, but the entire window. The SetWindowRgn() function allows you to create a window that includes only part of the title bar, or to eliminate one or more of the window borders, as well as many other effects. The function's general form is as follows:

```
int SetWindowRgn(
        HWND hWnd,      // 1
        HRGN hRgn,      // 2
        BOOL bRedraw    // 3
        );
```

The first parameter is the handle to the window whose region is to be changed. The second parameter is the handle to the region that is to be used in redefining the window area. If this parameter is NULL then the window has no window area; therefore, becoming invisible. The third parameter is a redraw flag. If set to TRUE, the operating system automatically redraws the window to the new output area. If the window is visible the redraw flag is usually TRUE. The function returns nonzero if it succeeds and zero if it fails.

The function GetWindowRgn() is used to obtain the window area of a window, which has been usually set by SetWindowRgn(). The function's general form is as follows:

```
int GetWindowRgn(
        HWND hWnd, // 1
        HRGN hRgn  // 2
        );
```

The first parameter is the handle to the window whose region is to be obtained. The second parameter is the handle to a region that receives a copy of the window region. The return value is one of the constants listed in Table 24.8.

The OffsetRgn() function is used to move a region to another location. The function's general form is as follows:

```
int OffsetRgn(
        HRGN hrgn,       // 1
        int nXOffset,    // 2
        int nYOffset     // 3
        );
```

The first parameter is the handle to the region that is to be moved. The second parameter is the number of logical units that the region is to be moved along the x-axis. The third parameter is the number of logical units along the y-axis. The function returns one of the constants listed in Table 24.8.

Sometimes the OffsetRgn() function does not perform as expected. It appears that when a region is moved by means of this function, some of the region attributes are not preserved. For example, assume a region that has been filled red is moved to a new location that does not overlap the old position. If we now call InvertRgn() on the translated window, the result is not a cyan-colored window, but one that is the reverse of the background color. In this case the red fill attribute of the original window was lost as it was translated into a new position, and the translated window has no fill. If the translated window partially overlaps the original one, however, then the overlap area's original color is negated when the InvertRgn() function is called on the translated region. Figure 24.17 shows the result of inverting a region translated by means of OffsetRgn().

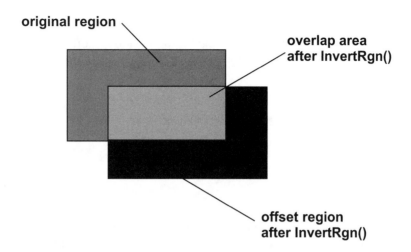

original region

**overlap area
after InvertRgn()**

**offset region
after InvertRgn()**

Figure 24.17 *Effect of OffsetRgn() on Region Fill*

Two functions, SelectClipRgn() and ExtSelectClipRgn(), refer to the use of regions in clipping. These functions, along with clipping operations, are discussed later in this chapter.

The InvalidateRgn() function adds the specified region to the current update region of the window. The invalidated region is marked for update when the next WM_PAINT message occurs. The function's general form is as follows:

```
BOOL InvalidateRgn(
        HWND hWnd,      // 1
        HRGN hRgn,      // 2
        BOOL bErase     // 3
        );
```

The first parameter is the handle to the window that is to be updated. The second parameter is the handle to the region to be added to the update area. If this parameter is NULL then the entire client area is added to the update area. The third parameter is an update flag for the background area. If this parameter is TRUE then the background is erased. The function always returns a nonzero value.

The ValidateRgn() function removes the region from the update area. It has the reverse effect as InvalidateRgn(). The function's general form is as follows:

```
BOOL ValidateRgn(
        HWND hWnd,      // 1
        HRGN hRgn       // 2
        );
```

The first parameter is the handle to the window. The second parameter is the handle to the region to be removed from the update area. If this parameter is NULL then the entire client area is removed from the update area. The function returns TRUE if it succeeds and FALSE if it fails.

24.4.5 Obtaining Region Data

A few region-related functions are designed to provide region data to application code. The GetRegionData() function is used mainly in relation to the ExtCreateRegion() function.

PtInRegion() tests if a point defined by its coordinates is located within a region. The function's general form is as follows:

```
BOOL PtInRegion(
        HRGN hrgn,    // 1
        int X,        // 2
        int Y         // 3
        );
```

The first parameter is the handle to the region to be examined. The second and third parameters are the x- and y-coordinates of the point. If the point is located within the region, the function returns TRUE. If not, the function returns FALSE.

The RectInRegion() function determines if any portion of a given rectangle is within a specified region. The function's general form is as follows:

```
BOOL RectInRegion (
                HRGN,          // 1
                CONST RECT *   // 2
                );
```

The first parameter is the handle to the region to be examined. The second parameter is a pointer to a RECT structure that holds the coordinates of the rectangle. If any part of the specified rectangle lies within the region, the function returns TRUE. If not, the function returns FALSE.

EqualRgn() tests if two regions are identical in size and shape. The function's general form is as follows:

```
BOOL EqualRgn(
        HRGN hSrcRgn1,    // 1
        HRGN hSrcRgn2     // 2
        );
```

The first parameter identifies one of the regions and the second parameter the other one. If the two regions are identical, the function returns TRUE. If not, the function returns FALSE.

The GetRgnBox() function retrieves the bounding rectangle that encloses the specified region. The function's general form is as follows:

```
int GetRgnBox(
        HRGN hrgn,    // 1
        LPRECT lprc   // 2
        );
```

The first parameter is the handle to the region. The second parameter is a pointer to a RECT structure variable that receives the coordinates of the bounding rectangle. The function returns one of the first three constants listed in Table 24.8. If the first parameter does not identify a region then the function returns zero.

24.5 Clipping Operations

One of the fundamental graphics manipulations is clipping. In Windows programming, clipping is associated with regions, since the clip action is defined by a region. In practice, a clipping region is often of rectangular shape, which explains why some clipping operations refer specifically to rectangles. Figure 24.18 shows the results of a clipping operation.

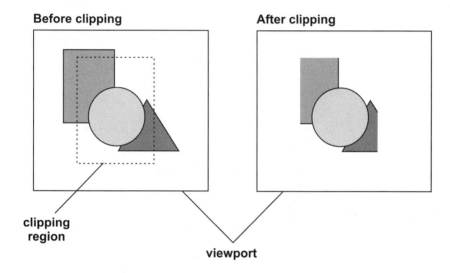

Figure 24.18 *Results of Clipping*

A clipping region is an object of the device context. The default clipping region is the client area. Not all device contexts support a predefined clipping region. If the device context is supplied by a call to BeginPaint(), then Windows creates a default clipping region and assigns to it the area of the window that needs to be repainted. If the device context was created with a call to CreateDC() or GetDC(), as is the case with a private device context, then no default clipping region exists. In this case an application can explicitly create a clipping region. Table 24.9, on the following page, lists the functions that relate to clipping.

Note that Metaregions were introduced in Windows NT and are supported in Windows 95 and later. However, very little has been printed about their meaning or possible uses. Microsoft documentation for Visual C++, up to the May prerelease of version 6.0, has nothing on metaregions beyond a brief mentioning of the two related functions listed in Table 24.9. For this reason, it is impossible to determine at this time if a metaregion is a trivial alias for a conventional region, or some other concept not yet documented. Metaregions are not discussed in the text.

Table 24.9

Windows Clipping Functions

FUNCTION	ACTION
CREATING OR MODIFYING A CLIPPING REGION:	
SelectClipRgn()	Makes a region the clipping region for a specified device context
ExtSelectClipRgn()	Combines a specified region with the clipping region according to a predefined mode
IntersectClipRect()	Creates a new clipping region from the interception of a rectangle and the current clipping region in a device context
ExcludeClipRect()	Subtracts a rectangle from the clipping region
OffsetClipRgn()	Moves the clipping region horizontally or vertically
SelectClipPath()	Appends the current path to the clipping region of a device context, according to a predefined mode
OBTAIN CLIPPING REGION INFORMATION:	
GetClipBox()	Retrieves the bounding rectangle for the clipping region
GetClipRgn()	Retrieves the handle of the clipping region for a specified device context
PtVisible()	Determines if a specified point is within the clipping region of a device context
RectVisible()	Determines whether any part of a rectangle lies within the clipping region
METAREGION OPERATIONS:	
GetMetaRgn()	Retrieves the metaregion for the specified device context
SetMetaRgn()	Creates a metaregion, which is the intersection of the current metaregion and the clipping region

24.5.1 Creating or Modifying a Clipping Region

In order for a region to be used to clip output, it must be selected as such in a device context. The SelectClipRgn() function is the primary method of achieving clipping. The region must first be defined and a handle for it obtained. Then the handle to the device context and the handle to the region are used to enforce the clipping. The function's general form is as follows:

```
int SelectClipRgn(
        HDC hdc,      // 1
        HRGN hrgn     // 2
        );
```

The first parameter is the handle to the device context that is to be clipped. The second parameter is the handle to the region used in clipping. This handle is obtained by any of the region-creating calls listed in Table 24.6. The function returns one of the values in Table 24.8.

Once the call is made, all future output is clipped; however, the existing screen display is not automatically changed to reflect the clipping area. Some unexpected or undesirable effects are possible during clipping. Once a clipping region is defined for a device context, then all output is limited to the clipping region. This requires that clipping be handled carefully, usually by installing and restoring clipping regions as necessary.

When a call is made to SelectClipRgn(), Windows preserves a copy of the previous clipping region. The newly installed clipping region can be removed from the device context by means of a call to SelectClipRgn() specifying a NULL region handle.

The ExtSelectClipRgn() function allows combining the current clipping region with a new one, according to one of five predefined modes. The function's general form is as follows:

```
int ExtSelectClipRgn(
        HDC hdc,            // 1
        HRGN hrgn,          // 2
        int fnMode          // 3
);
```

The first parameter is the handle to the device context that is to be clipped. The second parameter is the handle to the region used in clipping. The third parameter is one of the constants listed in Table 24.10.

Table 24.10

Clipping Modes

VALUE	ACTION
RGN_AND	The resulting clipping region combines the overlapping areas of the current clipping region and the one identified in the call, by performing a logical AND between the two regions
RGN_COPY	The resulting clipping region is a copy of the region identified in the call. The result is identical to calling SelectClipRgn(). If the region identified in the call is NULL, the new clipping region is the default clipping region
RGN_DIFF	The resulting clipping region is the difference between the current clipping region and the one identified in the call
RGN_OR	The resulting clipping region is the result of performing a logical OR operation on the current clipping region and the region identified in the call
RGN_XOR	The resulting clipping region is the result of performing a logical XOR operation on the current clipping region and the region identified in the call

The clipping regions that result from these selection modes are the same as those used in the CombineRgn() function, as shown in Figure 24.15. The ExtSelectClipRgn() function returns one of the values in Table 24.8.

The IntersectClipRect() function creates a new clipping region by performing a logical AND between the current clipping region and a rectangular area defined in the call. The function's general form is as follows:

```
int IntersectClipRect(
        HDC hdc,               // 1
        int nLeftRect,         // 2
        int nTopRect,          // 3
        int nRightRect,        // 4
        int nBottomRect        // 5
        );
```

The second and third parameters are the coordinates of the upper-left corner of the rectangle. The fourth and fifth parameters are the coordinates of the lower-right corner. The function returns one of the values in Table 24.8.

The function ExcludeClipRect() subtracts a rectangle specified in the call from the clipping region. The function's general form is as follows:

```
int ExcludeClipRect(
        HDC hdc,               // 1
        int nLeftRect,         // 2
        int nTopRect,          // 3
        int nRightRect,        // 4
        int nBottomRect        // 5
        );
```

The second and third parameters are the coordinates of the upper-left corner of the rectangle. The fourth and fifth parameters are the coordinates of the lower-right corner. The function returns one of the values in Table 24.8.

The function OffsetClipRgn() translates the clipping region along the horizontal or vertical axes. The function's general form is as follows:

```
int OffsetClipRgn(
        HDC hdc,               // 1
        int nXOffset,          // 2
        int nYOffset           // 3
        );
```

The second parameter is the amount to move the clipping region along the x-axis. The third parameter is the amount to move along the y-axis. Positive values indicate movement to the right or down. Negative values indicate movement to the left or up. The function returns one of the values in Table 24.8.

The SelectClipPath() function appends the current path to the clipping region of a device context, according to a predefined mode. The function's general form is as follows:

```
BOOL SelectClipPath(
        HDC hdc,         // 1
        int iMode        // 2
        );
```

The second parameter is one of the constants listed in Table 24.10. The function returns TRUE if it succeeds and FALSE if it fails. Paths are discussed later in this chapter.

24.5.2 Clipping Region Information

Code that uses clipping often needs to obtain information about the clipping region. Several functions are available for this purpose. The GetClipBox() function retrieves the bounding rectangle for the clipping region. This rectangle is the smallest one that can be drawn around the visible area of the device context. The function's general form is as follows:

```
int GetClipBox(
        HDC hdc,        // 1
        LPRECT lprc     // 2
        );
```

The second parameter is a pointer to a RECT structure that receives the coordinates of the bounding rectangle. The function returns one of the values in Table 24.8.

The GetClipRgn() function retrieves the handle of the clipping region for a specified device context. The function's general form is as follows:

```
int GetClipRgn(
        HDC hdc,            // 1
        HRGN hrgn           // 2
        );
```

The first parameter is the handle to the device context whose clipping region is desired. The second parameter is the handle to an existing clipping region that holds the results of the call. The function returns zero if there is no clipping region in the device context. The return value 1 indicates that there is a clipping region and that the function's second parameter holds its handle. A return value of –1 indicates an error. The function refers to clipping regions that result from SelectClipRgn() of ExtSelectClipRgn() functions. Clipping regions assigned by the system on calls to the BeginPaint() function are not returned by GetClipRgn().

The PtVisible() function is used to determine if a specified point is within the clipping region of a device context. The function's general form is as follows:

```
BOOL PtVisible(
        HDC hdc,    // 1
        int X,      // 2
        int Y       // 3
        );
```

The first parameter is the handle to the device context under consideration. The second and third parameters are the x- and y-coordinates of the point in question. The function returns TRUE if the point is within the clipping region, and FALSE otherwise.

The function RectVisible() is used to determine whether any part of a rectangle lies within the clipping region of a device context. The function's general form is as follows:

```
BOOL RectVisible(
        HDC hdc,            // 1
        CONST RECT *lprc    // 2
        );
```

The first parameter is the handle to the device context under consideration. The second parameter is a pointer to a structure variable of type RECT that holds the coordinates of the rectangle in question. The function returns TRUE if any portion of the rectangle is within the clipping region, and FALSE otherwise.

24.6 Paths

In previous chapters we have discussed paths rather informally. The project folder Text Demo No 3, in the book's on-line software package, contains a program that uses paths to achieve graphics effects in text display. We now consider revisit paths in a more rigorous manner, and apply paths to other graphics operations.

Paths were introduced with Windows NT and are also supported by Windows 95 and later. As its name implies, a path is the route the drawing instrument follows in creating a particular figure or set of figures. A path, which is stored internally by the GDI, can serve to define the outline of a graphics object. For example, if we start at coordinates 100, 100, and move to the point at 150, 100, then to 150, 200, from there to 100, 200, and finally to the start point, we have defined the path for a rectangular figure. We can now stroke the path to draw the rectangle's outline, fill the path to produce a solid figure, or both stroke and fill the path to produce a figure with both outline and fill. In general, there are path-related functions to perform the following operations:

- To draw the outline of the path using the current pen.
- To paint the interior of the path using the current brush.
- To draw the outline and paint the interior of a path.
- To modify a path converting curves to line segments.
- To convert the path into a clip path.
- To convert the path into a region.
- To flatten the path by converting each curve in the path into a series of line segments.
- To retrieve the coordinates of the lines and curves that compose a path.

The path is an object of the device context, such as a region, a pen, a brush, or a bitmap. One characteristic of a path is that there is no default path in the device context. Another one is that there is only one path in each device context; this determines that there is no need for a path handle. Every path is initiated by means of the BeginPath() function. This clears any old path from the device context and prepares to record the drawing primitives that create the new path, sometimes called the path bracket. Any of the functions listed in Table 24.11 can be used for defining a path in Windows NT. The subset of functions that can be used in paths in Windows 95 and later versions are listed in Table 24.12.

Since paths are mostly utilized in clipping operations, the CloseFigure() function is generally used to close an open figure in a path. After all the figures that form the path have been drawn into the path bracket, the application calls EndPath() to select the path into the specified device context. The path can then be made into a clipping region by means of a call to SelectClipPath().

Table 24.11

Path-Defining Functions in Windows NT

AngleArc()	LineTo()	Polyline()
Arc()	MoveToEx()	PolylineTo()
ArcTo()	Pie()	PolyPolygon()
Chord()	PolyBezier()	PolyPolyline()
CloseFigure()	PolyBezierTo()	Rectangle()
Ellipse()	PolyDraw()	RoundRect()
ExtTextOut()	Polygon()	TextOut()

Table 24.12

Path-Defining Functions in Windows 95 and Later

ExtTextOut()	PolyBezierTo()	PolyPolygon()
LineTo()	Polygon()	PolyPolyline()
MoveToEx()	Polyline()	TextOut()
PolyBezier()	PolylineTo()	

Notice that the term clip path, or clipping path, sometimes found in the Windows documentation, can be somewhat confusing. It is better to say that the SelectClipPath() function converts a path to a clipping region, thus eliminating the notion of a clip path as a separate entity.

Table 24.13 lists the paths-related functions.

Table 24.13

Path-Related Functions

FUNCTION	ACTION
PATH CREATION, DELETION, AND CONVERSION:	
BeginPath()	Opens a path bracket
EndPath()	Closes the path bracket and selects the path into the device context
AbortPath()	Closes and discards any open path bracket on the device context
SelectClipPath()	Makes the current path into a clipping region for a specified device context. Combines the new clipping region with any existing one according to a predefined mode
PathToRegion()	Closes an open path and converts is to a region

(continues)

Table 24.13

Path-Related Functions (continued)

FUNCTION	ACTION
PATH RENDERING OPERATIONS:	
StrokePath()	Renders the outline of the current path using the current pen
FillPath()	Closes and opens figure in the current path and fills the path interior with the current brush, using the current polygon fill mode
StrokeAndFillPath()	Renders the outline of the current path using the current pen and fills the interior with the current brush
CloseFigure()	Draws a line from the current pen position to the figure's start point. The closing line is connected to the figure's first line using the current line join style
PolyDraw()	Draws lines and curves that result from GetPath() (Windows NT only)
PATH MANIPULATIONS:	
FlattenPath()	Converts curves in the current path into line segments
WidenPath()	Redefines the current path in a given device context as the area that would be painted if the path were stroked with the current pen
SetMiterLimit()	Sets the length of the miter joins for the specified device context
OBTAIN PATH INFORMATION:	
GetPath()	Retrieves the coordinates of the endpoints of lines and control points of curves in a path
GetMiterLimit()	Returns the limit for the length of the miter joins in the specified device context
GetPolyFillMode()	Returns the current polygon fill mode

24.6.1 Creating, Deleting, and Converting Paths

A path is initiated by calling the BeginPath() function. The call discards any existing path in the device context and opens a path bracket. The function's general form is as follows:

```
BOOL BeginPath (HDC hdc);
```

The only parameter is the handle to the device context. The function returns TRUE if it succeeds and FALSE if it fails. After the call to BeginPath() is made an application can call, any of the functions in Table 24.11 or 24.12, according to the operating system platform.

The EndPath() function closes a path bracket and selects the path into the specified device context. The function's general form is as follows:

```
BOOL EndPath (HDC hdc);
```

The only parameter is the handle to the device context. The function returns TRUE if it succeeds and FALSE if it fails.

The AbortPath() functions closes and discards any open path bracket on the specified device context. The function's general form is as follows:

```
BOOL AbortPath (HDC hdc);
```

The only parameter is the handle to the device context.

A path bracket is created by calling BeginPath(), followed by one or more of the drawing functions listed in Table 24.11 and 24.12, and closed by a call to EndPath(). At this point applications usually proceed to stroke, fill, or stroke-and-fill the path or to install it as a clipping region. Two possible methods can be followed for converting a path into a clipping region. One method is to use PathToRegion() to create a region and then call ExtSelectClipRgn() to make the region a clipping region. Alternatively, code can call SelectClipPath() and perform both functions in a single call.

SelectClipPath() makes the current path into a clipping region for a specified device context, according to a predefined combination mode. The function's general form is as follows:

```
BOOL SelectClipPath(
        HDC hdc,      // 1
        int iMode     // 2
        );
```

The second parameter is one of the values listed in Table 24.7. The function returns TRUE if it succeeds and FALSE if it fails.

The PathToRegion() function closes an open path and converts is to a region. The function's general form is as follows:

```
HRGN PathToRegion (HDC hdc);
```

The function's only parameter is the handle to the device context. The call assumes that the path in the device context is closed. PathToRegion() returns the handle to the created region. Since there are no path handles, this function provides a way of identifying a particular path, although it must be first converted into a region. Unfortunately, there is no method for converting a region into a path.

24.6.2 Path-Rendering Operations

After a path is created it is possible to render it as an image by stroking it, filling it, or both. In Windows NT it is also possible to directly draw line segments and Bezier curves that form a path, whose end and control points are stored in an array of type POINT.

The StrokePath() function renders the outline of the current path using the current pen. The function's general form is as follows:

```
BOOL StrokePath (HDC hdc);
```

The only parameter is the handle to the device context that contains a closed path. Since a device context can only have a single path, there is no need for further specification. The path is automatically discarded from the device context after it is stroked. The function returns TRUE if it succeeds and FALSE if it fails. Notice that Microsoft Visual C++ documentation does not mention that StrokePath() discards the path automatically. What is worse, the remarks on the StrokeAndFillPath() function suggest that it is possible to first stroke and then fill the same path by making separate calls to the StrokePath() and FillPath() function. In reality, the StrokePath() function destroys the path before exiting execution. A subsequent call to FillPath() has no effect, since there is no longer a path in the device context. This is the reason why the StrokeAndFillPath() function exists. Without this function it would be impossible to both stroke and fill a path.

The FillPath() function closes an open figure in the current path and fills the path interior with the current brush, using the current polygon fill mode. The function's general form is as follows:

```
BOOL FillPath (HDC hdc);
```

The only parameter is the handle to the device context that contains a valid path. Since a device context can only have a single path, there is no need for further specification. The path is automatically discarded from the device context after it is filled. The function returns TRUE if it succeeds and FALSE if it fails.

The StrokeAndFillPath() function closes an open figure in the current path, strokes the path outline using the current pen, and fills the path's interior with the current brush, using the current polygon fill mode. The function's general form is as follows:

```
BOOL StrokeAndFillPath (HDC hdc);
```

The only parameter is the handle to the device context that contains a valid path. The path is automatically discarded from the device context after it is stroked and filled. StrokeAndFillPath() provides the only way for both stroking and filling a path in Windows, since StrokePath() and FillPath() destroy the path after they execute. The function returns TRUE if it succeeds and FALSE if it fails.

The CloseFigure() function draws a line from the current pen position to the figure's start point. The closing line is connected to the figure's first line using the current line join style. The function's general form is as follows:

```
BOOL CloseFigure (HDC hdc);
```

The only parameter is the handle to the device context that contains a valid path. A figure in a path is open unless the CloseFigure() call has been made, even if the figure's starting point and the current point coincide. Usually, the starting point of the figure is the one in the most recent call to MoveToEx().

The effect of closing a figure using the CloseFigure() function is not the same as using a call to a drawing primitive. For example, when the figure is closed with a call to the LineTo() function, end caps are used at the last corner, instead of a join. If the figure is drawn with a thick, geometric pen, the results can be quite different. Figure 24.19 shows the difference between closing a figure by calling LineTo() or by calling CloseFigure().

The triangles in Figure 24.19 are both drawn with a pen style that has a miter join and a round end cap. One of the figures is closed using the LineTo() drawing function and the other one with CloseFigure(). The apex of the triangle closed using the LineTo() function is rounded while the one closed using the CloseFigure() function is mitered. This is due to the fact that two segments drawn with LineTo() do not have a join at a common end point. In this case, the appearance of the apex is determined by the figure's round end cap. On the other hand, when the figure is closed with the CloseFigure() function, the selected join is used in all three vertices.

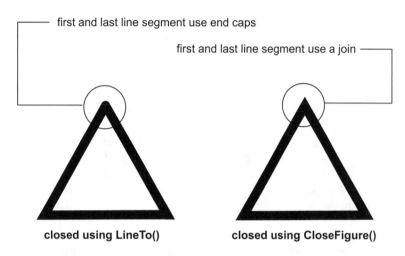

closed using LineTo()   **closed using CloseFigure()**

Figure 24.19 *Figure Closing Differences*

The PolyDraw() function, available only in Windows NT, draws lines segments and Bezier curves. Because of its limited portability we do not discuss it here.

24.6.3 Path Manipulations

Several functions allow modifying existing paths or determining the path characteristics. FlattenPath() converts curves in the current path into line segments. The function's general form is as follows:

```
BOOL FlattenPath (HDC hdc);
```

The only parameter is the handle to the device context that contains a valid path. The function returns TRUE if it succeeds and FALSE if it fails. There are few documented uses for the FlattenPath() function. The screen appearance of a flattened path is virtually undetectable. The documented application of this function is to fit text on a curve. Once a curved path has been flattened, a call to GetPath() retrieves the series of line segments that replaced the curves of the original path. Code can now use this information to fit the individual characters along the line segments.

The WidenPath() function redefines the current path in a given device context as the area that would be painted if the path were stroked with the current pen. The function's general form is as follows:

```
BOOL WidenPath (HDC hdc);
```

The only parameter is the handle to the device context that contains a valid path. Any Bezier curves in the path are converted to straight lines. The function makes a difference when the current pen is a geometric pen or when it has a width or more than one device unit. WidenPath() returns TRUE if it succeeds and FALSE if it fails. This is another function with few documented uses. The fact that curves are converted into line segments suggests that it can be used in text fitting operations, such as the one described for the FlattenPath() function.

The SetMiterLimit() function sets the length of the miter joins for the specified device context. The function's general form is as follows:

```
BOOL SetMiterLimit(
        HDC hdc,              // 1
        FLOAT eNewLimit,      // 2
        PFLOAT peOldLimit     // 3
        );
```

The first parameter is the handle to the device context. The second parameter specifies the new miter limit. The third parameter is a pointer to a floating-point variable that holds the previous miter limit. If this parameter is NULL the value of the previous miter limit is not returned. The function returns TRUE if it succeeds and FALSE if it fails.

The miter length is the distance from the intersection of the line walls on the inside of the join to the intersection of the line walls on the outside of the join. The miter limit is the ratio between the miter length, to the line width. Figure 24.20 shows the miter length, the line width, and the miter limit.

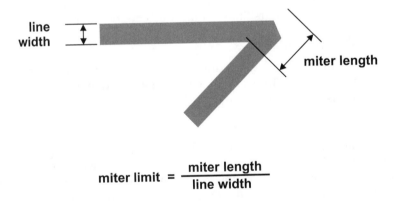

Figure 24.20 *Miter Length, Line Width, and Miter Limit*

The miter limit determines if the vertex of a join that was defined with the PS_JOIN_MITER style (see Figure 23.3 in the previous chapter) is drawn using a miter or a bevel join. If the miter limit is not exceeded, then the join is mitered. Otherwise, it is beveled. Mitered and beveled joins apply only to pens created with the ExtCreatePen() function and to stroked paths. The following code fragment shows the creation of two joins.

```
static HPEN     fatPen;    // Handle for pen
static FLOAT    oldMiter;  // Storage for miter limit
.
.
.
// Create a special pen
fatPen = ExtCreatePen (PS_GEOMETRIC | PS_SOLID |
        PS_ENDCAP_ROUND | PS_JOIN_MITER,
        15,
        &fatBrush, 0, NULL);
SelectBrush (hdc, GetStockObject (LTGRAY_BRUSH));
SelectPen (hdc, fatPen);
// Draw first angle
BeginPath (hdc);
MoveToEx (hdc, 100, 100, NULL);
LineTo (hdc, 250, 100);
LineTo (hdc, 150, 180);
EndPath (hdc);
StrokePath (hdc);
// Draw second angle
GetMiterLimit (hdc, &oldMiter);
SetMiterLimit (hdc, 2, &oldMiter);
BeginPath (hdc);
MoveToEx (hdc, 300, 100, NULL);
LineTo (hdc, 450, 100);
LineTo (hdc, 350, 180);
EndPath (hdc);
StrokePath (hdc);
SetMiterLimit (hdc, oldMiter, NULL);
```

Figure 24.21 is a screen snapshot of the execution of the preceding code fragment. Notice in Figure 24.21 that the image on the left, in which the default miter limit of 10 is used, is drawn with a miter join. In the right-hand figure the miter limit was changed to 1, therefore, the figure is drawn using a bevel join.

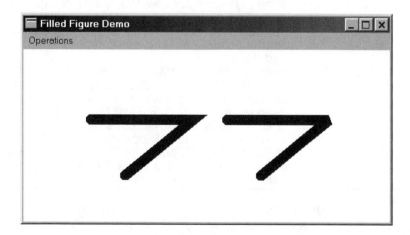

Figure 24.21 *Effect of the SetMiterLimit() Function*

24.6.4 Obtaining Path Information

Several functions provide information about the path, or about GDI parameters that affect the path. The GetPath() function retrieves the coordinates of the endpoints of lines and control points of curves in a path. The function's structure is quite similar to the PolyDraw() function discussed in Chapter 20. GetPath() is related to the PolyDraw() function mentioned earlier. The function's general form is as follows:

```
int GetPath(
        HDC hdc,              // 1
        LPPOINT lpPoints,    // 2
        LPBYTE lpTypes,      // 3
        int nSize            // 4
        );
```

The first parameter identifies the device context. The second parameter is a pointer to an array of POINT structures that contains the endpoints of the lines and the control points of the curves that form the path. The third parameter is an array of type BYTE which contains identifiers that define the purpose of each of the points in the array. The fourth parameter is the count of the number of points in the array of points. The function returns TRUE if it succeeds and FALSE otherwise. Table 24.14 lists the constants used to represent the identifiers entered in the function's third parameter.

Table 24.14

Constants for the GetPath() Vertex Types

TYPE	MEANING
PT_MOVETO	This point starts a disjoint figure. The point becomes the new current pen position.
PT_LINETO	A line is to be drawn from the current position to this point, which then becomes the new current pen position.
PT_BEZIERTO	This is a control point or end node for a Bezier curve. This constant always occurs in sets of three. The current position defines the start node for the Bezier curve. The other two coordinates are control points. The third entry (if coded) is the end node.
PT_CLOSEFIGURE	The figure is automatically closed after the PT_LINETO or PT_BEZIERTO type for this point is executed. A line is drawn from the end point to the most recent PT_MOVETO or MoveTo() point. The PT_CLOSEFIGURE constant is combined by means of a bitwise OR operator with a PT_LINETO or PT_BEZIERTO constant. This indicates that the corresponding point is the last one in a figure and that the figure is to be closed.

The GetMiterLimit() function, which was mentioned in regards to SetMiterLimit(), returns the limit for the length of the miter join in the specified device context. The function's general form is as follows:

```
BOOL GetMiterLimit(
```

```
HDC hdc,          // 1
PFLOAT peLimit    // 2
);
```

The first parameter is the handle to the device context. The second parameter stores the current miter limit. The function returns TRUE if it succeeds and FALSE if it fails.

The GetPolyFillMode() returns the current polygon fill mode. The only parameter is the handle to the device context. The value returned is either ALTERNATE or WINDING. The fill more affects the operation of the FillPath() and StrokeAndFillPath() functions.

24.7 Filled Figures Demo Program

The program named FIL_DEMO, located in the Filled Figure Demo project folder of the book's on-line software package, is a demonstration of the graphics functions and operations discussed in this chapter. The first entry in the Operations menu shows the offset of the hatch origin to visually improve a filled rectangle. The main menu contains several pop-up menus that demonstrate most of the graphics primitives discussed in the text. These include drawing solid figures, operations on rectangles, regions, clipping, and paths. Another menu entry demonstrates the use of the SetMiterLimit() function. Many of the illustrations used in this chapter were taken from the images displayed by the demonstration program.

Chapter 25

Displaying Bit-Mapped Images

Chapter Summary

This chapter is about bitmaps. A bitmap is a digitized image in which each dot is represented by a numeric value. Bitmap images are used in graphics programming at least as frequently as vector representation. The high resolution and extensive color range of current video display systems allows encoding bitmapped images with photo-realistic accuracy. The powerful storage and processing capabilities of the modern day PC make possible for software to rapidly and effectively manipulate and transform bitmaps. Computer simulations, virtual reality, artificial life, and electronic games are fields of application that rely heavily on bitmap operations.

25.0 Raster and Vector Graphics

The two possible ways of representing images in a computer screen, or a digital graphics device, are based on vector and raster graphics technologies. All of the graphics primitives discussed in previous chapters are based on vector techniques. Commercially speaking, vector graphics are associated with drawing programs, while raster graphics are associated with painting programs. The vector representation of a line consists of its start and end points and its attributes, which usually include width, color, and type. The raster representation of the same line is a mapping of adjacent screen dots. Most current computer systems are raster based, that is, the screen is a two-dimensional pixel grid and all graphics objects are composed of individual screen dots, as described in Chapter 16. Vector graphics are a way of logically defining images, but the images must be rasterized at display time.

Vector and raster representations have their advantages and drawbacks. Vector images can be transformed mathematically (as you will see later in the book); they can also be scaled without loss of quality. Furthermore, vector images are usually more compact. Many images can be conveniently represented in vector form, such as an engineering drawing composed of geometrical elements that can be mathematically defined. The same applies to illustrations, and even to artwork created by combining geometrical elements.

25.1 The Bitmap

An image of Leonardo's Mona Lisa, or a photograph of the Crab nebulae, can hardly be vectorized. When geometrical elements are not present, or when the image is rich in minute details, vector representations cease to be practical. In these cases it is better to encode the image as a data structure containing all the individual picture elements. This pixel-by-bit encoding is called a bitmap.

A bitmap is a form of raster image. A raster image can be defined as pixel-by-pixel enumeration, usually in scan-line order. A bitmap is a formatted raster image encoded according to some predefined standard or convention. A raster image, on the other hand, can be in raw format. For example, a scanning instrument onboard a satellite or space craft acquires and transmits image data in raster form. Once received, the raster data can be processed and stored as bitmaps that can be easily displayed on a computer screen. Television images are in raster form.

A bitmap is a memory object, not a screen image. It is the memory encoding of an image at the pixel level. Although bitmaps are often represented in image form, it is important to remember that a bitmap is a data construct. Bitmaps cannot be easily transformed mathematically, as is the case with vector images, nor can they be scaled without some loss of information. However, bitmaps offer a more faithful reproduction of small details than is practical in vector representations.

In bit-mapping, one or more memory bits are used to represent the attribute of a screen pixel. The simplest and most compact scheme is that in which a memory bit represents a single screen pixel: if the bit is set, so is the pixel. This one-bit-to-one-pixel representation leaves no choice about pixel attributes, that is, a pixel is either set or not. In a monochrome video system a set bit can correspond to a bright pixel and a reset bit to a black one. Figure 25.1 shows a bit-to-pixel image and bitmap.

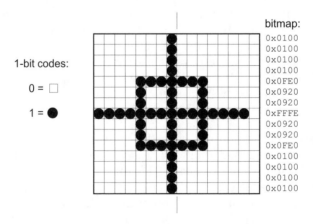

Figure 25.1 *One Bit-Per-Pixel Image and Bitmap*

Most current video systems support multiple attributes per screen pixel. PC color video systems are usually capable of representing 16, 256, 65,535, and 16.7 million colors. Shades of gray can also be encoded in a bitmap. The number of bits devoted to each pixel determines the attribute range. Sixteen colors or shades of gray can be represented in four bits. Eight bits can represent 256 colors. Sixteen bits encode 65,535 colors. So-called true color systems, which can display approximately 16.7 million colors, require 24 data bits per screen dot. Some color data formats use 32-bits per pixel, but this representation is actually 24 bits of pixel data plus 8 bits of padding. Figure 25.2 shows an image and bitmap in which each pixel is represented by two memory bits.

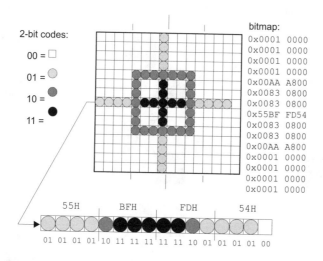

Figure 25.2 *Two Bits-Per-Pixel Image and Bitmap*

In Figure 25.2, each screen pixel can be in one of four attributes: background, light gray, dark gray, or black. In order to represent these four states it is necessary to assign a 2-bit field for each screen pixel. The four bit combinations that correspond with the attribute options are shown on the left side of Figure 25.2. At the bottom of the illustration is a map of one of the pixel rows, with the corresponding binary codes for each pixel, as well as the hexadecimal digits of the bitmap.

25.1.1 Image Processing

The fact that raster images cannot be transformed mathematically does not mean that they cannot be manipulated and processed. Image Processing is a major field of computer graphics. It deals exclusively with the manipulation and analysis of two-dimensional pictorial data in digital form. In practice, this means processing raster data. Although digital image processing originated in the science programs of the National Aeronautics and Space Administration (NASA) it has since been applied to many other technological fields, including biological research, document processing, factory automation, forensics, medical diagnostics, photography, prepress publishing operations, space exploration, and special effects on film and video.

25.1.2 Bitblt Operations

The fundamental bitmap operation is the bit block transfer, or bitblt (pronounced bit-blit). In the bitblt a rectangular block of memory bits, representing pixel attributes, is transferred as a block. If the destination of the transfer is screen memory, then the bitmapped image is displayed. At the time of the transfer, the source and destination bit blocks can be combined logically or arithmetically, or a unary operation can be performed on the source or the destination bit blocks. Figure 25.3 shows several binary and unary operations on bit blocks.

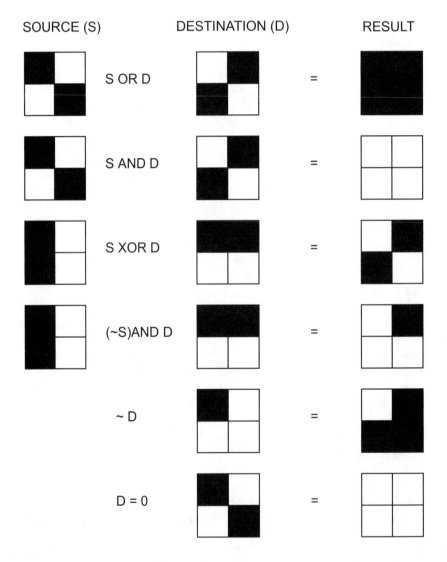

Figure 25.3 *Binary and Unary Operations on Bit Blocks*

25.2 Bitmap Constructs

Several Windows data structures and concepts relate to bitmaps:

- Bitmap formats
- Structures for bitmap operations
- Creating bitmap resources

You must understand these concepts in order to be able to manipulate and display bitmaps.

25.2.1 Windows Bitmap Formats

The computer establishment has created hundreds of bitmap image formats over the past two or three decades; among the best known are GIF, PCX, Targa, TIFF, and Jpeg. Windows does not provide support for manipulating image files in any commercial format. The only bitmap formats that can be handled are those specifically created for Windows. These are the original device-specific bitmaps (extension BMP) and the newer device-independent bitmaps (extension DIB).

The device-specific bitmap format was created in Windows 3.0. It can store images with up to 24-bit color. Files in BMP format are uncompressed, which makes them quick to read and display. The disadvantage of uncompressed files is that they take up considerable memory and storage space. Another objection to using the device-specific bitmap is that the BMP header cannot directly encode the original colors in the bitmap. For this reason, the format works well for copying parts of the screen to memory and pasting them back to other locations. But problems often occur when the device-specific bitmap must be saved in a disk file and then output to a different device, since the destination device may not support the same colors as the original one. For this reason it is usually preferable to use the device-independent format, although its data structures are slightly larger.

Note that many Windows bitmap functions operate both on device-specific and device-independent bitmaps. The Bitmap Demo project on the book's on-line software package displays bitmaps in BMP and DIB format using the same Windows functions. In the remainder of this chapter we deal exclusively with device-independent bitmaps encoded in DIB format, although device independent bitmaps can also be stored in .BMP files.

25.2.2 Windows Bitmap Structures

There are eleven structures directly or indirectly related to Windows bitmaps, listed in Table 25.1, on the following page.

25.2.3 The Bitmap as a Resource

In Visual C++ a bitmap can be a program resource. Developer Studio includes a bitmap editor that can be used for creating and editing bitmaps that do not exceed a certain size and color range. Bitmaps in BMP or DIB format created with other applications can also be imported into a program. In this case you can select the Resource command in the Insert menu, and then select the bitmap resource type and click the

Import button. The dialog that appears allows you to browse through the file system in order to locate the bitmap file. It is usually better to copy the bitmap file into the project's folder, thus ensuring that it is not deleted or modified inadvertently.

Table 25.1

Bitmap-Related Structures

STRUCTURE	CONTENTS
BITMAP	Defines the width, height, type, format, and bit values of a bitmap.
BITMAPCOREHEADER	Contains information about the dimensions and color format of a DIB.
BITMAPCOREINFO	Contains a BITMAPCOREHEADER structure and a RGBTRIPLE structure array with the Bitmap's color intensities.
BITMAPFILEHEADER	Contains information about a file that holds a DIB.
BITMAPINFO	Defines the dimensions and color data of a DIB. Includes a BITMAPINFOHEADER structure and a RGBQUAD structure.
BITMAPINFOHEADER	Contains information about the dimensions and color format of a DIB.
COLORADJUSTMENT	Defines the color adjustment values used by the StretchBlt() and StretchDIBits() functions.
DIBSECTION	Contains information about a bitmap created by means of the CreateDIBSection() function.
RGBQUAD	Describes the relative intensities of the red, green, and blue color components. The fourth byte of the structure is reserved.
RGBTRIPLE	Describes the relative intensities of the red, green, and blue color components. The BITMAPCOREINFO structure contains an array of RGBTRIPLE structures.
SIZE	Defines the width and height of a rectangle, which can be used to represent the dimensions of a bitmap.

Some of the structures in Table 25.1 are listed and described in this chapter.

Once a bitmap has been imported in the project, it is listed in the Resource tab of the Project Workspace pane. At this point it is possible to use the MAKEINTRESOURCE macro to convert it into a resource type that can be managed by the application. Since the final objective is to obtain the handle to a bitmap, code usually proceeds as follows:

```
HBITMAP        aBitmap;
...
aBitmap = LoadBitmap (pInstance,
                MAKEINTRESOURCE (IDB_BITMAP1));
```

In this case, IDB_BITMAP1 is the resource name assigned by Developer Studio when the bitmap was imported. It is listed in the Bitmap section of the Resource tab in the Project Workspace pane and also by the Resource Symbols command in the View menu.

25.3 Bitmap Programming Fundamentals

The bitmap is an object of the device context, such as a brush, a pen, a font, or a region. Bitmap manipulations and display functions in Windows are powerful, but not without some limitations and complications. This overview presents a preliminary discussion of simple bitmap programming. Later in this chapter we get into more complicated operations.

25.3.1 Creating the Memory DC

The unique characteristic of a bitmap is that it can be selected only into a memory device context. The memory DC is defined as a device context with a display surface. It exists only in memory and is related to a particular device context. In order to use a memory device context you must first create it. The CreateCompatibleDC() function is used for this purpose. Its general form is as follows:

```
HDC CreateCompatibleDC (HDC hdc);
```

Its only parameter is the handle to the device context with which the memory device context is to be compatible. This parameter can be NULL if the memory device context is to be compatible with the video screen. If the function succeeds, it returns the handle to the memory device context. The call returns NULL if it fails. CreateCompatibleDC() assumes that the device context supports raster operations, which is true in all PC video systems, but not necessarily so for other graphics devices, such as a plotter. The GetDeviceCaps() function with the RASTERCAPS constant can be used to determine if a particular device supports raster operations (see Table 25.4).

Like the device context, a memory device context has a mapping mode attribute. Applications often use the same mapping mode for the memory device context and for the device context. In this case the SetMapMode() and GetMapMode() functions can be combined as follows:

```
HDC        hdc;      // Handle to device context
HDC        memDC;    // Handle to memory device context
.
.
.
SetMapMode (memDC, GetMapMode (hdc));
```

25.3.2 Selecting the Bitmap

When the memory device context is created, it is assigned a display surface of a single monochrome pixel. The single pixel acts as a placeholder until a real one is selected into the memory DC. The SelectObject() function, discussed previously, can be used to select a bitmap into a memory device context, but the SelectBitmap() macro, discussed in Chapter 23, serves the same purpose. Both, SelectObject() and SelectBitmap() have the same interface. For SelectBitmap() it is as follows:

```
HBITMAP SelectBitmap (HDC, HBITMAP);
                       |     -------
                       |         |
                       1         2
```

The first parameter must be the handle of a memory device context. The second parameter is the handle to the bitmap being installed. If the call succeeds, the macro returns the handle to the device context object being replaced. If the call fails, it returns NULL. Using the SelectBitmap() macro instead of the SelectObject() function produces code that is correct and the coding is made easier. Recall that programs that use the object selection macros must include the windowsx.h file.

The handle to the bitmap used in the SelectBitmap() macro is usually obtained with the LoadBitmap() function previously discussed.

25.3.3 Obtaining Bitmap Dimensions

Bitmap functions often require information about the dimensions and other characteristics of the bitmap. For example, the function most often used to display a bitmap is BitBlt(); it requires the width and height of the bitmap. If the bitmap is loaded as a resource from a file, the application must obtain the bitmap dimensions before blitting it to the screen. The GetObject() function is used to obtain information about a bitmap. The function's general form is as follows:

```
int GetObject(
        HGDIOBJ hgdiobj,   // 1
        int cbBuffer,      // 2
        LPVOID lpvObject   // 3
);
```

The first parameter is the handle to a graphics object; in this case, the handle to a bitmap. The second parameter is the size of the buffer that holds the information returned by the call. In the case of a bitmap, this parameter can be coded as sizeof (BITMAP). The third parameter is a pointer to the buffer that holds the information returned by the call. In the case of a bitmap, the buffer is a structure variable of type BITMAP. The BITMAP structure is defined as follows:

```
typedef struct tagBITMAP {
   LONG    bmType;        // Must be zero
   LONG    bmWidth;       // bitmap width (in pixels)
   LONG    bmHeight;      // bitmap height  (in pixels)
   LONG    bmWidthBytes;  // bytes per scan line
   WORD    bmPlanes;      // number of color planes
   WORD    bmBitsPixel;   // bits per pixel color
   LPVOID bmBits;         // points to bitmap values array
} BITMAP;
```

The structure member bmType specifies the bitmap type. It must be zero. The member bmWidth specifies the width, in pixels, of the bitmap. Its value must be greater than zero. The member bmHeight specifies the height, in pixels, of the bitmap. The height must be greater than zero. The bmWidthBytes member specifies the number of bytes in each scan line. Since Windows assumes that the bitmap is word-aligned, its value must be divisible by 2. The member bmPlanes specifies the number of color planes. The member bmBitsPixel specifies the number of bits required to indicate the color of a pixel. The member bmBits points to the location of the bit values for the bitmap. It is a long pointer to an array of character-size values.

When the target of the GetObject() call is a bitmap, the information returned is the structure members related to the bitmap width, height, and color format. The GetObject() function cannot be used to read the value of the pointer to the bitmap data in the bmBits structure member. GetDIBits() retrieves the bit data of a bitmap.

The mapping mode can also be a factor in regard to bitmap data. The GetObject() function returns the bitmap width and height in the BITMAP structure. The values returned are in pixels, which are device units. This works well if the mapping mode of the memory device context is MM_TEXT, but is not acceptable in any of the mapping modes that use logical units. The DPtoLP() function allows the conversion of device coordinates (pixels) into logical coordinates for a particular device context. The function's general form is as follows:

```
BOOL DPtoLP(
        HDC hdc,              // 1
        LPPOINT lpPoints,     // 2
        int nCount            // 3
);
```

The first parameter identifies the device context. In the case of a bitmap, it is a memory device context. The second parameter points to an array of POINT structures that holds the transformed values for the x- and y-coordinates. The third parameter holds the number of points in the array specified in the second parameter. The function returns TRUE if it succeeds and FALSE if it fails.

A bitmap size is defined by two values: the x-coordinate is the width and the y-coordinate the height. When a call to DPtoLP() is made to obtain the bitmap size, the third parameter is set to 1. This indicates that the coordinates to be transformed refer to a single point. By the same token, the second parameter is a pointer to a single POINT structure that holds the bitmap width in the x member and the bitmap height in the y member.

25.3.4 Blitting the Bitmap

Once a bitmap has been selected onto a memory device context, and the code has obtained the necessary information about its width and height, it is possible to display it at any screen position by blitting the memory stored bitmap onto the screen. The BitBlt() function is the simplest and most direct method of performing the bitmap display operation. The function's general form is as follows:

```
BOOL BitBlt(
        HDC hdcDest,  // 1
        int nXDest,   // 2
        int nYDest,   // 3
        int nWidth,   // 4
        int nHeight,  // 5
        HDC hdcSrc,   // 6
```

```
    int nXSrc,     // 7
    int nYSrc,     // 8
    DWORD dwRop    // 9
);
```

The first parameter identifies the destination device context. If the call to BitBlt() is made to display a bitmap, this will be the display context. The second and third parameters are the x- and y-coordinates of the upper-left corner of the destination rectangle. Which is also the screen location where the upper-left corner of the bitmap is displayed. The fourth and fifth parameters are the width and height of the bitmap, in logical units. The sixth parameter is the source device context. In the case of a bitmap display operation, this parameter holds the memory device context where the bitmap is stored. The seventh and eighth parameters are the x- and y-coordinates of the source bitmap. Since the blitted rectangles must be of the same dimensions, as defined by the fourth and fifth parameters, the seventh and eighth parameters are usually set to zero.

The ninth parameter defines the raster operation code. These codes are called ternary raster operations. They differ from the binary raster operation codes (ROP2) discussed in Chapter 22 in that the ternary codes take into account the source, the destination, and a pattern determined by the brush currently selected in the device context. There are 256 possible raster operations, fifteen of which have symbolic names defined in the windows.h header file. The raster operation code determines how the color data of the source and destination rectangles, together with the current brush, are to be combined. Table 25.2 lists the fifteen raster operations with symbolic names.

Table 25.2

Symbolic Names for Raster Operations

NAME	DESCRIPTION
BLACKNESS	Fills the destination rectangle using the color associated with index 0 in the physical palette. The default value is black.
DSTINVERT	Inverts the destination rectangle.
MERGECOPY	Merges the colors of the source rectangle with the specified pattern using an AND operation.
MERGEPAINT	Merges the colors of the inverted source rectangle with the colors of the destination rectangle using an OR operation.
NOTSRCCOPY	Inverts the bits in the source rectangle and copies it to The destination.
NOTSRCERASE	Combines the colors of the source and destination rectangles using an OR operation, and then inverts the result.
PATCOPY	Copies the specified pattern to the destination bitmap.
PATINVERT	Combines the colors of the specified pattern with the colors of the destination rectangle using an XOR operation.

(continues)

Table 25.2

Symbolic Names for Raster Operations (continued)

NAME	DESCRIPTION
PATPAINT	Combines the colors of the pattern with the colors of the inverted source rectangle using an OR operation. The result of this operation is combined with the colors of the destination rectangle using an OR operation.
SRCAND	Combines the colors of the source and destination rectangles using an AND operation SRCCOPY. Copies the source rectangle directly to the destination rectangle. This is, by far, the most-used mode in bitblt operations.
SRCERASE	Combines the inverted colors of the destination rectangle with the colors of the source rectangle using an AND operation.
SRCINVERT	Combines the colors of the source and destination rectangles using an XOR operation.
SRCPAINT	Combines the colors of the source and destination rectangles using an OR operation.
WHITENESS	Fills the destination rectangle using the color associated with index 1 in the physical palette. The default value is White.

25.3.5 A Bitmap Display Function

Displaying a bitmap is a multistage process that includes the following operations:

1. Creating a memory device context.

2. Selecting the bitmap into the memory device context.

3. Obtaining the bitmap dimensions and converting device units to logical units.

4. Blitting the bitmap onto the screen according to a ternary raster operation code.

Many graphics applications can make use of a function that performs all of the previous operations. The function named ShowBitmap() is used in the Bitmap Demo project on the book's software on-line. The function's prototype is as follows:

```
void ShowBitmap (HDC, HBITMAP, int, int, DWORD);
                  |   -------   |    |    -----
                  |      |      |    |      |
                  |      |      |    |      |
                  1      2      3    4      5
```

The first parameter is the handle to the device context. The ShowBitmap() function creates its own memory device context. The second parameter is the handle to the bitmap that is to be displayed. The third and fourth parameters are the screen coordinates for displaying the upper-left corner of the bitmap. The fifth parameter is the ternary ROP code used in blitting the bitmap. If this parameter is NULL then the bitmap is displayed using the SRCCOPY raster operation. The following is a listing of the ShowBitmap() function:

```
void ShowBitmap (HDC hdc, HBITMAP hBitmap, int xStart, int yStart,\
                 DWORD rop3) {

BITMAP      bm;                 // BITMAP structure
HDC         memoryDc;           // Handle to memory DC
POINT       ptSize;            // POINT for DC
POINT       ptOrigin;          // POINT for memory DC
int         mapMode;           // Mapping mode

    // Test for NULL ROP3 code
    if (rop3 == NULL)
        rop3 = SRCCOPY;

    memoryDc = CreateCompatibleDC (hdc); // Memory device
                                         // handle
    mapMode = GetMapMode (hdc);          // Obtain mapping
                                         // mode
    SetMapMode (memoryDc, mapMode);      // Set memory DC
                                         // mapping mode

    // Select bitmap into memory DC
    // Note: assert statement facilitates detecting invalid
    //       bitmaps during program development
    assert (SelectBitmap (memoryDc, hBitmap));

    // Obtain bitmap dimensions
    GetObject (hBitmap, sizeof(BITMAP), (LPVOID) &bm);

    // Convert device units to logical units
    ptSize.x = bm.bmWidth;
    ptSize.y = bm.bmHeight;
    DPtoLP (hdc, &ptSize, 1);

    ptOrigin.x = 0;
    ptOrigin.y = 0;
    DPtoLP (memoryDc, &ptOrigin, 1);

    // Bitblt bitmap onto display memory
    BitBlt( hdc, xStart, yStart, ptSize.x, ptSize.y, memoryDc,
            ptOrigin.x, ptOrigin.y, rop3);

    // Delete memory DC with bitmap
    DeleteDC (memoryDc);
}
```

25.4 Bitmap Manipulations

In addition to displaying a bitmap stored in a disk file, graphics applications often need to perform other bitmap-related operations. The following are among the most common operations:

• Creating and displaying a hard-coded bitmap

• Creating a bitmap in heap memory

• Creating a blank bitmap and filling it by means of GDI functions

• Creating a system-memory bitmap which applications can access directly

• Using a bitmap to create a pattern brush

25.4.1 Hard-Coding a Monochrome Bitmap

With the facilities available in Developer Studio for creating bitmaps, the programmer is seldom forced to hard-code a bitmap. Our rationale for discussing this option is that hard-coding bitmaps is a basic skill for a programmer, and that it helps to understand bitmaps in general.

A monochrome bitmap has one color plane and is encoded in a one-bit-per-pixel format. In Windows, a monochrome bitmap is displayed by showing the 0-bits in the foreground color and the 1-bits in the background color. If the screen has a white foreground and a black background, 0-bits in the bitmap are displayed as white pixels, and vice versa. If the bitmap is to be blitted on the screen using the BitBlt() function, the action of the bits can be reversed by changing the raster operation code. This gives the programmer the flexibility of using either zero or one bits for background or foreground attributes. Figure 25.4 shows a hard-coded monochrome bitmap.

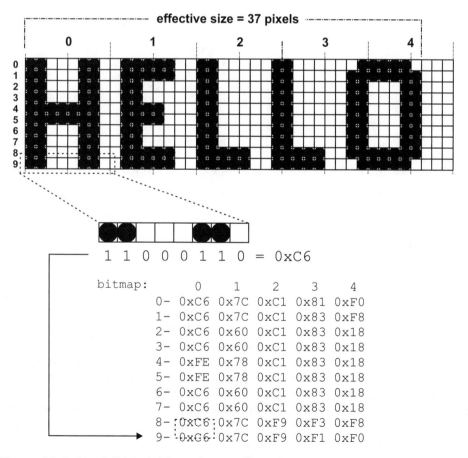

Figure 25.4 *Hard-Coded, Monochrome Bitmap*

In Figure 25.4 the dark pixels in the image are represented by 0-bits in the bitmap. The resulting data structure has five bytes per scan line and a total of 10 scan lines. The illustration shows how eight pixels in the bottom row of the image are represented by eight bits (one byte) of the bitmap. The dark pixels are encoded as 0-bits and the light pixels as 1-bits. In order to display this bitmap so that the letters are black on a white screen background, the black and white pixels have to be reversed by changing the raster operation mode from the default value, SCRCOPY, to NOTSCRCOPY, as previously explained.

Once the bit image of a monochrome bitmap has been calculated, we can proceed to store this bit-to-pixel information in an array of type BYTE. Windows requires that bitmaps be word-aligned; therefore, each scan line in the array must have a number of bytes divisible by 2. In regards to the bitmap in Figure 25.4, you would have to add a padding byte to each scan line in order to satisfy this requirement. The resulting array could be coded as follows:

```
//                               0     1     2     3     4     5
static BYTE hexBits[] = {0xc6, 0x7c, 0xc1, 0x81, 0xf0, 0x0,
                         0xc6, 0x7c, 0xc1, 0x83, 0xf8, 0x0,
                         0xc6, 0x60, 0xc1, 0x83, 0x18, 0x0,
                         0xc6, 0x60, 0xc1, 0x83, 0x18, 0x0,
                         0xfe, 0x78, 0xc1, 0x83, 0x18, 0x0,
                         0xfe, 0x78, 0xc1, 0x83, 0x18, 0x0,
                         0xc6, 0x60, 0xc1, 0x83, 0x18, 0x0,
                         0xc6, 0x60, 0xc1, 0x83, 0x18, 0x0,
                         0xc6, 0x7c, 0xf9, 0xf3, 0xf8, 0x0,
                         0xc6, 0x7c, 0xf9, 0xf1, 0xf0, 0x0};
//                                                       |
//                              padding byte -----------|
```

The CreateBitmap() function can be used to create a bitmap given the bit-to-pixel data array. The function's general form is as follows:

```
HBITMAP CreateBitmap(
        int nWidth,           // 1
        int nHeight,          // 2
        UINT cPlanes,         // 3
        UINT cBitsPerPel,     // 4
        CONST VOID *lpvBits   // 5
);
```

The first parameter is the actual width of the bitmap, in pixels. In relation to the bitmap in Figure 25.4, this value is 37 pixels, as shown in the illustration. The second parameter is the number of scan lines in the bitmap. The third parameter is the number of color planes. In the case of a monochrome bitmap this value is 1. The fourth parameter is the number of bits per pixel. In a monochrome bitmap this value is also 1. The fifth parameter is a pointer to the location where the bitmap data is stored. If the function succeeds, the returned value is the handle to a bitmap. If the function fails, the return value is NULL. The following code fragment initializes a monochrome bitmap using the bitmap data in the hexBits[] array previously listed:

```
    static HBITMAP      bmImage1;
    .
```

```
       .
       .
// Initialize monochrome bitmap
bmImage1 = CreateBitmap (37, 10, 1, 1, hexBits);
```

Alternatively, a bitmap can be defined using a structure of type BITMAP, listed previously in this chapter and in Appendix F. Before the bitmap is created, the structure members must be initialized, as in the following code fragment:

```
static BITMAP        monoBM;
       .
       .
       .
// Initialize data structure for a monochrome bitmap
monoBM.bmType = 0;        // must be zero
monoBM.bmWidth = 37;      // actual pixels used
monoBM.bmHeight = 10;     // scan lines
monoBM.bmWidthBytes = 6;  // width (must be word aligned)
monoBM.bmPlanes = 1;      // 1 for monochrome bitmaps
monoBM.bmBitsPixel = 1;   // 1 for monochrome bitmaps
monoBM.bmBits = (LPVOID) &hexBits; // address of bit field
```

When the bitmap data is stored in a structure of type BITMAP, the bitmap can be created by means of the CreateBitmapIndirect() function. The function's general form is as follows:

```
HBITMAP CreateBitmapIndirect(
        CONST BITMAP *lpbm     // 1
);
```

The function's only parameter is a pointer to a structure of type BITMAP. If the function succeeds, the returned value is the handle to a bitmap. If the function fails, the return value is NULL. The following code fragment uses CreateBitmapIndirect() to initialize a monochrome bitmap using the bitmap data in the hexBits[] array previously listed:

```
static HBITMAP        bmImage1;
       .
       .
       .
// Initialize monochrome bitmap
bmImage1 = CreateBitmapIndirect (&monoBM);
```

Whether the bitmap was created using CreateBitmap() or CreateBitmapIndirect(), it can now be displayed by means of the ShowBitmap() function developed and listed previously in this chapter.

25.4.2 Bitmaps in Heap Memory

A bitmap can take up a considerable amount of memory or storage resources. For example, a 1200-by-1200 pixel bitmap encoded in 32-bit color takes up approximately 5.7 Mb. Applications that store large bitmaps in their own memory space can run into memory management problems. One possible solution is to store large bitmaps in dynamically allocated memory, which can be freed as soon as the bitmap is no longer needed. Note that freeing memory where the bitmap is stored does not affect the screen image.

Several programming techniques can be used to allocate and release heap memory for a bitmap. The one most suitable depends on the particular needs of each particular programming problem. In one case you may allocate heap memory for a bitmap during WM_CREATE message processing, and then use this allocated space to create or copy different bitmaps during program execution. The allocated memory can then be freed during WM_DESTROY processing. Another option is to allocate memory at the time it is required, create or copy the bitmap to this allocated space, and deallocate the memory when the bitmap is no longer needed.

Many fables and fantastic theories have originated in the complications and misunderstandings of Windows memory management. Although most of these problems were corrected in Windows 3.1, an entire programming subculture still thrives on discussions related to these topics. A programmer encountering memory management in Windows finds that there are three separate sets of memory allocation and deallocation operators that serve apparently identical purposes: the C++ operators new and delete, the traditional C operators malloc and free, and the Windows kernel functions LocalAlloc(), GlobalAlloc(), LocalFree(), and GlobalFree().

In Win32 programming, the first simplification is a result of the fact that there is no difference between the global and the local heaps. Therefore, GlobalAlloc() and LocalAlloc(), as well as GlobalFree() and LocalFree(), actually perform virtually identical functions. Because of their greater flexibility we will use GlobalAlloc() and GlobalFree() instead of new and delete or malloc and free operators from the C++ and C libraries. Another reason for this preference is that most Windows compilers implement malloc and new in terms of GlobalAlloc(); therefore, the traditional operators offer no advantage.

Traditionally, three types of memory are documented as being available to applications: fixed memory, moveable memory, and discardable memory. The justification for preferring moveable memory disappeared with Windows 95, in which a memory block can be moved in virtual memory while retaining the same address. For the same reason, the use of fixed memory no longer needs to be avoided. Discardable memory is only indicated when the data can be easily recreated, which is not usually the case with image data structures. In conclusion, fixed memory is usually quite suitable for dynamically storing bitmaps and other image data.

Note that the terms moveable and movable are both accepted, although movable is preferred. However, the Windows API contains the constants GMEM_MOVEABLE and LMEM_MOVEABLE. For this reason we have used the former.

In this section we discuss the bare essentials of memory allocation and deallocation in Windows. The topic of Windows memory management can easily fill a good-size volume. Graphics applications are often memory-intensive and may require sophisticated memory management techniques. A graphics programmer should have a thorough knowledge of Win32 memory architecture, virtual memory, and heap management. The book Advanced Windows, by Richter (see Bibliography) has chapters devoted to each of these topics.

The GlobalAlloc() function is used to allocate memory from the default heap. The default heap is initially 1Mb, but under Windows, this heap grows as it becomes necessary. The function's general form is as follows:

```
HGLOBAL GlobalAlloc(
        UINT uFlags,      // 1
        DWORD dwBytes     // 2
);
```

The first parameter is a flag that determines how memory is allocated to the caller. Table 25.3 lists the most commonly used allocation flags. The second parameter is the number of bytes of memory to be allocated. If it succeeds, the function returns the handle to the allocated memory object. If the allocation flag is GMEM_FIXED, the handle can be directly cast into a pointer and the memory used. If the allocated memory is not fixed, then it must be locked using GlobalLock() before it can be used by code. The function returns NULL if the allocation request fails.

Table 25.3

Win-32 Commonly Used Memory Allocation Flags

FLAG	MEANING
GMEM_FIXED	Allocates fixed memory. This flag cannot be combined with the GMEM_MOVEABLE or GMEM_DISCARDABLE flag. The return value is a pointer to the memory block.
GMEM_MOVEABLE	Allocates moveable memory. This flag cannot be combined with the GMEM_FIXED flag. The return value is the handle of the memory object, which is a 32-bit quantity private to the calling process.
GMEM_DISCARDABLE	Allocates discardable memory. This flag cannot be combined with the flag. Win32-based operating systems ignore this flag.
GMEM_ZEROINIT	Initializes memory to 0.
GPTR	Combines the GMEM_FIXED and GMEM_ZEROINIT flags.
GHND	Combines the GMEM_MOVEABLE and GMEM_ZEROINIT flags.

The process of creating a bitmap in heap memory and displaying it on the video screen can be accomplished by the following steps:

1. Dynamically allocate the memory required for the bitmap and the corresponding data structures.

2. Store the bitmap data and the bitmap dimensions and color information, thus creating a DIB.

3. Convert the device independent bitmap (DIB) into a device dependent bitmap using CreateDIBitmap().

4. Display the bitmap using SetDIBitsToDevice().

Suppose you wanted to create a 200-pixel wide bitmap, with 255 scan lines, in 32-bit color. Since each pixel requires four bytes, each scan line consists of 800 bytes, and the entire bitmap occupies 204,000 bytes. A BITMAPINFO structure variable is used to hold the bitmap information. Notice that because this is a true-color bitmap, the RGBQUAD structure is not necessary. In this case the memory allocation operations can be coded as follows:

```
static PBITMAPINFO  pDibInfo;       // pointer to BITMAPINFO structure
static BYTE         *pDib;          // pointer to bitmap data
.
.
.
pDibInfo = (PBITMAPINFO) LocalAlloc(LMEM_FIXED, \
                          sizeof(BITMAPINFOHEADER));
pDib = (BYTE*) LocalAlloc(LMEM_FIXED, 204000);
```

At this point the code has allocated memory for both the bitmap and the BITMAPINFOHEADER structure variable that is to hold the bitmap format information. The pointers to each of these memory areas can be used to fill them in. First the bitmap information:

```
pDibInfo->bmiHeader.biSize = (LONG) sizeof(BITMAPINFOHEADER);
pDibInfo->bmiHeader.biWidth = (LONG) 200;       // pixel width
pDibInfo->bmiHeader.biHeight = (LONG) 255;      // pixel height
pDibInfo->bmiHeader.biPlanes = 1;               // number of planes
pDibInfo->bmiHeader.biBitCount = 32;            // bits per pixel
```

Assume that the bitmap is to represent a blue rectangle with 255 decreasing intensities of blue, along the scan lines. The code to fill this bitmap can be coded as follows:

```
int     i, j, k;        // counters
BYTE    shade;          // shade of blue
.
.
.
// Fill the bitmap using 32-bit color data
// <---------- 200 pixels (4 bytes each) -------->
// |
// | ... 255 scan lines
shade = 0;
  for (k = 0; k < 255; k++){        // Counts 255 scan lines
     for (i = 0; i < 200; i++){     // Counts 200 pixels
        for(j = 0; j < 4; j++) {    // Counts 4 bytes
              pDib[((k*800)+(i*4)+0] = shade; // blue
              pDib[((k*800)+(i*4)+1] = 0;     // green
              pDib[((k*800)+(i*4)+2] = 0;     // red
              pDib[((k*800)+(i*4)+3] = 0;     // must be zero
              };
        };
     shade++;
  };
```

Now that the bitmap data structures have been initialized and the bitmap data entered into the allocated heap memory, it is time to create the device-dependent bitmap that can be displayed on the display context. The function used for this purpose is named CreateDIBitmap(); this name is somewhat confusing since it actually creates a dependent device from a device-independent bitmap. The function's general form is as follows:

```
HBITMAP CreateDIBitmap(
        HDC hdc,                            // 1
CONST BITMAPINFOHEADER *lpbmih,   // 2
        DWORD fdwInit,                     // 3
        CONST VOID *lpbInit,               // 4
        CONST BITMAPINFO *lpbmi,           // 5
        UINT fuUsage                       // 6
);
```

The first parameter is the handle to the device context for which the device dependent bitmap is to be configured. The second parameter is a pointer to a BITMAPINFOHEADER structure variable that contains the bitmap data. The third parameter is a flag that determines how the operating system initializes the bitmap bits. If this parameter is zero the bitmap data is not initialized and parameters 4 and 5 are not used. If it is set to CBM_INIT, then parameters 4 and 5 are used as pointers to the data used in initializing the bitmap bits. The fourth parameter is a pointer to the array of type BYTE that contains the bitmap data. The fifth parameter is a pointer to a BITMAPINFO structure that contains the bitmap size and color data. The sixth parameter is a flag that determines whether the bmiColors member of the BITMAPINFO structure contains explicit color values in RGB format or palette indices. In the first case the constant DIB_RGB_COLORS is used for this parameter, and in the second case the constant is DIB_PAL_COLORS. The function returns the handle to the bitmap if it succeeds, or NULL if it fails.

In the example that we have been following, the device dependent bitmap is created as follows:

```
static HBITMAP      hBitmap;      // handle to a bitmap
.
.
.

hBitmap = CreateDIBitmap ( hdc,
            (LPBITMAPINFOHEADER) &pDibInfo->bmiHeader,
            CBM_INIT,
            (LPSTR) pDib,
            (LPBITMAPINFO) pDibInfo,
            DIB_RGB_COLORS );
```

Having obtained its handle, the bitmap can be displayed using the ShowBitmap() function developed earlier in this chapter. Alternatively, you can use SetDIBitsToDevice() to set the screen pixels. The function's general form is as follows:

```
int SetDIBitsToDevice(
        HDC hdc,                // 1
        int XDest,              // 2
        int YDest,              // 3
        DWORD dwWidth,          // 4
        DWORD dwHeight,         // 5
        int XSrc,               // 6
```

```
        int YSrc,                       // 7
        UINT uStartScan,                // 8
        UINT cScanLines,                // 9
        CONST VOID *lpvBits,            // 10
        CONST BITMAPINFO *lpbmi,        // 11
        UINT fuColorUse                 // 12
);
```

The first parameter is the handle to the display context to which the bitmap is to be output. The second and third parameters are the x- and y-coordinates of the destination rectangle, in logical units. This is the screen location where the bitmap is displayed. The fourth and fifth parameters are the width and height of the DIB. These values are often read from the corresponding members of the BITMAPINFOHEADER structure variable that defines the bitmap. The sixth and seventh parameters are the x- and y-coordinates of the lower-left corner of the DIB. The eighth parameter is the starting scan line of the DIB. The ninth parameter is the number of scan lines. The tenth parameter is a pointer to the bitmap data and the eleventh parameter is a pointer to the BITMAPINFO structure variable that describes the bitmap. The twelfth parameter is a flag that determines whether the bmiColors member of the BITMAPINFO structure contains explicit color values in RGB format or palette indices. In the first case the constant DIB_RGB_COLORS is used, and in the second case the constant DIB_PAL_COLORS. If the function succeeds the return value is the number of scan lines displayed. The function returns NULL if it fails.

In the current example, the bitmap can be displayed with the following call to SetDIBitsToDevice():

```
SetDIBitsToDevice (hdc, 50, 50,
        pDibInfo->bmiHeader.biWidth,
        pDibInfo->bmiHeader.biHeight,
        0, 0, 0,
        pDibInfo->bmiHeader.biHeight,
        pDib,
        (BITMAPINFO FAR*) pDibInfo,
        DIB_RGB_COLORS);
```

25.4.3 Operations on Blank Bitmaps

Sometimes an application needs to fill a blank bitmap using GDI functions. The functions can include all the drawing and text display primitives discussed in previous chapters. There are several programming approaches to creating a bitmap on which GDI operations can be performed. The simplest approach is to select a bitmap into a memory device context and then perform the draw operation on the memory device context. Note that all the drawing functions discussed previously require a handle to the device context. When the drawing takes place on a memory DC, the results are not seen on the video display until the memory DC is blitted to the screen. In this approach the following steps are required:

1. Select the bitmap into a memory device context using the SelectObject() function.

2. Clear or otherwise paint the bitmap using the PatBlt() function.

3. Perform drawing operations on the memory device context that contains the bitmap.

4. Display the bitmap by blitting it on the screen, typically with BitBlt().

The CreateCompatibleBitmap() function has the following general form:

```
HBITMAP CreateCompatibleBitmap(
        HDC hdc,            // 1
        int nWidth,         // 2
        int nHeight         // 3
);
```

The first parameter is the handle to the device context with which the created bitmap is to be compatible. If the bitmap is to be displayed this parameter is set to the display context. The second and third parameters are the width and height of the bitmap, in pixels. If the function succeeds it returns the handle to the bitmap. The function returns NULL if it fails.

In the following code sample we create a blank, 300-by-300 pixel bitmap, draw a rectangle and an ellipse on it, and then blit it to the screen. First we start by creating the blank bitmap:

```
static HDC        aMemDC;           // memory device context
static HBITMAP    bmBlank;          // handle to a bitmap
static HGDIOBJ    oldObject;        // storage for current object
.
.
.
// Preliminary operations
aMemDC = CreateCompatibleDC (NULL);   // Memory device handle
mapMode = GetMapMode (hdc);           // Obtain mapping mode
SetMapMode (aMemDC, mapMode);         // Set memory DC mapping mode
// Create the bitmap
bmBlank = CreateCompatibleBitmap (hdc, 300, 300);
oldObject = SelectObject (aMemDC, bmBlank);
```

Note that we use a generic handle (HGDIOBJ) to store the current handle in the device context.

There is no guarantee that the bitmap thus created and selected into the device is initialized. The PatBlt() function can be used to set all the bitmap bits to a particular attribute or to a predefined pattern. The function's general form is as follows:

```
BOOL PatBlt(
        HDC hdc,          // 1
        int nXLeft,       // 2
        int nYLeft,       // 3
        int nWidth,       // 4
        int nHeight,      // 5
        DWORD dwRop       // 6
);
```

The first parameter is the handle to the device context, which can be a memory device context. The second and third parameters are the x- and y-coordinates of the upper-left corner of the rectangle to be filled. The fourth and fifth parameters are the width and height of the bitmap, in logical units. The sixth parameter is one of the following constants: PATCOPY, PATINVERT, DSTINVERT, BLACKNESS, or WHITENESS. The constants are defined in Table 25.2. The call returns TRUE if it succeeds and FALSE if it fails.

Following the current example, the call clears the bitmap and sets all bits to the white attribute:

```
PatBlt (aMemDC, 0, 0, 300, 300, WHITENESS);
```

At this point in the code we can start performing drawing operations on the bitmap. The only requirement is that the drawing primitives reference the handle to the memory device context where the blank bitmap was selected, as in the following example:

```
Ellipse (aMemDC, 10, 10, 210, 110);
Polyline (aMemDC, rectangle, 5);
```

Once you have finished drawing on the blank bitmap, it can be displayed by means of a bitblt, as in the following example:

```
BitBlt(hdc, 50, 50, 300, 300, aMemDC, 0, 0, SRCCOPY);
```

In this case, the call references both the display context (hdc) and the memory device context containing the bitmap (aMemDC).

Clean-up operations consist of reselecting the original object to the memory device context, then deleting the device context and the bitmap.

25.4.4 Creating a DIB Section

The methods described in the preceding section are satisfactory when the bitmap area requires drawing operations that can be implemented by GDI functions, but code has no direct access to the bitmap itself. This is due to the fact that CreateCompatibleBitmap() does not return a pointer to the bitmap data area. The CreateDIBSection() function, first introduced in Win32 and formalized in Windows 95, allows creating a device-independent bitmap that applications can access directly.

Note that the original Windows documentation for Win32 contained incorrect information about the CreateDIBSection() function and the associated DIBSECTION structure. Some of the errors and omissions were later corrected so that current documentation is more accurate, although not very clear.

Before CreateDIBSection(), an application would access bitmap data by calling the GetDIBits() function, which copies the bitmap into a buffer supplied by the caller. At the same time, the bitmap size and color data is copied into a BITMAPINFO structure from which the application can read these values. After the bitmap is changed, the SetDIBits() function is used to redisplay the bitmap. Both functions, GetDIBits() and SetDIBits(), allow selecting the first scan line and the number of scan lines. When operating on large bitmaps, this feature makes it possible to save memory by reading and writing portions of it at a time.

There are several shortcomings to modifying bitmaps at run time using GetDIBits() and SetDIBits(). The most obvious one is that the system bitmap must be copied into the application's memory space, then back into system memory. The process is wasteful and inefficient. If the entire bitmap is read during the GetDIBits() call, there are two copies of the same data, thus wasting memory. If it is broken down into regions in order to reduce the waste, then processing speed suffers considerably. The solution offered by CreateDIBSection() is to create a bitmap that can be accessed by both the system and the application. Figure 25.5 shows both cases.

Although CreateDIBSection() provides a better alternative than GetDIBits() and SetDIBits(), it is by no means the ultimate in high-performance graphics. DirectDraw methods, not discussed in this book, provide ways of accessing video memory directly and of taking advantage of raster graphics hardware accelerators.

In the following example, we create a DIB section, using the pointer returned by the CreateDIBSection() call to fill the bitmap, and the bitmap handle to perform GDI drawing functions on its memory space. The bitmap is 50 pixels wide and 255 scan lines long. It is encoded in 32-bit true color format. The code starts by defining the necessary data structures, initializing the variables, and allocating memory.

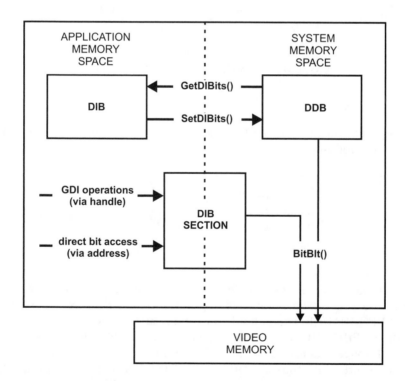

Figure 25.5 *Memory Image of Conventional and DIB Section Bitmaps*

```
HDC                 aMemDC;          // Memory DC
static HBITMAP      aBitmap;
static BYTE*        lpBits;

BITMAPINFOHEADER    bi;
BITMAPINFOHEADER*   lpbi;
HANDLE              hDIB;
int                 shade;
static int BMScanLines;      // Bitmap y-dimension
static int BMWidth;          // Bitmap x-dimension
.
.
.
// Initialize size variables
BMScanLines = 255;
BMWidth = 50;
// Initialize BITMAPINFOHEADER structure
bi.biSize = sizeof(BITMAPINFOHEADER);
bi.biWidth = BMWidth;
bi.biHeight = BMScanLines;
bi.biPlanes = 1;
bi.biBitCount = 32;
bi.biCompression = BI_RGB;
bi.biSizeImage = 0;
bi.biXPelsPerMeter = 0;
bi.biYPelsPerMeter = 0;
bi.biClrUsed = 0;
bi.biClrImportant = 0;

// Allocate memory for DIB
hDIB = GlobalAlloc (GMEM_FIXED, sizeof(BITMAPINFOHEADER));
// Initialize bitmap pointers
lpbi = (BITMAPINFOHEADER*) hDIB;
*lpbi = bi;
```

At this point everything is ready to call CreateDIBSection(). The function's general form is as follows:

```
HBITMAP CreateDIBSection(
            HDC hdc,                         // 1
            CONST BITMAPINFO *pbmi,          // 2
            UINT iUsage,                     // 3
            VOID *ppvBits,                   // 4
            HANDLE hSection,                 // 5
            DWORD dwOffset                   // 6
);
```

The first parameter is the handle to a device context associated with the DIB section. The second parameter is a pointer to a structure variable of type BITMAPINFOHEADER, which holds the bitmap attributes. The first five members of the BITMAPINFOHEADER structure are required; the other ones can often be omitted, although it is usually a good idea to fill in the entire structure. The third parameter is either the constant DIB_PAL_COLORS or DIB_RGB_COLORS. In the first case the bmiColors array member of the RGBQUAD structure in BITMAPINFO is a set of 16-bit palette color indices. In the second case the bmiColors member is not used and the colors are encoded directly in the bitmap. The fourth parameter is a pointer to a pointer to type VOID that contains the location of the bitmap values. This parameter is incorrectly documented in the Windows help files as a pointer to type VOID. If the parameter is not correctly typecast to (VOID**) the CreateDIBSection() call fails.

The fifth parameter is a handle to a file-mapping object. In file mapping, a physical file on disk is associated with a portion of the virtual address space of a process. The file-mapping object is the mechanism that maintains this association. Its main purpose is to share data between applications and to facilitate access to files. Although file mapping is a powerful mechanism, it is outside the scope of this book and is not discussed any further. If no file mapping is used, the fifth parameter is set to NULL, and the sixth one, which sets the offset of the file mapping object, is set to zero.

Following the current example, the call to CreateDIBSection() is coded as follows:

```
aBitmap = CreateDIBSection (hdc,
          (LPBITMAPINFO)lpbi,  // Pointer to
                               // BITMAPINFOHEADER
          DIB_RGB_COLORS,      // True color in RGB format
          (VOID**) &lpBits,    // Pointer to bitmap data
          NULL,                // File mapping object
          (DWORD) 0);          // File mapping object offset

assert (aBitmap);
assert (lpBits);
```

The two assertions that follow the call ensure that a valid bitmap and pointer are returned. If the call succeeds we now have a handle to a bitmap and its address in system memory. Using the address, we can fill the bitmap. The following code fragment uses the soft-coded bitmap parameters to fill the entire bitmap, scan line by scan line, with increasing intensities of blue. The access to the bitmap is by means of the pointer (lpBits) returned by the previous call.

```
// Fill the bitmap using 32-bit color data
// <--------- BMWidth * 4 -------->
// |
// |  ... BMScanLines

shade = 0;
for (k = 0; k < BMScanLines; k++){    // Counts 255 lines
   for (i = 0; i < BMWidth; i++){     // Counts 50 pixels
      for(j = 0; j < 4; j++) {  //Counts 4 bytes per pixel
         lpBits[(k*(BMWidth*4))+(i*4)+0] = shade; // blue
         lpBits[(k*(BMWidth*4))+(i*4)+1] = 0x0;   // green
         lpBits[(k*(BMWidth*4))+(i*4)+2] = 0x0;   // red
         lpBits[(k*(BMWidth*4))+(i*4)+3] = 0;     // zero
         };
      };
   shade++;
};
```

Since we have also acquired the handle to the bitmap, we can use GDI functions to perform drawing operations on its surface. As described earlier in this chapter, the GDI functions require that the bitmap be first selected into a memory device context. The following code fragment shows one possible processing method:

```
aMemDC = CreateCompatibleDC (NULL); // Memory device handle
mapMode = GetMapMode (hdc);          // Obtain mapping mode
SetMapMode (aMemDC, mapMode);        // Set memory DC
                                     // mapping mode
// Select the bitmap into the memory DC
oldObject = SelectObject (aMemDC, aBitmap);
```

Drawing operations can now take place, as follows:

```
// Draw on the bitmap
blackPenSol = CreatePen (PS_SOLID, 2, 0);
redPenSol = CreatePen (PS_SOLID, 2, (RGB (0xff, 0x0, 0x0)));
SelectPen (aMemDC, blackPenSol);
Polyline (aMemDC, rectsmall, 5);        // Draw a rectangle
SelectPen (aMemDC, redPenSol);
Ellipse (aMemDC, 4, 4, 47, 47);         // Draw a circle
```

You may be tempted to display the bitmap at this time; the display operation, however, cannot take place until the memory device context has been deleted. In the following instructions we re-select the original object in the memory device context and then delete it. We also delete the pens used in the drawing operations.

```
// Erase bitmap and free heap memory
SelectObject (aMemDC, oldObject);
DeleteDC (aMemDC);
DeleteObject (redPenSol);
DeleteObject (blackPenSol);
```

Displaying the bitmap can be performed by any of the methods already discussed. In this code fragment we use the ShowBitmap() function developed earlier in the chapter. A necessary precaution relates to the fact that some versions of Windows NT place GDI calls that return a boolean value in a batch for later execution. In this case, it is possible to attempt to display a DIB section bitmap before all the calls in the GDI batch have been executed. In order to prevent this problem, it is a good idea to flush the GDI batch buffer before displaying a DIB section bitmap, as shown in the following code:

```
GdiFlush();          // Clear the batch buffer
ShowBitmap (hdc, Abitmap, 50, 50, SRCCOPY);
```

Now that you have finished displaying the bitmap, a tricky problem arises: how to free the system memory space allocated by CreateDIBSection(). The solution is easy. Since the bitmap resides in system memory, all we have to do in application code is delete the bitmap; Windows takes care of freeing the memory. On the other hand, if the BITMAPINFOHEADER structure was defined in heap memory, your code must take care of freeing this memory space in the conventional manner. Processing is as follows:

```
// Erase bitmap and free heap memory
// Note: deleting a DIB section bitmap also frees
//       the allocated memory resources
DeleteObject (aBitmap);    // Delete the bitmap
GlobalFree (hDIB);
```

Figure 25.6 is a screen snapshot of a program that executes the listed code. The listing is found in the Bitmap Demo project folder on the book's software on-line.

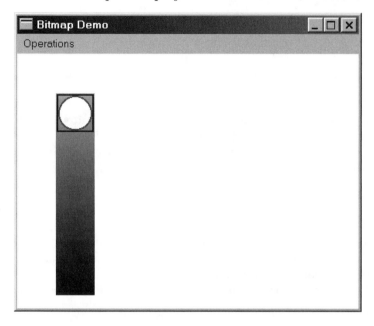

Figure 25.6 *Screen Snapshot Showing a DIB Section Bitmap Manipulation*

25.4.5 Creating a Pattern Brush

In Chapter 21 we mentioned that applications can create a brush with a hatch pattern different than the ones predefined in Windows. This is done by using a bitmap to define the brush pattern. In Windows 95 and later the size of the bitmap cannot exceed 8-by-8 pixels, but there is no size restriction in Windows NT. The function's general form is as follows:

```
HBRUSH CreatePatternBrush (HBITMAP hbitmap);
```

The function's only parameter is a handle to the bitmap that defines the brush. The bitmap can be created with CreateBitmap(), CreateBitmapIndirect() or CreateCompatibleBitmap() functions. It can also be a bitmap drawn using Developer Studio bitmap editor, or any other similar utility, and loaded with the LoadBitmap() function. The one type of bitmap that is not allowed is one created with the CreateDIBSection() function. CreatePatternBrush() returns the handle to the brush if it succeeds, and NULL if it fails.

Once the handle to the brush has been obtained, the pattern brush is selected into the device context. Thereafter, all GDI drawing functions that use a brush use the selected pattern brush. The following code fragment shows the creation of a pattern brush from a bitmap resource named IDC_BITMAP5. The pattern brush is then used to draw a rectangle.

```
static HBITMAP    brushBM1; // Handle to a bitmap
```

```
static HBRUSH       patBrush;  // Handle to a brush
  .
  .
  .
brushBM1 = LoadBitmap (pInstance,
                       MAKEINTRESOURCE (IDB_BITMAP5);
patBrush = CreatePatternBrush (brushBM1);
SelectBrush (hdc, patBrush);
Rectangle (hdc, 10, 10, 110, 160);
DeleteObject (patBrush);
```

In displaying solid figures that use a pattern brush, Windows sets the origin of the brush bitmap to the origin of the client area. The SetBrushOrgEx() function is used to reposition the brush bitmap in relation to the origin of the client area. This matter was discussed in Chapter 23 in relation to brush hatch patterns.

25.5 Bitmap Transformations

In addition to manipulating bitmaps, Windows provides functions that transform the bitmaps themselves. You have already seen that the BitBlt() function allows you to define a ternary raster operation code that determines how the source bitmap and a pattern are combined to form the destination bitmap. In this section we discuss the following transformations that are useful in bitmap programming:

- Painting a bitmap using a raster operation based on the brush selected in the device context

- Stretching or compressing a bitmap according to the dimensions of a destination rectangle, a predefined stretch mode, and the selected ternary raster operation code

Windows NT provides two powerful bitmap transforming functions named MaskBlt() and PlgBlt(). Since the scope of this chapter includes functions that are available only in Windows 95 and later, these functions are not discussed.

25.5.1 Pattern Brush Transfer

A pattern brush transfer consists of transferring the pattern in the current brush into a bitmap. The PatBlt() function is used in this case. If the PATCOPY raster operation code is selected, as is usually the case, the brush pattern is copied to the destination bitmap. If the PATINVERT raster operation code is used, then the brush and the destination bitmap are combined by performing a boolean XOR operation. The remaining raster operation codes that are documented for the PatBlt() function with symbolic names (DSTINVERT, BLACKNESS, and WHITENESS) ignore the brush and are useless in a pattern block transfer. The raster operation performed by PatBlt() is a binary one since it combines a pattern and a destination. In theory, any of the raster operations codes that do not have a source operand can be used in PatBlt(), although it may be difficult to find a useful application for most of them.

Note that there is a not-so-subtle difference between a rectangle filled with a pattern brush, and a bitmap created by means of a pattern transfer. Although the results can be made graphically identical by drawing the rectangle with a NULL pen, the possibilities of further manipulating and transforming a bitmap are not possible with a figure created by means of a GDI drawing function.

The following code fragment creates a blank bitmap in a memory device context and fills it with a pattern brush. Since the processing is based on functions already discussed, the code listing needs little comment.

```
static HBITMAP     brushBM1;   // Handle to a bitmap
static HBRUSH      patBrush;   // Handle to a brush
.
.
.
// Create the brush pattern bitmap from a resource
brushBM1 = LoadBitmap (pInstance,
                       MAKEINTRESOURCE (IDB_BITMAP5);
// Create a pattern brush
patBrush = CreatePatternBrush (patBM1);
// Create a memory device context
aMemDC = CreateCompatibleDC (NULL);   // Memory DC
mapMode = GetMapMode (hdc);           // Obtain mapping mode
SetMapMode (aMemDC, mapMode);         // Set memory DC
                                      // mapping mode
// Create the bitmap
bmBlank = CreateCompatibleBitmap (hdc, 300, 300);
oldObject = SelectObject (aMemDC, bmBlank);
// Select the pattern brush into the memory DC
SelectBrush (aMemDC, patBrush);
// Blit the pattern onto the memory DC
PatBlt (aMemDC, 0, 0, 300, 300, PATCOPY);
// Display the bitmap
BitBlt(hdc, 50, 50, 300, 300, aMemDC, 0, 0, SRCCOPY);
// Clean-up
SelectObject (aMemDC, oldObject);
DeleteDC (aMemDC);
DeleteObject (bmBlank);
```

The demonstration program named Bitmap Demo, in the book's on-line software package, displays a pattern bitmap using code very similar to the one listed.

25.5.2 Bitmap Stretching and Compressing

Occasionally, an application must fit a bitmap into a destination rectangle that is of different dimensions, and even of different proportions. In order to do this, the source bitmap must be either stretched or compressed. One possible use of bitmap stretching or compressing is adapting imagery to a display device that has a different aspect ratio that the one for which it was created. The method can also be used to accommodate a bitmap to a resizable window, as well as for producing intentional distortions, such as simulating the effect of a concave or convex mirror, or other special visual effects.

The StretchBlt() function, one of the more elaborate ones in the API, allows stretching or compressing a bitmap if this is necessary to fit it into a destination rectangle. StretchBlt() is a variation of BitBlt(); therefore, it is used to stretch or compress and later display the resulting bitmap. StretchBlt() is also used to reverse (vertically) or invert (horizontally) a bitmap image. The stretching or compressing is done according to the stretching mode attribute selected in the device context. The stretch mode is selected by means of the SetStretchBltMode() function, which has the following general form:

```
int SetStretchBltMode(
        HDC hdc,            // 1
        int iStretchMode    // 2
);
```

The first parameter is the handle to the device context to which the stretch mode attribute is applied. The second parameter is a predefined constant that corresponds to one of four possible stretching modes. All stretching modes have an old and a new name. Table 25.4 lists and describes the stretching modes. The new, preferred names are listed first.

Table 25.4

Windows Stretching Modes

NAME	DESCRIPTION
STRETCH_ANDSCANS BLACKONWHITE	Performs a logical AND operation using the color values for the dropped pixels and the retained ones. If the bitmap is a monochrome bitmap, this mode preserves black pixels at the expense of whiteones.
STRETCH_DELETESCANS COLORONCOLOR	Deletes the pixels. This mode deletes all dropped pixel lines without trying to preserve their information. This mode is typically used to preserve color in a color bitmap.
STRETCH_HALFTONE HALFTONE	Maps pixels from the source rectangle into blocks of pixels in the destination rectangle. The average color over the destination block of pixels approximates the color of the source pixels. Windows documentation recommends that after setting the HALFTONE stretching mode, an application must call the SetBrushOrgEx() function in order to avoid brush misalignment.
STRETCH_ANDSCANS WHITEONBLACK	Performs a logical OR operation using the color values for the dropped and preserved pixels. If the bitmap is a monochrome bitmap, this mode preserves white pixels at the expense of black ones.

Note that, on many systems, the entire discussion on stretch modes is purely academic, since Microsoft has reported a Windows 95 bug in which the StretchBlt() function always uses the STRETCH_DELETESCANS mode, no matter which one has been selected by means of SetStretchBltMode(). The Microsoft Knowledge Base article describing this problem is number Q138105. We have found no other Microsoft Knowledge Base update regarding this matter.

The actual stretching or compression of the bitmap is performed by means of the StretchBlt() function. The function's general form is as follows:

```
BOOL StretchBlt(
        HDC hdcDest,         // 1
        int nXOriginDest,    // 2
```

```
        int nYOriginDest,  // 3
        int nWidthDest,    // 4
        int nHeightDest,   // 5
        HDC hdcSrc,        // 6
        int nXOriginSrc,   // 7
        int nYOriginSrc,   // 8
        int nWidthSrc,     // 9
        int nHeightSrc,    // 10
        DWORD dwRop        // 11
);
```

The first parameter is the destination device context and the sixth parameter is the source device context. The second and third parameters are the x- and y-coordinates of the upper-left corner of the destination rectangle. The fourth and fifth parameters are the width and height of the destination rectangle. The seventh and eighth parameters are the x- and y-coordinates of the upper-left corner of the source rectangle. The ninth and tenth parameters are the width and height of the source rectangle. The eleventh parameter is one of the ternary raster operation codes listed in Table 25.2.

Although the function's parameter list is rather large, it can be easily simplified. Parameters 1 through 5 are the handle to the device context and the location and size of the destination rectangle. Parameters 6 through 10 contain the same information in regards to the source rectangle. The last parameter defines the raster operation code, which is usually set to SRCCOPY.

If the source and destination width parameters have opposite signs, the bitmap is flipped about its vertical axis. In this case the left side of the original bitmap is displayed starting at the right edge. If the source and destination height parameters have opposite signs the image is flipped about its horizontal axis. If both, the width and the height parameters have opposite signs, the original bitmap is flipped about both axes.

The following example takes an existing bitmap and stretches or compresses it to fit the size of the client area. Code assumes an existing bitmap resource named IDB_BITMAP2. The code starts by creating a bitmap from the resource and storing its dimensions in a structure variable of type BITMAP. The dimensions of the client area, which serves as a destination bitmap, are also retrieved and stored in a structure variable of type RECT.

```
BITMAP           bm;       // Storage for bitmap data
RECT             rect;     // Client area dimensions
static  HBITMAP  hstScope; // Handle for a bitmap
.  .  .

// Create bitmap from resource

hstScope = LoadBitmap (pInstance, MAKEINTRESOURCE (IDB_BITMAP2);
// Get bitmap dimensions into BITMAP structure variable
GetObject (hstScope, sizeof(BITMAP), &bm);
// Get client area dimensions
GetClientRect (hwnd, &rect);
```

Figure 25.7 shows the image changes in each case.

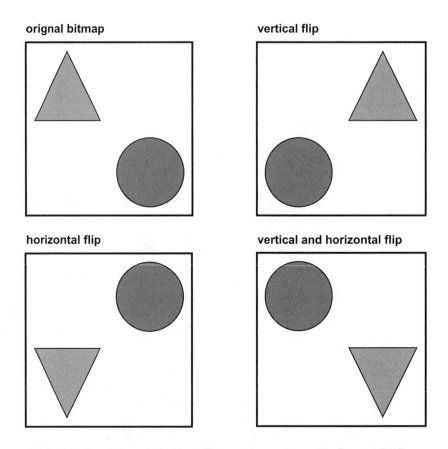

Figure 25.7 *Horizontal and Vertical Bitmap Inversion with StretchBlt()*

The StretchBlt() function requires two device contexts: one for the source bitmap and another one for the destination. In this case we create a memory device context and select the bitmap into it. This device context is the source rectangle. The code also sets the stretch mode.

```
aMemDC = CreateCompatibleDC (hdc);
SelectObject (aMemDC, hstScope);
SetStretchBltMode (hdc, STRETCH_DELETESCANS);
```

All that is left is to display the bitmap using StretchBlt(). Once the bitmap is blitted, the destination device context can be deleted. If the bitmap is not necessary, it can also be erased at this time.

```
StretchBlt(hdc,                        // Destination DC
          0, 0, rect.right, rect.bottom,  // dest. dimensions
          aMemDC,                      // Source DC
          0, 0, bm.bmWidth, bm.bmHeight,  // Source dimensions
          SRCCOPY);
DeleteDC (aMemDC);
```

25.6 Bitmap Demonstration Program

The program named BMP_DEMO, located in the Bitmap Demo project folder of the book's on-line software package, is a demonstration of the bitmap operations and functions discussed in this chapter. The Operations menu contains commands that correspond to all the bitmap programming primitives, manipulations, and transformations discussed in the text.

Part III

Project Engineering

Chapter 26

Fundamentals of Systems Engineering

Chapter Summary

In this chapter we provide a technical overview of some topics from the field of software engineering, stressing those that would be most useful to the engineer/programmer. The contents are an arbitrary selection of the topics that would be most useful in the context of a smaller software project, with little speculative discussions on the respective merits of the various software engineering paradigms.

26.0 What Is Software Engineering

Software engineering was first introduced in the 1960s in an effort to treat more rigorously the often frustrating task of designing and developing computer programs. It was around this time that the computer community became increasingly worried about the fact that software projects were typically over budget and behind schedule. The term software crisis came to signify that software development was the bottleneck in the advancement of computer technology.

During these initial years it became evident that we had incurred grave fallacies by extending to software development some rules that were true for other fields of human endeavor, or that appeared to be common sense. The first such fallacy states that if a certain task takes a single programmer one year of coding, then four programmers would be able to accomplish it in approximately three months. The second fallacy is that if a single programmer was capable of producing a software product with a quality value of 25, then four programmers could create a product with a quality value considerably higher than 25, perhaps even approaching 100. A third fallacy states that if we have been capable of developing organizational tools, theories, and techniques for building bridges and airplanes, we should also be capable of straightforwardly developing a scientific methodology for engineering software.

The programmer productivity fallacy relates to the fact that computer programming, unlike ditch digging or apple harvesting, is not a task that is easily partitioned into isolated functions that can be performed independently. The different parts of a computer program interact, often in complicated and hard-to-predict ways. Considerable planning and forethought must precede the partitioning of a program into individually executable tasks. Furthermore, the individual programmers in a development team must frequently communicate to ensure that the separately developed elements will couple as expected and that the effects of individual adjustments and modifications are taken into account by all members of the group. All of which leads to the conclusion that team programming implies planning and structuring operations as well as interaction between the elements of the team.

The cumulative quality fallacy is related to some of the same interactions that limit programmer productivity. Suppose the extreme case of an operating system program that was developed by ten programmers. Nine of these programmers have performed excellently and implemented all the functions assigned to them in a flawless manner, while one programmer is incompetent and has written code that systematically crashes the system and damages the stored files. In this case it is likely that the good features of this hypothetical operating system will probably go unnoticed to a user who experiences a destructive crash. Notice that this situation is different from what typically happens with other engineered products. We can imagine a dishwasher that fails to dry correctly, but that its other functionalities continue to be useful. Or a car that does not corner well, but that otherwise performs as expected. Computer programs, on the other hand, often fail catastrophically. When this happens it is difficult for the user to appreciate any residual usefulness in the software product.

This leads to the conclusion that rather than a cumulative quality effect, software production is subject to a minimal quality rule, which determines that a relatively small defect in a software product can impair its usability, or, at best, reduce its appraised quality. The statement that "the ungrateful look at the sun and see its spots" justifiably applies to software users. Also notice that the minimal quality rule applies not only to defects that generate catastrophic failures, but even to those that do not affect program integrity. For example, a word processor performs perfectly except that it hyphenates words incorrectly. This program may execute much more difficult tasks extremely well; it may have excellent on-screen formatting, a high-quality spelling checker, an extensive dictionary of synonyms and antonyms, and many other important features. However, very few users will consider adopting this product since incorrectly hyphenated text is usually intolerable.

Finally, there is the unlikely assumption that we can engineer software programs in much the same way that we engineer bridges, buildings, and automobiles. The engineering methods fallacy is a consequence of the fact that software does not deal with tangible elements, subject to the laws of physics, but with purely intellectual creations. A program is more a thought than a thing. Things can be measured, tested, stretched, tempered, tuned, and transformed. A construction engineer can take a sample of concrete and measure its strength and resistance. A mechanical engineer can determine if a certain motor will be sufficiently powerful to operate a given machine. Regarding software components these determinations are not unequivocal. At the present stage-of-the-art a software engineer cannot look up a certain algorithm or data structure in a reference manual and determine rigorously if it is suitable for the problem at hand.

26.0.1 The Programmer as an Artist

Donald Knuth established in the title of his now classic work that programming is an art. He starts the preface by saying:

> *"The process of preparing programs for a digital computer is especially attractive, not only because it can be economically and scientifically rewarding, but also because it can be an aesthetic experience much like composing poetry or music."*

Therefore, it is reasonable to deduce that any effort to reduce the art of programming to the following of a strict and scientifically defined rule set is likely to fail. What results is like comparing the canvas by a talented artist with one produced by using a paint-by-the-numbers toy. Without talent the programmer's productions will consist of the dull rehashing of the same algorithms and routines that are printed in all the common textbooks. With genius and art the code comes alive with imaginative and resourceful creations that make the program a beautiful conception.

The fact that computer programming is an art and that talented programmers are more artists than technicians does not preclude the use of engineering and scientific principles in pursuing this art. Software engineering is not an attempt to reduce programming to a mechanical process, but a study of those elements of programming that can be approached technically and of the mechanisms that can be used to make programming less difficult. However, we must always keep in mind that technique, no matter how refined, can never make an artist out of an artisan.

26.1 Software Characteristics

From an engineering viewpoint a software system is a product that serves a function. However, one unique attribute makes a computer program much different from a bridge or an airplane: a program can be changed. This malleability of software is both an advantage and a danger. An advantage because it is often possible to correct an error in a program much easier than it would be to fix a defect in an airplane or automobile. A danger because a modification in a program can introduce unanticipated side effects that may impair the functionality of those components that were executing correctly before the change.

Another notable characteristic of programs relate to the type of resources necessary for their creation. A software product is basically an intellectual commodity. The principal resource necessary for producing it is human intelligence. The actual manufacturing of programs is simple and inexpensive compared to its design, coding, testing, and documenting. This contrasts with many other engineered products in which the resources used in producing it are a substantial part of the product's final cost. For example, a considerable portion of the price of a new automobile represents the cost of manufacturing it, while a less significant part goes to pay for the engineering costs of design and development. In the case of a typical computer program the proportions are reversed. The most important element of the product cost is the human effort in design and development while the cost of manufacturing is proportionally insignificant.

26.1.1 Software Qualities

An engineered product is usually associated with a list of qualities that define its usability. For example, in performing its functions a bridge supports a predetermined weight and withstands a given wind force. An airplane is capable of transporting a specific load, at a certain speed and altitude. By the same token, a software product is associated with a given set of qualities that define its functionality. The principal goals of software engineering is to define, specify, and measure software qualities and to describe the principles that can be applied to achieve them.

The classification of software qualities can be based on the relation with the software product. In this sense we can speak of qualities desirable to the user, to the developer, or to the manager. Table 26.1 lists some qualities according to this classification.

Table 26.1

Software Qualities

USER	DEVELOPER	MANAGER
reliable	verifiable	productive
easy to use	maintainable	controllable
efficient	extensible	lucrative
	portable	

In this sense we can also talk about software qualities internal and external to the product. The internal ones are visible to developers and managers, while the external ones are visible to the user. It is easy to see that reliability from the user's view point implies verifiability from the developer's view point. On the other hand, this distinction is often not well defined. The following are some of the most important qualities usually associated with software products.

Correctness

A term sometimes incorrectly used as a synonym for reliability or robustness, relates to the fact that a program behaves as expected. More technically we can say that a program is correct if it behaves according to its functional specifications (described later in this chapter). Correctness can be verified experimentally (by testing) or analytically. Certain programming tools and practices tend to improve correctness while others tend to diminish it.

Reliability

Relates to a program's trustworthiness or dependability. More formally reliability can be defined as the statistical probability that a program will continue to perform as expected over a period of time. We can say that reliability is a relative measure of correctness.

The immaturity of software engineering as a technical discipline is made evident by the fact that we expect defects in software products. Without much embarrassment, developers often release programs accompanied by lists of known bugs. Who would purchase a car that came with a lists of known defects? What is more, programs are often sold with disclaimers that attempt to void any responsibility on the part of the producer for program defects. Warnings state that breaking the shrink-wrap cancels any accountability on the part of the producer. Not until we are willing to assume accountability for our products will software engineering become a technical field. We must be liable for the correctness of products that we claimed have been technically engineered.

Robustness

This software quality attempts to measure program behavior in circumstances that exceed those of the formal requirements. In other words, it is a measure of the program's reaction to unexpected circumstances. For example, a mathematical program assumes that input values are within the legal ranges defined for the operations. A user should not input a negative value to the square root function since this operation is undefined. A more robust program will recover from this error by posting an error message and requesting a new input, while a less robust one may crash.

Although robustness and correctness are related qualities they are often quite distinct. Since correctness is a measure of compliance with the formal specifications, the mathematical program that crashes on an invalid input value may still be correct, however, it will not be robust. Obviously, robustness is a quality difficult to measure and specify. Reliability is also related to robustness since robustness usually increases reliability.

Efficiency

This quality is a measure of how economically a system uses available resources. An efficient system is one that makes good use of these resources. In software engineering efficiency is equated with performance, so the terms are considered equivalent. A slow application or one that uses too much disk space reduces user productivity and increases operational cost.

Often software efficiency is difficult and costly to evaluate. The conventional methods rely on measurement, analysis, and simulation. Measurement consists of timing execution of a program or a part thereof. The stopwatch is sometimes used, but a more sensible approach consists of three timing routines: one starts an internal timer, another one stops it, and a third one displays the time elapsed between the timer start and stop calls. In this case we let the computer measure the execution time between two points in the code, an operation that is quite complicated when using an external timing device. Notice that this method requires access to the source code. Favorable outgrowth of execution time measurements is locating possible processing bottlenecks. In this case the execution time of different program sections are measured separately, using a progressively finer grid, until the code section that takes longest to execute is isolated. One problem with measurements of execution time as a way of determining efficiency is that the values are often meaningless without a point of reference. However, when comparing two or more processing routines no other method is as simple and effective.

The analytical method consist of determining efficiency by analyzing the complexity of the algorithms used by the code. The field of algorithm analysis defines theories that determine the worst, best, and average case behavior in terms of the use of critical resources such as time and space. But algorithm analysis is often too costly to be used in the smaller application development project.

The simulation method of determining efficiency is based on developing models that emulate the product so that we can analyze and quantify its behavior. Although this method may occasionally be useful in determining efficiency of medium-size applications, more typically its cost would be prohibitive.

Verifiability

A software product is verifiable if its properties can be easily ascertained by analytical methods or by testing. Often verifiability is an internal quality of interest mainly to the developers. At other times the user needs to determine that the software performs as expected, as is the case regarding program security. Some programming practices foster verifiability while other ones do not, as will be shown often in this book.

Maintainability

Programmers discover rather early in their careers that a software product is never finished: a finished program is an oxymoron. Programs evolve throughout their life span due to changes that correct newly detected defects or modifications that implement new functionalities as they become necessary. In this sense we refer to software maintenance as a program upkeep operation. The easier that a program is to upkeep, the greater its maintainability. If the modification refers to correcting a program defect we speak of repairability. If it refers to implementing a new function, we speak of evolvability.

User Friendliness

This is perhaps the least tangible software property since it refers to a measurement of human usability. The main problem is that different users may consider the same program feature as having various degrees of friendliness. In this sense, a user accustomed to a mouse device may consider that a program that uses this interface is friendly, while one unaccustomed to the mouse and may consider the program unfriendly.

The user interface is often considered the most important element in a program's user friendliness. But here again user preference often varies according to previous level of expertise. Also notice that several other attributes affect a program's user friendliness. For example, a program with low levels of correctness and reliability and that performs poorly could hardly be considered as user friendly. We revisit the topic of user friendliness on a later discussion of human engineering topics.

Reusability

The maturity of an engineering field is characterized by the degree of reusability. In this sense the same electric motor used in powering a washing machine may also be used in a furnace fan and a reciprocating wood saw. However, notice that in the case of the electric motor reusability is possible due to a high degree of standardization. For instance, different types of belts and pulleys are available off-the-shelf for adapting the motor to various devices. Also, the electric power is standardized so that the motor generates the same power and revolutions per minute in the various applications.

Perhaps the most important reason for promoting software reusability is that it reduces the cost of production. On the other hand, reusability of software components is often difficult to achieve because of lack of standardization. For example, the search engine developed for a word processor may not be directly usable in another one due to variations in the data structures used for storing and formatting the text file. Or a mathematical routine that calculates the square root may not be reusable in another program due to variations in required precision or in numeric data formats.

Another interesting feature of reusability is that it is usually introduced at design-time. Reusable software components must be designed as such, since reusability as an afterthought is rarely effective. Another consideration is that certain programming methodologies, such as object orientation, foster and promote reusability through specific mechanisms and techniques. Throughout this book we discuss reusability as it relates to many aspects of software production.

Portability

The term is related to the word port, which is a computer connection to the outside world. Software is said to be portable if it can be transmitted through a port, usually to another machine. More generally, a program of part thereof is portable if it can execute in different hardware and software environments. Portability is often an important economic issue since programs often need to be transferred to other machines and software environments.

One of the great advantages of developing applications in high-level languages is their greater portability. A program that uses no machine-specific features, written entirely in C++, on the PC, can probably be converted to run on a Macintosh computer with no great effort. On the other hand, program developers often need to use low-level machine features, related to either specific hardware or software properties, to achieve a particular functionality for their products. This decision often implies a substantial sacrifice in portability.

Other Properties

Many other properties of software are often mentioned in software engineering textbooks, among them are timeliness, visibility, productivity, interoperability, and understandability. In special circumstances these and other properties may take on special importance. Nevertheless, in general application development, those specifically listed appear to be the most significant ones.

26.1.2 Quality Metrics

One of the greatest challenges of software engineering is the measurement of the software attributes. It is relatively easy to state that a program must be robust and reliable and another matter to measure its robustness and reliability in some predetermined units. If we have no reliable way of measuring a particular quality it will be difficult to determine if it is achieved in a particular case, or to what degree it is present. Furthermore, in order to measure a quality precisely we must first be capable of accurately defining it, which is not always an easy task.

Most engineering fields have developed standard metrics for measuring product quality. For example, we can compare the quality of two car batteries by means of the cold cranking amps which they are capable of delivering. On the other hand, nonengineered products are typically lacking quality measurements. In this sense we cannot determine from the label on a videocassette what is the entertainment value of the movie that it contains, nor are units of information specified in the jacket of a technical book. Software is also a field in which there are no universally accepted quality metrics, although substantial work is this direction is in progress. The verification of program correctness, discussed later in the book, directly relates to software quality metrics.

26.2 Principles of Software Engineering

We started this chapter on the assumption that software development is a creative activity and that programming is not an exact science. From this point of view even the term software engineering may be considered unsuitable since we could preferably speak of software development technique, which term does not imply the rigor of a formal engineering approach. In our opinion it is a mistake to assume that programs can be mechanically generated by some mechanical methodology, no matter how sophisticated. When software engineering falls short of producing the expected results it is because we over-stressed the scientific and technical aspects of program development over those that are artistic or aesthetic in nature or that depend on talent, personal endowments, or know-how. Nevertheless, as there is technique in art, there is technique in program development. Software engineering is the conventional name that groups the technical and scientific aspects of program development.

Smaller software projects usually take place within the constraints of a limited budget. Often financial resources do not extend to hiring trained software project managers or specialists in the field of software engineering. The person in charge of the project usually wears many hats, including that of project manager and software engineer. In fact, it is not unusual that the project manager/engineer is also part-time designer, programmer, tester, and documentation specialist. What this all means is that the formality and rigor used in engineering a major project may not apply to one of lesser proportions. In other words, the strictness and rigidity of software engineering principles may have to be scaled down to accommodate the smaller projects.

In this sense we must distinguish between principles, techniques, and tools of software engineering. Principles are general guidelines that are applicable at any stage of the program production process. They are the abstract statements that describe desirable properties, but that are of little use in practical software development. For example, the principle that encourages high program reliability does not tell us how to make a program reliable. Techniques or methods refer to a particular approach to solving a problem and help ensure that a product will have the desirable properties. Tools are specific resources that are used in implementing a particular technique. In this case we may state as a principle that floating-point numbers are a desirable format for representing decimals in a digital machine. Also that the floating-point techniques described in the ANSI standard 754 are suitable for our application and should be followed. Finally, that a particular library of floating-point routines, which complies with ANSI 754, would be an adequate tool for implementing the mathematical functions required in our application. Figure 26.1 graphically shows the relationship between these three elements.

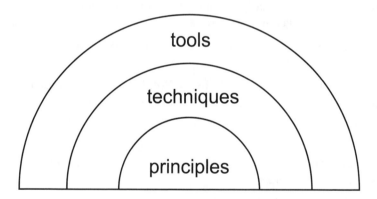

Figure 26.1 *Relation Between Principles, Techniques, and Tools*

26.2.1 Rigor

One of the drawbacks of considering a program as an art form is that it places the emphasis on inspiration rather than on accuracy and preciseness. But programming is an applied rather than a pure art. We may find it charming that Michelangelo planned his statue of David rather carelessly and ended up with insufficient marble for sculpturing the feet. But a client may not be willing to forgive an artistically-minded programmer who did not find inspiration to implement hyphenation in developing a word processor program. We must conclude that the fact that programming is often a creative activity does not excuse us from pursuing program design and development with the necessary rigor.

Some authors distinguish between degrees of rigor. The highest degree, called formality, requires that development be made strictly according to laws of mathematics and logic. In this sense formality is a high degree of rigor. One field in which the issue of rigor and formality gain particular importance is in software specifications. A logical formalism has been developed which allows the precise specification of software. The mathematical expressions, usually based on the predicate calculus, allow the representation of complete programs, or of program fragments, in symbolic expressions that can be manipulated according to mathematical laws. Some of the advantages of this methodology claimed by it advocates are the following:

1. Formal methods allow the software engineer to specify, develop, and verify a software system using a formal specification language so that its correctness can be assessed systematically.

2. They provide a mechanism for the elimination of program ambiguity and inconsistency through mathematical analysis, thereby discovering errors that could otherwise go undetected.

3. They allow a software problem to be expressed in terms of an algebraic derivation, thus forcing the specifier to think in more rigorous terms.

4. These methods will eventually make possible the mechanization of programming thereby revolutionizing software development.

On the other hand, the detractors of formal methods of specification also have their own counter arguments:

1. Formal specifications are difficult to learn. Nothing short of a graduate course is required to obtain even a cursory understanding of this subject.

2. So far these methods have not been used successfully in practical program development.

3. Most programmers, even most software engineers working today, are unfamiliar with formal specifications.

At this point we could declare that because the scope of our book is the development of smaller size program we will be satisfied with rigor and exclude formal methodologies from its contents. But the fact is that some smaller programs that deal with a particular subject matter, can (perhaps should) be specified formally. Therefore we exclude the consideration of formal methods of specification based purely on expediency. Our excuse is that this is a rather intricate subject, and that most program development done today still takes place without the use of formal methods of specifications.

Therefore we settle for the lowest level of rigor, which we define as methodology for program development based on following a sequence of well-defined and precisely stated steps. In the programming stage of development rigor is unavoidable. In fact, programming is performed in a formal context of well-defined syntax and semantics. But rigor should also apply in every stage of the program development process, including program design, specification, and verification. A program developed according to a rigorous methodology should contain the desirable attributes previously mentioned. In other words, the results should be reliable, understandable, verifiable, maintainable, and reusable.

26.2.2 Separation of Concerns

It is common sense that when dealing with complex issues we must look separately at its various facets. Since software development is an inherently complex activity, separation of concerns becomes a practical necessity. A coarse-grain observation of any construction project immediately reveals three concerns or levels of activity: technical, managerial, and financial. Technical concerns deal with the technological and scientific part of the project. The managerial concerns refer to project administration. The financial concerns relate to the monetary and fiscal activities.

An example shows this notion of separation of concerns into technical, managerial, and financial. Suppose homeowners who are considering the possibility of building a new garage. The project can be analyzed by looking at the three separate activities required for its completion. Technically, the homeowners must determine the size, type, and specifications of the garage to be built and its location within the property. A finer level of technical details could include defining the preparatory groundwork, the concrete work necessary for the foundation, the number and type of doors and windows, the siding, the electrical installation, roofing, as well as other construction parameters. The managerial aspects of the project may include supervising of the various subcontractors, purchasing materials, obtaining necessary building permits, and the timely disbursing of funds. Financial activities include evaluating how much the project would add to the value of the property, selecting the most desirable bids, and procuring a construction loan.

It is important to realize that separation of concerns is a convenience for viewing the different aspects of a project. It is an analytical tool which does not necessarily imply a division in the decision authority. In the previous example, the homeowners may decide to perform the financial activities themselves, while they delegate the managerial activities to a general contractor, who in turn, leaves the technical building details to the subcontractors. Nevertheless, it is a basic rule of capitalism that whoever pays the bills gives the orders. Therefore the homeowners would retain the highest level of authority. They would have the right to request managerial or even technical information from the general contractor, who they may override as they see fit, or even fire if they estimate it necessary.

By the same token, the participants of a software development project should always be aware of the fact that separation of concerns does not imply any permanent delegation of authority. The fact that a software designer is assigned the task of defining and specifying the program does not mean that whoever is in charge cannot step in at any point in the project's development and override the designer. In this respect two extremes should be avoided. In one case the participants in the project lose sight of where the final authority resides and are misled by the assumption that it has been totally and permanently delegated in their favor. This misconception often generates a take-it-or-leave-it attitude on the part of the project's technical staff who seem to state: "this is what is best, and disagreement only proves your technical ignorance." The other extreme occurs when authority is held so closely that the participants are afraid of showing their knowledge or expertise for fear of being ridiculed or overridden.

It is the task of those in charge of a project to make their subordinates aware of where final authority resides and of the fact that this authority will be used if necessary, perhaps even in a form that may seem arbitrary. Also, of what are the delegated levels of decision and the boundaries within which each participant can be freely creative. This initial definition of the various levels of concern, responsibility, and authority is a most important element in the success of a software development project.

We have seen that the notion of separation of concerns is related to separation of responsibilities, furthermore, it is related to specialization of labor. At the coarser level this determines the separation of activities into financial, managerial, and technical. Each of the groups can be further subdivided. For example, the technical staff can be divided into specialists in program design, coding, and testing. The coding personnel can be further specialized in low- and high-level languages, and the high-level language ones into C++, Pascal, and Cobol programmers. By the same token, the program designers can be specialists in various fields, and so can the managers and financial experts.

26.2.3 Modularization

Modularization is a general-purpose mechanism for reducing complexity which transcends programming languages or environments. Modules should be conceptualized as units of program division, but not necessarily equated to subroutines, subprograms, or disk files. For smaller software development projects modularization is a most effective organizational tool.

Two modularization methods are usually identified: one called top down decomposes a problem into sub-problems that can be tackled independently; another one, called bottom up builds up a system starting with elementary components. In this same context it is often stated that the main advantage of modularization is that it makes the system understandable. However, this understandability is usually achieved in stages during program development. Modularization typically takes place progressively. Since it is often impractical to wait until a system is fully understood to start its development, the most practical approach is to define an initial modular division and progressively refine it as the project progresses and the participants become more acquainted with its features.

The term modular cohesion refers to the relationship between the elements of a module. For example, a module in which the processing routines, procedures, and data structures are strongly related is said to have high cohesion. Plugging into a module some unrelated function just because we can find no better place for it reduces the module's cohesion. It is usually preferable to allocate a catch-all module, which we can somewhat euphemistically call the general-support module, in which we can temporarily place those functions or data structures for which no ideal location is immediately apparent. Later in the development we may rethink the program's modularization and move elements from the catch-all into other modules.

The term modular coupling refers to the relationship between modules in the same program. For example, if one module is heavily dependent on the procedures or data in another module, then the modules are said to be tightly coupled. Modules that have tight coupling are difficult to understand, analyze, and reuse separately, therefore tight coupling is usually considered an undesirable property. On the other hand all the modules in a program must be coupled in some way since they must interact in producing a common functionality. It is difficult to imagine a module that is entirely uncoupled from the system and still performs some related function.

It is often stated, as a general principle, that a good modular structure should have a high cohesion and a loose coupling. The elements must be closely related within the module and there should be a minimum of intermodular dependency.

26.2.4 Abstraction and Information Hiding

It is a unique property of the human mind that it can reduce a problem's complexity by concentrating on a certain subset of its features while disregarding some of the details. For example, we can depict an automobile by its component parts: motor, transmission, chassis, body, and wheels. This functional abstraction, which is based on the tasks performed by each component, ignores the details of certain parts that are included in the abstraction. For example, the abstraction may assume that the trunk lid need not be mentioned specifically since it is part of the body, or the starter, since it is part of the motor.

Since an abstraction is a particular representation of reality, there can be many abstractions of the same object. The previous automobile abstraction into motor, transmission, chassis, body, and wheels may be useful for a general description. However, a car-wash operation may not be interested in the components with which it does not have to deal. Therefore a car-wash project may define that an automobile abstraction composed of hood, trunk, fenders, roof, windshield, and windows is more useful to its purpose. Regarding software we often find that the abstraction that could be useful in describing a system to a user may not be the most suitable one for the developers. By the same token, those charged with the task of testing a software system may prefer an abstraction that is different from the one adopted by the users, designers, or programmers.

Since abstraction helps define the entities that comprise a software product it is a modularization tool. We modularize by abstraction. Information hiding and encapsulation are terms that appeared during the 1970s in relation to object-oriented design and programming. The notion was proposed by David Parnas who states that each program component should encapsulate or hide a design decision in such a way that it will reveal as little as possible about its inner workings. In other words, information hiding is one of the most desirable properties of abstraction. Its main usefulness, according to Coad and Yourdon, is that it localizes volatility thus facilitating changes that are less likely to have side effects. The notions of encapsulation and information hiding will be revisited in the chapters related to object oriented analysis and design.

26.2.5 Malleability and Anticipation of Change

A program should not be envisioned as a finished product but as a stage or phase in a never-ending development process. We can speak of a program as satisfying a certain set of requirements, as being suitable for a given purpose, or a certain version being ready for market, but to a software developer the notion of a "finished program" is an oxymoron. For a program to be finished we must assume that is has no remaining defects, that all present or future user requirements are perfectly satisfied, and that no improvements or modifications will ever be necessary. Since none of these conditions can ever be proved, a program can never be considered finished.

It is all a matter of semantics, but the ever-changing nature of software makes it preferable that we consider a program as a category more than as a product. In this sense we can say that the program WordMaster consists of the various implementations named WordMaster 1.0, WordMaster 2.1, and WordMaster Deluxe. In this same sense we speak of the automobile category named Chevrolet which includes the models '56 Corvette, '76 Station Wagon, and '94 Pickup.

The fact that software products are fundamentally fragile encourages the previous view. Knowing this, we must design software so it is malleable, which means that we must anticipate change. The reason for change comes from two possible causes: program repair and evolution. It is indeed a naive software developer who does not know that program defects will be detected after its release, or who cannot anticipate that new requirements will be identified and old ones will change. It is perhaps this inherent fragility that distinguishes software from many other industrial products.

The two desirable features of program maintainability and evolvability, previously mentioned, are closely related to program malleability and anticipation of change. Also reusability, since a software component is reusable if it can be easily adapted to perform a different function or to execute in a new context. This usually requires that the component be malleable, which in turn implies that the original designers anticipated change.

26.2.6 Maximum Generalization

The more generalized a software product, the larger the audience to which it appeals. The more generalized a program routine or procedure, the greater its possibilities for reusability. This implies the convenience of maximum generalization at the level of the application's design and of the individual routines, procedures, or modules. In these cases the maximum generalization approach is based on trying to discover a more general problem that is equivalent to the one presented, or a more general solution.

At the product development level the principle of maximum generalization advises that we search for more general problems hidden behind the one at hand. For example, a DOS image processing application is required to execute in a true-color graphics mode. The initial approach may be to incorporate into the application low-level drivers for a VESA true-color mode. However, a careful analysis of the problem shows that a more general approach would be to develop a library of support functions for several true-color VESA modes. The library could be used by the application under development and also furnished as a commercial toolkit. In this case the greater development effort required for creating the library may prove to be well justified.

26.2.7 Incremental Development

Sculpturing is a creative activity that resembles programming. How would a sculptor approach building a realistic marble statue of Franklin D. Roosevelt? One conceivable method would be to start at the top and attempt to carve out and finish the hat, then the head, then the torso, then the arms, and so forth until reaching the shoes, at which point the statue would be finished. But this method is not very reasonable since it assumes that each part of the anatomy of FDR would be finished to its finest details before the next one is started. An actual sculptor would probably approach the problem by progressive approximations. He or she would start by making a drawing of what the final statue is to look like. Then the basic traces of the drawing would be transferred to the marble block in order to determine which parts of the block could be roughly chipped away. The process of chipping away the unnecessary material would continue progressively on the whole figure, which would begin to approximate the final result by stages, rather than by parts.

An inexperienced sculptor could possibly err in two ways: he or she may be tempted to skip the drawing stage (program design) and proceed directly into the carving (coding), or may devote too much time to the details of the drawing and delay the carving unnecessarily. On the other hand, the experienced artist knows that the drawing is a basic guide but that the finer details of the statue come during the creative process. If the drawing shows FDR with his hands in his pockets, later the sculptor will not be able to change his or her mind and show the president waving at the crowd, since the necessary material would no longer be available. Therefore, FDR's general form and pose must be decided before the carving starts. However, the exact dimension of the cigarette holder, his facial expression, or the texture of his coat, are details that can be decided at a later stage.

Regarding software, incremental development sometimes consists of the creation of product subsets that can be delivered to the customer for early evaluation and feedback. In reality, very few software projects can be exactly defined at the starting point. Therefore it is often reasonable to divide the program specification into progressive stages, perhaps associating each stage with a test version of the final product. One advantage of this method is that the customer gets to examine and evaluate several prototypes.

However, the iterative development method introduces new management and organizational problems due to the fact that the project itself is defined "on the fly." One issue that must be specifically addressed is the management of the documentation for each of the various development stages as well as for the corresponding prototypes. A badly managed incremental development project can easily turn into disorder and anarchy.

26.3 Software Engineering Paradigms

Some software engineering authors, including Pressman, speak of a series of steps in program development that encompass the use of specific methods, tools, and policies. In software engineering the term methods refers to technical procedures used in the development of a project. It includes project planning, systems and requirements analysis, data structure design, algorithm selection and evaluation, coding, estimation of program correctness, and maintenance procedures.

Software engineering tools are software aides to program development. These tools, usually called computer-aided software engineering, or CASE tools, create a development environment that tends to automate development and provides support for the methods.

The policies (sometimes called procedures) bind together the methods and tools into a practical methodology of program development. For example, the policies adopted for a particular project define the methods that will be applied, the deliverables that will be generated during the development process, the controls that will help assure the quality of the final product, and the various control points in the development schedule.

A particular series of steps (methods, tools, and policies) gives rise to specific paradigms. Three of these paradigms have been extensively discussed in the literature: the waterfall model, the prototype methods, and the spiral development model. A fourth one, sometimes called a fourth-generation technique, is based on the use on nonprocedural languages and specialized software tools. Because of its unconventionality we will not discuss fourth-generation techniques, although the reader should note that many authors envision a promising future for this development methodology.

26.3.1 The Waterfall Model

This classical model of a software engineering project is based on the notion of a system life-cycle. Although the name waterfall model is rather recent, and so is the notion of a life-cycle paradigm, the notion of a flow of development activities can be found in the early software engineering literature. Tausworthe talks about a flow of activities in top-down development and provides a graph in which one activity follows another one in a waterfall-like fashion. Figure 26.2 is a representation of the classic waterfall model for a software project.

The specification phase of the system life cycle consists of a requirements gathering process through analysis and systems engineering. Whenever the project must interface with existing software or hardware elements the specification phase must include a systems requirements definition. During this phase customer and developer work very closely: the customer provides the requirements and the developer reflects these requirements in a formal specification that is, in turn, reviewed by the customer. The requirements/specification cycles continue until both parties agree that the project has been clearly and unambiguously defined.

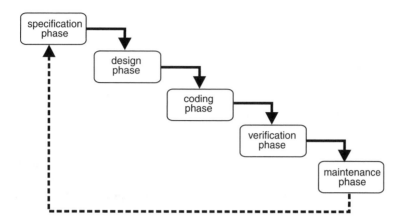

Figure 26.2 *The Waterfall Model*

The design phase of the software life-cycle focuses on four elements: data structures, program architecture, procedures, and interfaces. This phase concludes with a representation of the software that can be evaluated and assessed on its own. The maxim that states that engineering is design is also applicable to software engineering, since the design stage is often the most critical and difficult one. During the coding phase we convert the design into a machine-executable product.

In the life cycle paradigm it is conventionally accepted that the verification phase starts at the conclusion of the coding phase. Once the code executes in a machine it must be evaluated for correctness. This means that we must ascertain that it meets the requirements developed during the specifications phase and that it is free from defects. Although this phase is sometimes associated with debugging, it should also include all formal and experimental verifications of program correctness.

Although the maintenance phase was not included in the original model, today it is generally accepted that programs will undoubtedly undergo change and that these changes are part of the software life-cycle. The reason for changes include errors detected, modifications in the program's hardware or software environment, or the implementation of new functional requirements identified by the customer. Maintenance procedures require revisiting all the stages of the software life-cycle, as depicted by the dotted arrow in Figure 26.2.

Although the waterfall model has been used successfully in many software engineering projects and is still considered an effective method for program development, several shortcomings have been pointed out. The most notable one is that software development projects can rarely be specified and defined in one sitting. A more typical scenario is a series of iterative development cycles such as those mentioned in regards to the incremental development principle. Another difficulty in applying this model is that projects do not often follow the sequential flow described in the paradigm. For example, program verification is often done concurrently with the coding phase, as are the design and specification phases. Another criticism is that maintenance operations often require a repetition of the program life cycle. For instance, developing a new version of an existing program in order to correct known defects and to incorporate recently defined features requires revisiting all the life-cycle stages, starting with the specification phase. Therefore it is difficult to consider program maintenance as a typical development phase since it has unique effects that are not evident in the other phases of the model.

In spite of these and other weaknesses of the waterfall model, this paradigm is the one most widely used in software engineering practice. Although other more complicated models are usually judged to be more rigorous and accurate, the waterfall model remains the most popular and simplest methodology for software development.

26.3.2 Prototyping

Many software development projects are of an experimental or speculative nature. Consider the following examples:

- A research group wishes to determine if it is possible to develop an expert system that uses data obtained by remote-sensing satellites in order determine pollution levels in the lakes and streams of the United States.

- An entrepreneur wishes to determine if it is feasible to develop a word processing program in which the user is equipped with foot pedals that activate some of the program functions.

In either of these cases we can see that the software development project can hardly be stated a priori. The objectives are described so generally that it is difficult to define specific program requirements that could serve as a base for a detailed design. In both cases, as well as in many others in which an initial detailed design is not possible or practical, a prototyping approach could be a feasible alternative.

In prototyping the developer is able to create a model of the software. This model can later be used to better define the final product or to ascertain its feasibility. The prototype can be a simple paper model of the software, which can be produced with little or no coding, a working prototype that implements a subset of the program functions, or a complete program in which some functions are not implemented.

The purpose of the prototype is to allow both customer and developer to make decisions regarding the feasibility and practicality of the project, and, if judged feasible and practical, to better define the final product. Prototyping is often depicted as a development cycle with the sequence of steps shown in Figure 26.3.

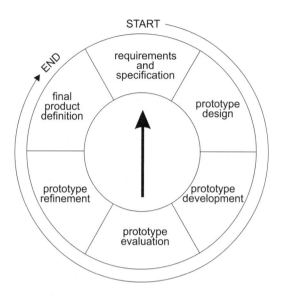

Figure 26.3 *The Prototyping Model*

Prototype development begins by collecting requirements and specifications. Then the prototype is designed, usually by following an abbreviated process which produces results quicker than conventional program design procedures. The prototype is built, also shortening the development processes by skipping all processing steps that are not strictly necessary for the purpose at hand. The prototype is finally evaluated, first by the developer and later by the customer. If necessary, it is further refined and tuned in an iterative cycle. The finished prototype is used to further define the final software product.

One major risk of prototype development relates to customers that misunderstand its purposes. This misconception can give rise to the following:

1. A customer not aware that the prototype is an unfinished product, usually with less functionality and lower performance that can be achieved in the final version, may make wrong decisions regarding the suitability of the final product. In extreme cases a promising project may be cancelled by the customer due to misconceptions that originated in the prototype.

2. The customer, in a rush to get a product to market or to put a product to practical use, decides that the prototype itself is sufficiently adequate. The customer then requests that the developer perform some final adjustments and modifications and deliver the prototype as a final product. One incentive this customer could have is the anticipated lower cost of using the prototype itself rather than following the development process as originally planned.

In most cases the possibility of a misunderstanding regarding the prototype phase can be eliminated by clearly explaining these risks to the customer. Alternatively, it may be advisable to exchange a memorandum of agreement between developer and customer regarding the reduced functionality of the prototype or the fact that it will not be suitable as a finished product. Unfortunately, in some cases, the risks of the prototype being misinterpreted by the client have made developers decide against adopting this paradigm.

26.3.3 The Spiral Model

This model, first suggested by Boehm in 1988, proposes to merge the best features of the life-cycle and the prototyping paradigm with the principle of incremental development. Figure 26.4 (based on Pressman) shows a spiral progression through four different stages.

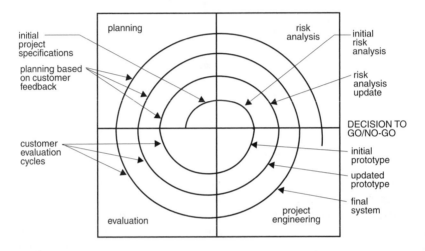

Figure 26.4 *The Spiral Model*

Notice that the drawing in Figure 26.4 is meant as a general illustration of the method and is not be interpreted literally. For example, the number of cycles around the spiral will vary from project to project.

A unique feature of the spiral model is the introduction of a risk analysis stage, which culminates in a go or no-go decision at the conclusion of each development cycle. However, this phase is also its most controversial feature. In the first place, risk analysis requires a particular expertise, and is trivialized when performed by un-trained personnel. In fact, the risk analysis phase is undesirable if it can lead to invalid interpretation of results. In addition, customers often believe that they performed a risk analysis of the project before deciding to undertake it, and that further consideration of this matter is unnecessary. Furthermore, the possibility that at the conclusion of each development cycle the entire project could be scrapped by a no-go decision may lead to apprehensions on the part of the customer.

On the other hand, if the difficulties and perils associated with the risk analysis phase can be conjured, then the spiral model constitutes the most satisfactory paradigm for the development of large software systems. The incremental development approach proposed by this model, with its repeated prototyping and risk evaluation phases, provides a realistic framework for program development. Both customer and developer have repeated opportunities in which to identify possible defects and shortcomings and make the necessary adjustments. Note that this model is relatively new and that not much experience regarding its application has been collected yet.

26.3.4 A Pragmatic Approach

The practicing software developer must decide, in each case, which model or combinations of models are most suitable to the project at hand. The decision is often based on practical risks and limitations rather than on theoretical applicability. For example, a developer may consider that the most adequate paradigm is the prototyping model, but a substantial risk that the customer will misinterpret the results advises against its use. In another case a project may fit quite well within the spiral model but the fact that personnel trained in risk analysis will not be available suggests a modification of the paradigm or the adoption of the more conventional waterfall model.

A wise man once said: "All systems and no system: behold the best system." This maxim is quite applicable to the various software engineering paradigms mentioned in the preceding sections, in particular when they concern smaller projects in which development time and resources are limited. The most common scenario is a combination of paradigms. Often the most useful one is a spiral model in which the risk evaluation function is scaled down to accommodate two factors. The first one is customer nervousness that results from the possibility of a project cancellation. The second one is the short supply of experts in the risk analysis field. Since the spiral model is in itself a combination of the waterfall and the prototyping model, all of the mentioned paradigms will actually be used. A possible scaling of the risk analysis phase could consist of limiting it to the first or the first and second cycles around the model. Thereafter it will be assumed that a go decision is permanently on. This modification certainly compromises one of the best features of the spiral model, but in many cases practical considerations will make the adjustment necessary.

Other circumstances may also require modifications of the theoretical models. For instance, most software engineering and system analysis textbooks assume that the roles of customer and developer are undertaken by separate individuals or entities. In real life many software development projects take place in circumstances where customer and developer are the same person or commercial firm. This is the case when a software company develops a project that will be marketed directly, or when a scientific research group develops its own project. In either case one possible remedy it to artificially emulate the customer function.

For example, the marketing department of the software company can play the customer role, or the head of the research project can pretend to be the client for which the product is being developed. The dual roles of developers and customer are so important in the software development process that any solution that will preserve it, even fictitiously, is an attractive option. However, we should always keep in mind that a virtual customer is a fictional creation, which may be overridden by higher authority at any time. When this happens, the integrity of the paradigm is seriously compromised, since the division between judgment and authority ceases to exit. The project then becomes like a democracy in which the executive branch takes on the power of firing the members of the judicial branch. In either case, although appearances may be preserved, the actual check-and-balances mechanism ceases to exist.

26.4 Concurrent Documentation

One of the most important lessons of software engineering refers to the need for adequate and rigorous project documentation. By the same token, the most notable difference between a correctly engineered development project and a haphazard effort is the documentation. Too often the tendency has been to consider program documentation as a secondary problem, one that can be addressed once the project is finished. This tendency is probably traceable to the same human fallacy that makes some programmers believe that comments can be inserted into the code after the programming has concluded. As writing comments is part of the chore of programming, documentation is part of the task of program development. Either one cannot be approached as an afterthought, at risk of writing spaghetti code or of developing undecipherable projects.

Tausworthe, who is responsible for the notion of concurrent documentation, states:

> *"The second principle guiding this work is that the definition, design, coding, and verification phases of development cannot be regarded as complete until the documentation is complete and certified by some form of correctness audit... good documentation is inextricably bound up in each facet of the project, from conception, to design, to coding, testing, etc., and because the formalization enforces a discipline, creating a program methodology."*

It has been the information systems community that has stressed the importance of documenting the software development project. Often a book on software engineering will devote little space to this topic, on the other hand, one on system analysis and design is more likely to contain a detailed discussion of the topic. Capron's *System Analysis and Design* contains a thorough treatment of documentation. Perhaps the awareness of the importance of documentation is more natural to the business community than to programmers and program designers, who often show distaste for paperwork.

In regards to software project development the following types of documentation can be clearly identified:

1. Written reports that mark the conclusion of a phase of the development cycle. These documents are sometimes called the deliverables, since they are often presented to the client as each development phases concludes. Typical deliverables are the feasibility study, the analysis and requirements document, and the detailed design document.

2. User manuals and training guides, which can be printed or online.

3. Operations documents, more often found in large computer environments, include run-time schedules, input and output forms and media, delivery, routing, and distribution charts, data file specifications, update schedules, recovery procedures, and security controls.

4. The project scrapbook is used to collect memos, schedules, meeting minutes, and other communications generated during the project.

26.4.1 Objections and Excuses

Software projects are developed by individuals who often resent having to spend time in what some consider useless paperwork. Even when the project managers are aware of the importance of documentation, they often have to overcome considerable resistance on the part of their technical staff. The following are excuses used to rationalize the documentation aversion:

1. Our time is better spent in programming and other technical matters than in useless paperwork.

2. If we attempt to document this activity now, the results will shortly be outdated. Therefore it is preferable to wait until the project is closer to its completion before we start on the documentation.

3. People do not read manuals, so why bother writing them.

4. We are programmers and designers, not writers. We should hire technical writers to do the manuals and paperwork.

26.4.2 Advantages of Good Documentation

The reasons why the excuses mentioned in the previous section are effective is that they have some truth value. It is only by realizing that documentation is one of the fundamental results of a development project, and that it should never be an afterthought at project conclusion time, that we can invalidate these arguments. The project manager must be conscious of the fact that program documentation is at least as important as any other technical phase of development. The following are undeniable advantages of concurrently documenting the development project:

1. A well-documented project is better able to resist personnel changes since new programmers can catch up by studying the project documents.

2. Documentation can serve as a management tool by requiring that each development stage conclude in a document, which must be approved before the next stage can proceed.

3. Concurrent documentation establishes a project history and can serve, among other things, as a progress report. Documents can be used as scheduling landmarks and therefore serve to measure and evaluate project progress.

The most important principle of project documentation is that of concurrency. Documentation must be a substantial part of the development effort and must take place simultaneously with each development phase. At the same time, documentation is often the development activity most easily postponed or even sacrificed. When time is running short it is tempting to defer documentation. At this point the project manager must be aware that when documentation loses its concurrency it also loses a great portion of its usefulness.

Chapter 27

Description and Specification

Chapter Summary

This chapter is about one an essential preliminary step in the design and coding of a software system: its specification. These preparatory steps include the feasability study and requirements analysis and specifications. It also covers the use of data flow and data modeling tools.

27.0 System Analysis Phase

The engineer's first task is understanding the system, the second one is specifying it, the third one is building it. Analysis must precede specification since it is impossible to define or describe what is unknown. Specification must precede construction, since we cannot build what has not been defined. The first two tasks (understanding and defining the system) can be quite challenging in the software development field, particularly regarding small- to medium-size projects.

The system analysis phases refer to the evaluation of an existing system. Therefore it may be arguable that this phase can be skipped when implementing a system which has no predecessor. In any case, we should not assume that the analysis phase can be circumvented simply because a system has been judged obsolete or because a decision has already been reached regarding its replacement. Outdated, obsolete, and even totally unsuitable systems may contain information that is necessary or convenient for designing and implementing a new one. Very rarely would it be advisable to completely discard an existing system without analyzing its operation, performance, and implementation characteristics. At the same time, the feasibility of a new system must often be determined independently of the one being replaced. In fact, a system's feasibility should be evaluated even when it has no predecessor.

27.0.1 The System Analyst

In the information systems world the specialist attempting to understand and define a software system is usually called the system analyst. In computer science-oriented environments this person is often referred to as a systems engineer. Other terms used to describe this computer professional are programmer analyst, systems designer, systems consultant, information system analyst, and information system engineer.

Under whatever name, the systems analyst's job description is usually quite different from that of a programmer. While the responsibility of a programmer ends with the computer program, the analyst's duties extend into management and administration. Typically the analyst is responsible for equipment, procedures, and databases. Furthermore, the system analyst is usually in charge of the software development team. Therefore, the analyst's job requires management skills in addition to technical competence. Oral and written communications skills are often necessary since the analyst usually has to perform presentations to clients, lead specifications discussions, and oversee the generation of multiple documents and reports.

Programming and system-building projects are usually directed by a lead or principal analyst, who may take on the responsibility of managing and overseeing the activities of a team of analysts, programmers, program designers, software testers, and documentation specialists. The ideal profile for the principal system analyst is that of an experienced programmer and program designer (preferably holding a graduate degree in computer science or information systems) who has experience and training in business. In addition, the lead analysist should be a person who has successfully directed the analysis and development of software or systems projects of comparable complexity. The job description includes gathering and analyzing data, executing feasibility studies of development projects, designing, developing, installing, and operating computer systems, and performing presentations, recommendations, and technical specifications.

However, the smaller software project can rarely count on the services of an ideally-qualified lead analyst. Quite often a person with less-than-optimal skills is appointed to the task. In this respect it is important for the project directors to realize that it is the lead system analyst, or project manager, who should be the most highly qualified member of the development team. One common mistake is to think of the principal analyst as a chore that can be undertaken by a professional administrator with little or no technical knowledge. Or to judge that the personnel with higher technical qualifications is best employed as programmers or program designers. Experience shows that appointing someone who is not the most highly qualified and technically competent member of the development team to the task of lead system analyst is an invitation to disaster.

Although the lead system analyst often wears many hats, the most necessary skills for this job are technical ones. A competent programmer and designer can be coached or supported in administrative and business functions and helped regarding communication skills. But it would be indeed astonishing that a person without the necessary technical skills could analyze, specify, and direct a systems development project.

27.0.2 Analysis and Project Context

The conventional treatment of systems analysis, as found in most current text and reference books, assumes that a typical project consists of developing a commercial system to be implemented in a particular business enterprise. In this context the system analysis phase consists of evaluating and describing the existing system and its bottom line is determining whether the new system will be a realistic and profitable venture. This business system orientation is typical of computer information systems and management information systems programs as taught in many of our universities. On the other hand, the software engineering textbooks (sometimes considered as the counterpart of systems analysis in a computer science curricula) view the analysis phase more from a system modeling point and usually stress the requirements analysis element in this development phase.

In other words, the term system analysis has different connotations when seen from the information systems viewpoint than when seen from a computer science perspective. Furthermore, sometimes neither perspective exactly suits the problem at hand. For example, to assume that a typical systems development project consists of replacing an outdated business payroll program (information systems viewpoint) or designing a new, more efficient operating system (computer science viewpoint) leaves out many other real-life scenarios. For example, a development project could be an experimental, scientific, or research enterprise for which there is no precedent and in which profit is not a factor. Or it may consist of creating a software product that will be marketed to the public, such as a word processor application, a computer game, or a toolkit for performing engineering calculations. In either case, many of the system analysis operations and details, as described in books on systems analysis and design or on software engineering, are not pertinent.

Another conceptualization that can vary with project context relates to the components of a computer system. Normally we think of a computer system as an integration of hardware and software elements. Therefore systems engineering consists both of hardware systems and software systems engineering. Therefore, the systems analysis phase embodies an investigation and specification of the hardware and the software subsystems. In reality it often happens that either the hardware or the software elements of a computer system take on a disproportionate importance in a particular project. For instance, in a project consisting of developing a computer program, to be marketed as a product to the user community, the hardware engineering element becomes much less important than the software engineering ingredient. Under different circumstances the relative importance of hardware and software elements could be reversed.

Consequently, the details of the systems analysis phase often depend on project context. What is applicable and pertinent in a typical business development scenario, or in a technical development setting, may have to be modified to suit a particular case. Two elements of this phase are generally crucial to project success: the feasibility study and the project requirements analysis. Each of these activities is discussed in the following sections.

27.1 The Feasibility Study

Often a software development project starts with a sketchy and tentative idea whose practicality has not been yet determined. Other projects are initially well-defined and enjoy the presumption of being suitable and justified. In the first case the system analyst must begin work by performing a technical study that will determine if the project is feasible or not. According to Pressman, given unlimited time and resources, all projects are feasible. But in the real world resources are limited and often a project's feasibility is equated with the economic benefits that can be expected from its completion. This excludes systems for national defense, investigative and research projects, or any development endeavor in which financial returns are not a critical consideration.

In most cases the first project activity is called the feasibility or preliminary study. It can be broken down into the following categories:

1. Technical feasibility is an evaluation that determines if the proposed system is technologically possible. Intended functions, performance, and operational constraints determine the project's technical feasibility.

2. Economic feasibility is an evaluation of the monetary cost in terms of human and material resources that must be invested in its development, compared with the possible financial benefits and other advantages which can be realistically anticipated from the project's completion.

3. Legal feasibility is an evaluation of the legal violations and liabilities that could result from the system's development.

Before the feasibility study phase begins, intermediate and high-level management (often in consultation with the lead analyst) must determine if the effort will be advantageous and how much time and resources will be invested in this phase. It is difficult to define arbitrary rules in this respect since the convenience of a feasibility study as well as the time and human resources devoted to it are frequently determined by financial and practical considerations. It is reasonable to expect that a minor development project, anticipated to generate large revenues and requiring a relatively small investment, will often be undertaken with a less rigorous feasibility study than a major project involving greater complexity and uncertainty.

Two business management techniques are often used in the feasibility study phase of a system analysis: risk analysis and cost-benefits analysis. By means of risk analysis we attempt to establish the dangers associated with the project and to define the points at which it is preferable to cancel the project than to continue it. By means of cost-benefit analysis we attempt to measure the gains that will result from developing the system and the expenses associated with it. A feasible project is one in which the risks and costs are acceptable in consideration of its expected benefits.

Some books on systems analysis and design propose that risk analysis and cost-benefits analysis are activities that precede the main system analysis effort. Therefore, they should be considered as parts of a phase sometimes called the preliminary analysis. The rationale for this approach is that if a project is judged to be not feasible or financially not justifiable, then no further analysis or design effort is necessary. This is consistent with the accepted definition of risk analysis as a technique that serves to ascertain a project's technical feasibility, and cost-benefits analysis as a method to determine the project's economic feasibility. In reality, the technical and economic feasibility of many projects are taken for granted. In these cases risk analysis and cost-benefit analysis become either a formality, or as a justification of the original assumptions. The analyst should keep in mind that loss of objectivity in risk analysis or cost-benefit analysis renders these activities virtually useless.

A unique problem-set of the smaller-size software project is that technically qualified personnel for performing feasibility studies and cost-benefit analysis is often not available. Therefore, linking the feasibility study to risk analysis and cost-benefit analysis techniques may result in no feasibility study being performed. Since in most system development projects it is important that some form of feasibility study be executed, to the highest degree of rigor that circumstances permit, then it is preferable to consider risk analysis and cost-benefit analysis as alternative phases. The project feasibility flowchart in Figure 27.1, on the following page, illustrates this view.

A word of caution regarding the feasibility study relates to the fact that those performing it may have a vested interest in the results. For example, we may suspect that the feasibility study could turn into a self-fulfilling prophesy if it is performed by a software company who is likely to be the eventual project developer. In such cases it may be advisable to contract an independent consultant to either perform or evaluate the feasibility study, thus assuring the objectivity of the results.

27.1.1 Risk Analysis

One of the factors considered in the feasibility study is the project's risk. For example, a large project, with a great potential for financial gain, could be a big risk if it could lead to a costly lawsuit. In this case a detailed analysis of this risk factor is indicated. By the same token, a small project, with questionable technical and economic feasibility, could be a desirable risk if its successful completion represents a substantial gain. In this case the potential for gain may make it acceptable to undertake the project even if the feasibility study has determined that the possibilities for its completion are doubtful. In other words, the project is judged to be a good gamble.

Risk analysis is usually associated with the following activities:

1. risk identification

2. risk estimation

3. risk assessment

4. risk management

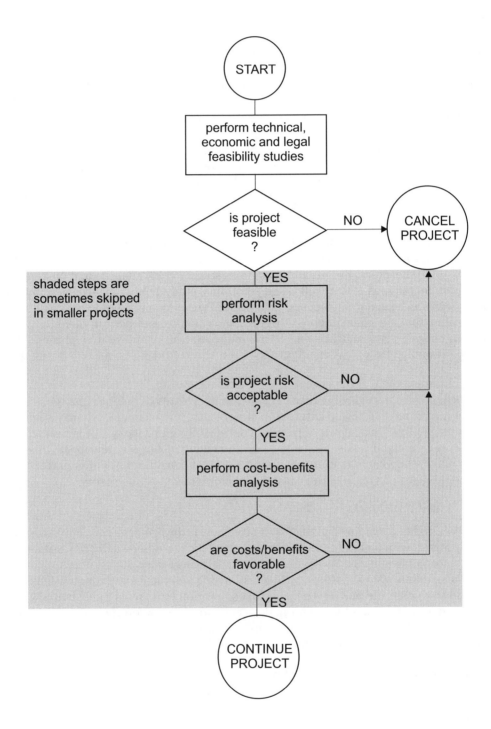

Figure 27.1 *Project Feasibility Flowchart*

We discuss each of these risk-related operations separately.

Risk Identification

Consists of categorizing risk into particular fields. In this sense project risk relates to budget, schedule, and staffing problems that may impact development. In this sense project type and complexity may pose additional risk factors. Technical risk is related to the possibility of encountering difficulties that were not initially anticipated. For example, an unexpected development in the field may make a project technically obsolete. Business risk, perhaps the most deluding one, relates to market changes, company strategy changes, sale difficulties, loss of management support, and loss of budgetary support.

Risk Estimation

In this phase of risk analysis we attempt to rate the likelihood and consequences of each risk. The first step is to establish the risk probability, the second one to describe the consequences of the risk, the third one to estimate the risk's impact on the project, and finally to determine the accuracy of the risk estimate itself. These four items can be quantified with a "yes" or "no" answer, but a better approach is to establish a probability rating for risk likelihood, consequence, impact, and projection accuracy. Historical data collected by the developers can often serve to quantify some of the project risks. As a general rule we can assume that the higher the probability of occurrence and impact of a risk the more concern that it should generate.

Risk Assessment

Is based on the aforementioned impact and probability of occurrence of a risk, as determined during the risk estimation phase. The accuracy of the risk estimate is also a factor that must be taken into account. The first task at this stage is the establishment of acceptable risk levels for the project's critical activities.

For system development projects three typical critical activities are cost, schedule, and performance. A degradation of one or more of these factors will determine the project's termination. Risk analysts talk about a referent point (or break point) at which the decision to terminate or continue a project is equally acceptable. Since risk analysis is not an exact science, the break point can better be visualized as a break area.

Figure 27.2 shows the effects of cost and schedule overruns on a software project. The dark regions bound by the curves (the uncertainty or break area) represents the region in which the decisions to stop or continue the project are approximately of equal weight. Below this area is the go region, in which the project continues without question. The area above the curve (no-go region) represents conditions that determine the cancellation of the project.

Note that in Figure 27.2 we have considered only two project risks, often there are others that can influence the decision to terminate. The reader should not assign any meaning to the particular shape of the curve in Figure 27.2. The fact that it is not uniform simply indicates that the uncertainty factor in risk assessment may not be a smooth function.

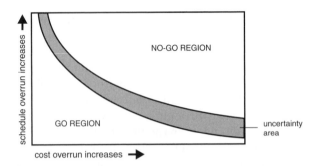

Figure 27.2 *Referent Level in Risk Assessment*

The following steps are usually followed during the risk assessment process:

1. The referent level is defined for each significant risk in the project.

2. The relationship between the risk, its likelihood, and its impact on the project is established for each referent level.

3. The termination region and the uncertainty area are defined.

4. An attempt is made to anticipate how a compound risk will affect the referent level for each participating risk.

Risk Management

Once the elements of risk description, likelihood, and impact have been established for each project risk, they can be used to minimize the particular risk or its effects. For example, assume that the analysis of historical data and the risk estimation process determine that for the particular project, there is a substantial risk of schedule overruns due to low programmer productivity. Knowing this, management can attempt to locate alternative manpower sources which could be tapped if the identified risk is realized. Or perhaps steps can be taken to improve productivity by identifying the causes of low programmer output.

Risk management, in itself, can become costly and time-consuming. Scores of risks can be identified in a large development project. Each risk may be associated with several risk management measures. Keeping track of risks and remedies can become a major project in itself. Therefore, managers and directors must sometimes perform cost-benefit analysis of the risk management activities to determine how it will impact project cost. The Pareto principle, sometimes called the 80/20 rule, states that 80 percent of project failures are related to 20 per cent of identified risks. The risk analysis phase should help recognize this critical 20 per cent, and eliminate the less significant risks from the risk management plan.

27.1.2 Risk Analysis in a Smaller Project

Risk analysis is a specialty field on which numerous monographs and textbooks have been published. A frivolous or careless attempt at risk analysis and management is often counterproductive. In the smaller software project it is common that specialized personnel is unavailable to perform accurate risk analysis. Furthermore, even if technically qualified personnel could be found, it is doubtful that a formal risk investigation would be economically justified. Consequently, most smaller projects take place with no real risk analysis and without a risk management plan. In this case the feasibility study or other project documents should clearly state that risk analysis and risk management efforts have not been performed, the reasons why this step was skipped, and the dangers that can result from the omission.

Perhaps a greater danger than the total omission of a formal risk analysis is its trivialization. When risk analysis has been performed superficially, with little rigor, and by untrained personnel, the most likely results are the identification of unreal risks, or worse, a false feeling of confidence resulting from a plan that has failed to identify the real perils associated with the project. If the risk analysis and management have been performed with less-than-adequate resources then the resulting low accuracy of the risk projection should be clearly noted in the project documentation. Occasionally, it may be considered valid to attempt some form of perfunctory risk analysis for the purpose of gaining experience in this field or for experimenting with a specific methodology. This should also be clearly noted so that management does not place undue confidence in the resulting plan.

Table 27.1 is a variation of the risk management and monitoring plan as presented by Charette. We have simplified the plan to include only those elements that, in our judgment, should never be omitted.

Table 27.1

Simplification of a Risk Management Outline

```
  I.   Introduction
      II.  Risk Analysis
          1. Risk Identification
              a. Risk Survey
              b. Risk Item Checklist
          2. Risk Estimation
              a. Probability of Risk
              b. Consequence of Risk
              c. Estimation Error
          3. Risk Evaluation
              a. Evaluation of Risk Analysis Methods
              b. Evaluation of Risk Referents
              c. Evaluation of Results
      III. Risk Management
              1. Recommendations
              2. Options for Risk Aversion
              3. Risk Monitoring Procedures
```

27.1.3 Cost-Benefit Analysis

Most computer systems development projects demand a tangible economic benefit. This does not mean that every system must generate a profit, but rather that the expected benefits must justify the cost. Exceptions to this rule are systems that are legally mandated or those whose purpose is beyond economic analysis, as could be the case with a project for national defense, or a scientific or academic experiment. Under normal circumstances the system's economic justification is the most important result of the feasibility study since it is a critical factor in deciding project support.

Cost-benefit analysis is a standard business evaluation instrument, but not all project results can be exactly measured in dollars and cents. Some systems bring about less tangible benefits that relate to factors such as the company's long-term plans, a higher quality of product or services, or increased employee morale. It could be a taxing task to attempt to attach a dollar value to some of the less concrete project benefits. It is often preferable to include a category of intangible benefits in the cost-benefit analysis document rather than to attempt to quantify these gains.

The reference point for evaluating the benefits of a new system is usually the existing system. For example, a manufacturer called Acme Corporation plans to upgrade its present inventory management system with a new one based on the use of bar codes to identify products in stock. The new system requires an investment of $500,000 in software and computer equipment but has the following advantages:

1. The number of employees used in inventory management functions will be reduced by 4. This will bring about a saving of $90,000 per year.

2. The flow of materials and supplies to the assembly line will be improved, with an estimated reduction in lost time of 200 work hours per month. Since an hour of lost time at the assembly line costs the company $12, this will bring about a saving of $2400 per month.

3. The better control of stocks will allow the company to reduce the size of its inventory. The smaller investment and the reduction in warehousing costs is estimated to bring the company a saving of $20,000 per year.

In this case the cost-benefit analysis could appear as follows:

```
Acme Corporation
Cost-benefit analysis for new inventory management system
over a 5-year period
1. COST
        a. Computer and peripherals ......................... $300,00
        b. Software ........................................   110,00
        c. Installation and technical support ..............    90,00
                                                             =========
                                total investment ..........$500,00
2. BENEFITS
        a. Personnel reductions .......... $450,000
        b. Reductions in down time .......  144,000
        c. Savings in inventory costs ....  100,000
                                          ========
            total savings .............. $694,000
            Balance ....................$194,000
```

In the case of Acme Corporation the benefits over a five-year period are relatively modest. Therefore it is likely that Acme's higher management would like to see a more detailed and accurate analysis before committing to a major overhaul of the company's existing system. On the other hand, some systems can be economically justified clearly and definitively. Under different circumstances the development and implementation of a computerized inventory control would bring about savings that leave no doubt regarding the project's benefits, even without considering less-tangible factors such as improvements in distribution operations.

27.2 Requirements Analysis and Specification

Once the feasibility of a proposed system has been established, the next task that the analyst must undertake is defining the system's requirements. Requirements analysis, or requirements engineering as it is sometimes called, has been considered a critical phase of computer systems development since the original conception of an engineering methodology for software development. The fundamental notion consists of generating an unambiguous and consistent specification that describes the essential functions of the proposed system. In other words, the specifications phase of the system development process must describe what the system is to do. How it is to do it is addressed during the design phase.

27.2.1 The Requirements Analysis Phase

Before the system requirements can be specified, the interface between the system and the real world must be analyzed. Based on this interface, a set of specifications of the software requirements is developed. The documents containing the program specification play a decisive role during the entire development process. For example, the development of a software system to remote-control the operation of a probe vehicle to operate on the surface of the planet Mars should be based on a detailed analysis of the interface requirements. In this case the following elements may be considered critical:

1. What information is returned by the sensors on the Mars probe.

2. What vehicle operation controls installed on the probe are governed by the software.

3. What is the time lapse between perception by the probe's sensors and reception by the earth system.

4. What is the time-lapse between a command issued by the earth system and its execution by the probe.

The analysis of this data helps determine the operational specification of the system. In this case the delay in receiving information and transmitting commands may require that upon sensing certain type of obstacles, the vehicle come to a stop until a command signal is received. In this case the requirements analysis phase serves to define the system specifications.

Concrete system specification guidelines result from the requirements analysis phase. In particular this could refer to difficulties, tradeoffs, and conflicting constraints; also from the functional and nonfunctional requirements. Functional requirements are those which are absolutely essential while the nonfunctional ones are merely desirable features. In this sense a functional requirement of the Martian probe could be that the vehicle not be damaged by obstacles in its path. A nonfunctional requirement could be that given several possible safe paths to a destination point, the software selects the one that can be traversed in the shortest time.

Customer/User Participation

An ideal scenario is that the customer, client, or eventual user of a system provides the detailed specification. In this "neat" world all the developer has to do is implement a system according to these specifications thus ensuring that the client's needs are thoroughly satisfied. In reality the customer/user often has little more than a sketchy idea of the intended system; its specification must be developed through a series of interviews, proposals, models, and revisions. Communications skills and system analysis experience are important factors in the success of this difficult and critical phase.

The developer of systems specification must not assume that a customer does not really know what he or she wants. It is the client's system, not the developer's. New systems often encounter considerable resistance from employees who must devote time and effort to learning its operation, often with little or no compensation. Furthermore, employees may feel threatened by a new system which could make their jobs unnecessary. The more participation that clients and users have in the system specification and definition the less resistance that they will present to its implementation.

If there is a single secret to being a successful analyst it is to maximize the client's participation in the specification, design, development, and implementation phases, thus making sure that the participants include those who will be the effective users of the new system. A regrettable situation is that the developer initially deals with an administrative elite, while the actual system operators, administrators, and users do not become visible until much later in the development process.

The Virtual Customer

While in many development projects there is a clearly identifiable customer, in others the client role in not quite visible. For example, a software company developing a program that is to be marketed to the public may not have a visible client who can participate in the system's specification, design, and development phases. Another example: a research group at a university or a private firm is often confronted with the task of developing a program for which no user or client is immediately at hand.

The customer/user element plays such an important role in the development process that in these cases it may be advisable to invent some sort of a virtual participant to play this role. In the context of research and development, the team's director sometimes takes on this function. Or perhaps a higher level of management will play the devil's advocate for the duration. Like separation of powers insures the operation of a democracy, separation of interests between clients and developers promotes the creation of better systems. When this duality does not naturally exist, creating it fictitiously may be an option.

27.2.3 The Specifications Phase

Once the system requirements have been determined the analyst creates a list of program specifications. These specifications serve to design, code, and test the proposed program and, eventually, validate the entire development effort; their importance can hardly be overstressed.

Since specifications are derived directly from the requirement analysis documents, a preliminary step to composing a list of specifications is to validate the requirement analysis phase. Whatever method or notation is used in requirements analysis, the following aims should be achieved:

1. Information contents and flow are determined.

2. Processes are defined.

3. System interfaces are established.

4. Layers of abstraction are drawn and processing tasks are partitioned into functions.

5. A diagram of the system essentials and its implementation is made.

We can adopt varying degrees of formality and rigor in composing the actual specifications. Many difficulties in system design and coding can be traced to imprecise specifications. The English language (in fact, any natural language) contains terms and expressions that require interpretation. For example: the statement that a routine to calculate mathematical functions must do so with the "highest possible degree of precision" may seem a strict specification. However, at system design and coding time the expression "the highest possible degree of precision" has to be interpreted to mean one of the following:

1. The precision of the most accurate computer in the world.

2. The precision of the most accurate machine of a certain type.

3. The highest precision that can be achieved in any high-level language.

4. The highest precision that can be achieved in a particular programming language.

5. The highest precision that can be achieved in a certain machine.

6. The highest precision that will be required by the most demanding user or application.

7. The highest degree of precision that will be required by the average user or application.

This list could be easily expanded to include many other alternatives. On the other hand, the specifications could have stated that mathematical functions will be calculated "so that the resulting error will not exceed 20 units of machine Epsilon." In this case very little interpretation is possible at design or coding time. Furthermore, the specification could later be used to validate the resulting product by performing tests that ascertain that the numerical error of the mathematical calculations is within the allowed error range.

In recent years the software engineering and programming communities have moved towards stricter and more formal methods of specification, some of which are discussed later in this section. The analyst should be wary of imprecise terms and expressions in the specification documents and should aim at describing the problems and tasks at hand in the least unambiguous terms possible.

The Software Specifications Document

The Software Specifications Document (SSD), also called the Software Requirements Specification, is the culmination and the conclusion of the system analysis. The elements always present in the SSD include the following:

1. An information description

2. A functional description

3. A list of performance requirements

4. A list of design constraints

5. A system testing parameters and a criterium for validation

The National Bureau of Standards, the IEEE, and others have proposed formats for the SSD. Table 27.2 is a simplified outline of the fundamental topics that must be visited in this document.

Table 27.2

Outline of the Software Specifications Document

I.   Introduction
    1. Purpose and Scope
    2. System Overview
    3. General Description and Constraints
II. Standards and Conventions
III. Environment Elements
    1. Hardware Environment
    2. Software Environment
    3. Interfaces
IV. Software Specifications
    1. Information Flow
    2. Information Contents
    3. Functional Organization
    4. Description of Operations
    5. Description of Control Functions
    6. Description of System Behavior
    7. Database and Data Structures
V.   Testing and Validation
    1. Performance Parameters
    2. Tests and Expected Response
    3. Validation Protocol
VI. Appendices
    1. Glossary
    2. References

Headington and Riley propose a six-step program specification plan consisting of the following parts:

1. Problem title

2. Description

3. Input specifications

4. Output specification

5. Error handling

6. Sample program execution (test suite)

Although this simplification is intended for the specification of small projects within the context of an undergraduate course in C++ programming, it does cover the fundamental elements. Even if it is not used as a model, it may be a good idea to check the actual specification against this plan.

27.3.4 Formal and Semiformal Specifications

The proposers of formal specifications methods postulate that software development will be revolutionized by describing and modeling programs using strict syntax and semantics based on a mathematical notation. The fundamentals of formal specification are in set theory and the predicate calculus. The main advantages of formal specifications are their lack of ambiguity and their consistency and completeness, thus eliminating the many possible interpretations that often result from problem descriptions in a natural language. Of these attributes, lack of ambiguity results from the use of mathematical notation and logical symbolism. Consistency can be ensured by proving mathematically that the statements in the specification can be derived from the initial facts. However, completeness is difficult to ensure even when using formal methods of specification, since program elements can be purposefully or accidentally omitted from the specification.

The original notion of applying the methods of the predicate calculus to problems in computing, and, in particular to the formal development of programs, is usually attributed to Professor Edsger W. Dijkstra (University of Texas, Austin TX, USA) who achieved considerable fame in the field of structured programming, first exposed his ideas in a book titled A Discipline of Programming. Professor David Gries (Cornell Univeristy, Ithaca NY, USA) and Edward Cohen have also made major contributions to this field.

One of the most interesting postulates made by the advocates of formal specifications is that once a program is mathematically described it can be coded automatically. In this manner it is claimed that formal specifications will eventually lead into automated programming, and thus, to the final solution of the software crisis.

In this book we have opted not to discuss the details of formal specifications for the following reasons:

1. So far, the viability of this method has been proven only for rather small programs dealing with problems of a specific and limited nature.

2. The use of this method assumes mathematical maturity and training in formal logic. It is considered difficult to learn and use.

3. Although formal specifications, program proving, and automated coding have gained considerable ground in recent years, these methods are not yet established in the system analysis and programming mainstream. Many still considered them more an academic refinement than a practical procedure.

27.3.5 Assertions Notation

Although formal specification methods have not yet achieved the status of a practical methodology applicable in general system analysis and programming, there is little doubt that reducing the ambiguity and inconsistency of natural languages is a desirable achievement in the software specifications stage. Some current authors of programming textbooks (including Headington and Riley) have adopted a method of program documentation which can also serve as a model for semiformal specifications. This technique, sometimes referred to as an assertion notation, consists of a specific style of comments that are inserted into the code so as to define the state of computation. Although assertions were originally proposed as a code documentation technique, the notation can also be used as a semiformal method of specification during the analysis phase. The following types of assertion are described:

ASSERT

This comment header describes the state of computation at a particular point in the code. Before this notation was suggested, some programmers inserted comments in their code with headers such as AT THIS POINT: that performed a functionally equivalent purpose. The idea of the ASSERT: header is to state what is certainly known about variables and other program objects so that this information can be used in designing and testing the code that follows.

Consider the following C++ example: a function called Power() is implemented so that it returns a float, the result of raising a float-type base to an integer power. The resulting code could be documented as follows:

```
Power(base, index);
// ASSERT:
//      Power() == base raised to index
//      (variables base and index are unchanged by call)
```

Certain symbols and terms are often used in the context of assertions. For example, the term Assigned is used to represent variables and constants initialized to a value. This excludes the possibility of these elements containing "garbage." In C++ assertions the symbols && are often used to represent logical AND, while the symbol || represents logical OR regarding the terms of the assertion. The symbol == is used to document a specific value and the symbol ?? indicates that the value is unknown or undeterminable at the time. Finally, the symbol —, or the word Implies, are sometimes used to represent logical implication. The programmer should also feel free to use other symbols and terms from mathematics, logic, or from the project's specific context, as long as they are clear, unambiguous, and consistent.

INV

This comment header represents a loop invariant. It is an assertion which is used to describe a state of computation at a certain point within a loop. The intention of the loop invariant is to document the fundamental tasks to be performed by the loop. Since the invariant is inserted inside the loop body, its assertion must be true before the body executes, during every loop iteration, and immediately after the loop body has executed for the last time. The following example in C++ is a loop to initialize to zero all the elements of an array of integers.

```
// ASSERT:
//        All set1[0 .. sizeof (Set1)/ sizeof (int)] == ??
for (int i = 0, i (sizeof (Set1) / sizeof (int)), i++)
                            // INV:
                            //      0 i elements in array Set1[]
    set1(i) = 0;
// ASSERT:
//        All set1[0 .. sizeof (Set1) / sizeof *int)] == 0
```

In this example the number of elements in an array is determined at compile time using the C++ sizeof operator. The expression:

```
        sizeof (Set1)
```

returns the number of bytes in array Set1. This value is divided by the number of bytes in the array data type to determine the number of elements. Although this way of determining the number of elements in an array is generally valid in C++, it should be used with caution since the calculation is only permissible if the array is visible to sizeof at the particular location in the code where it is used. The loop invariant states that at this point in the loop body the index to the array (variable i) is in the range 0 to number of array elements, which can be deduced from the loop statement.

PRE and POST

These comment headers are assertions associated with subroutines. In C++ they are used to establish the conditions that are expected at the entry and exit points of a function. Before the assertion notation was proposed some programmers used the terms ON ENTRY: and ON EXIT:.

The PRE: assertion describes the elements (usually variables and constants) that must be supplied by the caller and the POST: assertion describes the state of computation at the moment the function concludes its execution. Thus, the PRE: and POST: conditions represent the terms of a contract between caller and function, stating that if the caller ensures that the PRE: assertion is true at call time, the function guarantees that the POST: assertion is satisfied at return time. For example, the following function named Swap() exchanges the values in two integer variables.

```
void Swap( int& var1, int& var2)
// PRE:    Assigned var1 && var2
// POST:   var1 == var2 && var2 == var1
{
  int temp = var1;
  var1 = var2;
  var 2 = temp;
  return;
}
FCTVAL
```

This assertion is also used in the context of subroutines to indicate the returned value. In C++ FCTVAL represents the value returned by a function and associated with the function name. Therefore it should not be used to represent variables modified by the function. For example, the following trivial function returns the average of three real numbers passed by the caller.

```
float Avrg( float num1, float num2, float num3)
// PRE:     Assigned num1 && num2 && num3
//                      num1 + num2 + num3
// POST:    FCTVAL == ------------------
//                             3
{
  return ((num1 + num2 + num3) / 3));
}
```

27.4 Tools for Process and Data Modeling

Technical methodologies for system development propose that the documents that result from the analysis phase serve as a base for the design phase. Therefore the more detailed, articulate, and clear the analysis documents, the more useful they will be in program design. One way to accomplish this is through the use of modeling tools that allow representing the essence of a system, usually in graphical terms. Many such tools have been proposed over the years and two have achieved widespread acceptance, namely, data flow diagrams and entity-relationship diagrams.

Data flow diagrams originated in, and are often associated with, structured analysis. Entity-relationship diagrams are linked with database systems and with object-oriented methods. Nevertheless, we believe that either method is useful independently of the context or analysis school in which it originated. In fact, data flow diagrams and entity relationship diagrams have been used in modeling processes and data flow in both the structured analysis and the object-oriented paradigms.

System modeling can be undertaken using varying degrees of abstraction. One type of model, sometimes called a physical or implementation model, addresses the technical details as well as the system's essence. Program flowcharts are often considered as an implementation modeling tool since a flowchart defines how a particular process can be actually coded. Essential or conceptual models, on the other hand, are implementation independent. Their purpose is to depict what a system must do without addressing the issue of how it must do it.

Although implementation models are useful in documenting and explaining existing systems, it is a general consensus among system analysis that conceptual models are to be preferred in the analysis phase. The following arguments are often listed in favor of conceptual models:

1. Some incorrect solutions that result from implementation-dependent models can be attributed to bias or ignorance on the part of the designers. A conceptual model, on the other hand, fosters creativity and promotes an independent and fresh approach to the problem's solution.

2. Excessive concern with implementation details at system specification and analysis time takes our attention away from the fundamentals. This often leads to missing basic requirements in the specification.

3. Implementation-dependent models require a technical jargon that often constitutes a communications barrier between clients and developers.

The target of this modeling effort can be either an existing or a proposed system. In this respect we should keep in mind that the fundamental purpose of modeling an existing system is to learn from it. The learning experience can be boiled down to avoiding defects and taking advantage of desirable features.

27.4.1 Data Flow Diagrams

In his work on structured analysis DeMarco suggests that one of the additions required to the analysis phase is a graphical notation to model information flow. He then proceeds to create some of the essential graphical symbols and suggests a heuristic to be used with these symbols. A few years later Gane and Sarson refined this notation calling it a data flow diagram or DDF.

A data flow diagram is a process modeling tool that graphically depicts the course of data thorough a system and the processing operations performed on this data. Figure 27.3 shows the basic symbols and notation used in data flow diagrams.

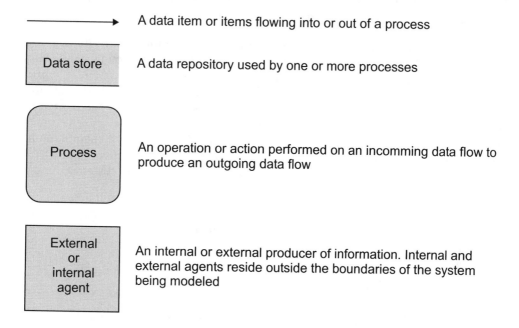

Figure 27.3 *Gane and Sarson Symbols for Data Flow Diagrams*

One variation often used in the Gane and Sarson notation is to represent processes by means of circles (sometimes called bubbles). A valid justification of this alternative is that circles are easier to draw than rounded rectangles.

Processes are the core operation depicted in data flow diagrams. A process is often described as an action performed on incoming data in order to transform it. In this sense a process can be performed by a person, by a robot, or by software. Processes are usually labeled with action verbs, such as Pay, Deposit, or Transform, followed by a clause that describes the nature of the work performed. However, on higher level DFDs we sometimes see a general process described by a noun or noun phrase, such as accounting system, or image enhancement process. Routing processes that perform no operation on the data are usually omitted in essential DFDs. The operations performed by processes can be summarized as follows:

1. Perform computations or calculations

2. Make a decision

3. Modify a data flow by splitting, combining, filtering, or summarizing its original components

Internal or external agents or entities reside outside the system's boundaries. They provide input and output into the system. In this sense they are often viewed as either sources or destinations. An agent is external if it is physically outside the system, for example, a customer or a supplier. Internal agents are located within the organization, but are logically outside of the system's scope. For example, another department within the company can be considered as an internal agent.

It is possible to vary the level of abstraction and magnification of details in a DFD. Some authors propose the designation of specific abstraction levels. In this sense a level 0 DFD, called a fundamental system model, consists of a single process bubble which represents the entire software system. A level I DFD could depict five or six processes which are included in the single process represented in the level 0 bubble. Although the notion that several DFDs can be used to represent different levels of abstraction is useful, the requirement that a specific number of processes be associated with each level could be too restrictive. Figure 27.4 shows an initial level of detail for representing a satellite imaging system.

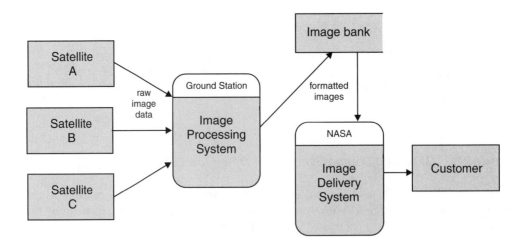

Figure 27.4 *Initial Model of a Satellite Imaging System*

Notice that in Figure 27.4 the process bubble has been refined with a header label that adds information regarding the process. In this case the header refers to who is to perform the process. Also notice that the processes are represented by noun phrases, which is consistent with the general nature of the diagram. Although this initial model provides some fundamental information about the system, it is missing many important details. For example, one satellite may be equipped with several sensing devices that generate different types of image data, each one requiring unique processing operations. Or a particular satellite may produce images that are secret, therefore they should be tagged as such so that the delivery system does not make them available to unqualified customers. Figure 27.5 shows a second level refinement of the image processing system.

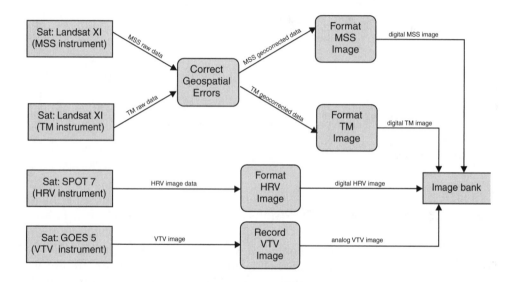

Figure 27.5 *Refined Model of the Satellite Image Processing System*

The level of detail shown by the DDF in Figure 27.5 is much greater than that in Figure 27.4. From the refined model we can actually identify the various instruments on board the satellites and the operations that are performed on the image data produced by each instrument. However, the actual processing details are still omitted from the diagram, which are best postponed until the system design and implementation phases. Note that in Figure 27.5 the process names correct, format, and record are used as verbs, consistent with the greater level of detail shown by this model. Note that the names of satellites, instruments, and image formats do not exactly correspond to real ones.

Event Modeling

Many types of systems can be accurately represented by modeling data flow, but some cannot. For example, a burglar alarm system is event-driven, as is the case with many real-time systems. First, Ward and Mellor, and later Hatley and Pribhai introduced variations of the data flow diagram which are suitable for representing real-time systems or for the event-driven elements of conventional systems. The Ward and Mellor extensions to DFD symbols are shown in Figure 27.6.

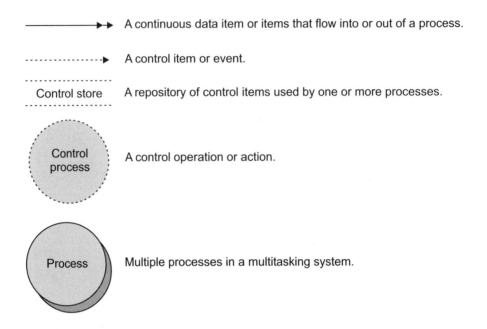

———————▸▸ A continuous data item or items that flow into or out of a process.

·······················▸ A control item or event.

Control store   A repository of control items used by one or more processes.

Control process   A control operation or action.

Process   Multiple processes in a multitasking system.

Figure 27.6 *Ward and Mellor Extensions to Data Flow Diagrams*

The resulting model, called a control flow diagram or CFD, includes the symbols and functions of a conventional DFD plus others that allow the representation of control processes and continuous data flows required for modeling real-time systems. Equipped with this tool we are now able to model systems in which there is a combination of data and control elements. For example, a burglar alarm system as installed in a home or business includes a conventional data flow component as well as a control flow element. In this case the conventional data flow consists of user operations, such as activating and deactivating the system and installing or changing the password. The control flow elements are the signals emitted by sensors in windows and doors and the alarm mechanism that is triggered by these sensors. Figure 27.7 represents a CFD of a burglar alarm system.

In depicting the burglar alarm scheme of Figure 27.7 we have merged the Gane and Sarson symbol set of conventional DFDs with the extensions proposed by Ward and Mellor. By combining both symbol sets in a control flow diagram we are able to model a system that contains both conventional data flow and real-time elements. Notice that the sensors and the alarm mechanism are a real-time component, since the status of the sensing devices must be monitored continuously. The control panel and user interface, on the other hand, are activated by a user request and do not require continuous data flow. However, the on-demand and real-time elements sometimes interact. For example, when the system is turned on or off the corresponding signals must be sent to the control panel as well as to the sensors and sensor control logic. Another interaction between subsystems results from the fact that the sensor status is displayed on the control panel. Therefore the sensor monitoring logic sends a continuous signal (double-headed arrow) to the update control panel process so that the sensor status is displayed in real-time.

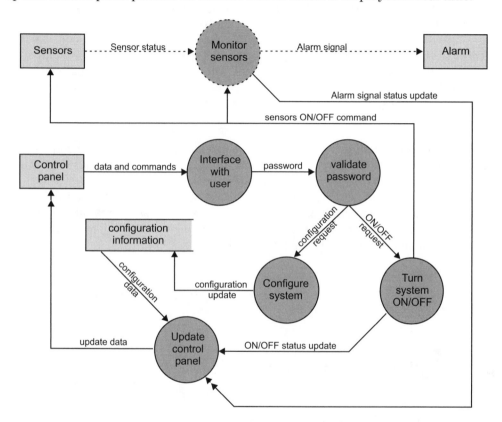

Figure 27.7 *Control Flow Diagram of a Burglar Alarm System*

27.4.2 Entity-Relationship Diagrams

Currently the most used tool in data modeling is the entity-relationship (E-R) diagram. This notation, first described in the early 1980s by Martin and later modified by Chen and Ross, is closely related to the entity-relationship model used in database design. E-R diagram symbols have been progressively refined and extended so that the method can be used to represent even the most minute details of the schema of a relational database. However, in the context of general system analysis we will restrict our discussion to the fundamental elements of the model.

Since one of the fundamental operations of most computer systems is to manipulate and store data, data modeling is often a required step in the system analysis phase. E-R diagrams are often associated with database design, however, their fundamental purpose is the general representation of data objects and their relationship. Therefore the technique is quite suitable to data analysis and modeling in any context. Furthermore, E-R notation can be readily used to represent objects and object-type hierarchies, which makes this technique useful in object-oriented analysis.

The entity-relationship model is quite easy to understand. An entity, or entity type, is a "thing" in the real world. It may be a person or a real or a conceptual object. For example, an individual, a house, an automobile, a department of a company, and a college course are all entity types. Each entity type is associated with attributes that describe it and apply to it. For example, the entity type company employee has the attributes name, address, job title, salary, and social security number. While the entity type college course has attributes name, number, credits, and class times. Notice that the entity employee is physical, while the entity college course is abstract or conceptual.

The entity-relationship model also includes relationships. A relationship is defined as an association between entities. In this sense we can say that the entity types employee and department are associated by the relationship works for since in the database schema every employees work for a department. By the same token we can say that the works on relationship associates an employee with a project, thus associating the entity types employee and project. Figure 27.8 shows the elementary symbols used in E-R diagrams.

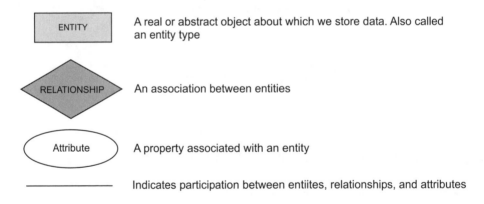

Figure 27.8 *Symbols Used in Entity-Relationship Diagrams*

As shown in Figure 27.8, it is conventional practice to type entity and relationships in uppercase letters while attribute names have initial caps. Singular nouns are usually selected for entity types, and verbs for relationships. Attributes are also nouns that describe the characteristics of an entity type. An entity type usually has a key attribute which is unique and distinct for the entity. For example, the social security number would be a key attribute of the entity type employee, since no two employees can have the same social security number. By the same token, job title and salary would not be key attributes since several employees could share this characteristic. Notice that while some entities can have more than one key attribute, others have none. An entity type with no key attribute is sometimes called a weak entity. Key attributes are usually underlined in the E-R diagram.

In regards to relationships we can distinguish two degrees of participation: total and partial. For example, if the schema requires that every employee work for a department, then the works for relationship has a total participation. Otherwise, the degree of participation is a partial one. Cardinality constraints refer to the number of instances in which an entity can participate. In this sense we can say that the works for relationship for the types DEPARTMENT:EMPLOYEE have a cardinality ratio of 1:N; in this case an employee works for only one department but a department can have more than one employee. Figure 27.9 shows an entity-relationship diagram for a small company schema.

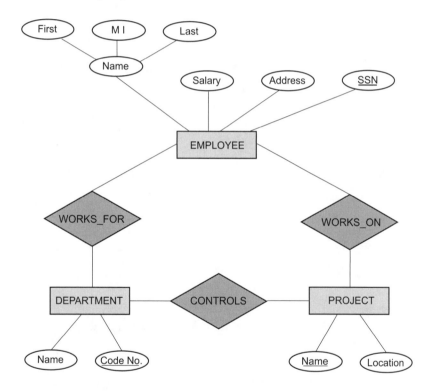

Figure 27.9 *Entity-Relationship Diagram for a Company*

The diagrams are read left-to-right or top to bottom. This convention is important: for example, when the diagram in Figure 27.9 is read left-to-right it is the department that controls the project, not vice versa.

Chapter 28

The Object-Oriented Approach

Chapter Summary

This chapter is an introduction to a programming and software development paradigm known as object orientation. It describes the history and evolution of this approach, discusses its fundamental features, and provides guidelines on when it is suitable.

28.0 History and Chronology

Although the object-oriented approach appeared in the computing mainstream during the early 1980s, its origin dates back to the 1960s. Three concepts are associated with object-oriented systems: data abstraction, inheritance, and dynamic binding.

The first notions of data abstraction are due to Kristen Nygaard and Ole-Johan Dahl working at the Norwegian Computing Center. Nygaard and Dahl developed a computer language named SIMULA I, intended for use in simulations and operations research. A few years later SIMULA I was followed by SIMULA 67, which was actually an extension of ALGOL 60. Although SIMULA 67 had little impact on programming, its historical importance was the introduction of the class construct.

The basic idea of a class is a form or template that packages together data and processing routines. Note that this definition of class as a template implies that a class is a formal construct, rather than a concrete entity. The programmer creates instances of a class as they become necessary. In modern-day terminology the instances of a class are called objects. It is interesting to note that although SIMULA 67 actually implemented data abstraction by classes it was not until 1972 that this connection was formally recognized.

But although data abstraction is clearly an ingredient of SIMULA 67, inheritance is implemented in a limited way, and dynamic binding is altogether missing. It was Alan Kay who first described a fully operational object-oriented language. Kay's vision was that desktop computers would become very powerful machines, with megabytes of memory and executing millions of instructions per second; since these

machines would be used mostly by nonprogrammers, a powerful graphical interface would have to be developed to replace the awkward teletype terminals and
batch processing methods of the time. It actually took over ten years for these machines to become available.

Kay's original solution to a friendly and easy-to-use interface was an information processing system named Dynabook. Dynabook was based on the visualization of a desk in which some documents were visible and others were partially
covered. By means of the keyboard a user would select documents. A touch-sensi-
tive screen allowed moving the documents on the desk. The result was very similar to some of our present day windowing environments. Processing was to be
performed by the Flex programming language which Kay had helped design and
which was based on SIMULA 67. Kay eventually went to work at the Xerox Palo
Alto Research Center where these ideas evolved into a mouse controlled, windowing interface and the Smalltalk programming language. Both of these notions,
windows and object-oriented systems, were destined to become major forces in
computing technology.

The first fully operational object-oriented programming language to become
available was Smalltalk 80. The Smalltalk 80 design team at Xerox PARC was led
by Adele Goldberg and the language soon became the de facto description of object-oriented programming. Several object-oriented languages have since been developed. Perhaps the most notable ones are Eiffel, CLOS, ADA 95, and C++.
Smalltalk 80 adopted a dynamically typed approach in which there is no typing
system outside the class structure. Eiffel is also a pure implementation but with
static typing. The other ones are hybrid languages that use a mixture of static and
dynamic typing and have type systems that extend beyond the class structures.

28.1 Object-Oriented Fundamentals

Different object-oriented languages implement the paradigm differently and to
varying degrees of conceptual purity. In recent years the development and marketing of object-oriented products has become a major commercial venture. Thus, to
the natural uncertainties of an elaborate and sophisticated concept, we must add
the hype of commercial entrepreneurs and promoters. The resulting panorama for
object-oriented methods is a confusing mixture of valid claims and unjustified
praise, some of it originating in the real virtues of a new and useful model, some of it
in pure mercantilism.

In the following sections we will attempt to provide the reader with a description of the fundamentals of the object-oriented approach and to point out what, in
our opinion, are the strengths of this model. Because object-oriented programming languages vary in the purity of their treatment of the underlying paradigm,
we have focused our description on a particular one: C++. This does not imply a
judgment on the suitability of the C++ approach, but is determined by the fact
that C++ programming is the central topic of this book.

28.1.1 Problem-Set and Solution-Set

A computer program, whether it be a major operating system or a small application, is a machine-code solution to a real-world problem. Therefore, at the beginning of any programming project is a problem-set that defines it and at its conclusion there is a solution-set in the form of a group of instructions that can be executed by a digital computing machine. The art and science of programming are a transit from the real-world problem-set to the machine-code solution-set. In this wide conceptualization, programming includes the phases of analysis, design, coding, and testing of a software product. Figure 28.1 graphically represents this concept.

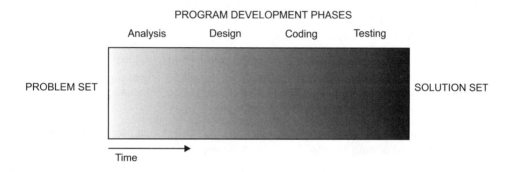

Figure 28.1 *Evolution of a Programming Project*

The smooth, white-to-black gradation in Figure 28.1 indicates that the various program development phases (analysis, design, coding and testing) have no discrete boundaries. It is difficult to determine exactly where one phase ends and the next one begins. Furthermore, in a particular project some of the phases (except for the coding phase) could be missing altogether. In one extreme case a program is developed by coding alone, with no analysis, design, or testing. In fact, some programs are developed in this manner; sometimes successfully but more often unsuccessfully.

Many methodologies have been developed to facilitate the transition from a real world problem to a digital solution in machine instructions. Assemblers, high-level programming languages, CASE tools, analysis and design methodologies, formal specifications, and scientific methods of program testing are all efforts in this direction. Perhaps the most significant simplifications are those based on generalized abstractions. In this sense structured programming can be considered as an abstraction that focuses on the solution-set, while the object-oriented approach focuses on the problem-set. Therefore, a project engineered in terms of structured programming is based on a model of the solution set, while a project analyzed and designed using object-oriented methods closely focuses on the problem set.

There is an additional factor to be considered in this comparison of object-oriented versus structured techniques. The development of high-level, object-oriented programming languages serves to close the gap between the problem-set and the solution-set. It is the simultaneous use of object-oriented analysis, object-oriented design, and object-oriented programming that makes a smooth transition possible. In reality, although the use of a uniform model can be a considerable advantage, there is also a price to pay in terms of program performance and code size, and very much so in programming complications. In order to use a real-world model we are forced to introduce several additional levels of complexity to the coding phase.

28.1.2 Rationale of Object Orientation

The result of object-oriented methods can be summarized by stating that we have complicated the coding phase in order to simplify the analysis and design phases of program development. The defenders of the object-oriented approach make the following reasonable claims:

1. Object-oriented-analysis and design methods facilitate communications with clients and users since the model does not require technical knowledge of programming.

2. Many real-world problems are easier to model in object-oriented terms than when using structured analysis and design techniques.

3. Object-oriented programming languages promote and facilitate code reuse, which eventually increases programmer productivity.

4. Object-oriented programs have internal mechanisms that make them resilient to change, therefore they are better able to accommodate the natural volatility of the problem-domain.

A question often raised is not whether these claims are valid but whether they are worth the complications. One answer can be found in the fact that the programming community appears to be leaning in favor of the object-oriented approach.

28.2 Classes and Objects

In the context of object orientation an object is a conceptual entity related to the problem domain. It is an abstraction, not a thing in the real world. A fallacy found in some books is the claim that a conceptual object, in the context of the paradigm, can be considered equivalent to an object in the conventional sense. This misconception can turn into a major stumbling block to understanding object-orientation. In reality the OO object is a hybrid that shares some characteristics of common objects with features of a computer construct. Instead of giving a formal definition for an object, we start with a rather simplified, and stereotyped, listing of its properties:

1. Every object belongs to an object class. An object cannot exist without a class that defines it. In this sense we say that an object is an instance of a class. To understand the difference between a class and an object we can visualize the class as a cookie cutter and the object as the cookie. In programming terms a class relates to a type definition and an object to variable declaration.

2. An object (and the class that contains it) is an encapsulation that includes data and its related processing operations. The object's data elements are called its attributes, and the processing operations are called its methods. Attributes and methods are the class members: the attributes are the data members and the methods are the member functions.

3. The object's attributes serve to store and preserve the object's state. An object's methods are the only way of accessing its data or modifying its state. This is accomplished by sending a message to the object.

28.2.1 Classes and Data Abstraction

In conventional programming we often think of a data type as mere grouping. In object-oriented terms we should think in terms of both the data type and its associated operations. For example, in a certain hardware environment the type integer includes the whole numbers in the range +32,767 to –32,768, as well as the operations that can be performed on integers, such as addition, subtraction, multiplication, division, remainder, and comparison. In this same context we can say that a variable is an instance of a type. For example, the variables num1 and num2 could be regarded as instances (objects) of the integer type. In C++ we declare and initialize these variables with the statements:

```
int num1 = 12;
int num2 = 6;
```

We can now perform any of the allowed operations on the instances (declared variables) of type int. For example:

```
int num3 = (num1 / num2) + num2;
```

which would result in assigning the value 8 ((12/6)+6) to the newly declared variable num3.

What happens if we need to operate on an entity that is not defined as a data type in our language? For example, suppose we needed to manipulate several integer values so that the last value stored is the first one retrieved. This mechanism corresponds to a data structure known as a stack, but C++ does not support the stack data type.

To solve this limitation, object-oriented languages have facilities for implementing programmer-defined data types, also called abstract data types or ADTs. In some languages (such as C++) an abstract data type is implemented in terms of one or more built-in data types plus a set of allowable operations. In the case of the stack the built-in data type could be an array of integers and the operations could be called Initialize(), Push(), and Pop().

In C++ the program structures used for implementing abstract data types are classes and objects. In the previous example we could define a class stack, with the attribute array of int, and the methods Initialize(), Push(), and Pop(). This class would serve as a template for creating as many stacks as necessary. Each instantiation of the class would be an object of type stack. This is consistent with the notion of a class being equivalent to a type, and an object to a variable.

28.2.2 Classes and Encapsulation

In object-oriented terms encapsulation is related to information hiding, which is a fundamental notion of the paradigm. Instead of basing a programming language on subprograms that share global data or that receive data passed by the caller, the object-oriented approach adopts the notion that data and functions (attributes and methods) are packaged together by means of the class construct. The methods that are visible to the user are called the object's interface. The fundamental mechanism of encapsulation consists of hiding the implementation details while stressing the interface. The goal is to create an abstraction that forces the programmer to think conceptually. The first golden rule of object-oriented programming, as stated by Gamma et. al, is:

> *program to an interface, not to an implementation.*

In a typical class the data members are invisible to the user. In C++ this is achieved by means of the keyword private (called an access specifier). In general, class members are either private, public, or protected. Public members are visible to the client and constitute the class interface. Private members are visible only to the other members of the class. Protected members are related to inheritance, which is discussed later in this chapter. If a data member must be made accessible to the client, then it is customary to provide a function that inspects it and returns its value. In C++ the const keyword can also be used to preclude the possibility of an unauthorized change in the value of the variable.

28.2.3 Message Passing

In structured programming, subroutines or subprograms are accessed in various manners: by means of a jump, a call, a trap, an interrupt, or, in distributed systems, an interprocess communication. Object-oriented systems make the actual access procedure invisible by means of a mechanism called message passing.

One of the advantages of message passing is that it eliminates hard-coded program structures such as jump tables, cascaded if statements, or case constructs. One disadvantage of hard-coded components is that they must be updated with every program modification or change. For example, suppose a conventional graphics package with functions to draw lines, rectangles and circles. In a non-object-oriented language these operations would be executed by individual subprograms. Typically, there would be a procedure named DrawLine(), another one named DrawRectangle(), and a third one named DrawCircle(). Somewhere in the program a selection logic would direct execution to the corresponding procedure. In C this logic may consist of cascaded ifs or a case construct. If we wanted to extend the functionality of the graphics package by providing a new function to draw an ellipse, we would not only need to code the ellipse-drawing routine, but also to modify the selection logic so that execution would be directed to this new function as necessary, which implies changing the existing code.

The message-passing mechanism used in object-oriented systems automatically directs execution to the corresponding member function. The object-oriented approach for an equivalent graphics package would be to create classes named Rectangle, Line, and Circle, each with a member function named Draw(). When we wish to display a rectangle we send a draw message to an object of the class rectangle, and so forth. If we now want to extend the functionality to draw an ellipse, we can create an Ellipse class, which has a Draw() function. Then we can draw an ellipse by passing a draw message to an object of type Ellipse. By using preprocessor services and run-time libraries we can add the new functionality without modifying existing code.

Evidently this method of adding functionality brings about two major advantages: first, we can avoid introducing new defects into code that has already been tested and debugged; second, systems can be expanded by supplying relatively small modules that contain the new functions.

28.2.4 Inheritance

The notion of inheritance, which originated in the field of knowledge representation, actually refers to the inheritance of properties. Knowledge is usually organized into hierarchies based on class relationships. (Note that now we are using the word class in its more common connotation). Thus, we say that an individual inherits the properties of the class to which it belongs. For example, animals breathe, move, and reproduce. If fish belongs to the animal class, then we can infer that fish breathe, move, and reproduce since these properties are inherited from the base class. Figure 28.2 is an inheritance diagram for some animal classes.

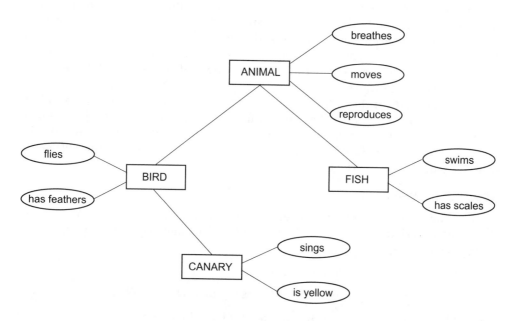

Figure 28.2 *Inheritance Diagram*

Inheritance systems are used in knowledge bases to ensure the highest levels of abstraction. For example, a knowledge base about birds first defines the traits that are common to all birds (fly and have feathers) and then the traits of the particular species (canary sings and is yellow). In this manner the species canary inherits the properties of its parent class, bird, which in turn inherits the properties of its parent class, animal. This reduces the size of the knowledge base by requiring that common properties be asserted only once. Inheritance also serves to maintain the consistency of the knowledge base and to simplify programming.

In object-oriented systems inheritance refers to the possibility of one class acquiring the public members of its parent class. As in knowledge bases, class inheritance promotes the highest level of abstraction, reduces code size, and simplifies programming. This allows the building of a class hierarchy, from the most general to the most specific. In the diagram of Figure 28.2 we could say that animal is a base class, and that bird, fish, and canary are derived classes. The derived class usually incorporates all the features of its parent classes and adds some unique ones of its own. Also, a derived class has access to all the public members of its base class. The terms parent, child, and sibling are sometimes used to express degrees of inheritance between classes. As in biological systems, sibling classes do not inherit from each other. Therefore the properties of fish (in Figure 28.2) are not inherited by its sibling class, bird.

28.2.5 Polymorphism

Etymologically the word polymorphism means "many forms." In object-oriented terms polymorphism relates to several methods that share the same name. One way to implement polymorphism is by overloading conventional functions or operators. For example, the standard C library contains three functions that return the absolute value of the operand: abs() refers to the absolute value of an integer, labs() to the absolute value of a long, and fabs() the absolute value of a float. The programmer must take into consideration the data type when calculating the absolute value. In object-oriented systems it is possible to overload the function abs() in order to use the same name for three different calculating routines. The compiler examines the data type of the operand to determine which of the abs() functions is pertinent to the call. Thus, the programmer need not remember three different names, nor be concerned with the operand's data type. Standard operators, such as the arithmetic symbols, can also be overloaded in C++.

Abstract Classes

Polymorphism is more than just overloading functions or operators. When combined with inheritance, polymorphism provides a mechanism for achieving the following desirable properties:

1. Allows the creation of extensible programs

2. Helps localize changes and hence prevents the propagation of bugs

3. Provides a common interface to a group of related functions

4. Allows the creation of libraries that are easy to extend and reuse, even when the source code is not available to the programmer

All of this is made possible through the use of a special type of functions called virtual functions and a construct called an abstract class. An abstract class is one in which a method's interface is defined but it is not implemented. This empty method serves as a template for the interface of a family of functions that share the same name. The concept of abstract classes requires that the polymorphic functions in the base and the derived classes be virtual functions and that they have the same name and identical interfaces. This makes abstract classes different from function overloading, which requires that either the number of the type of the parameters be different. In the context of virtual functions the word overriding is used to designate the redefinition of a virtual function by a derived class.

In C++ there are two types of virtual functions: regular and pure virtual functions. Both regular and pure virtual functions support polymorphism as well as early and late binding. For example, a graphics package has a base class called GeometricalFigure and three derived classes called Rectangle, Line, and Circle, all of which have a member function named Draw(). When we wish to display a rectangle we send a Draw() message to an object of type rectangle. Because each object knows the class to which it belongs, there is no doubt where the Draw() message will be directed, even though there are other classes with equally-named methods. Since the destination method is determined according to the object type, known at compile time, this type of polymorphism is said to use early binding.

However, when a virtual function is accessed by means of a pointer, the binding is deferred until run time. This type of late binding, or run-time polymorphism, is one of the most powerful mechanisms of object-oriented systems. A discussion of the details of implementation of polymorphism and virtual functions in C++ are not discussed in this book.

28.4 A Notation for Classes and Objects

It seems that every author of a book on object-oriented topics feels the need to develop a new notation or to introduce a modification to an existing one. In this manner we have the Coad and Yourdon notation, the Booch notation, the Rumbaugh notation, as well as half a dozen refinements and modifications. These variations are not always due to simple matters of form or style. Each notation is designed to model a particular view of the problem and to do it with the desired degree of detail. Furthermore, each of the various notations is often associated with a particular phase of the development process or with a design methodology. For example, the Coad and Yourdon notation is often linked with the analysis phase, the Booch notation is associated with the design phase, and Rumbaugh's with models and design patterns. There is no easy way out of this notational labyrinth.

We will start by using the object and class notation proposed by Coad and Yourdon because we will be closely following these authors in our treatment of object-oriented analysis. Later on we will add refinements consistent with Rumbaugh's symbology as they become necessary for the discussion of models and patterns. Unable to resist the aforementioned tendency to modify, we will also introduce some minor variations to the Coad and Yourdon symbols. Our main objection to Coad and

Yourdon notation is that it requires shades of gray, which is difficult to achieve with pencil and paper. Since the class and object diagrams are often necessary during brainstorming sessions and informal meetings, it is better if the components can be easily drawn by longhand. Figure 28.3 shows the symbols for classes and objects.

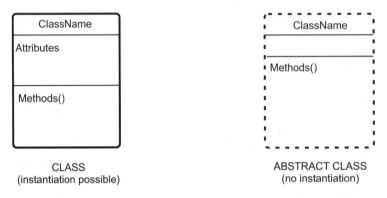

Figure 28.3 *Notation for Classes and Objects*

In Figure 28.3 we notice that a normal class can be instantiated while an abstract class cannot. In their book on object-oriented analysis Coad and Yourdon refer to classes that can be instantiated as a Class-&-Object element in the diagram. Therefore, their class designation always refers to an abstract class. The methods in an abstract class constitute an interface with no implementation, as will be shown in the example of the following section.

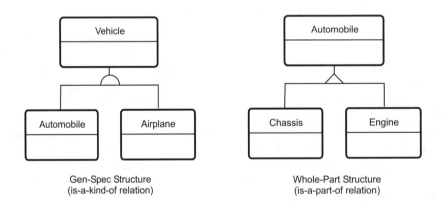

Figure 28.4 *Notation for Inheritance and Class Relations*

Figure 28.4 shows the Coad and Yourdon notation to represent inheritance and types of associations between classes. These types of associations are called structures. The Generalization-Specification structure (Gen-Spec) corresponds to the "is-a-kind-of" relation. For example, in a classification in which vehicle is the base class, then automobile and airplane are specializations of the base class, since automobile "is a kind of" vehicle and so is airplane. On the other hand, classification may be based on the Whole-Part association, depicted with a triangle symbol in Figure 28.4. In this case the chassis is a part of an automobile and so is the engine.

28.5 Example Classification

The concepts and notation covered in this chapter can be illustrated with an example. Suppose the following project:

> *We are to design and code a graphics toolkit that can display geometrical figures at any screen position. The figures are straight lines, rectangles, circles, ellipses, and parabolas.*

A first reading of the problem description shows that the program is to draw geometrical figures of four different types: line, rectangle, circle, ellipse, and parabola. We also notice that circle, ellipse, and parabola belong to a family of curves called the conic sections. These are obtained by sectioning a right-circular cone, as shown in Figure 28.5.

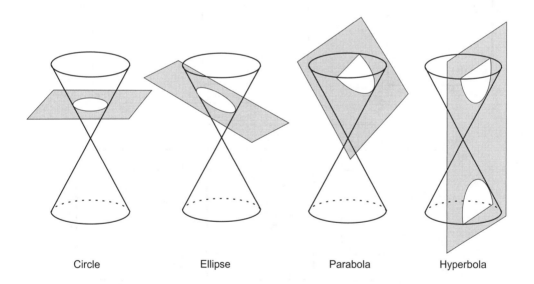

| Circle | Ellipse | Parabola | Hyperbola |

Figure 28.5 *Visualization of the Conic Sections*

A rough and instinctive classification can be based on using a class for each of the figure elements to be displayed. The notion of a geometrical figure could serve as a base class as shown in Figure 28.6, on the following page.

Since GeometricalFigure is an abstraction, it is represented by an abstract class, with no attributes and no possible instantiation. Line and Rectangle are concrete classes since it is possible to draw a line or a rectangle, therefore they have attributes and can be instantiated. Circle, Ellipse, and Parabola are concrete classes of the abstract class ConicSection, which is also a subclass of GeometricalFigure. Observe that the interface, which is the Draw() method, is defined in the abstract classes (GeometricalFigure and ConicSection) and implemented in the concrete classes. We have used the C++ convention of equating the function to zero to indicate that the methods in the abstract classes are pure virtual functions and that no implementation is provided. Also note that virtual functions in the abstract and concrete classes are in italics.

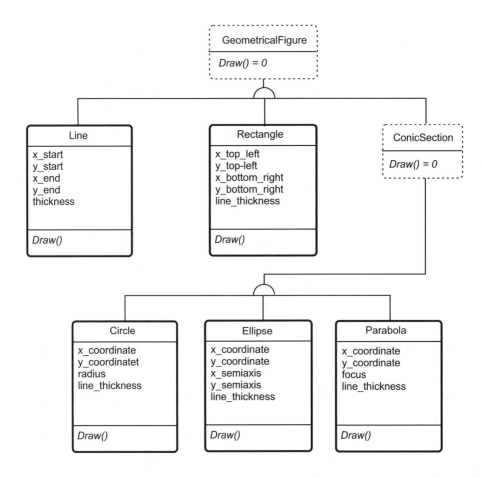

Figure 28.6 *Classification of a Graphics Toolkit*

Since the concrete classes can be instantiated, we can create objects of Line, Rectangle, Circle, Ellipse, and Parabola. There can clearly be no object of the abstract classes ConicSection or GeometricalFigure since there can be no possible instantiation of these concepts. For these reasons abstract classes have no attributes. We can now proceed to draw on the video display using objects. For example:

```
Object:              Attributes
Line_A               x_start = 10
                     y_start = 5
                     x_end = 40
                     y_end = 20
                     thickness = 0.005
Rectangle_A          x_top_left = 5
                     y_top_left = 20
                     x_bottom_right = 25
                     y_bottom_right = 30
                     line_thickness = 0.010
Rectangle_B          x_top_left = 30
                     y_top_left = 5
                     x_bottom_right = 45
                     y_bottom_right = 10
                     line_thickness = 0.030
Circle_A             x_coordinate = 40
                     y_coordinate = 30
                     radius = 5
                     line_thickness = 20
```

Figure 28.7 represents these objects on the video display. Note that each object is identified by its name and preserved by its attributes. Also that there are two objects of the class Rectangle, one is designated as Rectangle_A and the other one as Rectangle_B. The fact that there are no objects of the class Ellipse or Parabola means only that these classes have not been instantiated.

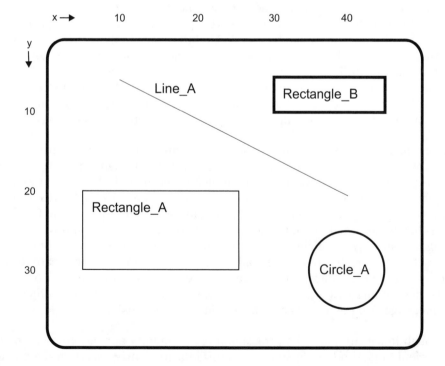

Figure 28.7 *Objects Displayed Using the Graphics Toolkit*

In Figure 28.5 we notice that there is a type of conic curve, the hyperbola, that is not defined in the toolkit. The class structure in the object-oriented paradigm allows to later implement this subclass without introducing modifications to the existing code. The mechanics of how a library can be dynamically extended are discussed later in the book.

28.6 When to Use Object Orientation

The fact that this book's title includes C++ as a programming language does not mean that we assume that every programming project will benefit from an object-oriented approach. One advantage of hybrid languages, such as C++, is that they make object-orientation optional. Furthermore, in a hybrid language environment we have the possibility of adopting some features of an object-oriented system while not others. For instance, we can design a program that uses classes and objects without inheritance, or one that uses inheritance but not run-time polymorphism. Or we may use some forms of inheritance and not others. In other words, the systems designer is free to decide how much object-orientation, if any, is suitable for a particular project. This is often a crucial and significant decision which may determine a project's success or failure.

28.6.1 Operational Guidelines

Object-oriented methods have permeated programming activities to such depth that sometimes we forget that its fundamental objective is to provide a model of the problem domain. In Figure 28.1 we showed programming as a bridge from a real-world problem-set to a machine-coded solution-set. In this representation of programming we see that the object-oriented paradigm emphasizes modeling the problem domain while structured programming models the solution domain. Modeling the solution domain facilitates the coding phase (which it approximates). Modeling the problem domain facilitates analysis. In one case we concentrate on the tail-end of the process and in the other one we concentrate on its start.

By emphasizing the problem domain in the analysis phase we are able to achieve two undisputable benefits: first, we are able to analyze more complicated problems; second, the analyst can communicate better with the domain experts since many technical elements are removed from the model. However, these benefits are considerably dimished if the object-oriented model that results from the analysis has to be translated into the code of a non-object-oriented programming language. In spite of the claims made by some hard-core defenders of object-orientation, this approach is of doubtful advantage when it does not permeate the entire development process. Therefore, it is usually undesirable to perform object-oriented analysis and design of a project to be coded in a non-object-oriented language. Mixing models and development tools is usually a bad idea.

Although in recent years the programming community appears to be favoring object-oriented methods, it is sometimes difficult to clearly ascertain the advantages of object-orientation over conventional structured programming for a particular project. In order to make a valid judgment we would have to follow through a host of major software projects developed using both methods, measure and keep track of the development effort, and critically evaluate the results. Even then, many subjective considerations would muddle our conclusions. In any case, we must not forget that the advantages of object orientation come at a high price in programming complications. Few will argue that object-oriented languages are more difficult to master and use, and that object-oriented analysis and design techniques are elaborate and complicated.

The decision whether to use or not to use an object-oriented approach should be given careful consideration by the analyst. The following elements favor the use of object-oriented methods:

1. Projects in which modeling the problem domain is a major issue

2. Projects in which communication with domain experts can be expected to be a difficult and critical part of the development activity

3. Volatile projects in which changes can be anticipated or in which frequent updates will be required

4. Projects with many elements with potential for reuse

5. Toolkits and frameworks are often good candidates for the object-oriented approach

On the other hand there are elements that often advise against object-orientation:

1. If the data structures and processing operations resulting from object-orientation are not compatible with existing data bases. In other words, it may be a bad idea to develop an object-oriented database management system if the new system forces the re-design or restructuring of the existing databases.

2. If the project, or a substantial part thereof, must be developed in a non-object-oriented programming language. Thus, object-oriented methods may be of doubtful use in developing a short, low-level device driver for a hardware accessory.

3. If the problem domain model is already well defined in non-object-oriented terms.

4. If the problem-domain is better expressed using a conventional, non-object-oriented model. Projects in which software elements are the subjects to be modeled sometimes fall into this category.

In general, major business and industrial applications are suitable for object orientation, while small, technical, or support utilities are not.

Chapter 29

Object-Oriented Analysis

Chapter Summary

Object-oriented programming is mostly a software design methodology. In this chapter we discuss the elements of object-oriented analysis which is the first and most important step in the design of an object-oriented application. The topics include modelling the problem domain, decomposition into classes and objects, testing object validity, systems and sub-systems, attributes, services, instance and message connections.

29.0 Elements of Object-Oriented Analysis

It has been observed by several authors that both major software engineering paradigms (structured programming and object orientation) started with programming languages, were later extended to software design, and finally to analysis. This implies that the advent of a system analysis methodology marks the maturity of a particular paradigm. Regarding object-oriented systems we have seen that full-featured programming languages were already available in the late seventies and early eighties. System design discussions started at about that time, while the object-oriented analysis methodologies are a product of the late eighties and early nineties.

Observe that the object-oriented approach to analysis does not attempt to replace every technique and tool of conventional system analysis. The feasibility study, risk analysis, and cost-benefit analysis can be performed independently of whether the project is modeled using structured programming or an object-oriented approach. Object-oriented analysis provides a practical methodology for project modeling during the analysis phase. Its purpose is to complement, not to replace.

We mentioned in Section 3.6 that the analyst should critically evaluate if a project will gain from using an object-oriented approach. Also that the greatest benefits of the object-oriented paradigm result when the methodology permeates the entire development process. In most cases, our decision to use object-oriented analysis implies that we will also follow object orientation in the design phase and that the program will be coded in an object-oriented programming language. Paradigm mixing is theoretically possible, and sometimes recommended by hard-core defenders of object orientation. However, regarding the smaller project, the complications brought about by simultaneously using more than one model far outweigh the uncertain advantages. In particular regarding the smaller development project, it is our opinion that the decision to proceed with object-oriented analysis should presuppose that object orientation will also be used during the design and coding phases of development.

In this chapter we follow Coad and Yourdon, although not very strictly. These authors are a principal force in the field of object-oriented analysis. In addition, their approach is elegant and simple. The Coad and Yourdon notation was described in Section 28.4.

29.0.1 Modeling the Problem-Domain

One of object orientation's principal characteristics, as well as its main justification, is that it provides a reasonable way of modeling the problem-domain. Other advantages often mentioned are that it makes modeling complex systems possible, facilitates building systems that are more flexible and adaptable, and promotes reusability.

Object-oriented analysis is most useful in problem-set modeling. One of the predicaments of software development is that the programmer must become an expert in the domain field. Thus, someone contracted to develop an air-traffic control system has to learn about radar, about air-to-ground communications, about emergency response systems, about flight scheduling, and about a multitude of other technical and business topics related to the activity at hand. How much knowledge must the analyst acquire, and to what technical level must this knowledge extend, is difficult to determine beforehand. Many projects have failed because the developers did not grasp important details of the problem set. The urge to "get on with the coding" often works against us during this stage.

Because the analyst needs to quickly understand the problem-domain, any tool that facilitates this stage of the process is indeed valuable. Once the analyst has grasped the necessary knowledge, then this information has to be transmitted to other members of the development team. Here again, any tool that assists in communicating the knowledge is greatly appreciated. Finally, the model of the proposed solution-set must be presented to the clients or users for their validation, and to obtain feedback. A model that facilitates this communication between clients and developers is an additional asset.

29.0.2 Defining System Responsibilities

In addition to providing a model of the problem-domain, object-oriented analysis must define the system's responsibilities. For example, the analysis of an air-traffic control system includes determining what functions and operations are within the system's burden. Does the air-traffic control system have the obligation of informing commercial airlines of delays in arrivals or departures of aircraft? How does the air-traffic control system interface with the emergency response system at the airport? At what point does the tracking of an aircraft become the responsibility of a particular air-traffic control system and when does this responsibility cease? A system's responsibilities refer to what a system should do. Answering this question is one of the fundamental tasks of object-oriented analysis. Questions related to how a system operates are left for the design and coding phases.

29.0.3 Managing Complexity

Analysis is necessary because natural systems are often so elaborate and complicated that they are difficult to understand or manage. Object-oriented analysis provides a toolset for managing complexity. This toolset is based on abstraction, encapsulation, inheritance, and message passing.

Abstraction

Abstraction consists of eliminating what is superfluous or trivial and concentrating on what is fundamental. Any definition is an exercise in abstraction. We can state that a fountain pen is a hand-held writing instrument that uses liquid ink, an ink discharge mechanism, and a fine tracing point. This description attempts to gather the fundamental attributes of a fountain pen that distinguish it from a typewriter, a ball-point pen, and a pencil. However, it ignores the less important features such as the pen's color, the color of the ink, the instrument's exact dimensions, and its design or style.

In the context of object-oriented analysis, abstraction refers to the class mechanism which simultaneously provides procedural and data abstraction. In other words, the object-oriented notion of a class is an abstraction that represents both processing and data elements of the problem set.

Encapsulation

Encapsulation is one of the fundamental notions of object orientation. In the object-oriented approach encapsulation is related to the notion of data and functions (attributes and methods) being packaged together by means of the class construct. Those methods visible outside the class are its interface. The principal purpose of encapsulation is hiding the implementation details while stressing the interface.

Inheritance

In the object-oriented paradigm inheritance refers to the possibility of one class accessing the public and protected members of its parent class. Class inheritance promotes the highest level of abstraction, reduces code size, and simplifies programming. The result is a class hierarchy that goes from the most general to the most specific. Typically, a derived class incorporates all the features of its parent classes and adds some unique ones of its own.

Message Passing

Object-oriented systems access processing functions by means of a mechanism called message passing. One of the disadvantages of hard-coded program structures such as jump tables, cascaded if statements, or case constructs, is that they must be updated with every program modification or change. The message passing mechanism, on the other hand, automatically directs execution to the appropriate member function. This functionality brings about two major advantages: first, we avoid introducing new defects into code that have already been tested and debugged. Second, a system can be expanded by supplying relatively small modules that contain the new functions.

29.1 Class and Object Decomposition

The first discipline of object-oriented analysis is learning to think in terms of classes and objects. At this stage we must free our minds of concerns regarding algorithms, programming structures, or any other implementation issues. Our task during analysis is to model the problem-domain and to define the system's responsibilities. Both of these purposes are accomplished by means of class and object decomposition.

In this sense an object can be considered an abstraction of a discrete element in the problem domain. The object-oriented encapsulation is a matter of defining attributes and methods. An object is characterized by the possibility of its identity being preserved by the system. Therefore each object must have a uniqueness that allows its identification.

Every object belongs to a class of objects. A class is a description of a set of unique attributes and methods associated with an object type. An object is an instance of a class. An abstract class is a description of an interface which lacks implementation. Its fundamental purpose is to define an inheritance path. It is not possible to instantiate objects from an abstract class.

Do we start modeling by thinking of objects or of classes? It is a matter of semantics, but, strictly speaking, an object is a run-time construct while a class is an abstraction that groups common attributes and methods for an object type. Therefore it appears that we should think of object types or classes, rather than of possible instantiations. On the other hand, the mental concept of a class of objects requires that we visualize a typical object. For example, suppose we are attempting to model a system which uses a viewport window to display a text message. In order to define the text message window type we must first imagine what the object would look like. Figure 29.1 shows what may be our initial visualization of a text window object.

From the object's visualization we can deduce the attributes and methods of the object itself and the class that represents it. For example, the screen location and the object size are defined by the start and end coordinates. The object also has a border thickness and a color. Its text attributes are the window's title and its text. A control button on the top left corner of the object allows the user to erase the text message window. There are three methods associated with this object: one to display the window, one to report its current status, and one to erase it. Using class diagrams we can now represent the object type TextWindow to any desired degree of detail. The left-hand diagram in Figure 29.2 merely states that there is an object class named TextWindow, while its attributes and methods are yet undefined. The diagram on the right shows the specific attributes and methods.

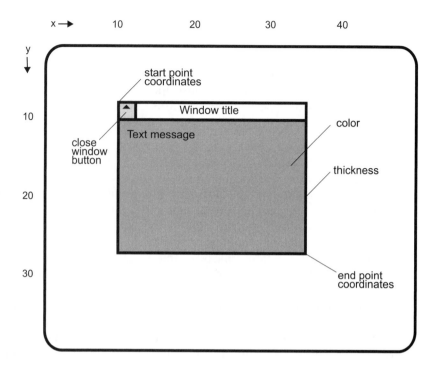

Figure 29.1 *Visualization of a Text Window Object*

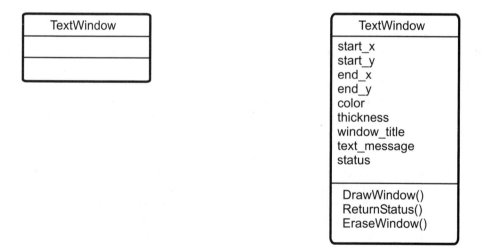

Figure 29.2 *Class Diagrams for a TextWindow Object*

During classification it is common practice to make the class name a singular noun or an adjective followed by a noun. It is also a good idea to use the client's standard vocabulary when naming classes. Suppose we are modeling a home alarm system with a rather elaborate video display and command control. For historical reasons the client calls the video display and command control a "push-button panel." During analysis we may be tempted to rename this element so that its designation describes its actual functionality more exactly. Clearly the term "command console" would be more forceful and descriptive than "push-button panel." However, this new name would serve only to confuse clients, who would have to mentally translate "command console" for "push-button panel" every time they encounter the new term.

During the analysis phase the class name could be entered using any reasonable style of spacing and capitalization. However, if our class names are not compatible with the syntax to be used in the design and coding phases, then these names would have to be modified at a later date. Since any modification could be a source of errors, it is better to name classes, attributes, and methods using a style that is consistent with the later phases of the development environment. In the context of this book decisions regarding style are facilitated by the fact that we have committed to C++ as a programming language. Therefore, to improve readability, and in conformance with one popular C++ programming style, we capitalize the first letter of every word in class names and methods. Methods are followed by parenthesis to indicate processing. Attributes, which later become variables, are typed in lowercase, and the underscore symbol is used as a separator. Note that any other style is equally satisfactory as long as it can be consistently maintained throughout the development process.

While methods are depicted by verbs and verb phrases, attributes are represented by nouns or noun phrases, as in Figure 29.2. C++ allows us to use rather simple naming schemes for attributes and methods since these names have class scope, where the same names can be reused in other classes. In this sense the DrawWindow method of the TextWindow class is not in conflict with other methods called DrawWindow that could be later defined for other classes. By the same token, the thickness attribute can also be reused in reference to other types of graphic objects.

Simplified class diagrams in which attributes and methods are not explicitly listed can be useful in showing class relations and inheritance. Consider a system with two types of window objects, one representing windows that contain nothing but text, a second one for windows that include a bitmap. The first type could be designated as a TextWindow and the second one as a GraphicWindow. The class structure would be depicted in the class diagram in Figure 29.3.

Note that in Figure 29.3 TextWindow is a subclass of Window, as is GraphicWindow. Also that Window is an abstract class, which implies that its methods are not implemented. Also that at run time we could instantiate objects of type TextWindow or of type GraphicWindow, however, Window objects could not exist since the abstract nature of the base class precludes this possibility. A more detailed diagram of the classes is shown in Figure 29.4.

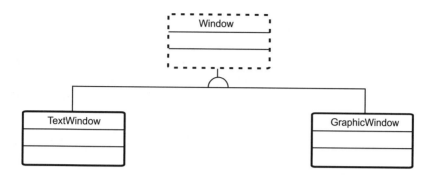

Figure 29.3 *Sample Inheritance Diagram for Window_Type Classes*

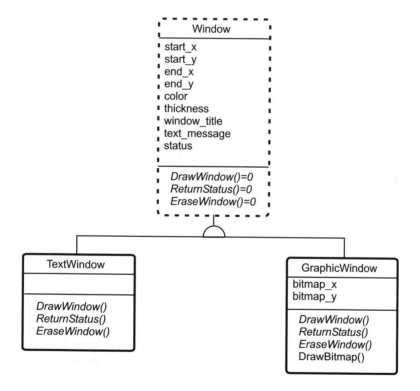

Figure 29.4 *Refined Diagram for Window-Type Classes*

In Figure 29.4 a GraphicWindow class is a subclass of Window which adds functionality to the base class. The new functionality is in the form of three new attributes which locate the bitmap on the viewport and a new method to draw the bitmap. The basic coordinates and characteristics of the window itself are defined in the base class, since these attributes apply to any window object.

29.1.1 Searching for Objects

The search for objects should not begin until the analyst has gained some familiarity with the problem domain itself. An analyst that is unfamiliar with an air-traffic control system should not be searching for air-traffic control objects. Instead, the first step should be an attempt to understand the system's fundamental characteristics. It is futile to attempt to identify objects and classes in a technology that is unfamiliar.

The following are the main sources of information about the problem domain:

1. Published materials and the Internet

2. First hand observation

3. Consultation with domain experts

4. Client documents

The search for published material usually leads to libraries, bookstores, and the Internet. First-hand observation consists of familiarizing ourselves with a system by participating in its operation. For example, to analyze an air-traffic control system we may begin by sitting in the air-traffic control tower for several days while observing the operation of the air-traffic controllers. Domain experts are often the source of valuable and unique information. Many systems are totally or partially unpublished, part of trade or government secret, or are considered confidential. In these cases published materials may be of little use. Our only option is to locate individuals with knowledge and expertise and to have them describe or explain the system to us. Client documentation can also be a valuable source of information. Expert analysts make it a standard practice to request a detailed document from the client. This document should identify the system and its purpose to the greatest possible detail.

29.1.2 Neat and Dirty Classes

In some contexts classification is a clear and neat process, but not in others. When conceptual objects correspond to real-world objects the classification process is often simple and straightforward. For example, when the task is to define a mini CAD program, classification could turn out to be relatively uncomplicated. In this case the screen objects give rise to classes, such as drawings composed of lines, curves, geometrical figures, and text.

On the other hand, classification may be somewhat less clear when the task is to design an editor or word processing program. In this case we may start thinking of program operations as objects. But is it reasonable to consider operations such as inserting, replacing, and deleting text as physical objects? What about the user's text file? However, since the text file is stored as an array of characters we may be tempted to think of it as an attribute. One possible classification is shown in Figure 29.5.

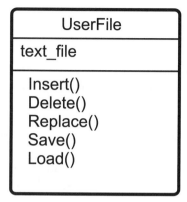

Figure 29.5 *Possible Class Diagram for a Simple Text Editor*

Other projects may be even more difficult to classify. For example, assume the task is designing and coding a virus-detection utility. In this case are the viruses the objects? Would this mean that if no viruses are found the program executes without instantiation? Or is the search itself an object? In this case whatever kind of classification is developed will probably be unsatisfactory, and the classes themselves will be untidy and unreal. What happens is that this is probably one of those cases, mentioned in Section 3.6, that is best handled outside of object orientation.

29.2 Finding Classes and Objects

The skills of an object-oriented analyst include the individual's ability to locate objects and to derive valid and useful classifications. In this respect there is no substitute for experience with various types of problem domains. However, there are some circumstances, patterns, and constructs that the analyst can use to locate possible objects. One proposed methodology consists of searching through the elements of class associations, mechanisms and devices, related systems, preserved data, roles played, sites, and organizational units.

29.2.1 Looking at Class Associations

In Section 27.4 we described a notation for representing two types of class associations: the Gen-Spec and the Whole-Part structure. This first type refers to a generalization-specialization relation that corresponds to one class being in a "kind-of" association with a parent or base class. The second type refers to an association where a class is "a-part-of" a parent or base class. The notation for class associations is often taken to imply an inheritance configuration whereby the public elements of the base class become accessible to the derived class. Although this assumption is valid during analysis, we should mention that Whole-Part associations are often implemented without using class inheritance.

Associations (also called structures) can be used to identify objects in the problem domain and to uncover levels of complexity in the class structures. The method consists of looking at object associations within the problem domain. For example, in Figure 29.3 we found that two types of windows, called GraphicWindow and

TextWindow, were subclasses of a general Window type. A detailed class diagram was later derived from these associations.

Gen-Spec Structures

The name of a specialization is usually derived from its generalization. In this manner TextWindow and GraphicWindow are specializations of the Window type (see Figure 29.4). In a Gen-Spec structure the bottommost classes are concrete classes, from which objects can be instantiated. The higher-level classes can be concrete or abstract. Often class structures are flexible and can be expressed in different ways. For example, we could redo the class diagram in Figure 29.4 so that a concrete class Window serves as a generalization with a possible specialization in the form of a GraphicWindow. This scheme is shown in Figure 29.6.

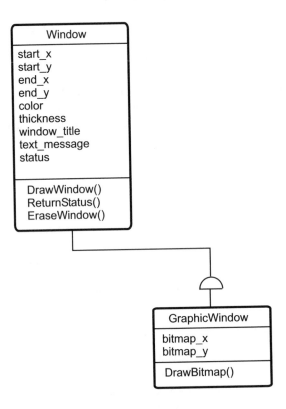

Figure 29.6 *Alternative Structure for Window-Type Classes*

Note that Figure 29.6 depicts a class structure in which the methods DrawWindow(), ReturnStatus(), and EraseWindow() are actually implemented in the base class. Therefore they need not be re-implemented in the derived class, except if these methods need update or modification.

Once we have identified a generalization-specialization structure, perform tests to ascertain whether the structure is both valid and necessary.

1. Does the subclass respond to the "is-a-kind-of" relation regarding the parent class?

2. Does the subclass add attributes or methods to the parent class?

3. Is the structure actually within the problem domain as it pertains to proposed system? In other words: is this structure really necessary to represent the problem at hand?

4. Is the structure within the system's responsibilities? Often a structure that actually exists in the problem domain turns out to be unnecessary in the model since the proposed system does not perform any function related to it. For example, we may recognize that a graphic window (see Figures 29.3 and 29.6) could contain vector drawings as well as bitmaps. However, if the system to be developed deals only in bitmap graphics, then the introduction of vector graphics would be an unnecessary complication since it falls outside of the system's responsibilities.

5. Are there lines of inheritance between the subclass and the parent class? A generalization-specialization structure in which no methods are inherited from the parent class should be questioned since it is inheritance that makes the structure useful.

6. Would it be simpler to use an attribute that qualifies a type than a Gen-Spec structure? For example, a system that deals with vehicles has to distinguish between autos, motorcycles, and trucks. This could possibly be accomplished by a Gen-Spec structure in which Vehicle is a base class and AutoVehicle, MotorcyleVehicle, and TruckVehicle are subclasses. Alternatively, the system could be modeled using a Vehicle class with a vehicle_type attribute. Both alternatives are shown in Figure 29.7.

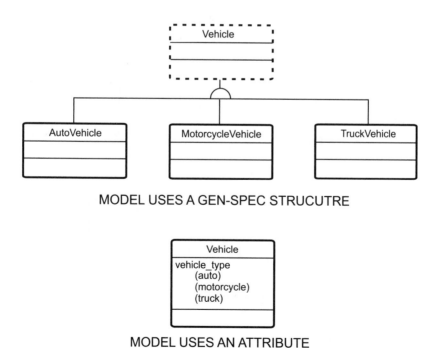

MODEL USES A GEN-SPEC STRUCUTRE

MODEL USES AN ATTRIBUTE

Figure 29.7 *Two Modeling Alternatives*

The general rule used in determining which model is to be preferred is that a generalization-specialization structure should not be used for extracting a common attribute.

Multiple Inheritance in Gen-Spec Structures

The common case in a Gen-Spec structure is that a class inherits from a single ancestor. However, Coad and Yourdon distinguish between an Gen-Spec structure used as a hierarchy and one used as a lattice. The hierarchy is the most common form of the structure and corresponds to the examples used previously in this chapter. In this sense the top-most model in Figure 29.7 represents a hierarchy of the Vehicle class. The notion of a hierarchy corresponds to single inheritance in object-oriented languages. A lattice, on the other hand, corresponds with multiple inheritance, which occurs when a class inherits attributes or methods from more than one ancestor.

For example, in the Vehicle class structure of Figure 29.7 we could define two additional subclasses. The class PickupVehicle could have some of the properties of an AutoVehicle (carries passengers) and some of the properties of a TruckVehicle (transports cargo). Or an ATVVehicle may inherit some properties from the AutoVehicle class and others from MotorcycleVehicle. Figure 29.8 shows the resulting lattice-type diagram.

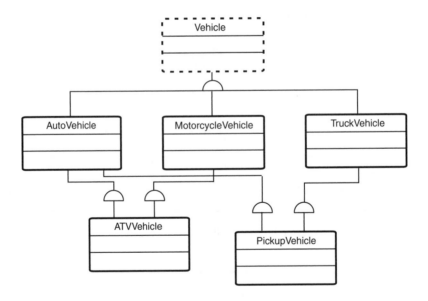

Figure 29.8 *Gen-Spec Structure Lattice*

Whole-Part Structures

The triangle markings in Figure 28.4 represent a Whole-Part structure which corresponds to the concept of an "is-a-part-of" class relation. Sometimes the Whole-Part notation is supplemented with digits or codes to represent the number of elements that take part in the relation. For example, it may help the model if we clarify that a vehicle can have four wheels but only one engine, or that a wheel can belong to only one vehicle, but a person can own more than one vehicle. A comma separates the low from the high numerical limits. The letter "m" is often used as a shorthand for "many." Figure 29.9 shows a Whole-Part structure regarding some components of an AutoVehicle class.

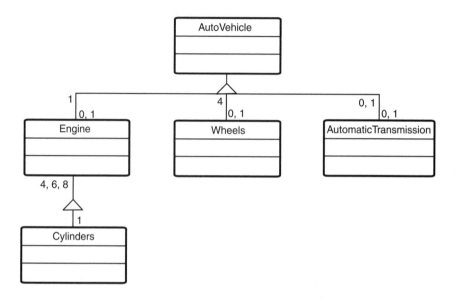

Figure 29.9 *Whole-Part Structure*

Note that the digits in Figure 29.9 add information to the model. For example, we can tell that an auto vehicle has one engine, four wheels, and zero or one automatic transmission. Also that engine, wheel, and automatic transmission can exist independently of the automobile, but that cylinders do not exist outside of the engine. Finally that an engine can have either four, six, or eight cylinders.

There are certain common constructs and relationships that can be investigated when searching for Whole-Part associations. In the example of Figure 29.9 the construct can be described as an assembly-components relationship. Another common case can be described as a container-contents relationship. For example, the driver of an automobile is not one of its components, but may be considered to be in a container-contents relationship. A third common construct is the collection-member relationship. For example, the collection coins has the members quarter, dime, nickel, and penny. Or the collection hospital has the members patient, nurse, doctor, and administrator.

The following tests can be performed to ascertain whether a Whole-Part structure is valid and necessary:

1. Does the subclass correspond to the "is-a-part-of" relation regarding the parent class?

2. Does the subclass add attributes or methods to the parent class?

3. Is the structure actually within the problem domain as it pertains to the proposed system? Is this structure really necessary to represent the problem at hand?

4. Is the structure within the system's responsibilities?

5. Would it be simpler to use an attribute that qualifies a type rather than a Whole-Part structure?

Compound Structures

It often happens that a structure contains both Gen-Spec and Whole-Part class relationships. For example, the structures in Figures 29.8 and 29.9 would give rise to the compound structure shown in Figure 29.10.

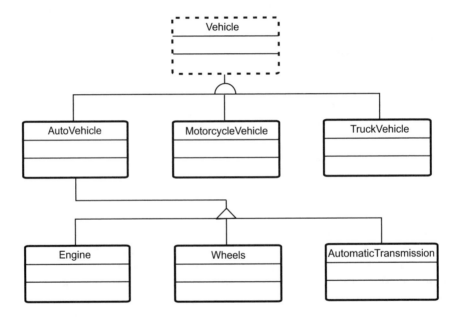

Figure 29.10 *Compound Structure*

29.2.2 Looking at Mechanisms and Devices

Another element of the search for classes and objects is the location of mechanisms and devices within the problem domain. For example, in modeling a burglar alarm system we may notice that it includes devices such as sensors to be placed in possible entry points, audible warnings that sound when a security breach is detected, switches for enabling and disabling the system, and input/output hardware such as a keypad for entering passwords and displays for echoing input and reporting system status. In this case the devices and mechanisms that are within the system's responsibilities are all candidates for classification. In the case of the burglar alarm system we should examine the possibility of having a class named Sensor, a class named AudibleAlarm, a class named OnOffSwitch, and so on.

29.2.3 Related Systems

At the fringes of a system's responsibilities we can sometimes detect other systems with which it interacts. The interaction could be hard-wired (if both systems are mechanically connected), through a communications link, or can result from human-computer interaction. Any associated or related system should be considered as a possible candidate for classification if it is within the base system's responsibilities and if it forms part of the problem domain. For example, if the burglar alarm system mentioned in Section 29.2.2 is connected to a local police alert system, then perhaps some elements of the police alert system should be included in the model.

29.2.4 Preserved Data

Another possible source of class and objects are the data or events that must be preserved or remembered by the system. Events preserved for historical purposes, modifications to databases, and legal documents, are all included in this category. For example, a computer system that uses a telescope to scan the sky for unknown celestial objects must somehow record and remember the position of a new comet in a possible collision path with Earth. The file used to store the time, date, celestial coordinates, and photographic image of the newly detected object is a possible candidate for classification. A more mundane example would be the transaction data stored by an automated teller machine.

29.2.5 Roles Played

Individuals play roles within organizations and these roles are another possible source of classes that should be examined. For example, a Health Maintenance Organization (HMO) has individuals playing the roles of physicians, nurses, patients, administrators, and staff. It is likely that a model of this HMO will consider these roles as classes. Figure 29.11 shows one possible (rather loose) interpretation of the interaction of the class Physician within a fictitious HMO.

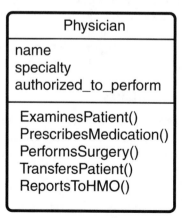

Figure 29.11 *Class Depicting Human Interaction*

29.2.6 Operational Sites

Operational sites or physical locations within the system's responsibilities are often a source of classes. For example, a hypothetical satellite image processing system includes a site in New Mexico where the image data is downloaded from the satellites, a second site in South Dakota receives this raw data and reformats it into specific image files, which are then transmitted to a third site in California which distributes the image files to the system clients. In this case the New Mexico, South Dakota, and California sites perform different functions and could therefore be considered as candidates for classification.

Another example: the quality control department of a food product manufacturer consists of an office and a laboratory. The food product samples are collected in eight manufacturing plants throughout the state. In modeling this system we may have to decide if the classification is based on ten sites, one of which is the lab, the other one the office, and eight sites are manufacturing plants, or if the model can be rationalized to three classes by combining all eight manufacturing plants sites into a single class. In this last option an attribute could be used to designate each one of the individual plants. This case is shown in Figure 29.12.

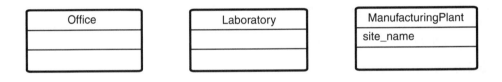

Figure 29.12 *Example of Site-Based Classification*

29.2.7 Organizational Units

Organizations are divided into units that perform different tasks and assume distinct responsibilities within the system. At modeling time these organizational units are also strong candidates for classification. For example, a manufacturing division includes departments called engineering, production, quality control, and shipping. These organizational units should be considered as possible classes.

29.3 Testing Object Validity

Once we have developed a basic classification scheme for the problem at hand it is a good idea to challenge its validity and soundness. There are certain requirements usually associated with objects which can serve to test their legitimacy and significance. In challenging proposed classes and objects the following questions are pertinent:

1. Is there information that the system needs to remember regarding this object?

2. Does this object provide some method or processing operation?

3. Is there more than one attribute associated with this object?

4. Will more than one object be instantiated from the class?

5. Do the attributes always apply to the object?

6. Do the methods always apply to the object?

7. Are the objects related to the problem domain or to a particular implementation?

8. Are the results a simple calculation or derivation?

In the following sections we proceed to examine each of these questions separately.

29.3.1 Information to Remember

For an object to be individually remembered by the system it must be associated with one or more related attributes. In this sense an instance of the class Clerk must have a name, and instance of the class Office must have an address on a set of specific functions, an instance of the class AutoVehicle must have a model and a serial number, and so on. An object that is not associated with any attribute should be questioned since it could be irrelevant to the system, even though it may be an object in the real world. Although an object could have methods but no attributes, this case should always be investigated before it is accepted as an exception to the general rule.

29.3.2 Object Behavior

A class with no methods is suspect of a classification error. Even an abstract class has methods, although these methods are not implemented. If we accept that, in principle, attributes should not be accessible from outside the class, then the fact that an object is affiliated with one or more attributes implies that there are one or more methods to access these attributes. A careful investigation is necessary if an object has attributes but no methods. Some authors consider that classes with methods but no attributes are a result of functional-style thinking and should be reconsidered.

29.3.3 Multiple Attributes

While a class with no attributes is suspect of incorrect classification, one with a single attribute may indicate a poorly conceived classification. In this case the analyst could use an attribute (typically in the parent class) instead of a full-fledged object.

29.3.4 Multiple Objects

The fundamental notion of classification implies that a class is a type which usually consists of more than one individual. In this sense we can correctly conceive the class FootballPlayer or the class Quarterback. Multiple objects of either class can be instantiated. However, the class JohnSmith, of which only one object can be instantiated, should set off an alarm indicating the need for further investigation. On the other hand, often we must create a class with only one possible object. For example, a satellite image processing system may have a class ReceivingStation with a single object. However, in this case we may decide to validate the class considering that another receiving station could be built. The decision about the validity and suitability of a class with a single object must be made considering the problem domain.

29.3.5 Always-Applicable Attributes

The general rule is that the attributes in a class apply to all objects instantiated from the class. For example, suppose that the class AutoVehicle includes the attributes manufacturer, model_number, and size-of-bed. All objects generated from the class include the first two attributes, however, the size-of-bed attribute does not apply to those vehicles that are not trucks. The fact that an attribute is not applicable to all objects instantiated from the class suggests that there is a possible class structure yet undiscovered. In this example perhaps we could use the Gen-Spec structure in Figure 29.7.

29.3.6 Always-Applicable Methods

The fact that there are methods in a class that do not apply to all possible instantiations indicates that there is room for improvement in the classification. Perhaps there is a class structure that has not yet been considered. This is unrelated to how simple or elaborate a particular method is, but it is linked to the fact that methods are expected to be applicable to all possible objects. For example, consider a class named HospitalEmployee that includes doctors, nurses, and administrators. This class has a method named PerformMajorSurgery that is applicable only to doctors, but not to nurses or administrators. The fact that a method does not apply to all objects that can be instantiated from the class strongly suggests a missing class structure.

29.3.7 Objects Relate to the Problem Domain

Many beginning analysts find it difficult to isolate themselves from implementation concerns during the analysis phases. Due to inattention, matters that pertain to implementation often permeate the model. Suppose a system that deals with licensing of automobile drivers. Because the system must digitally store each driver's personal data we may be tempted to create a class called DriverDiskFile. But, in reality, the concept of a disk file pertains to the solution set. From the problem set's viewpoint the class that stores driver data could be better named DriverInfo. How this information is stored or preserved in the computer hardware is outside the scope of the model at the analysis stage.

An important conclusion is that the choice of processing operations, of hardware devices, of file formats, and of other computer-related factors should not be made during the analysis phase. The specific methods, systems, or devices used in implementing the solution are inconsequential at this stage. Therefore, a class that represents a solution-related element should be carefully investigated. Exceptionally, we may find that project specifications require the use of a particular algorithm or method of computing. For example, the client of a graphics package requests in the specification documents that lines be drawn using Bresenham's algorithm. In this case the algorithm becomes a domain-based requirement and its inclusion in the model is valid and appropriate. On the other hand, if no mention of a specific algorithm is made in the specification, then it would be premature to decide what method is to be used for drawing lines at the analysis phase. A good general rule is that all implementation matters are best postponed until design or implementation, except if the problem domain specifically mandates it.

29.3.8 Derived or Calculated Results

Finally, we should investigate those objects that represent a mere calculation or that are derived by manipulating stored data. For example, suppose a payroll system which must issue a printed report of all salaries and wages disbursed during the period. In this case we may be tempted to create a class, perhaps called WeeklyPayrollReport, to represent this document. However, since this document is directly derived from stored data it could be an improper classification. An alternative and more satisfactory solution is to make the generation of a payroll report a method in the class that stores the pertinent information, rather than an object by itself.

29.4 Subsystems

The complexity of a model sometimes gets out of hand. According to Coad and Yourdon a system containing 35 classes is of average size, while one with about 110 classes is considered large. It is also common for the problem domain to include several subdomains, each one with up to 100 classes. With this level of complexity the fundamental features of the problem domain may become difficult to grasp, and therefore, to explain to clients and users. A convenient mechanism for reducing model complexity is its division into subsystems, sometimes called subjects in object-oriented analysis literature.

Subsystems do not reduce complexity by simplifying the model but rather by progressively focusing on its different components. For example, suppose a model that depicts a full-featured graphics library, shown in Figure 29.13.

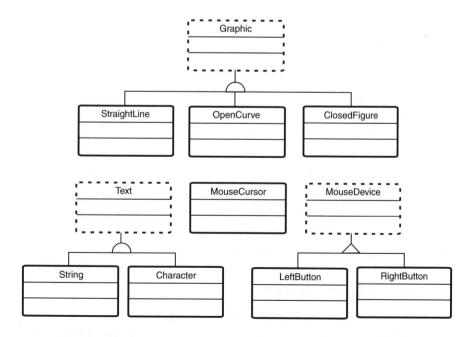

Figure 29.13 *Classes and Structures of a Graphics Toolkit*

In such cases the resulting model can be elaborate and complicated. However, by noticing that the graphics operations refer to drawing geometrical figures, to displaying text, and to interacting with a mouse, we are able to decompose it into three independent subsystems. Each of these subsystems can then be treated as a logical entity, even though they may share functions or interact with each other. In Figure 29.14 we have repositioned some of the classes and added rectangles to enclose the subsystems. Each rectangle is labeled with the subsystem's name. Some authors use numbers to designate subsystems. In either case the result is a more manageable model.

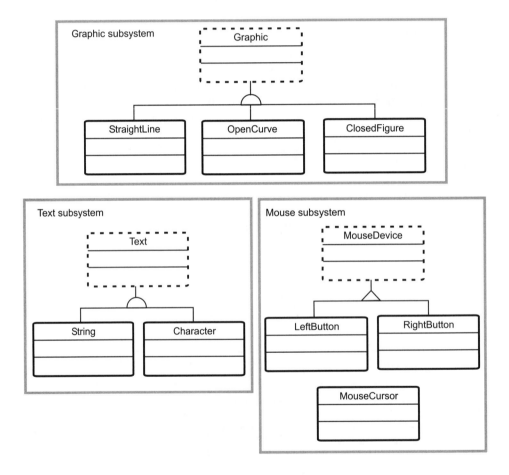

Figure 29.14 *Graphics Toolkit Subsystems*

29.4.1 Subsystems as Modules

In Section 26.2.3 we discussed the concept of a program module as a logical unit of division based on structural or functional considerations. The concept of a program module should not be equated with that of a procedure, a subroutine, or a disk file, although occasionally they may coincide. This viewpoint determines that modularization should not be considered as an implementation issue, but as one related to system analysis and design.

In this sense the notion of a subsystem is quite compatible with that of a module. Therefore subsystems identified during the analysis stage can become modules during the design stage. But for this to happen the subsystem must meet the same requirements demanded of modules.

Subsystem Cohesion

Regarding modules, cohesion or binding refers to the relationship between its internal elements; the same applies to subsystems. A subsystem in which the contained classes are strongly related has high cohesion. Sometimes we include a class into a subsystem simply because there is no better place for it. Inclusion of an unrelated class reduces subsystem cohesion. However, this does not mean that all classes in a subsystem must be intimately linked, but rather that they must be strongly related in the problem domain. For example, in Figure 29.14 the Mouse subsystem contains the class MouseCursor which is not part of any class structure. However, because this class is closely related to the problem domain of mouse functions it fits well in the subsystem.

Using conventional terminology we can say that a subsystem in which the classes are related loosely can show only coincidental cohesion, while one in which the classes are related logically show logical cohesion. Cohesion levels are difficult to quantify, but the functional strength of a subsystem depends on it having high cohesion.

Subsystem Coupling

The concept of coupling is a measure of the degree of interconnection between subsystems. In this sense two closely coupled subsystems have a high degree of interaction and interdependence, while loosely coupled subsystems stand independently of each other. Subsystems with high coupling are difficult to understand, analyze, and reuse separately, therefore high coupling is generally considered an undesirable property. On the other hand all subsystems are usually coupled in some way since they must interact in producing a common functionality. It is difficult to imagine a system in which all its component subsystems are totally uncoupled.

In general, good subsystem structure should have a high cohesion and a low coupling. The classes must be closely related within the subsystem and there should be a minimum interdependency between subsystems.

29.5 Attributes

An attribute is a data element within a class, and each different object instantiated from a class is associated with a unique set of values for its attributes. For example, in Figure 29.11 each object instantiated from the class Physician has a unique name, specialty, and authorized_to_perform attribute values. This unique set of values is sometimes referred to as the object's state.

One of the fundamental rules of object orientation is that attributes should not be directly accessible from outside the class. The only way to obtain or change the value of an attribute is by using a method that performs this function. However, many programming languages do not impose this rule; for example, in C++ it is possible to declare an attribute to be of public access type and thus make it visible to other classes and to client code. Nevertheless, making attributes visible outside the class violates the principles of encapsulation and data abstraction, which, in turn, defeats one of the fundamental purposes of the object-oriented paradigm.

While classes and structures are relatively stable over the lifetime of a system, attributes are likely to change. In refining or expanding the functionality of a class we often have to introduce new attributes and methods. For example, the Physician class in Figure 29.11 may be refined by introducing attributes to reflect the individual's social security number and home address. In this case encapsulation requires that new methods be developed to access the new attributes.

29.5.1 Attribute Identification

During the initial stage of analysis we often draw diagrams of classes and structures excluding attributes and methods, whose definition is postponed until we become more familiar with the basic elements of the model. However, the analysis phase is not complete until we have identified attributes and methods for every class.

The fundamental notion of an attribute relates to what the containing class is responsible for knowing. For example, an analyst defining an Employee class may be tempted to include sex as an attribute. However, if the system does not use the employee's sex in any of the processing operations performed by the class, and if this attribute is not required in defining an object's state, then it is outside of the system's responsibilities and should not be included. If the attribute later becomes necessary it can be added at that time. Attempting to predict, at analysis time, the attributes or methods that may become necessary in the future is usually a bad idea. The result is often a system overloaded with useless elements which serve only to complicate it and add to its cost. It is much more practical to make generalized allowances that will accommodate growth or modification, rather than attempt to guess all the requirements that could become necessary.

In identifying attributes we may ask the following questions:

1. How is the object described?

2. What properties are associated with the object in the problem domain?

3. What properties associated with the object are necessary for implementing the system's responsibilities?

4. What properties of the object need to be remembered over time?

5. What states can the object be in?

Not all of these questions are pertinent for all objects and, in some cases, more than one question can have the same answer. At the same time some objects are conceptually very simple, while others are complex. For example, the LeftButton object of a mouse device (see Figure 29.14) is a rather simple one. It appears that questions number 3, 4, and 5 are pertinent in this case. Specifically: the system's responsibilities may require that it be known if the left mouse button has been pressed (button down) or released (button up). The properties to be remembered over time are the button's current status, which are also the object's possible states. Therefore, it appears that one attribute that encodes the state of the left mouse button as currently being down or up may be sufficient in this case.

On the other hand, an Employee class that keeps track of the individuals in the employment of an organization may require attributes such as name, address, social_sec_number, date_of_employment, hourly_wage/salary, hours_worked, taxable_income, and many others. Several ethical and legal issues may be considered at this time. For example, if it is illegal to base any employment decision on the employee's age, should the individual's date-of-birth be stored as an attribute? If a company has no other reason for knowing an employee's age, could the fact that this information is stored with other employee data be used against the company in an age discrimination dispute? The following argument could be made in this case: why would a firm incur in the expense of storing data that it does not intend to use?

The decision to include or exclude an attribute is usually made jointly by client and analyst. The safest advice is not to include attributes that are outside of the system's current responsibilities. Crystal-ball-based system analysis can be counterproductive.

29.5.2 Attributes and Structures

Occasionally we may be uncertain regarding which class of a Gen-Spec structure should contain a particular attribute. For example, when assigning attributes to the classes depicted in the Text subsystem (shown in Figure 29.14) we may decide that each text string or character will have a particular letter style and point size. Should these attributes appear in the classes String and Character, or on the parent class Text? The general rule in this case is to position the attribute at the highest point in the class structure applicable to all specializations. Therefore the attributes should appear in the class Text, as shown in Figure 29.15.

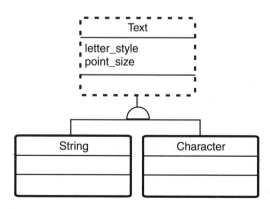

Figure 29.15 *Attribute Positioning in a Gen-Spec Structure*

29.6 Methods or Services

A system is functional and useful by virtue of the processing operations that it performs: a system must do something. A perfect structure of classes and attributes, with no methods, is worthless. The specific behavior that an object must exhibit is called a service. Service is synonymous with processing operations and coincides with a class' methods.

29.6.1 Identifying Methods

The analyst can follow several routines in order to identify the methods associated with each class. In this respect the following questions could be useful:

1. What are the object states?

2. What are the required services?

Object States

When an object is created it is associated with a set of attribute values that define its state. These attributes can change from the time the object is created until it is released. Every change in attribute values defines a new object state. We can determine an object's state by examining the range of values that can be assigned to each attribute in the light of the system's responsibilities. From this analysis we can construct a simple diagram showing the various states which are defined for the object as well as the transitions between states. For example, regarding the MouseCursor object in Figure 29.14 we may define that the system's responsibilities include knowing if the cursor is on or off; if on, its current screen position. The class can now be refined by drawing the state transition diagram. Later on we can use the state transition diagram to deduce the class' methods. The preliminary steps are shown in Figure 29.16.

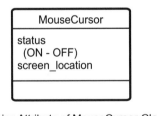

Defining Attribute of MouseCursor Class

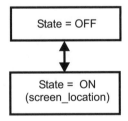

State Transition Diagram

Figure 29.16 *Attributes and Object State Diagram*

Required Services

From the object states and the object attributes it is possible to deduce the required services or methods. Here again, the services to be considered should be those within the system's responsibilities. Several categories of algorithmically-simple services should always be investigated. These are usually called Create, Connect, Access, and Release:

1. Create. This service is used to create and optionally initialize an object. When no particular initialization is required the Create service is often implicit. However, if an object's creation requires the initialization of attributes it should be distinctly defined. This approach is consistent with the operation of constructor functions in C++.

2. Connect. This type of service is used to establish or break a connection between objects. It corresponds to the notion of an instance connection described in Section 29.7.

3. Access. This type of service assigns values to attributes or returns attribute values. Its existence results from the principles of encapsulation and data abstraction, which require that attributes not be accessed from outside the class.

4. Release or Destroy. This service destroys the object. It corresponds to destructors in C++.

Much of the required behavior of an object can be performed by the Create, Connect, Access, or Destroy services. In many systems these algorithmically-simple methods account for 80 to 95 percent of the required functionality. However, algorithmically-complex methods are also often required. They usually correspond to the types Calculate and Monitor.

• Calculate. This type of service uses the attribute values in the object to perform a calculation.

• Monitor. This type of service is usually associated with an external device or system. Its functions relate to input, output, data acquisition, and control.

The methodology to follow in defining the required services consists in first examining the object for the four algorithmically-simple service types (Create, Connect, Access, and Destroy) and then for the two algorithmically-complex types (Calculate and Monitor). All object methods fall into one of these categories. For example, examining the MouseCursor class in Figure 29.17, we can identify two algorithmically-simple services. An access-type service is used to determine whether the cursor is in the ON or OFF state. A Connect-type service is used to establish a link between the MouseCursor object and the service that monitors the cursor movement on the screen. Figure 29.17 shows the corresponding class diagram.

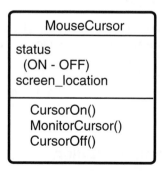

Figure 29.17 *Defining Services for the MouseCursor Object*

29.7 Instance Connections

While structures depict associations between classes, instance connections, as the name suggests, show associations between objects. Since objects cannot be instantiated from abstract classes, there can be no instance connection to or from an abstract class. In the analysis phase an instance connection reflects a mapping between objects in the problem domain. Note that while instance connections denote association they do not imply inheritance since instance connections are a weaker type of association than a Whole-Part or Gen-Spec structure. Like all other elements of system analysis, all instance connections should be within the system's responsibilities.

29.7.1 Instance Connection Notation

Instance connections are shown in class diagrams as solid lines between objects. Optionally, instance connection lines can be labeled to show the nature of the association. Coad and Yourdon consider that these labels are usually unnecessary. The notation includes numbers that indicate the upper and lower bounds of the object association. Figure 29.18 shows an example of instance connection notation.

Figure 29.18 *Example of Instance Connection Notation*

In Figure 29.18 we note that each driver can have zero or one license, and that each license must correspond to a single driver. In the first case the numerical constraint is separated by a comma to indicate a range. When there is no explicit upper bound the letter m can be used to replace the word "many".

29.8 Message Connections

While an instance connection represents a mapping between objects in the problem domain, a message connection refers to messages being sent between objects. In this context the term "sender" and "receiver" are often used. The purpose of the message is to obtain a processing operation which must be defined in both the sender's and the receiver's specifications. Trivial and literal messages are not included in this convention. Message connections occur when an object instantiates an object of another class, or when an object uses methods defined in another class.

29.8.1 Message Connection Notation

Coad and Yourdon notation uses gray lines to indicate message connections. In hand-drawn diagrams we could substitute them by dashed lines. To illustrate a message connection suppose a graphics package with a class StraightLine and a class

Polygon. Each StraightLine object has a start point, an end point, and a thickness attribute. A Polygon is drawn by instantiating several objects of the class StraightLine. Figure 29.19 shows the objects and the corresponding message connection notation.

Figure 29.19 *Example of Message Connection Notation*

Note that the arrow in Figure 29.19 points from the sender to the receiver of a message. This implies that the sender takes some specified action to which the receiver replies by returning a result to the sender or by performing a predefined function. Although typically the message connection takes place between objects, it is also possible to make the receiver a class, in which case it responds by instantiating a new object. An alternate notation uses a multipointed arrow in the case where more than one object is a receiver. Note that human interactions with a system (usually in the form of commands or requests) can often be depicted by means of message connection symbols.

29.9 Final Documentation

Once the analysis has concluded we must put together the corresponding documents. The most important one is normally the class diagram. However, in substantial projects it is often necessary to reduce complexity by layering the class diagram, as follows:

1. A layer indicating the various subsystems

2. A layer for each subsystem indicating inheritance in the form Gen-Spec and Whole-Part structures

3. A layer showing attributes and services for each class

4. A layer depicting instance and message connection for each object

In addition to the class and object diagrams, other documents are usually associated with the analysis phase, namely:

1. A description of class and object specifications

2. State transition diagrams

3. Documents describing system constraints and critical threads of execution

PART IV

Appendices

Appendix A

C++ Math Unit Programming

Summary

This appendix is about programming the Intel math unit which is part of all Pentium and later CPUs. The contents of this chapter relates to techniques for programming the Intel math unit using low-level code (Assembly Language) from C++.

AA.0 Programming the Math Unit

High-level languages, such C++, provide no facilities for programming the Intel math unit. Usually, a programmer's only choice consists of selecting a particular library that is documented to use the math unit, and to hope for the best. In regards to the math unit programmer control is limited to selecting rounding modes and levels of precision within C++ input and output functions. Numerical calculations take place inside a black box; all you see are input and results. The difficulty is that math unit programming requires Assembly Language. Fortunately, some C++ development systems for the PC, such as Microsoft's Visual C++, make it relatively easy to use Assembly Language code from C++ programs.

The Intel processor architecture supports a segmented, 16-bit memory model, and a flat, 32-bit model. Throughout the book we adopt the later one. Using the 32-bit, flat memory model ensures that the resulting code is compatible with Windows applications. At the same time, Visual C++ console applications can also be developed in flat memory model. The one objection to this approach is that 32-bit code is not compatible with MS DOS programs. There are two possible routes for developing the low-level component of a C++ application:

1. Develop the low-level code with assembler-based environment, such as Microsoft's MASM. After the source is assembled, the resulting object files are included in the Visual C++ project and referenced at link time.

2. Develop low-level routines using inline assembly.

Although both methods work well, here are a few points to keep in mind:

1. Inline code is usually easier to develop and debug. The program is created and tested in a single environment, which often leads to faster development times.

2. On the other hand, inline code cannot use assembler directives to define data objects. The program must use C++ variables for all data, but not all math unit data types can be created in C++.

3. Developing math unit code in MASM results in separate low-level object files. These files can be later converted into a static library or into dynamic-link library (DLL) which can be shared by several applications.

The approach we have followed in this book is as follows:

1. We use pure assembly language modules, developed with MASM version 6.14, for the core math unit routines. The routines use a flat, 32-bit, memory model that is compatible with both Windows programs and Visual C++ console-based applications. These routines are contained in the modules named UN32_?.ASM that are part of MATH32.LIB.

2. Each directly accessible assembly language procedure is furnished with a C++ interface function. The C++ interface often contains inline assembly code. The interface functions are located in header files for the corresponding modules of MATH32.LIB.

For example, UN32_4 defines the following public procedures:

```
PUBLIC  _ASCII_TO_EXP      ;.... ASCII string to exponential form
PUBLIC  _MATH_UNIT_OUTPUT  ;.... ST(0) register to ASCII decimal
PUBLIC  _MATH_UNIT_INPUT   ;.... ASCII exponential string to ST(0)
PUBLIC  _GET_TAG           ;.... Get math unit register tag code
```

These four low-level procedures were developed in MASM and the resulting object file (UN32_4) is one of the modules of MATH32.LIB.

The C++ interface functions to these four procedures are located in the header file Un32_4.h. They are prototyped as follows:

```
int MathunitInput(char[]);
void AsciiToExp(char[], char[]);
int MathunitOutput(char[]);
int GetTag(int, char[]);
```

The header file also declares the external references to the corresponding procedures in MATH32.LIB. The C++ application developer uses the interface functions to access the core procedures.

AA.1 MASM Sources in C++ Programs

Suppose an assembly language module named TESTX.ASM that contains a procedure named _ID_CPU. The module uses the following MASM directives:

```
PUBLIC    _ID_CPU        ; The PUBLIC declaration makes the procedure
                ; accessible from outside the module
.486            ; Use math unit instruction set
.MODEL FLAT                 ; Generate 32-bit code for a flat memory
                ; architecture
.DATA               ; Module data follows
```

```
...
.CODE               ; Module code follows
...
```

To assemble the module you will use the command:

```
MASM  TESTX /Zi;
```

Note that the /Zi switch ensures Codeview compatibility for easy debugging. During successful assembly MASM produces the file TESTX.OBJ. After assembly you copy both files, TESTX.ASM and TESTX.OBJ, to the Visual C++ project directory. Now, from Visual C++, you use the Add to Project button to add both files.

At this point both the source and object files for the low-level module are available to the Visual C++ environment. In order to gain access to the low-level procedure named _ID_CPU (located in the TESTX.OBJ module) you must declare it as an external reference, as follows:

```
extern "C" void ID_CPU();
```

Notice the C++ adds a leading underscore symbol to all external references automatically. For this reason the assembly language function (_ID_CPU) must contain the leading _. By the same token, the C++ extern statement does not use a leading _ in referencing the procedure.

The C++ program is compiled and linked normally, whether it be a Windows or a Console application.

AA.1.1 Sample Code

The following example is from the C++ project Test Un32_1 that is developed elsewhere in the book. The project contains three source files:

1. The C++ main source, which is the program's entry point. This file is named Un32_1.cpp.

2. A C++ header file, named Un32_1.h, which contains the interface routines to the low-level function.

3. The Assembly Language source file containing the low-level code for the processing routines.

The C++ source file, Test Un32_1.cpp, is coded as follows:

```
//*****************************************************************
//    Test and exercise for the interface functions to the
//    routines in the Un32_1 module of the MATH32 library
//*****************************************************************
// Reference:   SOFTWARE SOLUTIONS FOR SCIENTISTS AND ENGINEERS
//              By Sanchez and Canton
// Workspace  : \Test Un32_1
// Entry point: Test Un32_1.cpp
// Other files: Un32_1.asm
//              Un32_1.obj
//              Un32_1.h
// Development: Visual C++ 6.0 Console Application
// Dates:       December 23/2000
//              April 4/2001
```

```cpp
// Description:
//                Driver module for exercising procedures in the
//                Un32_1 module of MATH32.LIB
//***************************************************************
#include <iostream.h>
#include "Un32_1.h"

int main()
{
    // Variables for program
    int menuOption = 1;
    char buf5[] = "       ";
    int num1;
    char buf4[] = "     ";
    char asc10[10];
    int res1;

 while(menuOption)
 {
    // Display menu
     cout << "C++ INTERFACE FUNCTIONS TO THE UN32_1 MODULE OF";
        cout << " MATH32.LIB\n"
                << "  0 = END TEST\n"
                << "  1 = IntToAsc10()"
                << "  2 =  IntToAsc16()\n"
                << "  3 = AscToInt()\n"
                << " SELECT: ";
    cin >> menuOption;

    switch(menuOption)
    {
    case 1:
    // IntToAsc10()
        cout << "Testing IntToAsc10()\n";
        cout << "Enter value: ";
        cin >> num1;
        IntToAsc10(num1, buf5);
        cout << "Result = " <  buf5;
        cout << "\n\n";
        break;

    case 2:
    // IntToAsc16()
        cout << "Testing IntToAsc16()\n";
        cout << "Enter value: ";
        cin >> num1;
        IntToAsc16(num1, buf4);
        cout << "Result = " <  buf4;
        cout << "\n\n";
        break;

    case 3:
    // AscToInt()
        cout << "Testing AscToInt\n";
        cout << "Enter ASCII string (10 digits or less): ";
        cin >> asc10;
        res1 = AscToInt(asc10);
        cout << "Result = " <  res1;
        cout << "\n\n";
        break;
```

```
      }
    }
  return 0;
}
```

The C++ header file, Un32_1.h, is coded as follows:

```
//****************************************************************
//    Interface functions to the Un32_1 module of MATH32.LIB
//****************************************************************
// Reference:    SOFTWARE SOLUYTIONS FOR SCIENTISTS AND ENGINEERS
//               by Sanchez and Canton
// Workspace  : \Test Un32_1
// Entry point: Test Un32_1
// Other files: Un32_1.asm
//              Un32_1.obj
//              Test Un32_1.cpp
// Development: Visual C++ 6.0 Console Application
// Dates:       December 23/2000
//              April 4/2001
// Description:
//         C++ interface functions for some of the procedures
//         in the Un32_1 module of the MATH32 library
// Note:
//         Interface to the other procedures in the Un32_1
//         module of MATH32.LIB is not provided since the
//         book includes equivalent functions in C++. If
//         other interfaces are required, the reader can use
//         the listed ones as a pattern.
//****************************************************************
// Listing of all procedures in the Un32_1 module of MATH32.LIB
extern "C" BIN_TO_ASC10(); // 16-bit binary into 5-digit ASCII decimal
extern "C" BIN_TO_ASC16(); // 16-bit binary into 4-digit ASCII hex
extern "C" ASCII_TO_EDX(); // 4-digit ASCII decimal to binary in EDX
extern "C" BINF_TO_DECF(); // 8-bit binary to 8-digit decimal fraction
extern "C" DECODE_SINGLE();// Isolate elements of single precision
extern "C" BIN_TO_BCD();   // 16-bit binary to 5 BCD digits
extern "C" BCD_TO_ASCII(); // 5 unpacked BCD to 5 ASCII digits
extern "C" BCD_INTO_REG(); // 5-digit BCD into a binary in DX
extern "C" BCD12_TO_ASCSTR();// BCD12 number in ASCII string
extern "C" ASCII_TO_BCD12(); // ASCII decimal string into BCD12 format
extern "C" ASCII_TO_EDX(); // ASCII to binary in EDX register

//****************************************************************
//                    Prototypes
//****************************************************************
void IntToAsc10(int, char[]);
void IntToAsc16(int, char[]);
int AscToInt(char[]);

//****************************************************************
//      C++ interface functions to some procedures in
//            the Un32_1.asm module of MATH32.LIB
//****************************************************************
void IntToAsc10(int value, char *buffer)
{
// Function to convert a 16-bit binary number into 5 ASCII
// decimal digits
// On entry:
//         parameter value holds binary number
```

```
//          parameter buffer[] is and array of char for result
// Uses the BIN_TO_ASC10 procedure in the Un32_1 module of
// the MATH32 library
_asm
    {
        MOV       EAX,value
        MOV       EDI,buffer
        CALL      BIN_TO_ASC10
    }
  return;
}

void IntToAsc16(int value, char *buffer)
{
// Function to convert a 16-bit binary number into 4 ASCII
// hexadecimal digits
// On entry:
//          parameter value holds binary number
//          parameter buffer[] is and array of char for result
// Uses the BIN_TO_ASC16 procedure in the Un32_1 module of
// the MATH32 library
_asm
    {
        MOV       EAX,value
        MOV       EDI,buffer
        CALL      BIN_TO_ASC16
    }
return;
}

int AscToInt(char *buffer)
// Convert an ASCII string to binary
// On entry:
//          parameter buffer[] is an array of char holding
//          10-digit ASCII decimal string in the range
//          0 to 4,294,967,295
// On exit:
//          returns binary value in an int type
// Uses the ASCII_TO_EDX procedure in the Un32_1 module of
// the MATH32 library
{
int aVal = 0;
_asm
    {
        MOV       ESI,buffer
        CALL      ASCII_TO_EDX
        MOV       aVal,EDX
    }
return aVal;
}
```

The Assembly Language source file is named UN32_1.ASM. A partial listing of the code follows:

```
;******************************************************************
;******************************************************************
;                          UN32_1.ASM
;******************************************************************
;******************************************************************
```

```
; Library of math routines from the book
;
; Creation date:  January 9, 2001
; Updated: March 31/2001
; Libary name: MATH32.LIB
; Reference: Chapter 1
; Module name: UN32_1.ASM
; Description:
;         Procedures for number system conversions.
;
;******************************************************************
;                    About the MATH32 library
;******************************************************************
; Environment:
;     All modules in this library were assembled with MASM Version
;     6.11 and used in console applications, under Visual C++ 6.0.
;     32-bit code cannot be used in DOS.
; Procedure names:
;     Procedures referenced by C++ code have names that start with
;     the _ symbol. Local support procedure names do not use the
;     _ symbol as the first character.
;
;******************************************************************
;                         publics
;******************************************************************
PUBLIC _BIN_TO_ASC10   ; .. 16-bit binary into 5-digit ASCII decimal
PUBLIC _BIN_TO_ASC16   ; .. 16-bit binary into 4-digit ASCII hex
PUBLIC _ASCII_TO_DX    ; .. 4-digit ASCII decimal to binary in DX
PUBLIC _BINF_TO_DECF   ; .. 8-bit binary to 8-digit decimal fraction
PUBLIC _DECODE_SINGLE  ; .. Isolate elements of single precision
PUBLIC _BIN_TO_BCD     ; .. 16-bit binary to 5 BCD digits
PUBLIC _BCD_TO_ASCII   ; .. 5 unpacked BCD to 5 ASCII digits
PUBLIC _BCD_INTO_REG   ; .. 5-digit BCD into a binary in DX
PUBLIC _BCD12_TO_ASCSTR;.. BCD12 number in ASCII string
PUBLIC _ASCII_TO_BCD12 ;.. ASCII decimal string into BCD12 format
PUBLIC _ASCII_TO_EDX   ; .. ASCII to binary in EDX register
;******************************************************************
;            definitions for 32-bit flat model
;******************************************************************

          .486
        .MODEL flat


        .DATA
;***************************************
;      data for this module
;***************************************
; BCD conversion factors
BCD_FACTORS     DB      50H,00,00,00    ; .50000000
                DB      25H,00,00,00    ; .25000000
                DB      12H,50H,00,00   ; .12500000
                DB      06H,25H,00,00   ; .06250000
                DB      03H,12H,50H,00  ; .03125000
                DB      01H,56H,25H,00  ; .01562500
                DB      00H,78H,12H,50H ; .00781250
                DB      00H,39H,06H,25H ; .00390625


; Code segment temporary data
ASC_TEMP_BUF    DB      5 DUP (20H)
                DB      0               ; Terminator
```

```
OFFSET_ASC           DD        0          ; Entry offset of ASCII buffer
OFFSET_BCD           DD        0          ; Entry offset of BCD buffer

OFFSET_ASCBUF        DD        0          ; Entry offset of ASCII buffer
BCD_BUF              DB        5 DUP (0)

;********************************************************************
;                               code
;********************************************************************
            .CODE

_BIN_TO_ASC10        PROC
; Procedure to convert a 16-bit binary number into 5 ASCII decimal
; digits
; On entry:
;           AX = binary number
;           EDI -> 5 byte ASCII buffer
; On exit:
;           Buffer holds ASCII decimal digits
;
;**********************|
;   clear ASCII buffer |
;**********************|
            MOV       ECX,5                ; Five digits to clear
CLEAR_5:
            MOV       BYTE PTR [EDI],20H        ; Clear digit
            INC       EDI                  ; Bump pointer
            LOOP      CLEAR_5              ; Repeat for 5 digits
            DEC       EDI                  ; Adjust buffer pointer to last
                                           ; digit
            MOV       ECX,10               ; Decimal divisor to CX
;**********************|
;   obtain ASCII decimal |
;         digit        |
;**********************|
GET_ASC10:
            MOV       DX,0                 ; Clear for word division
            DIV       CX                   ; Perform division AX/CX
; Quotient is in AX and remainder in DL
; Convert decimal to ASCII decimal
            ADD       DL,30H               ; Add 30H to bring to ASCII range
;**********************|
;     store digit      |
;   bump buffer pointer |
;**********************|
            MOV       BYTE PTR [EDI],DL         ; Store digit in buffer
            DEC       EDI                  ; Buffer pointer to next digit
; Note: the binary quotient is left in AX by the DIV CX instruction
;**********************|
;   is quotient = 0 ?  |
;**********************|
            CMP       AX,0                 ; Test for end of binary
            JNZ       GET_ASC10            ; Continue if not 0
;**********************|
;   end of conversion  |
;**********************|
            CLD
            RET
_BIN_TO_ASC10        ENDP
```

```
;
;
_BIN_TO_ASC16     PROC    USES edi esi ebx ebp
; Procedure to convert a 16-bit binary number into 4 ASCII
; hexadecimal digits
; On entry:
;          AX = binary number
;          EDI —> 4 byte ASCII buffer
; On exit:
;          Buffer holds ASCII decimal digits

; CPU compatibilty chart:
;                          | 80386           | YES |
;                          | 80486/PENTIUM   | YES |
;
. . . and so on
```

SOFTWARE ON-LINE

The example mentioned in this chapter is contained in the TestUn32_1 project located in the Chapter 26 folder in the book's on-line software.

Appendix B

Accuracy of Exponential Functions

Summary

In this Appendix we compare the accuracy and performance of the three core methods for calculating exponential functions mentioned in the text. Although rigorous algorithm analysis is beyond the scope of this book, we have attempted some empirical evaluations of the various methods. The implementation of the exponent factoring algorithm used for the tests for the case 10^y is the procedure named TEN_TO_EXP. This procedure is contained in the Un32_4 module of the MATH32 library. The implementations of the binary powering and the log approximation algorithms for the 10^y case were based on the procedures for obtaining x^y listed in Chapter 8.

AB.0 Accuracy Calculations

Average performance for the various algorithms was determined by timing the calculation of all possible 4932 values of the exponent of 10^y. The results of this test are shown in Table AB.1.

Table AB.1
Average Execution Time in the Calculation of 10^y and x^y

CASE	LOGARITHMIC APPROXIMATION	EXPONENT FACTORING	BINARY POWER
10^y	1	1.08	0.80
x^y	1	0.43	1.28

The performance tests in Table AB.1 allow us to conclude that the best overall performance is by the exponent factoring algorithm, which was slightly better than by log approximation.

Accuracy tests, which were also empirical, were based on the assumption that the accumulated error increases with the number of arithmetic operations. Therefore calculation errors were examined for the highest values of the exponent within the limit of the testing environment (10^{4932}). The reference point for the accuracy tests were memory constants defined to machine epsilon precision. Table AB.2 shows the results of this accuracy test.

Table AB.2

Accuracy Tests for the Calculation of 10^y

EXPONENT	CONSTANT TO E_{MACH}	EXPONENT FACTORING	BINARY POWER	LOG APPROXIMATION
4926	D11CCH	D11CCH 0	D11E0H 5	CFBC3 12
4927	82B1FH	82B1FH 0	82B2CH 4	82501 10
4928	A35E7H	A35E8H 1	A35CCH 5	A37CEH 8
4929	CC361H	CC361H 0	CC33FH 6	CAB86H 12
4930	FFA1DH	FFA1DH 0	FFA07H 5	FF2ACH 10
4931	BF8A4H	BF8A5H 1	BF889H 5	BF8ABH 3
4932	EF6CDH	EF6CDH 0	EF6AAH 6	EDD3AH 12
4922	96D32H	96D32H 0	96D3FH 4	95317H 12

Note:
  Values are 5 low-order hex digits of result/highest binary digit level of error propagation.

The entries in Table AB.2 are the 20 low-order bits (5 hexadecimal digits) of the result over the highest binary level of error propagation. A value of 0 in the error propagation field indicates that no error was detected, compared to the value of a constant defined to machine epsilon. A value of i indicates that the error propagated to the ith binary digit, counted low-to-high, the lowest order binary digit being number 1.

We also compared the results of calculating 10^y for all values of y in the range of the FPU temporary real format (1 to 4932) using log approximation and exponent factoring. The comparison showed that the largest absolute difference of the calculation of 10^y was for an exponent of 4922. In this case the level of error propagation by the logarithmic method was the 12th binary digit. The result of this accuracy test for the exponent value 4922 is shown in the last row of Table AB.2.

From the test in Table AB.2 we conclude that exponent factoring showed the highest accuracy for the values tested, which never differed by more than one binary digit from $å_{mach}$ precision. Recall that the implementation of exponent factoring used in the test was the TEN_TO_EXP procedure that is contained in the

Un32_4 module of MATH32 library. Since, in this case, the result is obtained by performing a maximum of 4 multiplication operations on values defined at machine epsilon precision, a higher accuracy result was to be expected. We also tested other variants of the exponent factoring algorithm (not shown in Table AB.2) and found that the error never propagated beyond the 3rd binary digit. The binary power algorithm showed errors that propagated up to the 6th binary digit. The highest error was in the log approximation, which propagated to the 12th binary digit in 4 of the 8 cases tested.

The tests for the x^y algorithms were based on the procedures named X_TO_Y_BYFAC, X_TO_Y_BYBP, and X_TO_Y_BYLOG discussed in Chapter 8. Table AB.3 shows the absolute and relative size of the various implementations.

Table AB.3
Code Image Cost of x^y Algorithms

	LOG APPROXIMATION	EXPONENT FACTORS	BINARY POWER
absolute	214	519	179
relative	1	2.43	0.84

Average cost in execution time for the various implementations of the x^y algorithms was determined by timing the calculation of all possible 4932 values of the exponent for the FPU temporary format. The results of this test are shown in Table AB.3. Conclusion of the execution time tests for the x^y case is that the lowest execution time was by binary factoring.

We also compared the results of calculating x^y for all values of y in the range of the FPU temporary real format (1 to 4932) using logarithmic approximation and the exponent factoring. The comparison showed that the largest absolute difference of the calculation of x^y was for an exponent of 4923. In this case the level of error propagation for the log approximation was the 12th binary digit. The results of the accuracy tests for this y^{4923} are shown in the last row of Table AB.4.

The conditions and assumptions adopted for the accuracy tests of the x^y case were similar to those mentioned in relation to the 10^y case. Table AB.4 shows the results of this accuracy test for the x^y case. The conclusion of the tests performed is that performance and accuracy in the calculation of exponential functions can be considerably improved by exponent factoring and other methods (such as binary powering) that avoid the use of logarithms. In the case of C^y the accuracy obtained by exponent factoring is several orders of magnitude better than that obtained by logarithmic approximation and by binary powering. Cost of the exponent factorization algorithms in code image size and execution time can be considered acceptable for many applications. In the case of x^y the advantages of exponent factorization over binary powering are less evident, although the results are more accurate by exponent factoring. In both cases, C^y and x^y where y is an integer, there seems to be no justification in code size cost, performance, or accuracy for the use of logarithmic approximations. Mixed methods can be used to improve accuracy when y is a rational or mixed number.

Table AB.4

Accuracy Tests of the Calculation of xʸ

EXPONENT	CONSTANT TO E_MACH	EXPONENT FACTORING	BINARY POWER	LOG APPROXIMATION
4926	D11CCH	D11B8H 5	D11E0H 5	D22DCH 12
4927	82B1FH	82B13H 4	82B2CH 4	82500H 10
4928	A35E7H	A35D8H 4	A35CCH 5	A37CFH 8
4929	CC361H	CC34DH 5	CC33FH 6	CD1B5H 8
4930	FFA1DH	FFA11H 4	FFA07H 5	FF2ACH 10
4931	BF8A4H	BF895H 4	BF889H 5	BF8ACH 4
4932	EF6CDH	EF6BAH 5	EF6AAH 6	F0283H 11
4923	3C87FH	3C86AH 5	3C88FH 5	3B463H 12

Note:
Values are 5 low-order hex digits of restult/highest binary
digit level of error propagation.

Notice that the calculations performed in the exponent factoring algorithms take place in data-independent stages. This makes the method suitable for execution in a multi-processor system, such as the ones described by Akl (1989, 3-28-29). In the case of a parallel processing environment, algorithm performance would be three to four times that of a single processor environment. On the other hand, logarithmic and binary factoring methods have no inherent parallelism.

Appendix C

C++ Indirection

Summary

In this appendix we discuss the topic of indirection in general. Its purpose is to serve as a review and to focus the reader's attention on certain aspects of indirection in C and C++ that make it required for the complete integral use of the language.

AC.0 Indirection in C++

The extensive use of addresses and indirection is one of the most unique and powerful features of C, and of its descendant language C++. Although several imperative languages implement pointer variables, as is the case with Pascal, Ada, Fortran 90, and PL/I, few do it with comparable flexibility and power. The uniqueness of C and C++ in regards to pointers relates mainly to how the languages handle pointer arithmetic, as well as to some unique features, such as pointers to void, pointers to functions, pointers to objects, and pointers to pointers.

In C the use of pointers includes accessing variables and arrays, passing strings, implementing data structures such as stacks and linked lists, and facilitating heap management. In C++ the use of pointers is imperative in inheritance and run-time polymorphism. For this reason the C++ programmer needs much more than a superficial understanding of pointers and indirection. Only with thorough understanding of pointers will the C++ programmer be able to take advantage of all the power and flexibility of object orientation.

AC.0.1 Pointer Phobia

Most beginning programmers agree that there is something intimidating about pointers. Those of us who teach programming can attest to the difficulty with which the notion of indirection is grasped by beginning students of high-level languages. On the other hand, students of assembly language usually master the same notions with little

833

effort. The reasons why similar concepts are easily learned in one environment, and
the cause of great consternation in another one, perhaps relates to how indirection
is implemented in code. The fact that low-level indirection is not difficult to learn
has led Dr. Robert Gammill of North Dakota State University to propose teaching C
and assembly together.

In x86 assembly language indirection is achieved in code by using bracket symbols to enclose the machine register that serves as a pointer. The student quickly
learns to associate brackets with pointers. The fact that brackets are used for
nothing else and pointers can be implemented in no other manner facilitates the
learning process. In addition, the programming terminology and the instruction
mnemonics are self-explanatory and consistent. In this manner the LEA instruction (Load Effective Address) can be used to initialize a pointer register. The mnemonics indicate that the register has been loaded with an address. Later on, the
indirection symbols [] show that the code is using the register as a pointer. The
following code fragment serves as an example:

```
DATA    SEGMENT
; A memory area named BUFFER is reserved and initialized with a text
; string, terminated in NULL
BUFFER1      DB       'This is a test',0
; A second buffer is created and left uninitialized
BUFFER2      DB     20 (DUP '?')

DATA    ENDS
;
CODE    SEGMENT
       .
       .
       .


; The SI register is first setup as a pointer to the memory area
; named BUFFER1
        LEA    SI,BUFFER1    ; SI holds the address of the first
                             ; memory cell in the area named BUFFER1
        LEA    DI,BUFFER2    ; A second pointer register is set up
                             ; to the area named BUFFER2
MOVE_STRING:
        MOV    AL,[SI]       ; The pointer register SI is used indirectly
                             ; to access the first byte stored in
                             ; BUFFER1, which goes into the AL register
; Assert: AL holds a character or a NULL
        MOV    [DI],AL       ; The character is stored in BUFFER2
        CMP    AL,0          ; AL is tested for the string terminator
        JE     END_OF_MOVE   ; Go if AL holds the string terminator
        INC    SI            ; First pointer is bumped to next character
        INC    DI            ; and so is the second pointer
        JMP    MOVE_STRING   ; Execution returns to the loop start
; When execution reaches the END_OF_MOVE label the string in BUFFER1
; has been copied into BUFFER2
END_OF_MOVE:
       .
       .
       .
```

The following C++ program uses pointers to transfer a string from one buffer into another one. The processing is similar to the one performed by the assembly language code previously listed.

```cpp
//********************************************************************
// C++ program to illustrate data transfer using pointers
// Filename: SAM06-01.CPP

//********************************************************************
#include <iostream.h>

main(){

  char buffer1[] = "This is a test";   // buffer1 is initialized
  char buffer2[20];                     // buffer2 is reserved
  char* buf1_ptr;                       // One pointer variable
  char* buf2_ptr;                       // A second pointer variable

// Set up pointers to buffer1 and buffer2. Note that since array
// names are pointer constants, we can equate the pointer variables
// to the array name. However, we cannot say: buf1_ptr = &buffer1;
  buf1_ptr = buffer1;
  buf2_ptr = buffer2;

// Proceed to copy buffer1 into buffer2
  while (*buf1_ptr) {
    *buf2_ptr = *buf1_ptr;       // Move character using pointers
    buf1_ptr++;                  // Bump pointer to buffer1
    buf2_ptr++;                  // Bump pointer to buffer2
    }
  *buf2_ptr = NULL;              // Place string terminator

// Display both buffers to check program operation
  cout < "\n\n\n"
       < buffer1 < "\n"
       < buffer2 < "\n\n";
  return 0;
}
```

The C code has several peculiarities, for example, the statements

```cpp
char* buf1_ptr;
char* buf2_ptr;
```

declare that buf1_prt and buf2_ptr are pointer variables to variables of type char. If the statements had been:

```cpp
char *buf1_ptr;
char *buf2_ptr;
```

the results would have been identical. Students are often confused by these syntax variations. Placing the asterisk close to the data type seems to emphasize that the pointer is a pointer to a type. However, if we were to initialize several variables simultaneously we would have to place an asterisk before each variable name, either in the form:

```
char* buf_ptr1, * buf_ptr2;
```

or in the form:

```
char *buf_ptr1, *buf_ptr2;
```

either syntax seems to favor the second style.

Once a pointer variable has been created, the next step is to initialize the pointer variables (in this case buf_ptr1 and buf_ptr2) to the address of the first byte of the data areas named buffer1 and buffer2. A student familiar with the use of the & operator to obtain the address of a variable would be tempted to code:

```
buf1_ptr = &buffer1;     // INVALID FOR ARRAYS
```

However, in C and C++ an array name is an address constant. Therefore, this expression is illegal for arrays, but legal and valid for any other data type. In the case of an array we must initialize the pointer variable with an expression such as:

```
buf1_ptr = buffer1;
```

We again overload the * symbol when we need to address the characters pointed to by the pointer variable. As is the case in the expressions:

```
while (*buf1_ptr) {
*buf2_ptr = *buf1_ptr;
buf1_ptr++;
buf2_ptr++;
}
*buf2_ptr = NULL;
```

The result is that even in the simplest cases, such as the one discussed, the treatment of pointers and indirection by C and C++ can be confusing and inconsistent. The fact that these concepts are much clearer in assembly language may tempt us to use assembler in order to explain indirection in C++. The problem with this approach is that those that are fluent in assembler are not likely to have problems in understanding indirection in C++, which means that using assembler to explain C pointers may be like preaching to the choir.

AC.1 Indirect Addressing

In the x86 family of Intel microprocessors indirect addressing is supported by the hardware. The assembly language code listed in Section AC.0.1 shows instructions to load the effective address of a memory operand (LEA), and the use of the [] symbols (sometimes called the indirection indicators) in accessing memory via a pointer register. The registers SI, DI, BX, and BP, as well as their 32-bit counterparts ESI, EDI, EBX, and EBP, can be used as pointers. In addition to indirect addressing of memory operands, the x86 CPUs support indirect jumps and calls that transfer execution to the address contained in any general register.

The x86 assemblers also support various refinements to indirection, such as based, indexed, and based-indexed addressing. In these cases the processor calculates the effective address using the contents of more than one register, as well as optional immediate data contained in the instruction opcode. In addition, the so-called string instructions implicitly use indirect addressing. In conclusion, x86 hardware supports indirect addressing quite well.

AC.2 Pointer Variables

In high-level languages a pointer is a variable used for storing a memory address or a special value, sometimes called "nil." The value nil indicates that the stored value is invalid and cannot be legally dereferenced. A pointer variable can be used to access the contents of the memory cell or cells whose address it contains. This way of accessing memory through an address stored in another memory location is called indirect addressing. Assembly languages rely heavily on indirect addressing for accessing memory operands. Figure AC.1 is a symbolic representation of indirect addressing by means of a pointer variable.

Figure AC.1 *Visualization of Indirect Addressing*

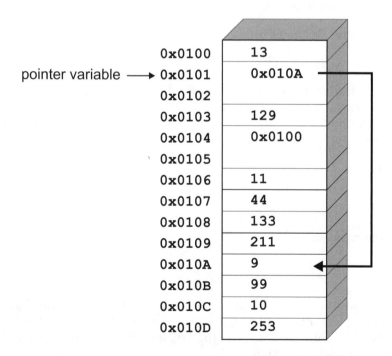

In Figure AC.1 the pointer variable holds the value 0x010A, the address at which the value 9 is stored. By using the pointer variable we can access this value stored at memory location 0x010A indirectly. However, in the x86 architecture, pointer variables have no special tag to identify them. Therefore, memory location 0x0104 in Figure AC.1, which holds the value 0x100, could be a pointer variable or not. The same can be said of any other memory location in the system's address space. It is up to the programmer and the compiler to keep track of pointer variables.

AC.2.1 Dangling Pointers

A dangling pointer is a pointer variable that contains an invalid or deallocated address. Dangling pointers are a danger because the referenced memory area can contain a constant, invalid data, or garbage. Since pointers can be used to direct program execution, a dangling pointer could be the cause of a crash, especially in programs that execute in real mode. In x86 architecture the addresses at the beginning of memory are reserved for the interrupt vector table. Therefore it would be possible, in the x86 machine environment, to assign a zero value to any un-initialized or invalid pointer and check for this value before using it. However, C++ provides no internal checks for dangling pointers, which are the programmer's responsibility.

AC.2.2 Pointers to Variables

One use of pointers is in indirectly accessing the contents of variables. Three different operations are available in C++: pointer declaration, pointer assignment or initialization, and pointer dereferencing.

Pointer Variable Declaration

The first step in implementing C++ indirection is usually declaring the variables that will hold pointers. C probably inherited the * as a pointer symbol from the syntax of some older assemblers. C++ is concerned with two elements of pointers: the address of the target variable and its data type. The address is necessary to locate it in the system's memory space. The data type is used in implementing pointer arithmetic, discussed later in this chapter.

The statement

```
int* ptr1;
```

or its alternative syntax

```
int *ptr1;
```

indicate that the variable ptr1 is a pointer to a variable of int type. Note that this does not refer to the pointer but to the type of the target variable. The pointer to void is a special case of a pointer that can refer to a target variable of any data type. Pointers to void are discussed in Section AC.6.

Pointer Variable Assignment

Once a pointer variable is declared, it can be initialized to the address of a target variable that corresponds to the one declared for the pointer. For example, if ptr1 is a pointer variable to the type int, and if ivar1 is an integer variable, then we assign the address of the target variable to the pointer variable with the statement:

```
ptr1 = &ivar1;
```

The ampersand symbol (&), which in this case corresponds to the address-of operator, extracts the address of the target variable and stores so it is in the pointer. One important exception is the address extraction of arrays. In C and C++ the array name represents the address of the first element of the array, which corresponds to the notion of an address constant. Therefore, if arr1 is an array of type int, we initialize the pointer variable ptr1 to the address of the array with the statement:

```
ptr1 = arr1;
```

In this case the use of the address of operator is not only unnecessary but illegal.

Pointer Variable Dereferencing

The process of accessing the value of the target variable by means of a pointer is called dereferencing. For example, dereferencing the pointer variable in Figure AC.1 results in the value 9, which is stored in the target variable. The asterisk symbol (*) is also used in dereferencing a pointer. In this case it precedes the name of the pointer variable and is sometimes called the indirection operator. For example, if prt1 is a pointer to the integer variable var1, which holds the value 22, then the following statement displays this value:

```
cout < *ptr1;
```

```
The following program shows the three operations previously described:
//*****************************************************************
// C++ program to illustrate the use of a pointer variable by means
// of its declaration, initialization, and dereferencing.
// Filename: SAM06 02.CPP
//*****************************************************************

#include <iostream.h>

main(){
// Declare pointer and target variable
  int* ptr1;              // Pointer to type int
  int var1 = 99;          // Variable of type int

// Initialize pointer to target variable
  ptr1 = &var1;

// Dereference of var1 using the pointer
  cout < "the value of var1 is " < *ptr1;
  return 0;
}
```

AC.3 Pointers to Arrays

An array name represents the address of the first element of the array. When we use the array name as an address we remove the brackets used to enclose the array index. Since an array name is an address constant, the address-of operator is not applicable. For example, in the program SAM06-01.CPP listed at the beginning of this chapter, we first created an array and the corresponding pointer variable:

```
char buffer1[] = "This is a test";   // buffer1 is initialized
char* buf1_ptr;                      // One pointer variable
```

then we initialized the pointer variable using the array name

```
buf1_ptr = buffer1;
```

A word of caution can be inserted at this point: because an array name is an address constant its value cannot be changed by code. In other words, it can be used to initialize a pointer but it cannot serve as a pointer itself. Therefore, note the following operations:

```
++buf1_ptr;              // legal
++buffer1;               // illegal
```

A final point to note is that access to array elements can also be achieved by means of the conventional subscript notation, although most compilers internally substitute subscript notation with pointers.

AC.4 Pointers to Structures

Some of the lexical elements used in structures, characteristic of C, have been reused in C++ in relation to classes and objects. For this reason it is useful to revisit structures and structure pointers before discussing classes and object pointers.

While all the elements of an array must belong to the same data type, the elements of a structure can be of different types. Accordingly, structures are useful in creating data constructs in which several, often different, data types, are related. A typical example of a structure is one used for storing employee data files, as in the following declaration:

```
struct employee_data {
  char emp_name[30];        // array for storing employee name
  char SS_num[12];          // array for SSN
  float wage;               // storage for hourly wage
  float hours_worked;       // storage for time worked
  unsigned int dependents;  // number of dependents
};
```

Once the structure has been declared we can create variable of the structure type. For example, we create instances of the structure employee_data with the following structure variable declarations:

```
struct employee_data welder_1;
struct employee_data welder_2;
struct employee_data shop_foreman;
```

Now we can access the structure members by means of the dot or member-of operator, as follows:

```
strcpy (welder_1.emp_name, "Joseph W. Smith);
strcpy (welder_1.SS_num, "255-47-0987");
welder_1.wage = 18.50;
welder_1.hours_worked = 45.6;
welder_1.dependents = 3;
```

Note that in initializing arrays within structures we used the string copy function strcpy(), defined in the string.h header, since aggregate array assignment is illegal in C and C++.

C's treatment of structure pointers is based on several unique concepts. One fundamental notion is that a structure is a class declaration which creates a template for instantiating structure variables, but is not a memory object in itself. Therefore, we define a pointer to a structure type, but initialize it to a structure variable, for example:

```
struct employee_data *ed_ptr;   // Pointer to type
ed_ptr = &welder_1;             // Pointer initialization
```

At this point we have created a pointer to the structure variable welder_1, of the structure type employee_data. We can now proceed to access the structure members by means of the pointer-member operator, which consists of a dash and greater-than symbol combined to simulate an arrow. Therefore we can say

```
float wage_of_w1 = ed_ptr -> wage;
```

Furthermore, we can pass the pointer of a local structure to a function, as shown in the following program:

```
//********************************************************************
// C++ program to illustrate the use of pointers to structures
// Filename: SAM06 03.CPP
//********************************************************************
#include <iostream.h>
#include <string.h>

// Structure type declaration
struct employee_data {
  char name[30];
  char SS_num[12];
  float wage;
  float hours;
  unsigned int dependents;
};

// Function prototype
void Show_employee(struct employee_data *);

//***************************
//          main()
//***************************
main(){

// Structure variable declaration
```

```
      struct employee_data welder_1;

// Initialize structure variables
    strcpy (welder_1.name,"Joseph W. Smith");
    strcpy (welder_1.SS_num, "263 98 0987");
    welder_1.wage = 18.5;
    welder_1.hours = 22.3;
    welder_1.dependents = 3;

// Pointer operations
    struct employee_data *edata_ptr;      // pointer type defined
    edata_ptr = &welder_1;                // pointer initialized to
                                          // variable

// Call function show_employee passing pointer
    Show_employee(edata_ptr);
    return 0;
}

//***************************
// function Show_employee()
//***************************
void Show_employee (struct employee_data *e_ptr) {
// Note that the function receives a copy of the pointer variable
// edata_ptr
    cout < "\n\nEmployee data: ";
    cout < "\n" < e_ptr  > name;
    cout < "\n" < e_ptr  > SS_num;
    cout < "\n" < e_ptr  > wage;
    cout < "\n" < e_ptr  > hours;
    return;
}
```

AC.5 Pointer Arithmetic

In handling pointers the assembly language programmer must take into account the stored data type. For example, if a machine register is set as a pointer to a memory area containing an array of characters, stored at the rate of one byte per element, then the programmer adds one to the pointer which accesses the successive elements in the array. However, if the array is of integers (two bytes per element), then the code must add two to the value of the pointer register in order to access the consecutive array entries. By the same token, we would add four to access doubleword entries, eight to access quadwords, and ten to access successive ten-bytes. This manipulation is, of course, machine dependent; in a machine in which integers are stored in 4 bytes, the code would have to add four during each successive increment in order to access adjacent values.

The machine dependency of low-level pointer handling compromises the portability of a high-level language and complicates programming. C and C++ automate the scaling of pointers in order to make the size of the stored elements transparent to the code. This automatic pointer scaling is sometimes called pointer arithmetic. Three types of operations are permitted in C and C++ pointer arithmetic: addition, subtraction, and comparison. Operations that are meaningless in pointer scaling are also illegal. Therefore multiplication, division, remainder, as well as all other mathematical functions, are prohibited.

Addition and subtraction of pointer variables are typically used in accessing successive array elements. In the program SAM06-01.CPP listed in Section AC.0.1, we bumped an array pointer with the statement:

```
buf1_ptr++;
```

In this case C++ adds 1 to each successive value of the pointer since the array is of type char. By the same token, subtraction operations can be used to access preceding entries in the array.

Comparison is a pointer operation with less obvious applications. In comparison we usually relate two or more pointers to the same data construct. For example, in a stack structure we can set a pointer to the stack top and compare it to the current stack position in order to detect an overflow. The following program shows pointer addition and comparison operations:

```
//***************************************************************
// C++ program to illustrate pointer arithmetic by adding all the
// elements in an int array. The end of the array is detected
// by means of a pointer comparison
// Filename: SAM06 04.CPP
//***************************************************************

#include <iostream.h>

main(){
// Variable declarations
  const int ENTRIES = 5;
  int num_array[ENTRIES] = {12, 8, 10, 5, 3};
  int total = 0;
  int *num_ptr;


  num_ptr = num_array;          // Initialize array pointer

// The following loop continues while the address of the pointer
// is less than the address of the end of the array
  while (num_ptr < num_array + ENTRIES)
    total = total + *num_ptr++;
  cout < "\nThe total is: " < total;

  return 0;
}
```

AC.6 Pointers to Void

C and C++ provide a mechanism for implementing a generic pointer that is independent of the object's data type. This mechanism is called a pointer to void. Most C programmers first encountered pointers to void in the context of memory allocation using malloc(). However, pointers to void are powerful tools that allow solving problems which are difficult to approach without them. In C++ some of the principal application of pointers to void are in accessing system memory areas, in passing addresses to functions that perform operations independently of the data type, in dynamic memory allocation, and in accessing system services.

When initializing a conventional pointer in C++, the programmer must make sure that the pointer type coincides with the target data type. For example:

```
double dbl_var = 22.334455;
int* i_ptr = &dbl_var;              // ILLEGAL EXPRESSION
```

The preceding code fragment would produce a compile-time error since we have declared a pointer to int and tried to initialize it with the address of a variable of type double. One possible solution is a type cast of the pointer to a double:

```
int* i_ptr;
(double*) i_ptr = &dbl_var;
```

Type casting pointers is allowed in C and C++. After the previous type cast the compiler treats i_ptr as a pointer to type double, and it is up to the programmer to ensure that all operations that follow the type cast are consistent and valid. An alternative approach is to define a pointer to type void, which can then be initialized with the address of any variable, array, or structure type. For example:

```
double dbl_var = 22.334455;
int_var = 12;

void* v_ptr;
v_ptr = &dbl_var;
  .
  .
v_ptr = &int_var;
```

A pointer to void is capable of holding the address of a variable of type int or of type double. In fact, a void pointer can hold the address of any variable, independently of its data type. One question that arises is how can C or C++ perform type checking and pointer arithmetic with void pointers? Another one is that if void pointers work, why do we need typed pointers? The answer to both questions is that void pointers have a limited functionality since they cannot be dereferenced. For example, if v_ptr is a pointer of type void, then the expression:

```
*v_ptr = 12;                 // ILLEGAL EXPRESSION
```

is illegal. Since C++ has no knowledge of the data type of the pointer it cannot use it to change the contents of a memory variable. However, like conventional pointers, void pointers can be type cast:

```
*(int*) v_ptr = 12;          // v_ptr a pointer to type void, pointing
                             // to a variable of type int
```

Although the preceding expression is valid, the fact that pointers to void must be type cast before dereferencing somewhat limits their practical use.

A historical note is in order. The first high-level language to implement pointer variables was PL/I. In PL/I the object of a pointer variable can be a static or dynamic variable of any type, therefore all PL/I pointers are equivalent to the notion of a C++ pointer to void. The result is a very flexible but dangerous construct which is the cause of several PL/I pointer problems, mainly related to type checking, dangling pointers, and lost objects. Language designers took into account the problems in PL/I and attempted to solve them by means of typed pointer variables, which simplify type checking and make code safer. Pointers to void are implemented for those cases in which a generic pointer is necessary.

AC.6.1 Programming with Pointers to Void

Pointers to void can be useful in many programming situations. For example, the malloc() function is used in C to dynamically allocate heap memory. The prototype for malloc is:

```
void *malloc(size_t size);
```

where size_t is defined in STDLIB.H as an unsigned integer and size is the number of memory bytes requested. Note that malloc() returns a pointer to void since the function need not know the data type to be used in formatting the allocated memory. The caller assigns the allocated memory to any valid unit of storage. The following program shows the use of malloc(), and its associated function free(), in dynamically allocating and deallocating an array.

```
//*************************************************************
// C++ program to illustrate void pointers and dynamic memory
// allocation using malloc() and free()
// Filename: SAM06 05.CPP
//*************************************************************

#include <iostream.h>
#include <string.h>
#include <alloc.h>

main() {
   char* str_array;                  // Declare pointer to char

// allocate memory for char array   Post error message and exit
// if malloc() fails
   str_array = (char *) malloc(20);
   if (str_array == NULL){
     cout < "\nNot enough memory to allocate array\n";
     return(1);
    }

   // copy "Hello World!" string into allocated array
   strcpy(str_array, "Hello World!");

   cout < str_array;                 // display array
   free(str_array);                  // free allocated memory
   return 0;
}
```

In the program SAM06-05.CPP, malloc() that returns a pointer to void. Our code type casts this pointer into a pointer to an array of char with the statement:

```
str_array = (char *) malloc(20);
```

Another frequent use for pointers to void is in passing pointers to functions that perform operations independently of data types. For example, a routine to perform a hexadecimal dump, such as is commonly found in disassemblers and debuggers, can receive the address of the data element in a pointer to void. This makes the code transparent to specific data types. The following program is an example of a hexadecimal dump function that uses a pointer to void.

```
//*************************************************************
// C++ program to illustrate use of pointers to void in a
// function that performs a hexadecimal dump
// Filename: SAM06 06.CPP
//*************************************************************

#include <iostream.h>

void Show_hex(void *p, int);        // Function prototype

//****************************
//          main()
//****************************
main(){
// Declare and initialize sample arrays and variables
   int int_array[] = {21, 33, 44, 55, 66};
   char char_array[] = {"abcdef"};
   float float_array[] = {22.7, 445.6, 111.0};
   long int long_var = 123456789;
   double dbl_var = 267.889911;

   cout < "\nHexadecimal dump: \n";

// Arrays and variables are passed to hex_dump() function
   Show_hex(int_array, sizeof(int_array));
   Show_hex(char_array, sizeof(char_array));
   Show_hex(float_array, sizeof(float_array));
   Show_hex(&long_var, sizeof(long_var));
   Show_hex(&dbl_var, sizeof(dbl_var));
   return 0;
}

//****************************
//    show_hex() function
//****************************
void Show_hex(void *void_ptr, int elements) {
unsigned char digit, high_nib, low_nib;

   for(int x = 0; x < elements; x++) {
      digit = *((char *)void_ptr + x);
      high_nib = low_nib = digit;          // copy in high and low nibble
      high_nib = high_nib & 0xf0;          // Mask out 4 low order bits
      high_nib = high_nib > 4;             // Shift high bits left
      low_nib = low_nib & 0x0f;            // Mask out 4 high order bits
```

```
// Display in ASCII hexadecimal digits
    if(low_nib > 9)
       low_nib = low_nib + 0x37;
    else
       low_nib = low_nib + 0x30;
// Same thing for high nibble
    if(high_nib > 9)
       high_nib = high_nib + 0x37;
    else
       high_nib = high_nib + 0x30;
    cout < high_nib < low_nib < " ";
}
  cout < "\n";
  return;
}
```

AC.7 Reference Variables

C++ introduced a new way of achieving indirection by means of a mechanism called a reference variable or a reference type. The syntax for creating a reference is as follows:

```
int var_1 = 456;
int& var_1A = var_1;
```

Now var_1A is a reference variable to var_1. From this point on we can access var_1 by means of var_1A at any time. This fact leads some authors to refer to a reference as an alias, which implies that it is merely another name for the same entity.

Several unique features of references make them useful to the programmer. In the first place, once created, the use of a reference variable does not require the dereference (*) or the address-off (&) operators. In addition, the compiler treats a reference variable as a constant pointer that cannot be reassigned, as is the case with array names. After a reference variable has been initialized its address value cannot be again accessed. For example, if var_1A is a reference variable, then

```
var1A = 26;
var_1A++;
```

The last statement increments the stored value, so that now var_1A holds the value 27. This notational convenience saves the programmer from having to continuously use the * operator to refer to the contents of a variable referenced by the pointer. The fact that reference variables cannot be reassigned serves as a safety device.

One frequent use of reference variables is in passing a scalar to functions so that the function has access to the original variable. The following program illustrates this operation in a rather trivial example:

```
//*************************************************************
// C++ program to illustrate use of reference variables
// Filename: SAM06 07.CPP
//*************************************************************
```

```
#include <iostream.h>

void Swap(int&, int&);                  // Prototype for function

//****************************
//         main()
//****************************
main(){
  int var_1 = 11;
  int var_2 = 22;

  Swap(var_1, var_2);

  // Testing function Swap()
  cout < "\nvar_1 now holds " < var_1;
  cout < "\nvar_2 now holds " < var_2 < "\n";
  return 0;
}
//****************************
//       Swap() function
//****************************
  void Swap(int& x, int& y) {
  // This function swaps the contents of two variables in
  // the caller's memory space
  int temp = x;
  x = y;
  y = temp;
  return;
}
```

Note that by using reference variables we can simplify the call to the function which appears identical to a call by value. The reference variables are declared in the header line of the function, as follows:

```
void Swap(int& x, int& y)
```

From this point on variable x can access the contents of variable var_1 and the variable y can access the contents of var_2. The fact that the call statement using reference variables does not show that an address is being passed to the function is sometimes mentioned as a disadvantage to their use.

AC.8 Dynamic Memory Allocation in C++

In Section AC.6 we described the use of malloc() and free() in implementing dynamic memory allocation in C. C++ provides more powerful and easier ways of handling dynamic allocation, although malloc() and free() are still available.

In C, program data can be of three storage classes: static, automatic, and dynamic. The storage class determines a variable's lifetime, which can be defined as the period of time during which an identifier holds a memory allocation. Static data refers to global variables and to those declared using the static keyword. These variables retain their value throughout program execution. Therefore, the storage space remains allocated until execution terminates. In more precise terms we can say that the lifetime of a static variable is the entire program. An automatic variable is one whose storage is allocated within its activation frame. The activation frame is either a block or a function. If no storage class specifier is given, automatic storage class is assumed for all local variables (those declared inside a function or block) and static is assumed for all global variables.

There is a difference between a global variable and a local one declared with the static keyword. Both have the same lifetime, but the local scope makes the variable invisible to other functions in the program. In some cases this restricted visibility can safeguard against undesirable changes. The visibility rules for identifiers are similar in C and C++. For example, an identifier can be local to a block:

```
if (var_1 > 10) {
   int input_var;
   cin > input_var;
   .
   .
   .
}
```

In this case the variable input_var is visible within the compound statement block of the if construct, but not outside of it. Because a function body is a syntactical block, the notion of block scope also applies to functions.

AC.8.1 Dynamic Data

The main objection to the use of static and automatic variables is that it often leads to wasted memory space. The worse case is with static variables, in which the memory storage persists throughout the program's execution time. In the case of static variables the storage assigned is never released or reused. The storage assigned to automatic variables is usually located in the stack, therefore it is reused during execution. But a program's control over automatic variables depends upon the activation frame of the function or block. This forces the programmer to declare all variables that could be needed at execution time, which also leads to wasted storage.

Suppose that we were designing the data structure to hold the text file of a word processor or editor. One possible option would be to make the text file an external array, visible to all code. However, how large would we make this array? If too small, the user could run out of storage space at execution time. If too large, the system may not find enough free memory to allocate at run time. Another example, this one mentioned by Headington and Riley (see Bibliography), is an array to hold seating assignments in an airline reservation system. If the largest airplane in the fleet had 400 seats, then we would have to define and allocate an array with 400 elements. The remainder of the allocated array would be wasted storage if only 40 of these seats were booked for a particular flight. At the same time, the program would have to be recoded if a new model of aircraft which contained 410 seats were added to the fleet. Problems such are these may lead to a design decision in which static and automatic data structures are assigned the largest possible memory space. The results are often the waste of a valuable resource.

Dynamic data provides a mechanism whereby storage is assigned as needed, at run time. Specific instructions are used to allocate and deallocate memory according to occasional requirements. In this case the array for storing seating assignments in the airline reservation system of the preceding paragraph would be allocated as needed. If only 40 seats were booked, the array would have 40 entries. If a new aircraft were put into service, the array could be configured to hold any number of items as long as there was storage available.

AC.8.2 The *new* and *delete* Operators

In C++ the unary operators new and delete can be used to control allocation and deallocation of dynamic data. These operators are C++ counterparts to malloc() and free(). The new operator returns a pointer to the allocated memory. If there is insufficient memory to satisfy the request then new returns NULL. The delete operator frees the memory allocated by new. Their general forms are as follows:

```
ptr_var = new type;
delete ptr_var;
```

where ptr_var is a pointer variable, and type is a C++ data type. The delete operator must refer to a valid pointer created with new. One of the differences between new and malloc() is that new automatically calculates the size of the operand. A second one is that new does not require an explicit type cast. Finally, both new and delete can be overloaded. Figure AC.2 is a visualization of the operation of the new and delete.

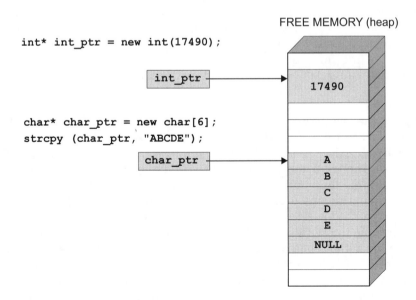

Figure AC.2 *Visualization of C++ Dynamic Allocation*

Another feature of new is that it allows the optional initialization of the allocated storage. For example:

```
int *ptr_var;
*ptr_var = new int (25);        // dynamic variable is initialized
```

When allocating arrays using the new operator the general format is as follows:

```
ptr_var = new array_type [array_size];
```

For example:

```
char *ptr_var;
ptr_var = new char[10];         // Allocate array to hold 10 elements
                                // of type char
```

To deallocate this array use the form:

```
delete []ptr_var;
```

Dynamically allocated arrays cannot be initialized.

The following program uses new and delete to dynamically allocate and deallocate memory for an array. Comparing it to SAM06-05.CPP, which performs the identical function using malloc() and free(), shows how the coding is simplified by using the new and delete operators.

```
//***************************************************************
// C++ program to illustrate dynamic memory allocation using
// new and delete
// Filename: SAM06 08.CPP
//***************************************************************

#include <iostream.h>
#include <string.h>

main() {

   char* str_array;                  // Declare pointer to char

// allocate memory for char array    Post error message and exit
// if new() fails
   str_array = new char(19);
   if (str_array == NULL){
      cout < "\nNot enough memory to allocate array\n";
      return(1);
    }
   // copy "Hello New World!" string into allocated array
   strcpy(str_array, "\nHello New World!\n");

   cout  < str_array;               // display array

   delete [] str_array;             // free allocated memory

   return 0;
}
```

AC.9 Pointers to Functions

In Section AC.1 we mentioned that the Intel x86 microprocessor family used in the PC supports indirect jumps and calls. This indirect access to code is achieved either through a register or memory operand, or through both simultaneously. One commonly used technique of indirect access to code is by means of a memory resident jump or call table, which holds a set of addressed to various processing routines. A offset value is added to the address of the start of the table to determine the destination for a particular jump or call. The following trivial program in x86 assembly language shows the use of a jump table in accessing one of four possible routines:

```
;*******************************************************************
;*******************************************************************
;                            IND_JMP.ASM
;*******************************************************************
;*******************************************************************
; Program description:
;     Test program for indirect jumps in x86 assembler language
;
;*******************************************************************
;                          stack segment
;*******************************************************************
STACK    SEGMENT stack
         DB       1024 DUP ('?')   ; Default stack is 1K
STACK    ENDS
;*******************************************************************
;                          data segment
;*******************************************************************
DATA    SEGMENT
;
; Address table is defined in memory
ADD_TABLE        DW       ACTION1, ACTION2, ACTION3, ACTION4
                 DW       0
;
;*************************|
;      text messages      |
;*************************|
USER_PROMPT      DB       'Enter a number from 1 to 4: $'
BAD_ACTION       DB       0AH,0DH,'Invalid input',0AH,0DH,'$'
;
AC1_MSG          DB       0AH,0DH,'Action 1 executed',0AH,0DH,'$'
AC2_MSG          DB       0AH,0DH,'Action 2 executed',0AH,0DH,'$'
AC3_MSG          DB       0AH,0DH,'Action 3 executed',0AH,0DH,'$'
AC4_MSG          DB       0AH,0DH,'Action 4 executed',0AH,0DH,'$'
;
DATA    ENDS
;*******************************************************************
;                          code segment
;*******************************************************************
CODE    SEGMENT
        ASSUME  CS:CODE
;******************|
; initialization   |
;******************|
ENTRY_POINT:
; Initialize the DATA segment so that the program can access the
; stored data items using the DS segment register
```

```
        MOV     AX,DATA         ; Address of DATA to AX
        MOV     DS,AX           ; and to DS
        ASSUME  DS:DATA         ; Assume directive so that
                                ; the assemble defaults to DS
;************************|
;  prompt user for input |
;************************|
        LEA     DX,USER_PROMPT  ; Pointer to message
        MOV     AH,9            ; MS DOS service number
        INT     21H             ; MS DOS interrupt
; Input character and echo
        MOV     AH,1            ; MS DOS service number
        INT     21H
; At this point AL holds ASCII of user input
; Test for invalid value
        CMP     AL,'4'          ; 4 is highest allowed
        JBE     OK_HIGH_RANGE   ; Go if in range
        JMP     ERROR_EXIT      ; Error if out of range
OK_HIGH_RANGE:
        CMP     AL,'1'          ; Test low range
        JAE     OK_RANGE        ; Go if in range
; Display error message and exit
ERROR_EXIT:
        LEA     DX,BAD_ACTION   ; Pointer to message
        MOV     AH,9            ; MS DOS service number
        INT     21H
        JMP     DOS_EXIT
; Convert to binary and scale
OK_RANGE:
        SUB     AL,30H          ; AL now holds binary
        SUB     AL,1            ; Reduce to range 0 to 3
        ADD     AL,AL           ; Double the value
        MOV     BX,0            ; Clear BX
        MOV     BL,AL           ; Offset to BX
        JMP     ADD_TABLE[BX]   ; Jump to table address plus offset
;*****************|
;   exit to DOS   |
;*****************|
DOS_EXIT:
        MOV     AH,76           ; MS DOS service request code
        MOV     AL,0            ; No error code returned
        INT     21H             ; TO DOS
;
;*********************************************************************
;                      ACTION PROCEDURES
;*********************************************************************
ACTION1:
        LEA     DX,AC1_MSG      ; Pointer to message
        MOV     AH,9            ; MS DOS service number
        INT     21H
        JMP     DOS_EXIT

ACTION2:
        LEA     DX,AC2_MSG      ; Pointer to message
        MOV     AH,9            ; MS DOS service number
        INT     21H
        JMP     DOS_EXIT

ACTION3:
        LEA     DX,AC3_MSG      ; Pointer to message
```

```
        MOV     AH,9                ; MS DOS service number
        INT     21H
        JMP     DOS_EXIT

ACTION4:
        LEA     DX,AC4_MSG          ; Pointer to message
        MOV     AH,9                ; MS DOS service number
        INT     21H
        JMP     DOS_EXIT

CODE    ENDS
        END     ENTRY_POINT         ; Reference to start label
```

The IND_JMP program operates as follows:

1. A jump table (named ADD_TABLE) is defined as data. At assembly time the table is filled with the offset of the four routines referenced by name. The corresponding routines are found toward the end of the code segment.

2. The user is prompted to enter a value in the range 1 to 4. After testing for a valid input, the code converts the ASCII character to binary, scales to the range 0 to 3, doubles its value so that it serves as an index to the word-size offsets in the jump table, and stores the result in the BX register.

3. An indirect jump instruction (JMP ADD_TABLE[BX]) accesses the address in the jump table that corresponds to the offset in the BX register. Thus, if BX = 0 the routines labeled ACTION1 executes, if BX = 2 then the routine named ACTION2 executes.

One point to note is that the IND_JMP program code performs a check for an input in the valid range. This check is advisable, since, in this type of execution, an invalid offset sends the program into a garbage address which produces an almost certain crash.

AC.9.1 Simple Dispatch Table

C++ implements code indirection by means of pointers to functions. Since a function address is its entry point, this address can be stored in a pointer and used to call the function. When these addresses are stored in an array of pointers, then the resulting structure is a call table similar to the one discussed in the preceding section. Jump and call tables are sometimes called dispatch tables by C and C++ programmers.

The implementation of pointers to functions and dispatch tables in C and C++ requires a special syntax. First, a pointer to a function has a type which corresponds to the data type returned by the function and is declared inside parentheses. For example, to declare a function pointer named fun_ptr, which receives two parameters of int type in the variables named x and y, and returns void, we would code:

```
void (*fun_ptr) (int x, int y);
```

The parentheses have the effect of binding to the function name not to its data type. If in this statement we were to remove the parentheses, the result would be a pointer to a function that returns type void. Note that the previous line creates a function pointer that is not yet initialized. This pointer can be set to point to any function that receives two int-type parameters and returns void. For example, if there was a function named Fun1 with these characteristics we could initialize the function pointer with the statement:

```
fun_ptr = Fun1;
```

Note that the C and C++ compiler assumes that a function name is a pointer to its entry point, thus the address of (&) operator is not used. Once the function pointer is initialized, we can access the function Fun1 with the statement:

```
(*Fun1)(6, 8);
```

In this case we are passing to the function the two integer parameters, in the conventional manner.

The declaration and initialization of a dispatch table also need a special syntax. Since a dispatch table should not be changed by code, it is a good idea to declare it of static type. For example, the following statement creates an array named dispatch that holds the addresses of the functions named Fun1, Fun2, Fun3, and Fun4, all of which receive no arguments and return void.

```
static void (*dispatch[]) (void) = {Fun1, Fun2, Fun3, Fun4};
```

To access a particular function whose address is stored in the dispatch table we must create a variable to store this offset. For example:

```
unsigned int off_var = 1;
```

Since off_var is now initialized to the second entry in the array (recall that array indexes are zero-based) we can access Fun2 with the statement:

```
(*dispatch[off_var]) ();
```

The following program shows the use of a dispatch table in C++ to achieve the same purpose as the assembly language program named IND_JMP listed in Section AC.9.

```
//****************************************************************
// C++ program to illustrate pointers to functions and the use
// of dispatch tables
// Filename: SAM06 09.CPP
//****************************************************************

#include <iostream.h>
#include <string.h>

// Function prototypes
void Action1(void);
void Action2(void);
```

```
void Action3(void);
void Action4(void);

//****************************
//          main()
//****************************
main() {
// Create dispatch table
static void (*add_table[]) (void) = {Action1, Action2, Action3, Action4};

// Create variable for offset into table
unsigned int table_offset = 0;

// Prompt user for input
   cout < "\nEnter a number from 1 to 4: ";
   cin > table_offset;
// Test for invalid user input
   if (table_offset < 1 || table_offset > 4) {
     cout < "\nInvalid input\n";
     return 1;
   }

   table_offset  ;                    // Adjust offset to range
   (*add_table[table_offset]) ();     // Access function using table
                                      // and offset entered by user

   return 0;
}

//****************************
//          functions
//****************************
void Action1(void) {
  cout < "\nAction 1 executed\n";
  return;
}

void Action2(void) {
  cout < "\nAction 2 executed\n";
  return;
}

void Action3(void) {
  cout < "\nAction 3 executed\n";
  return;
}

void Action4(void) {
  cout < "\nAction 4 executed\n";
  return;
}
```

AC.9.2 Indexed Dispatch Table

The program SAM06-09.CPP, previously listed, shows the simplest possible way of implementing a dispatch table. The processing can be modified so as to display a menu of options, from which the user selects the one to be executed. The code then manipulates the user input so that it serves as the offset into the jump table. However, there are several objections to this simple mechanism. Perhaps the most important one is the code assumes that the address of the various processing routines are stored in a single memory word. Therefore it uses an unsigned int as an index to the various entries. One possible risk of this approach is that it may be true in some memory models, machines, or compilers, but not in others.

A more elegant and rigorous solution, although slightly more complicated, is based on an indexed dispatch table. The indexed dispatch table approach allows ignoring the size of the stored addresses, since the correct one is found not by addition of an offset to a base, but by performing a table look-up operation. The table itself relates the input codes entered by the user (or other identification codes or strings) to the addresses of the various selectable functions. Program SAM06-10.CPP uses an indexed dispatch table to obtain the pointer to the corresponding function.

```cpp
//****************************************************************
// C++ program to illustrate pointers to functions and the use
// of indexed dispatch tables
// Filename: SAM06 10.CPP
//****************************************************************

#include <iostream.h>

// Prototypes
void Action1(void);
void Action2(void);
void Action3(void);
void Action4(void);

// Structure for creating an indexed dispatch table
struct index_cmd {
  int user_code;
  void (*function) (void);
};

//***************************
//          main()
//***************************
main() {

// Setup indexed dispatch table using the index_cmd structure
struct index_cmd add_table[] = {
  { 1,    Action1 },
  { 2,    Action2 },
  { 3,    Action3 },
  { 4,    Action4 }
};

// Create variable to store user input
int user_input = 0;

// Prompt user for input
   cout < "\nEnter a number from 1 to 4: ";
   cin > user_input;
// Test for invalid input
   if(user_input < 1 || user_input > 5){
     cout < "\ninvalid input\n" ;
```

```
      return 1;
   }

// Search lookup table for match
for (int i = 0; i < 5; i++) {
  if (user_input == add_table[i].user_code) {
     (*add_table[i].function) ();
     break;
     }
   }
   return 0;
}
//***************************
//        functions
//***************************
void Action1(void) {
  cout < "\nAction 1 executed\n";
  return;
}

void Action2(void) {
  cout < "\nAction 2 executed\n";
  return;
}

void Action3(void) {
  cout < "\nAction 3 executed\n";
  return;
}

void Action4(void) {
  cout < "\nAction 4 executed\n";
  return;
}
```

Program SAM06-10.CPP uses an integer value entered by the user to search into a structure that contains all four legal codes (1 to 4) matched with the addresses of the corresponding functions. The search is performed by comparing the value of the user input with the stored codes. Once the correct code is found, the function is called with the statement

```
(*add_table[i].function) ();
```

An alternative approach, based on a string, can be implemented using strcmp() to perform the compare operation.

AC.10 Compounding Indirection

C allows the creation of pointers to pointers, thus formulating a mechanism for compounding indirection. For example, first we create and initialize an integer variable:

```
in value1 = 10;
```

then we create and initialize a pointer to the variable value1:

```
int *int_ptr = &value1;
```

at this point we can create a pointer to the pointer variable int_ptr with the statement

```
int **ptr_ptr = &int_ptr;
```

Figure AC.3 is a representation of compounded indirection.

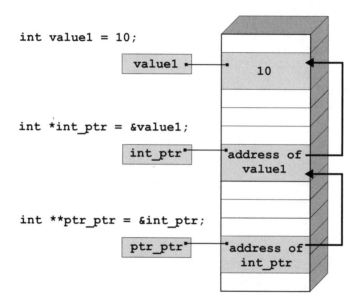

Figure AC.3 *Visualization of Compounded Indirection*

In Figure AC.3 the variable ptr_ptr points to the pointer int_ptr, which in turn points to the integer variable named value1. To access the variable value1 and change its value you can code:

```
value1 = 20;
```

or by simple indirection:

```
*int_ptr = 20;
```

also by compounded indirection:

```
**ptr_ptr = 20;
```

Appendix D

Multiple File Programs

Summary

This appendix is about the convenience of partitioning a program into several files or components. Applications soon grow to unmanageable dimensions and the partitioning into several files simplifies and helps organize development and maintenance.

AD.0 Partitioning a Program

There are many reasons for dividing a program into several files, the most obvious one, mentioned previously, is that the source code has grown to such size that it has become unmanageable. In this case partitioning into files is based on physical reasons or a storage and handling convenience. The resulting files are simple physical divisions of the program with no other logical or organizational base. However, even when the fundamental reason for dividing a program into files is a simple matter of size, a programmer or designer usually attempts to find some other more reasonable criteria by which to perform the actual division. For example, one option may be to place all functions and support routines into a separate file, while the main function and other central program components are stored in another one. Although this partitioning rationale is far from ideal, it is better than just arbitrarily slicing the code into two or more parts as if it were a loaf of bread.

In most cases the partitioning of a program's source code into several files is motivated by reasons other than its physical size, for instance, to facilitate development by several programmers or programmer groups. In this case program partitioning is usually done according to the modules defined during the analysis and design stages.

The fact that dividing into modules facilitates project organization, as well as the distribution of tasks in the coding phase, does not imply that modularization is nothing more than a project management tool. Modules are logical divisions of a program, and, in themselves, have a legitimate reason for being. The fundamental notion of a program module should be related to the program's conceptual design and not to an administrative convenience. Once the modules are established, placing each one of them in separate program files (or in a group of related files if the module's size so requires) is usually a good idea in itself since it promotes the preservation of the program's logical structure.

The linker programs that are furnished with all current versions of C++ for the PC support multiple source files. How C++ multifile programs are actually produced is discussed in Section AD.1.

AD.0.1 Class Libraries

Object code libraries are supported by most procedural languages. The fundamental notion is to provide a mechanism for code reuse that does not require scavenging or cut-and-paste operations. Utility programs, sometimes called library managers, allow the creation of program files that can be referenced at link time. The implementation of libraries is usually operating-system dependent. In the PC world, DOS libraries are in different format than Windows libraries and library code cannot be easily ported across the two environments. The library is usually organized into files or modules according to its logical design. Libraries for implementing GUI functions, graphics, and to provide special mathematical support such as statistics or vector analysis, have been available for many years.

The structure of a conventional DOS library for a non-object-oriented language usually consists of a collection of functions and associated data structures. Class libraries became possible after the advent of object-oriented programming languages. Since classes provide a mechanism for encapsulating data and operations, and because object orientation facilitates modeling the problems of the real world, class libraries are an effective way of providing a specific functionality. In this case the interface between the client code and the library itself is both flexible and clean. The one disadvantage of class libraries is that the client must be relatively fluent in object-oriented programming in order to access its services. This explains why, for many years, developers of Windows libraries have stayed away from object orientation. It is only recently that object-oriented development tools for Windows programming have gained prevalence.

AD.0.2 Public and Private Components

The components of class libraries are usually divided into two groups: public and private. The public elements of a class library are files and declarations necessary to document its interface requirements. In C++ the public components are typically furnished in header files that use the .h extension. Client code can then use the #include directive to physically incorporate the header file into its own source. In object-oriented programs the header files usually contain class declarations. These declarations make it possible for the client to instantiate the corresponding objects in order to access the class' methods, or to inherit from these classes. An interesting possibility is to use abstract classes to define the library's interface and then collect all the abstract classes in a header file.

The private elements of the declared classes usually contain program variables and constants, data structures, and member functions that need not be made visible to client code. Making these program elements private ensures two practical objectives:

- The client does not have to deal with information that is not necessary for using the class library.

- The confidentiality of the source code is protected.

According to this division into public and private elements, class libraries for the PC usually contain the following file types.

1. Header files. These are used to encode the public class elements that describe and document the interface and provide the necessary access hooks. These files can be collections of simple predeclarations or prototypes. They can contain actual implementation of the class that provides the library's interface, or they can contain a combination of both declarations and implementations. In C and C++ the header files have the extension .h. Assembly language header files for the PC are usually called include files, and have the extension .INC.

2. Object files. These contain executable code for the private elements of the library. These object files are usually referenced in the header files previously mentioned. Object files are in binary format, therefore the source code is not accessible by the client.

3. Library files. These are a collection of object files organized and formatted by a library management utility.

One advantage of libraries over individual objects files is that the client can deal with a single program element, instead of several individual components which often interact. Another one is that the library manager checks for consistency in all the component modules and detects any duplicate names or structural errors.

Whenever it is practical and logical to equate files and modules no other partition rationale could be preferable. However, practical considerations often force the designers to include several modules in the same file, or to distribute a single module into several files.

AD.0.3 Object-Oriented Class Libraries

Inheritance is a powerful mechanism for code reuse. Its core notion is that a useful type is often a variant of another type. Instead of producing the same code for each variant, a derived class inherits the functionality of a base class. The derived class can then be modified by adding new members, by modifying existing functions, or by changing access privileges. The idea that a taxonomic classification can be used to summarize a large body of knowledge is central to all science.

Both types of polymorphism (compile-time and run-time) promote code reuse. Compile time polymorphism, either by classification or by overloading of functions and operators, allows extending functionality without replacing exiting code. The name-mangling machinery in the compiler generates different internal names for functions that appear with the same external desgination, as long as the passed parameters are different. Run-time polymorphism requires a pointer to a base class which is redirected to a derived object that contains a virtual function. In this case the type of object determines the version of the function that is called. In the case of run-time polymorphism no variations in the parameter list are necessary; in fact, any variation in the parameter list generates an overloaded function and its virtual nature is lost.

In a programming language such as C++, a class library can take advantage of all the mechanisms and constructs of object orientation to create a product that is powerful and easy to access. By allowing class inheritance the library permits the client to modify or extend its functionalities without changing what has already been tested and tried. Furthermore, a library may use object composition to free the client from having to perform many elaborate or complicated tasks, and to hide other details of implementation, as shown in some of the patterns described in Chapter 12.

AD.1 Multifile Support in C++

In the reality of commercial programming all but the simplest applications require multiple files. All commercial C++ compilers and development platforms provide support for multifile projects. Some environments (such as the project managers in Borland C++ and Microsoft Visual C++) include development tools that automate the creation and maintenance of multifile projects. In these environments the programmer defines a list of sources and object files, as well as other structural constraints, and the project manager handles the compiling and linking operations that are necessary to produce executable code.

Alternatively, each source file can be compiled separately, then a linker can be called from the command line with a list of the object files and libraries that must be used to produce the executable. Or the files can be listed in a special text file that is read by the linker. In either case the result is a program that is composed of more than one source file. Figure AD.1 shows the various components of a C++ multifile program.

Multifile program architecture promotes reusability by locating the potentially reusable components in easily referenced individual program elements. Such is the case with header files, class, and object file libraries.

AD.2 Multilanguage Programming

So far we have considered a multifile program as one composed of several files coded in the same programming language. However, the use of more than one language in the creation of a program is a powerful development technique. When we refer to multilanguage programming we mean the mixing of two or more high- or low-level languages. Mixing high-level languages can be a way of reusing code developed in another programming environment. For example, we may have available a powerful mathematical library developed in FORTRAN that we wish to use in an application coded in C++.

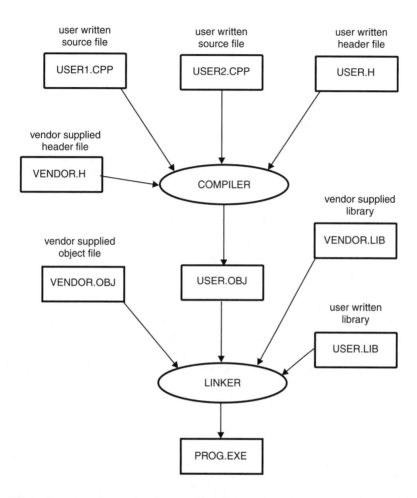

Figure AD.1 *Components of a C++ Multifile Program*

In many language implementations it is possible to access a procedure, function, or subroutine coded in another high level language. Thus, we may be able to create C++ code that calls the functions in a FORTRAN mathematical library. In most development environments the interface is implemented at the object file level, not at the source code level. Three issues must be considered in multilanguage programming: naming, calling, and parameter-passing conventions.

AD.2.1 Naming Conventions

This refers to how a compiler or assembler alters names as they are read from the source file into the object file, as well as to the number of upper- or lowercase characters and symbols that are recognized. For example, a much used FORTRAN compiler sold by Microsoft translates all lowercase characters to uppercase and admits at most six characters per name, while a recent C++ compiler recognizes thirty-two characters, which can be in upper- or lowercase, and appends an underscore symbol in front of the name.

Not taking into account the naming conventions of each language may lead to the linker reporting an unresolved external symbol. For example, the routine named Add in FORTRAN source code will appear as ADD in the object file. The same routine is labeled _Add in C++. Since ADD and _Add are not the same name, a link-time error takes place. Therefore, the multilanguage programmer often has to consider the calling convention of the participating languages and make the necessary adjustments. For example, an assembly language routine that is to be referenced as ADD in C++ code must be coded as _ADD in the assembly source, since the C++ compiler automatically appends the underscore to the name. Specific adjustments depend on the particular languages and implementations being interfaced.

AD.2.2 Calling Conventions

Calling conventions refer to the specific protocol used by a language in implementing a function, procedure, or subroutine call. In this sense we may often speak of the C, BASIC, or Pascal calling conventions.

The Microsoft languages BASIC, FORTRAN, and Pascal use the same calling convention. When a call is made in any of these languages, the parameters are pushed onto the stack in the order in which they appear in the source. On the other hand, the C and C++ calling convention is that the parameters are pushed onto the stack in reverse order of appearance in the source. The result is that in a BASIC CALL statement to a subroutine named Add(A, B) the parameter A is pushed on the stack first, and B is pushed second. However, in a C function Add(A, B) the reverse is true.

A language's calling convention determines how and when the stack is cleaned up. In the BASIC, FORTRAN, and Pascal calling conventions, the stack is restored by the called routine, which has the responsibility of removing all parameters before returning to the caller. The C calling convention, on the other hand, is that the calling routine performs the stack clean-up after the call returns.

An advantage of the BASIC, FORTRAN, and Pascal calling convention is that the generated code is more efficient. The C and C++ calling convention has the advantage that the parameters show the same relative position in the stack. Since, in the C convention, the first parameter is the last one pushed, it always appears at the top of the stack, the second one appears immediately after the first one, and so on. This allows the creation of calls with a variable number of parameters since the frame pointer register always holds the same relative address independently of the number of parameters stored in the stack.

Some programming languages allow changing the default calling convention to that of another language. For example, most C and C++ compilers allow using the Pascal calling convention whenever the programmer desires. In this case, the C compiler imitates the behavior of Pascal by pushing the parameters onto the stack in the same order in which they are declared and by performing a stack clean-up before the call returns. In C and C++ the calling convention keyword can be included in the function declaration.

AD.2.3 Parameter-Passing Conventions

In C and C++, parameters can be passed to a function by reference or by value. When a parameter is passed by reference, use the address of a scalar variable or a pointer variable. When passed by value, a copy of the variable's value is stored in the stack so that it can be removed by the called function. In passing by reference, it is the variable's address that is placed in the stack.

In the segmented memory architecture of systems based on the Intel x86 processors (such as the PC) we have two possible representations of the memory address of a parameter. If the parameter is located in the same segment, then its address is a word offset from the start of the segment. However, if it is located in a different segment, then the address must contain two elements: one identifies the segment within the machine's address space and the other one the offset within the segment. When a parameter is in the same segment we sometimes speak of a near reference. When in another segment the we say that it is a far reference. A third option is when the architecture is based on a flat memory space, in which case the address is a linear, 32-bit value which locates the element within a 4 gigabyte memory area.

For these reasons the calling and the called routines must agree on whether parameters are passed by reference or by value as well as on the memory model. If a parameter is passed by value, the called routine removes the parameter directly from the stack. If it is passed by reference, then the called routine obtains the address of the parameter from the stack and uses this address to dereference its value. Additionally, when making a call by reference in an environment with segmented memory architecture, such as a PC under DOS, the called routine must know if the address stored in the stack is a near or far reference and proceed accordingly.

Programming languages sometimes allow selecting the parameter passing convention. In C and C++, the memory model selected at compile time determines the default address type for functions and pointers. The language also allows the use of the near and far keywords as address type modifiers to force pointers and addresses to be of a type different from the default.

AD.2.4 Difficulties and Complications

One of the difficulties of mixed-language programming is that the interface between language components changes in the different language combinations and in products from different software vendors. Furthermore, different implementations or versions of the same language combinations may interface differently. Such is the case in the languages developed by Borland and Microsoft.

In an effort to facilitate mixed-language programming Borland and Microsoft have both published directives and conventions that attempt to simplify the interface requirements. The main difficulty is that these directives and conventions are not identical. Therefore, multilanguage code is often not portable between development environments from different vendors.

Another difficulty related to mixed-language programming is insufficient technical information. Although software manuals sometimes include a few paragraphs on how to interface program elements in different languages, the discussion often does not contain all the necessary data. For example, consider the rather simple problem of passing an array from Borland C++ to x86 assembly language. Borland documentation states that in C++, array variables are passed by reference. What the manual does not say is that if the array is declared public in the C++ code, then it is assigned space in the program's data area. Otherwise, the array is allocated dynamically in the stack. In both cases the calling routine passes the array by reference, as mentioned in the manual, but in one case the called code finds the segment base of the address in the SS segment register and in another case in DS.

From these observations we must conclude that multilanguage programming, although a powerful and versatile technique, usually compromises the portability of the code, even to other versions of the languages or development environment. A decision to use several languages in the development process should always consider this possible loss of portability.

AD.3 Mixing Low- and High-Level Languages

Although many mixtures of programming languages are possible, by far the most likely and useful one is that of a low- and a high-level language. For a C++ programmer working in the PC environment this means mixing x86 assembly language with the C++ code.

Objections to assembly language programming are usually related to difficulties in learning the language and with lower programmer productivity. Both points are debatable, but even if we were to grant their validity, there are still many other valid reasons for using assembly language in PC programming. The following list covers some of the more important ones:

1. Power. Assembly language allows access to all machine functions. The language itself places no limitations on the functionalities that can be achieved. In other words, every operation which the machine is capable of performing can be achieved in assembly language. It is valid to state that if it cannot be done in assembly language code, then it cannot be done at all.

2. Compactness. Programs coded in assembly language are more compact than those coded in any high-level language since the code is customized to contain only those instructions that are necessary for performing the task at hand. Compilers, on the other hand, must generate substantial amounts of additional code in order to accommodate the multiple alternatives and options offered by the language.

3. Speed. Programs coded in assembly language can be optimized to execute at the fastest possible speed allowed by the processor and the hardware. Here again, code generated by high-level languages has to be more complicated and intricate, which determines its slower speed of execution.

4. Control. In order to locate program defects, or to discover processing bottlenecks, assembly language code can be traced or single-stepped one machine instruction at a time. The fact that there is no software layer between the program and the hardware determines that all defects can be found and corrected. The only exceptions

are those bugs caused by defects in the hardware or in the microcode. Defects in high-level language programs are sometimes caused by the compiler generating incorrect code. These bugs are very difficult to locate and impossible to fix.

These advantages of assembly language code are summarized in four words: power, speed, compactness, and control. Whenever one of these four elements is at a premium, then the possible use of assembly language should be investigated.

AD.3.1 C++ to Assembly Interface

Of all high-level languages, C and C++ provide the easiest interface with assembly language routines. The reason is that C (and its descendant C++) call an assembly language procedure in the same manner as they would a C or C++ function.

C++ passes variables, parameters, and addresses to the assembly language routine on the stack. As previously mentioned, these values appear in opposite order as they are listed in the C++ code. A unique feature of the C++ language interface is that it automatically restores stack integrity. For this reason the assembly language routine can return with a simple RET or RETF instruction, without keeping count of the parameters that were pushed on the stack by the call. One consequence of this mode of operation is that assembly language routines can be designed to operate independently of the number of parameters passed by the C++ code, as long as the stack offset of those that are necessary to the routine can be determined.

Not all implementations and versions of C++ follow identical interface conventions. In the Microsoft versions of C and C++, the memory model adapted by the C program determines the segment structure of the assembly language source file. Table AD.1, on the following page, shows the segment structure of the various memory models used by the Microsoft C and C++ compilers.

To the programmer, Table AD.1 means that if the C++ source file was compiled using the small or compact model, then the assembly language procedure should be defined using the NEAR directive. On the other hand, if the C++ source was compiled using the medium or large models, then the assembly language procedure should be defined with the FAR directive.

AD.4 Sample Interface Programs

In this section we provide two examples of interfacing C++ and assembly language code. The first one is based on the use of C or C++ compilers by Microsoft or IBM, and the second one on Borland compilers. In the first example a parameter is passed by value to a routine coded in assembly language. In this case the assembly language code also changes a variable declared in the C++ source file. In the second example we use a Borland C or C++ compiler to pass the address of a structure to assembly language code. The assembly code modifies the structure variables in the C++ source.

In either case it is our intention to furnish the reader with a general example that illustrates the interface, but by no means are these examples to be taken as universally valid. When interfacing C++ and assembler language each implementation must be carefully analyzed for its interface requirements. The elements of naming, calling, and parameter-passing conventions must all be taken into account. Even different versions of the same language often differ from each other. In general, it is safer to use language development tools from the same vendor since it is reasonable to expect that these tools have been designed to interact with each other. In this sense when using a Microsoft C++ compiler, use a Microsoft assembler, such as MASM. The same applies to products from Borland or any other vendor.

Table AD.1

Segment Structures in Microsoft C++ Memory Models

MEMORY MODEL	SEGMENT TYPE	NAME	ALIGN TYPE	COMBINE TYPE	CLASS NAME	GROUP
Small	CODE	_TEXT	word	PUBLIC	'CODE'	
	DATA	_DATA	word	PUBLIC	'DATA'	DGROUP
	CONST		word	PUBLIC	'CONST'	DGROUP
		_BSS	word	PUBLIC	'BSS'	DGROUP
	STACK	STACK	para	STACK	'STACK'	DGROUP
Compact	CODE	_TEXT	word	PUBLIC	'CODE'	
	DATA	_DATA	word	PUBLIC	'DATA'	DGROUP
		CONST	word	PUBLIC	'CONST'	DGROUP
		_BSS	word	PUBLIC	'BSS'	DGROUP
		FAR_DATA	para	private	'FAR_DATA'	
		FAR_BSS	para	private	'FAR_BSS'	
	STACK	STACK	para	STACK	'STACK'	DGROUP
Medium	CODE	xx_TEXT	word	PUBLIC	'CODE'	
	DATA	_DATA	word	PUBLIC	'DATA'	DGROUP
		CONST	word	PUBLIC	'CONST'	DGROUP
		_BSS	word	PUBLIC	'BSS'	DGROUP
	STACK	STACK	para	STACK	'STACK'	DGROUP
Large	CODE	xx_TEXT	word	PUBLIC	'CODE'	
	DATA	_DATA	word	PUBLIC	'DATA'	DGROUP
		CONST	word	PUBLIC	'CONST'	DGROUP
		_BSS	word	PUBLIC	'BSS'	DGROUP
		FAR_DATA	para	private	'FAR_DATA'	
		FAR_BSS	para	private	'FAR_BSS'	
	STACK	STACK	para	STACK	'STACK'	DGROUP

Both the Microsoft and the Borland versions of C and C++ have an additional interface constraint: the SI and DI machine registers are used to store the values of variables defined with the "register" keyword. Therefore, the assembly language procedure should always preserve these registers. The Microsoft implementations also assume that the processor's direction flag (DF) is clear when C regains control. Therefore, if there is any possibility of this flag being changed during processing, the assembly language code should contain a CLD instruction before returning. Although the Borland manuals do not document this requirement, it is probably a good idea to clear the direction flag in any case even though it may not be strictly necessary. Even if a compiler does not use the direction flag, as appears to be the case with those from Borland, the CLD instruction produces no undesirable effects.

C and C++ compilers automatically add an underscore character in front of every public name. This is true for both the Microsoft and the Borland C and C++ compilers. For this reason, a C++ language reference to a function named RSHIFT is compiled so that it refers to one named _RSHIFT. In coding the assembly language program element it is necessary to take this into account and add the required underscore character in front of the name.

This convention applies not only to functions but also to data items that are globally visible in C++ code. For example, if the C++ source contains a variable named ODDNUM this variable is visible to an assembly language source under the name _ODDNUM. The assembly language code can gain access to the variable by using the EXTRN statement. Note the spelling in this case, which is different from the extern keyword of C and C++.

AD.4.1 Microsoft C-to-Assembly Interface

The first interface example uses code generated by a Microsoft C or C++ compiler and the MASM assembler. The example consists of a C++ program that calls an assembly language routine. The caller passes an integer variable to assembly language code and the assembly language returns the value with all bits shifted right by one position. In addition, the assembly language routine modifies the value of a C++ variable named NUMB2, which is defined globally.

The example is implemented in two source files. The C++ file (TESTC.CPP) was created using a standard editor and compiled using a Microsoft C++ compiler. The assembly language source was created with an ASCII text editor and assembled with a DOS MASM assembler. The assembled file is named CCALLS.OBJ. Note that the examples work with all the Microsoft fullfeatured C and C++ compilers, but not with the Microsoft QuickC programming environment.

The following is a listing for the C++ source file in this interface example.

```
/* File name: TESTC.CPP                                         */


/* C++ source to show interfacing with assembly language
   code. This sample assumes the use of Microsoft C or C++
   compilers, excluding QuickC */

#include <iostream.h>
extern int RSHIFT(int);
/* Variables visible to the assembly source must be declared
   globally in the C++ source */

int numb1, NUMB2;

main() {
      cout << "Enter the variable numb1: ";
      cin >> numb1;
      cout << "The shifted number is: %d\n"
          <<   RSHIFT(numb1));
      cout << "The variable NUMB2 has the value: %u\n"
          << NUMB2;
}
```

Note the following points in the C++ source file:

1. The assembly language source is declared external to the C code with the extern keyword. The name in this declaration is not preceded with an underscore character, since it is automatically appended by the compiler.

2. Once declared external, the RSHIFT routine is called as if it were a C++ function.

3. The variables numb1 and NUMB2 are declared to be of int type. Both are global in scope. The value of numb1 is transferred by the C++ code to the assembly language source file in the statement RSHIFT(numb1). The assembler code recovers the value of the variable numb1 from the stack. During execution the assembly language code accesses the variable NUMB2 directly, using its name. Note that the variable to be accessed by name is declared using capital letters. This is for compatibility with assembler programs that automatically convert all names to uppercase.

The assembly language code is as follows:

```
;******************************************************************
;                              CCALLS.ASM
;******************************************************************
; Example of C++ to assembly language interface for Microsoft
; C and C++ compilers, excluding QuickC
;
PUBLIC    _RSHIFT
EXTRN     _NUMB2: WORD

_TEXT     SEGMENT word    PUBLIC  'CODE'
          ASSUME  CS:_TEXT
;********************************|
;  C language interface routine  |
;********************************|
_RSHIFT         PROC    NEAR
; Function to rightshift the bits in an integer passed by
; rightshifted value returned to caller
        PUSH       BP              ; Save caller's BP
        MOV        BP,SP           ; Set pointer to stack frame
        MOV        AX,[BP+4]       ; Value passed by value
                                   ; which is C language default mode
        CALL       INTERNAL_P      ; Procedure to perform shift and
                                   ; to access C variable NUMB2
        CLD                        ; Clear direction flag
        POP        BP              ; Restore caller's BP
        RET                        ; Simple return. C repairs stack
_RSHIFT         ENDP
;
INTERNAL_P      PROC       NEAR
; value of arbitrary new value assigned to C variable
        SHR        AX,1            ; Shift right AX 1 bit
                                   ; Return value is left in AX
        MOV        CX,12345        ; Arbitrary value to CX
        MOV        _NUMB2,CX       ; and to C variable NUMB2
        RET
INTERNAL_P      ENDP
_TEXT      ENDS
        END
```

Note the following points of the assembly language source file:

1. The procedure _RSHIFT is declared PUBLIC to make it visible at link time. An underscore character is affixed at the start of the procedure's name so that it matches the name used by the C++ compiler.

2. The segment structure is compatible with the Microsoft small memory model as shown in Table AD.1. The memory model of the C program is determined at link time by the library selected. Also note that the code segment is defined with the statement:

```
_TEXT      SEGMENT word    PUBLIC  'CODE'
```

which are the same as those listed for the C and C++ small and compact models in Table AD.1.

3. Procedures to be linked with C++ programs of the small or compact model must be declared with the NEAR operator, as follows:

```
_RSHIFT      PROC    NEAR
```

If the C++ program used the medium, large, or huge models then the assembly language procedure would have been declared using the FAR operator.

4. The _RSHIFT procedure in the assembly language code can serve as a general interface template for Microsoft C++/assembly programming. However, note that this template requires modifications in order to adapt it to C++ programs that use other memory models or that pass or return parameters of other data types.

5. The structure of the stack frame at call time also depends on the adopted memory model. In this case, since the chosen model is small, the parameter passed is located at BP+4 and the return address consists of only the offset component.

6. The variable declared as NUMB2 and initialized as int in the C source is declared as EXTRN _NUMB2: WORD in the assembly language code. The underscore character is added in the assembly source file because C++ appends this character at the beginning of the variable name. The variable is assigned an arbitrary value in the INTERNAL_P procedure.

7. Upon exiting, the assembly language routine clears the direction flag (CLD), restores the BP register (POP BP), and returns with a simple RET instruction. Note that the processing routine in this example does not use the SI or DI registers; hence they were not preserved. However, if SI or DI is used in processing, the entry value must be saved and restored by the assembler code.

The actual configuration of the stack frame at call-time varies with the different memory models. One of the determining factors is that the return address requires two bytes in the small and compact models and four bytes in the other ones. Also, each parameter passed by value occupies a storage space equivalent to its type. Therefore, a char variable takes up one byte, an int takes up two bytes, and so forth. When parameters are passed by reference, the amount of space taken up by each parameter depends exclusively on whether it is of NEAR or FAR type. In the PC environment, FAR addresses require four bytes and NEAR address two bytes. Figure AD.2 shows a stack frame in which two parameters of int type were passed using the small or compact memory models (NEAR reference).

high addresses

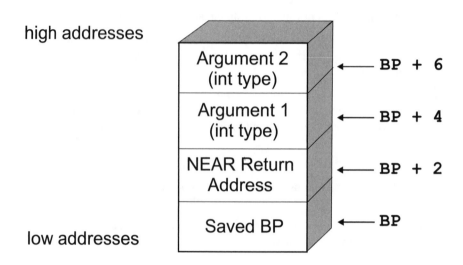

low addresses

Figure AD.2 *C++ Stack Frame Example*

Return Values in Microsoft C++ Compilers

In the Microsoft implementations of C++ for the PC, parameters are returned to the calling routine in the AX register, or in the AX and the DX registers. Table AD.2 shows the Microsoft language conventions for returning values.

Table AD.2

Microsoft C++ Returned Values

DATA TYPE OR POINTER	REGISTER OR REGISTER PAIR
Char short int unsigned char unsigned short unsigned int	AX
Long unsigned long	DX = high-order word AX = low-order word
struc or union float or double	AX = address value
near pointer	AX = offset value of address
far pointer	DX = segment value of address AX = offset value of address

Another way of interchanging values between the C++ and assembly language code is to access the C++ variables by name (as demonstrated in the previous example). This method requires that the assembly code use the same storage format as the C++ source. Since variable storage formats change in different implementations of C++, this access technique must be adapted to each particular implementation of the language.

AD.4.2 Borland C++ to Assembly Interface

The following example, coded in Borland C++ and in x86 assembly language, also implements a C++ to assembly language interface. The C++ code can be compiled using Borland C++ Version 3.1 and later. The assembler code is crated using either Borland's TASM or Microsoft MASM. The example consists of passing a structure to an assembly language source. The call is made using the FAR access type and the address of the structure is a FAR reference.

```
/* Program name: TEST1.CPP

   Other files:
      Project: ASMTEST.PRJ (in C:\BORLANDC)
      Assembly Source File: CCALLS.OBJ (in C:\BORLANDC)

*/

#include <iostream.h>
#include <string.h>

extern "C" {
      int SETSTR(struct pix_params *t);  /* Passing a structure by
      |       |                      |             reference
      |       |                      |_____ structure tag name
      |       |_____ Procedure name (_SETSTR in assembler
      |                                    source file)
      |_____ Return type
                           Procedure SETSTR sets new values in the
                           pix_params structure variable passed to
                           it by reference */
}

// STRUCTURE
{
    struct pix_params point1;       // Declare one structure variable

    point1.x_coord = 0x20;          // Initialize structure variables
    point1.y_coord = 0x40;
    point1.blue = 0x30;
    point1.green = 0x31;
    point1.red = 0x32;
    strcpy ( point1.text_string, "This is a test message" );

    SETSTR[&point1];

  return (0);
}
struct pix_params
{
  int x_coord;                      // point coordinate on x axis
```

```
  int y_coord;                    // point coordinate on y axis
  unsigned char blue;             // color components
  unsigned char green;
  unsigned char red;
  char text_string[80];
};

main()
     .
     .
     .
```

The following points should be observed regarding the C++ program:

1. Which sources are linked is determined by the files referenced in the project.

2. The names of the assembly procedures to be called from C++ are defined with the extern "C" directive.

3. Assembly procedures must take into account the memory model when accessing parameters in the stack. If the procedure is accessed with a NEAR call then the stack frame holds a 2 byte return address and the first parameter is located at BP+4. If the procedure is accessed with a FAR call, then the stack frame holds a 4 byte return address and the first parameter is located at BP+6. In the Borland compilers, the C++ model is determined by the setting of the Options/Compiler menu item. In this case the memory model is LARGE and the call is FAR.

4. C language code prefixes the function name with a leading underscore character (_) at call time. The assembler code must take this into account by inserting the corresponding _ symbol in front of the public name.

The following is the assembly language source that receives the C++ call:

```
;******************************************************************
;                          CCALLS.ASM
;******************************************************************
; Example C++ to assembly language interface routine for Borland
; compilers
;
;
PUBLIC  _SETSTR

;******************************************************************
;                          local code
;******************************************************************
_TEXT    SEGMENT word    PUBLIC  'CODE'
         ASSUME  CS:_TEXT
;*****************************|
;  C language interface routine  |
;*****************************|
_SETSTR           PROC    FAR
; Function to access a structure located in the C++ code and change
; the value of some of its variables
; On entry:
;          structure address passed by reference

         PUSH    BP                ; Save caller's BP
         MOV     BP,SP             ; Set pointer to stack frame
```

```
; The C++ register operator uses SI and DI for temporary storage
; therefore, it is usually a good idea to save these registers
        PUSH    SI                  ; Saved in stack
        PUSH    DI
; The FAR model locates the first word size parameter at +6
        MOV     BX,[BP+6]           ; Offset of address
        MOV     ES,[BP+8]           ; Segment of address
; At this point ES:BX holds address of structure in C++ source
; The following code accesses the variables within the C++ structure
; and modifies them
        MOV     WORD PTR ES:[BX],80         ; change x_coord
        MOV     WORD PTR ES:[BX+2],90       ; change y_coord
        MOV     BYTE PTR ES:[BX+4],0A0H     ; change red
        MOV     BYTE PTR ES:[BX+5],0B0H     ; change blue
        MOV     BYTE PTR ES:[BX+6],0C0H     ; change green
; C exit code
        POP     DI                  ; Restore DI and SI from stack
        POP     SI
        CLD                         ; Clear direction flag
        POP     BP                  ; Restore caller's BP
        MOV     AX,0                ; Return 0
        RET                         ; Simple return. C repairs stack
_SETSTR         ENDP
_TEXT   ENDS
        END
```

AD.4.3 Using Interface Headers

Note in the above examples that the assembly language code is designed to accommodate C++ calling conventions. Therefore, the procedure _SETSTR does not work with another high-level language. Furthermore, the C++ memory model determines where in the stack are the passed parameters placed. If the calling routine is compiled with a different memory model, the interface fails.

One way to accommodate varying calling conventions between languages or between implementations of the same language is by means of interface headers. These are usually short routines that receive the parameters passed by one language and locate them in the machine registers or data structures expected by the other language. This scheme allows portability while maintaining maximum coding efficiency in both languages. It also allows reusing code that is not designed to interface with a particular language. For example, an assembly language library of graphics primitives can be used from C++ by means of interface headers, even though the assembly source was not originally designed to be called from a high-level language. The one drawback of using interface header routines is some loss of performance, since each call has the added load of traversing the header code.

In C++ programming, interface header routines can be coded in line and placed in include files that are selected by means of preprocessor conditional directives. This simple technique allows choosing a different header routine according to the memory model of the C++ program, thus solving a major interface limitation of the language.

Appendix E

The MATH32 Library

Summary

This Appendix contains a listing of all the low-level procedures in the MATH32 library furnished in the book's on-line software package.

Table AE.1

Procedures in MATH32.LIB

PROCEDURE NAME	OPERATION
UN32_1 MODULE	
_BIN_TO_ASC10	16-bit binary into 5-digit ASCII decimal
_BIN_TO_ASC16	16-bit binary into 4-digit ASCII hex
_ASCII_TO_DX	4-digit ASCII decimal to binary in DX
_BINF_TO_DECF	8-bit binary to 8-digit decimal fraction
_DECODE_SINGLE	Isolate elements of single precision
_BIN_TO_BCD	16-bit binary to 5 BCD digits
_BCD_TO_ASCII	5 unpacked BCD to 5 ASCII digits
_BCD_INTO_REG	5-digit BCD into a binary in DX
_ASCII_TO_EDX	ASCII to binary in EDX register
BCD12 conversions:	
_BCD12_TO_ASCSTR	BCD12 number in ASCII string
_ASCII_TO_BCD12	ASCII decimal string into BCD12 format
BCD20 conversions:	
_BCD20_TO_ASCSTR	BCD20 number in ASCII string
_ASCII_TO_BCD20	ASCII decimal string into BCD20 format
UN32_2 MODULE	
_SIGN_DIV_BCD12	Signed BCD12 arithmetic
_SIGN_MUL_BCD12	
_SIGN_ADD_BCD12	
_SIGN_SUB_BCD12	
UN32_3 MODULE	
_SIGN_DIV_BCD20	Signed BCD20 arithmetic
_SIGN_MUL_BCD20	

(continues)

Table AE.1

Procedures in MATH32.LIB (continued)

PROCEDURE NAME	OPERATION
_SIGN_ADD_BCD20	
_SIGN_SUB_BCD20	
UN32_4 MODULE	
_ASCII_TO_EXP	ASCII decimal string to exponential form
_FPU_OUTPUT	ST(0) register to ASCII decimal string
_FPU_INPUT	ASCII exponential string to ST(0)
_GET_TAG	Get math unit register tag code
UN32_5 MODULE	
_TEN_TO_Y_BYFAC	10**y by exponent factoring
_X_TO_Y_BYFAC	x**y by exponent factoring
_X_TO_Y_BYBP	x**y by binary powering
_X_TO_Y_BYLOG	x**y by logarithms
_X_TO_Y_MIXED	x**y by logs and binary powering
_LOG_10	base 10 logarithm
_LOG_E	base e logarithm
_ALOG_10	base 10 antilogarithm
_ALOG_E	base e antilogarithm
_ALOG_2	base 2 antilogarithm
UN32_6 MODULE	
_DEG_TO_RAD	degrees-to-radian conversion
_RAD_TO_DEG	radians-to-degrees conversion
_CIRCLE_SCALE	scale angle to unit circle
_OCTANT_SCALE	scale angle to first octant
_TANGENT	tangent function
_SINE	sine
_COSINE	cosine
_COTANGENT	cotangent
_SECANT	secant
_COSECANT	cosecant
_ARC_SINE	arcsine function
_ARC_COSINE	arccosine
_ARC_TANGENT	arctangent
_SINH	hyperbolic sine
_COSH	hyperbolic cosine
_TANH	hyperbolic tangent
_ARC_SINH	hyperbolic arcsine
_ARC_COSH	hyperbolic arcosin
_ARC_TANH	hyperbolic arctangent
UN32_7 MODULE	
_FACTORIAL	x!
_ORDER	organize x > y
_MAX	find larger or two
_MIN	find smaller of two

(continues)

Table AE.1

Procedures in MATH32.LIB (continued)

PROCEDURE NAME	OPERATION
_MODULUS	remainder of x/y
_PYTHAGORAS	solve triangle using Pythagoras
_SOLVE_FOR_A	solve triangle using tangent
_INT_FRAC	separate integer and fractional parts
_QUAD_FORM_REAL	solve equation for real roots
_QUAD_FORMULA	solve equation for all roots
_ADD_CMPLX	add complex numbers
_SUB_CMPLX	subtract complex numbers
_MUL_CMPLX	multiply complex numbers
_DIV_CMPLX	divide complex numbers
_ABS_CMPLX	find absolute value of complex number
_POLAR_TO_CART	convert polar to Cartesian coordinates
_CART_TO_POLAR	convert Cartesian to polar coordinates
_PAYMENTS	periodic payments on a loan
_PRESENT_VALUE	present value of an investment
_FUTURE_VALUE	future value of an investment
_SL_DEPRECIATION	straight-line depreciation
_COMP_PERIODS	number of compounding periods
UN32_8 MODULE	
_PAYMENTS	periodic payments on a loan
_ANNUITY_PV	present value of an annuity
_ANNUITY_FV	future value of an investment
_ANNUITY_DUE	annuity due at start of period
_COMP_PERIODS	number of compounding periods
_SINKING_FUND	periodic payment on annuity
_SIMPLE_INTEREST	simple interest
_FUTURE_VALUE	future value at compound interest
_PRESENT_VALUE	present value at compound interest
_EFFECTIVE_YIELD	effective annual yield
_APR	annual Percentage Rate
_AOI_PAYMENT	add-on interest payment
_SL_DEPRECIATION	straight-line depreciation
UN32_9 MODULE	
_PERMUTATIONS	number of permutations
_COMBINATIONS	number of combinations
_BINOMIAL_PROB	binomial probability
_NORMAL_CURVE_Y	f(x) for normal dist. curve
_LOADE	load constant e
UN32_13 MODULE	
_ROW_TIMES_SCALAR	matrix row multiplied by matrix row
_ROW_PLUS_SCALAR	scalar added to matrix row
_ROW_DIV_SCALAR	matrix row divided by scalar
_ROW_MINUS_SCALAR	scalar subtracted from matrix row
_MAT_TIMES_SCALAR	matrix multiplied by scalar

(continues)

Table AE.1

Procedures in MATH32.LIB (continued)

PROCEDURE NAME	OPERATION
_MAT_DIV_SCALAR	matrix divided by scalar
_MAT_PLUS_SCALAR	scalar added to matrix
_MAT_MINUS_SCALAR	scalar subtracted from matrix
_ADD_MATRICES	matrix addition
_MUL_MATRICES	matrix multiplication
_GAUSS_SOLVE	solve linear system using Gauss-Jordan
UN32_14 MODULE	
_CALCULATE_Y	calculate value or dependent variable
_EVALUATE	evaluate non-parenthetical element of equation string

Appendix F

Windows Structures

Summary

This appendix contains the structures mentioned in the text. Structures are listed in alphabetical order.

```
BITMAP
    typedef struct tagBITMAP {   /* bm */
        int       bmType;
        int       bmWidth;
        int       bmHeight;
        int       bmWidthBytes;
        BYTE      bmPlanes;
        BYTE      bmBitsPixel;
        LPVOID    bmBits;
    };

BITMAPCOREHEADER
    typedef struct tagBITMAPCOREHEADER { // bmch
        DWORD     bcSize;
        WORD      bcWidth;
        WORD      bcHeight;
        WORD      bcPlanes;
        WORD      bcBitCount;
    } BITMAPCOREHEADER;

BITMAPCOREINFO
    typedef struct _BITMAPCOREINFO {     // bmci
        BITMAPCOREHEADER  bmciHeader;
        RGBTRIPLE         bmciColors[1];
    } BITMAPCOREINFO;

BITMAPFILEHEADER
    typedef struct tagBITMAPFILEHEADER { // bmfh
        WORD      bfType;
        DWORD     bfSize;
        WORD      bfReserved1;
        WORD      bfReserved2;
        DWORD     bfOffBits;
    } BITMAPFILEHEADER;
```

```
BITMAPINFO
    typedef struct tagBITMAPINFO { // bmi
        BITMAPINFOHEADER bmiHeader;
        RGBQUAD          bmiColors[1];
    } BITMAPINFO;

BITMAPINFOHEADER
    typedef struct tagBITMAPINFOHEADER{ // bmih
        DWORD   biSize;
        LONG    biWidth;
        LONG    biHeight;
        WORD    biPlanes;
        WORD    biBitCount
        DWORD   biCompression;
        DWORD   biSizeImage;
        LONG    biXPelsPerMeter;
        LONG    biYPelsPerMeter;
        DWORD   biClrUsed;
        DWORD   biClrImportant;
    } BITMAPINFOHEADER;

CHOOSECOLOR
    typedef struct {    // cc
        DWORD           lStructSize;
        HWND            hwndOwner;
        HWND            hInstance;
        COLORREF        rgbResult;
        COLORREF*       lpCustColors;
        DWORD           Flags;
        LPARAM          lCustData;
        LPCCHOOKPROC    lpfnHook;
        LPCTSTR         lpTemplateName;
    } CHOOSECOLOR;

COLORADJUSTMENT
    typedef struct  tagCOLORADJUSTMENT {     /* ca */
        WORD   caSize;
        WORD   caFlags;
        WORD   caIlluminantIndex;
        WORD   caRedGamma;
        WORD   caGreenGamma;
        WORD   caBlueGamma;
        WORD   caReferenceBlack;
        WORD   caReferenceWhite;
        SHORT  caContrast;
        SHORT  caBrightness;
        SHORT  caColorfulness;
        SHORT  caRedGreenTint;
    } COLORADJUSTMENT;

CREATESTRUCT
    typedef struct tagCREATESTRUCT { // cs
        LPVOID     lpCreateParams;
        HINSTANCE  hInstance;
        HMENU      hMenu;
        HWND       hwndParent;
        int        cy;
        int        cx;
        int        y;
        int        x;
```

```
        LONG        style;
        LPCTSTR     lpszName;
        LPCTSTR     lpszClass;
        DWORD       dwExStyle;
    } CREATESTRUCT;

DDBLTFX
    typedef struct _DDBLTFX{
        DWORD   dwSize;
        DWORD   dwDDFX;
        DWORD   dwROP;
        DWORD   dwDDROP;
        DWORD   dwRotationAngle;
        DWORD   dwZBufferOpCode;
        DWORD   dwZBufferLow;
        DWORD   dwZBufferHigh;
        DWORD   dwZBufferBaseDest;
        DWORD   dwZDestConstBitDepth;
    union
    {
            DWORD               dwZDestConst;
            LPDIRECTDRAWSURFACE lpDDSZBufferDest;
    };
        DWORD   dwZSrcConstBitDepth;
    union
    {
            DWORD               dwZSrcConst;
            LPDIRECTDRAWSURFACE lpDDSZBufferSrc;

    };
        DWORD   dwAlphaEdgeBlendBitDepth;
        DWORD   dwAlphaEdgeBlend;
        DWORD   dwReserved;
        DWORD   dwAlphaDestConstBitDepth;
    union
    {
            DWORD               dwAlphaDestConst;
            LPDIRECTDRAWSURFACE lpDDSAlphaDest;
    };
        DWORD   dwAlphaSrcConstBitDepth;
    union
    {
            DWORD               dwAlphaSrcConst;
            LPDIRECTDRAWSURFACE lpDDSAlphaSrc;
    };
    union
    {
            DWORD               dwFillColor;
            DWORD               dwFillDepth;

            LPDIRECTDRAWSURFACE lpDDSPattern;
    };
    DDCOLORKEY  ddckDestColorkey;
    DDCOLORKEY  ddckSrcColorkey;
    } DDBLTFX,FAR* LPDDBLTFX;

DDCAPS
    typedef struct _DDCAPS{
        DWORD   dwSize;
        DWORD   dwCaps;
```

```
        DWORD       dwCaps2;
        DWORD       dwCKeyCaps;
        DWORD       dwFXCaps;
        DWORD       dwFXAlphaCaps;
        DWORD       dwPalCaps;
        DWORD       dwSVCaps;
        DWORD       dwAlphaBltConstBitDepths;
        DWORD       dwAlphaBltPixelBitDepths;
        DWORD       dwAlphaBltSurfaceBitDepths;
        DWORD       dwAlphaOverlayConstBitDepths;
        DWORD       dwAlphaOverlayPixelBitDepths;
        DWORD       dwAlphaOverlaySurfaceBitDepths;
        DWORD       dwZBufferBitDepths;

        DWORD       dwVidMemTotal;
        DWORD       dwVidMemFree;
        DWORD       dwMaxVisibleOverlays;
        DWORD       dwCurrVisibleOverlays;
        DWORD       dwNumFourCCCodes;
        DWORD       dwAlignBoundarySrc;
        DWORD       dwAlignSizeSrc;
        DWORD       dwAlignBoundaryDest;
        DWORD       dwAlignSizeDest;
        DWORD       dwAlignStrideAlign;
        DWORD       dwRops[DD_ROP_SPACE];
        DDSCAPS     ddsCaps;
        DWORD       dwMinOverlayStretch;
        DWORD       dwMaxOverlayStretch;
        DWORD       dwMinLiveVideoStretch;

        DWORD       dwMaxLiveVideoStretch;
        DWORD       dwMinHwCodecStretch;
        DWORD       dwMaxHwCodecStretch;
        DWORD       dwReserved1;
        DWORD       dwReserved2;
        DWORD       dwReserved3;
        DWORD       dwSVBCaps;
        DWORD       dwSVBCKeyCaps;
        DWORD       dwSVBFXCaps;
        DWORD       dwSVBRops[DD_ROP_SPACE];
        DWORD       dwVSBCaps;
        DWORD       dwVSBCKeyCaps;
        DWORD       dwVSBFXCaps;
        DWORD       dwVSBRops[DD_ROP_SPACE];
        DWORD       dwSSBCaps;
        DWORD       dwSSBCKeyCaps;

        DWORD       dwSSBCFXCaps;
        DWORD       dwSSBRops[DD_ROP_SPACE];
        DWORD       dwReserved4;
        DWORD       dwReserved5;
        DWORD       dwReserved6;
    } DDCAPS,FAR* LPDDCAPS;

DDCOLORKEY
    typedef struct _DDCOLORKEY{
        DWORD   dwColorSpaceLowValue;
        DWORD   dwColorSpaceHighValue;
    } DDCOLORKEY,FAR* LPDDCOLORKEY;
```

```
DDPIXELFORMAT
   typedef struct _DDPIXELFORMAT{
        DWORD   dwSize;
        DWORD   dwFlags;
        DWORD   dwFourCC;
   union
   {
            DWORD   dwRGBBitCount;
            DWORD   dwYUVBitCount;
            DWORD   dwZBufferBitDepth;
            DWORD   dwAlphaBitDepth;
   };
   union
   {
            DWORD   dwRBitMask;
            DWORD   dwYBitMask;
   };
   union
   {
            DWORD   dwGBitMask;
            DWORD   dwUBitMask;
   };
   union
   {
            DWORD   dwBBitMask;
            DWORD   dwVBitMask;
   };
   union
   {
            DWORD   dwRGBAlphaBitMask;

            DWORD   dwYUVAlphaBitMask;
   };
   } DDPIXELFORMAT, FAR* LPDDPIXELFORMAT;


DDSCAPS2
   typedef struct _DDSCAPS2 {
        DWORD     dwCaps;    // Surface capabilities
        DWORD     dwCaps2;   // More surface capabilities
        DWORD     dwCaps3;   // Not currently used
        DWORD     dwCaps4;   // .
   } DDSCAPS2, FAR* LPDDSCAPS2;


DDSURFACEDESC2
   typedef struct _DDSURFACEDESC2 {
        DWORD          dwSize;
        DWORD          dwFlags;
        DWORD          dwHeight;
        DWORD          dwWidth;
        union
        {
            LONG       lPitch;
            DWORD      dwLinearSize;
        } DUMMYUNIONNAMEN(1);
        DWORD          dwBackBufferCount;
        union
        {
            DWORD      dwMipMapCount;
```

```
            DWORD       dwRefreshRate;
        } DUMMYUNIONNAMEN(2);
        DWORD           dwAlphaBitDepth;
        DWORD           dwReserved;
        LPVOID          lpSurface;
        DDCOLORKEY      ddckCKDestOverlay;
        DDCOLORKEY      ddckCKDestBlt;
        DDCOLORKEY      ddckCKSrcOverlay;
        DDCOLORKEY      ddckCKSrcBlt;
        DDPIXELFORMAT ddpfPixelFormat;
        DDSCAPS2        ddsCaps;
        DWORD           dwTextureStage;
    } DDSURFACEDESC2, FAR* LPDDSURFACEDESC2;

DIBSECTION
    typedef struct tagDIBSECTION {
        BITMAP              dsBm;
        BITMAPINFOHEADER    dsBmih;
        DWORD               dsBitfields[3];
        HANDLE              dshSection;
        DWORD               dsOffset;
    } DIBSECTION;

DIDATAFORMAT
    typedef struct {
        DWORD dwSize;
        DWORD dwObjSize;
        DWORD dwFlags;
        DWORD dwDataSize;
        DWORD dwNumObjs;
        LPDIOBJECTDATAFORMAT rgodf;
    } DIDATAFORMAT;

DIDEVCAPS
    typedef struct {
        DWORD dwSize;
        DWORD dwDevType;
        DWORD dwFlags;
        DWORD dwAxes;
        DWORD dwButtons;
        DWORD dwPOVs;
    } DIDEVCAPS;

DIDEVICEINSTANCE
    typedef struct {
        DWORD dwSize;
        GUID  guidInstance;
        GUID  guidProduct;
        DWORD dwDevType;
        TCHAR tszInstanceName[MAX_PATH];
        TCHAR tszProductName[MAX_PATH];
    } DIDEVICEINSTANCE;

DIDEVICEOBJECTDATA
    typedef struct {
        DWORD dwOfs;
        DWORD dwData;
        DWORD dwTimeStamp;
        DWORD dwSequence;
    } DIDEVICEOBJECTDATA;
```

```
DIJOYSTATE
   typedef struct DIJOYSTATE {
       LONG     lX;
       LONG     lY;
       LONG     lZ;
       LONG     lRx;
       LONG     lRy;
       LONG     lRz;
       LONG     rglSlider[2];
       DWORD    rgdwPOV[4];
       BYTE     rgbButtons[32];
   } DIJOYSTATE, *LPDIJOYSTATE;

DIJOYSTATE2
   typedef struct DIJOYSTATE2 {
       LONG     lX;
       LONG     lY;
       LONG     lZ;
       LONG     lRx;
       LONG     lRy;
       LONG     lRz;
       LONG     rglSlider[2];
       DWORD    rgdwPOV[4];
       BYTE     rgbButtons[128];
       LONG     lVX;
       LONG     lVY;
       LONG     lVZ;
       LONG     lVRx;
       LONG     lVRy;
       LONG     lVRz;
       LONG     rglVSlider[2];
       LONG     lAX;
       LONG     lAY;
       LONG     lAZ;
       LONG     lARx;
       LONG     lARy;
       LONG     lARz;
       LONG     rglASlider[2];
       LONG     lFX;
       LONG     lFY;
       LONG     lFZ;
       LONG     lFRx;
       LONG     lFRy;
       LONG     lFRz;
       LONG     rglFSlider[2];
   } DIJOYSTATE2, *LPDIJOYSTATE2;

DIMOUSESTATE
   typedef struct {
       LONG lX;
       LONG lY;
       LONG lZ;
       BYTE rgbButtons[4];
   } DIMOUSESTATE;

DIPROPDWORD
   typedef struct {
       DIPROPHEADER      diph;
       DWORD             dwData;
```

```
        } DIPROPDWORD;

DIPROPHEADER
    typedef struct {
        DWORD    dwSize;
        DWORD    dwHeaderSize;
        DWORD    dwObj;
        DWORD    dwHow;
    } DIPROPHEADER;

DIPROPRANGE
    typedef struct {
        DIPROPHEADER diph;
        LONG         lMin;
        LONG         lMax;
    } DIPROPRANGE;

    DISPLAY_DEVICE
    typedef struct _DISPLAY_DEVICE {
        DWORD   cb;
        WCHAR   DeviceName[32];
        WCHAR   DeviceString[128];
        DWORD   StateFlags;
    } DISPLAY_DEVICE, *PDISPLAY_DEVICE, *LPDISPLAY_DEVICE;

DSETUP_CB_UPGRADEINFO
    typedef struct _DSETUP_CB_UPGRADEINFO
    {
        DWORD UpgradeFlags;
    } DSETUP_CB_UPGRADEINFO;

LOGBRUSH
    typedef struct tag LOGBRUSH { /* lb */
        UINT     lbStyle;
        COLORREF lbColor;
        LONG     lbHatch;
    } LOGBRUSH;

LOGPEN
    typedef struct tagLOGPEN {  /* lgpn */
        UINT     lopnStyle;
        POINT    lopnWidth;
        COLORREF lopnColor;
    } LOGPEN;

LV_KEYDOWN
    typedef struct tagLV_KEYDOWN {
        NMHDR hdr;
        WORD wVKey;
        UINT flags;
    } LV_KEYDOWN;

MONITORINFO
    typedef struct tagMONITORINFO {
        DWORD   cbSize;
        RECT    rcMonitor;
        RECT    rcWork;
        DWORD   dwFlags;
    } MONITORINFO, *LPMONITORINFO;
```

```
MONITORINFOEX
    typedef struct tagMONITORINFOEX {
        DWORD   cbSize;
        RECT    rcMonitor;
        RECT    rcWork;
        DWORD   dwFlags;
        TCHAR   szDevice[CCHDEVICENAME]
    } MONITORINFOEX, *LPMONITORINFOEX;

MSG
    typedef struct tagMSG {        // msg
        HWND    hwnd;
        UINT    message;
        WPARAM  wParam;
        LPARAM  lParam;
        DWORD   time;
        POINT   pt;
    } MSG;

NMHDR
    typedef struct tagNMHDR {
        HWND hwndFrom;
        UINT idFrom;
        UINT code;
    } NMHDR;

PAINTSTRUCT
    typedef struct tagPAINTSTRUCT { // ps
        HDC  hdc;
        BOOL fErase;
        RECT rcPaint;
        BOOL fRestore;
        BOOL fIncUpdate;
        BYTE rgbReserved[32];
    } PAINTSTRUCT;

POINT
    typedef struct tagPOINT {
        LONG x;
        LONG y;
    } POINT;

    RECT
    typedef struct tagRECT {
        LONG left;
        LONG top;
        LONG right;
        LONG bottom;
    } RECT;

RGBQUAD
    typedef struct tagRGBQUAD { // rgbq
        BYTE    rgbBlue;
        BYTE    rgbGreen;
        BYTE    rgbRed;
        BYTE    rgbReserved;
    } RGBQUAD;

RGBTRIPLE
    typedef struct tagRGBTRIPLE { // rgbt
```

```
        BYTE rgbtBlue;
        BYTE rgbtGreen;
        BYTE rgbtRed;
    } RGBTRIPLE;


RGNDATA
    typedef struct _RGNDATA { /* rgnd */
        RGNDATAHEADER rdh;
        char          Buffer[1];
    } RGNDATA;

RGNDATAHEADER
    typedef struct _RGNDATAHEADER { // rgndh
        DWORD dwSize;
        DWORD iType;
        DWORD nCount;
        DWORD nRgnSize;
        RECT  rcBound;
    } RGNDATAHEADER;

SCROLLINFO
    typedef struct tagSCROLLINFO {  // si
        UINT cbSize;
        UINT fMask;
        int  nMin;
        int  nMax;
        UINT nPage;
        int  nPos;
        int  nTrackPos;
    }    SCROLLINFO;
    typedef SCROLLINFO FAR *LPSCROLLINFO;

SIZE
    typedef struct tagSIZE {
        int cx;
        int cy;
    } SIZE;

TBBUTTON
    typedef struct _TBBUTTON { \\ tbb
        int iBitmap;
        int idCommand;
        BYTE fsState;
        BYTE fsStyle;
        DWORD dwData;
        int iString;
    } TBBUTTON, NEAR* PTBBUTTON, FAR* LPTBBUTTON;
    typedef const TBBUTTON FAR* LPCTBBUTTON;

TEXTMETRICS
    typedef struct tagTEXTMETRIC {  /* tm */
        int  tmHeight;
        int  tmAscent;
        int  tmDescent;
        int  tmInternalLeading;
        int  tmExternalLeading;
        int  tmAveCharWidth;
        int  tmMaxCharWidth;
        int  tmWeight;
```

```
        BYTE tmItalic;
        BYTE tmUnderlined;
        BYTE tmStruckOut;
        BYTE tmFirstChar;
        BYTE tmLastChar;
        BYTE tmDefaultChar;
        BYTE tmBreakChar;
        BYTE tmPitchAndFamily;
        BYTE tmCharSet;
        int  tmOverhang;
        int  tmDigitizedAspectX;
        int  tmDigitizedAspectY;
   } TEXTMETRIC;


TOOLINFO
   typedef struct {  // ti
        UINT      cbSize;
        UINT      uFlags;
        HWND      hwnd;
        UINT      uId;
        RECT      rect;
        HINSTANCE hinst;
        LPTSTR    lpszText;
   } TOOLINFO, NEAR *PTOOLINFO, FAR *LPTOOLINFO;


WNDCLASSEX
   typedef struct _WNDCLASSEX {     // wc
        UINT    cbSize;
        UINT    style;
        WNDPROC lpfnWndProc;
        int     cbClsExtra;
        int     cbWndExtra;
        HANDLE  hInstance;
        HICON   hIcon;
        HCURSOR hCursor;
        HBRUSH  hbrBackground;
        LPCTSTR lpszMenuName;
        LPCTSTR lpszClassName;
        HICON   hIconSm;
   } WNDCLASSEX;
```

Bibliography

Abdi, Herve, and Dominique Valentin, and Betty Edelman. *Neural Networks*. California. SAGE Publications, Inc., 1999.

Akl, Selim G. *The Design and Analysis of Parallel Algorithms*. Prentice-Hall. NJ., 1989.

Angermeyer, John, and Kevin Jaeger. *MS DOS Developer's Guide*. Indianapolis, IN.: Howard W. Sams and Co., 1986.

Anton, Howard. *Elementary Linear Algebra*. Third Edition. New York: John Wiley, 1981.

Anton, Howard, and Chris Rorres. *Elemetary Linear Algebra with Applications*. New York: John Wiley, 1987.

Berkeley, Edmund C. *Symbolic Logic and Intelligent Machines*. New York: Reinholt Publishing Corporation, 1959.

Bliss, Elizabeth. *Data Processing Mathematics*. Englewood Cliffs, NJ.: Prentice-Hall, 1985.

Bradley, David J. *Assembly Language Programming for the IBM Personal Computer*. Englewood Cliffs, NJ.: Prentice-Hall, 1984.

Brassard, Giles, and Paul Bratley. *Fundamentals of Algorithmics*. Englewood Cliffs, NJ., Prentice-Hall, 1996.

Brumm, Peter, and Don Brumm. *80386 A Programming and Design Handbook*. Blue Ridge Summit, PA.: Tab Books, 1987.

Burden, Richard L., and J. Douglas Faires. *Numerical Analysis. Fourth Edition*. Boston: PWS-KENT Publishing Company. 1989.

Byrne, Richard J. *Number Systems: An Elementary Approach*. New York: McGraw-Hill, 1967.

Capron, H. L. *System Analysis and Design.* Benjamin Cummins. Menlo Park, CA., 1986.

Cavanagh, Joseph J. F. *Digital Computer Arithmetic: Design and Inplementation.* New York: McGraw-Hill, 1984.

Chester, Michael. *Neural Networks: A Tutorial.* Engelwood Cliffs, NJ.: Prentice-Hall, Inc., 1993.

Chien, Chao C. *Programming the IBM Personal Computer: Assembly Language.* New York: CBS College Publishing, 1984.

Cleaves, Cheryl, and Margie Hobbs. *Business Math, Fifth Edition.* Upper Saddle River, N.J.: Prentice-Hall, Inc. 1999.

Cluts, Nancy Winnick. *Programming the Windows 95 User Interface.* Microsoft Press. Redmond, WA., 1995.

Cody, William J, and W. Waite. *Software Manual for Elementary Func ions.*Prentice-Hall, NJ., 1980.

Cooke, D. J., and H. E. Bez. *Computer Mathematics.* Cambridge, England: Cambridge University Press, 1984.

Conrad, Clifford L. with Nancy J Conrad and Harry B. Higley. *Computer Mathematics.* Rochelle Park, NJ.: Hayden Book Company, 1975.

Dayhoff, Judith. *Neural Network Architectures: An Introduction.* New York: Van Nostrand Reinhold, 1990.

Dewar, Robert B. K. and Mathew Smosna. *Microprocessors: A Programmer's View.* New York: McGraw-Hill, 1990.

Dromey, R. G. *How to Solve it by Computer.* Englewood Cliffs, NJ.: Prentice-Hall, 1982.

Duncan, Ray. *The MS DOS Encyclopedia.* Redmond, WA.: Microsoft Press, 1988.

Edgars, William J. *The Elements of Logic for Use in Computer Science, Mathematics, and Philosophy.* Chicago, Ill.: Science Research Associates, 1989.

Fausett, Laurene. *Fundamentals of Neural Networks.* Englewood Cliffs, NJ.: Prentice Hall, 1994.

Flores, Ivan. *The Logic of Computer Arithmetic.* Englewood Cliffs, NJ.: Prentice-Hall, 1963.

Forsythe, George E., Michael A. Malcolm, and Cleve B. Moler. *Computer Methods for Mathematical Computations.* Englewood Cliffs, NJ.: Prentice-Hall, Inc., 1977.

Franklin, Mark. *Using the IBM PC: Organization and Assembly Language Programming.* New York: CBS Publishing, 1984.

Froberg, Carl-Erik, *Numerical Mathematics. Theory and Computer Applications.* Menlo Park, CA.: The Benjamin/Cummings Publishing Company, Inc., 1985.

Gallant, Stephen I. *Neural Network Learning and Expert Systems.* Cambridge, MA.: The MIT Press, 1993.

Gamma, Erich, and R. Helm, and R. Johnson, and J.Vlissides. *Design Patterns: Elements of Reusable Object-Oriented Software.* Addison Wesley Professional Computing Series. NJ., 1995.

Gibson, Carol. Editor. *The Facts on File Dictionary of Mathematics.* New York: Facts on File Inc., 1982.

Gonnet, G.H., and R. Baeza-Yates. *Handbook of Algorithms and Data Structures. Second Edition.* Reading, MA.: Addison-Wesley, 1991.

Hager, William H. *Applied Numerical Linear Algebra.* Englewood Cliffs, NJ.: Prentice-Hall, Inc., 1988.

Hamming, R.W. *Numerical Methods for Scientists and Engineers. Second Edition.* New York: Dover Publications, Inc., 1973.

Hardy, F. Lane, and Bevan K. Youse. *College Mathematics for the Managerial, Social, and Life Sciences.* St. Paul, MN.: West Publishing Co., 1984.

Haykin, Simon. *Neural Networks: A Comprehensive Foundation.* New York: Macmillan College Publishing Company, Inc., 1994.

Headington, Mark R., and David D. Riley. *Data Abstraction and Structures Using C++.* D.C. Heath and Company, 1994

Hogan, Thom. *The Programmer's PC Sourcebook.* Redmond, WA.: Microsoft Press, 1988.

Hoggar, S.G. *Mathematics for Computer Graphics.* Cambridge, England: Cambridge University Press, 1992.

Hungerford, Thomas W., and Richard Mercer. *Algebra and Trigonometry.* New York: CBS College Publishing, 1982.

IBM Corporation. *Technical Reference, Personal Computer.* Boca Raton, Fla.: IBM, 1984.

—— *Technical Reference, Personal Computer PCjr.* Boca Raton, Fla.: IBM, 1983.

—— *Technical Reference, Personal System/2, Model 30.* Boca Raton, Fla.: IBM, 1987.

—— *Technical Reference, Personal System/2, Model 50 and 60.* Boca Raton, Fla.: IBM, 1987.

—— *Personal System/2 and Personal Computer BIOS Interface Technical Reference.* Boca Raton, Fla.: IBM, 1987.

—— *Technical Reference, Options and Adapters, vol. 2.* Boca Raton, Fla.: IBM, 1986.

—— *Technical Reference, Options and Adapters.* Boca Raton, Fla.: IBM, 1986.

—— *Technical Reference, Supplements for the PS/2 Model 70, Hardware Interface, and BIOS Interface Technical References.* Boca Raton,Fla.: IBM, 1988.

Institute of Electrical and Electronics Engineers, *IEEE Standard for Binary Floating-Point Arithmetic.* New York,: IEEE, 1985.

—— IEEE Standard for Radix-Independent Floating-Point Arithmetic. New York, NY: IEEE, 1987.

Intel Corporation, *iAPX 86/88, 186/188 User's Manual (Programmer's Reference).* Santa Clara, Calif.: Reward Books, 1983.

—— *iAPX 86/88, 186/188 User's Manual (Programmer's Reference).* Santa Clara, Calif.: Intel, 1987.

—— *80286 and 80287 Programmer's Reference Manual.* Santa Clara, Calif.: Intel, 1987.

—— *80386 Programmer's Reference Manual.* Santa Clara, Calif.: Intel, 1986.

—— *i486 Microprocessor Programmer's Reference Manual.* Santa Clara, Calif.: 1989.

—— *Pentium Processor Unser's Manual. Volume 3: Architecture and Programming Manual.* Santa Clara, Calif. 1993.

Irwin, Wayne C. *Digital Computer Principles.* New York: D. Van Nostrand Company, 1960.

Isaacson, Eugene, and Herbert Bishop Keller. *Analysis of Numerical Methods.* New York: Dover Publications, Inc., 1994.

Kartalopoulos, Stamatios V. *Understanding Neural Networks and Fuzzy Logic.* Piscataway, NJ.: IEEE Press., 1996.

Kernighan, Brian W. and Dennis M. Ritchie: *The C Programming Language.* Prentice-Hall, NJ., 1988.

Kohn, Michael C. *Practical Numerical Methods: Algorithms and Programs.* New York: Macmillan Publishing Company, 1987.

Kolmand, Bernard, and Robert C. Busby. *Discrete Mathematical Structures for Computer Science.* Second Edition. Englewood Cliffs, NJ.: Prentice-Hall, 1987.

Koren, Israel. *Computer Arithmetic Algorithms.* Englewood Cliffs, NJ.: Prentice-Hall, 1993.

Donald E. Knuth: *The Art of Computer Programming (Semi-numerical Algorithms)* Addison Wesley, MA.: 1969.

Jourdain, Robert. *Programmer's Problem Solver for the IBM PC, XT & AT.* New York: Brady Communications, 1986.

Knuth, Donald E. *The Art of Computer Programming. Volume 1/ Fundamental Algorithms.* Second Edition. Reading, MA.: Addison-Wesley, 1973

—— *The Art of Computer Programming. Volume 2 / Seminumerical Algorithms.* Second Edition. Reading, MA.: Addison-Wesley, 1981.

Kovach, Ladis D. *Computer-oriented Mathematics.* San Francisco: Holden-Day, Inc., 1964.

Lay, Clark L. *The Study of Arithmetic.* New York: Macmillan Publishing Company, 1966.

Layton, W.I. *Essential Business Mathematics.* New York: John Wiley & Sons, Inc. 1965.

Lew, Art. *Computer Science: A Mathematical Introduction.* Englewood Cliffs, NJ.: Prentice-Hall, 1985.

Luger, George F., and William A. Stubblefield. Artificial Intelligence: Third Edition. Reading, MA.: Addison Wesley Longman, Inc., 1998.

Mak, Ronal. *Writing Compilers & Interpreters. An Applied Approach.* New York: John Wiley, 1991.

Malvino, Albert Paul, and Donald P. Leach. *Digital Principles and Applications.* New York: McGraw-Hill, 1986.

Mathews, John H. *Numerical Methods for Computer Science, Engineering, and Mathematics.* Englewood Cliffs, NJ.: Prentice-Hall, 1987.

Mayer, Joerg. *Assembly Language Programming: 8086/8088,8087.* New York: John Wiley and Sons, 1988.

Mehrotra, Kishan, and Chilukuri K. Mohan and Sanjay Ranka. *Elements of Artificial Neural Networks.* Cambridge, MA.: MIT Pres+s, 1997.

Merris, Russell. *Introduction to Computer Mathematics.* Rockville, MD.: Computer Science Press, 1985.

Microsoft. *Windows Interface Guidelines for Software Design* Microsoft Corporation, Redmond, WA., 1995.

Morgan, Christopher L. *Bluebook of Assembly Language Routines for the IBM PC & XT.* New York: Waite Group, 1984.

Morse, Stephen P. *The 80386/387 Architecture.* New York: John Wiley, 1987.

Mott, Joe L., with Abraham Kandel and Theodore P. Baker. *Discrete Mathematics for Computer Scientists.* Reston, VA.: Prentice-Hall, 1983.

Nakamura, Shoichiro. *Applied Numerical Methods with Software.* Englewood Cliffs, NJ.: Prentice-Hall, 1991.

Omondi, Amos R. *Computer Arithmetic Systems: Algorithms, Architecture, and Implementation.* New York: Prentice-Hall International Limited, 1994.

Palmer, John F., and Stephen P. Morse. *The 8087 Primer.* New York: Wiley, 1984.

Polya, G. *How to Solve It: A New Aspect of Mathematical Method.* Princeton, NJ.: Princeton University Press, 1945.

Ralston, Anthony, and Chester L. Meek. *Encyclopedia of Computer Science.* New York: Mason and Charter, 1983.

Rash, Bill. *Getting Started with the Numeric Data Processor.* Santa Clara, CA.: Intel, 1981.

Roberts, A. Wayne. *Elementary Linear Algebra.* Menlo Park, CA..: The Benjamin/Cummings Publishing Company Inc., 1982.

Rojas, Raul. *Neural Networks: A Systematic Introduction.* Berlin: Springer-Verlag, 1996.

Rosch, Winn L. *The Winn Rosch Hardware Bible.* New York: Prentice-Hall, 1989.

Roueche, Nelda W. Business Mathematics Second Edition: A Collegiate Approach. Englewood Cliffs, NJ.: Prentice-Hall, Inc., 1973.

Russell, Bertrand. *Wisdom of the West.* London: Rathbone Books Limited, 1977.

Sanchez, Julio. *Assembly Language Tools and Techniques for the IBM Microcomputers.* Englewood Cliffs, NJ.: Prentice-Hall, 1990.

—— *Graphics Design and Animation on the IBM Microcomputers.* Englewood Cliffs, NJ.: Prentice-Hall, 1990.

—— and Maria P. Canton. *IBM Microcomputers: A Programmer's Handbook.* New York: McGraw-Hill, 1990.

—— and Maria P. Canton. *Programming Solutions Handbook for the IBM Microcomputers.* New York: McGraw-Hill, 1990.

—— *Logical and Mathematical Methods for IBM Microcomputer.* Boca Raton, FL., CRC Press, 1991.

—— and Maria P. Canton. *Graphics Programming Solutions.* New York: McGraw-Hill, 1993.

—— and Maria P. Canton. *High-Resolution Video Graphics.* New York: McGraw-Hill, 1994.

—— and Maria P. Canton. *Numerical Programming the 387, 486, and Pentium.* New York: McGraw-Hill, 1995.

Sargent, Richard, III, and Richard L. Shoemaker. *The IBM Personal Computer from the Inside Out.* Reading, MA.: Addison-Wesley, 1984.

Scanlon, Leo J. *Assembly Language Subroutines for MS-DOS Computers.* Blue Ridge Summit, PA.: Tab Books, 1987.

Scott, Norman R. *Computer Number Systems and Arithmetic.* Engelwood Cliffs, NJ.: Prentice-Hall, 1985.

Shoup, Terry E. *Numerical Methods for the Personal Computer.* Englewood Cliffs, NJ.: Prentice-Hall, 1983.

Smith, Bud E., and Mark T. Johnson. *Programming the Intel 80386.* Glenview, IL.: Scott, Foresman, 1987.

Smith, Karl J. *The Nature of Mathematics, Sixth Edition.* Pacific Grove, CA.: Brooks/Cole Publishing Company, 1991.

Snyder, Llewellyn R., and William F. Jackson. *Essential Business Mathematics.* New York: McGraw-Hill Inc, 1972.

Startz, Richard. *8087 Applications and Programming for the IBM PC, XT, and AT.* New York: Brady Communications, 1985.

Stanat, Donald F., and David F. McAllister. *Discrete Mathematics in Computer Science.* Englewood Cliffs, NJ.: Prentice-Hall, Inc., 1977.

Sullivan, Richard L. *Modern Electronics Mathematics.* Albany, NY.: Delmar Publishers Inc., 1985.

Swokowski, Earl W. *Functions and Graphs, Fourth Edition.* Boston: Prindle, Weber & Schmidt, 1984.

Tausworthe, Robert C. *Standardized Development of Computer Software: Methods and Standards.* Prentice-Hall, NJ., 1977.

Venit, Stewart, and Wayne Bishop. *Elementary Linear Algebra. Second Edition.* Boston, MA.: Prindle, Weber and Schmidt, 1984.

Volder, Jack E. *The CORDIC Trigonometric Computing Technique.* IRE Transactions on Electronic Computers, NY., September 1959.

Walsh, J. E., with N. E. Williams, E. E. Barnett, and R. H. Collyer. *Principles of Data Processing.* New York: Sir Isaac Pitman Limited, 1968.

Williams, Gerald E. *Boolean Algebra with Computer Applications.* New York: McGraw-Hill, 1970.

Woram, John. *The PC Configuration Handbook.* New York: Bantam Books, 1987.

Yakowitz, Sidney, and Ferenc Szidarovszky. *An Introduction to Numerical Computations.* New York: Macmillan Publishing Company, 1986.

Zill, Dennis G. *Calculus with Analytical Geometry.* Boston, MA.: PWS Publishers, 1985.

Index

.NET 425

_
_ABS_CMPLX 237,883
_ADD_CMPLX 234,883
_ADD_MATRICES 361-363,884
_ALOG_10 212,882
_ALOG_2 212-213,882
_ALOG_E 212-213,882
_ANNUITY_DUE 267-268,883
_ANNUITY_FV 263-264,883
_ANNUITY_PV 265-266,883
_AOI_PAYMENT 260,883
_APR 261-262,883
_ARC_COSH 882
_ARC_COSINE 210,882
_ARC_SINE 209-210,882
_ARC_SINH 882
_ARC_TANGENT 207,209,246,882
_ARC_TANH 882
_ASCII_TO_BCD12 827,881
_ASCII_TO_BCD20 881
_ASCII_TO_DX 827,881
_ASCII_TO_EDX 22-23,827,881
_ASCII_TO_EXP 133,822,882
_BCD_INTO_REG 827,881
_BCD_TO_ASCII 827,881
_BCD12_TO_ASCSTR 827,881
_BCD20_TO_ASCSTR 881
_BIN_TO_ASC10 17-18,827-828,881
_BIN_TO_ASC16 18-19,827,829,881
_BIN_TO_BCD 827,881
_BINF_TO_DECF 827,881
_BINOMIAL_PROB 285-286,883
_CALCULATE_Y 382,386-388,884
_CART_TO_POLAR 246-247,883
_CIRCLE_SCALE 201-202,205,882
_COMBINATIONS 284-286,883
_COMP_PERIODS 271,883
_COSECANT 206,882
_COSH 219,882

_COSINE 205,210,246,882
_COTANGENT 206-207,882
_DECODE_SINGLE 48,827,881
_DEG_TO_RAD 200,229,245,247,882
_DIV_CMPLX 235,237,883
_EFFECTIVE_YIELD 256,883
_EVALUATE 382-387,884
_FACTORIAL 220-221,283-285,882
_FPU_INPUT 131,133,882
_FPU_OUTPUT 132-133,882
_FUTURE_VALUE 252-253,883
_GAUSS_SOLVE 372-373,884
_GET_TAG 153-154,822,882
_INT_FRAC 226,883
_LOADE 298,883
_LOG_10 211-212,882
_LOG_E 211-212,271,882
_MAT_DIV_SCALAR 884
_MAT_MINUS_SCALAR 884
_MAT_PLUS_SCALAR 884
_MAT_TIMES_SCALAR 358,883
_MAX 222,882
_MIN 222-223,354,357,882-884
_MODULUS 224-225,883
_MUL_CMPLX 234-235,883
_MUL_MATRICES 363,366,884
_NORMAL_CURVE_Y 299-302,883
_OCTANT_SCALE 203-204,882
_ORDER 222,882
_PAYMENTS 258,883
_PERMUTATIONS 283,883
_POLAR_TO_CART 245-246,883
_PRESENT_VALUE 254-255,883
_PYTHAGORAS 227-228,247,883
_QUAD_FORM_REAL 230-231,240,883
_QUAD_FORMULA 240,243,883
_RAD_TO_DEG 201,246-247,882
_ROW_DIV_SCALAR 353-354,883
_ROW_MINUS_SCALAR 354,883
_ROW_PLUS_SCALAR 353-355,883
_ROW_TIMES_SCALAR 352-355,883
_SECANT 207,882

_SIGN_ADD_BCD12 881
_SIGN_ADD_BCD20 882
_SIGN_DIV_BCD12 881
_SIGN_DIV_BCD20 881
_SIGN_MUL_BCD12 881
_SIGN_MUL_BCD20 881
_SIGN_SUB_BCD12 881
_SIGN_SUB_BCD20 882
_SIMPLE_INTEREST 249-250,260,883
_SINE 205,209-210,246,882
_SINH 217,219,882
_SINKING_FUND 269-270,883
_SL_DEPRECIATION 883
_SOLVE_FOR_A 228-229,883
_SUB_CMPLX 234,883
_TANGENT 205,207,209,229,246,882
_TANH 219,882
_TEN_TO_Y_BYFAC 192,882
_X_TO_Y_BYBP 188-189,198,882
_X_TO_Y_BYFAC 192,197,286,882
_X_TO_Y_BYLOG 184-186,198,216-217,882
_X_TO_Y_MIXED 198-199,253-256,258,
       263,265,267,269,300,882

2 to an integer power 174
32-bit true color format 711
486
    DX 95-96,100
    SX 100,102
80286 33,99-101,104,118,141,147
80287 95-96,98-103,105-106,111,117,
       141,143-144,150,155,163,166-167,
       170-172,174,178-179,181,183,186,
       199,201,204-205
80386 33,95,99-102,104-105,118,124,
       199,829
80387 95-96,99-100,102-103,105-107,
       111,113,117,140-144,150,155,163,
       167,170-171,173-175,178-179,
       181-183,186,199,204-205,217-219
8086 33,75,95-96,99-100,104,118,147
8087 60,95-96,98-103,105-107,111,117,
       123,140-141,143-145,147,150,155,
       162-164,166-167,170-172,174,178,
       181-184,186,199,201-205
8088 33,95-96,100,104,118,147
80x86
    family. 55,95
    flags 149,222-223
80x87 Numeric Support Libraries 128
8259A 141

A
Abacists and algorists 5
Abacus 5,31
Abort macro 540-541
Absolute value 34-35,82,87-88,97-98,115,
       138,145,168,171,199,215,233,237,
       239,371,782,883
Abstract classes 783,786,816,864
Accelerator table 572
Access key 568-569,571
Accuracy 176,831-834
    of exponential functions 831,833
Accuracy tests 831
ACM 106,176
ACM Transactions on Mathematical Soft-
    ware 176
Activation function 391-392,395,400,406,413
Adaline networks 389,410
Add-on interest 257,259-262,883
Add-subtract rule 4
Affine closure 110-111,140
ALGOL family of programming languages
    423
Algorists 5
Algorithm
    analysis 831
    for BCD addition 82
Al-Khowarizmi 5
Allocated heap memory 706
Alternate mode 641
American National Standards Institute
    32,42,80,107
Amortization 249,257
    problems 249
Anding
    with a 0 bit 58
    with a 1 bit 58
Animation controls 552,583
Annual percentage rate 257,261-262
Annuities 249,263,266,271
Annuity due 266-268,883
    future value 263
    present value 265
ANSI (see: American National Standards In-
    stitute)
ANSI/IEEE 854 80
ANSI/IEEE
    floating point formats 42
    single format 42-44
    standard 754-1985 42,107
Antilogarithm 181,212-214,216,324,882
    to any base 212
Anton, Howard 368

API
    drawing functions 600
    programming 463
    services 459,463
Application's message queue 528,602
APR 261-262,883
Arc
    direction 599,620,638
    -cosine 181,209,882
    -sine 209
    -tangent 181,199,207,210
Area under a curve 332,334
Arithmetic
    instructions 51,60,118,138,161
    mean 287
Array processors 349
Arrays of arrays 345
Artificial
    intelligence 389,391
    neural network 391,394
ASCII
    numbers 62
    to machine register 21
    to bcd12 51-52
    to-binary conversions 25
    to-exponential conversion 133
Assert macro 540-541
Assertions notation 447,457
Association for Computing Machinery (see
    also: ACM) 106
Augmented matrix 344,368-369,372
Average
    execution time 831
    value 286
Axon 389-390,865
Axonic ending 389-390

B
Background mode 599,619,638
Backpropagation 389,412-419
Backward and forward difference cases of
    the derivative 328
Base 2 logarithm 211
Basic 108,345
    controls 558,565
Batch file 441-442
Bayliss, John 96
BCD
    addition 82-83

arithmetic 31,49-51,56,63,79-82,86,
    90,93,249
    conversion 49,51
    division 85
    multiplication 83-84,274
    subtraction 82
BCD12 48-53,80-83,85-91,120,825,827,
    881
    format 49-53,80,82-83,85-87,89-91,
        825,827,881
BCD20 49,51,80-82,90-93,120,274,
    881-882
    format 49,51,81,90-91,120
BCPL 423
Bell Laboratories 423
Bell Labs (see Bell Laboratories)
Bernoulli
    conditions 285
    Jacob 285
Best-fitting line 307-308
Bias 42-43,45,47-48,81,91,108,112,114,
    120,122,140,166
    form 43,120
Binary
    arithmetic 36,39,63,380
    integer formats 118-119
    integers 117
    numbers 6,11,34,70,117,120,143
    powering 186-190,198,254,831,
        833,882
    raster operation 617,698
    raster operation codes 698
    real 117,120,125,138
    resources 440
    system 6-7,10,14-15,35-36
    -coded decimal 6,31,61,117
    -coded decimals 6,31,117
    -to-ascii conversions 23
    -to-ascii-decimal 15,17-18
    -to-hexadecimal 18
Binomial probability 285-286,883
Biocomputer 390
Biological neuron 389
BIOS 141
Bi-stable components 32
Bit manipulation instructions 55-57
Bit scan 63,69
    forward 69
    reverse 69
    test opcodes 63,69
Bitblt 692,699,710
Bitmap
    dimensions 696,699-700,705,719

Bitmap *(continued)*
  in heap memory 700,705
  is flipped 719
  programming 695,716,721
  stretching 717
BITMAPINFO structure 706-708,710
Bit-mapping 690
Blank bitmap 700,708-710,717
Blitting the bitmap 699
Boolean operators 617
Bootstrap sequence 431
Borland C++ 425,440,866,870,877
Borrow operations 35
Bradley 184
Break action 529
Brush hatch patterns 617,716
  origin 638-639
Brushes 452,525,599,604-606,613,
  615-616,619, 642,644
Bubble sort 280
Built-in cursors 545-549

C
C++
  console application 821
  conversion routines 23
Calculating
  powers and roots 254
  the error sum 409
  the tangent 202,205,384
Calculation
  errors 831
  of powers 182,198
  of powers and roots 182,198
Calculator operations 215
California Institute of Technology 277
Calling conventions 444,469,868,879
Cambridge University 423
Cancellation error 275
Caption bar 552
Capturing the mouse 545
Caret
  processing 537
  shape and size 537
  screen position 537
Carry flag 56,60-62,64,66-67,69-71,130,
  132,149,241
Cascading menus 569
Central difference formulas 331
Character
  codes 529,532-533

  recognition 396
Characteristic 3,6,12,32,34,41-42,44,82,
  117,137,158,296,368
  of significand 44
Chebyshev's theorem 297
Check
  boxes 455
  button 587
Checkbox 492,559,561-562,566
Child
  menus 568,570,574
  windows 473,479,500,528,545,
  551-554,556,558,560,566,601
Chop-off operation 110
Class
  and object decomposition 794
  associations 799
  libraries 463,864-865
Classes and objects 779,784,788,791,794,
  804,806,842
Client area
  coordinates 506
  into screen coordinates 575
  area mouse messages 542,544
Clip path 678-679
Clipping
  operations 671,673,679
  region 502,523,637,663,669,
  673-679,681
  region handle 669
Clock
  arithmetic 163-165
  modulus 164
Closed figures 604-605,633,637-638,
  644,665
Cluts, Nancy 593
Code
  generator 378
  image cost 833
Coding the parser 377
Coefficient of linear correlation 305
Coefficients matrix 344
COFF 441-442
Color palettes 452
Color planes 606,613,696-697,702
COLORREF Bitmap 612
Column vector 349-350
Column-major order 345
Combinations 283,285
Combo box 437,455,476,551,557,
  559-560,593-594
  in a toolbar 593

Common
  antilog 212
  controls 455,551,583-584,591-592
  controls library 583,591-592
  dialog boxes 576,580-581,583
  elements in a menu 569
  exponentials 183
  log 181,212
  summations 278
Compare operands 56
Comparison instructions 171
Complementary numbers 35
Complete pivoting 370
Complex
  addition and subtraction 233
  arithmetic 233-234,237
  division 233
  multiplication 233
  numbers 10,215,232-235,237-241,
    244,883
  region 663,666
  roots 240-243
Compound interest 249,251-257,883
Compounding periods 252-257,271-272,883
Computational error 120,160,272,274,355
Computationally-intensive applications 352
Computer
  arithmetic 41
  errors in financial calculations 272
Concurrent documentation 746
Condition
  code bits 105,144-145,147,149,154,
    165-166,202-203
  codes 145-147,150,165,171,203
Conditional jumps 71,150
Console-based applications 499,507,822
Constant instructions 177-178
Context-sensitive submenu 573
Control events 437
Control
  instructions 98,100,178
  register 178-179
  unit 137
Conversion
  algorithm 20-21,131
  errors 80,200,273
  routines 3,23,25,30-31,51,86,98,
    117,128,130-132,168,184
Converting a region into a path 681
Coonen, Jerome T. 106
Coordinate systems 244-245,505-506
Coprocessor state 156
CORDIC 174,176

Cornet, Jean Claude 96
Correctness 381,383,728-730,732,734,
    740-741,746
Cosines 102,173
Cosmetic pen 614-615
Cost in execution time 833
Cost-benefit analysis 752-753,756,758,
    791
Counting 3,6,8,12-14,220,249,272,
    281,768
  techniques 281
Covariance 305
CPU
  flags 150
  identification 75
CPUID instruction 75
Create a bitmap 702,711
Creating
  a toolbar 586,588
  custom buttons 589
CUI 499
Current pen position 619
Cursor manipulations 547
Curve fitting 305,312-313,319
  by minimizing residuals 319
Curvilinear regression 307
Custom
  brushes 644
  pens 613
Customized cursor 546,548

D
Dangling pointers 840,847
Data
  dispersion 295,297
  flow diagrams 766-767
  manipulation primitives 278
  pointer 137,151-152,154-155
  transfer instructions 160-161
  variables 126,363
DD and DQ Directives 124
Decimal
  arithmetic 36,80
  integers 117,121
  numbers 6,11,14,34-35,42,49,55-56,
    61,80-81,95,97-98,109,128,133,
    160,272-274
  rounding 129
  -to-binary 20,80,272
Decoding floating-point numbers 45
Default
  mapping mode 505,507,605

rounding mode 109
    windows procedure 542
Deferred payment 251
Definition of a permutation 282
Degrees to radians 200,228-229,245
Delta rule 405
Dendrites 389-391
Dennis Ritchie 423
Denormal
    exception 143
    numbers 43,142
    operand. 143
Denormalized
    numbers 43
    representations 115
Denormalizing the significand 116,142
Denormals 43-44,48,112,115,121-123,143
Derivative approximated by the slope of a
    secant 327
Descriptive statistics 277
Desktop surface 501
Destination device context 698,719-720
Developer Studio menu editor 570,572
Developers Studio 423
Device context
    attributes 500
    capabilities 609
Device
    coordinates 504-507,697
    unit 505-507,521,607,638,684,697,
        699-700
    -dependent scan code 529
    -independent bitmap 693,707,710
    -independent graphics 605
    -specific bitmaps 693
Diacritical characters 533
Dialog box editor 576,578-579,584
DIB 607,617,643,693-694,697,705,
        707-708,710-715,890
    section 711-712,714
Differentiation 311,313,315,317,319,321,
        323,325,327,329,331,333,335,
        337,339
Diminished radix representation 36
Discardable memory 704-705
Discriminant 230-231,240-242
Dispatch table 856-859
Display;
    context
    context types 502,504
    device 453,484,500-502,504,599-600,
        603,717
    mode parameters 481

Dithering 644
Division by zero 110,112,114,141,143
Division by zero exception 141,143
DLL 442,444,456,584,822
DOS
    cursor 536
    interrupts 435
Double precision
    formats 45,120,122
    precision shift 63,67-68
Double-click
    intercept 544
    processing 544
Drawing
    operations on the bitmap 710
    primitives 678,710
DT directive 125
Dynamic
    linking 444
    memory allocation 845,850,853

E
E flags register 104-105
Edit box 455,476,556,559,584,
        589-590,593
Effective rate 256
Emach 190,832,834
Empirical rule 297
Emulator software 96,182
End cap style 614-615
Engelbart, Douglas 432
Entity-relationship diagrams 437,766
Environment area 150,152,155-156
Equation
    grammar 376,384-385
    parsing 375
    solving engine 382
    string 376,378,381-382,385-387,884
    syntax 380-381
    -solving algorithm 382
    -solving routine 376-377
Error
    exception interrupt 141
    in trapezoidal integration 336
    sum 144,150,320-326,409-411,419
    in Gaussian elimination 370
    in the derivative 330
    of syntax 376
Euclidean geometry 227
Evaluating
    an expression 386
    user equations 379

Event
    chain 436-437
    handler 436
    manager 436
    -driven program 431,434,436,438,448,481
    -driven programming 431,434,438,481
Exact remainder 164,167
Exception
    bit 141
    flag 98,141,178
    pointers 151,154-155
Excess-n notation 43
Executable files 275
Exponent
    factoring 191-192,198,213,254,
        831-834,882
    factoring algorithm 192,831,834
    methods 198
    field 43,51,119-120,163,168
    range 43,90,182
Exponential
    form 41,52-53,130,132-133,135,382,
        386,822,882
    functions 131,172,184,198,831,833
    notation 124-125,130-133,135
Exponentials 183,190,199,211,272-273,341
    in financial formulas 253
Exponentiation 174,186,376,379,384-385
    function 174,379
Expressions of central difference 328
Extended
    key 531
    precision 42,117,120-123,125-126,129,
        132,138,140,160,168,177,190,192,
        205,216,220,298,355,370,372,381

F
Factorial 215,220-221
Fast multiplication and division 162
FCTVAL 458,765-766
Feasibility study 746,752-753,757,791
Feed-forward network 413,417
File-mapping 713
    object 713
Financial
    calculations 249,251,253,255,257,259,
        261,263,265,267,269,271-273,275
    formulas 211,254
    management 249
    primitives 199
    software 275
Finite algebra 163,190,224

First derivative of a function 327
Fixed
    memory 704-705
    point representations 40
Fixed-size mapping 504,507
Flags register 56,76-77,104-105,222-223
Flat address space 33,150
Floating-point
    BCD 49,80-81,86
    emulator 182
    exponent 81
    hardware 79-80,90,96,162,182
    representation 31,41,44,115,182,
        274,370
    significand 44
Flowchart
    for mode calculation 290
    symbols 426
Flowcharts 423,427,766
Flush to zero 115
Foreground mix mode 617
FORTRAN 345
Forward and backward difference 328-329
FPREM and FPREM1 166-167,201
FPU 99,103,105-106,130-137,140,149,
        152-153,155-156,158,168,178,184,
        204,2 15,345,370,377,382,386,832,
        882
Fractional
    numbers 95,344
    part 40-42,44,46,97,120,183,185-186,
        198,225-226,254,883
Framing regions 668
Function mapping 375
Functional expression 375
Fundamental assumption for interpolation
    312
Future value 251-256,261,263-265,
        267-271,883
    of an account 263
    of an annuity 263
    of an investment 252-253,264,883

G
Gauss, Carl Friedrich 368
Gaussian
    elimination 368,370
    methods of elimination 371
Gauss-Jordan
    algorithm 370,371
    elimination 368-372

method 368
Gauss-Siedel 368
GDI 444,453,499-501,504,508,520,
    524,599,604-606,613,619,628,
    637-640,662-663,678,686,696,700,
    708-711,713- 716
General purpose controls 551
Geometric pen 615,682,684
GIF 693
Gonnet and Baeza-Yates 186
Gradual
    overflow 115
    underflow 115
Graphic
    interpretation of the derivative 327
    interpretation of the second derivative
    329
Graphical user interface
    427,432,434,447-448,453,540,551
Graphics-based applications. 499
Groupbox 559
GUI 432,468,499,540,551,864,890

H
Half-carry flag 56
Handle
    to a bitmap 537,694,696,702-703,707,
        709,713
    to the device context 500,510,517,617,
        674-675,677-678,681-684,695-696,
        707,709,718-719
    to the instance 469,473,480,548
Hard-coded bitmap 700
Heap memory 700,704-706,714,847
    for a bitmap 704
Hello World program 447-448,497
Hewlett-Packard calculators 377
High-precision BCD arithmetic 79,80,90
Hindu-Arabic numerals 5-6
Historical computer systems 32
Hoff, Marcian 405
Horizontal skip factor 348,352-358,
    361-362,367
Hungarian notation 450
Hyperbolic
    cosine 216,218-219,882
    functions 210,215-217
    sine 216-217,219,882
    tangent 216,218,882

I

IBM 704 96
Icon
    bitmap 494
    resource 475,494
Identifiers 376-378,381,451,551,632-633,
    643,686,851
IEEE 32,42-49,79-81,91-92,95,98,101-
    102,106-115,117,121-122,125,128-
    129,131-132,140,142-143,161,164,
    166-167,171,184,274,355,451,762
    754 single format 43
    Computer Society 80,107
Image
    editor 547
    processing 691,738,769,806-807
Imaginary
    and complex numbers 233
    number 232-233
    unit 10,232
Implicit
    1-bit 79,112,123,125-126
    operands 159,162,172
Indirect addressing 838-839
Indirection in C++ 838
Inexact 116,143,272-274
    result exception 116
Inferential statistics 277
Infinity 44,102,109-112,114-115,
    122-123,131,133-134,136,
    140-143,146-147,200,413
    arithmetic 112
Infix form 376
Inheritance 424,775,780-782,784,788,
    793-794,796,799,801-802,816-817,
    835,866
Initializing data 124-125
Input focus 482,527-528,537,542,559,585
Input/output controls 453-455
Instance connection notation 816
Institute for Advanced Study and Research
    31
Institute of Electrical and Electronics Engi-
    neers (see: IEEE)
Instruction
    and data pointers 151-152,155
    patterns 158
    pointer 137,155
    set 137,139,141,143,145,147,149,
        151,153,155,157,159
Integer
    and fractional parts 97,225-226,883
    conversion routines 3,23
    encodings 33

identifier 553
Integration 302,311,313,315,317,319,
    321,323,325,327,329,331-333,
    335-337,339
    by Simpson's Rule 336
    methods 332
Intel
    80x86 family 37,46,55,61
    mnemonics 61

I
Interactive controls 447
Intercept routine 531,571,581,586
Interest calculations 249,251
Interpolation 311-317,319,321,323,325,
    327,329,331,333,335,337,339
    techniques 312
Interrupt
    enable bit 141
    request 144
    request bit 144
Interval arithmetic 107,110
Invalid
    operation 86,102,112-113,142-144,
        171,173
    operation exception 102,113,142,171
    rectangle 601
Inverse trigonometric functions
    181,199,206-207
Iomanip operators 92
Irrational numbers 9,120,225

J
Jacobi 368
Jet Propulsion Laboratory 277
Jordan, Camille 368
Jpeg 693

K
Kahan, William 96,106
Kay, Allan 432
Kernighan and Ritchie 447
Key state 529
Keyboard
    and mouse programming 527
    data 128,528
    input 425,527-528,532,536,542,
        560,576
    LED 531
Keystroke

message 528
    processing 530
Knuth, Donald 186,188,280
Koehler, Bob 96

L
Lagrange
    expansion 317
    Joseph Louis 317
    interpolation 312,317-319
Lagrange's formula 317
Large bitmaps 703,710
Least squares
    line 308
    interpolation 321,324,409
    linear interpolation function 321
Leibnitz, G.W. 6
Library files 439,442,444
Line join attributes 614
Linear
    classifiers 396
    correlation 305-307
    correlation coefficient 305,306
    dependency 305
    equations 323,341-345,368-369,372
    interpolation 313,315,317,321
    interpolation by similar triangles 314
    regression analysis 307
    regression coefficient 305,307
    system 341,343-345,347,349,351,353,
        355,357,359,361,363,365,367-369,
        371-373
Line-drawing functions 621-622,631
List
    and tree view controls 552
    box 437,441,443,455,476,556-557,559-560
    box control 559
Little-endian
    byte ordering scheme 46
    format 46
Least mean squares 405
Loan
    amortization 257
    period 251-252,261,263
Locating a matrix entry 348
Log approximation algorithms 831
Logarithm base 2 of e 177,216
Logarithmic
    approximation 131,183-184,198,833
        of exponentials 184
    constants 211
    function 98,181-182,211,326,341

instructions 183,211
method 184,190,254,832
opcodes 181
transcendental primitives 174
Logarithms
41,98,131,174,181,183,211-213,254,
272,324,833,882
to bases other than 2 211
LOGBRUSH Structure Members 617,643
Logical
AND 58,69,396,458,658,675-676,718,
764
brush 642-643
coordinates 489,504-505,507,517,
619,697
instructions 57,425
left 64
NOT 37,57,60
OR 394,458,524,659,675,718,764
XOR 675
Long
double 127
real 97,117
Longhand
algebraic method 377
decimal arithmetic 36
multiplication 65
Loss of numeric precision 143
Lukasiewicz, Jan 376

M
Machine epsilon 190-191,832-833
Macintosh computer 432,731
Main window 453-454,470,476,481,483,
485,552-553,558,562,568
Mak 48,104-105,220,246,358,377
Make action 529
MAKEINTRESOURCE macro 474,494,
548,694
Managing complexity 793
Manipulating
bitmaps 716
long bit strings 67
instructions 55-57
Mantissa 41,44
Mapping mode 499,504-505,507,521,605,
695,697,700,709,714,717
Masked responses 141
MASM 100,120,123,821-823,827
sources 822
Math
coprocessor 96

Math unit
data formats 117
instruction patterns 158
instruction set 137,156,159,161,199,822
limitations 98
programming 821,823,825,827,829
real formats 121
register stack 137-138
stack 139
state area 156-157
trigonometry 199
versions 100
Mathematical coprocessor 49,60,79,
95-96,100-103,111,113,123,182,
199
Matrices 341,344-345,347-350,355,
359-365,367-368,372
in C++ 347
Matrix
addition 341,359-362,367,884
dimensions 346,348
entries 348-349,355,362,368,370
mathematics 349
multiplication 359-360,363-364,367,884
of coordinates 345
representations of linear systems 344
-by-matrix operations 359
-by-scalar multiplication 357
-by-scalar operations 357
-like arrays 345
McCulloch, Warren S. 391
McCulloch-pitts neuron 391,396,398-400
Mean 282,287,293,296
Measurement errors 312
Measures
of central tendency 277,286,294
of dispersion 277,294-295
Median 287-288
Memory
addressing 12,33,158-159
device context 695-699,708-710,714,
717,720
management techniques 704
operands 158,839
storage 12,49,160,179,381,851
variables 118,123,126,177,231
Menu
bar 453,568,585,593
commands 454,568-569,585-586,602,
609,636
title 568-570,574
Menus 440,452-453,473,486,491-492,
504,553,568-571,573-574,653,687

Message
    box 506,540-541,551,558,575-578,592
    connections 791,816
    loop 482,484,528,533
    parameters 483,487
    passing 481,560,780,793
    processing 461,485,496,503,553-555,
        566,575,581,586,602,704
    queues 481
    -driven architecture 527
    -type structure 471
Metafile 604-605
Metaregions 673
MFC 428,459,463,494,571,586
Microsoft
    Developers Studio 423
    Foundation Classes (see also: MFC)
    428,463
Midrange 288
Minsky and Papert. 412
Miter
    length 684
    limit 684-685,687
Mixed
    exponent 183-184,187
    methods 198
MMX 79,96
Modal dialog boxes 576,583
Mode 15,33,151,162,172,289-292,319,323
Model of a neuron 391,395,398
Modeless dialog boxes 576
Modular arithmetic 163,201
Modularization 736-737,811,863
Modulo division 163
Modulus 153,163-167,170-171,174,202,
    224-225,638
Monochrome bitmap 547,701-703,718
Monospaced fonts 509
Monotonicity 129,132,176
Mouse
    handling operations 548
    messages 541-542,544
    programming 527,545
Moveable memory 704-705
Multi-digit arithmetic 60,81
Multi-dimensional arrays 345,347
Multilanguage programming 866-867,870
Multiple
    attributes 691
    inheritance 424,802
    modes 289
    regression 307
    text lines 518

Multitasking environment 101,156,453,
    469,501,510,599

N
Naming conventions 447,868
NAN 107,112-114,121-123,129,131,
    142,145,171,173-174
NASA 277,691
National Aeronautics and Space Administra-
    tion (see also: NASA) 691
Natural
    and common logarithms 211
    antilog 212,324
    log 211,271-272,323-325
    numbers 8-9
Nave, Rafi 96
Networks 389,391,396,398,410,
    412-413,417
NEU 137
Neural networks 389,391,398,412
Neuron 389-392,394-396,398-400,404,
    406-407,413-418
    output 392
Neurotransmitter 390
Newton-Cotes formulae 334
Non-linear
    equations 323
    models 323
Non-logarithmic algorithms 186
Nontranscendental instructions 161,168-171
Normal
    curve 297-299
    distribution 277,282,297-302
    probability distribution 301,338
Normalized
    form 44
    significand 44-45,52-53,87,115
Normals 43-44,48,112,115,121-123,
    143,174
Not-a-number 112
NPX 96
Number
    crunching 3,174
    of combinations 281,283-284,286,883
    of compounding periods
    252-257,271-272,883
    system 3,5-8,10-14,18,24-25,32,39,
        65,110,163,827
Numeric
    constants 387,447,451
    data conversions 127

data coprocessor 96
data encoding 107-108
   in matrix form 345
   in memory 31,33,35,37,39,41,43,45,
      47,49,51,53,123-124
   execution unit 137
   processor extension 96
   unit 102
   analysis 311
Numerical
   approximations of the derivative 330
   differentiation 327
   errors 272
   integration 332
   methods 272,278,311,370
   parsing 375

O
Object files 443,821-823,865-866
Octal encoding 11
One-byte tokens 383
Operand modes 162,172
Operating system/2 432
Operations
   on matrix entries 349
   on rectangles 687
Operators 56-57,92,126,376-382,386-
      387,424,617,704,760,782,849,852-
      853,865
OR and NOT neurons 394
Ordered compare 102,171,173
Ordering numeric data 221
Ordinary annuity 263
ORing with a 1 bit 59
OS/2 432
   warp version 4.0 432
Output precision 608
Overflow 36-38,56-57,62,70-71,106,111-
      112,114-115,129,131,142-144,194,
      516,845
   exception 142
   flag 57,70-71

P
Packed
   BCD 49-51,79,81,91,117-120,125-127,
      131,160-161,168,825,827,881
   exponent 81
   BCD format 49,119,125,127
   decimal 49,72,117
   format 49

Palette colors 612
Palmer
   and Morse 96,166,174,182,184,202
   John F. 96,106
Parameter-passing conventions 867,872
Parity flag 56,70,149
Parser 377-379,381-383
Partial
   arc-tangent 172,181,199
   pivoting 370
   remainder 102,161,163-166
   tangent 172,181,199,201
Pascal 345
   calling convention 469,868
Paths 12,520,525,599,604,613-614,621,
      678-679,683,685,687,759
   -related functions 679
Pattern
   brush 617,643,701,715-717
      transfer 716
PCX 693
PDP 11 33,42,423
Pen
   position 599,619-624,626,630-633,680,
      682,686
   type 614
Pentium math unit 31,47,49-50,103,105
Perceptron 389,399-406,409-410,412
   learning 400
   training 402
Periodic payment 252,257-259,26- 270,
      883
Periodicity identities 201
Permutations 282-283
Pitch and family 525
Pittman, Tom 106
Pitts, Walter 391
Pivotal condensation 370
Planetary Data System 277
Pohlman, Bill 95
Pointer
   arithmetic 63,835,840,844-846
   to type VOID 712
   variables 835,837-840,845,847
Pointers
   to functions 835,848,856-857,859
   to structures 843
   to VOID 278,835,845-846,848
Polar
   and Cartesian Coordinates 244
   coordinate 244-245,883
      system 244
Polish notation 376-377

Polygon fill mode 640-642,649,665,680,
    682,687
Polymorphism 782-783,788,835,865-866
Polynomial interpolation 317
Pop-up menus 573-574,687
Positional 5-8,13-15,18-20,25,39-40
    feature 6-7
    weight 7,15,18-20,25,39-40
Power function 183,190,323
PRE and POST 458,765
Precision
    and Computation Errors 274
    exception 116,143,163
Predefined
    child windows 556
    classes 551-552,556,558
    control classes 558
    controls 455
Preemptive multitasking 434
Prefix form 377
Present value 164,249-255,257-259,
    265-266,271,883
    at compound interest 254
    of an annuity 265,883
Private display context 502-504
Probability of normal distribution 301
Processor control instructions 98,100,178
Processor/coprocessor
    interface 99
    synchronization 99
Program
    events 434,437
    resource 440,475,491,571-572,576,
        578,693
Programming style 429,447,796
Progress bars 455,552,583
Project workspace pane 571-572,579-580
Projective closure 110-111
Proof of Correctness 176
Property sheets 455,552,583
Proportionally spaced font 509-510,513
Prototyping 742,744-745
Pure stack machine 162
Pushbutton control 560
Pythagorean theorem 9

Q
QNAN 113
quadratic
    equation 10,215,229-231,240,243,305
    formula 229-230,240
Quality metrics 732

Quantification of data relationships 305
Quiet NAN 112-113,122-123,129,171

R
Radians to degrees 200-201
Radio buttons 428,561,566
Radix 7-8,10,35-36,38-45,80,174,181
    complement 35-36,38,42-44
    complement representation 36,42-44
    -minus-one form 36
Ramon y Cajal, Santiago 389
Range 127,129,170,201,294
    -scaling operations 201
Raster
    fonts 523
    graphics 689,711
    operation 617,695,698-699,701-702,
        716,719
        code 698-699,701,716,719
    operations 695,698,716
Rational
    and irrational numbers 9,120
    numbers 9,120,225,273
Ravenel, Bruce 96
Real
    and complex roots 240
    numbers 9,40,43,110,114-115,117,
        120-121,125,184,241,243
Rectangle structure 518,534,538,550,
    611,656
Recursive
    definition 220
    factorial function 220
Redefinition of symbol error 571
Redraw responsibility 600
Reduction
    to the first octant 165,174
    to the unit circle 201
Reference variables 849-850

Region
    combination modes 667
    manipulations 669
    -based operations 663
    -related functions 663,672
Register
    operands 158
    stack 137-139,142,152-153,158,160
Regression 304,307
    analysis 277,307,319
Representation error 274

Representations for zero 34
Requirements analysis phase 759
Resource
    editors 440,491-493
    files 429,440,442,586,588
    header file 492
    id 572,592,594
Response to underflow 143
Reverse polish notation, or rpn 377
RGB format 612,707-708,713
Rich edit controls 552,583
Richards, Martin 423
Risk
    analysis 744-745,752-753,755-757,791
    assessment 753,755-756
    estimation 753,755-756
    identification 753
    management 753,756-757
Roman numerals 4-5
Root instruction 138
ROP2 617-618,698
Rosenblatt, Frank 399,403,406
Round
    to nearest 109,114,225
    to nearest even 109,225
    to negative infinity 110,114
    to positive infinity 109,114
Rounding
    and computational errors 160
    control field 132,140,168,225
    to nearest 129
    to the nearest even 109,129,166
    -off 109
Round-off errors 80
Row vector 349-350,352-358
Row-major order 345
RPN (see also: Reverse Polish Notation)
    377

S
Scalable mapping modes 504,507,605
Scaling factor 163
Scan
    code 529
    line 690
Scanner 377-378,381
    module 381
Scatter diagrams 304
        and correlation 304
Scientific notation 41,133,136
Screen
    coordinates 506,573,575,699

erasing flag 603
    pixel 118,620,690-691,707
    resizing operation 600
Script file 440,492,494,548,570-571,579
Scroll bar controls 557,562-563
Scroll bars 437,454-455,500,504,552,557,
    560,562-564
GUI 468,499,540,551,864
Selection sort 280
Semi-conductor 32
Separation of concerns 735-736
Separators 25,27-28,535-536,568,
    585-587,589-590,592-593,595
Sequential programming model 434
Stack fault 144
Shift
    and rotate bits 56
    arithmetic left 64
Short real 117
Shortcut keys 568,570,572
Side-angle formula 228
Sign
    bit field 120
    extension instructions 72
    flag 57,70-71
Signaling
    and quiet NANs 113
    NAN 112-113,122-123,142,171
Signed numbers 8,32,34-35,43,57,60-61,
    64,70
Significand field
    48,50,91,107,120,126,131,168
Sign-magnitude representation 34
Similar triangles 314
Simonyi, Charles 450
Simple
    interest 249-251,256,259,883
    message box 578
Simpson's rule 302,336,338-339
        approximation 337
single precision 48,117,122,124-126,130,
    132,184,355,825,827,881
sinking fund 269-270
slope of a secant 327
    -intercept form 320,324,409-410
SNaN 113
Software
    emulators 128,182
    qualities 728
Solution
    of a linear equation 343
    of a linear system 368
Solving

and parsing equations 375,377,379, 381,383,385,387
   triangles 227
Soma 389-390
Sorting 280
   algorithms 280
Source files 429,438-440,442-444, 456,494,586,588,823,864,873
Spaghetti code 456,746
Special encodings 121-123,125-126
Specifications phase 741,759
Spiral model 744-745
Square root 98,107,112,114,138,142, 161-162,170,182,185,210,215,228, 296,349,379,729,731
Stack top register 130-132,134,138, 146-147,207
Standard
   deviation 296
      of a data set 296
      of a population 296
   form 32,341-342,578,614
   normal curve 299
   push button 587
   summations 278
   toolbar buttons 591-592
Stanford University 405
Startz 184
Static controls 476,560,565
Statistical
   data 277,292
   notation 282,284
Status
   bar 454-455,500,504,551-552,583-584
   flags 56-57,61
   register 56,179
Stevenson, David 106-107
Stock
   brushes 615-616,644
   pens 613
Storage of numerical data 32
Stretching or compressing a bitmap 717
Stroking the path 637
Stroustrup, Bjarne 423
Structure of type BITMAP 703
Subsystem cohesion 811
Sudden overflow 114
Summation
   operations 279
   primitives 307
Sum-of-products calculation 395
Symbol table 381
Synapse 390-391

Synaptive cleft 390
Syntax rules 380

System
   analysis phase 749,751,771
   analyst 750,752
   fixed font 556
   font 510,512,556
   memory 711,713-714,845
   of linear equations 341,343-344,368-369
   queue 481-482,528
   -level graphics information 599
   -level message 530
   memory bitmap 700
Systems of linear equations 343

T
Tab
   controls 455
   codes 152
Tag word 137,151-153,155-156
   register 152
Tally system 3-4
Temporary real 97,117,131,832-833
Ternary raster operations 698
Text
   box 464-465,551
   dimensions 512-513
   display 454,499,505,508,515,517, 519-520,525,678,708
   formatting 489,513
   graphics 508,520
   processing 512-513
Textmetric structure 510-513,521
Theory of counting 12
Thompson, Ken 423
TIFF 693
Tokenization of keywords 383
Tokens 377,383
Toolbar
   bitmap 588,590-592,595
   buttons 586-588,591-592,594
Toolbars 455,492,551-552,576,583-585, 587,590-591
Tooltip 601
Transcendental
   algorithms 176-177
   functions 102,176,181
   instructions 138,142,161,168,172-174
   primitives 174,215
Transcendentals 172,174,176
Trapezoidal

approximation of the integral 333
rule 332,334-336,338
Tree views 455
Triangles 314,683
Trigonometric
    arcfunctions 207
    functions 97-98,163-164,166,172-174,
        181,199-200,202,206-207,216,349
    opcodes 199
    transcendentals 172
Truncate 91,114,142
Truth-in-Lending Act 261
TTY 431,508
Turbo Assembler 120,123
Twips 507
Two-button mouse 541
Typeface 509-510,524-525
    family 509
Typematic action 528-529
Types of numbers 8,121

U
Unclassed child windows 552
Underflow 71,112,115-116,121,129,
        131-132,142-144,272
    exception 115,132,143
Units of memory 12
UNIVAC I 31
UNIX operating system 423,441
Unnormals 174
Unrepresentable number 273
Unsigned multiplication 61
Update region 601,603,671

V
Vector 350,352
    fonts 523,609
    graphics 508,689,801
    operations 341,350
    -by-Scalar Operations 350,352
Very small numbers 115,274
Viewport 506-507,794,797
    and window 506-507
Virtual-key 529,531-534,539
    code 529,531-534,539
von Neumann, John 6,31
von Neumann computer 31
von Neumann, Burks, and Goldstine 6

W
WAIT 99-101,105,147,149,153,
        178-179,185,202-203,208,222-223,
        225,242,549
Ward and Mellor 769-770
Waterfall model 740-742,745
Weighted
    measures of central tendency 292
    median 293
    mode 293
Weitek chip 96
Whole numbers 8-9,33,779
Widrow
    Bernard 405
    -Hoff learning 405,414
Win32 51,82,355
Winding mode 640,642,665
Window
    class 471-472,474-476,479,482-483,
        502,544,552,556,558
    class name 474-475,479
    display context 506
    function 482,579
    procedure 473,483,495,529,546-547,
        549,552-553,555,578-579,586,601
    tyles 551
Windows
    client area 452
    explorer 474,491,544,553
    GUI 468,551
    procedure 470,472,480-487,503,542,
        545,549,556,566,572
Wizards 463,552,583
WNDCLASSEX structure 469,471,475,
        483,491,493,502,546-547,549,551-
        552,555,558
Word size 32-33,38,67,879
Wozniak, Steve 432

X
Calculating f 299
Xerox PARC 432,540,776
XORing with a 1-bit 59

Z
Zero divisor 81,86
Zero flag 57,60,69,71,74,149
Zero-based system 13